AF291728

Signaling by Receptor Tyrosine Kinases

A subject collection from *Cold Spring Harbor Perspectives in Biology*

OTHER SUBJECT COLLECTIONS FROM *COLD SPRING HARBOR PERSPECTIVES IN BIOLOGY*

Mitochondria

DNA Repair, Mutagenesis, and Other Responses to DNA Damage

Cell Survival and Cell Death

Immune Tolerance

DNA Replication

Endoplasmic Reticulum

Wnt Signaling

Protein Synthesis and Translational Control

The Synapse

Extracellular Matrix Biology

Protein Homeostasis

Calcium Signaling

The Golgi

Germ Cells

The Mammary Gland as an Experimental Model

The Biology of Lipids: Trafficking, Regulation, and Function

Auxin Signaling: From Synthesis to Systems Biology

The Nucleus

Neuronal Guidance: The Biology of Brain Wiring

Cell Biology of Bacteria

Cell–Cell Junctions

Generation and Interpretation of Morphogen Gradients

Immunoreceptor Signaling

NF-κB: A Network Hub Controlling Immunity, Inflammation, and Cancer

Symmetry Breaking in Biology

The Origins of Life

The p53 Family

SUBJECT COLLECTIONS FROM *COLD SPRING HARBOR PERSPECTIVES IN MEDICINE*

Cystic Fibrosis: A Trilogy of Biochemistry, Physiology, and Therapy

Hemoglobin and its Diseases

Addiction

Parkinson's Disease

Type 1 Diabetes

Angiogenesis: Biology and Pathology

HIV: From Biology to Prevention and Treatment

The Biology of Alzheimer Disease

Signaling by Receptor Tyrosine Kinases

A subject collection from *Cold Spring Harbor Perspectives in Biology*

EDITED BY

Joseph Schlessinger

Yale University School of Medicine

Mark A. Lemmon

*University of Pennsylvania
Perelman School of Medicine*

COLD SPRING HARBOR LABORATORY PRESS
Cold Spring Harbor, New York • www.cshlpress.org

Signaling by Receptor Tyrosine Kinases

A Subject Collection from *Cold Spring Harbor Perspectives in Biology*
Articles online at www.cshperspectives.org

Executive Editor	Richard Sever
Managing Editor	Maria Smit
Project Manager	Barbara Acosta
Permissions Administrator	Carol Brown
Production Editor	Diane Schubach
Production Manager/Cover Designer	Denise Weiss
Publisher	John Inglis

Front cover artwork: Stylized representations of the 20 different receptor tyrosine kinase (RTK) families found in humans—accounting for a total of 58 different receptors. The common intracellular tyrosine kinase domain (lower part of each receptor) is shown as a red rectangle. Domains in the extracellular region (upper part of each receptor) are much more variable across families and include immunoglobulin domains (blue), fibronectin type III domains (orange), and many others. The chapters in this volume describe mechanisms by which ligand binding to the extracellular region controls activity of the intracellular kinase domain, which vary substantially across the family and drive a variety of intracellular signaling pathways. (Cover illustration created by J. Schlessinger and M. Lemmon.)

Library of Congress Cataloging-in-Publication Data

Signaling by receptor tyrosine kinases : a subject collection from Cold Spring Harbor perspectives in biology / edited by Joseph Schlessinger, Yale University School of Medicine and Mark A. Lemmon, University of Pennsylvania Perelman School of Medicine.
    pages cm
 Includes bibliographical references and index.
 ISBN 978-1-936113-33-0 (hardcover : alk. paper)
1. Protein kinases. 2. Cell receptors. 3. Cellular signal transduction. I. Schlessinger, Joseph. II. Lemmon, Mark A.

 QP606.P76S54 2013
 572'.696--dc23

                              2013004466

10 9 8 7 6 5 4 3 2 1

Contents

Preface , ix

Contents

Preface

SINCE THEIR DISCOVERY IN THE LATE 1970s, the receptor tyrosine kinases (RTKs) have been the subject of intensive study in both the basic and clinical arenas. With 58 RTKs in the deduced human proteome, this family of receptors accounts for around two-thirds of all tyrosine kinases and $\sim$10% of all protein kinases. Recognition of their importance in disease, notably the insulin receptor in diabetes and several growth factor receptors in cancer, predated sophisticated mechanistic understanding. As time has passed, all (or almost all) RTK families have been associated with one disease or another—most predominantly cancers—and they have become major therapeutic targets. In this arena, RTKs have the advantage that approaches to their inhibition (which is typically desired in cancers) have benefited greatly from development of small molecule kinase inhibitors. Moreover, because RTKs are cell surface molecules they can be targeted with antibodies that may act as inhibitors or mediate immunotherapy.

In putting together this volume, we were acutely aware that the majority of literature on RTKs is driven by disease context, clinical evaluation, and efforts to identify and target cancer drivers. This can make an overview of RTKs as a class of molecules difficult to discern. One of our primary goals, therefore, was to draw together the extraordinary recent advances in mechanistic, structural, and biochemical understanding of RTKs. The collection also highlights several new areas of biology in which RTKs play crucial roles. Our goal was to emphasize conceptual developments and mechanistic insights into RTK activation and signaling that will stand the test of time.

In Section I, Joseph Schlessinger and Tony Hunter give conceptual and historical perspectives on the discovery of RTKs and of tyrosine phosphorylation itself, respectively, setting the stage for understanding of the distinct biologies and mechanisms of the individual RTK families. Chapters in Section II describe the conceptual basis for common signaling and regulatory mechanisms employed by most RTKs. These include "canonical" mechanisms through which RTK activation is linked to downstream signaling via SH2 domains (chapter by Tony Pawson and colleagues), as well as more receptor-specific (and emerging) mechanisms of nuclear signaling (chapters by Graham Carpenter and Hong-Jun Liao and by Gabriel Corfas and colleagues). Section II also considers trafficking and internalization of RTKs (chapters by Alexander Sorkin and Lai Kuan Goh and by Marta Miaczynska)—a crucial part of their regulation—and the related (and burgeoning) area of RTK ligand processing (chapter by Matthew Freeman and Colin Adrain). Finally, in this section, Boris Kholodenko and Natalia Volinsky provide a systems perspective on RTK signaling.

Our intention in the remaining sections of the book was—with this stage set—to provide a useful resource for comparing the characteristics of different RTK families. There are 20 different families of human RTKs, and few (if any) resources with which they can be compared and contrasted at the level of biology and mechanism. In the remaining 18 chapters, we have tried to be as comprehensive as possible in covering these families. Some families (such as the EGFR, insulin receptor, and VEGFR families) are represented in more than one chapter—where the volume and diversity of work warrants this. Others have a single chapter. Only a few families—where work remains quite preliminary or RTK function is ill-defined—are not represented. With this organization, we believe that this book represents a unique resource for comparing the features of different RTKs, which we hope will provide the inspiration for new approaches to therapeutic targeting and the discovery of new RTK biology. We are deeply indebted to all of the authors who wrote superb chapters in their areas of expertise. We also greatly appreciate their excellent scientific contributions to this field, and their patience in the process of completing this project.

We thank Richard Sever at CSHL Press for initiating this project and, in particular, Barbara Acosta, who has shown a level of patience and support that ought not to have been necessary and without whom this volume would never have been completed.

JOSEPH SCHLESSINGER
MARK A. LEMMON

In memory of Tony Pawson (1952–2013)

Receptor Tyrosine Kinases: Legacy of the First Two Decades

Joseph Schlessinger

Department of Pharmacology, Yale University School of Medicine, New Haven, Connecticut 06520

Correspondence: joseph.schlessinger@yale.edu

Receptor tyrosine kinases (RTKs) and their cellular signaling pathways play important roles in normal development and homeostasis. Aberrations in their activation or signaling leads to many pathologies, especially cancers, motivating the development of a variety of drugs that block RTK signaling that have been successfully applied for the treatment of many cancers. As the current field of RTKs and their signaling pathways are covered by a very large amount of literature, spread over half a century, I am focusing the scope of this review on seminal discoveries made before tyrosine phosphorylation was discovered, and on the early days of research into RTKs and their cellular signaling pathways. I review the history of the early days of research in the field of RTKs. I emphasize key early findings, which provided conceptual frameworks for addressing the questions of how RTKs are activated and how they regulate intracellular signaling pathways.

The family of cell-surface receptors designated receptor tyrosine kinases (RTK) received their name more that a decade after the same molecules were already known as the cell-surface receptors for insulin (insulin receptor), epidermal growth factor (EGFR), and many other growth factor receptors. Following the pioneering discoveries of nerve growth factor and epidermal growth factor (EGF; Levi-Montalcini and Booker 1960; Cohen 1962) and the establishment of the important roles of these two growth factors in the control of neuronal differentiation and cell proliferation in vivo and in vitro, it became clear that these cytokines bind specifically to cell-surface receptors. Insulin had already been discovered by this time, and had been applied successfully to treat diabetes patients since the early twentieth century. The resulting homogenous preparations of pure insulin enabled the quantitative characterization of insulin binding to its receptor on intact cells or to solubilized insulin receptor preparations using radiolabeled insulin (De Meyts et al. 1973). These studies greatly advanced understanding of the ligand binding characteristics of insulin receptor and, later on EGFR (Carpenter et al. 1975), including the establishment of negative cooperativity in insulin binding to its receptor expressed on the surface of living cells (De Meyts et al. 1973). Moreover, these studies shed important light on the dynamic nature of the cellular behavior of these receptors. The capacities of insulin receptor and EGFR to undergo ligand-dependent down-regulation and desensitization through receptor-mediated internalization and degradation (Carpenter and

Cite this article as *Cold Spring Harb Perspect Biol* doi: 10.1101/cshperspect.a008912

Cohen 1976; Gordon et al. 1978; Schlessinger et al. 1978a,b; Carpentier et al. 1979; Haigler et al. 1979) were also established well before the realization that growth factors receptors are endowed with intrinsic protein tyrosine kinase activities (Fig. 1).

Progress was also made in elucidating the role of growth factors in normal embryonic development, wound healing, and pathological conditions such as cancer. Early studies in the 1960s and 1970s showed that growth factors play an important role in oncogenesis induced by retroviruses and in the proliferation of tumor-derived cancer cells. Pioneering studies performed by Howard Temin (1966, 1967) showed that cancer cells need less insulin and serum growth factors for cell proliferation compared with normal cells, suggesting that cancer cells

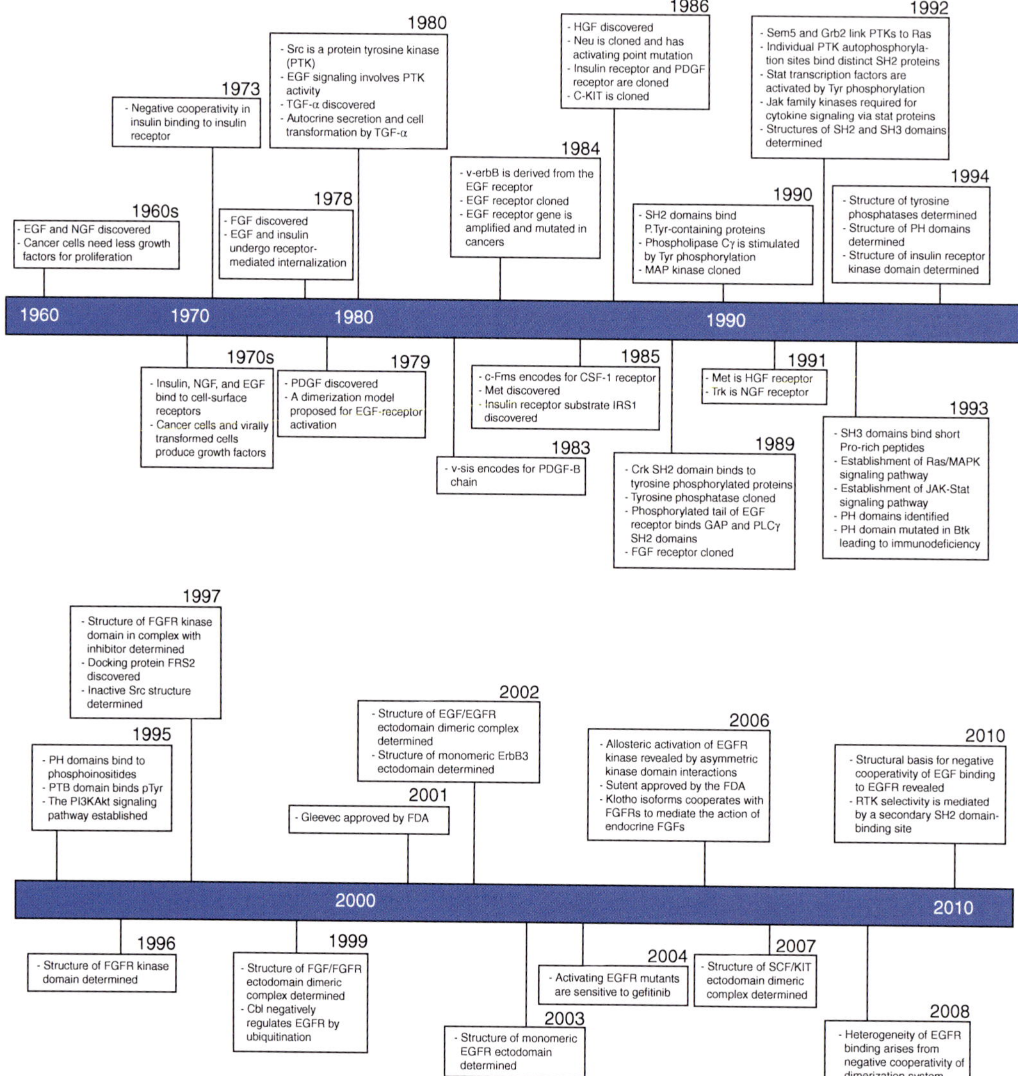

Figure 1. A time line of key findings during the history of RTKs, with emphasis on findings and discoveries that produced the conceptual framework in the development of the RTK field and its application for cancer therapy. References for the key findings are also presented in the text (Lee et al. 1985; Libermann et al. 1985; Margolis et al. 1990; Bottaro et al. 1991; Bae et al. 2009).

produce and use their own growth factors and/or use cellular processes that in normal cells are regulated by exogenously supplied growth factors; both predictions were subsequently confirmed. A variety of new polypeptide growth factors that stimulate cell proliferation by binding to receptors at the cell surface were subsequently discovered. Those include a growth factor isolated from human platelets designated platelet-derived growth factor (PDGF; Antoniades et al. 1979; Heldin et al. 1979), a growth factor isolated from bovine brain designated fibroblast growth factor (FGF; Gospodarowicz et al. 1978), a growth factor isolated from rat platelets that stimulates the proliferation of mature hepatocytes, designated hepatocyte growth factor (HGF; Nakamura et al. 1986). In addition to EGF, another growth factor that binds selectively to cells expressing EGFR was isolated from virally and chemically transformed cells, suggesting that this growth factor—designated transforming growth factor α—may play a role in oncogenesis by an autocrine mechanism (Roberts et al. 1980, 1982). This discovery provided further support to the earlier finding that transformation by murine and feline sarcoma viruses selectively interferes with EGF binding to EGFR in transformed cells (Todaro et al. 1976). Together with many other studies published since the 1980s, this work showed that growth factors and their receptors play numerous important roles during development and in many normal cellular processes as well as in pathologies such as cancer, diabetes, atherosclerosis, severe bone disorders, and tumor angiogenesis.

Visualization of dynamic cellular redistribution of ligand/receptor complexes, and rapid receptor-mediated internalization of growth factors such as insulin or EGF, led to the proposal that cell-surface receptors for these ligands may play a passive role in delivering them to intracellular compartments in which internalized EGF or insulin molecules exert their actions (Vigneri et al. 1978; Podlecki et al. 1986; Jiang and Schindler 1990). In other words, according to this hypothesis, the biological signals induced by insulin or EGF were thought to be mediated by binding of the ligands themselves to intracellular target(s) in the cytoplasm or nucleus, with the role of the cell-surface receptor being to act as a "carrier" that delivers them directly to these targets. An alternative hypothesis was that insulin or EGF activates their cognate receptors at the cell surface, which in turn stimulate the production of an intracellular second messenger molecule analogous to cAMP in signaling by the G-protein-activating β-adrenergic receptor. Indeed, several potential second messengers that are generated in cells on stimulation with insulin or other growth factors were proposed before (and even after) it became clear that insulin receptor, EGFR, and other RTKs are endowed with intrinsic tyrosine kinase activity (Larner et al. 1979; Das 1980; Saltiel and Cuatrecasas 1986).

A demonstration that anti-insulin receptor antibodies from the serum of certain diabetic patients could mimic cellular responses of insulin (Flier et al. 1977; Van Obberghen et al. 1979) provided the first conclusive answer to the question of whether the biological activity of growth factors is mediated directly or indirectly through their membrane receptors. This experiment ruled out the possibility that insulin receptor functions as a passive carrier that delivers insulin to an intracellular target to induce cellular responses. Studies showing that intact, bivalent antibodies against the insulin receptor can activate its signaling, whereas monovalent Fab fragments of the same antibodies cannot further argued that ligand-induced receptor dimerization or stimulation of a particular arrangement between two receptor molecules in a dimer can activate the insulin receptor (Kahn et al. 1978).

A similar conclusion was reached using certain monoclonal antibodies that bind to the extracellular region of EGFR and block ligand binding (Schreiber et al. 1981). Whereas intact antibodies were able to mimic EGF in stimulating a variety of EGF-like responses including cell proliferation, monovalent Fab fragments of the same monoclonal EGFR antibodies failed to do so—and acted instead as EGFR antagonists (Schreiber et al. 1981, 1983). These experiments provided strong evidence both that EGFR plays a crucial role in mediating EGF-induced cellular responses and that EGFR is activated by ligand-induced receptor dimerization (Schreiber 1981, 1983).

GROWTH FACTOR RECEPTORS ARE ENDOWED WITH PROTEIN TYROSINE KINASE ACTIVITY

Valuable insights into the mode of action of growth factors and their receptors were obtained from advancement in molecular characterization of the modes of action of retroviral oncogenic proteins. An interesting convergence of two different fields of biomedical research was taking place in the late seventies and early eighties, leading to the elucidation of mechanisms responsible for transformation of cells by a family of retroviral oncogenes. These studies led to the discovery of a common enzymatic activity used in controlling cell growth, and revealed the key features of how growth factors stimulate cell proliferation by binding to their cell-surface receptors.

The first key insight was provided by molecular characterization of the oncogene product, Src, from Rous sarcoma virus (Brugge and Erikson 1977). It was reported that Src is a protein kinase that phosphorylates primarily threonine residues (Collett and Erikson 1978). Subsequent experiments performed using lysates of EGF-stimulated human epidermoid carcinoma cells A431, which express very high levels of EGFR (more than two million copies) on their cell surface, similarly revealed EGF stimulation of threonine phosphorylation (Carpenter et al. 1978). A more thorough examination by Hunter and Sefton (1980) of the phosphorylation induced by the Src oncogenic product revealed that this was a unique kinase that phosphorylates tyrosines rather than serines or threonines. Subsequent studies revealed that EGFR stimulation also leads to phosphorylation of tyrosines, the mistake arising because phosphotyrosine (P-Tyr) and phosphothreonine (P-Thr) comigrate in the electrophoresis experiments performed at pH 1.9 (Ushiro and Cohen 1980). It was also shown that the Abelson Leukemia virus protein, Abl, promotes tyrosine phosphorylation (Witte et al. 1980). Moreover, subsequent studies showed that treatment of cultured cells with PDGF or insulin-induced tyrosine phosphorylation. The fact that this was dependent on expression of platelet derived

growth factor receptor (PDGFR) or insulin receptor, respectively, in the stimulated cells suggested that insulin receptor and PDGFR possess tyrosine kinase activity (Ek et al. 1982; Kasuga et al. 1982). Becasue tyrosine phosphorylation was seen with the oncogenic proteins Src and Abl, as well as with growth factor-stimulated surface receptors such as EGFR, PDGFR, and insulin receptor strongly suggested that it plays an important role in normally regulated cell proliferation as well as in aberrantly stimulated proliferation of cancer cells, and that these proteins all activate a common enzymatic activity (i.e., protein tyrosine kinases).

Biochemical analysis of EGFR expressed in A-431 or other cultured cells showed that EGFR is a glycosylated, transmembrane receptor of molecular weight of 170 kDa, which is comprised of a large extracellular ligand binding region connected via a transmembrane domain to a cytoplasmic region that becomes phosphorylated on tyrosines on stimulation with EGF. However, in the absence of any direct information about the primary structure of EGFR and other members of the growth factor receptor family, it was not obvious whether the tyrosine kinase activity is encoded within the same gene that endows EGFR with specific EGF binding or whether the observed tyrosine kinase activity arises from a closely associated protein encoded by a separate gene. Other EGFR-associated activities were also reported. For example, a topoisomerase activity associated with purified EGFR—which leads to ATP-stimulated DNA-nicking activity—was proposed to represent an intrinsic activity of EGFR or to become associated with EGFR during membrane solubilization (Miskimins et al. 1983; Mroczkowski et al. 1984; Basu et al. 1985). Further characterization of EGFR isolated from A-431 cells showed that the isolated EGFR can be recognized specifically by antibodies generated against certain synthetic peptides from the tyrosine kinase domain of Src, suggesting that the cytoplasmic region of EGFR may indeed contain a tyrosine kinase domain (Lax et al. 1984).

With the partial sequencing of peptides derived from human EGFR revealing a link between the retroviral oncogene v-erbB and the

Cite this article as *Cold Spring Harb Perspect Biol* doi: 10.1101/cshperspect.a008912

EGFR (Downward et al. 1984), and subsequent cloning of human EGFR cDNA that allowed the receptor's amino acid sequence to be deduced (Ullrich et al. 1984), the molecular nature of EGFR became clear. These findings showed that the EGFR contains a large extracellular ligand binding domain ($\sim$620 aa), a single transmembrane helix, and a large ($\sim$540 aa) cytoplasmic region that contains a tyrosine kinase domain. The tyrosine kinase domain is preceded by a short juxtamembrane region of $\sim$30 aa that was later shown to play an important regulatory role, and is followed by a long carboxy-terminal tail of $\sim$230 aa that contains most of the tyrosine autophosphorylation sites (Downward et al. 1984; Ullrich et al. 1984). This molecular characterization also showed that the v-erbB oncogene of the avian erthryoblastosis virus encodes a truncated avian EGFR that has lost most of the extracellular ligand-binding region as well as part of the carboxy-terminal tail containing tyrosine autophosphorylation sites. Importantly, these studies revealed that the retroviral oncogene v-erbB induces cell transformation by encoding an EGF-independent variant of avian EGFR that has constitutively stimulated tyrosine kinase activity (Downward et al. 1984; Ullrich et al. 1984; Kris et al. 1985).

This finding added to earlier studies that had revealed the first link between a retroviral oncogene and a growth factor, namely, the demonstration that the *v-sis* oncogene of simian sarcoma virus encodes the growth factor PDGF (Doolittle et al. 1983; Waterfield et al 1983). It was concluded that this virus had captured/transduced the gene for a growth factor and the product of the retroviral *v-sis* oncogene induces cell transformation by autocrine stimulation of PDGFRs expressed on the surface of the same cells that secrete PDGF molecules (Doolittle et al. 1983; Waterfield et al. 1983).

RTKs ARE ACTIVATED BY DIMERIZATION

As previously mentioned, the central role of receptor dimerization as a mechanism of activation of insulin receptor and EGFR was revealed before these two receptors were defined as RTKs. A wealth of information accumulated during the late 1970s and early 1980s demonstrating that insulin or EGF stimulation leads to the clustering of insulin receptor or EGFR, respectively, on the cell surface. Moreover, the cellular responses of EGF or insulin could be stimulated by bivalent, but not by monovalent, receptor antibodies. Biochemical studies showed that, before ligand activation, EGFR is expressed on the cell surface as a monomer, whereas insulin receptor and the closely related IGF1 receptor are both expressed on the cell surface as preexisting disulfide linked dimers. It was also shown that the homodimeric structure of the inactive insulin receptor is essential for the negative cooperativity observed in binding of insulin to its receptor (De Meyts 2008).

The establishment of the primary structures of EGFR, PDGFR, and other RTKs showed that the ligand-binding region of RTKs is separated from the cytoplasmic region by a single transmembrane helix. The identification of the structural topology of EGFR and other monomeric RTKs triggered an intense debate about how ligand binding to an extracellular region transfers a stimulatory cue across the cell membrane to stimulate tyrosine kinase activity in the cytoplasmic region.

The two main mechanisms proposed for how EGF binding to the extracellular region stimulates the tyrosine kinase activity in the EGFR cytoplasmic region are presented in Figure 1. The top panel in Fig. 2A depicts an "intramolecular mechanism" in which EGF induces a conformational change in the extracellular region that is transmitted via the transmembrane helix to activate the tyrosine kinase domain in the cytoplasmic region. Accordingly, ligand-stimulated EGFR monomers transmit a structural change across the single transmembrane helix to activate the catalytic tyrosine kinase activity in *cis* and to mediate autophosphorylation on multiple tyrosine residues. In other words, according to this model, both tyrosine kinase activation and autophosphorylation are mediated by an intramolecular process (Bertics et al. 1985; Gill et al. 1987; Koland and Cerione 1988; Northwood and Davis 1988).

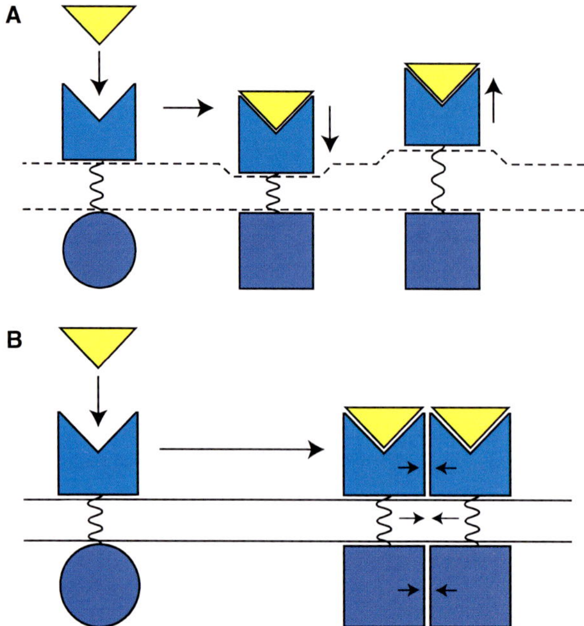

Figure 2. Two contrasting views during the early days for how EGF binding to the extracellular region of EGFR stimulates the tyrosine kinase domain in the cytoplasm. (*A*) An "intramolecular" mechanism. Ligand-stimulated EGFR monomers transmit a conformational change through the transmembrane helix to activate the tyrosine kinase domain and mediate autophosphorylation by an intramolecular process. (*B*) An "intermolecular" mechanism. Ligand binding stimulates lateral contacts between a pair of EGFR molecules, resulting in EGFR dimerization mediated by interactions between extracellular regions, transmembrane domains, and cytoplasmic regions resulting in stimulation of tyrosine kinase activity and autophosphorylation by an intermolecular process.

The bottom panel (Fig. 2B) depicts an "intermolecular mechanism" for information transfer from the extracellular region to the cytoplasmic region, in which ligand binding induces lateral contacts between two receptors, resulting in their dimerization, which is responsible for stimulation of the tyrosine kinase activity and for autophosphorylation. Accordingly, both tyrosine kinase activation (Kashles et al. 1991) and tyrosine autophosphorylation are mediated in *trans* by an intermolecular process (Honegger et al. 1989).

It is now well established that dimerization or formation of larger oligomeric structures is required for activation of most, if not all, RTKs (Lemmon and Schlessinger 2010). Numerous cellular studies using advanced microscopic methods, as well as biochemical studies in cells and in vitro have shown that EGF promotes dimerization of its receptor. However, the preexisting insulin receptor dimer provides an example in which ligand-induced dimerization per se is not required for signaling. Insulin activates its receptor by promoting an allosteric transition within this preexisting receptor homodimer (De Meyts 2008)—effectively stabilizing an active configuration of the dimeric receptor. An allosteric/dimerization model for ligand-induced stimulation of EGFR is shown in Figure 3A, which is modified from the original model described by Yarden and Schlessinger (1987). The model proposes that monomeric receptors are in equilibrium with receptor dimers. It is proposed that the monomeric receptor shows weak ligand-binding affinity and low or inactive tyrosine kinase activity. The dimeric receptor shows high ligand-binding affinity and elevated tyrosine kinase activity. K_D is the dissociation constant for EGF binding to EGFR monomers and K_R is the dissociation constant for EGF

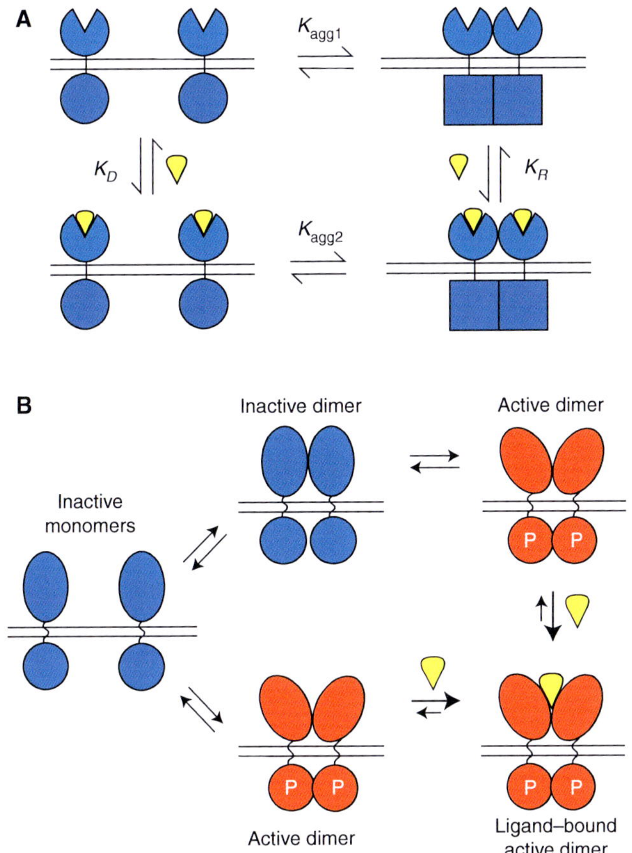

Figure 3. Models for ligand stimulation of RTK dimerization and activation. The scheme presented in (*A*) is based on data from Yarden and Schlessinger (1987) with minor modifications. The scheme presented in (*B*) is based on data presented in Schlessinger et al. (2000). This is a generalized allosteric/dimerization model for RTK activation that applies equally to EGFR and insulin receptor. Full details are presented in the body of the article.

binding to EGFR dimers. K_{agg1} is the dimerization constant of EGFR monomers and K_{agg2} is the dimerization constant for EGFR dimers. It was previously shown that $K_{agg1}/K_{agg2} = K_R^2/K_D^2$ (Yarden and Schlessinger 1987). According to this model, the enhanced binding affinity of EGF to dimeric EGFR confers the dimeric state on EGFR resulting in stimulated tyrosine kinase activity. For EGFR and many other RTKs, before ligand-binding monomers predominate—although this depends on the receptor expression level (Chung et al. 2010). For the insulin receptor, this equilibrium is shifted almost entirely to dimer even in the absence of ligand. The fact that the dimeric insulin receptor is not active until

it binds ligand argues that dimerization—although required—is not necessarily sufficient for receptor activation. Instead, it is likely that "inactive" dimers are in equilibrium with "active" dimers as shown in Figure 3B. Ligand binding to an active dimer will stabilize it, whereas binding to an inactive dimer will induce a conformational change to promote activation (as seen for insulin and its receptor). In both cases, this ligand binding will pull the equilibrium to the right in Figure 3—constituting ligand-induced dimerization. The original model from 1987 assumed that the ligand-binding sites in the dimeric EGFR do not interact. More recent studies suggest that interactions between the ligand-binding sites are responsible for the negative cooperativity and for the curvilinear Scatchard plots of EGF binding to EGFR in living cells (Alvarado et al. 2010; Pike 2012)—as is also well documented for the insulin receptor (De Meyts 2008). A generalized allosteric/dimerization model can be drawn for RTK activation that applies equally to EGFR and insulin receptor described by Schlessinger (2000) is presented in Figure 3B.

Although ligand stimulation of receptor by specific dimerization is a universal mechanism for RTK activation, different ligands use different strategies for stabilizing structural changes that are required for stimulation of tyrosine kinase activity.

Dimeric RTKs Ligands

The ligands of all type-III RTKs, including PDGFR, KIT, and other members of the family, are dimeric molecules. Structural studies have shown that binding of stem cell factor to the ligand-binding region in the ectodomain effectively "crosslinks" two KIT molecules, resulting in the formation of homotypic contacts between the membrane proximal regions of the ectodomains of receptor pairs (Yuzawa et al. 2007).

Ligand-Induced Receptor-Mediated Dimerization

Dimerization of EGFR and other members of the family is entirely mediated by receptor-mediated interactions. Ligand binding to individual protomers induces a conformational change in the extracellular region of EGFR that exposes a dimerization site that is occluded in the absence of ligand by autoinhibitory intramolecular interactions (Burgess et al. 2003).

Ligand and Accessory Molecule-Mediated Dimerization

FGF acts in concert with heparin sulfate proteoglycans to crosslink two fibroblast growth factor receptors (FGFRs). Direct contacts between FGF and heparin, FGF and FGFR, and heparin with FGFR mediate the formation of a stable activated (2:2:2) FGFR-containing complex (Schlessinger et al. 2000).

Structural analysis of the free or ligand-occupied extracellular regions of EGFR, KIT, FGFR, and other RTKs as well as the analysis of the catalytic kinase domains of the insulin receptor, EGFR, and other RTKs provided valuable insights into the allosteric nature of the interactions that regulate them (Lemmon and Schlessinger 2010). Examples include asymmetric interactions between the tyrosine kinase domains of EGFR that provide a direct allosteric mechanism for tyrosine kinase activation (Zhang et al. 2006), asymmetric arrangement and partial ligand occupation of dimers of the *Drosophila* EGFR extracellular region, which provide a mechanism for negative cooperativity of EGF binding to EGFR (Alvarado et al. 2010; Pike 2012), asymmetric contacts between the tyrosine kinase domains of FGFR providing a potential mechanism for the ordered tyrosine autophosphorylation of FGFR (Bae et al. 2010), and structural insights into the role played by juxtamembrane domain of EGFR in the control of tyrosine kinase activity (Jura et al. 2009; Red Brewer et al. 2009).

HOW TYROSINE PHOSPHORYLATION ACTIVATES CELLULAR SIGNALING PATHWAYS

The mid 1990s saw a dramatic convergence of experimental insights obtained from biochemical, structural, and genetic studies that provided a molecular view of both how tyrosine phosphorylation regulates RTK activity and how tyrosine phosphorylation activates multiple signaling pathways to relay information from the cell membrane to the nucleus and other intracellular compartments. Intriguingly, regulation of the tyrosine kinases themselves (through autophosphorylation) is one of the most important roles of tyrosine phosphorylation of RTKs and nonreceptor protein tyrosine kinases (Hubbard et al. 1994). All tyrosine kinases contain 1–3 tyrosines in the activation loop of the catalytic core that, in most known cases, are the first tyrosine(s) to be phosphorylated following stimulation. Enzymatic analyses and

     Cite this article as *Cold Spring Harb Perspect Biol* doi: 10.1101/cshperspect.a008912

structural studies have shown that autophosphorylation of these tyrosines maintains the activation loop in an "open" configuration permitting ATP binding to the nucleotide binding site and (by stabilizing the "active" conformation of the kinase) enabling transfer of the γ phosphate to tyrosines of bound substrate molecules (Hubbard et al. 1994; Lemmon and Schlessinger 2010).

Inspection of the pattern of tyrosine phosphorylated cellular proteins following growth factor stimulation showed that, in most cases, the most strongly phosphorylated protein is the receptor of the stimulating growth factor. Most of the tyrosines that become autophosphorylated on RTKs are located in noncatalytic regions of the cytoplasmic domain, such as the juxtamembrane region (e.g., PDGFR and KIT), the kinase insert region (e.g., FGFR and KIT), or the carboxy-terminal tail. EGFR and other members of the ErbB family of RTKs become phosphorylated on multiple tyrosines located primarily in the carboxy-terminal tail of the receptor. One of the first clues as to how this autophosphorylation serves to propagate RTK signaling came from the finding that one substrate molecule that becomes phosphorylated in response to EGF stimulation, phospholipase Cγ (PLCγ), also forms a stable complex with the activated receptor (Margolis et al. 1989; Meisenhelder et al. 1989). It was subsequently shown that complex formation between activated EGFR and PLCγ is mediated by the Src homology 2 (SH2) domain of PLCγ that binds directly to a P-Tyr site in the carboxy-terminal tail of EGFR (Margolis et al. 1990). The SH2 domain is a small protein module of ∼100 aa, initially discovered as regulatory region in Src kinases responsible for maintaining tyrosine kinase activity of Src kinases in an inactive configuration by mediating intramolecular autoinhibitory interactions (Sadowski et al. 1986). Src, PLCγ, and many signaling proteins contain an additional small protein module designated the SH3 (for Src homology 3) domain, which binds specifically to short, proline-rich regions in target proteins (Pawson 1995, 2004). Interestingly, a viral oncogenic protein, designated Crk, does not contain any enzymatic activity, but only a single

SH2 domain followed by two SH3 domains (Matsuda et al. 1990). It was shown that in virally transformed cells, Crk forms complexes with several tyrosine-phosphorylated proteins (Matsuda et al. 1990). Importantly, it was shown that the SH2 domains of Crk bind directly to P-Tyr-containing proteins and peptides, suggesting that the biological role of SH2 domain of signaling molecules is to recognize P-Tyr sites within specific sequence contexts (Mayer and Hanafusa 1990; Matsuda et al. 1991; Pawson 2004; Waksman and Kuriyan 2004).

On the basis of these and other studies that are described in several review articles (Pawson 1995, 2004; Lemmon and Schlessinger 2010), it was established that a variety of SH2 domain-containing signaling molecules, including enzymes such as PLCγ, the tyrosine phosphatase Shp2, and the GTPase-activating protein GAP for Ras, form complexes with activated EGFR and other RTKs by binding to specific P-Tyr sites in the receptor molecule. Complex formation with an activated RTK leads to efficient tyrosine phosphorylation of PLCγ, a step necessary for stimulation of its phospholipase activity (Lemmon and Schlessinger 2010). Additional proteins that form complexes with activated receptors include the small adaptor proteins Crk, Grb2, and Nck. These adapter proteins are composed entirely of SH2 and SH3 domains with architectures of SH2-SH3-SH3 for Crk, SH3-SH2-SH3 for Grb2, and SH3-SH3-SH3-SH2 for Nck (Pawson 1995). Biochemical studies together with genetic screening in *Drosophila* and Nematodes have shown that Grb2 provides a link between RTKs and the Ras/Map kinase (MAPK) signaling pathway. Grb2 uses its SH2 domain for binding to the phosphorylated tail of activated EGFR and its two SH3 domains for recruitment of the nucleotide exchange factor, SOS, to provide a direct link between EGFR stimulation and the GTPase Ras (Pawson 2004; Lemmon and Schlessinger 2010). Activated Ras in turn stimulates a kinase cascade, ultimately leading to MAP kinase stimulation. Another example is the regulatory subunit, p85, of PI-3 kinase, which uses its SH2 domain to form a complex with activated RTKs leading to PI-3K activation (Pawson 1995, 2004). It was concluded that RTKs function not only as enzymes but also as platforms for recruitment of a variety of signaling molecules that stimulate the activities of a variety of intracellular signaling pathways including the RAS/MAPK signaling pathway, the PI-3K/Akt signaling pathway, and the Jak2/STAT signaling pathways, among others.

Another mechanism for recruitment and activation of signaling molecules by activated RTKs involves docking proteins such as insulin receptor substrate 1 (IRS1) and other member of the IRS family and FGF receptor substrate 2 (FRS2). It was shown that docking proteins bind via their PTB domains to the insulin receptor or FGFRs, respectively, and become phosphorylated on numerous tyrosine. The P-Tyr sites of docking proteins provide a platform for the recruitment and activation of an additional complement of signaling molecules that bind to the P-Tyr sites of the docking proteins via their own SH2 domains (Lemmon and Schlessinger 2010), and are responsible for mediating many of the known insulin-induced cellular responses (see Boucher et al. 2014). The signaling pathways that are initiated at the cell membrane by activated RTKs and their specifically associated docking proteins are responsible for regulating many critical cellular processes essential for cell proliferation, cell differentiation, cell survival, and metabolism during development and normal homeostasis. Dysfunction in the activation or regulation of RTK stimulated signaling networks plays a critical role in a variety of human pathologies.

ACKNOWLEDGMENTS

I thank Mark Lemmon for his excellent suggestions for this work.

REFERENCES

*Reference is also in this collection.

Alvarado D, Klein DE, Lemmon MA. 2010. Structural basis for the negative cooperativity in growth factor binding to an EGF receptor. *Cell* **142:** 568–579.

Antoniades HN, Scher CD, Stiles CD. 1979. Purification of human platelet-derived growth factor. *Proc Natl Acad Sci* **76:** 1809–1813.

Bae JH, Boggon TJ, Tomé F, Mandiyan V, Lax I, Schlessinger J. 2010. Asymmetric receptor contact is required for tyrosine autophosphorylation of fibroblast growth factor receptor in living cells. *Proc Natl Acad Sci* **107:** 2866–2871.

Bae JH, Lew ED, Yuzawa S, Tomé F, Lax I, Schlessinger J. 2009. The selectivity of receptor tyrosine kinase signaling is controlled by a secondary SH2 domain binding site. *Cell* **138:** 514–524.

Basu M, Frick K, Sen-Majumdar A, Scher CD, Das M. 1985. EGF receptor-associated DNA-nicking activity is due to a Mr-100,000 dissociable protein. *Nature* **316:** 640–641.

Bertics PJ, Weber W, Cochet C, Gill GN. 1985. Regulation of the epidermal growth factor by phosphorylation. *J Cell Biochem* **29:** 195–208.

Bottaro DP, Rubin JS, Faletto DL, Chan AM, Kmiecik TE, Vande Woude GF, Aaronson SA. 1991. Identification of the hepatocyte growth factor receptor as the c-met proto-oncogene product. *Science* **251:** 802–804.

* Boucher J, Kleinridders A, Kahn CR. 2014. Insulin receptor signaling in normal and insulin-resistant states. *Cold Spring Harb Perspect Biol* doi: 10.1101/cshperspect.a009191.

Brugge JS, Erikson RL. 1977. Identification of a transformation-specific antigen induced by an avian sarcoma virus. *Nature* **269:** 346–348.

Burgess AW, Cho HS, Eigenbrot C, Ferguson KM, Garrett TP, Leahy DJ, Lemmon MA, Sliwkowski MX, Ward CW, Yokoyama S. 2003. An open-and-shut case? Recent insights into the activation of EGF/ErbB receptors. *Mol Cell* **12:** 541–542.

Carpenter G, Cohen S. 1976. ^{125}I-labeled human epidermal growth factor. Binding, internalization, and degradation in human fibroblasts. *J Cell Biol* **71:** 159–171.

Carpenter G, King L Jr, Cohen S. 1978. Epidermal growth factor stimulates phosphorylation in membrane preparations in vitro. *Nature* **276:** 409–410.

Carpenter G, Lembach KJ, Morrison MM, Cohen S. 1975. Characterization of the binding of ^{125}I-labeled epidermal growth factor to human fibroblasts. *J Biol Chem* **250:** 4297–4304.

Carpentier JL, Gorden P, Barazzone P, Freychet P, Le Cam A, Orci L. 1979. Intracellular localization of ^{125}I-labeled insulin in hepatocytes from intact rat liver. *Proc Natl Acad Sci* **76:** 2803–2807.

Chung I, Akita R, Vandlen R, Toomre D, Schlessinger J, Mellman I. 2010. Spatial control of EGF receptor activation by reversible dimerization on living cells. *Nature* **464:** 783–787.

Cohen S. 1962. Isolation of a mouse submaxillary gland protein accelerating incisor eruption and eyelid opening in the new-born animal. *J Biol Chem* **237:** 1555–1562.

Collett MS, Erikson RL. 1978. Protein kinase activity associated with the avian sarcoma virus src gene product. *Proc Natl Acad Sci* **75:** 2021–2024.

Das M. 1980. Mitogenic hormone-induced intracellular message: Assay and partial characterization of an activator of DNA replication induced by epidermal growth factor. *Proc Natl Acad Sci* **77:** 112–116.

De Meyts P. 2008. The insulin receptor: A prototype for dimeric, allosteric membrane receptor? *Trends Biochem Sci* **33**: 376–384.

De Meyts P, Roth J, Neville DM Jr, Gavin JR III, Lesniak MA. 1973. Insulin interactions with its receptors: Experimental evidence for negative cooperativity. *Biochem Biophys Res Commun* **55**: 154–161.

Doolittle RF, Hunkapiller MW, Hood LE, Devare SG, Robbins KC, Aaronson SA, Antoniades HN. 1983. Simian sarcoma virus *onc* gene, v-*sis*, is derived from the gene (or genes) encoding a platelet-derived growth factor. *Science* **221**: 275–277.

Downward J, Yarden Y, Mayes E, Scrace G, Totty N, Stockwell P, Ullrich A, Schlessinger J, Waterfield MD. 1984. Close similarity of epidermal growth factor receptor and v-*erb*-B oncogene sequences. *Nature* **307**: 521–527.

Ek B, Westermark B, Wasteson A, Heldin CH. 1982. Stimulation of tyrosine-specific phosphorylation by platelet-derived growth factor. *Nature* **295**: 419–420.

Flier JS, Kahn CR, Jarrett DB, Roth J. 1977. Autoantibodies to the insulin receptor. Effect of the insulin-receptor interaction in IM-9 lymphocytes. *J Clin Invest* **60**: 784–794.

Gill GN, Santon FB, Bertics PJ. 1987. Regulatory features of the epidermal growth factor receptor. *J Cell Physiol* **5**: 35–41.

Gordon P, Carpentier JL, Freychet P, LeCam A, Orci L. 1978. Intracellular translocation of iodine-125 labeled insulin: Direct demonstration in isolated hepatocytes. *Science* **200**: 782–785.

Gospodarowicz D, Bialecki H, Greenburg G. 1978. Purification of the fibroblast growth activity from bovine brain. *J Biol Chem* **253**: 3736–3743.

Haigler HT, McKanna JA, Cohen S. 1979. Direct visualization of the binding and internalization of a ferritin conjugate of epidermal growth factor in human carcinoma cells A-431. *J Cell Biol* **81**: 382–395.

Heldin CH, Westermark B, Wasteson A. 1979. Platelet-derived growth factor: Purification and partial characterization. *Proc Natl Acad Sci* **76**: 3722–3726.

Honegger AM, Kris RM, Ullrich A, Schlessinger J. 1989. Evidence that autophosphorylation of solubilized receptors for epidermal growth factor is mediated by intermolecular cross-phosphorylation. *Proc Natl Acad Sci* **86**: 925–929.

Hubbard SR, Wei L, Ellis L, Hendrickson WA. 1994. Crystal structure of the tyrosine kinase domain of the human insulin receptor. *Nature* **372**: 746–754.

Hunter T, Sefton BM. 1980. Transforming gene product of Rous sarcoma virus phosphorylates tyrosine. *Proc Natl Acad Sci* **77**: 1311–1315.

Jiang LW, Schindler M. 1990. Nucleocytoplasmic transport is enhanced concomitant with nuclear accumulation of epidermal growth factor (EGF) binding activity in both 3T3-1 and EGF receptor reconstituted NR-6 fibroblasts. *J Cell Biol* **110**: 559–568.

Jura N, Endres NF, Engel K, Deindl S, Das R, Lamers MH, Wemmer DE, Zheng X, Kuriyan J. 2009. Mechanism for activation of the EGF receptor catalytic domain by the juxtamembrane segment. *Cell* **137**: 1293–1307.

Kahn CR, Baird KL, Jarrett DB, Flier JS. 1978. Direct demonstration that receptor crosslinking or aggregation is important to insulin action. *Proc Natl Acad Sci* **75**: 4209–4213.

Kashles O, Yarden Y, Fischer R, Ullrich A, Schlessinger J. 1991. A dominant negative mutation suppresses the function of normal epidermal growth factors by heterodimerization. *Mol Cell Biol* **11**: 1454–1463.

Kasuga M, Zick Y, Blithe DL, Crettaz M, Kahn CR. 1982. Insulin stimulates tyrosine phosphorylation of the insulin receptor in a cell-free system. *Nature* **298**: 667–669.

Klein R, Jing SQ, Nanduri V, O'Rourke E, Barbacid M. 1991. The trk proto-oncogene encodes a receptor for nerve growth factor. *Cell* **65**: 189–197.

Koland JG, Cerione RA. 1988. Growth factor control of epidermal growth factor receptor kinase activity via an intramolecular mechanism. *J Biol Chem* **263**: 2230–2237.

Kris RM, Lax I, Gullick W, Waterfield MD, Ullrich A, Fridkin M, Schlessinger J. 1985. Antibodies against a synthetic peptide as a probe for the kinase activity of the avian EGF receptor and v-erbB protein. *Cell* **40**: 619–625.

Larner J, Galasko G, Cheng K, DePaoli-Roach AA, Huang L, Daggy P, Kellogg J. 1979. Generation by insulin of a chemical mediator that controls protein phosphorylation and dephosphorylation. *Science* **206**: 1408–1410.

Lax I, Bar-Eli M, Yarden Y, Libermann TA, Schlessinger J. 1984. Antibodies of two defined regions of the transforming protein pp60src interact specifically with the epidermal growth factor receptor kinase system. *Proc Natl Acad Sci* **81**: 5911–5915.

Lee DC, Rose TM, Webb NR, Todaro GJ. 1985. Cloning and sequence analysis of a cDNA for rat transformation growth factor-α. *Nature* **313**: 489–491.

Lemmon MA, Schlessinger J. 2010. Cell signaling by receptor tyrosine kinases. *Cell* **141**: 1117–1134.

Levi-Montalcini R, Booker B. 1960. Excessive growth of sympathetic ganglia evoked by a protein isolated from mouse salivary gland. *Proc Natl Acad Sci* **46**: 373–384.

Libermann TA, Nusbaum HR, Razon N, Kris R, Lax I, Soreg H, Whittle N, Waterfield MD, Ullrich A, Schlessinger J. 1985. Amplification, enhanced expression and possible rearrangement of EGF receptor gene in primary human brain tumours of glial origin. *Nature* **313**: 144–147.

Margolis B, Bellot F, Honegger AM, Ullrich A, Schlessinger J, Zilberstein A. 1990. Tyrosine kinase activity is essential for the association of phospholipase C-γ with the epidermal growth factor receptor. *Mol Cell Biol* **10**: 435–441.

Margolis B, Li N, Kock A, Mohammadi M, Hurwitz DR, Zilberstein A, Ullrich A, Pawson T, Schlessinger J. 1990. The tyrosine phosphorylated carboxyterminus of the EGF receptor is a binding site for GAP and PLC-γ. *EMBO J* **9**: 4375–4380.

Margolis B, Rhee SG, Felder S, Mervic M, Lyall R, Levitzki A, Ullrich A, Zilberstein A, Schlessinger J. 1989. EGF induces tyrosine phosphorylation of phospholipase C-II: A potential mechanism for EGF receptor signaling. *Cell* **57**: 1101–1107.

Matsuda M, Mayer BJ, Fukui Y, Hanafusa H. 1990. Binding of transforming protein, P47$^{gag-crk}$, to a broad range of phosphotyrosine-containing proteins. *Science* **248**: 1537–1539.

Matsuda M, Mayer BJ, Hanafusa H. 1991. Identification of domains of the v-*crk* oncogene product sufficient for association with phosphotyrosine-containing proteins. *Mol Cell Biol* **11:** 1607–1613.

Mayer BJ, Hanafusa H. 1990. Association of the v-*crk* oncogene product with phosphotyrosine-containing proteins and protein kinase activity. *Proc Natl Acad Sci* **87:** 2642–2648.

Meisenhelder J, Suh PG, Rhee SG, Hunter T. 1989. Phospholipase C-γ is a substrate for the PDGF and EGF receptor protein-tyrosine kinases in vivo and in vitro. *Cell* **57:** 1109–1122.

Miskimins R, Miskimins WK, Bernstein H, Shimizu N. 1983. Epidermal growth factor-induced topoisomerase(s). Intracellular translocation and relation to DNA synthesis. *Exp Cell Res* **146:** 53–62.

Mroczkowski B, Mosig G, Cohen S. 1984. ATP-stimulated interaction between epidermal growth factor receptor and supercoiled DNA. *Nature* **309:** 270–273.

Nakamura T, Teramoto H, Ichihara A. 1986. Purification and characterization of a growth factor from rat platelets for mature parenchymal hepatocytes in primary cultures. *Proc Natl Acad Sci* **83:** 6489–6493.

Northwood IC, Davis RJ. 1988. Activation of the epidermal growth factor receptor tyrosine protein kinase in the absence of receptor oligermerization. *J Biol Chem* **263:** 7450–7453.

Pawson T. 1995. Protein modules and signaling networks. *Nature* **373:** 573–580.

Pawson T. 2004. Specificity in signal transduction: From phosphotyrosine-SH2 domain interactions to complex cellular systems. *Cell* **116:** 191–203.

Pike LJ. 2012. Negative co-operativity in the EGF receptor. *Biochem Soc Trans* **40:** 15–19.

Podlecki DA, Smith RM, Kao M, Tsai P, Huecksteadt T, Brandenburg D, Lasher RS, Jarett L, Olefsky JM. 1986. Nuclear translocation of the insulin receptor. *J Biol Chem* **262:** 3362–3368.

Red Brewer M, Choi SH, Alvarado D, Moravcevic K, Pozzi A, Lemmon MA, Carpenter G. 2009. The juxtamembrane region of the EGF receptor functions as an activation domain. *Mol Cell* **34:** 641–651.

Roberts AB, Anzano MA, Lamb LC, Smith JM, Frolik CA, Marquadt H, Todaro GJ, Sporn MB. 1982. Isolation from murine sarcoma cells of novel transforming growth factors potentiated by EGF. *Nature* **295:** 417–419.

Roberts AB, Lamb LC, Newton DL, Sporn MB, De Larco JE, Todaro GJ. 1980. Transforming growth factors: Isolation of polypeptides from virally and chemically transformed cells by acid/ethanol extraction. *Proc Natl Acad Sci* **77:** 3494–3498.

Sadowski I, Stone JC, Pawson T. 1986. A noncatalytic domain conserved among cytoplasmic protein-tyrosine kinases modifies the kinase function and transforming activity of Fujinami sarcoma virus P130$^{gag\text{-}fps}$. *Mol Cell Biol* **6:** 4396–4408.

Saltiel AR, Cuatrecasas P. 1986. Insulin stimulates the generation from hepatic plasma membranes of modulators derived from an inositol glycolipid. *Proc Natl Acad Sci* **83:** 5793–5797.

Schlessinger J. 2000. Cell signaling by receptor tyrosine kinases. *Cell* **103:** 211–225.

Schlessinger J, Plotnikov AN, Ibrahimi OA, Eliseenkova AV, Yeh BK, Yayon A, Linhardt RJ, Mohammadi M. 2000. Crystal structure of a ternary FGF-FGFR-heparin complex reveals a dual role for heparin in FGFR binding and dimerization. *Mol Cell* **6:** 743–750.

Schlessinger J, Shechter Y, Cuatrecasas P, Willingham MC, Pastan I. 1978a. Quantitative determination of the lateral diffusion coefficients of the hormone-receptor complexes of insulin and epidermal growth factor on the plasma membrane of cultured fibroblasts. *Proc Natl Acad Sci* **75:** 5353–5357.

Schlessinger J, Shechter Y, Willingham MC, Pastan I. 1978b. Direct visualization of binding, aggregation and internalization of insulin and epidermal growth factor on living fibroblastic cells. *Proc Natl Acad Sci* **75:** 2659–2663.

Schreiber AB, Lax I, Yarden Y, Eshhar Z, Schlessinger J. 1981. Monoclonal antibodies against receptor for epidermal growth factor induce early and delayed effects of epidermal growth factor. *Proc Natl Acad Sci* **78:** 7535–7539.

Schreiber AB, Libermann TA, Lax I, Yarden Y, Schlessinger J. 1983. Biological role of epidermal growth factor-receptor clustering. Investigation with monoclonal anti-receptor antibodies. *J Biol Chem* **258:** 846–853.

Temin HM. 1966. Studies on carcinogenesis by avian sarcoma viruses. III: The differential effect of serum and polyanions on multiplication of uninfected and converted cells. *J Natl Cancer Inst* **37:** 167–175.

Temin HM. 1967. Studies on carcinogenesis by avian sarcoma viruses. VI. Differential multiplication of uninfected and of converted cells in response to insulin. *J Cell Physiol* **69:** 377–384.

Todaro GJ, De Larco JE, Cohen S. 1976. Transformation by murine and feline sarcoma viruses blocks binding of epidermal growth factor to cells. *Nature* **264:** 26–31.

Ullrich A, Coussens L, Hayflick JS, Dull TJ, Gray A, Tam AW, Lee J, Yarden Y, Libermann TA, Schlessinger J, et al. 1984. Human epidermal growth factor receptor cDNA sequence and aberrant expression of the amplified gene in A431 epidermoid carcinoma cells. *Nature* **309:** 418–425.

Ushiro H, Cohen S. 1980. Identification of phosphotyrosine as a product of epidermal growth factor-activated protein kinase in A-431 cell membranes. *J Biol Chem* **255:** 8363–8365.

Van Obberghen E, Spooner PM, Kahn CR, Chernick SS, Garrison MM, Karlsson FA, Grunfeld C. 1979. Insulin-receptor antibodies mimic a late insulin effect. *Nature* **280:** 500–502.

Vigneri R, Goldfine ID, Wong KY, Smith GJ, Pezzino V. 1978. The nuclear envelope. The major site of insulin binding in rat liver nuclei. *J Biol Chem* **253:** 2098–2103.

Waksman G, Kuriyan J. 2004. Structure and specificity of the SH2 domain. *Cell* **116:** S45–S48.

Waterfield MD, Scrace FT, Whittle N, Stroobant P, Johnsson A, Wasteson A, Westermark B, Heldin CH, Huang JS, Deuel TF. 1983. Platelet-derived growth factor is structurally related to the putative transforming protein p28sis of simian sarcoma virus. *Nature* **304:** 35–39.

Witte ON, Dasgupta A, Baltimore D. 1980. Abelson murine leukaemia virus protein is phosphorylated in vitro to form phosphotyrosine. *Nature* **283:** 826–831.

Yarden Y, Schlessinger J. 1987. Self-phosphorylation of epidermal growth factor receptor: Evidence for a model of intermolecular allosteric activation. *Biochemistry* **26:** 1434–1442.

Yuzawa S, Opatowsky Y, Zhang Z, Mandiyan V, Lax I, Schlessinger J. 2007. Structural basis for activation of the receptor tyrosine kinase KIT by stem cell factor. *Cell* **130:** 323–334.

Zhang X, Gureasko J, Shen K, Cole PA, Kuriyan J. 2006. An allosteric mechanism for activation of the kinase domain of epidermal growth factor receptor. *Cell* **125:** 1137–1149.

The Genesis of Tyrosine Phosphorylation

Tony Hunter

Salk Institute for Biological Studies, La Jolla, California 92037

Correspondence: hunter@salk.edu

Tyrosine phosphorylation of proteins was discovered in 1979, but this posttranslational modification had been "invented" by evolution more than a billion years ago in single-celled eukaryotic organisms that were the antecedents of the first multicellular animals. Because sophisticated cell–cell communication is a sine qua non for the existence of multicellular organisms, the development of cell-surface receptor systems that use tyrosine phosphorylation for transmembrane signal transduction and intracellular signaling seems likely to have been a crucial event in the evolution of metazoans. Like all types of protein phosphorylation, tyrosine phosphorylation serves to regulate proteins in multiple ways, including causing electrostatic repulsion and inducing allosteric transitions, but the most important function of phosphotyrosine (P.Tyr) is to serve as a docking site that promotes a specific interaction between a tyrosine phosphorylated protein and another protein that contains a P.Tyr-binding domain, such as an SH2 or PTB domain. Such docking interactions are essential for signal transduction downstream from receptor tyrosine kinases (RTKs) on the cell surface, which are activated on binding a cognate extracellular ligand, and, as a consequence, elicit specific cellular outcomes.

The first eukaryotic tyrosine kinases (TKs) were discovered through studies of animal tumor virus transforming proteins, such as polyoma virus middle T antigen and the Rous sarcoma virus v-Src protein (Eckhart et al. 1979; Hunter and Sefton 1980). When the v-Src TK sequence was reported in 1983, this immediately led to the revelation that despite their unique amino acid specificity, the TKs are related to the Ser/Thr kinases, exemplified by the cAMP-dependent protein kinase. Bioinformatic analysis and targeted cDNA cloning quickly revealed the existence of a surprisingly large number of related protein kinases, now known as the eukaryotic protein kinase (ePK) family. Based on bioinformatic analysis of the completely sequenced human genome (Manning et al. 2002), the number of ePK genes stands at 478 (see Fig. 1), and the total number of protein kinase genes, which includes other protein kinases either distantly related to the ePKs or unrelated to the ePKs, is 566. Surprisingly, given the scarcity of P.Tyr in cellular proteins, 90 kinases are classified as TKs (Fig. 1), although a small number of these lack significant kinase activity, but have conserved noncatalytic functions.

The first hint that a growth factor receptor might have an intrinsic protein kinase activity came from Stanley Cohen's 1978 report that EGF stimulated protein phosphorylation in a membrane preparation from A431 cells (Carpenter et al. 1978), which have an extraordinari-

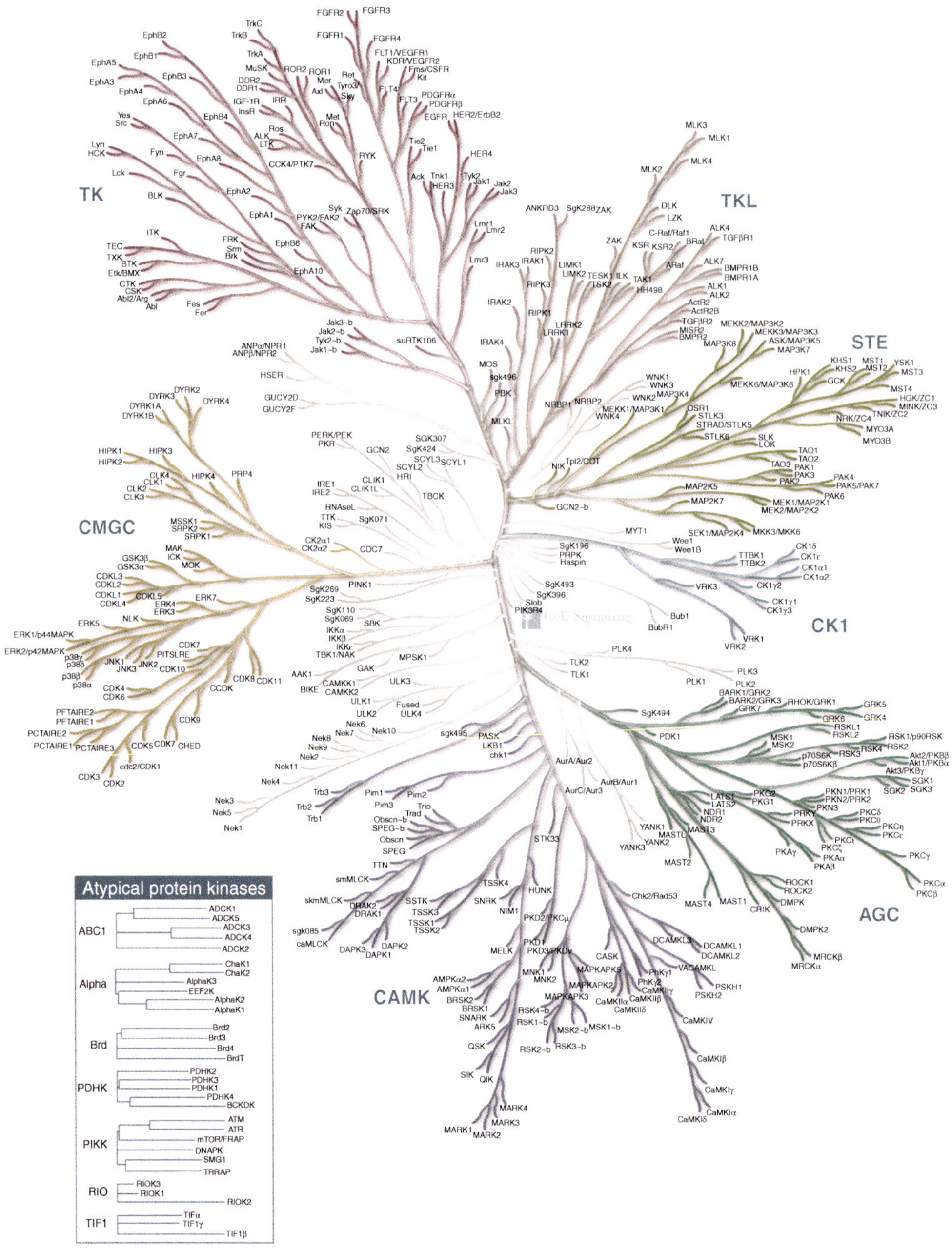

Figure 1. The human kinome. Based on the catalog of human protein kinases compiled by Manning et al. (2002), an unrooted relatedness tree was constructed using the catalytic domain sequences of the 478 eukaryotic protein kinases (ePKs). The seven major branches of the kinome are indicated: AGC, CAMK, CMGC, TK, TKL, STE, and CK1. The ends of the branches representing individual kinases are labeled with the names of each protein kinase. The TK (tyrosine kinase) branch at the top of the tree has 90 members. The RTKs are present in four major branches: EPH, INSR/TRK/AXL, FGFR/PDGFR/CSF-1R, and EGFR. The atypical protein kinases, shown in the *inset* at the *bottom left*, fall into seven small families, which are either distantly related to the ePKs or else unrelated in sequence. (Illustration reproduced courtesy of Cell Signaling Technology, Inc., www.cellsignal.com.)

ly high number of surface EGF receptors. This group's subsequent July 1979 paper concluded that EGF stimulated threonine phosphorylation of proteins in membranes (Carpenter et al. 1979). However, with the realization that phosphothreonine (P.Thr) and P.Tyr comigrate on electrophoresis at pH 1.9, they reevaluated this conclusion, and in September 1980 the Cohen group published that EGF actually stimulated Tyr phosphorylation (Ushiro and Cohen 1980). By June 1981, purified EGF receptor preparations had been shown to have TK activity (Chinkers and Cohen 1981), and EGF had been shown to stimulate Tyr phosphorylation in the cell, resulting in Tyr phosphorylation of specific proteins within minutes of EGF treatment (Hunter and Cooper 1981). The 1984 cloning of the EGF receptor revealed that it has a catalytic domain related to that of the c-Src TK (Downward et al. 1984; Lin et al. 1984), confirming the intrinsic nature of the TK activity. In quick succession, several additional growth factor receptors were shown to have TK activity, starting with the PDGF receptor in 1982. Further RTKs were added through directed cloning and sequence analysis combined with biochemical testing, and by the end of the decade >10 RTKs had been reported. By this time, it was clear that ligand-induced Tyr phosphorylation was a major mechanism for the transmission of signals across the plasma membrane. This conclusion was reinforced by the discovery of two component receptors, like the antigen receptors and the cytokine receptors, in which the ligand-binding subunit of the receptor complexes with a cytoplasmic TK, which is activated upon ligand binding. We now know that the human genome encodes 58 RTKs grouped in 20 distinct families (Lemmon and Schlessinger 2010).

In general, RTKs are type 1 transmembrane proteins with an extracellular ligand-binding domain linked by a transmembrane domain to an intracellular domain that includes a TK catalytic domain, and, usually, an unstructured carboxy-terminal tail that possesses autophosphorylation sites (Lemmon and Schlessinger 2010). RTK kinase activity is increased in response to binding of a cognate ligand, such as a growth factor, to the extracellular domain. Ligand-induced dimerization and intermolecular phosphorylation was originally proposed as an RTK activation mechanism by Yarden and Schlessinger (1987), and based on much subsequent work, this is now accepted as the general mechanism of RTK activation by ligands, although the specific details differ between different RTK subfamilies. Ligand binding either alters the conformation of a preexisting RTK dimer or induces RTK dimer formation, which results in juxtaposition of the catalytic domains and activation in *trans*, either through an induced conformational change or via transphosphorylation of activating residues in the activation loop or the cytoplasmic juxtamembrane domain. Once activated, RTKs autophosphorylate at additional sites, enabling recruitment of SH2 and PTB domain P.Tyr-binding proteins, and also directly phosphorylate substrates to propagate downstream signals. In this regard, although we have structures of dimerized ligand-bound RTK extracellular domains, and cytoplasmic domains, structures of an intact liganded RTK dimer are still needed to understand exactly how ligand-induced dimerization of the extracellular domain results in catalytic activation through juxtaposition of the cytoplasmic kinase domain (Arkhipov et al. 2013; Endres et al. 2013).

TYROSINE PHOSPHORYLATION AND DISEASE

The discovery that transforming proteins of tumor viruses had TK activity immediately suggested that unbridled Tyr phosphorylation might be a potent transforming mechanism. Analysis of temperature-sensitive transforming mutants of Rous sarcoma virus showed that v-Src TK activity correlated precisely with transforming potential, providing direct evidence that Tyr phosphorylation was required for transformation (Sefton et al. 1980). A search for human tumor oncogenes quickly revealed that chronic myelogenous leukemia (CML) results from the fusion of the *BCR* gene with the c-*ABL* TK gene, yielding the BCR-ABL fusion protein, a constitutively activated TK that is encoded by the t22:9 Philadelphia chromosomal fusion (Hunter 2007). BCR-ABL TK activity is necessary for transformation of myeloid cells in culture and leukemia in animals. Subsequently, many additional oncogenically activated human TK mutants have been reported in cancer. Several of these are mutant forms of RTKs, including many instances in which a chimeric protein is made as a result of the fusion of a dimerization domain from one protein with the cytoplasmic catalytic domain of an RTK, resulting in a constitutively activated TK.

The finding that an activated TK was causal in human disease spurred efforts to develop protein kinase inhibitors as cancer therapeutics. The first effort to develop selective protein kinase inhibitors began in the 1980s with the goal of using these to study protein kinase function in the cell (Hidaka et al. 1984), and ultimately to develop them as therapeutics. The importance of elevated Tyr phosphorylation in cancer triggered efforts to develop selective inhibitors against individual TKs known to be activated by mutation or overexpression in different types of cancer. In 1998, the first drug antagonizing a TK was approved for cancer therapy; trastuzumab (Herceptin), an inhibitory monoclonal antibody directed against the extracellular domain of the HER2 RTK, is used for therapy of HER2-positive breast cancer. This was quickly followed in 2001 by the approval of a small-molecule TK inhibitor (TKI), imatinib (Gleevec), an inhibitor of the activated BCR-ABL TK responsible for CML (Hunter 2007). Imatinib has proved to be remarkably successful in treating CML, and most patients who are put on treatment during the chronic indolent phase of the disease go into long-term remission, provided they continue taking the drug. Since 2001, several additional small-molecule TKIs and protein drugs have been approved for cancer therapy, targeting a wide range of mutationally activated or overexpressed TKs characteristic of different human cancers. A striking recent example is crizotinib, which is a selective inhibitor of the ALK RTK that is activated by chromosomal translocation in about 4% of non-small-cell lung cancers. Within 4 years of the discovery of ALK fusion genes in NSCLC in 2007, crizotinib had been approved for treatment of NSCLC patients diagnosed with an ALK mutation (Hallberg and Palmer 2010). As of August 2013, 19 TKIs have been approved for cancer therapy, and, notably, 12 of these target activated RTKs; several more TKIs are in phase III trials, and are likely to be approved in the near term. In addition, five approved protein antibody drugs are directed against RTK extracellular domains or their ligands, with more in the pipeline.

WHY WAS TYROSINE PHOSPHORYLATION MISSED FOR SO LONG?

Ser/Thr kinase activities were first identified in 1954 (Burnett and Kennedy 1954), but, even though as Phoebus Levene, who reported the synthesis of P.Tyr in 1933 (Levene and Shcormuller 1933) had realized that phosphorylation of the tyrosine hydroxyl is theoretically possible, phosphorylation of Tyr in proteins was not described until nearly 25 years later (Eckhart et al. 1979; Hunter and Sefton 1980). This long gap was despite the fact that, as we now know, a large number of TKs exists (Manning et al. 2002). Why was Tyr phosphorylation missed for so long? Unlike many Ser and Thr phosphorylations, most Tyr phosphorylations are very short-lived owing to the presence of extremely active P.Tyr-specific phosphatases (PTPs) that rapidly dephosphorylate any P.Tyr residue that is not protected through binding to an SH2 or PTP domain or via an intramolecular interaction. A good example of how rapidly phosphate on Tyr turns over is the finding that the EGF-induced P.Tyr residues on the EGF receptor have half-lives of only a few seconds, and are turned over >100 times during the early phase of the cellular EGF response when the EGF receptor is maximally phosphorylated (Kleiman et al. 2011). As a result, P.Tyr constitutes $<1\%$ of the total phosphohydroxy-amino acids in proteins in a typical mammalian cell, even when stimulated with growth factors to activate RTKs or other types of receptor systems that signal through TKs. To compound this scarcity, for many years the routine method for identifying phosphoamino acids in proteins involved electrophoretic separation of partial acid hydrolysates of ^{32}P-labeled proteins at pH 1.9, and, at this pH, P.Thr and

Cite this article as *Cold Spring Harb Perspect Biol* doi: 10.1101/cshperspect.a020644

P.Tyr comigrate (Ushiro and Cohen 1980), meaning that the weak P.Tyr signal is masked by the much more abundant P.Thr signal (5% of total cellular phosphoamino acids). This meant that even when kinase activity was assayed in vitro, a protein kinase that phosphorylated Tyr would have been mistakenly identified as a Thr kinase, as was the case with the EGF receptor RTK (Carpenter et al. 1979) and v-Src TK (Collett et al. 1979). Once two-dimensional electrophoretic methods had been developed to separate P.Tyr from P.Thr and P.Ser (Hunter and Sefton 1980), it was possible to show that P.Tyr was present in proteins isolated from cells, and that its levels increased significantly when an activated TK, such as v-Src (Hunter and Sefton 1980), was expressed, or when the cells were stimulated with EGF to activate the EGF receptor RTK (Cooper and Hunter 1981). Moreover, individual cellular proteins containing an increased level of P.Tyr were identified in such cells by two-dimensional gel electrophoresis (Radke et al. 1980; Cooper and Hunter 1981), and through physical association with the TKs themselves (Hunter and Sefton 1980).

WHAT IS SPECIAL ABOUT TYROSINE PHOSPHORYLATION?

From a chemical perspective there is nothing particularly unusual about the chemical properties of the O^4-phenolic phosphate ester bond of P.Tyr, which, like those of phosphoserine (P.Ser) and P.Thr, is a relatively high energy bond (8–10 kcal). However, because the phosphate on Tyr is linked to the O^4 position of the phenolic ring, it lies much further away from the peptide backbone than the phosphate on the β-OH groups of Ser and Thr, and, in consequence, this in itself provides an element of binding specificity. In addition, the phenolic ring of P.Tyr is unique in providing significant additional binding energy for phosphospecific-binding domains that are mediated by hydrophobic or π bond-ring interactions, which P.Ser/Thr cannot make. These properties allowed the evolution of selective P.Tyr-binding domains, which have much deeper binding pockets than those for P.Ser and P.Thr-binding domains. In addition,

the greater distance of the O^4 hydroxyl from the peptide backbone allowed evolution of Tyr-specific kinases, and also P.Tyr-specific phosphatases. Fortuitously, the distinct chemical properties of P.Tyr also allow the immune system to generate antibodies that selectively recognize P.Tyr over P.Ser and P.Thr, and such anti-P.Tyr antibodies have been extraordinarily useful in studying Tyr phosphorylation.

WHERE DID TKs AND PHOSPHATASES COME FROM?

Prokaryotes made early use of the reactive tyrosine hydroxyl group in proteins for regulatory purposes. For example, adenylylation of a specific Tyr residue in glutamine synthetase decreases its activity (Shapiro and Stadtman 1968). True Tyr phosphorylation is also used in prokaryotes, and was apparently "invented" separately from the process in eukaryotes, because the small family of bacterial receptor-like TKs, known as BY kinases, is unrelated in sequence to the large family of Tyr TKs in eukaryotes. BY-catalyzed Tyr autophosphorylation is used in bacteria to regulate biosynthesis and export of extracellular polysaccharide (Grangeasse et al. 2012). P.Tyr-specific phosphatases counteract the phosphorylation of BY kinases, but no specific P.Tyr-binding proteins have been found in bacteria, and the consequences of Tyr phosphorylation are exerted primarily through allosteric/electrostatic effects.

Based on the sequence similarities between their catalytic domains, it is likely that conventional TKs evolved from Ser/Thr kinases, which in turn appear to have been derived from bacterial small-molecule kinases, like the eukaryotic-like kinases (ELKs), such as the aminoglycoside kinases (Kannan et al. 2007), which have a distantly related catalytic domain that has a very similar three-dimensional fold. Dual-specificity kinases that could phosphorylate Ser/Thr and Tyr may have been an intermediate step in the evolution of the first Tyr-specific kinase. For example, the MAP2K family dual-specificity kinases, which phosphorylate both a Thr and a Tyr in target MAP kinases are of ancient origin, and are found in all eukaryotes. The main requirement for converting a Ser/Thr kinase, which phosphorylates the β-OH group of Ser/Thr that lies close to the peptide backbone, into a TK is to redesign the active site so that it can accommodate a Tyr residue, whose phenolic OH group is ~6 Å further away from the peptide backbone than the β-OH group of Ser/Thr. The conventional TKs are characterized by a difference in the conserved catalytic domain motif on the amino-terminal side of the activation loop (HRDLAARN in TKs versus HRDLKPEN in Ser/Thr kinases; in both cases the Asp residue serves as the catalytic base). Both the Arg residues in this motif hydrogen bond to the target Tyr OH group, and the Tyr phenolic ring makes van der Waals interactions with a conserved Pro in the P + 1 loop, which adopts a different configuration in the TKs than in the Ser/Thr kinases and appears to be a major determinant in conferring specificity for Tyr versus Ser/Thr (Hubbard et al. 1994).

One can never know with certainty in which species the first TK appeared, but almost certainly it was a unicellular organism, because extant unicellular choanoflagellate species, which lie at the base of the metazoan branch of the evolutionary tree, have a well-developed Tyr phosphorylation system, with many TK genes (Manning et al. 2008; Pincus et al. 2008). Indeed, an unexpectedly complex repertoire of TKs has been found in the genomes of two unicellular organisms that are close relatives of metazoans, namely, the choanoflagellate *Monosiga brevicollis* and the filasterian *Capsaspora owczarzaki*. These two holozoan genomes each encode >100 TKs, more than the total human count, although both organisms are predominantly unicellular in lifestyle. Almost nothing is known about the functions of these TKs, but their sequences do shed light on TK evolution and the diversity of domain contexts in which TKs can operate. The majority of TKs in both species is predicted to be membrane-spanning cell-surface receptors (RTKs) (88/128 in *Monosiga* and 92/103 in *Capsaspora*), but for the most part these RTKs are not orthologous to metazoan RTKs. Possible homologs of EPH and IGF-1R may be present in choanoflagellates, but no other examples of human RTK families are evident.

Interestingly, no additional RTK families are shared between *Monosiga* and *Capsaspora*, suggesting that the large numbers of RTKs in both clades evolved independently. Not only is the overall number of TKs in choanoflagellates greater than that of a typical vertebrate, but they also possess all the machinery needed for Tyr phosphorylation-based signaling, including an abundance of PTPs and SH2 domain proteins. These include homologs of several specific metazoan PTP and SH2 proteins, as well as many novel proteins with unique domain arrangements. Overall, the analysis of TKs in holozoa suggests that an early holozoan had a mature set of 6−8 nonreceptor TKs (CTKs), including SRC, CSK, ABL, TEC, FER, and FAK, several fast-evolving RTKs, and an extensive network of SH2 and PTB P.Tyr-binding domains, and PTP domain proteins to reverse the action of TKs, and presumably downstream TK target substrates. In all unicellular and multicellular organisms that have TKs, they constitute 10%−20% PKs, underscoring their fundamental importance to cellular physiology.

At what point in the emergence of Tyr phosphorylation-based signaling did RTKs arise? Because of their key role in intercellular communication, RTKs were originally proposed to have evolved in parallel with multicellularity in animals, and it was suggested that the development of Tyr phosphorylation-based signaling may have played a vital part in the emergence of metazoans by providing the essential means of coordinating function between different cells in a multicellular organism. The lack of Tyr phosphorylation in single-celled organisms, such as the yeasts, reinforced this idea. Nevertheless, because of the unexpectedly complex repertoire of the holozoan RTKs, it seems likely that RTKs did indeed evolve in single-celled organisms, and that they were used as a means of sensing extracellular stimuli, perhaps nutrients or toxic compounds. Subsequently, they adapted to a new function in signaling cell−cell interactions either directly or through paracrine factors. Whether RTKs arose through the fusion of a gene encoding a receptor-like protein and a gene for a TK catalytic gene, or perhaps from a receptor-serine kinase, and whether all RTKs

 Cite this article as *Cold Spring Harb Perspect Biol* doi: 10.1101/cshperspect.a020644

arose from a single progenitor RTK or whether RTKs arose multiple times are unanswered questions.

The existence of a large family of P.Tyr-specific protein phosphatases (PTPs), which oppose the actions of the TKs, underscores the importance of Tyr phosphorylation as an intracellular signaling system. Like the TKs, PTPs developed early in eukaryotic evolution, apparently being derived from a family of dual-specificity phosphatases (DSPs), whose evolutionary origins are obscure, but which are found in all extant eukaryotes, serving to dephosphorylate MAP kinases at the activating P.Tyr and P.Thr sites. Unlike the protein kinases, which are predominantly in a single family, there are several distinct and unrelated protein phosphatase catalytic domain families. Among these, there are two main P.Tyr phosphatase families: the classical PTPs, which are selective for P.Tyr, and the DSPs, many of which can hydrolyze P.Tyr but also P.Ser and P.Thr. In humans, there are 38 classical PTPs, split into 21 receptor-like PTPs and 17 nonreceptor PTPs, 61 DSPs (although all DSPs hydrolyze phosphate ester linkages, not all of them act on protein substrates), four Asp-based PTPs, three Cdc25 DSPs that act primarily on the inhibitory P.Thr and P.Tyr residues in the CDKs, and the LMW PTP, for a total of 107 (Alonso et al. 2004). Taken together, this means that nearly as many P.Tyr phosphatases exist as TK catalytic entities. However, although there are genetic hints that specific RTK/PTP pairs exist, in general there do not appear to be one-to-one relationships.

NEW METHODS HAVE BEEN ESSENTIAL FOR UNDERSTANDING TYROSINE PHOSPHORYLATION

It is often said that progress in biology is dependent on advances in technology, and nowhere is this truer than in the field of Tyr phosphorylation. Historically, many of the advances in understanding Tyr phosphorylation have depended on the development of new methods and reagents that played and continue to play a vital part in our progress toward a complete understanding of Tyr phosphorylation-based signaling networks. Areas in which key methodological advances were made include detection of P.Tyr, three-dimensional structural analysis, degenerate library methods for defining P.Tyr-binding and TK phosphorylation consensus sequences, development of sensitive nonradioactive kinase assays, generation of specific TKIs and analog-sensitive protein kinase mutants, MS-based P.Tyr proteomics, the use of kinase and domain arrays, functional analysis of TK function by RNAi, genetic analysis in model organisms through knockout and mutant knock-in analysis, and live cell imaging techniques for spatiotemporal localization of TK signaling events using fluorescently tagged proteins and biosensors. I will touch on a few of these key advances.

Detection of P.Tyr

Initially, the detection of Tyr phosphorylation and the identification of the first TK substrates required labeling cells with massive doses of ^{32}P-orthophosphate, and in vitro kinase reactions with γ-^{32}P-ATP. However, within 2 years of the discovery of Tyr phosphorylation, antibodies that specifically recognize P.Tyr in proteins were developed (Ross et al. 1981; Frackelton et al. 1983), and the use of anti-P.Tyr polyclonal and then monoclonal antibodies quickly supplanted in vivo ^{32}P labeling, and greatly increased the number of identified TK substrates. Ironically, as it turned out, these were not the first man-made anti-P.Tyr antibodies. Immunologists had been using phenylarsonate as a hapten in antibody induction experiments since the 1940s, without realizing, as later analysis showed, that they were generating antibodies that cross-reacted with P.Tyr, albeit weakly! Anti-P.Tyr antibodies have been valuable for studying patterns of Tyr phosphorylation in response to specific stimuli by immunoblotting, and also for immunostaining cells to identify structures enriched in P.Tyr proteins (Marchisio et al. 1984; Maher et al. 1985). In addition to sequence-independent P.Tyr antibodies, site-specific P.Tyr antibodies raised against synthetic P.Tyr-containing peptides have been particularly useful for assessing TK activation states in the cell, and both types of P.Tyr antibody have been used for assaying TK activity in vitro.

In the last 10 years, phosphoproteomic analysis using high-throughput, high-sensitivity mass spectrometry (MS) instruments has revolutionized the identification of Tyr phosphorylation sites. Such global analysis of P.Tyr-containing proteins, using anti-P.Tyr monoclonal antibodies to enrich for P.Tyr-containing tryptic peptides from digests of cellular proteins followed by MS analysis, has revealed an unexpectedly complex repertoire of proteins and sites that can be phosphorylated on Tyr in metazoans (Rush et al. 2005; Rikova et al. 2007). Thousands of P.Tyr sites have been reported, and, in the vast majority of cases, the function of these Tyr phosphorylation events has not been investigated. In this regard, it is possible that a significant fraction of these sites do not have a functional output, and could be construed as "noise" Tyr phosphorylation.

Three-Dimensional Structural Analysis

Advances in the use of protein crystallography to define the structures of isolated protein domains, intact proteins, and protein complexes has played a key role in affording major insights into how Tyr phosphorylation and dephosphorylation is catalyzed, and how P.Tyr residues are recognized by P.Tyr-binding domains in a sequence-dependent fashion to promote protein–protein interactions and propagate signaling initiated by RTK activation at the cell surface upon binding a growth factor. The first crystal structure of a TK, namely, that of the insulin receptor catalytic domain, yielded the secrets underlying kinase specificity for Tyr versus Ser and Thr (Hubbard et al. 1994; Taylor et al. 1995) Likewise, structures of the SH2 domain bound to a synthetic P.Tyr peptide revealed that the P.Tyr side chain binds into a pocket, such that its O^4-phosphate interacts with an Arg at the base of the pocket, which is too deep for the phosphate on β-OH of Ser or Thr to reach (Waksman et al. 1992). Higher-order multidomain structures obtained by crystallography and cryo-EM tomography analysis have begun to reveal how receptor and nonreceptor TKs are

regulated, both positively and negatively. A nice recent example emerging from structural analysis is the unexpected stimulatory role of specific contacts made in *cis* between the SH2 domains of Fes and Abl and the amino lobes of their catalytic domains (Filippakopoulos et al. 2008), a mode of regulation that contrasts with the inhibitory role of the combined SH3-SH2 domain unit in autoinhibition of Src and Abl family kinase activity, which is exerted through a distinct set of SH3 and SH2 contacts with the amino and carboxy lobes. Recently, nuclear magnetic resonance (NMR) solution analysis of kinase catalytic domains and P.Tyr interaction domains has begun to emphasize the importance of dynamic motions within the catalytic domain. This concept has been extended by the use of microsecond-long molecular dynamic simulations of catalytic domain structures to define transitions between active and inactive conformations.

Degenerate Peptide Libraries and Target Identification

Twenty years ago, Cantley and Songyang's development of degenerate, position-oriented peptide libraries to delineate the sequence specificity of SH2 domain binding to P.Tyr sites (Zhou et al. 1993), and subsequently to define the primary sequence selectivity of TK catalytic domains (Songyang et al. 1994) revolutionized the field. Degenerate peptide library technology has subsequently been widely used to define protein kinase specificity in general, and such primary sequence preferences have been incorporated into programs such as Scansite; this algorithm enables the user to input a primary sequence, and obtain predictions as to which protein kinase(s) might phosphorylate a residue of interest, which can then be tested experimentally (Obenauer et al. 2003). Such prediction algorithms are continually being improved by combining sequence preference with other relevant contextual information, such as known protein–protein interactions obtained from systematic interactome analysis, subcellular localization, genetic epistasis relationships, and other functional connections (e.g., NetworKIN and NetPhorest) (Linding et al. 2008). In the case of SH2 domain specificity, this approach has been extended by using arrays of P.Tyr-containing peptides derived from physiological Tyr phosphorylation sites and recombinant SH2 domains (Machida et al. 2007; Tinti et al. 2013).

In an attempt to solve the thorny problem of which PK phosphorylates a particular site in the cell, a number of new methods for identifying targets of individual PKs have been developed. One such method involves cross-linking of a catalytic domain to substrate proteins bound to it in the cell; this is achieved by making a Cys substitution mutation at the substrate phosphorylation site, and then using a cross-linker designed to covalently couple the Cys to the Lys in the ATP-binding site, followed by MS identification of the kinase bound to the tagged substrate (Statsuk et al. 2008). In another method, an analog-sensitive PK mutant is used to thiophosphorylate substrates in vitro using γ-S-ATP, and then thiophosphorylated residues are modified with an adduct that can be recognized by an antibody, allowing the modified peptides to be immunoaffinity enriched and identified by MS (Allen et al. 2007). In another approach, spotted arrays of complete proteomes can be phosphorylated by a purified TK to identify potential substrates, and arrays of isolated catalytic domains can be tested for phosphorylation of a specific protein of interest. Many nonreceptor TKs interact with targets via their SH2 or SH3 domains, and a new method for identifying SH2 and SH3 domain binding partners in the cell has recently been developed. This method takes advantage of "unnatural amino acid" (UAA) technology, in which a UV photoactivatable UAA is incorporated into a protein domain at specific sites on its ligand-interacting surface in appropriately engineered cells, allowing cross-linking to bound proteins when cells are irradiated with UV, and recovery of the tagged domains for MS analysis (Okada et al. 2011; Uezu et al. 2012).

Bioinformatic Analysis of Tyrosine Phosphorylation

The wealth of global phosphoproteomics data now available allows cross-species comparisons to define phosphorylation sites that are conserved through evolution (Beltrao et al. 2012). Sites that are conserved are most likely to have important physiological functions. In addition, knowledge of the primary sequence preference for a PK, combined with its interaction partners, subcellular localization, signaling pathway information, and functional genetic analysis can be used to define potential new PK targets (Linding et al. 2008). Information obtained from biochemical analysis, si/shRNA screens, phosphoproteomic data sets, protein interactomes, and other sources, can be used to build phosphorylation networks. Such networks can then be subjected to computer simulation to predict input/output responses, and modulation by cross talk and feedback mechanisms. This is particularly important in diseases of aberrant signaling, such as cancer, because it may ultimately be possible to predict the consequences of using a specific inhibitor on signaling outcome based on the genotype of a tumor, and thereby determine which combinations of signal transduction inhibitor drugs might be most effective.

Protein Kinase Assays, Inhibitors, and Kinase Engineering

Traditionally, PK assays involved using a protein substrate or a short synthetic peptide corresponding to a known high-affinity phosphorylation site for the PK in question, combined with γ-^{32}P-ATP to monitor phosphorylation. However, short Tyr-containing peptides are generally poor substrates for TKs, although the synthetic polymer poly(Glu.Tyr) has proved useful as a generic TK substrate. The general trend toward nonradioactive assays has led to the use of the sequence-independent anti-P.Tyr MAbs or sequence-specific anti-P.Tyr antibodies to assay phosphorylation, using FRET, often in a high-throughput format for inhibitor screening. More recently, MS-based methods using mass-encoded substrate peptides have been developed, and these can be multiplexed to simultaneously assay multiple PKs in a mixture, such as a cell lysate, with high sensitivity and speed. Assays in which His-tagged TKs are bound to NTA-modified lipid vesicles have proved very useful in recapitulating Tyr phosphorylation on cell membranes, because of the large increase in local protein concentration afforded by using a two-dimensional surface (Zhang et al. 2006). Another important advance has been the development of FRET-based biosensors to measure PK activity in living cells. These genetically encoded reporters, such as AKAR (Zhang et al. 2001), have a short peptide substrate sequence, coupled by a linker to an appropriate phosphobinding domain flanked by YFP and CFP, such that phosphorylation of the substrate site on the biosensor results in its interaction with the phosphobinding domain intramolecularly, leading to a change in the YFP-CFP FRET signal that can be measured in living cells to report local activity of the PK in question (Ting et al. 2001). One of the first PKs this method was applied to was the EGF receptor RTK, in which the EKAR biosensor was used to assess activation of the EGF receptor in response to addition of EGF. Another fluorescence-based method for assaying substrate Tyr phosphorylation in cells makes use of a modified SH2 domain in which a Trp lying adjacent to the bound phosphate is replaced with a coumarin derivative using UAA technology; P.Tyr binding is read out as a change in fluorescence output (Lacey et al. 2011).

The development of selective TK inhibitors has been a primary focus of the pharmaceutical industry because of the importance of TKIs in cancer therapy. Methods for developing truly selective inhibitors are continually improving, and fragment/scaffold-based approaches, combined with structure-based refinement are generating inhibitors with exquisite selectivity, although there are some arguments for designing multitargeted TKIs for cancer therapy. Selective inhibitors are an important research tool, because they allow rapid and reversible inhibition of the target kinase in cells, which can avoid compensatory mechanisms often observed with knockout or knockdown studies. In this regard, it would undoubtedly benefit the research community enormously if pharma could be persuaded to make available without strings selective inhibitors for research purposes that

came out of the same series as a compound that was ultimately taken forward into the clinic. With the use of any inhibitor, however selective, the caveat is that the cellular response may be due to an off-target effect, and the expression of mutant forms of the target PK designed to be resistant to the inhibitor can provide a crucial control for inhibitor specificity, including allosteric inhibitors (Holt et al. 2009).

A major advance in studying PK function was Shokat's development of mutant PKs in which mutation of the "gatekeeper" residue at the base of the catalytic cleft to a residue with a smaller side chain allows the binding of base-modified forms of ATP with bulky groups at the N^6 position of the purine ring, which are excluded by the WT PK. This method was originally developed to identify PK substrates in cell lysates using γ-^{32}P-labeled N^6-modifed ATP analogs (Shah et al. 1997). Subsequently, Shokat extended this concept by developing cell-permeant, purine-based inhibitors with similar bulky groups at the N^6 position that can be used to specifically inhibit these "analog-sensitive" PKs, without affecting other PKs in the cell (Bishop et al. 2000). This strategy has been very powerful, particularly in organisms in which gene replacement technology can be used to replace the resident PK gene with the *as* mutant gene. Treatment of the *as* mutant PK-expressing cells with such analog inhibitors has allowed identification of substrate proteins of the PK of interest, characterized by a rapid decrease in their phosphorylation (Holt et al. 2009).

Genetic Analysis of Tyrosine Phosphorylation

Genetic screens in *Drosophila* and *Caenorhabditis elegans* afforded some of the first glimpses into the functions of RTKs when mutants defective in differentiation and development were shown to harbor mutations in RTK genes. Targeted knockout and knock-in studies in mice have also provided a wealth of information about RTK function in vivo. More recently, the advent of RNAi technology has enabled si/shRNA depletion of individual TKs and PTPs revealing which Tyr phosphorylation events are

 Cite this article as *Cold Spring Harb Perspect Biol* doi: 10.1101/cshperspect.a020644

affected by the loss of individual TKs and PTPs, as well as kinome-wide functional screens (MacKeigan et al. 2005), which have given us important insights into Tyr phosphorylation-based signaling networks.

Subcellular Localization of Tyrosine Phosphorylation Events in Fixed and Living Cells

Major advances in light microscopy, such as confocal and superresolution microscopy, have made it possible to study signaling processes in real time in living cells (and organisms) at increasing resolution, using genetically encoded GFP or RFP reporters or microinjected fluorescently labeled antibodies. This has made it possible to visualize protein movements in the cell in response to an external signal, and also use FRET-based reporters to determine where a protein kinase is active in the cell or where a protein–protein interaction occurs in the cell. Superresolution microscopy, which breaks the diffraction limit of light, now makes it possible to resolve the position of single signaling protein molecules to within a few nm, and this unprecedented advance affords insights into localized signaling structures, such as nanoclusters, which can act as signaling depots (Lillemeier et al. 2010).

FUTURE

What does the future of tyrosine phosphorylation hold? Given the unabated rate of progress since its discovery more than 30 years ago, we are certainly in for more surprises and insights, and undoubtedly, this will require the development of new technologies. It seems unlikely that additional dedicated TKs will be identified, but the recent report that PKM2 can phosphorylate STAT3 on Tyr705 (Gao et al. 2012) means that other enzymes that use ATP, or another substrate with an energy-rich phosphate, might also moonlight as TKs under special circumstances. Likewise, additional P.Tyr phosphatase activities may emerge; a recent example is STS-1, a member of the histidine phosphatase family that has been reported to be a P.Tyr phosphatase

(Mikhailik et al. 2007). Over the past few years, new P.Tyr-binding domains have been identified (e.g., the Hakai HYB domain and the PLC-δ variant C2 domain), and there may be other examples, but, in contrast to the SH2 domain, these will not be found in large families. A high priority should be efforts to determine the functions of the thousands of reported Tyr phosphorylation sites, bearing in mind the possibility that some fraction of them may be silent.

An important concept to consider is that P.Tyr not only signals directly, but can also couple with other posttranslational modifications (PTMs) in the same protein to provide a unique signaling output dependent on both PTMs being present simultaneously, thus creating an AND logic gate. In this regard, the phenolic hydroxyl group of Tyr itself is subject to additional PTMs, namely, sulfation and nitration, and also adenylylation (AMPylation) (Worby et al. 2009). Such PTMs might compete with phosphorylation of specific Tyr under defined circumstances, although this would depend on stoichiometry.

From a technology perspective, more sensitive and selective genetically encoded TK biosensors are needed. These will be particularly useful in studying nuclear signaling by TKs, where biosensors localized to the nucleus could help address the somewhat controversial issue of whether RTKs actively phosphorylate targets in the nucleus following their activation, either through intramembrane cleavage and trafficking of the released cytoplasmic domain into the nucleus, or through translocation of the intact RTK into the nucleus. TK biosensors used in conjunction with superresolution microscopy will also be helpful in studying membrane nanoclusters as signaling nodes.

Although structural analysis has already taught us a great deal about Tyr phosphorylation, we need to define the structures of higher-order signaling complexes and nanoclusters in the membrane. Single molecule analysis in T cells has indicated the importance of signaling protein clusters that preexist before activation (Lillemeier et al. 2010; Sherman et al. 2011), and this principle seems likely to hold true for other systems. We can expect to learn that higher-order signaling complexes are built up through multiple low-affinity interactions involving P.Tyr-binding domains, but also additional contacts between proteins. Given that many scaffolding proteins involved in building these complexes have flexible linkers, a combination of approaches will be required to define the structures of large signaling complexes at atomic resolution, including crystallography, small-angle X-ray scattering, and cryo-EM tomography combined with NMR solution analysis and molecular dynamic simulations. Such efforts will give us a better picture of how TK signaling is propagated from the plasma membrane through the cytoplasm to the nucleus at the molecular level.

We will need to confront the deluge of systems data pertinent to Tyr phosphorylation-based signaling networks, to find the best way to extract meaningful information, and gain functional insights. This will require new ways of interrogating the systems-level datasets of Tyr phosphorylation events emerging through phosphoproteomics and interactomics, combined with kinome-wide si/shRNA screens and whole genome sequencing analysis of large numbers of human tumors for mutations in TKs, which will continue to provide a deeper understanding of how cellular processes are regulated by Tyr phosphorylation and how they go awry in cancer.

If we gaze into the crystal ball and look ahead 30 years, there is every reason to be optimistic that our quest to obtain a full understanding of tyrosine phosphorylation and its multiple roles in eukaryotic biology will have been achieved.

ACKNOWLEDGMENTS

This article is dedicated to the memory of Tony Pawson, a pioneer in the field of signal transduction and tyrosine phosphorylation, who, through his discovery of the SH2 P.Tyr-binding domain, introduced us to the concept that protein interaction domains transmit signals initiated by tyrosine phosphorylation, and championed the field of protein–protein interactions through modular binding domains.

REFERENCES

Allen JJ, Li M, Brinkworth CS, Paulson JL, Wang D, Hubner A, Chou WH, Davis RJ, Burlingame AL, Messing RO, et al. 2007. A semisynthetic epitope for kinase substrates. *Nat Methods* **4:** 511–516.

Alonso A, Sasin J, Bottini N, Friedberg I, Friedberg I, Osterman A, Godzik A, Hunter T, Dixon J, Mustelin T. 2004. Protein tyrosine phosphatases in the human genome. *Cell* **117:** 699–711.

Arkhipov A, Shan Y, Das R, Endres NF, Eastwood MP, Wemmer DE, Kuriyan J, Shaw DE. 2013. Architecture and membrane interactions of the EGF receptor. *Cell* **152:** 557–569.

Beltrao P, Albanese V, Kenner LR, Swaney DL, Burlingame A, Villen J, Lim WA, Fraser JS, Frydman J, Krogan NJ. 2012. Systematic functional prioritization of protein posttranslational modifications. *Cell* **150:** 413–425.

Bishop AC, Ubersax JA, Petsch DT, Matheos DP, Gray NS, Blethrow J, Shimizu E, Tsien JZ, Schultz PG, Rose MD, et al. 2000. A chemical switch for inhibitor-sensitive alleles of any protein kinase. *Nature* **407:** 395–401.

Burnett G, Kennedy EP. 1954. The enzymatic phosphorylation of proteins. *J Biol Chem* **211:** 969–980.

Carpenter G, King L Jr, Cohen S. 1978. Epidermal growth factor stimulates phosphorylation in membrane preparations in vitro. *Nature* **276:** 409–410.

Carpenter G, King L Jr, Cohen S. 1979. Rapid enhancement of protein phosphorylation in A-431 cell membrane preparations by epidermal growth factor. *J Biol Chem* **254:** 4884–4891.

Chinkers M, Cohen S. 1981. Purified EGF receptor-kinase interacts specifically with antibodies to Rous sarcoma virus transforming protein. *Nature* **290:** 516–519.

Collett MS, Erikson E, Erikson RL. 1979. Structural analysis of the avian sarcoma virus transforming protein: Sites of phosphorylation. *J Virol* **29:** 770–781.

Cooper JA, Hunter T. 1981. Similarities and differences between the effects of epidermal growth factor and Rous sarcoma virus. *J Cell Biol* **91:** 878–883.

Downward J, Yarden Y, Mayes E, Scrace G, Totty N, Stockwell P, Ullrich A, Schlessinger J, Waterfield MD. 1984. Close similarity of epidermal growth factor receptor and v-erb-B oncogene protein sequences. *Nature* **307:** 521–527.

Eckhart W, Hutchinson MA, Hunter T. 1979. An activity phosphorylating tyrosine in polyoma T antigen immunoprecipitates. *Cell* **18:** 925–933.

Endres NF, Das R, Smith AW, Arkhipov A, Kovacs E, Huang Y, Pelton JG, Shan Y, Shaw DE, Wemmer DE, et al. 2013. Conformational coupling across the plasma membrane in activation of the EGF receptor. *Cell* **152:** 543–556.

Filippakopoulos P, Kofler M, Hantschel O, Gish GD, Grebien F, Salah E, Neudecker P, Kay LE, Turk BE, Superti-Furga G, et al. 2008. Structural coupling of SH2-kinase domains links Fes and Abl substrate recognition and kinase activation. *Cell* **134:** 793–803.

Frackelton AR Jr, Ross AH, Eisen HN. 1983. Characterization and use of monoclonal antibodies for isolation of phosphotyrosyl proteins from retrovirus-transformed cells and growth factor-stimulated cells. *Mol Cell Biol* **3:** 1343–1352.

Gao X, Wang H, Yang JJ, Liu X, Liu ZR. 2012. Pyruvate kinase M2 regulates gene transcription by acting as a protein kinase. *Mol Cell* **45:** 598–609.

Grangeasse C, Nessler S, Mijakovic I. 2012. Bacterial tyrosine kinases: Evolution, biological function and structural insights. *Philos Trans R Soc Lond B Biol Sci* **367:** 2640–2655.

Hallberg B, Palmer RH. 2010. Crizotinib—Latest champion in the cancer wars? *N Engl J Med* **363:** 1760–1762.

Hidaka H, Inagaki M, Kawamoto S, Sasaki Y. 1984. Isoquinolinesulfonamides, novel and potent inhibitors of cyclic nucleotide dependent protein kinase and protein kinase C. *Biochemistry* **23:** 5036–5041.

Holt LJ, Tuch BB, Villen J, Johnson AD, Gygi SP, Morgan DO. 2009. Global analysis of Cdk1 substrate phosphorylation sites provides insights into evolution. *Science* **325:** 1682–1686.

Hubbard SR, Wei L, Ellis L, Hendrickson WA. 1994. Crystal structure of the tyrosine kinase domain of the human insulin receptor. *Nature* **372:** 746–754.

Hunter T. 2007. Treatment for chronic myelogenous leukemia: The long road to imatinib. *J Clin Invest* **117:** 2036–2043.

Hunter T, Cooper JA. 1981. Epidermal growth factor induces rapid tyrosine phosphorylation of proteins in A431 human tumor cells. *Cell* **24:** 741–752.

Hunter T, Sefton BM. 1980. Transforming gene product of Rous sarcoma virus phosphorylates tyrosine. *Proc Natl Acad Sci* **77:** 1311–1315.

Kannan N, Taylor SS, Zhai Y, Venter JC, Manning G. 2007. Structural and functional diversity of the microbial kinome. *PLoS Biol* **5:** e17.

Kleiman LB, Maiwald T, Conzelmann H, Lauffenburger DA, Sorger PK. 2011. Rapid phospho-turnover by receptor tyrosine kinases impacts downstream signaling and drug binding. *Mol Cell* **43:** 723–737.

Lacey VK, Parrish AR, Han S, Shen Z, Briggs SP, Ma Y, Wang L. 2011. A fluorescent reporter of the phosphorylation status of the substrate protein STAT3. *Angew Chem Int Ed Engl* **50:** 8692–8696.

Lemmon MA, Schlessinger J. 2010. Cell signaling by receptor tyrosine kinases. *Cell* **141.** 1117–1134.

Levene PA, Shcormuller A. 1933. The synthesis of tyrosinephosphoric acid. *J Biol Chem* **100:** 583–587.

Lillemeier BF, Mortelmaier MA, Forstner MB, Huppa JB, Groves JT, Davis MM. 2010. TCR and Lat are expressed on separate protein islands on T cell membranes and concatenate during activation. *Nat Immunol* **11:** 90–96.

Lin CR, Chen WS, Kruiger W, Stolarsky LS, Weber W, Evans RM, Verma IM, Gill GN, Rosenfeld MG. 1984. Expression cloning of human EGF receptor complementary DNA: Gene amplification and three related messenger RNA products in A431 cells. *Science* **224:** 843–848.

Linding R, Jensen LJ, Pasculescu A, Olhovsky M, Colwill K, Bork P, Yaffe MB, Pawson T. 2008. NetworKIN: A resource for exploring cellular phosphorylation networks. *Nucleic Acids Res* **36:** D695–D699.

Machida K, Thompson CM, Dierck K, Jablonowski K, Karkkainen S, Liu B, Zhang H, Nash PD, Newman DK, Nollau P, et al. 2007. High-throughput phosphotyrosine profiling using SH2 domains. *Mol Cell* **26:** 899–915.

MacKeigan JP, Murphy LO, Blenis J. 2005. Sensitized RNAi screen of human kinases and phosphatases identifies new regulators of apoptosis and chemoresistance. *Nat Cell Biol* **7:** 591–600.

Maher PA, Pasquale EB, Wang JY, Singer SJ. 1985. Phosphotyrosine-containing proteins are concentrated in focal adhesions and intercellular junctions in normal cells. *Proc Natl Acad Sci* **82:** 6576–6580.

Manning G, Whyte DB, Martinez R, Hunter T, Sudarsanam S. 2002. The protein kinase complement of the human genome. *Science* **298:** 1912–1934.

Manning G, Young SL, Miller WT, Zhai Y. 2008. The protist, Monosiga brevicollis, has a tyrosine kinase signaling network more elaborate and diverse than found in any known metazoan. *Proc Natl Acad Sci* **105:** 9674–9679.

Marchisio PC, Di Renzo MF, Comoglio PM. 1984. Immunofluorescence localization of phosphotyrosine containing proteins in RSV-transformed mouse fibroblasts. *Exp Cell Res* **154:** 112–124.

Mikhailik A, Ford B, Keller J, Chen Y, Nassar N, Carpino N. 2007. A phosphatase activity of Sts-1 contributes to the suppression of TCR signaling. *Mol Cell* **27:** 486–497.

Obenauer JC, Cantley LC, Yaffe MB. 2003. Scansite 2.0: Proteome-wide prediction of cell signaling interactions using short sequence motifs. *Nucleic Acids Res* **31:** 3635–3641.

Okada H, Uezu A, Mason FM, Soderblom EJ, Moseley MA 3rd, Soderling SH. 2011. SH3 domain-based phototrapping in living cells reveals Rho family GAP signaling complexes. *Sci Signal* **4:** prs13.

Pincus D, Letunic I, Bork P, Lim WA. 2008. Evolution of the phospho-tyrosine signaling machinery in premetazoan lineages. *Proc Natl Acad Sci* **105:** 9680–9684.

Radke K, Gilmore T, Martin GS. 1980. Transformation by Rous sarcoma virus: A cellular substrate for transformation-specific protein phosphorylation contains phosphotyrosine. *Cell* **21:** 821–828.

Rikova K, Guo A, Zeng Q, Possemato A, Yu J, Haack H, Nardone J, Lee K, Reeves C, Li Y, et al. 2007. Global survey of phosphotyrosine signaling identifies oncogenic kinases in lung cancer. *Cell* **131:** 1190–1203.

Ross AH, Baltimore D, Eisen HN. 1981. Phosphotyrosine-containing proteins isolated by affinity chromatography with antibodies to a synthetic hapten. *Nature* **294:** 654–656.

Rush J, Moritz A, Lee KA, Guo A, Goss VL, Spek EJ, Zhang H, Zha XM, Polakiewicz RD, Comb MJ. 2005. Immunoaffinity profiling of tyrosine phosphorylation in cancer cells. *Nat Biotechnol* **23:** 94–101.

Sefton BM, Hunter T, Beemon K, Eckhart W. 1980. Evidence that the phosphorylation of tyrosine is essential for cellular transformation by Rous sarcoma virus. *Cell* **20:** 807–816.

Shah K, Liu Y, Deirmengian C, Shokat KM. 1997. Engineering unnatural nucleotide specificity for Rous sarcoma virus tyrosine kinase to uniquely label its direct substrates. *Proc Natl Acad Sci* **94:** 3565–3570.

Shapiro BM, Stadtman ER. 1968. 5′-Adenylyl-O-tyrosine. The novel phosphodiester residue of adenylylated glutamine synthetase from *Escherichia coli*. *J Biol Chem* **243:** 3769–3771.

Sherman E, Barr V, Manley S, Patterson G, Balagopalan L, Akpan I, Regan CK, Merrill RK, Sommers CL, Lippincott-Schwartz J, et al. 2011. Functional nanoscale organization of signaling molecules downstream of the T cell antigen receptor. *Immunity* **35:** 705–720.

Songyang Z, Blechner S, Hoagland N, Hoekstra MF, Piwnica-Worms H, Cantley LC. 1994. Use of an oriented peptide library to determine the optimal substrates of protein kinases. *Curr Biol* **4:** 973–982.

Statsuk AV, Maly DJ, Seeliger MA, Fabian MA, Biggs WH 3rd, Lockhart DJ, Zarrinkar PP, Kuriyan J, Shokat KM. 2008. Tuning a three-component reaction for trapping kinase substrate complexes. *J Am Chem Soc* **130:** 17568–17574.

Taylor SS, Radzio-Andzelm E, Hunter T. 1995. How do protein kinases discriminate between serine/threonine and tyrosine? Structural insights from the insulin receptor protein-tyrosine kinase. *FASEB J* **9:** 1255–1266.

Ting AY, Kain KH, Klemke RL, Tsien RY. 2001. Genetically encoded fluorescent reporters of protein tyrosine kinase activities in living cells. *Proc Natl Acad Sci* **98:** 15003–15008.

Tinti M, Kiemer L, Costa S, Miller ML, Sacco F, Olsen JV, Carducci M, Paoluzi S, Langone F, Workman CT, et al. 2013. The SH2 domain interaction landscape. *Cell Rep* **3:** 1293–1305.

Uezu A, Okada H, Murakoshi H, del Vescovo CD, Yasuda R, Diviani D, Soderling SH. 2012. Modified SH2 domain to phototrap and identify phosphotyrosine proteins from subcellular sites within cells. *Proc Natl Acad Sci* **109:** E2929–E2938.

Ushiro H, Cohen S. 1980. Identification of phosphotyrosine as a product of epidermal growth factor-activated protein kinase in A-431 cell membranes. *J Biol Chem* **255:** 8363–8365.

Waksman G, Kominos D, Robertson SC, Pant N, Baltimore D, Birge RB, Cowburn D, Hanafusa H, Mayer BJ, Overduin M, et al. 1992. Crystal structure of the phosphotyrosine recognition domain SH2 of v-src complexed with tyrosine-phosphorylated peptides. *Nature* **358:** 646–653.

Worby CA, Mattoo S, Kruger RP, Corbeil LB, Koller A, Mendez JC, Zekarias B, Lazar C, Dixon JE. 2009. The fic domain: Regulation of cell signaling by adenylylation. *Mol Cell* **34:** 93–103.

Yarden Y, Schlessinger J. 1987. Self-phosphorylation of epidermal growth factor receptor: Evidence for a model of intermolecular allosteric activation. *Biochemistry* **26:** 1434–1442.

Zhang J, Ma Y, Taylor SS, Tsien RY. 2001. Genetically encoded reporters of protein kinase A activity reveal impact of substrate tethering. *Proc Natl Acad Sci* **98:** 14997–15002.

Zhang X, Gureasko J, Shen K, Cole PA, Kuriyan J. 2006. An allosteric mechanism for activation of the kinase domain of epidermal growth factor receptor. *Cell* **125:** 1137–1149.

Zhou S, Shoelson SE, Chaudhuri M, Gish GD, Pawson T, Haser WG, King F, Roberts T, Ratnofsky S, Lechleider RJ, et al. 1993. SH2 domains recognize specific phosphopeptide sequences. *Cell* **72:** 767–778.

Cite this article as *Cold Spring Harb Perspect Biol* doi: 10.1101/cshperspect.a020644

Molecular Mechanisms of SH2- and PTB-Domain-Containing Proteins in Receptor Tyrosine Kinase Signaling

Melany J. Wagner[1,3], Melissa M. Stacey[1,3], Bernard A. Liu[1], and Tony Pawson[1,2]

[1]Lunenfeld Tanenbaum Research Institute, Mount Sinai Hospital, Toronto, Ontario M5G 1X5, Canada

[2]Department of Molecular Genetics, University of Toronto, Toronto, Ontario M5S 1A8, Canada

Correspondence: pawson@lunenfeld.ca

Intracellular signaling is mediated by reversible posttranslational modifications (PTMs) that include phosphorylation, ubiquitination, and acetylation, among others. In response to extracellular stimuli such as growth factors, receptor tyrosine kinases (RTKs) typically dimerize and initiate signaling through phosphorylation of their cytoplasmic tails and downstream scaffolds. Signaling effectors are recruited to these phosphotyrosine (pTyr) sites primarily through Src homology 2 (SH2) domains and pTyr-binding (PTB) domains. This review describes how these conserved domains specifically recognize pTyr residues and play a major role in mediating precise downstream signaling events.

Receptor tyrosine kinase (RTK) signaling is initiated on binding of soluble growth factors to growth factor receptors such as the insulin receptor (IR) or epidermal growth factor receptor (EGFR), or on binding of membrane-bound ephrins, as is the case for Eph receptors. Intracellular signaling is then propagated through PTMs, which commonly serve to regulate protein function by acting as docking sites for recruitment of modular protein interaction domains. Phosphorylation is the best studied PTM, and is a principle mechanism regulating intracellular signaling.

A common element in RTK signaling involves autophosphorylation of the intracellular portion of the receptor (Fig. 1). RTKs become activated as a result of ligand-stabilized dimerization or oligomerization. For instance, in the EGFR subfamily (which includes ErbB and EGF receptors), the formation of homo- or heterodimers is initiated by ligand binding and subsequent exposure of a dimerization domain (Hynes and Lane 2005). Dimerization of the RTKs allows autophosphorylation of the RTKs; EGFR is exceptional in that an allosteric interaction between the kinase domains of adjacent monomers is responsible for the receptor activation (Zhang et al. 2006). However, in the majority of cases dimerization enhances RTK catalytic activity through phosphorylation of the

---

[3]These authors contributed equally to this article.

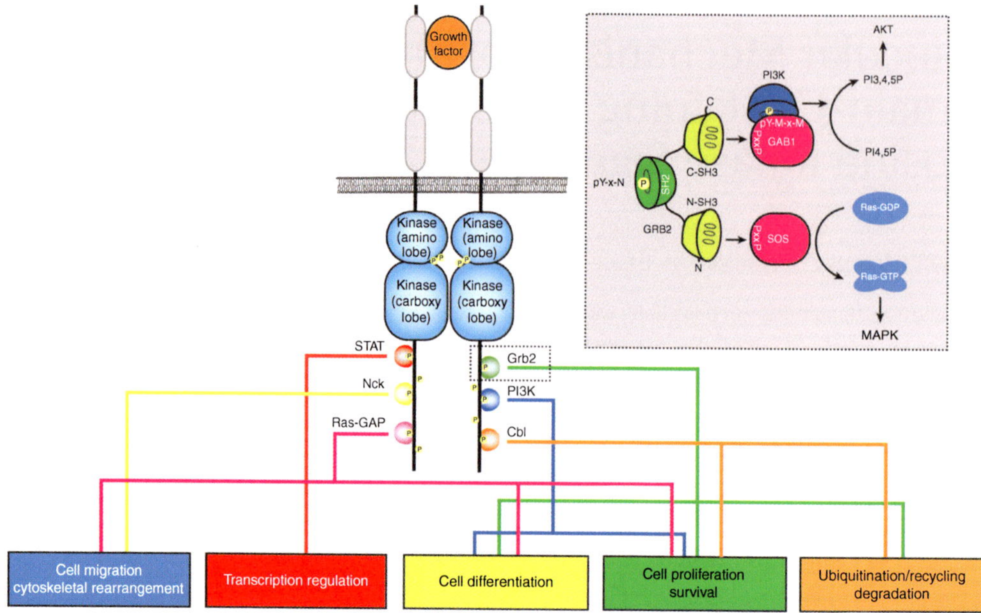

Figure 1. Receptor tyrosine kinases activate downstream pathways through recruitment of proteins containing pTyr-binding domains. Receptor tyrosine kinases are activated on growth factor binding to the extracellular domain of the receptor, leading to receptor dimerization and tyrosine phosphorylation (yellow circles labeled with a P) of their cytoplasmic tails, which act as docking sites for recruitment of PTB and SH2 domains. Various RTKs can mediate a diverse set of cellular processes (colored boxes) determined by the recruitment of specific SH2- and PTB-domain-containing proteins. The gray box displays how the adaptor Grb2 is recruited to an RTK through recognition of the pY-x-N (pY = pTyr, x = any natural amino acid) and activates cell growth and survival pathways such as MAPK and AKT, respectively, through complex formation via its SH3 domains.

kinase activation loop, and in some instances the juxtamembrane region, and recruits signaling effectors through the creation of pTyr docking sites. The specific interaction of signaling proteins with these pTyr-binding motifs activates signaling pathways, such as canonical signaling through the Ras-mitogen activated protein kinase (MAPK), phosphoinositide-3-kinase (PI3K)-Akt, and phospholipase C-gamma (PLC-γ) pathways. These RTK pathways can result in a variety of cellular processes, including differentiation, proliferation, survival, and migration (Fig. 1). The cellular context of signaling can dictate the biological outcome, and how each RTK initiates a given cellular process remains an area of active research.

Tyrosine phosphorylation mediates RTK signaling through the recruitment and activation of proteins involved in downstream signaling pathways, mediated through pTyr binding of the SH2 and PTB domains of signaling effectors. SH2 and PTB domains are found in an otherwise diverse set of proteins containing a range of distinct catalytic and interaction domains, and provide a degree of specificity through their recognition of both a pTyr residue and surrounding amino acids. Here we will discuss the properties of proteins that contain SH2 and PTB domains and their roles in signaling downstream of RTKs, as well as the mechanisms by which they regulate the activity of these signaling effectors.

SH2- AND PTB-DOMAIN-CONTAINING PROTEINS ARE DIVERSE IN NATURE

An SH2 or PTB domain typically recognizes a pTyr residue within the context of a specific amino acid sequence (Songyang et al. 1993). This characteristic allows for the distinct bind-

Cite this article as *Cold Spring Harb Perspect Biol* doi: 10.1101/cshperspect.a008987

ing of SH2- and PTB-containing proteins to activated RTKs and to other tyrosine phosphorylated signaling effectors (Sadowski et al. 1986; Anderson et al. 1990; Blaikie et al. 1994; Kavanaugh and Williams 1994). These domains are found in proteins of diverse function, and on RTK activation their recruitment to pTyr sites results in the activation of their host proteins, and the stimulation of downstream signaling pathways. As noted, these proteins usually contain other catalytic and noncatalytic domains that establish their functions, and can be used to categorize them into distinct groups, which are briefly discussed below (Fig. 2).

Adaptor proteins and docking/scaffolding proteins lack intrinsic catalytic activity, but act to assemble signaling complexes capable of selectively stimulating downstream pathways. Adaptors containing both SH2 and SH3 domains use the SH3 domain(s) to aggregate signaling effectors, whereas the SH2 domain mediates recruitment to the active RTK and assembly of the signaling complex. For example, the SH3 domains of Nck bind to effectors involved in the organization of the cytoskeleton (such as N-WASP and PAK) to link the active RTK to the cytoskeleton (Rivero-Lezcano et al. 1995; Zhao et al. 2000). Scaffolding proteins recruit signaling effectors through linear motifs, including pTyr-containing binding sites, and localized assembly of the signaling complex is directed by the scaffold's PTB or SH2 domain. Shc and IRS-1, for instance, localize to pTyr motifs in an active RTK via their PTB domains and bind the Grb2/Sos complex following phosphorylation of internal YXN motifs, allowing for subsequent activation of the Ras-MAPK pathway (Salcini et al. 1994; van der Geer et al. 1996; Kouhara et al. 1997; Saucier et al. 2004).

In combination with a catalytic domain, the SH2 domains of kinases, phosphatases, guanine nucleotide exchange factors (GEFs), GTPase activating proteins (GAPs), and the phospholipase PLC-γ generally recruit these proteins to their substrates. Such a substrate can be either a tyrosine-phosphorylated protein or a distinct substrate juxtaposed to a tyrosine-phosphorylated protein. Within an enzyme, an SH2 domain can also allow for intramolecular binding

and dynamic regulation of enzymatic activity, and this concept is discussed later in this article. In addition, enzymes containing domains capable of binding secondary messengers (such as PH domains binding phosphatidylinositol 3,4,5-triphosphate (PIP3) or C1 domains binding diacylglycerol (DAG)/phorbol esters) tend to show a higher level of specificity in both localization and activation.

Finally, in the STAT family of transcription factors, SH2 domains enable recruitment to active RTKs; on subsequent RTK-mediated phosphorylation, the SH2 domain mediates STAT dimerization and activation (Stahl et al. 1995; Darnell 1997; Levy and Darnell 2002; Schlessinger and Lemmon 2003). Thus, specific binding of SH2 and PTB domains plays a major role in the localized assembly and activation of signaling effectors. In the sections that follow, we describe in detail how the specific localization of a protein is achieved through the various modes of SH2 binding.

SH2 DOMAINS

The SH2 domain was initially discovered through the observation that an ∼100 amino acid sequence in the v-Fps/Fes oncoprotein was necessary for cellular transformation (Sadowski et al. 1986). This domain was named SH2 because of its homology with a corresponding region in Src family and Abl cytoplasmic tyrosine kinases. Since that time, more than 100 human proteins have been found to contain SH2 domains (Sadowski et al. 1986; Liu et al. 2006).

Early on, it was determined that the affinity of an SH2 domain for pTyr depends on the amino acid sequence surrounding the pTyr residue. A given SH2 domain binds to pTyr residues within a preferred peptide with a dissociation constant in the range of 0.2 to 5 μM (Songyang et al. 1993). By comparison, the dissociation constant for an SH2 domain bound to a pTyr-containing motif of random sequence is 4- to 100-fold lower (∼20 μM) (Songyang et al. 1993). This increase in affinity conveys specificity to each SH2 domain. Thus, domains bind specific pTyr motifs; the Src family kinases (SFKs) bind preferentially to a pYEEI motif,

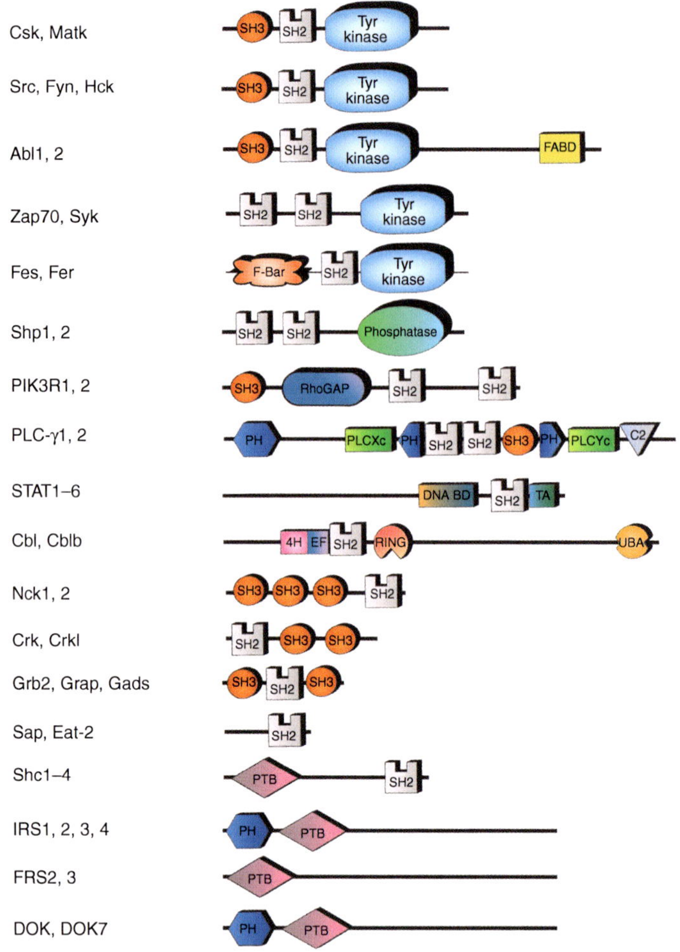

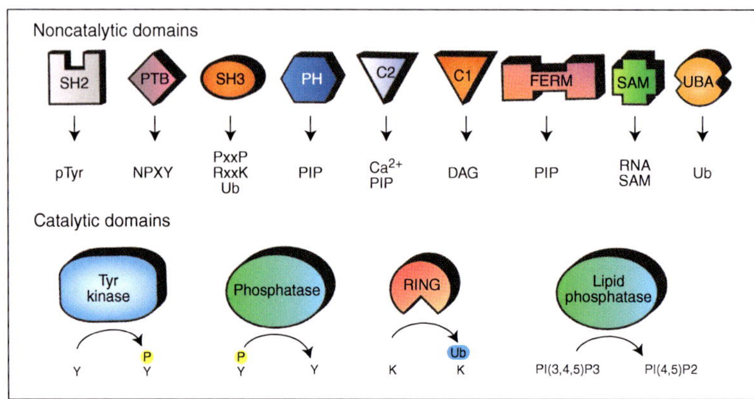

Figure 2. SH2- and PTB-containing proteins are diverse in nature. The modular domain organization of SH2- and PTB-domain proteins displays a diverse set of noncatalytic and catalytic domains for mediating protein–protein interactions and enzyme catalysis, respectively. See the box legend for description of the binding partners and function of these domains. More information on the individual domains portrayed can be found at www.pawsonlab.mshri.on.ca and www.smart.embl-heidelberg.de.

whereas the SH2 domains from PI3K or PLC-γ bind preferentially to pYφXφ (where φ is a residue with a hydrophobic side chain) (Songyang et al. 1993). The specificity of SH2 domains is therefore afforded to some extent by the binding of distinct pTyr-containing motifs, and by other modes of binding described below.

Canonical SH2 Binding

SH2 domains for which the crystal structures have been solved show a high degree of structural homology (reviewed in Kuriyan and Cowburn 1997 and Schlessinger and Lemmon 2003). The typical SH2 domain fold consists of three or four β strands, which make up an antiparallel β sheet, surrounded by two α helices. A positively charged binding pocket on the SH2 domain surface uses a critical Arg residue (within a very highly conserved FLVR motif) (Hidaka et al. 1991; Koch et al. 1991) to bind the pTyr of target ligands. In canonical binding, the residues surrounding the critical Arg typically engage amino acids from position +1 to +6 (carboxy terminal of the pTyr) of the ligand, and it is this sequence that dictates the specificity of a given SH2 domain (Songyang et al. 1993; Pawson 1995; Pawson et al. 2001). For example, binding of the preferred pYEEI motif involves regions on either side of the SH2 domain central β-sheet; between the β-sheet and the amino-terminal αA helix is a deep binding pocket that accommodates the pTyr, whereas on the opposite side of the β-sheet, adjacent to the carboxy-terminal αB helix, residues form a hydrophobic pocket into which the isoleucine chain extends (Fig. 3A,B) (Waksman et al. 1993). Therefore, the pTyr-binding pocket of an SH2 domain is adjacent to the region that dictates binding specificity and determines whether or not the signaling effector is recruited to a given RTK or scaffold protein. This in turn determines which pathway(s) is subsequently activated downstream of the receptor (reviewed in Songyang et al. 1993; Pawson 1995; Pawson et al. 2001).

The various forms of canonical binding give SH2 domains a complexity that contributes greatly to their selectivity. For example, the SH2 domain of Grb2 preferentially binds pYXNX (in which X represents any of the natural 20 amino acids) motifs present among its interaction partners including the Shc proteins (Fig. 3C) (Rozakis-Adcock et al. 1993). Grb2's preference for Asn at the +2 position is mediated by a Trp residue (W121) in the EF loop of the SH2 domain whereby mutation of this position (W121T) displays weak binding to pYXNX motifs (Marengere et al. 1994). The SLAM-associated protein (SAP) SH2 domain, which generally requires its target ligands to be phosphorylated, can also bind its target SLAM constitutively and independent of phosphorylation (Fig. 3D) (Sayos et al. 1998; Li et al. 1999; Poy et al. 1999; Ma and Deenick 2011). Tandem SH2 domains in proteins like PI3K, Shp2, and Zap-70 also enhance the specificity of enzyme recruitment to RTKs. Two closely-spaced tyrosine phosphorylated motifs bind to tandem SH2 domains with 20- to 50-fold greater affinity and specificity compared with the binding of a single SH2 domain with a single tyrosine phosphorylated motif (Ottinger et al. 1998). This enhanced selectivity is due in part to the spacing between the two pTyr motifs. In the case of Zap-70, this is known to be caused by the nature of binding; the carboxy-terminal SH2 domain binds one of the pTyr sites in a conventional manner, whereas the second pTyr activation motif is bound by residues from both the amino-terminal and carboxy-terminal SH2 domains (Hatada et al. 1995). To accommodate binding of one pTyr motif by both SH2 domains, a bisphosphorylated peptide would require specific spacing between the pTyr motifs. Similarly, on binding of one of the SH2 domains of Shp2, the SH2 domains maintain a defined relative position that is stabilized at an interface by a disulfide bond and a hydrophobic patch (Eck et al. 1996); this conformational rigidity is most likely responsible for the requirement of defined spacing between pTyr motifs in a Shp2 bisphosphorylated target (Pluskey et al. 1995).

Canonical SH2 binding may be subject to regulation during cell signaling. A recent publication examining phorbol ester stimulation showed that the PI3K SH2 domains are serine phosphorylated in a manner that occludes pTyr peptide binding (Lee et al. 2011), suggesting

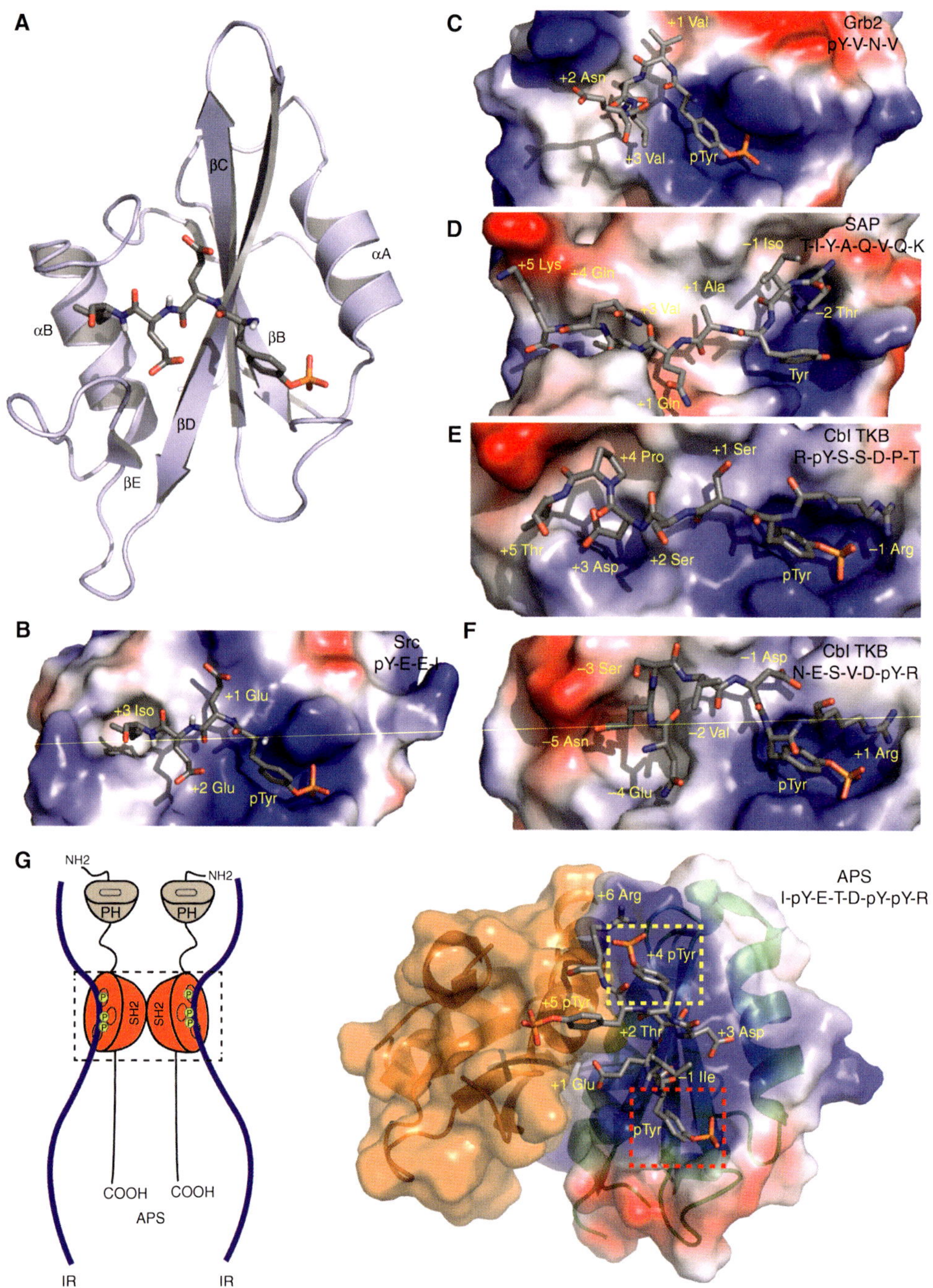

Figure 3. Src homology 2 domains recognize specific phosphotyrosine motifs. (*A*) Ribbon structure of the Src SH2 domain (light blue) bound to a pTyr-Glu-Glu-Ile peptide (PDB: 1SPS) (Waksman et al. 1993). The amino-terminal pTyr of the peptide (gray) occupies the pTyr-binding pocket. The peptide runs over the central β sheet of the SH2 domain, the +1 and +2 glutamates contact the surface of the domain, and the side chain of the +3 Ile (to the *left*) fits in a hydrophobic pocket. (*B*) The electrostatic surface of the SH2 domain reveals the positive charged pTyr-binding pocket (blue is positive, red is negative) and the ligand-binding pocket. (*C*) Grb2 SH2 domain in complex with pYVNV (red) (PDB: 1BMB). (*Legend continues on following page.*)

Cite this article as *Cold Spring Harb Perspect Biol* doi: 10.1101/cshperspect.a008987

that this phosphorylation may prevent SH2-mediated recruitment and activation. Further investigation is required to determine whether this represents a broad-spectrum mechanism of negative regulation.

Alternative Modes of Binding

Some SH2-containing proteins bind their targets in a noncanonical manner. Cbl is an E3 ubiquitin ligase that generally switches "off" protein tyrosine kinase (PTK) targets through ubiquitination and subsequent proteasomal or lysosomal degradation (Sanjay et al. 2001; Ryan et al. 2010). The Cbl SH2 domain has an unusual sequence compared with other SH2 domains (sharing only ~11% sequence homology), and its structure—although similar to other SH2 domains—lacks the conventional BG loop and secondary β-sheet of a typical SH2 domain (Meng et al. 1999). Nevertheless, the Cbl SH2 domain is functionally similar to other SH2 domains and is embedded within the TKB (tyrosine kinase-binding) domain, which includes an EF hand domain and a calcium-binding four-helix bundle (Hu and Hubbard 2005). Extensive interdomain contacts form within the TKB domain, which may contribute to the specificity of the SH2 domain in engaging target phosphopeptides. The extended Cbl SH2 domain recognizes residues both amino and carboxy terminal to the target pTyr, and in the case of binding APS it has been shown that amino acids amino terminal to the pTyr also associate with the four-helix bundle (Hu and Hubbard 2005). Binding of Cbl to proteins including the EGFR, vascular endothelial growth factor receptor, ZAP-70, Src, and Syk involves the canonical (N/D)XpY(S/T)XXP consensus motif

(Fig. 3E) (Lupher et al. 1997; Meng et al. 1999; Schmidt and Dikic 2005), whereas the unrelated DpYR motif in the Met and Plexin families also binds Cbl (Tamagnone et al. 1999; Penengo et al. 2003; Peschard et al. 2004). Surprisingly, the crystal structure of the Cbl TKB domain binding to the pY ligand of Met revealed that Cbl can bind this motif in two orientations: the "canonical" forward orientation and the reverse orientation (amino to carboxyl terminus) (Fig. 3F) (Ng et al. 2008).

In addition, the APS adaptor protein has been shown to dimerize and stabilize the activation of RTKs including the IR through interactions with the activation loop (Ahmed et al. 1999; Moodie et al. 1999; Yokouchi et al. 1999). This dimerization may be important for trans-phosphorylation—and activation of the kinase domains—of these receptors (Dhe-Paganon et al. 2004; Nishi et al. 2005). The solved structure for the dimerized APS SH2 domains bound to the IR activation loop peptide revealed conventional pTyr binding to Y1158. The dimerization induces a conformational change that creates a second pTyr-binding pocket, and an unusual turn in the peptide ligand running parallel to the β-strands. This new conformation allows for charged interactions between the second pTyr site on IR Y1162 and two Lys residues in the β-D strand (Hu et al. 2003; Dhe-Paganon et al. 2004; Hu and Hubbard 2006).

In some proteins, secondary binding sites in the SH2 domain, distinct from the pTyr motif-binding site, also participate in specific interactions. For example, in the Fes and Abl kinases, an intramolecular interaction between a secondary SH2-binding site and the kinase domain enhances enzymatic activity, as discussed later in this article (Filippakopoulos et al. 2008).

Figure 3. (*Continued*) (*D*) The SAP SH2 domain can recognize nonphoshorylated SLAM peptide and residues amino terminal to the tyrosine, such as -2 Thr (PDB: 1M27). (*E*) pTyr-binding pocket of the Cbl SH2 is bound in the canonical fashion with EGFR peptide pYSSDP (gray) with the carboxyl terminus extended across the SH2 surface (PDB: 3BUO). (*F*) Cbl TKB in complex with the MET peptide is oriented in the reverse direction with the amino acids amino terminal to the pTyr extended across the SH2 domain (PDB: 3BUX). (*G*) *Left* panel: A graphical representation of the dimerized APS molecules bound to the insulin receptor (IR). *Right* panel: The dimerized SH2 domain of APS bound to the activation loop peptide of the IR with the primary pTyr pocket (red box) and the second pocket (yellow box) indicated.

PTB DOMAINS

The pTyr-binding (PTB) domain was first identified in the scaffold protein Shc, which has a carboxy-terminal SH2 and amino-terminal PTB domain. This PTB domain was found to bind a pTyr residue in the EGF receptor (Blaikie et al. 1994; Kavanaugh and Williams 1994), and it was originally assumed that a characteristic function of all such domains was recruitment in a phospho-dependent manner.

To date, approximately 60 PTB domains have been identified in the human proteome; mutations in six of these have been associated with heritable diseases including Alzheimer's disease, diabetes, and coronary artery disease (reviewed in Uhlik et al. 2005). It turns out that many PTB domains do not in fact rely on ligand tyrosine phosphorylation for binding, but commonly PTB domains do bind phospholipid acidic head groups, which helps to localize them to membrane or juxtamembrane regions where they can easily bind RTKs and mediate downstream signaling. Binding to phospholipid head groups is mediated by a positively charged binding pocket composed of a cluster of basic residues, and is a distinct event from the PTB-domain-containing protein's interactions with pTyr motifs. This section discusses the pTyr-dependent PTB domains.

The 3D structure of the Shc PTB domain was published just 1 year after its identification (Zhou et al. 1995). Surprisingly, its structure resembles that of the pleckstrin homology (PH) domain, despite highly dissimilar sequences (Zhou et al. 1995). To date, all PTB domains for which the crystal structures have been solved appear to share the same folding pattern, including a β-sandwich composed of two virtually orthogonal β-sheets, which is capped by a carboxy-terminal α-helix (Fig. 4). Together, the β5 strand and α-helix make up the ligand-binding pocket (Schlessinger and Lemmon 2003). This binding pocket appears to be highly conserved, not only in its location and spatial arrangement but also in the binding mechanism: a cleft between the β5 strand and the carboxy-terminal α helix that accommodates the ligand (Harrison 1996; reviewed in Schlessinger and Lemmon 2003; Uhlik et al. 2005).

Modes of Peptide Binding

The motif NPXY is common among PTB-domain substrates and is considered the canonical binding motif for PTB-domain-containing proteins. Uhlik and others (Uhlik et al. 2005) conducted an extensive analysis of the PTB-containing proteins; using structural, functional, and evolutionary data, they classified PTB-domain-containing proteins into three groups: pTyr-dependent Shc-like; pTyr-dependent IRS-like; and pTyr-independent Dab-like (Fig. 4).

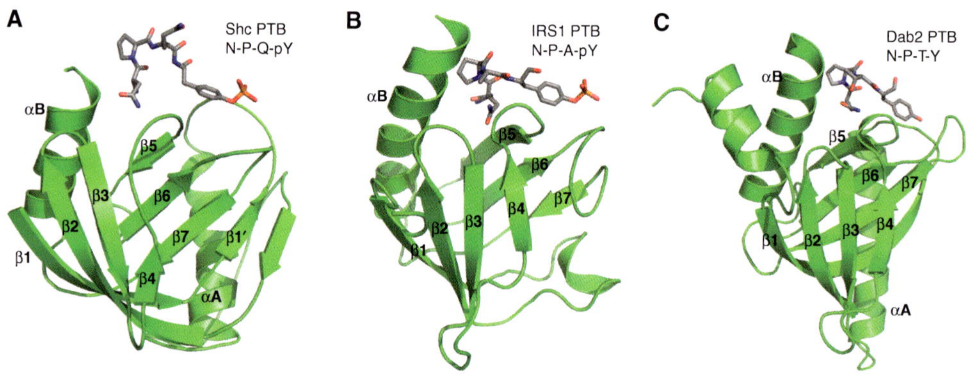

Figure 4. Structure of phosphotyrosine-binding domains. The ribbon structures shown in green of the PTB domains of Shc (*A*) (PDB: 1SHC), IRS-1 (*B*) (PDB: 1IRS), and Dab2 (*C*) (PTB: 1ME7) bound to their respective N-P-x-(p)Y ligand (gray). x Denotes any natural amino acid; pY represents phosphotyrosine. The mode of ligand binding to the PTB domains shown is similar.

These three classes are distinct in their ability to bind a phosphorylated NPXY motif. The Shc, IRS, and Dok scaffolds preferentially bind phosphorylated substrates, whereas the Dab-like PTB-domain-containing proteins bind preferentially to residues that are unphosphorylated or in which the tyrosine is replaced with phenylalanine (Zhang et al. 1997; Dho et al. 1998; Howell et al. 1999; Uhlik et al. 2005).

The three classes of PTB-containing proteins have three distinct modes of peptide binding described below. To accommodate the negatively charged phosphate moiety, the Shc-binding pocket is basic and positively charged, and forms a network of hydrogen bonding with the phosphate. Three key residues are involved in triangulating the oxygen atoms of the phosphate molecule, and these are conserved in all Shc family members: two arginines (Arg67, Arg 175 in Shc) and one lysine (Lys169 in Shc) (Uhlik et al. 2005). As with Shc, IRS-like PTB domains have a positively charged pocket, but only two key arginine residues bind the phosphoryl oxygen molecules: these are Arg212 and Arg227 in IRS-1 (Uhlik et al. 2005). The binding of target ligands with Dab-like PTB domains is non-phospho-dependent and the binding pocket is both less basic and more shallow, having a less important role in peptide-binding affinity. Even though the preferred binding motif is unphosphorylated, the tyrosine at position 0 is important; His136 (in Dab1) forms van der Waals contacts with this residue and Gly131 (Dab1) forms a hydrogen bond. A number of other residues in the β5 strand also form hydrogen bonds with the ligand (Borg et al. 1996; Zhang et al. 1997; Uhlik et al. 2005). A number of hydrophobic contacts and hydrogen bonds mediate extensive contacts between the Dab-like PTB domains and NPXY-containing ligands (Stolt et al. 2003; Yun et al. 2003; Uhlik et al. 2005).

PTB Domain Function

Proteins containing PTB domains lack inherent catalytic activity. Although a number of these proteins also lack any other distinguishable domains, many contain additional protein–protein interaction modules, such as PH, SH2, SH3, PDZ, and SAM domains (Uhlik et al. 2005). In these proteins, the domains allow for pTyr-dependent and -independent interactions, whereas the disordered regions can become phosphorylated to recruit additional signaling components.

Shc is the classical example of a dual SH2 and PTB-domain-containing protein. Shc becomes tyrosine phosphorylated on binding to a receptor, an association that is mediated by its PTB domain (Ravichandran 2001). This pathway is discussed in more detail elsewhere, but the Shc example illustrates how the combination of peptide-binding motifs within a single protein is capable of translating RTK activation into downstream signaling via specific protein–protein interactions.

Similar examples of docking proteins can be seen with IRS-like PTB domains, including IRS-1, FRS2, and Dok-2 (Guy et al. 2002). IRS-like members have been found to bind activated RTKs, including members of the IR, EGF, and RET families (reviewed in Uhlik et al. 2005). Signaling specificity is achieved by variations in binding affinity between recruited proteins (Uhlik et al. 2005). For example, the PTB domain of IRS binds strongly to Tyr960 of the IR juxtamembrane region (Gustafson et al. 1995; Sasaoka and Kobayashi 2000), whereas Shc binds poorly to the same position (van der Geer et al. 1999).

THE SH2 DOMAIN AS A MOLECULAR SWITCH

RTK signaling is initiated through SH2-mediated binding of active RTKs and phosphorylated scaffolding proteins. In this section, we discuss how this and other SH2-binding events act as "molecular switches" that "turn on" downstream signaling by giving a precise order to the recruitment and activation of signaling effectors.

SH2-Mediated Recruitment of Signaling Adaptors

Grb2 initiates two signaling pathways following RTK activation and SH2-mediated recruitment: the canonical Ras-MAPK pathway and the canonical PI3K-Akt pathway. Grb2 is recruited to the cell membrane by binding a pYXNX motif in the RTK itself or in an associated scaffold protein, like Shc or IRS (Fig. 3C). One of the SH3 domains of Grb2 facilitates signaling through the Ras-MAPK pathway by binding the GEF, and activator of Ras, Sos (Fig. 1, inset) (Li et al. 1993; Rozakis-Adcock et al. 1993). Translocation brings Sos proximal to its membrane-bound target Ras. Colocalization seems to be sufficient for Sos to activate Ras, stimulating Ras to exchange GDP for GTP. GTP-bound Ras then activates the serine–threonine kinase Raf, as well as binding to an allosteric regulatory site on Sos to stimulate catalytic activity (Margarit et al. 2003; Sondermann et al. 2004); this ultimately leads to activation of the serine/threonine kinase Erk (extracellular signal-regulated kinase). The outcome of this signaling varies from cell proliferation and survival to differentiation depending on the cellular context, which includes complex mechanisms of Erk regulation and a multiplicity of Erk substrates (Yoon and Seger 2006; Ramos 2008).

A second SH3 domain of Grb2 associates with the scaffold protein Grb2-associated binder 1/2 (Gab) to allow for PI3K signaling. PI3K is activated on recruitment to the cell membrane via binding of a phosphorylated activation motif (pYXXM) in a membrane-proximal protein such as Gab. Once proximal to its lipid substrate, PI3K converts phosphatidylinositol (4,5)-bisphosphate (PIP2) to PIP3. The formation of PIP3 allows for Akt activation because PIP3 specifically binds the PH domains of both Akt and the Akt activator, 3-phosphoinositide-dependent kinase 1 (PDK1) (Newton 2009). Colocalization and a conformational change in Akt (induced by PIP3 binding) allow PDK1 to phosphorylate Akt at threonine 308 (T308) (Stokoe et al. 1997; Stephens et al. 1998). Full Akt activity also requires phosphorylation at serine 473 (S473) by mammalian target of rapamycin complex 2 (mTORC2) (Sarbassov et al. 2005). The final biological outcome of Akt signaling includes promotion of cell growth, proliferation, and survival. Once again, the signaling outcome is thought to be dependent on which substrates are targeted by Akt (Manning and Cantley 2007).

Controlled Assembly of the Crk-Signaling Complex

The activity of an adaptor can be controlled by mechanisms aside from SH2-mediated recruitment. The Crk adaptor is autoinhibited through intramolecular interactions. Crk consists of an amino-terminal SH2 domain and two SH3 domains separated by a linker region containing a phosphorylation site (Y221). The potential conformations of this molecule include two structures that result in the inhibition of its binding activities. Under basal conditions, Crk is maintained in a compact structure with the inter-SH3 region forming contacts with the SH2 domain and each of its two SH3 domains (Kobashigawa et al. 2007). In this closed conformation, the binding site of the amino-terminal SH3 is occluded by the SH2 domain, leaving it unable to bind to signaling effectors. However, the binding site of the SH2 domain is unobstructed. This suggests that SH2-mediated recruitment to a phosphorylated activation motif may allow for rapid and specific binding of proximal signaling effectors at the amino-terminal SH3 domain. The second inhibitory conformation is thought to be induced as a negative feedback mechanism to limit signaling downstream of RTKs and other receptors. This conformation is induced by active Abl kinase, which phosphorylates Crk at Y221 in the SH3 linker region (Feller et al. 1994). The SH2 domain of Crk then binds to this pTyr and in doing so prevents its binding to intermolecular targets (Feller et al. 1994; Rosen et al. 1995).

Kinase Inhibition by Intramolecular SH2-Mediated Interactions

The pairing of a kinase domain and a preceding amino-terminal SH2 domain is a configuration that is highly conserved among nonreceptor tyrosine kinases. Whereas this configuration may have initially served to recruit an enzyme to its cognate substrate, it has evolved to play multiple roles in activating and deactivating kinases (Mayer et al. 1995; Li et al. 2008).

SH2 domains stabilize an inactive conformation in nonreceptor tyrosine kinases, including the SFKs and Abl. On a biochemical level,

Cite this article as *Cold Spring Harb Perspect Biol* doi: 10.1101/cshperspect.a008987

the activity of these kinases is enhanced as the enzyme switches from a "closed" inactive conformation to an "open" active configuration. For SFKs, this conformational change is regulated by the phosphorylation state of a carboxy-terminal inhibitory tyrosine residue (the equivalent of Y527 in Src). This is regulated by the kinase Csk (c-Src kinase) (Okada and Nakagawa 1989). Crystal structures of inactive SFKs indicate that the closed conformation is characterized by intramolecular interactions between the SH2 domain and the carboxy-terminal inhibitory pTyr motif, and between the SH3 domain and the linker region connecting the SH2 domain and the catalytic domain (Sicheri et al. 1997; Xu et al. 1997). This rigid structure is thought to limit the flexibility of the kinase active site (Young et al. 2001). Inactive Abl mirrors the rigid and closed conformation of inactive Src. However, Abl lacks a regulatory pTyr residue, and an amino-terminal myristoyl group is in part responsible for stabilizing or destabilizing the inhibitory interactions (Hantschel et al. 2003; Nagar et al. 2003).

Activation from a State of Autoinhibition

The intramolecular interactions of autoinhibition, such as those of SFKs, Abl, and the adaptor Crk, must be energetically favorable in the absence of a target peptide to prevent constitutive activation. However, they must also be suboptimal for the protein to be labile and able to respond to changes in cell signaling. Structural analysis of the inhibitory conformation of Crk supports this theory because the carboxy-terminal SH3 domain is in a thermodynamically unstable conformation (Cho et al. 2011). Similarly, "strained" conformations can also be observed in SFKs, in which dephosphorylation of the inhibitory tyrosines are sufficient to trigger an open and active conformation (Fig. 5A) (Cooper et al. 1986; Kmiecik and Shalloway 1987; Amrein and Sefton 1988; Xu et al. 1999). In addition, kinase activity can be triggered by the binding of intermolecular target ligands to the SH2 and/or SH3 domains (Liu et al. 1993; Lerner and Smithgall 2002). However, a mutant of the SFK Lck is constitutively inactive and can-

not be activated by T-cell receptor signaling because of a substitution of the native carboxy-terminal inhibitory motif for an alternative optimal SH2-binding sequence: pYEEI (Nika et al. 2007).

SH2-Mediated Allosteric Activation of Kinases

The SH2 domains of the nonreceptor tyrosine kinases Csk, Fes, and Abl participate in intramolecular interactions that enhance kinase activity. This is evidenced by the isolated kinase domains from Csk or Fes, which are up to 100-fold less efficient than the full-length kinases (Stone et al. 1984; Sadowski et al. 1986; Filippakopoulos et al. 2008). The isolated Abl kinase domain is also less efficient compared with an SH2-kinase construct of Abl (Fig. 5B) (Filippakopoulos et al. 2008), indicating that the adjacent SH2 domain directly activates the kinase through allosteric effects. In contrast, isolation of the SFK kinase domains (i.e., deletion of the amino-terminal SH2 and SH3 domains) does not hinder activity. Thus, some kinase domains require the SH2 domain for full activity, whereas those of the SFKs do not (Xu and Miller 1996).

The intramolecular mechanisms of activation have been extensively studied for Csk, Fes, and Abl. The crystal structure of full-length Csk shows extensive hydrophobic contacts between a regulatory αC helix of the kinase domain and the regions surrounding the SH2 domain, the SH3-SH2 linker, and the SH2-kinase domain linker (Ogawa et al. 2002). Similarly, in Fes, acidic residues in the amino-terminal region of the SH2 domain that are distinct from the pTyr-binding pocket form polar interactions with the αC helix of its kinase domain (Filippakopoulos et al. 2008). In each case, this interaction seems to stabilize the active configuration of the αC helix and allows for the formation of a lysine-glutamate salt bridge in the active site, and this is critical for orienting the γ-phosphate of ATP. In both structures, the SH2-kinase linker is flexible, suggesting that the SH2 domain can rotate to bind phosphopeptides. Thus, on RTK signaling, SH2-mediated binding of an activation motif may stabilize the active con-

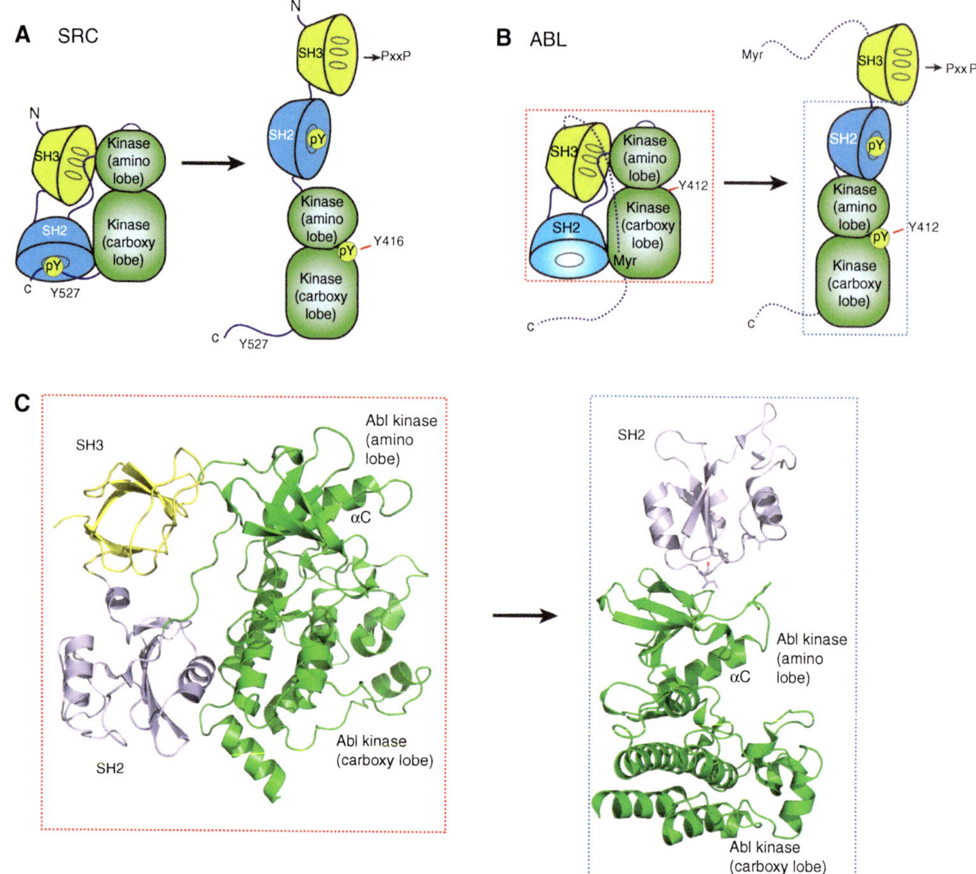

Figure 5. SH2 domains regulate tyrosine kinase activation. (*A*) In the autoinhibited state of Src, the SH2 domain recognizes an intramolecular pTyr site (Y572) and stabilizes the inactive kinase. Additional contacts with the SH3 and the SH2-kinase linker promote stabilization of the inactive conformation of Src. Activation of Src, initiated by dephosphorylation of Y527, frees the SH3 and SH2 to recognize other short linear motifs and further promote kinase activity. (*B*) The Abl tyrosine kinase remains in an inactive conformation analogous to Src but lacks the intramolecular phosphorylation site. Instead, an intramolecular amino-terminal myristolation stabilizes the inactive conformation of Abl. (*C*) For Abl kinase to be active, it requires the SH2 domain to stabilize the amino lobe of the kinase domain, thereby allowing it to couple pTyr ligand binding and substrate recognition (PDB:1OPL).

formation. This idea is supported by observations of increased activity of Csk on binding to its phosphorylated target in Cbp (Csk-binding protein) (Takeuchi et al. 2000), and reduced activity in Fes mutants containing a deactivated SH2 pTyr-binding pocket (Filippakopoulos et al. 2008).

Similar to Fes, Abl activity is driven by sites of contact between the SH2 domain (I164) and the amino lobe of the kinase domain (T291/ Y331) (Fig. 5C) (Nagar et al. 2006; Filippako-

poulos et al. 2008). This interaction may represent a promising avenue for novel therapeutics because it was shown to be strictly required for the formation of Bcr-Abl-driven leukemia in a mouse model (Grebien et al. 2011).

Tandem SH2 Domains Mediate SHP-2 Phosphatase Activity

Although phosphatases are often thought of as negative regulators of signaling, Shp2 is known to activate the Ras-MAPK pathway downstream of RTKs. In this pathway, the roles of Shp2 as a scaffold for Grb2 and as a phosphatase have yet to be fully elucidated. However, several studies have shown that catalytic activity is necessary for Ras-MAPK activation (Noguchi et al. 1994; Deb et al. 1998; Cunnick et al. 2000; Maroun et al. 2000; Yart et al. 2001).

Intramolecular binding of the amino-terminal SH2 domain of Shp2 limits the activity of its PTP (protein tyrosine phosphatase) domain. The crystal structure of Shp2 indicates that the amino-terminal SH2 domain interacts with the PTP domain at residues that are distinct from the pTyr-binding pocket (Hof et al. 1998). Mutation of the amino-terminal SH2 at its PTP interacting residues prevents autoinhibition, resulting in Shp2 catalytic activity (O'Reilly et al. 2000). Interaction between the SH2 domain and a monophosphorylated activation motif is also sufficient to disrupt the PTP-binding surface and activate the enzyme in vitro (Lechleider et al. 1993; Sugimoto et al. 1994; Hof et al. 1998). Collectively, these data suggest that SH2 target binding simultaneously localizes and disrupts the autoinhibition of Shp2.

Binding of both SH2 domains may further enhance the activity of Shp2. Whereas binding of a monophosphorylated activation motif from IRS-1 is sufficient to activate Shp2, catalytic activity is more potently induced by the corresponding bisphosphorylated motif (Lechleider et al. 1993; Sugimoto et al. 1994; Pluskey et al. 1995). This suggests that Shp2 activity may be enhanced by the engagement of both SH2 domains by IRS-1 in the IR pathway, and potentially by other targets in other RTK pathways. One study suggests that the amino-terminal and carboxy-terminal SH2 domains of Shp2 bind to intramolecular pTyrs Y542 and Y580 to enhance activity in vitro (Lu et al. 2001). As well, one of these pTyrs is necessary and sufficient to activate the Ras-MAPK pathway in vivo (Lu et al. 2001; Araki et al. 2003). Intramolecular interactions of the SH2 domain may therefore play a role in positively regulating Shp2 phosphatase activity. However, it is of note that some reports suggest that pY542 facilitates binding of

the Grb2 adaptor and that this represents an alternative mechanism of Ras activation (Bennett et al. 1994; Li et al. 1994; Vogel and Ullrich 1996; Araki et al. 2003). The role of these pTyr residues in activating Shp2 and Ras-MAPK signaling warrants further investigation.

SH2-Mediated Assembly of the PLC-γ Catalytic Domain

Active PLC-γ cleaves the membrane lipid PIP2 to give rise to the secondary messengers DAG and IP$_3$, and subsequently activates calcium signaling and protein kinase C to induce effects on cell proliferation, cell survival, and cytoskeletal movements. The amino-terminal and carboxy-terminal SH2 domains of PLC-γ act in concert to recruit PLC-γ to the cell membrane and assemble the two lipase subdomains into an active configuration. PLC-γ has an amino-terminal PH domain, a carboxy-terminal C2 domain, as well as a second PH domain and a catalytic domain, each of which is split into two subdomains located on either side of a stretch of amino acids containing two SH2 domains and an SH3 domain. The amino-terminal SH2 domain has been shown to play a dominant role in pTyr-mediated recruitment of PLC-γ to the cell membrane downstream of PDGFR (Ji et al. 1999). In the PLC-γ-FGFR interaction, the high-affinity binding of the amino-terminal SH2 domain is enhanced by a secondary binding site on the SH2, and this secondary site of contact is required for subsequent enzyme activation (Fig. 6) (Bae et al. 2009). Once bound to an RTK, PLC-γ becomes phosphorylated at a critical regulatory tyrosine (Y783) in the linker region between the carboxy-terminal SH2 and the SH3 domains (Kim et al. 1991). The carboxy-terminal SH2 domain binds this pTyr to enhance catalytic activity (Poulin et al. 2005), presumably through a conformational change that unifies the split catalytic domain. Thus the amino-terminal SH2 domain recruits PLC-γ for tyrosine phosphorylation and this allows the carboxy-terminal SH2 domain to restructure PLC-γ into an active state.

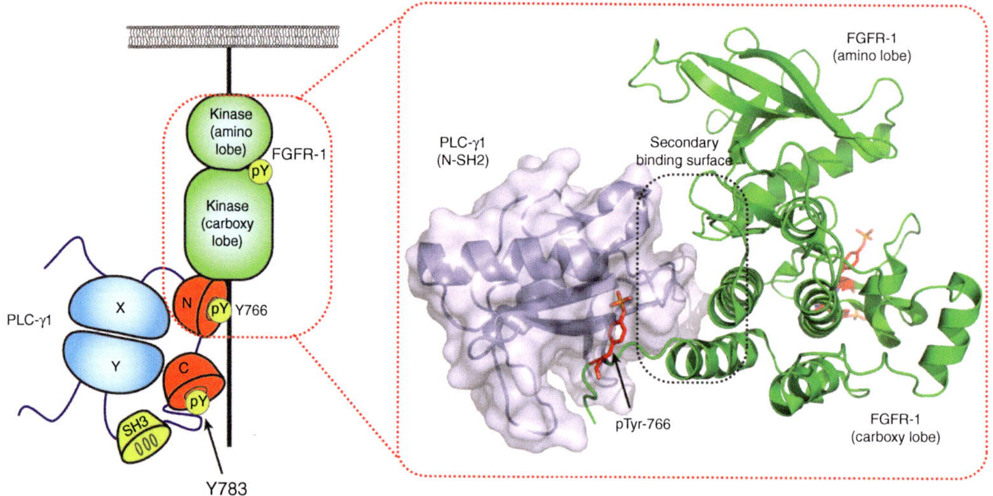

Figure 6. Activation of PLC-γ. Phospholipase C isoform γ (PLC-γ1) is a tandem SH2 domain-containing protein (see Fig. 2 for a complete view of the domain organization of PLC-γ1) that is recruited to activated FGFR-1. The amino-terminal SH2 domain of PLC-γ1 mediates primary pTyr binding to Y766 on FGFR-1 but also mediates secondary contacts through the B,C loop and D,E loops of the SH2 domain and the αE, I helix of the carboxy lobe of the kinase domain (PDB: 3GQI). In the activated state of PLC-γ1, the carboxy-terminal SH2 recognizes an intramolecular tyrosine phosphorylation site Y783 bringing together the PLC-γ-x (X) and PLC-γ-y (Y) lobes, activating the phospholipase enzyme.

SH2 DOMAINS AND PHOSPHOTYROSINE-MEDIATED SIGNALING IN DISEASE

As reviewed in this article, pTyr binding and other SH2-mediated molecular switches are critical for controlling basal levels of cell signaling and the level of activation following RTK ligation. Loss of these functions is known to cause disease in many instances.

Mutations abrogating pTyr binding of SH2 or PTB domains prevent specific recruitment of the signaling effector following RTK ligation and are sufficient to cause disease. For example, mutation of the c-Cbl SH2-binding site in the RTK c-Met is sufficient to induce malignant transformation of cells in vitro and in vivo (Park et al. 1986; Abella et al. 2005). Cbl binds RTKs and down-regulates signaling through receptor degradation (Peschard et al. 2001). Exposure to a chemical carcinogen causes loss of the exon that encodes for the juxtamembrane region, including the Cbl SH2-binding site (Park et al. 1986; Vigna et al. 1999; Peschard et al. 2001). This mutation prevents down-reg-

ulation of c-Met signaling and induces cellular transformation. Similarly, mutation of the PTB-binding site of the Dok-7 scaffold is sufficient to cause congenital myasthenic syndrome (CMS). CMS is characterized by malformed muscular junctions and manifests early in life as a muscular weakness that causes hypomobility (Engel 2012). The Dok-7 scaffold is essential for neuromuscular junction formation because it binds the RTK MUSK, and allows for MUSK dimerization and activation in the postsynaptic muscle (Okada et al. 2006; Bergamin et al. 2010). One of the known CMS mutations of Dok7 (Arg 158 to Glu) disrupts a critical salt bridge formed between the pTyr-binding pocket of the PTB domain and the pTyr-binding motif (NPXpY) in MUSK (Bergamin et al. 2010). The resulting disruption of MUSK dimerization and MUSK signaling leads to the malformed muscular junctions of CMS.

Mutation in the SH2 domain of Shp2 disrupts the inactive conformation of the enzyme and manifests in one of two distinct disease phenotypes: Noonan syndrome or myelomonocytic leukemia. Noonan syndrome is a disease characterized in part by congenital heart defects and reduced postnatal growth. Roughly one-half of Noonan syndrome patients have mutations in Shp2 at or close to the interacting residues of the phosphatase domain and the amino-terminal SH2 domain (Tartaglia et al. 2001). This mutation is predicted to result in a gain of function because this interaction surface is known to prevent Shp2-mediated activation of the Ras-MAPK pathway (Hof et al. 1998). Interestingly, less conservative mutations of the same region can cause juvenile myelomonocytic leukemia (Bentires-Alj et al. 2004), suggesting that autoinhibition of Shp2 plays multiple roles in controlling the development and maintenance of human tissues.

Loss of SH2-mediated autoinhibition leads to cellular transformation by the viral homolog of Src. Rous sarcoma virus contains an Src homolog known as v-Src (viral Src) (Martin 2001). Minor differences in the amino acid sequences differentiate v-Src and cellular Src. This includes the loss of the inhibitory pTyr motif. The result is a constitutively active form of Src that transforms the cell through activation of the Ras-MAPK pathway and the PI3K pathway (Cooper et al. 1986; Penuel and Martin 1999).

CONCLUDING REMARKS

The discovery of the SH2 domain allowed us to appreciate the importance of protein-interaction domains and pTyr-mediated signaling. We now have a strong understanding of the mechanisms involved in intracellular signaling and how it influences human development and disease. With this knowledge we have developed an array of therapeutics targeting nonreceptor kinases and RTKs in cancer and further advances in the field will hopefully allow us to develop novel therapeutics disrupting the intracellular interactions.

In recent years, there has been a shift from contextual studies of the individual components of signaling pathways to high-throughput methodologies that examine signaling in terms of the genetic requirements along with global changes in tyrosine phosphorylation and gene expression. Combining this approach with computational analysis of large data sets has the potential to rapidly advance the understanding of RTK signaling and other signaling systems.

REFERENCES

Abella JV, Peschard P, Naujokas MA, Lin T, Saucier C, Urbe S, Park M. 2005. Met/hepatocyte growth factor receptor ubiquitination suppresses transformation and is required for Hrs phosphorylation. *Mol Cell Biol* **25**: 9632–9645.

Ahmed Z, Smith BJ, Kotani K, Wilden P, Pillay TS. 1999. APS, an adapter protein with a PH and SH2 domain, is a substrate for the insulin receptor kinase. *Biochem J* **341**: 665–668.

Amrein KE, Sefton BM. 1988. Mutation of a site of tyrosine phosphorylation in the lymphocyte-specific tyrosine protein kinase, p56lck, reveals its oncogenic potential in fibroblasts. *Proc Natl Acad Sci* **85**: 4247–4251.

Anderson D, Koch CA, Grey L, Ellis C, Moran MF, Pawson T. 1990. Binding of SH2 domains of phospholipase C$_\gamma$1, GAP, and Src to activated growth factor receptors. *Science* **250**: 979–982.

Araki T, Nawa H, Neel BG. 2003. Tyrosyl phosphorylation of Shp2 is required for normal ERK activation in response to some, but not all, growth factors. *J Biol Chem* **278**: 41677–41684.

Bae JH, Lew ED, Yuzawa S, Tome F, Lax I, Schlessinger J. 2009. The selectivity of receptor tyrosine kinase signaling is controlled by a secondary SH2 domain binding site. *Cell* **138**: 514–524.

Bennett AM, Tang TL, Sugimoto S, Walsh CT, Neel BG. 1994. Protein-tyrosine-phosphatase SHPTP2 couples platelet-derived growth factor receptor β to Ras. *Proc Natl Acad Sci* **91**: 7335–7339.

Bentires-Alj M, Paez JG, David FS, Keilhack H, Halmos B, Naoki K, Maris JM, Richardson A, Bardelli A, Sugarbaker DJ, et al. 2004. Activating mutations of the Noonan syndrome-associated SHP2/PTPN11 gene in human solid tumors and adult acute myelogenous leukemia. *Cancer Res* **64**: 8816–8820.

Bergamin E, Hallock PT, Burden SJ, Hubbard SR. 2010. The cytoplasmic adaptor protein Dok7 activates the receptor tyrosine kinase MuSK via dimerization. *Mol Cell* **39**: 100–109.

Blaikie P, Immanuel D, Wu J, Li N, Yajnik V, Margolis B. 1994. A region in Shc distinct from the SH2 domain can bind tyrosine-phosphorylated growth factor receptors. *J Biol Chem* **269**: 32031–32034.

Borg JP, Ooi J, Levy E, Margolis B. 1996. The phosphotyrosine interaction domains of X11 and FE65 bind to distinct sites on the YENPTY motif of amyloid precursor protein. *Mol Cell Biol* **16**: 6229–6241.

Cho JH, Muralidharan V, Vila-Perello M, Raleigh DP, Muir TW, Palmer AG 3rd. 2011. Tuning protein autoinhibition by domain destabilization. *Nat Struct Mol Biol* **18**: 550–555.

Cooper JA, Gould KL, Cartwright CA, Hunter T. 1986. Tyr527 is phosphorylated in pp60c–src: Implications for regulation. *Science* 231: 1431–1434.

Cunnick JM, Dorsey JF, Munoz-Antonia T, Mei L, Wu J. 2000. Requirement of SHP2 binding to Grb2-associated binder-1 for mitogen-activated protein kinase activation in response to lysophosphatidic acid and epidermal growth factor. *J Biol Chem* 275: 13842–13848.

Darnell JE Jr. 1997. STATs and gene regulation. *Science* 277: 1630–1635.

Deb TB, Wong L, Salomon DS, Zhou G, Dixon JE, Gutkind JS, Thompson SA, Johnson GR. 1998. A common requirement for the catalytic activity and both SH2 domains of SHP-2 in mitogen-activated protein (MAP) kinase activation by the ErbB family of receptors. A specific role for SHP-2 in map, but not c-Jun amino-terminal kinase activation. *J Biol Chem* 273: 16643–16646.

Dhe-Paganon S, Werner ED, Nishi M, Hansen L, Chi YI, Shoelson SE. 2004. A phenylalanine zipper mediates APS dimerization. *Nat Struct Mol Biol* 11: 968–974.

Dho SE, Jacob S, Wolting CD, French MB, Rohrschneider LR, McGlade CJ. 1998. The mammalian numb phosphotyrosine-binding domain. Characterization of binding specificity and identification of a novel PDZ domain-containing numb binding protein, LNX. *J Biol Chem* 273: 9179–9187.

Eck MJ, Pluskey S, Trub T, Harrison SC, Shoelson SE. 1996. Spatial constraints on the recognition of phosphoproteins by the tandem SH2 domains of the phosphatase SH-PTP2. *Nature* 379: 277–280.

Engel AG. 2012. Current status of the congenital myasthenic syndromes. *Neuromuscul Disord* 22: 99–111.

Feller SM, Knudsen B, Hanafusa H. 1994. c-Abl kinase regulates the protein binding activity of c-Crk. *EMBO J* 13: 2341–2351.

Filippakopoulos P, Kofler M, Hantschel O, Gish GD, Grebien F, Salah E, Neudecker P, Kay LE, Turk BE, Superti-Furga G, et al. 2008. Structural coupling of SH2-kinase domains links Fes and Abl substrate recognition and kinase activation. *Cell* 134: 793–803.

Grebien F, Hantschel O, Wojcik J, Kaupe I, Kovacic B, Wyrzucki AM, Gish GD, Cerny-Reiterer S, Koide A, Beug H, et al. 2011. Targeting the SH2-kinase interface in Bcr-Abl inhibits leukemogenesis. *Cell* 147: 306–319.

Gustafson TA, He W, Craparo A, Schaub CD, O'Neill TJ. 1995. Phosphotyrosine-dependent interaction of SHC and insulin receptor substrate 1 with the NPEY motif of the insulin receptor via a novel non-SH2 domain. *Mol Cell Biol* 15: 2500–2508.

Guy GR, Yusoff P, Bangarusamy D, Fong CW, Wong ES. 2002. Dockers at the crossroads. *Cell Signal* 14: 11–20.

Hantschel O, Nagar B, Guettler S, Kretzschmar J, Dorey K, Kuriyan J, Superti-Furga G. 2003. A myristoyl/phosphotyrosine switch regulates c-Abl. *Cell* 112: 845–857.

Harrison SC. 1996. Peptide-surface association: The case of PDZ and PTB domains. *Cell* 86: 341–343.

Hatada MH, Lu X, Laird ER, Green J, Morgenstern JP, Lou M, Marr CS, Phillips TB, Ram MK, Theriault K, et al. 1995. Molecular basis for interaction of the protein tyrosine kinase ZAP-70 with the T-cell receptor. *Nature* 377: 32–38.

Hidaka M, Homma Y, Takenawa T. 1991. Highly conserved eight amino acid sequence in SH2 is important for recognition of phosphotyrosine site. *Biochem Biophys Res Commun* 180: 1490–1497.

Hof P, Pluskey S, Dhe-Paganon S, Eck MJ, Shoelson SE. 1998. Crystal structure of the tyrosine phosphatase SHP-2. *Cell* 92: 441–450.

Howell BW, Lanier LM, Frank R, Gertler FB, Cooper JA. 1999. The disabled 1 phosphotyrosine-binding domain binds to the internalization signals of transmembrane glycoproteins and to phospholipids. *Mol Cell Biol* 19: 5179–5188.

Hu J, Hubbard SR. 2005. Structural characterization of a novel Cbl phosphotyrosine recognition motif in the APS family of adapter proteins. *J Biol Chem* 280: 18943–18949.

Hu J, Hubbard SR. 2006. Structural basis for phosphotyrosine recognition by the Src homology-2 domains of the adapter proteins SH2-B and APS. *J Mol Biol* 361: 69–79.

Hu J, Liu J, Ghirlando R, Saltiel AR, Hubbard SR. 2003. Structural basis for recruitment of the adaptor protein APS to the activated insulin receptor. *Mol Cell* 12: 1379–1389.

Hunter T. 2007. The age of crosstalk: Phosphorylation, ubiquitination, and beyond. *Mol Cell* 28: 730–738.

Hynes NE, Lane HA. 2005. ERBB receptors and cancer: The complexity of targeted inhibitors. *Nat Rev Cancer* 5: 341–354.

Ji QS, Chattopadhyay A, Vecchi M, Carpenter G. 1999. Physiological requirement for both SH2 domains for phospholipase C-γ1 function and interaction with platelet-derived growth factor receptors. *Mol Cell Biol* 19: 4961–4970.

Kavanaugh WM, Williams LT. 1994. An alternative to SH2 domains for binding tyrosine-phosphorylated proteins. *Science* 266: 1862–1865.

Kim HK, Kim JW, Zilberstein A, Margolis B, Kim JG, Schlessinger J, Rhee SG. 1991. PDGF stimulation of inositol phospholipid hydrolysis requires PLC-γ 1 phosphorylation on tyrosine residues 783 and 1254. *Cell* 65: 435–441.

Kmiecik TE, Shalloway D. 1987. Activation and suppression of pp60c–src transforming ability by mutation of its primary sites of tyrosine phosphorylation. *Cell* 49: 65–73.

Kobashigawa Y, Sakai M, Naito M, Yokochi M, Kumeta H, Makino Y, Ogura K, Tanaka S, Inagaki F. 2007. Structural basis for the transforming activity of human cancer-related signaling adaptor protein CRK. *Nat Struct Mol Biol* 14: 503–510.

Koch CA, Anderson D, Moran MF, Ellis C, Pawson T. 1991. SH2 and SH3 domains: Elements that control interactions of cytoplasmic signaling proteins. *Science* 252: 668–674.

Kouhara H, Hadari YR, Spivak-Kroizman T, Schilling J, Bar-Sagi D, Lax I, Schlessinger J. 1997. A lipid-anchored Grb2-binding protein that links FGF-receptor activation to the Ras/MAPK signaling pathway. *Cell* 89: 693–702.

Kuriyan J, Cowburn D. 1997. Modular peptide recognition domains in eukaryotic signaling. *Ann Rev Biophys Biomol Struct* 26: 259–288.

Lechleider RJ, Sugimoto S, Bennett AM, Kashishian AS, Cooper JA, Shoelson SE, Walsh CT, Neel BG. 1993. Activation of the SH2-containing phosphotyrosine phosphatase SH-PTP2 by its binding site, phosphotyrosine 1009, on the human platelet-derived growth factor receptor. *J Biol Chem* **268:** 21478–21481.

Lee JY, Chiu YH, Asara J, Cantley LC. 2011. Inhibition of PI3K binding to activators by serine phosphorylation of PI3K regulatory subunit p85α Src homology-2 domains. *Proc Natl Acad Sci* **108:** 14157–14162.

Lerner EC, Smithgall TE. 2002. SH3-dependent stimulation of Src-family kinase autophosphorylation without tail release from the SH2 domain in vivo. *Nat Struct Biol* **9:** 365–369.

Levy DE, Darnell JE Jr. 2002. Stats: Transcriptional control and biological impact. *Nat Rev Mol Cell Biol* **3:** 651–662.

Li N, Batzer A, Daly R, Yajnik V, Skolnik E, Chardin P, Bar-Sagi D, Margolis B, Schlessinger J. 1993. Guanine-nucleotide-releasing factor hSos1 binds to Grb2 and links receptor tyrosine kinases to Ras signalling. *Nature* **363:** 85–88.

Li W, Nishimura R, Kashishian A, Batzer AG, Kim WJ, Cooper JA, Schlessinger J. 1994. A new function for a phosphotyrosine phosphatase: Linking GRB2-Sos to a receptor tyrosine kinase. *Mol Cell Biol* **14:** 509–517.

Li SC, Gish G, Yang D, Coffey AJ, Forman-Kay JD, Ernberg I, Kay LE, Pawson T. 1999. Novel mode of ligand binding by the SH2 domain of the human XLP disease gene product SAP/SH2D1A. *Curr Biol* **9:** 1355–1362.

Li W, Young SL, King N, Miller WT. 2008. Signaling properties of a non-metazoan Src kinase and the evolutionary history of Src negative regulation. *J Biol Chem* **283:** 15491–15501.

Liu X, Brodeur SR, Gish G, Songyang Z, Cantley LC, Laudano AP, Pawson T. 1993. Regulation of c-Src tyrosine kinase activity by the Src SH2 domain. *Oncogene* **8:** 1119–1126.

Liu BA, Jablonowski K, Raina M, Arce M, Pawson T, Nash PD. 2006. The human and mouse complement of SH2 domain proteins-establishing the boundaries of phosphotyrosine signaling. *Mol Cell* **22:** 851–868.

Lu W, Gong D, Bar-Sagi D, Cole PA. 2001. Site-specific incorporation of a phosphotyrosine mimetic reveals a role for tyrosine phosphorylation of SHP-2 in cell signaling. *Mol Cell* **8:** 759–769.

Lupher ML Jr, Songyang Z, Shoelson SE, Cantley LC, Band H. 1997. The Cbl phosphotyrosine-binding domain selects a D(N/D)XpY motif and binds to the Tyr^{292} negative regulatory phosphorylation site of ZAP-70. *J Biol Chem* **272:** 33140–33144.

Ma CS, Deenick EK. 2011. The role of SAP and SLAM family molecules in the humoral immune response. *Ann NY Acad Sci* **1217:** 32–44.

Manning BD, Cantley LC. 2007. AKT/PKB signaling: Navigating downstream. *Cell* **129:** 1261–1274.

Marengere LE, Songyang Z, Gish GD, Schaller MD, Parsons JT, Stern MJ, Cantley LC, Pawson T. 1994. SH2 domain specificity and activity modified by a single residue. *Nature* **369:** 502–505.

Margarit SM, Sondermann H, Hall BE, Nagar B, Hoelz A, Pirruccello M, Bar-Sagi D, Kuriyan J. 2003. Structural evidence for feedback activation by Ras.GTP of the Ras-

specific nucleotide exchange factor SOS. *Cell* **112:** 685–695.

Maroun CR, Naujokas MA, Holgado-Madruga M, Wong AJ, Park M. 2000. The tyrosine phosphatase SHP-2 is required for sustained activation of extracellular signal-regulated kinase and epithelial morphogenesis downstream from the met receptor tyrosine kinase. *Mol Cell Biol* **20:** 8513–8525.

Martin GS. 2001. The hunting of the Src. *Nat Rev Mol Cell Biol* **2:** 467–475.

Mayer BJ, Hirai H, Sakai R. 1995. Evidence that SH2 domains promote processive phosphorylation by protein-tyrosine kinases. *Curr Biol* **5:** 296–305.

Meng W, Sawasdikosol S, Burakoff SJ, Eck MJ. 1999. Structure of the amino-terminal domain of Cbl complexed to its binding site on ZAP-70 kinase. *Nature* **398:** 84–90.

Moodie SA, Alleman-Sposeto J, Gustafson TA. 1999. Identification of the APS protein as a novel insulin receptor substrate. *J Biol Chem* **274:** 11186–11193.

Nagar B, Hantschel O, Young MA, Scheffzek K, Veach D, Bornmann W, Clarkson B, Superti-Furga G, Kuriyan J. 2003. Structural basis for the autoinhibition of c-Abl tyrosine kinase. *Cell* **112:** 859–871.

Nagar B, Hantschel O, Seeliger M, Davies JM, Weis WI, Superti-Furga G, Kuriyan J. 2006. Organization of the SH3-SH2 unit in active and inactive forms of the c-Abl tyrosine kinase. *Mol Cell* **21:** 787–798.

Newton AC. 2009. Lipid activation of protein kinases. *J Lipid Res* **50:** S266–S271.

Ng C, Jackson RA, Buschdorf JP, Sun Q, Guy GR, Sivaraman J. 2008. Structural basis for a novel intrapeptidyl H-bond and reverse binding of c-Cbl-TKB domain substrates. *EMBO J* **27:** 804–816.

Nika K, Tautz L, Arimura Y, Vang T, Williams S, Mustelin T. 2007. A weak Lck tail bite is necessary for Lck function in T cell antigen receptor signaling. *J Biol Chem* **282:** 36000–36009.

Nishi M, Werner ED, Oh BC, Frantz JD, Dhe-Paganon S, Hansen L, Lee J, Shoelson SE. 2005. Kinase activation through dimerization by human SH2-B. *Mol Cell Biol* **25:** 2607–2621.

Noguchi T, Matozaki T, Horita K, Fujioka Y, Kasuga M. 1994. Role of SH-PTP2, a protein-tyrosine phosphatase with Src homology 2 domains, in insulin-stimulated Ras activation. *Mol Cell Biol* **14:** 6674–6682.

Ogawa A, Takayama Y, Sakai H, Chong KT, Takeuchi S, Nakagawa A, Nada S, Okada M, Tsukihara T. 2002. Structure of the carboxyl-terminal Src kinase, Csk. *J Biol Chem* **277:** 14351–14354.

Okada M, Nakagawa H. 1989. A protein tyrosine kinase involved in regulation of pp60c–src function. *J Biol Chem* **264:** 20886–20893.

Okada K, Inoue A, Okada M, Murata Y, Kakuta S, Jigami T, Kubo S, Shiraishi H, Eguchi K, Motomura M, et al. 2006. The muscle protein Dok-7 is essential for neuromuscular synaptogenesis. *Science* **312:** 1802–1805.

O'Reilly AM, Pluskey S, Shoelson SE, Neel BG. 2000. Activated mutants of SHP-2 preferentially induce elongation of *Xenopus* animal caps. *Mol Cell Biol* **20:** 299–311.

Ottinger EA, Botfield MC, Shoelson SE. 1998. Tandem SH2 domains confer high specificity in tyrosine kinase signaling. *J Biol Chem* **273:** 729–735.

Park M, Dean M, Cooper CS, Schmidt M, O'Brien SJ, Blair DG, Vande Woude GF. 1986. Mechanism of met oncogene activation. *Cell* **45:** 895–904.

Pawson T. 1995. Protein modules and signalling networks. *Nature* **373:** 573–580.

Pawson T, Gish GD, Nash P. 2001. SH2 domains, interaction modules and cellular wiring. *Trends Cell Biol* **11:** 504–511.

Penengo L, Rubin C, Yarden Y, Gaudino G. 2003. c-Cbl is a critical modulator of the Ron tyrosine kinase receptor. *Oncogene* **22:** 3669–3679.

Penuel E, Martin GS. 1999. Transformation by v-Src: Ras-MAPK and PI3K-mTOR mediate parallel pathways. *Mol Biol Cell* **10:** 1693–1703.

Peschard P, Fournier TM, Lamorte L, Naujokas MA, Band H, Langdon WY, Park M. 2001. Mutation of the c-Cbl TKB domain binding site on the Met receptor tyrosine kinase converts it into a transforming protein. *Mol Cell* **8:** 995–1004.

Peschard P, Ishiyama N, Lin T, Lipkowitz S, Park M. 2004. A conserved DpYR motif in the juxtamembrane domain of the Met receptor family forms an atypical c-Cbl/Cbl-b tyrosine kinase binding domain binding site required for suppression of oncogenic activation. *J Biol Chem* **279:** 29565–29571.

Pluskey S, Wandless TJ, Walsh CT, Shoelson SE. 1995. Potent stimulation of SH-PTP2 phosphatase activity by simultaneous occupancy of both SH2 domains. *J Biol Chem* **270:** 2897–2900.

Poulin B, Sekiya F, Rhee SG. 2005. Intramolecular interaction between phosphorylated tyrosine-783 and the C-terminal Src homology 2 domain activates phospholipase C-γ1. *Proc Natl Acad Sci* **102:** 4276–4281.

Poy F, Yaffe MB, Sayos J, Saxena K, Morra M, Sumegi J, Cantley LC, Terhorst C, Eck MJ. 1999. Crystal structures of the XLP protein SAP reveal a class of SH2 domains with extended, phosphotyrosine-independent sequence recognition. *Mol Cell* **4:** 555–561.

Ramos JW. 2008. The regulation of extracellular signal-regulated kinase (ERK) in mammalian cells. *Int J Biochem Cell Biol* **40:** 2707–2719.

Ravichandran KS. 2001. Signaling via Shc family adapter proteins. *Oncogene* **20:** 6322–6330.

Rivero-Lezcano OM, Marcilla A, Sameshima JH, Robbins KC. 1995. Wiskott-Aldrich syndrome protein physically associates with Nck through Src homology 3 domains. *Mol Cell Biol* **15:** 5725–5731.

Rosen MK, Yamazaki T, Gish GD, Kay CM, Pawson T, Kay LE. 1995. Direct demonstration of an intramolecular SH2-phosphotyrosine interaction in the Crk protein. *Nature* **374:** 477–479.

Rozakis-Adcock M, Fernley R, Wade J, Pawson T, Bowtell D. 1993. The SH2 and SH3 domains of mammalian Grb2 couple the EGF receptor to the Ras activator mSos1. *Nature* **363:** 83–85.

Ryan PE, Sivadasan-Nair N, Nau MM, Nicholas S, Lipkowitz S. 2010. The N terminus of Cbl-c regulates ubiquitin ligase activity by modulating affinity for the ubiquitin-conjugating enzyme. *J Biol Chem* **285:** 23687–23698.

Sadowski I, Stone JC, Pawson T. 1986. A noncatalytic domain conserved among cytoplasmic protein-tyrosine kinases modifies the kinase function and transforming activity of Fujinami sarcoma virus P130gag-fps. *Mol Cell Biol* **6:** 4396–4408.

Salcini AE, McGlade J, Pelicci G, Nicoletti I, Pawson T, Pelicci PG. 1994. Formation of Shc-Grb2 complexes is necessary to induce neoplastic transformation by overexpression of Shc proteins. *Oncogene* **9:** 2827–2836.

Sanjay A, Horne WC, Baron R. 2001. The Cbl family: Ubiquitin ligases regulating signaling by tyrosine kinases. *Sci STKE* **2001:** e40.

Sarbassov DD, Guertin DA, Ali SM, Sabatini DM. 2005. Phosphorylation and regulation of Akt/PKB by the rictor-mTOR complex. *Science* **307:** 1098–1101.

Sasaoka T, Kobayashi M. 2000. The functional significance of Shc in insulin signaling as a substrate of the insulin receptor. *Endocr J* **47:** 373–381.

Saucier C, Khoury H, Lai KM, Peschard P, Dankort D, Naujokas MA, Holash J, Yancopoulos GD, Muller WJ, Pawson T, et al. 2004. The Shc adaptor protein is critical for VEGF induction by Met/HGF and ErbB2 receptors and for early onset of tumor angiogenesis. *Proc Natl Acad Sci* **101:** 2345–2350.

Sayos J, Wu C, Morra M, Wang N, Zhang X, Allen D, van Schaik S, Notarangelo L, Geha R, Roncarolo MG, et al. 1998. The X-linked lymphoproliferative-disease gene product SAP regulates signals induced through the co-receptor SLAM. *Nature* **395:** 462–469.

Schlessinger J, Lemmon MA. 2003. SH2 and PTB domains in tyrosine kinase signaling. *Sci STKE* **2003:** RE12.

Schmidt MH, Dikic I. 2005. The Cbl interactome and its functions. *Nat Rev Mol Cell Biol* **6:** 907–918.

Sicheri F, Moarefi I, Kuriyan J. 1997. Crystal structure of the Src family tyrosine kinase Hck. *Nature* **385:** 602–609.

Sondermann H, Soisson SM, Boykevisch S, Yang SS, Bar-Sagi D, Kuriyan J. 2004. Structural analysis of autoinhibition in the Ras activator Son of sevenless. *Cell* **119:** 393–405.

Songyang Z, Shoelson SE, Chaudhuri M, Gish G, Pawson T, Haser WG, King F, Roberts T, Ratnofsky S, Lechleider RJ, et al. 1993. SH2 domains recognize specific phosphopeptide sequences. *Cell* **72:** 767–778.

Stahl N, Farruggella TJ, Boulton TG, Zhong Z, Darnell JE Jr, Yancopoulos GD. 1995. Choice of STATs and other substrates specified by modular tyrosine-based motifs in cytokine receptors. *Science* **267:** 1349–1353.

Stephens L, Anderson K, Stokoe D, Erdjument-Bromage H, Painter GF, Holmes AB, Gaffney PR, Reese CB, McCormick F, Tempst P, et al. 1998. Protein kinase B kinases that mediate phosphatidylinositol 3,4,5-trisphosphate-dependent activation of protein kinase B. *Science* **279:** 710–714.

Stokoe D, Stephens LR, Copeland T, Gaffney PR, Reese CB, Painter GF, Holmes AB, McCormick F, Hawkins PT. 1997. Dual role of phosphatidylinositol-3,4,5-trisphosphate in the activation of protein kinase B. *Science* **277:** 567–570.

Stolt PC, Jeon H, Song HK, Herz J, Eck MJ, Blacklow SC. 2003. Origins of peptide selectivity and phosphoinositide binding revealed by structures of disabled-1 PTB domain complexes. *Structure* **11:** 569–579.

Stone JC, Atkinson T, Smith M, Pawson T. 1984. Identification of functional regions in the transforming protein of Fujinami sarcoma virus by in-phase insertion mutagenesis. *Cell* **37:** 549–558.

Sugimoto S, Wandless TJ, Shoelson SE, Neel BG, Walsh CT. 1994. Activation of the SH2-containing protein tyrosine phosphatase, SH-PTP2, by phosphotyrosine-containing peptides derived from insulin receptor substrate-1. *J Biol Chem* **269:** 13614–13622.

Takeuchi S, Takayama Y, Ogawa A, Tamura K, Okada M. 2000. Transmembrane phosphoprotein Cbp positively regulates the activity of the carboxyl-terminal Src kinase, Csk. *J Biol Chem* **275:** 29183–29186.

Tamagnone L, Artigiani S, Chen H, He Z, Ming GI, Song H, Chedotal A, Winberg ML, Goodman CS, Poo M, et al. 1999. Plexins are a large family of receptors for transmembrane, secreted, and GPI-anchored semaphorins in vertebrates. *Cell* **99:** 71–80.

Tartaglia M, Mehler EL, Goldberg R, Zampino G, Brunner HG, Kremer H, van der Burgt I, Crosby AH, Ion A, Jeffery S, et al. 2001. Mutations in PTPN11, encoding the protein tyrosine phosphatase SHP-2, cause Noonan syndrome. *Nat Genet* **29:** 465–468.

Uhlik MT, Temple B, Bencharit S, Kimple AJ, Siderovski DP, Johnson GL. 2005. Structural and evolutionary division of phosphotyrosine binding (PTB) domains. *J Mol Biol* **345:** 1–20.

van der Geer P, Wiley S, Gish GD, Pawson T. 1996. The Shc adaptor protein is highly phosphorylated at conserved, twin tyrosine residues (Y239/240) that mediate protein–protein interactions. *Curr Biol* **6:** 1435–1444.

van der Geer P, Wiley S, Pawson T. 1999. Re-engineering the target specificity of the insulin receptor by modification of a PTB domain binding site. *Oncogene* **18:** 3071–3075.

Vigna E, Gramaglia D, Longati P, Bardelli A, Comoglio PM. 1999. Loss of the exon encoding the juxtamembrane domain is essential for the oncogenic activation of TPR-MET. *Oncogene* **18:** 4275–4281.

Vogel W, Ullrich A. 1996. Multiple in vivo phosphorylated tyrosine phosphatase SHP-2 engages binding to Grb2 via tyrosine 584. *Cell Growth Differ* **7:** 1589–1597.

Waksman G, Shoelson SE, Pant N, Cowburn D, Kuriyan J. 1993. Binding of a high affinity phosphotyrosyl peptide to the Src SH2 domain: Crystal structures of the complexed and peptide-free forms. *Cell* **72:** 779–790.

Xu B, Miller WT. 1996. Src homology domains of v-Src stabilize an active conformation of the tyrosine kinase catalytic domain. *Mol Cell Biochem* **158:** 57–63.

Xu W, Harrison SC, Eck MJ. 1997. Three-dimensional structure of the tyrosine kinase c-Src. *Nature* **385:** 595–602.

Xu W, Doshi A, Lei M, Eck MJ, Harrison SC. 1999. Crystal structures of c-Src reveal features of its autoinhibitory mechanism. *Mol Cell* **3:** 629–638.

Yart A, Laffargue M, Mayeux P, Chretien S, Peres C, Tonks N, Roche S, Payrastre B, Chap H, Raynal P. 2001. A critical role for phosphoinositide 3-kinase upstream of Gab1 and SHP2 in the activation of ras and mitogen-activated protein kinases by epidermal growth factor. *J Biol Chem* **276:** 8856–8864.

Yokouchi M, Wakioka T, Sakamoto H, Yasukawa H, Ohtsuka S, Sasaki A, Ohtsubo M, Valius M, Inoue A, Komiya S, et al. 1999. APS, an adaptor protein containing PH and SH2 domains, is associated with the PDGF receptor and c-Cbl and inhibits PDGF-induced mitogenesis. *Oncogene* **18:** 759–767.

Yoon S, Seger R. 2006. The extracellular signal-regulated kinase: Multiple substrates regulate diverse cellular functions. *Growth Factors* **24:** 21–44.

Young MA, Gonfloni S, Superti-Furga G, Roux B, Kuriyan J. 2001. Dynamic coupling between the SH2 and SH3 domains of c-Src and Hck underlies their inactivation by C-terminal tyrosine phosphorylation. *Cell* **105:** 115–126.

Yun M, Keshvara L, Park CG, Zhang YM, Dickerson JB, Zheng J, Rock CO, Curran T, Park HW. 2003. Crystal structures of the Dab homology domains of mouse disabled 1 and 2. *J Biol chem* **278:** 36572–36581.

Zhang Z, Lee CH, Mandiyan V, Borg JP, Margolis B, Schlessinger J, Kuriyan J. 1997. Sequence-specific recognition of the internalization motif of the Alzheimer's amyloid precursor protein by the X11 PTB domain. *EMBO J* **16:** 6141–6150.

Zhang X, Gureasko J, Shen K, Cole PA, Kuriyan J. 2006. An allosteric mechanism for activation of the kinase domain of epidermal growth factor receptor. *Cell* **125:** 1137–1149.

Zhao ZS, Manser E, Lim L. 2000. Interaction between PAK and nck: A template for Nck targets and role of PAK autophosphorylation. *Mol Cell Biol* **20:** 3906–3917.

Zhou MM, Ravichandran KS, Olejniczak EF, Petros AM, Meadows RP, Sattler M, Harlan JE, Wade WS, Burakoff SJ, Fesik SW. 1995. Structure and ligand recognition of the phosphotyrosine binding domain of Shc. *Nature* **378:** 584–592.

Receptor Tyrosine Kinases in the Nucleus

Graham Carpenter and Hong-Jun Liao

Department of Biochemistry, Vanderbilt University School of Medicine, Nashville, Tennessee 37232-0146

Correspondence: graham.carpenter@vanderbilt.edu

To date, 18 distinct receptor tyrosine kinases (RTKs) are reported to be trafficked from the cell surface to the nucleus in response to ligand binding or heterologous agonist exposure. In most cases, an intracellular domain (ICD) fragment of the receptor is generated at the cell surface and translocated to the nucleus, whereas for a few others the intact receptor is translocated to the nucleus. ICD fragments are generated by several mechanisms, including proteolysis, internal translation initiation, and messenger RNA (mRNA) splicing. The most prevalent mechanism is intramembrane cleavage by γ-secretase. In some cases, more than one mechanism has been reported for the nuclear localization of a specific RTK. The generation and use of RTK ICD fragments to directly communicate with the nucleus and influence gene expression parallels the production of ICD fragments by a number of non-RTK cell-surface molecules that also influence cell proliferation. This review will be focused on the individual RTKs and to a lesser extent on other growth-related cell-surface transmembrane proteins.

The localization of receptor tyrosine kinases (RTKs) in the nucleus is perhaps 20 years old and was for some time limited by the available techniques to descriptive experiments that suffered from potential artifacts. With improved technical approaches and clearer understanding of subcellular compartments more convincing and mechanistic data have appeared. The authors have previously reviewed this topic (Carpenter and Liao 2009) and will, therefore, concentrate on more recent and mechanistic advances.

Although there are several mechanisms involved in this trafficking pathway, the role of secretase-dependent processing of cell-surface molecules to the nucleus is the most clear and convincing in the case of Notch (Bray 2006). In that case, ligand binding initiates sequential proteolytic processing by α-secretase, which removes the ectodomain, and by γ-secretase, which cleaves within the transmembrane domain of the remaining cell-associated receptor fragment to release an intracellular domain (ICD) fragment into the cytosol. The ICD subsequently escorts a transcription activation factor into the nucleus to initiate cellular responses to the ligand. The Notch scenario is recapitulated to different extents by numerous RTKs, as indicated in Table 1, and other growth-related molecules as shown in Table 2.

In reviewing intramembraneous cleavage of RTKs, two points are emphasized. First, is the cleavage process or nuclear localization of the ICD fragment stimulated by a ligand? Although a number of receptors and other cell-surface molecules are cleaved following the addition of

Table 1. Receptor tyrosine kinase in the nucleus

RTK[a]	Mechanism
CSF-1	γ-Secretase-generated ICD
Eph B2	γ-Secretase-generated ICD
ErbB-1	Holoreceptor, mRNA splicing: rhomboid-generated ICD
ErbB-2	Holoreceptor, CTF
ErbB-3	Holoreceptor, RNA splicing
ErbB-4	γ-Secretase-generated ICD
FGFR1	Holoreceptor, granzyme-generated ICD
FGFR2	Holoreceptor
FGFR3	γ-Secretase-generated ICD
IGF-1	Holoreceptor, ICD
Insulin	ICD
Met	Holoreceptor, secretase- or caspase-generated ICD
PTK7	γ-Secretase-generated ICD
Ret	γ-Secretase-generated ICD
Ron	γ-Secretase-generated ICD
Ryk	γ-Secretase-generated ICD
Tie1	γ-Secretase-generated ICD
VEGFR1	γ-Secretase-generated ICD, holoreceptor
VEGFR2	γ-Secretase-generated ICD, holoreceptor

ICD, intracellular domain; CTF, carboxy-terminal fragment.

[a]References can be found in the text or in Carpenter and Liao (2009).

protein kinase C agonists, such as phorbol esters, it remains to be shown in those cases whether natural receptor ligands also influence receptor trafficking to the nucleus. Second, what is the evidence that the released ICD fragment produces a relevant biologic activity? These issues are important as it has been hypothesized that secretase processing of transmembrane proteins may be a cellular housekeeping mechanism to degrade these molecules, as the presence of a transmembrane domain presents a barrier to other proteolytic systems (Small 2002; Kopan and Ilagan 2004). These are not, however, necessarily mutually exclusive interpretations. For example, α- or β-secretase release of an ectodomain fragment may be biologically important, whereas the γ-secretase degradation of the remaining cell-associated fragment may proceed as a housekeeping function. However, when the cleavage is stimulated by a ligand, especially the cognate ligand, then it seems very likely that the nuclear trafficking also represents a signal transduction mechanism. Although the nuclear function(s) of ICD fragments is unknown in many cases, the lessons of ErbB-4 and Notch indicate that ICDs may activate transcription by transporting transcription factors from the cytoplasm to the nucleus.

SECRETASE PROCESSING OF RTKs

ErbB-4 Receptor

Within the family of RTKs that are processed to the nucleus, ErbB-4 is the most completely understood in terms of mechanism and biological function and will be discussed in more detail as a model for the secretase cleavage of other RTKs (Fig. 1). Ectodomain proteolytic processing of ErbB-4 includes a basal level, which can be increased by 12-O-tetradecanoylphorbol-13-acetate (TPA) in all cells or by the addition of its cognate ligand, neuregulin (heregulin), to certain cells (Vecchi et al. 1996; Zhou and Carpenter 2000). This cleavage results in the formation of two receptor fragments: a 120-kDa ectodomain fragment that is released into the media and an 80-kDa membrane-bound fragment, termed m80 or CTF (carboxy-terminal fragment). Cleavage requires ADAM 17 (TACE) and it is likely this is the enzyme that executes cleavage of ErbB-4 between His651 and Ser652 within the extracellular stalk or ecto-juxtamembrane region (Rio et al. 2000; Cheng et al. 2003). Hence, the m80 fragment includes eight ectodomain residues, the transmembrane domain, and entire ICD.

Sensitivity to ectodomain shedding is likely determined, at least in part, by the length of the stalk region as shown for the selectins (Migaki et al. 1995). There are two ErbB-4 isoforms termed Jm-a, in which the ectodomain is sensitive to cleavage, and Jm-b, which is not cleavable (Elenius et al. 1997). The presence of these two isoforms is unique to ErbB-4. Because ADAM-mediated cleavage events do not involve a defined sequence or cleavage site in the substrate, it seems that longer stalk regions in substrates may simply permit accessibility of the protease. Interestingly, the stalk region in Jm-b is much shorter (six residues) than the corresponding region of Jm-a (16 residues), and ErbB-1, -2,

Cite this article as *Cold Spring Harb Perspect Biol* doi: 10.1101/cshperspect.a008979

Table 2. Growth-related cell-surface molecules in the nucleus

Molecules	Mechanism	References[a]
ADAM10	γ-Secretase-generated ICD	Tousseyn et al. 2009
ADAM13	γ-Secretase-generated ICD	Cousin et al. 2011
CD44	γ-Secretase-generated ICD	Janiszewska et al. 2010; Miletti-González et al. 2012
EpCAM	γ-Secretase-generated ICD	Denzel et al. 2009; Maetzel et al. 2009; Ralhan et al. 2010; Huang et al. 2011b; Chaves-Pérez et al. 2013
Frizzled	γ-Secretase-generated ICD	Mathew et al. 2005
Growth hormone receptor	γ-Secretase-generated ICD	Wang et al. 2011a
IL-2R	γ-Secretase-generated ICD	Montes de Oca et al. 2010
Klotho	γ-Secretase-generated ICD	Bloch et al. 2009
LRP1	γ-Secretase-generated ICD	Rebeck 2009
Neogenin (DDC)	γ-Secretase-generated ICD	Goldschneider et al. 2008; Bai et al. 2011
p75NTR	γ-Secretase-generated ICD	Ceni et al. 2010; Parkhurst et al. 2010; Wang et al. 2010a; Le Moan et al. 2011
Patched	γ-Secretase-generated ICD	Kagawa et al. 2011
Pre-β-cellulin	γ-Secretase-generated ICD	Stoeck et al. 2010
Protocadherin	γ-Secretase-generated ICD	Buchanan et al. 2010
PTPμ	γ-Secretase-generated ICD	Burgoyne et al. 2009
TβRI	γ-Secretase-generated ICD, holoreceptor	Mu et al. 2011; Chandra et al. 2012
Ephrin B2	γ-Secretase-generated ICD	Tomita et al. 2006; Georgakopoulos et al. 2011
Notch	γ-Secretase-generated ICD	Fortini 2009

[a]Earlier references can be found in Carpenter and Liao (2009).

and -3 also have relatively short stalk regions (6–9 residues) and are not subject to a significant level of metalloprotease-mediated ectodomain cleavage (Vecchi et al. 1996). Hence the unique sensitivity of the Jm-a ErbB-4 isoform to secretase-dependent processing and signaling seems likely owing to the length of its stalk region.

It seems probable that the shed ErbB-4 ectodomain may function to block receptor activation by binding neuregulin. The function of the m80 fragment, however, is known. The capacity of γ-secretase to cleave substrates requires that the substrate have a short ectodomain region of 50 or fewer residues (Struhl and Adachi 2000). Hence, the ADAM-mediated removal of a large portion of the ErbB-4 ectodomain is a prerequisite step for subsequent γ-secretase cleavage of the m80 fragment (Ni et al. 2001).

γ-Secretase is a complex of at least four distinct transmembrane proteins of which presenilin is the catalytic protease (Selkoe and Wolfe 2007). The nicastrin subunit of this complex recognizes transmembrane proteins with short-

ened or nublike ectodomains and thereby acts as a targeting subunit for intramembrane cleavage by presenilin (Shah et al. 2005). Presenilin activity converts the ErbB-4 m80 fragment to a soluble s80 or ICD fragment that is found in the cytoplasm, nucleus, and mitochondria (Ni et al 2001; Naresh et al. 2006). Recently, a fifth protein (GASP) has been added to the γ-secretase complex (He et al. 2010). GASP, which was first identified as a target of the tyrosine kinase inhibitor imatnib, is a 16 kDa protein that is required for efficient γ-secretase cleavage of the cell-surface protein APP. This raises the issue of whether a similar targeting protein may participate in the cleavage of other γ-secretase substrates.

The carboxyl terminus of ErbB-4 encodes a PDZ domain recognition motif, which is required for presenilin cleavage of the m80 fragment (Ni et al. 2003). Deletion of this motif (TVV) does not influence ectodomain cleavage, but does attenuate presenilin association with the m80 fragment and production of the ICD

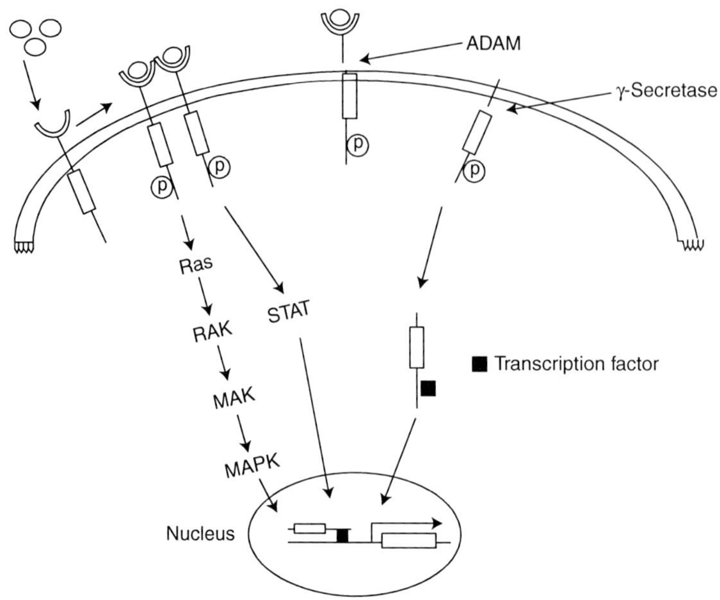

Figure 1. Depicted is the general mechanism for generation and nuclear localization of RTK ICD fragments. Included are examples of canonical signal transduction pathways to the nucleus (i.e., MAPK and STAT pathways) as contrasted to the noncanonical ICD mechanism.

fragment. Presenilin also contains a PDZ domain recognition motif, and it is possible that a scaffold of PDZ domain-containing proteins may be required for γ-secretase cleavage.

Presenilin cleavage of substrates occurs within the transmembrane domain and, based on APP and Notch processing, this may occur at multiple sites, producing several species of ICD fragments that may have differing levels of metabolic stability based on the N-end rule (Varshavsky 1997). Mutation within the transmembrane domain, which is often used to identify a cleavage site based on the lack of detectable ICD formation, can alter the metabolic stability of the ICD fragments. This is shown in the case of Notch where a transmembrane mutation appears to prevent cleavage, but actually results in a new ICD fragment that is very rapidly degraded owing to the presence of a metabolically destabilizing amino-terminal residue (Tagami et al. 2008).

It has been reported that the Val675Ala (Muraoka-Cook et al. 2006) or Val673Ile (Vidal et al. 2005) mutations within the ErbB-4 transmembrane domain abrogate γ-secretase cleav-

age, as judged by the inability to detect the ICD fragment. In view of the Notch mutagenesis data, it is not clear whether these mutations actually prevent cleavage or result in a less stable ICD fragment. Given the low level of ICD fragment normally detectable, a modest change in stability may render the fragment undetectable by the same methodology. Additional splice variants of ErbB-4 (Veikkolainen et al. 2011) produce two isoforms of the ICD fragment, CYT-1 and CYT-2. The two isoforms differ by 16 consecutive residues that are present in CYT-1, but absent in the CYT-2, and provide the CYT-1 isoform with overlapping docking sites for PI3 kinase and proteins containing WW domains. In a comparative study using Madin-Darby canine kidney (MDCK) cells, one group has shown that the CYT-1 ICD is less metabolically stable than the CYT-2 ICD (Zeng et al. 2009) and that the CYT-2 ICD is more readily detected in the nucleus (Zeng et al. 2007). These differences were due, at least in part, to association of the ubiquitin ligase Need 4, a WW-domain-containing protein, with selective sequence motifs present in the CYT-1 isoform.

 Cite this article as *Cold Spring Harb Perspect Biol* doi: 10.1101/cshperspect.a008979

Using different cell backgrounds, similar results have been reported by others (Omerovic et al. 2007; Feng et al. 2009).

In terms of the physiological relevance, it is now clear that endogenous generation of the ErbB-4 ICD by γ-secretase is required for control of astrogenesis in the developing mouse (Sardi et al. 2006). In this system, the ICD fragment interacts with TAB2, an adaptor protein, and thereby with N-CoR, a corepressor, and chaperones this complex into the nucleus. A similar chaperone mechanism between the ErbB-4 ICD and STAT5 has been proposed to be operative during mammary differentiation (Williams et al. 2004). Also, the nuclear interaction of the ErbB-4 ICD with HIF-1α is reported to attenuate metabolic degradation of HIF-1α and thereby promote gene expression in mammary tissue (Paatero et al. 2012).

It is clear that ErbB-4 is functionally involved in mammary development in the animal (Jones 2008). ErbB-4 nuclear localization has been observed in normal and tumor mammary tissue and exogenous ICD expression provokes differentiation events (Muraoka-Cook et al. 2006). However, as a cautionary note studies using exogenous expression of the Notch ICD have shown that different levels of expression can provoke distinct biologic responses (Mazzone et al. 2010; Han et al. 2011). Nevertheless, analysis of an ErbB-4 noncleavable mutant (Rokicki et al. 2010) and the intracellular distribution of the ICD fragment in breast tumor tissue (Thor et al. 2009) are consistent with the conclusion that endogenous ErbB-4 cleavage is physiologically relevant to differentiation in this tissue and a positive factor for breast cancer patients.

Also, consistent with a role of the ErbB-4 ICD fragment in various differentiation systems is the report that γ-secretase inhibition prevents neuregulin generation of the ErbB-4 ICD in oligodendrocytes and maturation of this cell type (Lai and Feng 2004). Interestingly, there is a genetic correlation between ErbB-4 and schizophrenia (Pan et al. 2011) and the involvement of the ICD fragment in controlling gene expression in relevant cell types has been reported (Wong and Weickert 2009; Allison et al. 2011). Also, the maturation of fetal lung cells is report-

ed to be dependent on secretase processing of ErbB-4 and ICD association with the transcription factors YAP (Hoeing et al. 2011), STAT 5a (Zscheppang et al. 2011a), and ERβ (Zscheppang et al. 2011b). It is suggested that these interactions allow ErbB-4 to regulate transcription of surfactant protein B and thereby promote lung maturation.

As mentioned above, the ErbB-4 ICD also has been localized in mitochondria and in that location may function as a proapoptotic protein. This is based on the capacity of the ICD to induce cell death, the presence of a BH3 domain in the ICD, the loss of apoptotic capacity following mutagenesis of this domain, and detection of an interaction with the antiapoptotic protein BCL-2, which, when overexpressed, abrogated ICD-induced cell death (Naresh et al. 2006).

Ephrin-B2

Addition of the ligand ephrin-B2 (Eph-B2) provokes the secretase cleavage of the EphB2 receptor releasing ectodomain and ICD fragments (Litterst et al. 2007; Lin et al. 2008). The cleavage events are also stimulated by ionomycin or by activation of the N-methyl-D-aspartate (NMDA) receptor, agents that mediate Ca^{2+} influx into cells. In this system, ephrin-mediated cleavage events require endocytosis, whereas cleavage mediated by ionomycin or NMDA receptor activation occurs on the cell surface. A similar endocytosis relationship between neuregulin- or TPA-mediated cleavage of ErbB-4 was noted (Zhou and Carpenter 2000). To date, it is unclear whether the Eph receptor ICD fragment is translocated from the cytoplasm to another organelle and there is no data related to its physiological function in mediating ligand responsiveness.

Colony Stimulating Factor-1 Receptor

Secretase cleavage of the colony-stimulating factor-1 receptor (CSF-1R) can be stimulated by CSF-1, LPS, or TPA (Wilhelmsen and van der Geer 2004; Glenn and van der Geer 2007, 2008). Lipopolysaccharide (LPS) is a ligand for the

Toll4 receptor and agonists for other Toll-like receptors also stimulate cleavage of CSF-1R. This heterologous stimulation of CSF-1R cleavage may be related to the fact that in macrophages both receptor systems are thought to be involved in producing innate immune responses. Although the CSF-1 ICD fragment does appear in both cytoplasm and nucleus, a physiologic function has not been identified.

Vascular Endothelial Growth Factor-1

Pigment epithelium-derived factor (PEDF) binds to an unknown receptor and promotes an antiangiogenic response that can oppose the capacity of vascular endothelial growth factor (VEGF) to promote endothelial cell proliferation. The addition of PEDF to endothelial cells promotes the γ-secretase-mediated cleavage of VEGFR1 (Flt1) with release of its ICD fragment (Cai et al. 2006, 2011a; Rahimi et al. 2009). The ICD fragment is only present when cells are treated simultaneously with PEDF and VEGF, and the fragment was detected in the cytoplasm, but not in the nucleus. In this system, the intact VEGFR1 molecule is found in the nucleus following the addition of VEGF (see below) and PEDF reduces VEGF-induced angiogenesis and the nuclear level of intact VEGFR in a manner dependent on γ-secretase activity. This implies that the PEDF-stimulated production of the VEGFR1 ICD fragment negatively regulates intact VEGFR1 levels in the nucleus and VEGF-induced angiogenesis. Based on sensitivity to γ-secretase inhibitors, it is reported that the capacity of PEDF to prevent VEGF-induced changes in vascular permeability requires production of the VEGFR1 ICD fragment (Cai et al. 2011b). In contrast, Ablonczy et al. (2009) report that it is VEGFR2 that is cleaved by γ-secretase in this system.

Tie 1

Tie 1 is an orphan receptor that forms a hetero-oligomeric complex with Tie 2, the receptor for angiopoietin 1 (Ang 1). Addition of Ang 1 activates Tie 2 and provokes tyrosine phosphorylation of Tie 1. Ectodomain cleavage of Tie 1 is stimulated by a variety of agents (TPA, VEGF, TNFα, sheer stress) and increases Ang 1 activation of Tie 2, apparently by allowing greater access of the ligand to its Tie 2 binding site. Following ectodomain release, the Tie 1 cell-associated 45-kDa cleavage fragment is processed by γ-secretase to produce a 42-kDa cytoplasmic ICD fragment (Marron et al. 2007). In this receptor system, the ectodomain secretase action is physiologically important; however, the significance of γ-secretase activity may be simply to remove the highly tyrosine phosphorylated 45 kDa fragment. Although the addition of Ang 1 promotes rapid endocytosis and degradation of Tie 2, Tie 1 is not cleared from the cell surface by this same route. Therefore, a secretase mechanism may provide the means by which Ang1-phosphorylated Tie 1 is inactivated.

Insulin-like Growth Factor-1 and Insulin Receptors

In preliminary reports, it has been shown that the insulin and IGF-1 receptors can be cleavaged by secretase action to produce ICD fragments (Kasuga et al. 2007; McElroy et al. 2007). However, although TPA stimulated formation of the ICD fragments neither cognate ligand was shown to do so.

Ryk

Wnt proliferative signaling is mainly mediated by Frizzled receptors and low-density lipoprotein-related proteins (LRP) through a canonical pathway involving β-catenin and T-cell factor (TCF). However, it is known that Ryk, a receptor tyrosine kinase that lacks kinase activity, binds Wnt and participates in Wnt signaling in certain biologic contexts. Evidence has appeared demonstrating that γ-secretase cleavage of Ryk is required for Wnt 3-dependent neuronal differentiation (Lyu et al. 2008). Release of the Ryk ICD fragment does not require Wnt, but nuclear translocation of the ICD does require Wnt. Because the Ryk ICD lacks a nuclear localization signal, the mechanism for nuclear translocation is unclear. Cdc 37, a cochaperone of Hsp 90, interacts with the Ryk ICD, attenuates metabolic degradation, and promotes nuclear ICD levels (Lyu et al. 2009). Because this interaction has not been shown to be a result of Wnt signaling, the mechanism by which Wnt promotes nuclear translocation of the ICD fragment is not known. It is interesting to note that in *Drosophila*, Wnt binding induces the γ-secretase cleavage of D Frizzled, a heptahelical receptor, with nuclear localization of a CTF (Mathew et al. 2005). Also, the RTKs Ror1/2 participate in signaling by certain Wnt ligands (Grumolato et al. 2010), but it is not known whether ICD fragments are produced.

Fibroblast Growth Factor 3

Activation of the fibroblast growth factor receptor 3 (FGFR3) by FGF-1 leads to an ErbB-4-like cleavage pathway (Degnin et al. 2011). Both FGFR3 and ErbB-4 can undergo regulated intramembrane proteolysis following addition of the homologous ligand or TPA, but in the former case autophosphorylation and endocytosis are required. Although internalized ErbB-4 is first cleaved by a metalloprotease, ectodomain cleavage of FGFR3 is catalyzed by a cathepsin protease. Subsequently, γ-secretase cleavage liberates a FGFR3 ICD fragment that translocates to the nucleus.

Met

Ligand-independent constitutive γ-secretase cleavage of Met in several cell types has been described (Foveau et al. 2009). Interestingly, an inhibitory antibody to Met (DN30), which induces receptor degradation and biologic activity of overexpressed Met, promotes the formation of Met ICD. The intracellular localization of this ICD fragment, however, was not reported.

Protein Tyrosine Kinase 7

Protein tyrosine kinase 7 (PRK7), a pseudokinase, is constitutively cleaved by ADAM metalloprotease and γ-secretase to produce an ICD fragment that can be found in the nucleus (Golubkov and Strongin 2012; Na et al. 2012). In these studies, exogenous expression of the ICD fragment promoted cell proliferation, migration, and colony formation. Although PTK7 functions as a coreceptor in the Wnt pathway, it has been implicated in a broad range of biologic processes. Whether the PTK7 ICD fragment has a role in these events is not known.

NONSECRETASE FORMATION OF RTK ICD FRAGMENTS

Caspase-Dependent Fragments

In the case of several RTKs (ErbB-2 [Tikhomirov and Carpenter 2001; Benoit et al. 2004; Strohecker et al. 2008], Ret [Bordeaux et al. 2000; Cabrera et al. 2011], ALK [Mourali et al. 2006; Racaud-Sultan et al. 2006], TrkC [Tauszig-Delamasure et al. 2007], and Met [Tulasne et al. 2004; Pozner-Moulis et al. 2006; Foveau et al. 2007; Deheuninck et al. 2009]) there is evidence that caspase activity cleaves the cytoplasmic domain to produce an ICD fragment. Because the fragment is often produced by two cleavage events within the cytoplasmic domain, the fragment is often considerably smaller than that produced by intramembrane proteolysis. In no reported case are these caspase proteolytic events stimulated by ligand binding or by TPA, and in some studies the presence of the cognate ligand prevents cleavage. The formation of caspase ICD fragments is functionally associated with the induction of apoptosis and in one instance (Strohecker et al. 2008) the fragment has been localized to the mitochondria.

Translation-Dependent Fragments

Metalloprotease activity (Codony-Servat et al. 1999) or internal translation initiation of ErbB-2 mRNA (Anido et al. 2006) leads to the production of ErbB-2 CTFs. The CTFs, which are generated at three methionine residues located on either side of the transmembrane domain, are found in the nucleus (Anido et al. 2006; Xia et al. 2011). These fragments promote cell migration (García-Castillo et al. 2009) and are reported to be oncogenic (Pedersen et al. 2009). Because two of the CTFs include the transmembrane domain, a mechanism for translocation of these CTFs to the nucleus is not clear. ErbB-2 is a major therapeutic target in breast cancer and the therapeutic agent used is an antibody to the ectodomain. Therefore the presence of these biologically active CTFs in tumor tissue represents a potential therapeutic problem (Arribas et al. 2011).

Splicing-Dependent Fragments

Another group has identified a spliced product of the epidermal growth factor (EGF) receptor (ErbB-1) that represents the splicing of exon 1 of the ectodomain to exon 23 of the cytoplasmic domain (Piccione et al. 2012). Hence, sequences required for ligand binding and tyrosine kinase activity are deleted. This variant, termed mLEEK, is constitutively produced in a variety of cell lines and tumor tissues, is present in the nucleus, and displays transcriptional activity. Because mLEEK retains a signal sequence and may be secreted, it remains to be shown how nuclear localization is achieved. Splicing of ErbB-3 produces at least 14 variants and a few of these encode sequences corresponding to an ICD. One of these encodes an ICD sequence encoding a protein 80 kDa, which is localized to the nucleus (Andrique et al. 2012). Data have been presented to implicate this protein in the regulation of cyclin D1 transcription. Another nuclear ErbB-3 variant has been reported to associate with active promoters in Schwann cells (Adilakshmi et al. 2011).

Granzyme-Dependent Fragments

The generation of an ICD fragment from FGFR1 in an FGF-dependent manner has been reported (Chioni and Grose 2012). The ICD fragment is localized to the nucleus and evidence is presented to argue that this fragment increases cell migration. Unexplained, however, was the mechanism by which the ICD fragment could be generated by granzyme B, a serine protease localized within cytoplasmic granules.

Rhomboid-Dependent Fragments

Rhomboids are intramembrane serine proteases that are best known for the regulation of the ectodomain shedding of transmembrane proteins, such as the precursor for TGF-α. The intramembrane cleavage of ErbB-1 (EGF receptor) by rhomboid activity has now been reported (Liao and Carpenter 2012). This cleavage is EGF-stimulated and produces an ICD-like fragment that is localized to membrane and nuclear fractions, but is not found in the cytosol. Because rhomboid cleavage occurs close to the ectodomain side of the transmembrane domain, this fragment may retain transmembrane domain residues at its amino terminus and therefore remain membrane associated.

INTACT RECEPTORS IN THE NUCLEUS

Listed in Table 1 are several receptors for which the data show that the intact or holoreceptor is present in the nucleus. That the nuclear species is the holoreceptor and not an ICD fragment is based on a nuclear fraction M_r value indicative of the mature receptor, which is significantly distinct from the size of an ICD fragment. Because cell fractionation can produce contamination of organelle preparations, this is a source of concern when nuclear extracts are used for size analysis. However, most localization studies are supplemented with imaging data of intact cells. In a few instances the data relies on immunohistochemistry alone and in those cases it is not clear that intact receptor is distinguishable from an ICD fragment unless both ectodomain and cytoplasmic domain antibodies are used. In nearly all cases, however, it does appear that the receptor is present in the nucleoplasm and not the nuclear envelope.

ErbB-1

The capacity of EGF to induce trafficking of the intact EGF receptor (ErbB-1) to the nucleus was first reported in 2001 and a nuclear target was identified (Lin et al. 2001). The nuclear receptor is reported to recognize the promoter of cyclin D1 and to transactivate this promoter in a reporter system. Other promoters are also reported to be recognized by the EGF receptor (Lo et al. 2005, 2010; Tao et al. 2005; Hanada et al. 2006). However, direct and specific binding to any promoter remains to be shown. Also, the nuclear receptor associates with proliferating cell nuclear antigen (PCNA) in the nucleus, modifying its stability (Wang et al. 2006), and with polynucleotide phosphorylase, increasing cellular radioresistance (Yu et al. 2012).

Trafficking from the cell surface to the nucleus requires receptor internalization through clathrin-coated pits (De Angelis Campos et al. 2011); retrograde transport from the Golgi to the endoplasmic reticulum (ER) (Wang et al. 2011b); phosphorylation of Ser 229, an ectodomain residue, by Akt (Huang et al. 2011a); phosphorylation of Thr 654 (Dittmann et al. 2010), a known cytoplasmic domain protein kinase C (PKC) phosphorylation site; and function of the lipid kinase PIKfyve (Kim et al. 2007). Importin β is required for ErbB-1 nuclear entry (Lo et al. 2006) and a nuclear localization sequence to facilitate this interaction is present in the cytoplasmic juxtamembrane region of the EGF receptor (Hsu and Hung 2007). The nuclear receptor has been identified by biochemical fractionation, immunohistochemistry, and live cell imaging methods and shown to be in a nonmembranous nuclear environment, based on its extractability with a high salt aqueous buffer.

The above trafficking steps, perhaps with the exception of Ser 229 phosphorylation, are reasonable given what is known about intracellular trafficking in general and more specifically about other molecules (e.g., bacterial toxins) (Sandvig and van Deurs 2002; Bonifacino and Rojas 2006), that are trafficked from the cell surface to the ER. The toxins are soluble molecules and after reaching the ER are translocated to the cytoplasm where they exert their biological activity. Therefore, a mechanism is required to facilitate movement of the membrane-bound EGF receptor in the ER to the nucleus as a nonmembrane localized molecule. Although no endocytic trafficking system was known to extract a transmembrane receptor from its lipid bilayer, it was suggested that an ER protein translocon could provide such a step (Carpenter 2003). The Sec61 translocon located in the ER is known to mediate the trafficking of extracellular toxins from the cell surface to the cytoplasm

and, as part of the ERAD pathway, to retrotranslocate malfolded transmembrane proteins in the ER to the cytoplasm. Subsequent experiments showed that EGF induced trafficking of the EGF receptor to the ER where it interacted with the Sec61 translocon that mediated receptor retrotranslocation to the cytoplasm and import into the nucleus (Liao and Carpenter 2007). Experimental data showed that knockdown of a Sec61 subunit attenuated both EGF nuclear localization of the EGF receptor and EGF induction of cyclin D1. This trafficking pathway is depicted in Figure 2. A subsequent report has confirmed the role of the Sec61 translocon in this pathway, but places the interaction between the EGF receptor and the translocon at the inner nuclear membrane (Wang et al. 2010b, 2012).

Additional reports have shown that nuclear localization of the EGF receptor is provoked by radiation (Dittmann et al. 2005; Liccardi et al. 2011) or treatment with the receptor antibody C225/Cetuximab (Liao and Carpenter 2009). The nuclear receptor is reported to interact with MUC1 and increase the level of chromatin-bound EGF receptor (Bitler et al. 2010; Merlin et al. 2011). MUC1 is a cell-surface protein that is known to interact with the receptor and promote receptor internalization. It is also known to be trafficked to the nucleus (Carson 2008).

ErbB-2 and ErbB-3

Using methodologies similar to those used to localize the EGF receptor to the nucleus (Lin et al. 2001), the same group showed that the intact ErbB-2 molecule translocated from the cell surface to the nucleus and thereby facilitates the expression of the Cox-2 mRNA (Wang et al. 2004; Giri et al. 2005) and ribosomal RNA (Li et al. 2011). The mechanism of ErbB-2 trafficking to the nucleus is reported to include a Sec61-dependent step analogous to that of ErbB-1 (Wang et al. 2012). Because ErbB-2 does not bind a known ligand, the studies represent a constitutive pathway. Progesterone, however, transactivates ErbB-2 and thereby initiates ErbB-2 translocation to the nucleus

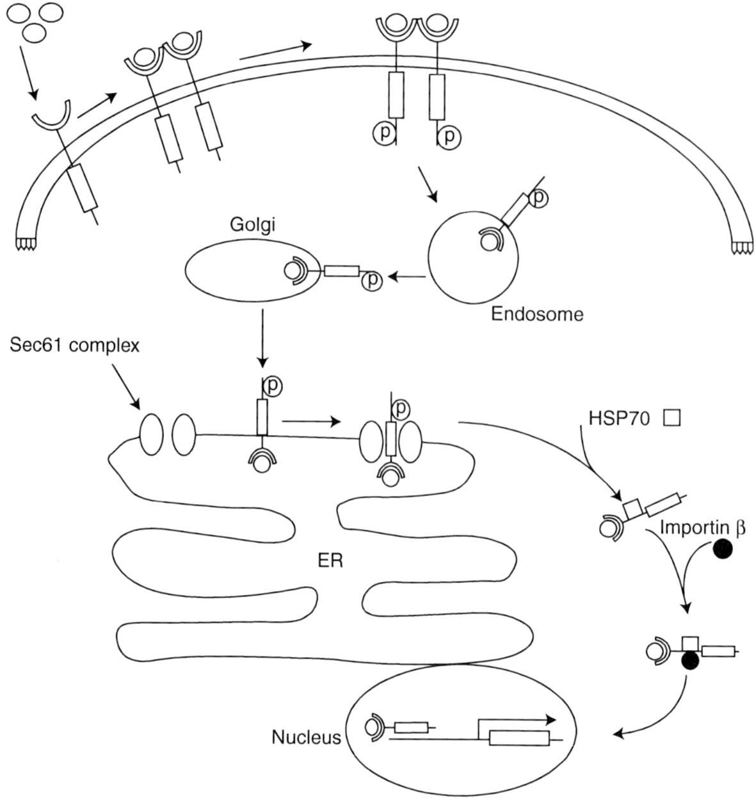

Figure 2. Illustrated is the Sec61-dependent trafficking pathway for intact RTKs, such as the EGF receptor, to translocate from the cell surface to the nucleus.

(Béguelin et al. 2010). This translocation of ErbB-2 chaperones STAT3 into the nucleus where the complex, together with the progesterone receptor, interacts with the cyclin D promoter to bring about progesterone-induced cyclin D expression. Less is known about the nuclear form of ErbB-3 (Offterdinger et al. 2002).

IGF-1

An initial report of intact IGF-1 receptors present in the nucleus (Chen and Roy 1996) has been confirmed and extended in recent reports from two groups (Aleksic et al. 2010; Deng et al. 2010; Sehat et al. 2010). Both groups indicate that this relocalization from plasma membrane to nucleus is IGF-1 dependent and that the nuclear receptor may function as a transcriptional coactivator. One report showed that nuclear lo-

calization and the capacity to increase transcription in reporter assays require sumoylation at three lysine residues in the β chain of the receptor. Chromatin-immunoprecipitation (ChIP) assays indicated the nuclear IGF-1R interacts with DNA sequences identified as predominately intergenic, with characteristics of enhancers and having the capacity to increase luciferase transcription in reporter assays.

Ron

Nuclear localization of the Ron RTK is reported to occur in bladder cancer cells and is not stimulatable by the Ron ligand MSP (Liu et al. 2010). Rather nuclear levels of Ron are increased during serum starvation. Interestingly, the increase in nuclear Ron requires the EGF receptor as a heterodimerization partner. ChIP assays have identified potential gene targets, including a number of stress-responsive genes.

Met

Nuclear localization of Met ICD fragments produced by either secretase or caspase cleavage are discussed above. One group has reported that the intact Met receptor is very rapidly translocated from the cell surface to the nucleus following the addition of HGF, and this translocation event is required for HGF induction of a nuclear Ca^{2+} signal (Gomes et al. 2008).

FGFR1 and 2

There is a large background of published reports describing the ligand-dependent translocation of intact FGFR1 to the nucleus (reviewed in Stachowiak et al 2007). There is a report detailing the dynamics of nuclear FGFR1 movement among receptor populations having different motilities taken to represent free receptor or receptor associated with chromatin or nuclear matrix (Dunham-Ems et al. 2009). The interaction of FGFR1 and the transcription factor Nurr 1 in the nucleus of dopaminergic neurons has been described (Baron et al. 2012). Morphologic studies have recognized FGFR2 in the nucleus of tissue samples (Lu et al. 2004; Schmahl et al. 2004; Giulianelli et al. 2008; Martin et al. 2011), but without analysis of the size it is not possible to distinguish whether this represents an intact receptor or a receptor fragment.

VEGFR1 and R2

Two reports have identified intact VEGFR1 in the nucleus and one indicates that the localization is increased by ligand (Cai et al. 2006; Lee et al. 2007). Several studies have found that intact VEGFR2 (KDR) is localized in the nucleus and in some reports this is ligand sensitive (Stewart et al. 2003; Fox et al. 2004; Santos and Dias 2004; Zhang et al. 2005; Santos et al. 2007). Domingues et al. (2001) conclude that nuclear VEGFR2 regulates its own transcription.

NUCLEAR LOCALIZATION OF OTHER GROWTH-REGULATING RECEPTORS

Although Notch is an obvious example of a cell-surface receptor that uses nuclear localization of the receptor to affect its biologic activity, there are a number of other examples for growth-related cell-surface proteins and these are presented in Table 2. This list is limited to recent additions; a more comprehensive list was previously published and should be consulted for earlier references (Carpenter and Liao 2009).

CONCLUDING REMARKS

The role of regulated intramembrane proteolysis in signal transduction by activated RTKs is reasonably well established biochemically and biologically. One open question revolves around the role of tyrosine kinase activity within the nuclear compartment. Are there nuclear substrates or does kinase activity not have a role in ICD function? A second issue is whether any ICD, regardless of its source, actually forms a direct and biologically important contact with DNA. To date, there is no convincing data to resolve either of those questions.

It might be important to note that the ligand-dependent trafficking of any RTK or its ICD to the nucleus would represent a noncanonical signaling pathway for that particular ligand. This would be unique as other signaling elements are canonical (i.e., common to multiple ligand–receptor systems).

Last, the reader is referred to other recent reviews on this topic (Ancot et al. 2009; Borlido et al. 2009; McCarthy et al. 2009; De Strooper and Annaert 2010; López-Otin and Hunter 2010; Wang et al. 2010c; Lal and Caplan 2011; Lemberg 2011).

ACKNOWLEDGMENT

The authors acknowledge support of the National Institutes of Health (grant RO1 CA 125649).

REFERENCES

Ablonczy Z, Prakasam A, Fant J, Fauq A, Crosson C, Sambamurti K. 2009. Pigment epithelium-derived factor maintains retinal pigment epithelium function by inhibiting vascular endothelial growth factor-R2 signaling through γ-secretase. *J Biol Chem* **284:** 30177–30286.

Adilakshmi T, Ness-Myers J, Madrid-Aliste C, Fiser A, Tapinos N. 2011. A nuclear variant of ErbB3 receptor tyrosine kinase regulates ezrin distribution and Schwann cell myelination. *J Neurosci* **31:** 5106–5119.

Aleksic T, Chitnis MM, Perestenko OV, Gao S, Thomas PH, Turner GD, Protheroe AS, Howarth M, Macaulay VM. 2010. Type 1 insulin-like growth factor receptor translocates to the nucleus of human tumor cells. *Cancer Res* **70:** 6412–6419.

Allison JG, Das PM, Ma J, Inglis FM, Jones FE. 2011. The ERBB4 intracellular domain (4ICD) regulates NRG1-induced gene expression in hippocampal neurons. *Neurosci Res* **70:** 155–163.

Ancot F, Foveau B, Lefebvre J, Leroy C, Tulasne D. 2009. Proteolytic cleavages give receptor tyrosine kinases the gift of ubiquity. *Oncogene* **28:** 2185–2195.

Andrique L, Fauvin D, El Maassarani M, Colasson H, Vannier B, Séité P. 2012. ErbB3(80 kDa), a nuclear variant of the ErbB3 receptor, binds to the *Cyclin D1* promoter to activate cell proliferation but is negatively controlled by p14ARF. *Cell Signal* **24:** 1074–1085.

Anido J, Scaltriti M, Bech Serra JJ, Santiago Josefat B, Todo FR, Baselga J, Arribas J. 2006. Biosynthesis of tumorigenic HER2 C-terminal fragments by alternative initiation of translation. *EMBO J* **25:** 3234–3244.

Arribas J, Baselga J, Pedersen K, Parra-Palau JL. 2011. p95HER2 and breast cancer. *Cancer Res* **71:** 1515–1519.

Bai G, Chivatakarn O, Bonanomi D, Lettieri K, Franco L, Xia C, Stein E, Ma L, Lewcock JW, Pfaff SL. 2011. Presenilin-dependent receptor processing is required for axon guidance. *Cell* **144:** 106–118.

Baron O, Förthmann B, Lee YW, Terranova C, Ratzka A, Stachowiak EK, Grothe C, Claus P, Stachowiak MK. 2012. Cooperation of nuclear fibroblast growth factor receptor 1 and Nurr1 offers new interactive mechanism in postmitotic development of mesencephalic dopaminergic neurons. *J Biol Chem* **287:** 19827–19840.

Béguelin W, Díaz Flaqué MC, Proietti CJ, Cayrol F, Rivas MA, Tkach M, Rosemblit C, Tocci JM, Charreau EH, et al. 2010. Progesterone receptor induces ErbB-2 nuclear translocation to promote breast cancer growth via a novel transcriptional effect: ErbB-2 function as a coactivator of Stat3. *Mol Cell Biol* **30:** 5456–5472.

Benoit V, Chariot A, Delacroix L, Deregowski V, Jacobs N, Merville MP, Bours V. 2004. Caspase-8-dependent HER-2 cleavage in response to tumor necrosis factor α stimulation is counteracted by nuclear factor κB through c-FLIP-L expression. *Cancer Res* **64:** 2684–2691.

Bitler BG, Goverdhan A, Schroeder JA. 2010. MUC1 regulates nuclear localization and function of the epidermal growth factor receptor. *J Cell Sci* **123:** 1716–1723.

Bloch L, Sineshchekova O, Reichenbach D, Reiss K, Saftig P, Kuro-o M, Kaether C. 2009. Klotho is a substrate for α-, β- and γ-secretase. *FEBS Lett* **583:** 3221–3224.

Bonifacino JS, Rojas R. 2006. Retrograde transport from endosomes to the *trans*-Golgi network. *Nat Rev Cell Mol Biol* **7:** 563–579.

Bordeaux MC, Forcet C, Granger L, Corset V, Bidaud C, Billaud M, Bredesen DE, Edery P, Mehlen P. 2000. The RET proto-oncogene induces apoptosis: A novel mechanism for Hirschsprung disease. *EMBO J* **19:** 4056–4063.

Borlido J, Zecchini V, Mills IG. 2009. Nuclear trafficking and functions of endocytic proteins implicated in oncogenesis. *Traffic* **10:** 1209–1220.

Bray SJ. 2006. Notch signalling: A simple pathway becomes complex. *Nat Rev Mol Cell Biol* **7:** 678–689.

Buchanan SM, Schalm SS, Maniatis T. 2010. Proteolytic processing of protocadherin proteins requires endocytosis. *Proc Natl Acad Sci* **107:** 17774–17779.

Burgoyne AM, Phillips-Mason PJ, Burden-Gulley SM, Robinson S, Sloan AE, Miller RH, Brady-Kalnay SM. 2009. Proteolytic cleavage of protein tyrosine phosphatase μ regulates glioblastoma cell migration. *Cancer Res* **69:** 6960–6968.

Cabrera JR, Bouzas-Rodriguez J, Tauszig-Delamasure S, Mehlen P. 2011. RET modulates cell adhesion via its cleavage by caspase in sympathetic neurons. *J Biol Chem* **286:** 14628–14638.

Cai J, Jiang WG, Grant MB, Boulton M. 2006. Pigment epithelium-derived factor inhibits angiogenesis via regulated intracellular proteolysis of vascular endothelial growth factor receptor 1. *J Biol Chem* **281:** 3604–3613.

Cai J, Chen Z, Ruan Q, Han S, Liu L, Qi X, Boye SL, Hauswirth WW, Grant MB, Boulton ME. 2011a. γ-Secretase and presenilin mediate cleavage and phosphorylation of vascular endothelial growth factor receptor-1. *J Biol Chem* **286:** 42514–42523.

Cai J, Wu L, Qi X, Li Calzi S, Caballero S, Shaw L, Ruan Q, Grant MB, Boulton ME. 2011b. PEDF regulates vascular permeability by a γ-secretase-mediated pathway. *PLoS ONE* **6:** e21164.

Carpenter G. 2003. Nuclear localization and possible functions of receptor tyrosine kinases. *Curr Opin Cell Biol* **15:** 143–148.

Carpenter G, Liao HJ. 2009. Trafficking of receptor tyrosine kinases to the nucleus. *Exp Cell Res* **315:** 1556–1566.

Carson DD. 2008. The cytoplasmic tail of MUC1: A very busy place. *Sci Signal* **1:** e35.

Chandra M, Zang S, Li H, Zimmerman LJ, Champer J, Tsuyada A, Chow A, Zhou W, Yu Y, Gao H, et al. 2012. Nuclear translocation of type I transforming growth factor β receptor confers a novel function in RNA processing. *Mol Cell Biol* **32:** 2183–2195.

Chaves-Pérez A, Mack B, Maetzel D, Kremling H, Eggert C, Harréus U, Gires O. 2013. EpCAM regulates cell cycle progression via control of cyclin D1 expression. *Oncogene* **32:** 641–650.

Ceni C, Kommaddi RP, Thomas R, Vereker E, Liu X, McPherson PS, Ritter B, Barker PA. 2010. The p75NTR intracellular domain generated by neurotrophin-induced receptor cleavage potentiates Trk signaling. *J Cell Sci* **123:** 2299–22307.

Chen CW, Roy D. 1996. Up-regulation of nuclear IGF-I receptor by short term exposure of stilbene estrogen, diethylstilbestrol. *Mol Cell Endocrinol* **118:** 1–8.

Cheng QC, Tikhomirov O, Zhou W, Carpenter G. 2003. Ectodomain cleavage of ErbB-4: Characterization of the cleavage site and m80 fragment. *J Biol Chem* **278:** 38421–38427.

Chioni AM, Grose R. 2012. FGFR1 cleavage and nuclear translocation regulates breast cancer cell behavior. *J Cell Biol* **197:** 801–817.

Codony-Servat J, Albanell J, Lopez-Talavera JC, Arribas J, Baselga J. 1999. Cleavage of the HER2 ectodomain is a pervanadate-activable process that is inhibited by the tissue inhibitor of metalloproteases-1 in breast cancer cells. *Cancer Res* **59:** 1196–11201.

Cousin H, Abbruzzese G, Kerdavid E, Gaultier A, Alfandari D. 2011. Translocation of the cytoplasmic domain of ADAM13 to the nucleus is essential for Calpain8-a expression and cranial neural crest cell migration. *Dev Cell* **20:** 256–263.

De Angelis Campos AC, Rodrigues MA, de Andrade C, de Goes AM, Nathanson MH, Gomes DA. 2011. Epidermal growth factor receptors destined for the nucleus are internalized via a clathrin-dependent pathway. *Biochem Biophys Res Commun* **412:** 341–346.

Degnin CR, Laederich MB, Horton WA. 2011. Ligand activation leads to regulated intramembrane proteolysis of fibroblast growth factor receptor 3. *Mol Biol Cell* **22:** 3861–3873.

Deheuninck J, Goormachtigh G, Foveau B, Ji Z, Leroy C, Ancot F, Villeret V, Tulasne D, Fafeur V. 2009. Phosphorylation of the MET receptor on juxtamembrane tyrosine residue 1001 inhibits its caspase-dependent cleavage. *Cell Signal* **21:** 1455–1463.

Deng H, Lin Y, Badin M, Vasilcanu D, Strömberg T, Jernberg-Wiklund H, Sehat B, Larsson O. 2010. Over-accumulation of nuclear IGF-1 receptor in tumor cells requires elevated expression of the receptor and the SUMO-conjugating enzyme Ubc9. *Biochem Biophys Res Commun* **404:** 667–671.

Denzel S, Maetzel D, Mack B, Eggert C, Bärr G, Gires O. 2009. Initial activation of EpCAM cleavage via cell-to-cell contact. *BMC Cancer* **9:** 402.

De Strooper B, Annaert W. 2010. Novel research horizons for presenilins and γ-secretases in cell biology and disease. *Annu Rev Cell Dev Biol* **26:** 235–260.

Dittmann K, Mayer C, Fehrenbacher B, Schaller M, Raju U, Milas L, Chen DJ, Kehlbach R, Rodemann HP. 2005. Radiation-induced epidermal growth factor receptor nuclear import is linked to activation of DNA-dependent protein kinase. *J Biol Chem* **280:** 31182–31189.

Dittmann K, Mayer C, Fehrenbacher B, Schaller M, Kehlbach R, Rodemann HP. 2010. Nuclear EGFR shuttling induced by ionizing radiation is regulated by phosphorylation at residue Thr654. *FEBS Lett* **584:** 3878–3884.

Domingues I, Rino J, Demmers JA, de Lanerolle P, Santos SC. 2001. VEGFR2 translocates to the nucleus to regulate its own transcription. *PLoS ONE* **6:** e25668.

Dunham-Ems SM, Lee YW, Stachowiak EK, Pudavar H, Claus P, Prasad PN. 2009. Stachowiak MK. Fibroblast growth factor receptor-1 (FGFR1) nuclear dynamics reveal a novel mechanism in transcription control. *Mol Biol Cell* **20:** 2401–2412.

Elenius K, Corfas G, Paul S, Choi CJ, Rio C, Plowman GD, Klagsbrun M. 1997. A novel juxtamembrane domain iso-form of HER4/ErbB4. Isoform-specific tissue distribution and differential processing in response to phorbol ester. *J Biol Chem* **272:** 26761–26768.

Feng SM, Muraoka-Cook RS, Hunter D, Sandahl MA, Caskey LS, Miyazawa K, Atfi A, Earp HS III. 2009. The E3 ubiquitin ligase WWP1 selectively targets HER4 and its proteolytically derived signaling isoforms for degradation. *Mol Cell Biol* **29:** 892–906.

Fortini ME. 2009. Notch signaling: The core pathway and its posttranslational regulation. *Dev Cell* **16:** 633–647.

Foveau B, Leroy C, Ancot F, Deheuninck J, Ji Z, Fafeur V, Tulasne D. 2007. Amplification of apoptosis through sequential caspase cleavage of the MET tyrosine kinase receptor. *Cell Death Differ* **14:** 752–764.

Foveau B, Ancot F, Leroy C, Petrelli A, Reiss K, Vingtdeux V, Giordano S, Fafeur V, Tulasne D. 2009. Down-regulation of the met receptor tyrosine kinase by presenilin-dependent regulated intramembrane proteolysis. *Mol Biol Cell* **20:** 2495–2507.

Fox SB, Turley H, Cheale M, Blázquez C, Roberts H, James N, Cook N, Harris A, Gatter K. 2004. Phosphorylated KDR is expressed in the neoplastic and stromal elements of human renal tumours and shuttles from cell membrane to nucleus. *J Pathol* **202:** 313–320.

García-Castillo J, Pedersen K, Angelini PD, Bech-Serra JJ, Colomé N, Cunningham MP, Parra-Palau JL, Canals F, Baselga J, Arribas J. 2009. HER2 carboxyl-terminal fragments regulate cell migration and cortactin phosphorylation. *J Biol Chem* **284:** 25302–25313.

Georgakopoulos A, Xu J, Xu C, Mauger G, Barthet G, Robakis NK. 2011. Presenilin1/γ-secretase promotes the EphB2-induced phosphorylation of ephrinB2 by regulating phosphoprotein associated with glycosphingolipid-enriched microdomains/Csk binding protein. *FASEB J* **25:** 3594–3604.

Giri DK, Ali-Seyed M, Li LY, Lee DF, Ling P, Bartholomeusz G, Wang SC, Hung MC. 2005. Endosomal transport of ErbB-2: Mechanism for nuclear entry of the cell surface receptor. *Mol Cell Biol* **25:** 11005–11018.

Giulianelli S, Cerliani JP, Lamb CA, Fabris VT, Bottino MC, Gorostiaga MA, Novaro V, Góngora A, Baldi A, Molinolo A, et al. 2008. Carcinoma-associated fibroblasts activate progesterone receptors and induce hormone independent mammary tumor growth: A role for the FGF-2/FGFR-2 axis. *Int J Cancer* **123:** 2518–2531.

Glenn G, van der Geer P. 2007. CSF-1 and TPA stimulate independent pathways leading to lysosomal degradation or regulated intramembrane proteolysis of the CSF-1 receptor. *FEBS Lett* **581:** 5377–5381.

Glenn G, van der Geer P. 2008. Toll-like receptors stimulate regulated intramembrane proteolysis of the CSF-1 receptor through Erk activation. *FEBS Lett* **582:** 911–915.

Goldschneider D, Rama N, Guix C, Mehlen P. 2008. The neogenin intracellular domain regulates gene transcription via nuclear translocation. *Mol Cell Biol* **28:** 4068–4079.

Golubkov VS, Strongin AY. 2012. Insights into ectodomain shedding and processing of protein-tyrosine pseudokinase 7 (PTK7). *J Biol Chem* **287:** 2009–42018.

Gomes DA, Rodrigues MA, Leite MF, Gomez MV, Varnai P, Balla T, Bennett AM, Nathanson MH. 2008. c-Met must translocate to the nucleus to initiate calcium signals. *J Biol Chem* **283**: 4344–4351.

Grumolato L, Liu G, Mong P, Mudbhary R, Biswas R, Arroyave R, Vijayakumar S, Economides AN, Aaronson SA. 2010. Canonical and noncanonical Wnts use a common mechanism to activate completely unrelated coreceptors. *Genes Dev* **24**: 2517–2530.

Han J, Allalunis-Turner J, Hendzel MJ. 2011. Characterization and comparison of protein complexes initiated by the intracellular domain of individual Notch paralogs. *Biochem Biophys Res Commun* **407**: 479–485.

Hanada N, Lo HW, Day CP, Pan Y, Nakajima Y, Hung MC. 2006. Co-regulation of B-Myb expression by E2F1 and EGF receptor. *Mol Carcinog* **45**: 10–17.

He G, Luo W, Li P, Remmers C, Netzer WJ, Hendrick J, Bettayeb K, Flajolet M, Gorelick F, Wennogle LP, et al. 2010. γ-Secretase activating protein is a therapeutic target for Alzheimer's disease. *Nature* **467**: 95–98.

Hoeing K, Zscheppang K, Mujahid S, Murray S, Volpe MV, Dammann CE, Nielsen HC. 2011. Presenilin-1 processing of ErbB4 in fetal type II cells is necessary for control of fetal lung maturation. *Biochim Biophys Acta* **1813**: 480–491.

Hsu SC, Hung MC. 2007. Characterization of a novel tripartite nuclear localization sequence in the EGFR family. *J Biol Chem* **282**: 10432–10440.

Huang WC, Chen YJ, Li LY, Wei YL, Hsu SC, Tsai SL, Chiu PC, Huang WP, Wang YN, Chen CH, et al. 2011a. Nuclear translocation of epidermal growth factor receptor by Akt-dependent phosphorylation enhances breast cancer-resistant protein expression in gefitinib-resistant cells. *J Biol Chem* **286**: 20558–20568.

Huang HP, Chen PH, Yu CY, Chuang CY, Stone L, Hsiao WC, Li CL, Tsai SC, Chen KY, Chen HF, et al. 2011b. Epithelial cell adhesion molecule (EpCAM) complex proteins promote transcription factor-mediated pluripotency reprogramming. *J Biol Chem* **286**: 33520–33532.

Janiszewska M, De Vito C, Le Bitoux MA, Fusco C, Stamenkovic I. 2010. Transportin regulates nuclear import of CD44. *J Biol Chem* **285**: 30548–30557.

Jones FE. 2008. HER4 intracellular domain (4ICD) activity in the developing mammary gland and breast cancer. *J Mammary Gland Biol Neoplasia* **13**: 247–258.

Kagawa H, Shino Y, Kobayashi D, Demizu S, Shimada M, Ariga H, Kawahara H. 2011. A novel signaling pathway mediated by the nuclear targeting of C-terminal fragments of mammalian Patched 1. *PLoS ONE* **6**: e18638.

Kasuga K, Kaneko H, Nishizawa M, Onodera O, Ikeuchi T. 2007. Generation of intracellular domain of insulin receptor tyrosine kinase by γ-secretase. *Biochem Biophys Res Commun* **360**: 90–96.

Kim J, Jahng WJ, Di Vizio D, Lee JS, Jhaveri R, Rubin MA, Shisheva A, Freeman MR. 2007. The phosphoinositide kinase PIKfyve mediates epidermal growth factor receptor trafficking to the nucleus. *Cancer Res* **67**: 9229–9237.

Kopan R, Ilagan MX. 2004. γ-Secretase: Proteasome of the membrane? *Nat Rev Mol Cell Biol* **5**: 499–504.

Lai C, Feng L. 2004. Implication of γ-secretase in neuregulin-induced maturation of oligodendrocytes. *Biochem Biophys Res Commun* **314**: 535–542.

Lal M, Caplan M. 2011. Regulated intramembrane proteolysis: Signaling pathways and biological functions. *Physiology (Bethesda)* **26**: 34–44.

Lee TH, Seng S, Sekine M, Hinton C, Fu Y, Avraham HK, Avraham S. 2007. Vascular endothelial growth factor mediates intracrine survival in human breast carcinoma cells through internally expressed VEGFR1/FLT1. *PLoS Med* **4**: e186.

Lemberg MK. 2011. Intramembrane proteolysis in regulated protein trafficking. *Traffic* **12**: 1109–1118.

Le Moan N, Houslay DM, Christian F, Houslay MD, Akassoglou K. 2011. Oxygen-dependent cleavage of the p75 neurotrophin receptor triggers stabilization of HIF-1α. *Mol Cell* **44**: 476–490.

Li LY, Chen H, Hsieh YH, Wang YN, Chu HJ, Chen YH, Chen HY, Chien PJ, Ma HT, Tsai HC, et al. 2011. Nuclear ErbB2 enhances translation and cell growth by activating transcription of ribosomal RNA genes. *Cancer Res* **71**: 4269–4279.

Liao HJ, Carpenter G. 2007. Role of the Sec61 translocon in EGF receptor trafficking to the nucleus and gene expression. *Mol Biol Cell* **18**: 1064–1072.

Liao HJ, Carpenter G. 2009. Cetuximab/C225-induced intracellular trafficking of epidermal growth factor receptor. *Cancer Res* **69**: 6179–6183.

Liao HJ, Carpenter G. 2012. Regulated intramembrane cleavage of the EGF receptor. *Traffic* **13**: 1106–1112.

Liccardi G, Hartley JA, Hochhauser D. 2011. EGFR nuclear translocation modulates DNA repair following cisplatin and ionizing radiation treatment. *Cancer Res* **71**: 1103–1114.

Lin SY, Makino K, Xia W, Matin A, Wen Y, Kwong KY, Bourguignon L, Hung MC. 2001. Nuclear localization of EGF receptor and its potential new role as a transcription factor. *Nat Cell Biol* **3**: 802–808.

Lin KT, Sloniowski S, Ethell DW, Ethell IM. 2008. Ephrin-B2-induced cleavage of EphB2 receptor is mediated by matrix metalloproteinases to trigger cell repulsion. *J Biol Chem* **283**: 28969–28979.

Litterst C, Georgakopoulos A, Shioi J, Ghersi E, Wisniewski T, Wang R, Ludwig A, Robakis NK. 2007. Ligand binding and calcium influx induce distinct ectodomain/γ-secretase-processing pathways of EphB2 receptor. *J Biol Chem* **282**: 16155–16163.

Liu HS, Hsu PY, Lai MD, Chang HY, Ho CL, Cheng HL, Chen HT, Lin YJ, Wu TJ, Tzai TS, et al. 2010. An unusual function of RON receptor tyrosine kinase as a transcriptional regulator in cooperation with EGFR in human cancer cells. *Carcinogenesis* **31**: 1456–1464.

Lo HW, Hsu SC, Ali-Seyed M, Gunduz M, Xia X, Wei Y, Bartholomeusz G, Shih JY, Hung MC. 2005. Nuclear interaction of EGFR and STAT3 in the activation of the iNOS/NO pathway. *Cancer Cell* **7**: 575–589.

Lo HW, Ali-Seyed M, Wu Y, Bartholomeusz G, Hsu SC, Hung MC. 2006. Nuclear-cytoplasmic transport of EGFR involves receptor endocytosis, importin β1 and CRM1. *J Cell Biochem* **98**: 1570–1583.

Lo HW, Cao X, Zhu H, Ali-Osman F. 2010. Cyclooxygenase-2 is a novel transcriptional target of the nuclear EGFR-STA3 and EGFRvIII-STAT3 signaling axes. *Mol Cancer Res* **8**: 232–245.

Cite this article as *Cold Spring Harb Perspect Biol* doi: 10.1101/cshperspect.a008979

López-Otín C, Hunter T. 2010. The regulatory crosstalk between kinases and proteases in cancer. *Nat Rev Cancer* **10:** 278–292.

Lu P, Ewald AJ, Martin GR, Werb Z. 2004. Genetic mosaic analysis reveals FGF receptor 2 function in terminal end buds during mammary gland branching morphogenesis. *Dev Biol* **321:** 77–87.

Lyu J, Yamamoto V, Lu W. 2008. Cleavage of the Wnt receptor Ryk regulates neuronal differentiation during cortical neurogenesis. *Dev Cell* **15:** 773–780.

Lyu J, Wesselschmidt RL, Lu W. 2009. Cdc37 regulates Ryk signaling by stabilizing the cleaved Ryk intracellular domain. *J Biol Chem* **284:** 12940–12948.

Maetzel D, Denzel S, Mack B, Canis M, Went P, Benk M, Kieu C, Papior P, Baeuerle PA, Munz M, et al. 2009. Nuclear signalling by tumour-associated antigen EpCAM. *Nat Cell Biol* **11:** 162–171.

Marron MB, Singh H, Tahir TA, Kavumkal J, Kim HZ, Koh GY, Brindle NP. 2007. Regulated proteolytic processing of Tie1 modulates ligand responsiveness of the receptor-tyrosine kinase Tie2. *J Biol Chem* **282:** 30509–30517.

Martin AJ, Grant A, Ashfield AM, Palmer CN, Baker L, Quinlan PR, Purdie CA, Thompson AM, Jordan LB, Berg JN. 2011. FGFR2 protein expression in breast cancer: Nuclear localization and correlation with patient genotype. *BMC Res Notes* **4:** 72.

Mathew D, Ataman B, Chen J, Zhang Y, Cumberledge S, Budnik V. 2005. Wingless signaling at synapses is through cleavage and nuclear import of receptor DFrizzled2. *Science* **310:** 1344–1347.

Mazzone M, Selfors LM, Albeck J, Overholtzer M, Sale S, Carroll DL, Pandya D, Lu Y, Mills GB, Aster JC, et al. 2010. Dose-dependent induction of distinct phenotypic responses to Notch pathway activation in mammary epithelial cells. *Proc Natl Acad Sci* **107:** 5012–5017.

McCarthy JV, Twomey C, Wujek P. 2009. Presenilin-dependent regulated intramembrane proteolysis and γ-secretase activity. *Cell Mol Life Sci* **66:** 1534–1555.

McElroy B, Powell JC, McCarthy JV. 2007. The insulin-like growth factor 1 (IGF-1) receptor is a substrate for γ-secretase-mediated intramembrane proteolysis. *Biochem Biophys Res Commun* **358:** 1136–1141.

Merlin J, Stechly L, de Beaucé S, Monté D, Leteurtre E, van Seuningen I, Huet G, Pigny P. 2011. Galectin-3 regulates MUC1 and EGFR cellular distribution and EGFR downstream pathways in pancreatic cancer cells. *Oncogene* **30:** 2514–2525.

Migaki GI, Kahn J, Kishimoto TK. 1995. Mutational analysis of the membrane-proximal cleavage site of l-selectin: Relaxed sequence specificity surrounding the cleavage site. *J Exp Med* **182:** 549–557.

Miletti-González KE, Murphy K, Kumaran MN, Ravindranath AK, Wernyj RP, Kaur S, Miles GD, Lim E, Chan R, Chekmareva M, et al. 2012. Identification of function for CD44 intracytoplasmic domain (CD44-ICD): Modulation of matrix metalloproteinase 9 (MMP-9) transcription via novel promoter response element. *J Biol Chem* **287:** 18995–19007.

Montes de Oca P, Malardé V, Proust R, Dautry-Varsat A, Gesbert F. 2010. Ectodomain shedding of interleukin-2 receptor β and generation of an intracellular functional fragment. *J Biol Chem* **285:** 22050–22058.

Mourali J, Bénard A, Lourenço FC, Monnet C, Greenland C, Moog-Lutz C, Racaud-Sultan C, Gonzalez-Dunia D, Vigny M, Mehlen P, et al. 2006. Anaplastic lymphoma kinase is a dependence receptor whose proapoptotic functions are activated by caspase cleavage. *Mol Cell Biol* **26:** 6209–6222.

Mu Y, Sundar R, Thakur N, Ekman M, Gudey SK, Yakymovych M, Hermansson A, Dimitriou H, Bengoechea-Alonso MT, Ericsson J, et al. 2011. TRAF6 ubiquitinates TGF-β type I receptor to promote its cleavage and nuclear translocation in cancer. *Nat Commun* **2:** 330.

Muraoka-Cook RS, Sandahl M, Husted C, Hunter D, Miraglia L, Feng SM, Elenius K, Earp HS. 2006. The intracellular domain of ErbB4 induces differentiation of mammary epithelial cells. *Mol Biol Cell* **17:** 4118–4129.

Na HW, Shin WS, Ludwig A, Lee ST. 2012. The cytosolic domain of protein-tyrosine kinase 7 (PTK7), generated from sequential cleavage by a disintegrin and metalloprotease 17 (ADAM17) and γ-secretase, enhances cell proliferation and migration in colon cancer cells. *J Biol Chem* **287:** 25001–25009.

Naresh A, Long W, Vidal GA, Wimley WC, Marrero L, Sartor CI, Tovey S, Cooke TG, Bartlett JM, Jones FE. 2006. The ERBB4/HER4 intracellular domain 4ICD is a BH3-only protein promoting apoptosis of breast cancer cells. *Cancer Res* **66:** 6412–6420.

Ni CY, Murphy MP, Golde TE, Carpenter G. 2001. γ-Secretase cleavage and nuclear localization of ErbB-4 receptor tyrosine kinase. *Science* **294:** 2179–2181.

Ni CY, Yuan H, Carpenter G. 2003. Role of the ErbB-4 carboxyl terminus in γ-secretase cleavage. *J Biol Chem* **278:** 4561–4565.

Offterdinger M, Schofer C, Weipoltshammer K, Grunt TW. 2002. c-erbB-3: A nuclear protein in mammary epithelial cells. *J Cell Biol* **157:** 929–939.

Omerovic J, Santangelo L, Puggioni EM, Marrocco J, Dall'Armi C, Palumbo C, Belleudi F, Di Marcotullio L, Frati L, Torrisi MR, et al. 2007. The E3 ligase Aip4/Itch ubiquitinates and targets ErbB-4 for degradation. *FASEB J* **21:** 2849–2862.

Paatero I, Jokilammi A, Heikkinen PT, Iljin K, Kallioniemi OP, Jones FE, Jaakkola PM, Elenius K. 2012. Interaction with ErbB4 promotes hypoxia-inducible factor-1α signaling. *J Biol Chem* **287:** 9659–9671.

Pan B, Huang XF, Deng C. 2011. Antipsychotic treatment and neuregulin 1-ErbB4 signalling in schizophrenia. *Prog Neuropsychopharmacol Biol Psychiatry* **35:** 924–930.

Parkhurst CN, Zampieri N, Chao MV. 2010. Nuclear localization of the p75 neurotrophin receptor intracellular domain. *J Biol Chem* **285:** 5361–5368.

Pedersen K, Angelini PD, Laos S, Bach-Faig A, Cunningham MP, Ferrer-Ramón C, Luque-García A, García-Castillo J, Parra-Palau JL, Scaltriti M, et al. 2009. A naturally occurring HER2 carboxy-terminal fragment promotes mammary tumor growth and metastasis. *Mol Cell Biol* **29:** 3319–3331.

Piccione EC, Lieu TJ, Gentile CF, Williams TR, Connolly AJ, Godwin AK, Koong AC, Wong AJ. 2012. A novel epidermal growth factor receptor variant lacking multiple domains directly activates transcription and is overexpressed in tumors. *Oncogene* **31:** 2953–2967.

Pozner-Moulis S, Pappas DJ, Rimm DL. 2006. Met, the hepatocyte growth factor receptor, localizes to the nucleus in cells at low density. *Cancer Res* **66:** 7976–7982.

Racaud-Sultan C, Gonzalez-Dunia D, Vigny M, Mehlen P, Delsol G, Allouche M. 2006. Anaplastic lymphoma kinase is a dependence receptor whose proapoptotic functions are ctivated by caspase cleavage. *Mol Cell Biol* **26:** 6209–6222.

Rahimi N, Golde TE, Meyer RD. 2009. Identification of ligand-induced proteolytic cleavage and ectodomain shedding of VEGFR-1/FLT1 in leukemic cancer cells. *Cancer Res* **69:** 2607–2614.

Ralhan R, He HC, So AK, Tripathi SC, Kumar M, Hasan MR, Kaur J, Kashat L, MacMillan C, Chauhan SS, et al. 2010. Nuclear and cytoplasmic accumulation of Ep-ICD is frequently detected in human epithelial cancers. *PLoS ONE* **5:** e14130.

Rebeck GW. 2009. Nontraditional signaling mechanisms of lipoprotein receptors. *Sci Signal* **2:** e28.

Rio C, Buxbaum JD, Peschon JJ, Corfas G. 2000. Tumor necrosis factor-α-converting enzyme is required for cleavage of erbB4/HER4. *J Biol Chem* **275:** 10379–10387.

Rokicki J, Das PM, Giltnane JM, Wansbury O, Rimm DL, Howard BA, Jones FE. 2010. The ERα coactivator, HER4/4ICD, regulates progesterone receptor expression in normal and malignant breast epithelium. *Mol Cancer* **9:** 150.

Sandvig K, van Deurs B. 2002. Transport of protein toxins into cells: Pathways used by ricin, cholera toxin and Shiga toxin. *FEBS Lett* **529:** 49–53.

Santos SC, Dias S. 2004. Internal and external autocrine VEGF/KDR loops regulate survival of subsets of acute leukemia through distinct signaling pathways. *Blood* **103:** 3883–3889.

Santos SC, Miguel C, Domingues I, Calado A, Zhu Z, Wu Y, Dias S. 2007. VEGF and VEGFR-2 (KDR) internalization is required for endothelial recovery during wound healing. *Exp Cell Res* **313:** 1561–1574.

Sardi SP, Murtie J, Koirala S, Patten BA, Corfas G. 2006. Presenilin-dependent ErbB4 nuclear signaling regulates the timing of astrogenesis in the developing brain. *Cell* **127:** 185–197.

Schmahl J, Kim Y, Colvin JS, Ornitz DM, Capel B. 2004. Fgf9 induces proliferation and nuclear localization of FGFR2 in Sertoli precursors during male sex determination. *Development* **131:** 3627–3636.

Sehat B, Tofigh A, Lin Y, Trocmé E, Liljedahl U, Lagergren J, Larsson O. 2010. SUMOylation mediates the nuclear translocation and signaling of the IGF-1 receptor. *Sci Signal* **3:** ra10.

Selkoe DJ, Wolfe MS. 2007. Presenilin: Running with scissors in the membrane. *Cell* **131:** 215–221.

Shah S, Lee SF, Tabuchi K, Hao HY, Yu C, LaPlant Q, Ball H, Dann CE, Sudhof T, Yu G. 2005. Nicastrin functions as a γ-secretase-substrate receptor. *Cell* **122:** 435–447.

Small DH. 2002. Is γ-secretase a multienzyme complex for membrane protein degradation? Models and speculations. *Peptides* **23:** 1317–1321.

Stachowiak MK, Maher PA, Stachowiak EK. 2007. Integrative nuclear signaling in cell development—A role for FGF receptor-1. *DNA Cell Biol* **26:** 811–826.

Stewart M, Turley H, Cook N, Pezzella F, Pillai G, Ogilvie D, Cartlidge S, Paterson D, Copley C, Kendrew J, et al. 2003. The angiogenic receptor KDR is widely distributed in human tissues and tumours and relocates intracellularly on phosphorylation. An immunohistochemical study. *Histopathology* **43:** 33–39.

Stoeck A, Shang L, Dempsey PJ. 2010. Sequential and γ-secretase-dependent processing of the β-cellulin precursor generates a palmitoylated intracellular-domain fragment that inhibits cell growth. *J Cell Sci* **123:** 2319–2331.

Strohecker AM, Yehiely F, Chen F, Cryns VL. 2008. Caspase cleavage of HER-2 releases a bad-like cell death effector. *J Biol Chem* **283:** 18269–18282.

Struhl G, Adachi A. 2000. Requirements for presenilin-dependent cleavage of notch and other transmembrane proteins. *Mol Cell* **6:** 625–636.

Tagami S, Okochi M, Yanagida K, Ikuta A, Fukumori A, Matsumoto N, Ishizuka-Katsura Y, Nakayama T, Itoh N, Jiang J, et al. 2008. Regulation of Notch signaling by dynamic changes in the precision of S3 cleavage of Notch-1. *Mol Cell Biol* **28:** 165–176.

Tao Y, Song X, Deng X, Xie D, Lee LM, Liu Y, Li W, Li L, Deng L, Wu Q, et al. 2005. Nuclear accumulation of epidermal growth factor receptor and acceleration of G_1/S stage by Epstein–Barr-encoded oncoprotein latent membrane protein 1. *Exp Cell Res* **303:** 240–251.

Tauszig-Delamasure S, Yu LY, Cabrera JR, Bouzas-Rodriguez J, Mermet-Bouvier C, Guix C, Bordeaux MC, Arumae U, Mehlen P. 2007. The TrkC receptor induces apoptosis when the dependence receptor notion meets the neurotrophin paradigm. *Proc Natl Acad Sci* **104:** 13361–13366.

Thor AD, Edgerton SM, Jones FE. 2009. Subcellular localization of the HER4 ntracellular domain, 4ICD, identifies distinct prognostic outcomes for breast cancer patients. *Am J Pathol* **175:** 1802–1809.

Tikhomirov O, Carpenter G. 2001. Caspase-dependent cleavage of ErbB-2 by geldanamycin and staurosporin. *J Biol Chem* **276:** 33675–33680.

Tomita T, Tanaka S, Morohashi Y, Iwatsubo T. 2006. Presenilin-dependent intramembrane cleavage of ephrin-B1. *Mol Neurodegener* **1:** 2.

Tousseyn T, Thathiah A, Jorissen E, Raemaekers T, Konietzko U, Reiss K, Maes E, Snellinx A, Serneels L, Nyabi O, et al. 2009. ADAM10, the rate-limiting protease of regulated intramembrane proteolysis of Notch and other proteins, is processed by ADAMS-9, ADAMS-15, and the γ-secretase. *J Biol Chem* **284:** 11738–11747.

Tulasne D, Deheuninck J, Lourenco FC, Lamballe F, Ji Z, Leroy C, Puchois E, Moumen A, Maina F, Mehlen P, et al. 2004. Proapoptotic function of the MET tyrosine kinase receptor through caspase cleavage. *Mol Cell Biol* **24:** 10328–10339.

Varshavsky A. 1997. The N-end rule pathway of protein degradation. *Genes Cells* **2:** 13–28.

Vecchi M, Baulida J, Carpenter G. 1996. Selective cleavage of the heregulin receptor ErbB-4 by protein kinase C activation. *J Biol Chem* **271:** 18989–18995.

Veikkolainen V, Vaparanta K, Halkilahti K, Iljin K, Sundvall M, Elenius K. 2011. Function of ERBB4 is determined by alternative splicing. *Cell Cycle* **10:** 2647–2657.

Vidal GA, Naresh A, Marrero L, Jones FE. 2005. Presenilin-dependent γ-secretase processing regulates multiple ERBB4/HER4 activities. *J Biol Chem* **280:** 19777–19783.

Wang SC, Lien HC, Xia W, Chen IF, Lo HW, Wang Z, Ali-Seyed M, Lee DF, Bartholomeusz G, Ou-Yang F, et al. 2004. Binding at and transactivation of the COX-2 promoter by nuclear tyrosine kinase receptor ErbB-2. *Cancer Cell* **6:** 251–261.

Wang SC, Nakajima Y, Yu YL, Xia W, Chen CT, Yang CC, McIntush EW, Li LY, Hawke DH, Kobayashi R, et al. 2006. Tyrosine phosphorylation controls PCNA function through protein stability. *Nat Cell Biol* **8:** 1359–1368.

Wang X, Cui M, Wang L, Chen X, Xin P. 2010a. Inhibition of neurotrophin receptor p75 intramembran proteolysis by γ-secretase inhibitor reduces medulloblastoma spinal metastasis. *Biochem Biophys Res Commun* **403:** 264–269.

Wang YN, Yamaguchi H, Huo L, Du Y, Lee HJ, Lee HH, Wang H, Hsu JM, Hung MC. 2010b. The translocon Sec61β localized in the inner nuclear membrane transports membrane-embedded EGF receptor to the nucleus. *J Biol Chem* **285:** 38720–38729.

Wang YN, Yamaguchi H, Hsu JM, Hung MC. 2010c. Nuclear trafficking of the epidermal growth factor receptor family membrane proteins. *Oncogene* **29:** 3997–4006.

Wang X, Cowan JW, Gerhart M, Zelickson BR, Jiang J, He K, Wolfe MS, Black RA, Frank SJ. 2011a. γ-Secretase-mediated growth hormone receptor proteolysis: Mapping of the intramembranous cleavage site. *Biochem Biophys Res Commun* **408:** 432–436.

Wang YN, Wang H, Yamaguchi H, Lee HJ, Lee HH, Hung MC. 2011b. COPI-mediated retrograde trafficking from the Golgi to the ER regulates EGFR nuclear transport. *Biochem Biophys Res Commun* **399:** 498–504.

Wang YN, Lee HH, Lee HJ, Du Y, Yamaguchi H, Hung MC. 2012. Membrane-bound trafficking regulates nuclear transport of integral epidermal growth factor receptor (EGFR) and ErbB-2. *J Biol Chem* **287:** 16869–16879.

Wilhelmsen K, van der Geer P. 2004. Phorbol 12-myristate 13-acetate-induced release of the colony-stimulating factor 1 receptor cytoplasmic domain into the cytosol involves two separate cleavage events. *Mol Cell Biol* **24:** 454–464.

Williams CC, Allison JG, Vidal GA, Burow ME, Beckman BS, Marrero L, Jones FE. 2004. The ERBB4/HER4 receptor tyrosine kinase regulates gene expression by functioning as a STAT5A nuclear chaperone. *J Cell Biol* **167:** 469–478.

Wong J, Weickert CS. 2009. Transcriptional interaction of an estrogen receptor splice variant and ErbB4 suggests convergence in gene susceptibility pathways in schizophrenia. *J Biol Chem* **284:** 18824–18832.

Xia W, Liu Z, Zong R, Liu L, Zhao S, Bacus SS, Mao Y, He J, Wulfkuhle JD, Petricoin EF III, et al. 2011. Truncated ErbB2 expressed in tumor cell nuclei contributes to acquired therapeutic resistance to ErbB2 kinase inhibitors. *Mol Cancer Ther* **10:** 1367–1374.

Yu YL, Chou RH, Wu CH, Wang YN, Chang WJ, Tseng YJ, Chang WC, Lai CC, Lee HJ, Huo L, et al. 2012. Nuclear EGFR suppresses ribonuclease activity of polynucleotide phosphorylase through DNAPK-mediated phosphorylation at serine 776. *J Biol Chem* **287:** 31015–31026.

Zeng F, Zhang MZ, Singh AB, Zent R, Harris RC. 2007. ErbB4 isoforms selectively regulate growth factor induced Madin-Darby canine kidney cell tubulogenesis. *Mol Biol Cell* **18:** 4446–4456.

Zeng F, Xu J, Harris RC. 2009. Nedd4 mediates ErbB4 JM-a/CYT-1 ICD ubiquitination and degradation in MDCK II cells. *FASEB J* **23:** 1935–1945.

Zhang Y, Pillai G, Gatter K, Blázquez C, Turley H, Pezzella F, Watt SM. 2005. Expression and cellular localization of vascular endothelial growth factor A and its receptors in acute and chronic leukemias: An immunohistochemical study. *Hum Pathol* **36:** 797–805.

Zhou W, Carpenter G. 2000. Heregulin-dependent trafficking and cleavage of ErbB-4. *J Biol Chem* **275:** 34737–34743.

Zscheppang K, Dörk T, Schmiedl A, Jones FE, Dammann CE. 2011a. Neuregulin receptor ErbB4 functions as a transcriptional cofactor for the expression of surfactant protein B in the fetal lung. *Am J Respir Cell Mol Biol* **45:** 761–767.

Zscheppang K, Konrad M, Zischka M, Huhn V, Dammann CE. 2011b. Estrogen-induced upregulation of Sftpb requires transcriptional control of neuregulin receptor ErbB4 in mouse lung type II epithelial cells. *Biochim Biophys Acta* **1813:** 1717–1727.

Biological Function of Nuclear Receptor Tyrosine Kinase Action

Sungmin Song[1,2,4], Kenneth M. Rosen[1,3,4], and Gabriel Corfas[1,2,3]

[1]F.M. Kirby Neurobiology Center, Children's Hospital Boston, Massachusetts 02115

[2]Department of Neurology, Harvard Medical School, Boston, Massachusetts 02115

[3]Department of Otolaryngology, Harvard Medical School, Boston, Massachusetts 02115

Correspondence: gabriel.corfas@childrens.harvard.edu

Receptor tyrosine kinases (RTKs) were believed until recently to act at the cell membrane in a singular fashion (i.e., binding of ligands on the extracellular domain would activate the intrinsic tyrosine kinase activity in the intracellular domain), which would then start a cascade involving other intracellular signaling molecules that would act as effectors. However, new evidence indicates that some RTKs can signal through a different modality; they can move into the nucleus where they directly exert their actions. Although some studies have showed that the proteolytically released intracellular domain of several RTKs can move to the nucleus where they influence gene expression and cell function, others suggest that RTKs can also move to the nucleus as holoproteins. The identification of this novel signaling mechanism calls for a critical reevaluation of the mechanisms of action of RTKs and their biological roles.

Receptor tyrosine kinases (RTKs) are type I transmembrane proteins that typically transduce signals originating at the cell surface, which then influence numerous cellular processes, in many cases by ultimately impacting on gene expression. For years, it had been believed that RTKs could only signal by what now should be referred to as "canonical RTK signaling" (Fig. 1), a multistep process involving: (1) the binding of their cognate ligands (most frequently growth factors), (2) hetero- or homodimerization of the receptor leading to activation of the intrinsic tyrosine kinase, (3) cytoplasmic domain auto- or transphosphorylation that generates binding sites for intracellular adaptor proteins, and (4) activation of intracellular signaling cascades that impact on the cell's biology (Lemmon and Schlessinger 2010). However, more recent studies have uncovered a novel signaling mechanism used by RTKs (i.e., direct nuclear signaling) (Fig. 2). In this modality, following receptor activation, instead of involving a multicomponent cytoplasmic signaling cascade, the cleaved intracellular domain transits to the nucleus to influence transcription. Some studies have also reported the translocation of full-length RTKs to the nucleus, but whether this actually occurs and whether it has a signaling role is still in debate (Schlessinger and Lemmon 2006). Here we summarize the current

---

[4]These authors contributed equally to this work.

Cite this article as *Cold Spring Harb Perspect Biol* doi: 10.1101/cshperspect.a009001

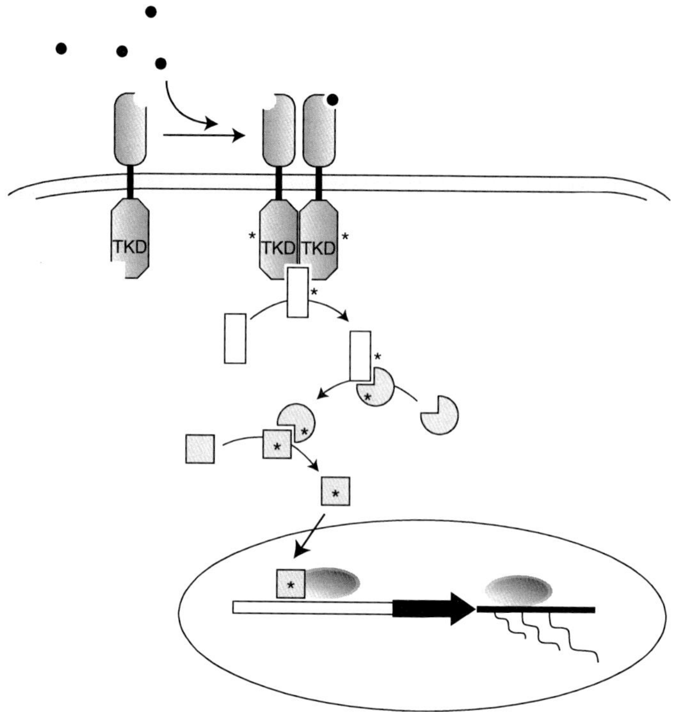

Figure 1. Canonical receptor tyrosine kinase signaling. Activation of a receptor tyrosine kinase at the cell surface leads first to dimerization, followed by activation and phosphorylation of the receptor itself in its intracellular domain. The activated receptor then recruits downstream cytoplasmic signaling molecules, in many cases other kinases. Activation of the downstream signaling cascade frequently leads to changes in gene expression. TKD, tyrosine kinase domain.

understanding of the wide array of mechanisms and biological roles that have been attributed to RTK nuclear signaling, with a particular emphasis on the ErbB receptors, the RTK family for which nuclear signaling has been characterized most extensively.

ErbB4 AND NUCLEAR SIGNALING

The ErbB4 RTK, as the name implies, was the fourth member of the EGF (or ErbB) receptor family to be identified (Plowman et al. 1993). Alternative splicing in the extracellular juxtamembrane (JM) and cytoplasmic domains of erbB4 generates several isoforms (Elenius et al. 1997, 1999). The splicing in both regions influences the biology of the receptor, but most germane to the discussion here is the variation that occurs in the juxtamembrane region. Two jux-

tamembrane isoforms exist, JMa and JMb (Elenius et al. 1997), which result in ErbB4 isoforms that differ in their susceptibility to targeted proteolytic cleavage. Stimulation of ErbB4-JMa by its ligand neuregulin 1 (NRG1) (Zhou and Carpenter 2000) or the activation of protein kinase C by TPA (12-O-tetradecanoylphorbol-13-acetate) (Vecchi et al. 1996; Elenius et al. 1997) promotes cleavage in the extracellular juxtamembrane region, causing the release of a 120-kD ectodomain fragment and leaving behind a transmembrane-tethered 80-kD intracellular domain fragment (mE4ICD). In contrast, ErbB4-JMb is uncleavable (Rio et al. 2000). ErbB4-JMa cleavage can be blocked by TAPI, an inhibitor of tumor necrosis factor-γ-converting enzyme (TACE), and does not occur in TACE-deficient cells (Rio et al. 2000) linking this metalloprotease (also known as α-secretase,

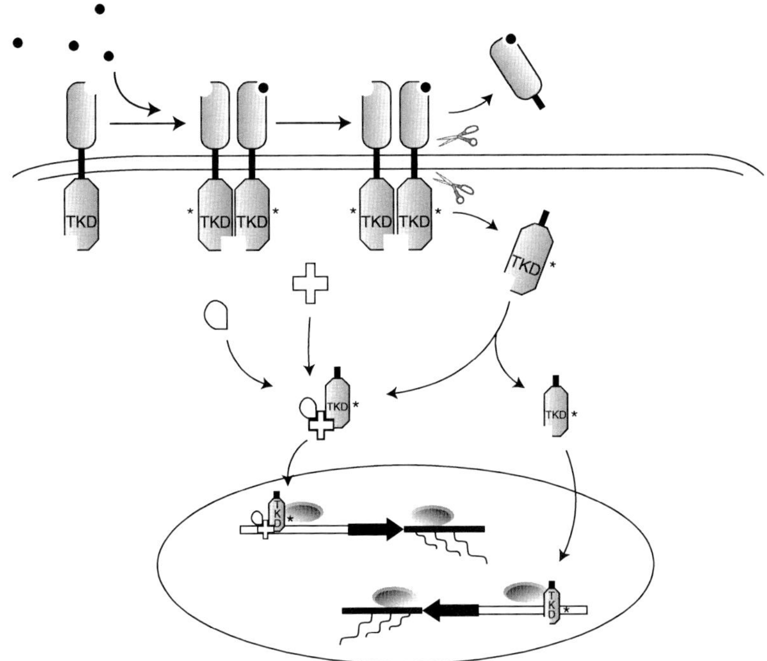

Figure 2. Direct nuclear signaling through a released RTK intracellular domain. Following activation by ligand, specific proteolytic cleavages lead to the release of an activated tyrosine kinase-containing intracellular domain (ICD). Known examples of nuclear signaling by RTK-ICDs include: (a) formation of a cytoplasmic complex containing the ICD and other effector molecules, which then translocates to the nucleus (e.g., ErbB4), and (b) direct nuclear translocation of a released RTK-ICD (e.g., Ryk). Both examples can lead to changes in gene expression.

ADAM17) to the ectodomain shedding of ErbB4. Following this cleavage event, the mE4ICD fragment serves as a substrate for regulated intramembrane proteolysis by the Presenilin-1 (PS1)/γ-secretase complex, releasing a soluble intracellular domain (sE4ICD), which has a nuclear localization signal that mediates its import to the nucleus (Ni et al. 2001). This adds ErbB4 JMa to a growing list of membrane proteins, including the β-amyloid precursor protein (βAPP) and Notch, that are cleaved sequentially by different secretase enzymes, specifically generating a soluble cytoplasmic domain, which can then enter the nucleus (reviewed by Fortini 2002).

Signaling via ErbB4 has been implicated in a number of both normal and pathological processes in multiple cell types; however, the functional importance of a nuclear-localized E4ICD is only beginning to be fully appreciated. Given that erbB4 is necessary for breast development

and lactation (Jones et al. 1999; Long et al. 2003; Tidcombe et al. 2003), and that it is involved in neoplastic breast tissue development (Sundvall et al. 2008), breast-derived tissue and cell lines have served as a rich area for investigation of the role of nuclear ErbB4 signaling. As milk production is a recognized sign of mammary epithelium differentiation, molecular events that regulate the expression of milk components are often viewed as proxies for mammary differentiation. In this context, Williams et al. (2004) found that stimulation of ErbB4 by neuregulin-1 (NRG1) in a breast cancer cell line leads to the association of sE4ICD and STAT5A with the promoter of the β-casein gene inducing its activity. Another study using breast cancer cells showed that activation of ErbB4 by NRG1 leads to the tyrosine phosphorylation of the largely nuclear localized c-Abl target Hdm2, and that this phosphorylation is sensitive to both inhibitors of ErbB receptor kinases and of TACE or γ-secretase, implying that sE4ICD generation is required (Arasada and Carpenter 2005). A recent study of resected breast tumors showed that the subcellular location of ErbB4 is a critical prognostic indicator, with tumors displaying membranous ErbB4 associated with better clinical outcomes, whereas the presence of nuclear-localized sE4ICD is associated with higher proliferative potential and greater transcription at genes bearing an estrogen response element (ERE) (Junttila et al. 2005). The association between E4ICD and estrogen responsive gene expression has been extended by a study showing that E4ICD is found along with ERα at the promoters for the progesterone receptor and SDF-1 genes, whereas the noncleavable mutant ErbB4 V673I fails to coactivate ERα-dependent gene transcription (Zhu et al. 2006).

Nuclear ErbB4 signaling has also emerged as an important signaling modality in the nervous system, one of the sites originally identified as expressing the cleavable JMa isoform (Elenius et al. 1997). A study examining oligodendrocytes showed that NRG1-induced changes in the maturation of these cells, including the induction of myelin basic protein expression, are dependent on both γ-secretase activity and sE4ICD nuclear translocation (Lai and Feng 2004). Sardi et al. (2006) found that, during brain development, ErbB4 nuclear signaling plays a critical role in the process of astrogenesis. This study showed that NRG1-induced ErbB4 activation in neural precursor cells induces the release of sE4ICD, which forms a transcriptional repressor complex with TAB2 and the nuclear corepressor N-CoR. After nuclear translocation, this complex binds to promoter elements of astrocytic-associated genes, repressing their expression. These data provide a mechanism for the precocious astrogenesis that was observed in mice lacking nervous system ErbB4 expression. More recently, Allison et al. (2011) have shown that NRG1 treatment of cultured hippocampal neurons leads to a dramatic increase in nuclear E4ICD in a presenilin-1/γ-secretase-dependent manner. Using global pathway analysis, this study also showed that nuclear sE4ICD signaling has a major effect on the expression of genes important to the morphogenesis of dendritic

spines, perhaps lending credence to the suspected involvement of the neuregulin-1/ErbB4 signaling pathway in the pathobiology of schizophrenia.

Ryk, NUCLEAR SIGNALING AND THE Wnt PATHWAY

The ligands of the Wnt family play critical roles in the establishment of the body map during embryogenesis in all animals (van Amerongen and Nusse 2009). Whereas a great deal is known about the canonical Wnt signaling pathway acting through the Frizzled/LRP receptor complex (Nusse 2005), much less is known about signals that are transduced by selected Wnt family members through the Ryk tyrosine kinase. Recent work has shown that this receptor controls aspects of neuronal differentiation in the developing brain (Lyu et al. 2008). In a manner that parallels the mechanism described above for ErbB4, once Ryk is activated by Wnt, its intracellular domain is released by a γ-secretase-dependent proteolytic cleavage. During cortical development, Ryk-ICD accumulates in the nucleus where it increases production of neurons from undifferentiated precursor cells. Suitably, the level of Ryk-ICD is highest at the peak of neurogenesis and decays thereafter (Lyu et al. 2008). It is interesting that both ErbB4 and Ryk presenilin-dependent nuclear signaling appear to enhance neurogenesis, the first by restricting astrogenesis, the latter by increasing neuronal differentiation.

TrkA NUCLEAR LOCALIZATION IN THE LIVER

The tropomyosin-related kinase (Trk A, B, C) family of RTKs is well known for its involvement in the transduction of signals induced by the neurotrophins (NGF, BDNF, NT-3, and NT-4) and for their effects on neuronal survival and differentiation in the developing and mature nervous system (Reichardt 2006). However, it is the liver where nuclear localization of TrkA ICD was observed by Bonacchi et al. (2008). Using immunostaining, this group found nuclear TrkA in hepatic stellate cells of the injured

 Cite this article as *Cold Spring Harb Perspect Biol* doi: 10.1101/cshperspect.a009001

liver. Because the signal was detected by antibodies against TrkA's intracellular domain but could not be detected with antibodies directed against its extracellular domain, they concluded that the released ICD was responsible for the labeling. Interestingly, a nuclear signal was also detected using phospho-TrkA-specific antibodies, suggesting that the nuclear TrkA-ICD might be in an active state. The mechanism by which TrkA is cleaved in this context remains to be determined. Previous research identified these same cells in the liver as a source of NGF that is modulated by injury (Cassiman et al. 2001). Together with the observation that NGF induces hepatic cell migration, these findings suggest the possibility that an autocrine loop involving NGF-mediated nuclear TrkA signaling might play a role in the response of the liver to injury (Bonacchi et al. 2008).

NUCLEAR TRANSLOCATION AND FUNCTION OF ErbB2

ErbB2 was originally identified as a cellular oncogene (Schechter et al. 1984) and has been found to be mutated or overexpressed in many human tumors. Unlike the other members of the ErbB family, ErbB2 remains an orphan receptor with no bona fide ligand. Nevertheless, ErbB2 becomes active in cell surface signaling through heterodimerization with other ErbB family members (Hynes and MacDonald 2009). Whereas nuclear ErbB2 has been detected in many tumors, there is limited information regarding what induces its translocation and what genes it may influence. Using both immunofluorescence and subcellular fractionation techniques, Wang et al. (2004) identified ErbB2 holoreceptor in the nucleus of both tumor tissue and cultured tumor cells and found that it interacts with a sequence in the promoter for the cyclooxygenase-2 gene, an important modulator of cytokine expression (Fig. 3). This interaction appears to depend on a transactivation domain in the carboxy-terminal cytoplasmic region of the receptor. Regarding how the receptor gains entry to the nucleus, Giri et al. (2005) suggested a mechanism involving a presumptive nuclear localization signal in ErbB2 that mediates inter-

action with importin β1. Similarly, Chen et al. (2005) identified a putative nuclear localization signal in the cytoplasmic domain of ErbB2 and showed that it was sufficient to drive the nuclear entry of a fusion construct created with green fluorescent protein. More recently, Béguelin et al. (2010) have shown that ErbB2, the progesterone receptor (PR), and the transcription factor Stat3 interact and that together they can regulate the expression of the cyclin D1 gene. Importantly, ErbB2 nuclear translocation depended not only on forming a complex with PR but also required the nuclear localization signal in the cytoplasmic domain of ErbB2. Understandably, the potential that nuclear localized ErbB2 regulates the expression of genes related to tumor progression has been a driving force for studying this process. However, a recent study suggests a potential for a much larger impact of nuclear ErbB2 signaling on cell behavior. Li et al. (2011) reported that nuclear ErbB2 regulates the expression of genes encoding ribosomal RNA by a mechanism involving RNA polymerase-I and that this interaction also affects cell translation and consequently cell size and overall growth, and that these effects occur independently of the canonical downstream activation of PI3 kinase and extracellular signal-regulated kinase (ERK).

FUNCTIONS OF NUCLEAR ErbB3

As a result of its identification (Kraus et al. 1989), ErbB3 has posed somewhat of a conundrum regarding the mechanism by which it signals. Because of the presence of specific amino acids at key sites in the presumed kinase domain, this RTK has largely been believed to be inactive and only able to function through heterodimerization with other ErbB receptors (Citri et al. 2003), but recent data showed that ErbB3 has tyrosine kinase activity (Shi et al. 2010). Studies of prostate cancer tissue samples as well as cell lines derived from prostate tumors have been a significant resource for characterizing the nuclear localization of ErbB3. A study that analyzed prostate neoplasms of varying pathological grades found that nuclear ErbB3 localization is associated with malignant growths but

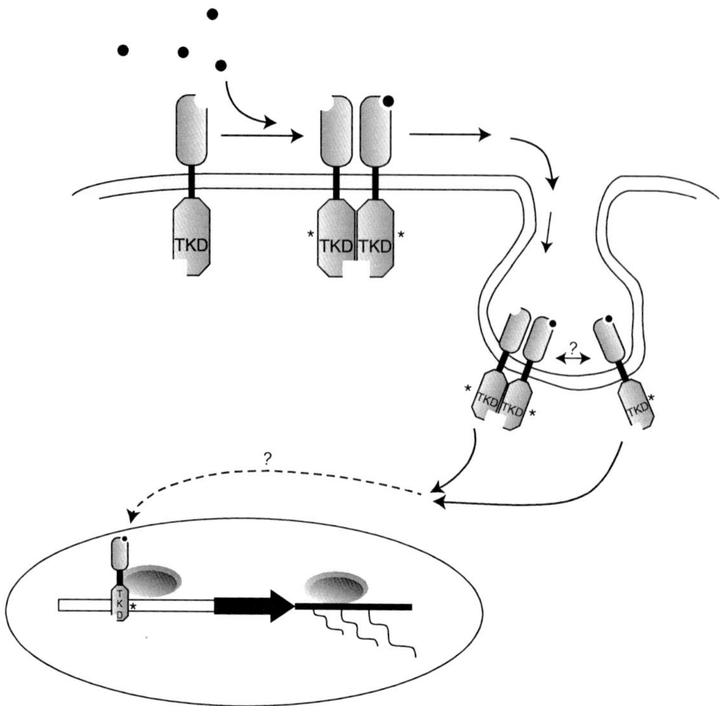

Figure 3. Nuclear signaling by holoreceptor tyrosine kinase. It has been proposed that following activation by its ligand, some intact activated RTKs can be internalized through endocytic or other yet unknown mechanisms, followed by their translocation to the nucleus. It is unknown whether this involves monomer or dimerized RTKs; some reports suggest that ligand-bound RTKs can be found in the nucleus.

not with normal tissue or tissue derived from benign prostatic hypertrophy, and that nuclear expression can also be used to differentiate between hormone-responsive versus nonresponsive growths (Koumakpayi et al. 2006). The implications of these data are that the presence of nuclear ErbB3 is associated with cells that are undergoing abnormally enhanced growth.

The nature of the ErbB3 molecule in the nucleus (holoreceptor vs. ICD) and how it gets there remains unclear. Koumakpayi et al. (2011), reported the presence of full-length ErbB3 holoprotein in the nuclei of prostate cancer cells and that it moves from the cell surface to the nucleus via a nonclathrin-, noncaveolin-dependent endocytic pathway that is sensitive to the effects of amiloride, a macropinocytosis inhibitor. In a line of immortalized, nonmalignant human mammary epithelial cells, Offterdinger et al. (2002) found ErbB3 nuclear localization to be mediated by its COOH terminus and an

apparent nuclear localization signal. Furthermore, the nuclear localization of ErbB3 was enhanced by allowing for the establishment of epithelial cell polarity. Additionally, treatment with NRG1 shifted ErbB3 localization from the nucleus into the nucleoplasm and ultimately into the cytoplasm, suggesting that both the extent of epithelial polarization and activation by NRG1 influence the subcellular localization of this RTK. Recently, Adilakshmi et al. (2011) detected in primary rat Schwann cells a nuclear variant of ErbB3 (nuc-ErbB3) that results from an alternative transcription initiation site. The nuc-ErbB3 isoform contains a functional nuclear localization signal sequence and it binds to chromatin, regulating transcriptional activity from the promoters of the ezrin and HMGB1 genes, and suppression of nuc-ErbB3 expression both reduces myelination and alters the distribution of ezrin in the nodes of Ranvier.

BIOLOGICAL FUNCTIONS OF NUCLEAR EGF RECEPTOR (ErbB1)

Numerous studies have reported the localization of the full-length epidermal growth factor receptor (EGFR, ErbB1) in the nucleus, where evidence suggests it can influence transcription, cell proliferation, and DNA repair. Activated nuclear EGFR has been shown to bind to AT-rich response sequences within the promoters of the genes for *cyclin D1*, *B-Myb*, *iNOS*, and *Aurora-A*, and nuclear EGFR has been shown to cooperate with several transcription factors including E2F1 and STAT5, ultimately modifying target gene expression (Lin et al. 2001; Lo et al. 2005; Hanada et al. 2006; Hung et al. 2008). More recently, it has been found that nuclear EGFR can interact with the STAT3 transcription factor to drive the expression of COX-2 in glioblastomas (Lo et al. 2010). Furthermore, nuclear localized EGFR also has been shown to phosphorylate the proliferating cell nuclear antigen (PCNA) at Tyr211, an event that increases its stability (Wang et al. 2006). Thus, nuclear EGFR signaling may contribute to cell proliferation independently of transcriptional regulation.

Exposure of cells to ionizing radiation leads to nuclear translocation of holo-EGFR in a ligand-independent manner, an event associated with an increase in the activity of DNA-dependent protein kinase and enhanced levels of DNA repair (Dittmann et al. 2005). This group also found that the nonsteroidal anti-inflammatory drug celecoxib increases radiosensitization of tumor cells by inhibiting ligand-independent nuclear EGFR transport in response to radiation exposure, and these effects are independent of the inhibitory effects of celecoxib on COX-2 activity as well as modulation of prostaglandin E2 levels (Dittmann et al. 2008). Furthermore, Hsu et al. (2009) showed that nuclear EGFR localization is necessary for tumor resistance to DNA damage caused by the alkylating agent, cisplatin. Whereas the exact molecular mechanisms as to how nuclear EGFR impacts on DNA repair in normal cells are still unclear, nuclear EGFR has been shown to associate with p53 and the MDC1 protein, essential proteins for the recruitment of DNA repair foci (Dittmann et al. 2008). These data suggest that nuclear EGFR may be beneficial for normal cells responding to DNA damage caused by radiation but it might negatively impact tumor-targeting therapies.

VASCULAR ENDOTHELIAL GROWTH FACTOR RECEPTOR 2 (VEGFR2 OR KDR) AND NUCLEAR ACTIVITY

The regulation of angiogenesis plays a key role throughout the course of development, and inappropriate vascularization is a frequent occurrence in multiple disease states, especially cancer. One of the best known positive modulators of angiogenesis is vascular endothelial growth factor (VEGF), which acts largely through its receptor VEGFR2, making these molecules important targets for cancer treatment (Folkman 2007). Investigation of the responses of endothelial cells to in vitro wounding identified nuclear translocation of an activated VEGFR2/VEGF complex as a key component of the cellular response to injury (Santos et al. 2007). Intriguingly, more recent work has shown that nuclear VEGFR2 may actually play a role in the regulation of expression of the VEGFR2 gene itself. This has been strengthened by the observation that a VEGFR2/Sp1 complex interacts with the VEGFR2 promoter (Domingues et al. 2011).

TYPE I IGF RECEPTOR, POSTTRANSLATIONAL MODIFICATION AND NUCLEAR TRANSLOCATION

Unlike all other receptors discussed here, the type 1 insulin-like growth factor receptor (IGF-1R) is not a single chain type I transmembrane protein. Like the insulin receptor, it exists as a heterodimer containing extracellular ligand-binding α subunits paired with transmembrane dimerized β subunits. Signaling via this RTK influences cell growth, size, and maintenance of many cell populations throughout the organism (LeRoith and Roberts 2003). A recent study showed that the complete receptor complex translocates to the nucleus in a manner that is dependent on the posttranslational and activation-dependent addition of the small ubiquitin-like modifier (SUMO) to the kinase-containing β-subunit (Sehat et al. 2010). Immunoprecipitation/Western blot analysis showed that, as occurs for many of the RTKs discussed here, the IGF-1R only enters the nucleus after activation, as judged by the presence of phosphorylated tyrosine residues. Using an unbiased approach with synthetic random sequence oligonucleotides in electrophoretic mobility shift assays, followed by assays using chromatin immunoprecipitation, Sehat et al. (2010) also find that the nuclear IGF-1R binds to chromatin and modifies transcriptional activity of many genes.

FIBROBLAST GROWTH FACTOR RECEPTOR 1

The fibroblast growth factor (FGF) family has many members that signal through four different high-affinity RTKs (FGFR1-4). The ligands also can interact with heparin sulfate, an association that can influence receptor signaling in a wide range of cells and tissues (Harmer 2006). The four FGF RTKs share similar domain structures (i.e., a single transmembrane domain, a split TK domain, and a carboxy-terminal domain). Currently, FGFR1 is the only FGF receptor reported to be present in the nucleus, where it is present in its full-length form and in association with FGF2 (Stachowiak et al. 1996a,b; Myers et al. 2003). Although it remains controversial as to how the ligand reaches the receptor, there is some evidence that this pathway has biological effects on neural precursors. Nuclear FGFR1 signaling has been implicated in the differentiation of neural precursor cells in culture, inducing expression of neuron-specific enolase and Neurofilament-L (Stachowiak et al. 2003) as well as tyrosine hydroxylase (Peng et al. 2002), possibly acting downstream of BMP signaling (Horbinski et al. 2002; Stachowiak et al. 2003). In the subventricular zone (SVZ) of the developing mouse brain, which contains proliferating progenitor cells, FGFR1 immunoreactivity is localized in a predominantly nonnuclear manner, whereas in differentiating cells outside of the SVZ, FGFR1 accumulates in the nuclei (Fang et al. 2005). Consequently, it has been proposed that nuclear FGFR1 signaling is involved in the differentiation of neuronal populations.

CONCLUDING REMARKS

Whereas early observations of nuclear translocation of RTKs, either as fragments or holoproteins, raised the possibility that RTKs had the ability to signal directly to the nucleus, the mechanism and biological roles for this type of signaling are just now beginning to emerge. As this new arena for RTK action is explored in greater detail, several important questions will need to be addressed: (1) If activated holoreceptors move from the plasma membrane to the nucleus, do they maintain the same conformation? (2) Do different pools of a specific RTK participate in cytoplasmic versus nuclear signaling, or can one molecule perform both? (3) Given the role of γ-secretase in the release of soluble RTK-ICDs, should their processing and subsequent nuclear signaling be evaluated in the context of Alzheimer's disease? (4) How do nuclear RTKs or RTK-ICDs interact with chromatin and what are their partners and targets? (5) And, of course, what are the biological roles of this new signaling modality? We believe that these studies will provide important insights into the basic mechanisms of cell biology, developmental biology, and disease.

ACKNOWLEDGMENTS

This work is supported in part by National Institute of Neurological Disorders and Stroke (NINDS) grant R01 NS35884 (to G.C.) and National Institute on Deafness and Other Communication Disorders (NIDCD) grant R01 DC04820 (to G.C.).

REFERENCES

Adilakshmi T, Ness-Myers J, Madrid-Aliste C, Fiser A, Tapinos N. 2011. A nuclear variant of ErbB3 receptor tyrosine kinase regulates ezrin distribution and Schwann cell myelination. *J Neurosci* **31:** 5106–5119.

Allison JG, Das PM, Ma J, Inglis FM, Jones FE. 2011. The ERBB4 intracellular domain (4ICD) regulates NRG1-induced gene expression in hippocampal neurons. *Neurosci Res* **70:** 155–163.

Arasada RR, Carpenter G. 2005. Secretase-dependent tyrosine phosphorylation of Mdm2 by the ErbB-4 intracellular domain fragment. *J Biol Chem* **280:** 30783–30787.

Beguelin W, Diaz Flaque MC, Proietti CJ, Cayrol F, Rivas MA, Tkach M, Rosemblit C, Tocci JM, Charreau EH,

Schillaci R, et al. 2010. Progesterone receptor induces ErbB-2 nuclear translocation to promote breast cancer growth via a novel transcriptional effect: ErbB-2 function as a coactivator of Stat3. *Mol Cell Biol* **30:** 5456–5472.

Bonacchi A, Taddei ML, Petrai I, Efsen E, Defranco R, Nosi D, Torcia M, Rosini P, Formigli L, Rombouts K, et al. 2008. Nuclear localization of TRK-A in liver cells. *Histol Histopathol* **23:** 327–340.

Cassiman D, Denef C, Desmet VJ, Roskams T. 2001. Human and rat hepatic stellate cells express neurotrophins and neurotrophin receptors. *Hepatology* **33:** 148–158.

Chen QQ, Chen XY, Jiang YY, Liu J. 2005. Identification of novel nuclear localization signal within the ErbB-2 protein. *Cell Res* **15:** 504–510.

Citri A, Skaria KB, Yarden Y. 2003. The deaf and the dumb: The biology of ErbB-2 and ErbB-3. *Exp Cell Res* **284:** 54–65.

Dittmann K, Mayer C, Fehrenbacher B, Schaller M, Raju U, Milas L, Chen DJ, Kehlbach R, Rodemann HP. 2005. Radiation-induced epidermal growth factor receptor nuclear import is linked to activation of DNA-dependent protein kinase. *J Biol Chem* **280:** 31182–31189.

Dittmann K, Mayer C, Kehlbach R, Rodemann HP. 2008. The radioprotector Bowman-Birk proteinase inhibitor stimulates DNA repair via epidermal growth factor receptor phosphorylation and nuclear transport. *Radiother Oncol* **86:** 375–382.

Domingues I, Rino J, Demmers JA, de Lanerolle P, Santos SC. 2011. VEGFR2 translocates to the nucleus to regulate its own transcription. *PLoS ONE* **6:** e25668.

Elenius K, Corfas G, Paul S, Choi CJ, Rio C, Plowman GD, Klagsbrun M. 1997. A novel juxtamembrane domain isoform of HER4/ErbB4. Isoform-specific tissue distribution and differential processing in response to phorbol ester. *J Biol Chem* **272:** 26761–26768.

Elenius K, Choi CJ, Paul S, Santiestevan E, Nishi E, Klagsbrun M. 1999. Characterization of a naturally occurring ErbB4 isoform that does not bind or activate phosphatidyl inositol 3-kinase. *Oncogene* **18:** 2607–2615.

Fang X, Stachowiak EK, Dunham-Ems SM, Klejbor I, Stachowiak MK. 2005. Control of CREB-binding protein signaling by nuclear fibroblast growth factor receptor-1: A novel mechanism of gene regulation. *J Biol Chem* **280:** 28451–28462.

Folkman J. 2007. Angiogenesis: An organizing principle for drug discovery? *Nat Rev Drug Discov* **6:** 273–286.

Fortini ME. 2002. γ-Secretase-mediated proteolysis in cell-surface-receptor signalling. *Nat Rev Mol Cell Biol* **3:** 673–684.

Giri DK, Ali-Seyed M, Li LY, Lee DF, Ling P, Bartholomeusz G, Wang SC, Hung MC. 2005. Endosomal transport of ErbB-2: Mechanism for nuclear entry of the cell surface receptor. *Mol Cell Biol* **25:** 11005–11018.

Hanada N, Lo HW, Day CP, Pan Y, Nakajima Y, Hung MC. 2006. Co-regulation of B-Myb expression by E2F1 and EGF receptor. *Mol Carcinog* **45:** 10–17.

Harmer NJ. 2006. Insights into the role of heparan sulphate in fibroblast growth factor signalling. *Biochem Soc Trans* **34:** 442–445.

Horbinski C, Stachowiak EK, Chandrasekaran V, Miuzukoshi E, Higgins D, Stachowiak MK. 2002. Bone morphogenetic protein-7 stimulates initial dendritic growth in sympathetic neurons through an intracellular fibroblast growth factor signaling pathway. *J Neurochem* **80:** 54–63.

Hsu SC, Miller SA, Wang Y, Hung MC. 2009. Nuclear EGFR is required for cisplatin resistance and DNA repair. *Am J Transl Res* **1:** 249–258.

Hung LY, Tseng JT, Lee YC, Xia W, Wang YN, Wu ML, Chuang YH, Lai CH, Chang WC. 2008. Nuclear epidermal growth factor receptor (EGFR) interacts with signal transducer and activator of transcription 5 (STAT5) in activating Aurora-A gene expression. *Nucleic Acids Res* **36:** 4337–4351.

Hynes NE, MacDonald G. 2009. ErbB receptors and signaling pathways in cancer. *Curr Opin Cell Biol* **21:** 177–184.

Jones FE, Welte T, Fu XY, Stern DF. 1999. ErbB4 signaling in the mammary gland is required for lobuloalveolar development and Stat5 activation during lactation. *J Cell Biol* **147:** 77–88.

Junttila TT, Sundvall M, Lundin M, Lundin J, Tanner M, Harkonen P, Joensuu H, Isola J, Elenius K. 2005. Cleavable ErbB4 isoform in estrogen receptor-regulated growth of breast cancer cells. *Cancer Res* **65:** 1384–1393.

Koumakpayi IH, Diallo JS, Le Page C, Lessard L, Gleave M, Begin LR, Mes-Masson AM, Saad F. 2006. Expression and nuclear localization of ErbB3 in prostate cancer. *Clin Cancer Res* **12:** 2730–2737.

Koumakpayi IH, Le Page C, Delvoye N, Saad F, Mes-Masson AM. 2011. Macropinocytosis inhibitors and Arf6 regulate ErbB3 nuclear localization in prostate cancer cells. *Mol Carcinog* **50:** 901–912.

Kraus MH, Issing W, Miki T, Popescu NC, Aaronson SA. 1989. Isolation and characterization of ERBB3, a third member of the ERBB/epidermal growth factor receptor family: Evidence for overexpression in a subset of human mammary tumors. *Proc Natl Acad Sci* **86:** 9193–9197.

Lai C, Feng L. 2004. Implication of γ-secretase in neuregulin-induced maturation of oligodendrocytes. *Biochem Biophys Res Commun* **314:** 535–542.

Lemmon MA, Schlessinger J. 2010. Cell signaling by receptor tyrosine kinases. *Cell* **141:** 1117–1134.

LeRoith D, Roberts CT Jr. 2003. The insulin-like growth factor system and cancer. *Cancer Lett* **195:** 127–137.

Li LY, Chen H, Hsieh YH, Wang YN, Chu HJ, Chen YH, Chen HY, Chien PJ, Ma HT, Tsai HC, et al. 2011. Nuclear ErbB2 enhances translation and cell growth by activating transcription of ribosomal RNA genes. *Cancer Res* **71:** 4269–4279.

Lin SY, Makino K, Xia W, Matin A, Wen Y, Kwong KY, Bourguignon L, Hung MC. 2001. Nuclear localization of EGF receptor and its potential new role as a transcription factor. *Nat Cell Biol* **3:** 802–808.

Lo HW, Hsu SC, Ali-Seyed M, Gunduz M, Xia W, Wei Y, Bartholomeusz G, Shih JY, Hung MC. 2005. Nuclear interaction of EGFR and STAT3 in the activation of the iNOS/NO pathway. *Cancer Cell* **7:** 575–589.

Lo HW, Cao X, Zhu H, Ali-Osman F. 2010. Cyclooxygenase-2 is a novel transcriptional target of the nuclear EGFR-STAT3 and EGFRvIII-STAT3 signaling axes. *Mol Cancer Res* **8:** 232–245.

Long W, Wagner KU, Lloyd KC, Binart N, Shillingford JM, Hennighausen L, Jones FE. 2003. Impaired differentiation and lactational failure of Erbb4-deficient mammary glands identify ERBB4 as an obligate mediator of STAT5. *Development* **130:** 5257–5268.

Lyu J, Yamamoto V, Lu W. 2008. Cleavage of the Wnt receptor Ryk regulates neuronal differentiation during cortical neurogenesis. *Dev Cell* **15:** 773–780.

Myers JM, Martins GG, Ostrowski J, Stachowiak MK. 2003. Nuclear trafficking of FGFR1: A role for the transmembrane domain. *J Cell Biochem* **88:** 1273–1291.

Ni CY, Murphy MP, Golde TE, Carpenter G. 2001. γ-Secretase cleavage and nuclear localization of ErbB-4 receptor tyrosine kinase. *Science* **294:** 2179–2181.

Nusse R. 2005. Wnt signaling in disease and in development. *Cell Res* **15:** 28–32.

Offterdinger M, Schofer C, Weipoltshammer K, Grunt TW. 2002. c-erbB-3: A nuclear protein in mammary epithelial cells. *J Cell Biol* **157:** 929–939.

Peng H, Myers J, Fang X, Stachowiak EK, Maher PA, Martins GG, Popescu G, Berezney R, Stachowiak MK. 2002. Integrative nuclear FGFR1 signaling (INFS) pathway mediates activation of the tyrosine hydroxylase gene by angiotensin II, depolarization and protein kinase C. *J Neurochem* **81:** 506–524.

Plowman GD, Culouscou JM, Whitney GS, Green JM, Carlton GW, Foy L, Neubauer MG, Shoyab M. 1993. Ligand-specific activation of HER4/p180erbB4, a fourth member of the epidermal growth factor receptor family. *Proc Natl Acad Sci* **90:** 1746–1750.

Reichardt LF. 2006. Neurotrophin-regulated signalling pathways. *Philos Trans R Soc Lond B Biol Sci* **361:** 1545–1564.

Rio C, Buxbaum JD, Peschon JJ, Corfas G. 2000. Tumor necrosis factor-α-converting enzyme is required for cleavage of erbB4/HER4. *J Biol Chem* **275:** 10379–10387.

Santos SC, Miguel C, Domingues I, Calado A, Zhu Z, Wu Y, Dias S. 2007. VEGF and VEGFR-2 (KDR) internalization is required for endothelial recovery during wound healing. *Exp Cell Res* **313:** 1561–1574.

Sardi SP, Murtie J, Koirala S, Patten BA, Corfas G. 2006. Presenilin-dependent ErbB4 nuclear signaling regulates the timing of astrogenesis in the developing brain. *Cell* **127:** 185–197.

Schechter AL, Stern DF, Vaidyanathan L, Decker SJ, Drebin JA, Greene MI, Weinberg RA. 1984. The neu oncogene: An erb-B-related gene encoding a 185,000-Mr tumour antigen. *Nature* **312:** 513–516.

Schlessinger J, Lemmon MA. 2006. Nuclear signaling by receptor tyrosine kinases: The first robin of spring. *Cell* **127:** 45–48.

Sehat B, Tofigh A, Lin Y, Trocme E, Liljedahl U, Lagergren J, Larsson O. 2010. SUMOylation mediates the nuclear translocation and signaling of the IGF-1 receptor. *Sci Signal* **3:** ra10.

Shi F, Telesco SE, Liu Y, Radhakrishnan R, Lemmon MA. 2010. ErbB3/HER3 intracellular domain is competent to bind ATP and catalyze autophosphorylation. *Proc Natl Acad Sci* **107:** 7692–7697.

Stachowiak MK, Maher PA, Joy A, Mordechai E, Stachowiak EK. 1996a. Nuclear accumulation of fibroblast growth factor receptors is regulated by multiple signals in adrenal medullary cells. *Mol Biol Cell* **7:** 1299–1317.

Stachowiak MK, Maher PA, Joy A, Mordechai E, Stachowiak EK. 1996b. Nuclear localization of functional FGF receptor 1 in human astrocytes suggests a novel mechanism for growth factor action. *Brain Res Mol Brain Res* **38:** 161–165.

Stachowiak MK, Fang X, Myers JM, Dunham SM, Berezney R, Maher PA, Stachowiak EK. 2003. Integrative nuclear FGFR1 signaling (INFS) as a part of a universal "feed-forward-and-gate" signaling module that controls cell growth and differentiation. *J Cell Biochem* **90:** 662–691.

Sundvall M, Iljin K, Kilpinen S, Sara H, Kallioniemi OP, Elenius K. 2008. Role of ErbB4 in breast cancer. *J Mammary Gland Biol Neoplasia* **13:** 259–268.

Tidcombe H, Jackson-Fisher A, Mathers K, Stern DF, Gassmann M, Golding JP. 2003. Neural and mammary gland defects in ErbB4 knockout mice genetically rescued from embryonic lethality. *Proc Natl Acad Sci* **100:** 8281–8286.

van Amerongen R, Nusse R. 2009. Towards an integrated view of Wnt signaling in development. *Development* **136:** 3205–3214.

Vecchi M, Baulida J, Carpenter G. 1996. Selective cleavage of the heregulin receptor ErbB-4 by protein kinase C activation. *J Biol Chem* **271:** 18989–18995.

Wang SC, Lien HC, Xia W, Chen IF, Lo HW, Wang Z, Ali-Seyed M, Lee DF, Bartholomeusz G, Ou-Yang F, et al. 2004. Binding at and transactivation of the COX-2 promoter by nuclear tyrosine kinase receptor ErbB-2. *Cancer Cell* **6:** 251–261.

Wang SC, Nakajima Y, Yu YL, Xia W, Chen CT, Yang CC, McIntush EW, Li LY, Hawke DH, Kobayashi R, et al. 2006. Tyrosine phosphorylation controls PCNA function through protein stability. *Nat Cell Biol* **8:** 1359–1368.

Williams CC, Allison JG, Vidal GA, Burow ME, Beckman BS, Marrero L, Jones FE. 2004. The ERBB4/HER4 receptor tyrosine kinase regulates gene expression by functioning as a STAT5A nuclear chaperone. *J Cell Biol* **167:** 469–478.

Zhou W, Carpenter G. 2000. Heregulin-dependent trafficking and cleavage of ErbB-4. *J Biol Chem* **275:** 34737–34743.

Zhu Y, Sullivan LL, Nair SS, Williams CC, Pandey AK, Marrero L, Vadlamudi RK, Jones FE. 2006. Coregulation of estrogen receptor by ERBB4/HER4 establishes a growth-promoting autocrine signal in breast tumor cells. *Cancer Res* **66:** 7991–7998.

Endocytosis of Receptor Tyrosine Kinases

Lai Kuan Goh[2] and Alexander Sorkin[1]

[1]Department of Cell Biology, University of Pittsburgh School of Medicine, Pittsburgh, Pennsylvania 15261
[2]Department of Pharmacology, University of Colorado Anschutz Medical Campus, Aurora, Colorado 80045
 and CryoCord Sdn Bhd, Cyberjaya, Selangor 63000, Malaysia

Correspondence: sorkin@pitt.edu

Endocytosis is the major regulator of signaling from receptor tyrosine kinases (RTKs). The canonical model of RTK endocytosis involves rapid internalization of an RTK activated by ligand binding at the cell surface and subsequent sorting of internalized ligand-RTK complexes to lysosomes for degradation. Activation of the intrinsic tyrosine kinase activity of RTKs results in autophosphorylation, which is mechanistically coupled to the recruitment of adaptor proteins and conjugation of ubiquitin to RTKs. Ubiquitination serves to mediate interactions of RTKs with sorting machineries both at the cell surface and on endosomes. The pathways and kinetics of RTK endocytic trafficking, molecular mechanisms underlying sorting processes, and examples of deviations from the standard trafficking itinerary in the RTK family are discussed in this work.

Functional activities of transmembrane proteins, including the large family of RTKs, are controlled by intracellular trafficking. RTKs are synthesized in the endoplasmic reticulum, transported to Golgi apparatus, and then delivered to the plasma membrane. At the cell surface RTKs undergo constitutive endocytosis (internalization) at a rate similar to that of other integral membrane proteins. Constitutive internalization of RTKs is much slower than their constitutive recycling from endosomes back to the cell surface. Therefore, RTKs are accumulated at the cell surface, which allows maximal accessibility to extracellular ligands. The rates of the constitutive internalization, recycling, and degradation determine the half-life of an RTK protein, which varies depending on the nature of the RTK and the cell type, and typically positively correlates with the expression level of the RTK. For example, the turnover rates range from $t_{1/2} < 1$ h for the colony stimulating factor 1 receptor (CSF-1R) in macrophages (Lee et al. 1999) to 24 h for the epidermal growth factor receptor (EGFR) overexpressed in carcinoma cells (Stoscheck and Carpenter 1984).

Activation of RTKs by growth factors elevates turnover and ultimately leads to downregulation of RTKs. First, at the cell surface, ligand binding increases the rate of RTK internalization by several folds (Fig. 1). Typically, surface RTKs can freely diffuse in the plane of the plasma membrane and, therefore, reach all endocytic sites. In some cases, however, a pool of RTKs is immobile owing to association with actin-based membrane protrusions (microvilli) or caveolae (Mineo et al. 1999; Foti et al. 2004; Hommelgaard et al. 2004). This pool is incapable of rapid internalization and retained at the

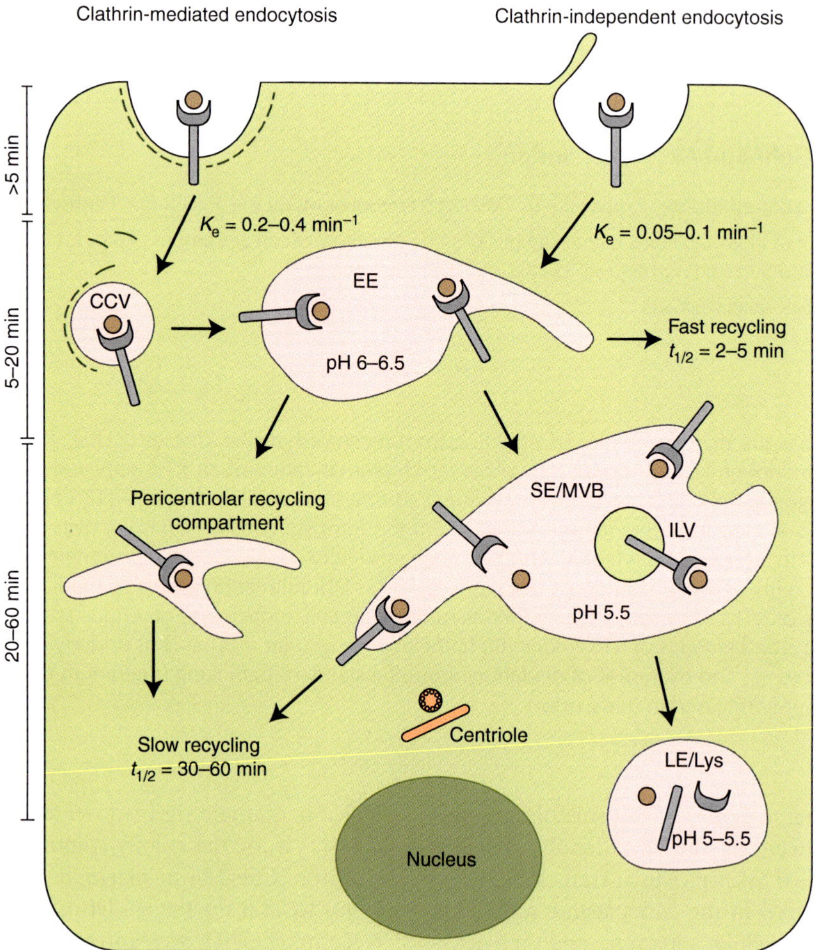

Figure 1. Pathways of RTK endocytosis. RTKs are endocytosed by clathrin-mediated and clathrin-independent mechanisms. Typical rate constants (K_e) for RTK internalization through both pathways are shown. Clathrin-coated vesicles are uncoated shortly after fission from the plasma membrane and fuse with early endosomes (EEs). In the most well-studied EGFR system, EGFR-ligand complexes are detected in these highly dynamic, morphologically heterologous compartments within 2–5 min after EGF stimulation (Haigler et al. 1979; Beguinot et al. 1984; Miller et al. 1986; Hopkins et al. 1990). Ligand-RTK complexes remain intact but certain ligands dissociate from the receptor in the acidic environment of the endosomal lumen. Released ligands remain in the vesicular parts of endosomes (most of the endosomal volume), whereas unoccupied receptors are found mainly in tubular extensions (most of the membrane area). Ligand-occupied and unoccupied RTKs can rapidly recycle from EEs through the process of back fusion of peripheral EEs with the plasma membrane or via tubular carriers derived from these endosomes (retroendocytosis). EEs mature into sorting endosomes (SE) or multivesicular bodies (MVBs) in which RTKs are incorporated into intraluminal vesicles (ILVs) by inward membrane invagination. RTKs can also be delivered to the pericentriolar Rab11-containing recycling compartment. Recycling of unoccupied and ligand-occupied RTKs is slower from the SE/MVBs and recycling compartment. SE/MVBs gradually lose early endosome components, such as Rab5 and EEA.1, and recycling cargo (such as transferrin receptors) while become enriched in resident late endosomal proteins (such as Rab7), thus maturing into late endosomes. Fusion of late endosomes with primary lysosomes carrying proteolytic enzymes results in degradation of receptors and growth factors. A typical time scale of RTK endocytosis, their accumulation in EEs, and SE/MVBs is shown in minutes.

Cite this article as *Cold Spring Harb Perspect Biol* doi: 10.1101/cshperspect.a017459

cell surface. EGFR and its closest homolog, ErbB2, can also be retained in the plasma membrane through interactions with Na^+-K^+ exchanger and erbin, respectively (Borg et al. 2000; Lazar et al. 2004). Thus, interplay of surface retention and endocytosis-promoting mechanisms determines the extent of activation-induced acceleration of RTK internalization. The outcomes of this interplay vary within the RTK family. For instance, endocytosis of ErbB2 is not significantly increased when ErbB2 is activated by homo- or heterodimerization with other ErbBs (Yarden and Sliwkowski 2001). In contrast, some RTKs display a relatively fast constitutive internalization that is not appreciably augmented by ligand binding (McClain 1992; Burke and Wiley 1999; Jopling et al. 2011).

The second trafficking step causing downregulation of activated RTKs is the efficient sorting of internalized RTKs to late endosomes and lysosomes for proteolytic degradation (Fig. 1). Members of all RTK subfamilies studied to date undergo accelerated lysosomal degradation on activation. Thus, activity-dependent acceleration of both sorting processes, at the cell surface and in endosomes, together resulting in RTK down-regulation, is a trademark of this receptor family. Significant advances have been made in recent years in understanding the molecular mechanisms of RTK endocytic trafficking. Yet, the specific components that mediate and regulate key internalization and sorting processes remain poorly defined. Because many conceptual findings have been made using the EGFR system and later reproduced in studies of other RTKs, EGFR has become the prototypic model system to study RTK endocytosis. Thus, we will rely on data derived in EGFR models for the broader discussion of other RTK families.

Endocytosis is the major regulator of signal transduction processes initiated by RTKs. Although the discussion of cross talk between endocytosis and signaling is outside of the scope of this article, we should emphasize that elucidation of the specific mechanisms of RTK endocytosis is essential for defining the role of endocytosis in regulation of RTK signaling. Hence our discussion will focus on various mo-

lecular mechanisms implicated in RTK endocytic trafficking. We regret that we are unable to cite many excellent studies on RTK endocytosis owing to the space constraints. We would also like to point out that a major limitation in the interpretation of RTK endocytosis literature is that methods used to analyze RTK internalization in the majority of studies fail to measure specific internalization rates because interference of receptor recycling is rarely considered. Therefore, changes in the apparent rates of RTK endocytosis observed in such studies could be owing to alterations in either internalization or recycling. We will discuss the literature in accordance to the authors' interpretation, but leave the detailed evaluation of each set of published data to reader discretion.

PATHWAYS OF RTK INTERNALIZATION

On ligand activation, increased localization of EGFR, neurotrophic tyrosine kinase receptor type 1 (TrkA), insulin receptor (IR), and a number of other RTKs within clathrin-coated pits (CCPs) has been observed (Gorden et al. 1978; Carpentier et al. 1982; Beattie et al. 2000; Bogdanovic et al. 2009). Further, internalization of EGFR and several other RTKs was blocked by small interfering RNAs (siRNAs) to clathrin heavy chain and chemical inhibitors of clathrin (Huang et al. 2004; Lampugnani et al. 2006; Zheng et al. 2008; von Kleist et al. 2011). Together, these studies show that clathrin-mediated endocytosis (CME) is the major pathway of internalization for ligand-occupied RTKs (Fig. 1).

CME is the fastest internalization pathway (rate constant $K_e \sim$ up to 0.6 min^{-1}). Kinetics analysis of EGFR endocytosis suggested that CME is saturated when a large number of surface EGFRs are activated by EGF, and the contribution of a slower clathrin-independent endocytosis (CIE) increases with the increase of EGF concentration and EGFR expression levels, leading to overall reduction in the apparent internalization rate (Wiley 1988). Indeed, siRNA depletion of clathrin significantly inhibited EGFR internalization only when low concentrations of EGF were used (Sigismund et al. 2005). Similar saturation of the rapid internalization pathway and reduced rates of internalization were observed when IR and insulin-like growth factor 1 receptor (IGF-1R) were activated with high ligand concentrations (Backer et al. 1991; Prager et al. 1994). CIE has also been shown for fibroblast growth factor receptor (FGFR) and vascular endothelial growth factor receptor (VEGFR) (Wiedlocha and Sorensen 2004; Lanahan et al. 2010; Haugsten et al. 2011).

Mechanisms of CIE for various RTK can be roughly divided into two groups. The macropinocytosis-like endocytosis involving actin cytoskeleton rearrangements and membrane ruffling has been observed in early studies of epidermoid carcinoma A-431 cells that express very high levels of EGFR (Haigler et al. 1979), and later in other cell types (Barbieri et al. 2000; Yamazaki et al. 2002; Orth et al. 2006; Valdez et al. 2007). Another mechanism implicated in the CIE of RTKs is defined by its sensitivity to inhibitors of caveolae and cholesterol-disrupting agents (Sigismund et al. 2005; Sehat et al. 2008; Salani et al. 2010). However, the lack of experimental tools to inhibit specific CIE pathways impedes analysis of the mechanisms and functional role of CIE. Given the major role of CME in the presence of physiological concentrations of growth factors, we therefore focus our discussion on the molecular mechanisms of this pathway for RTK endocytosis.

ROLE OF TYROSINE-BASED AND DILEUCINE MOTIFS IN RTK ENDOCYTOSIS

Sequence motifs in the cytoplasmic domains of transmembrane proteins are recognized by components of the clathrin coat to trap the endocytic cargo in CCP. Several such motifs have been implicated in RTK internalization (Fig. 2). IR uses the dileucine D(E)xxxLL(I) ("LL") internalization motif (Haft et al. 1994; Morrison et al. 1996; Hamer et al. 1997), which directly binds to the trunk domain of the clathrin adaptor protein complex AP-2 (Fig. 3). The NPxY motif (x, any amino acid residue) was shown to mediate internalization of IR, IGF-1R, and platelet-derived growth factor receptor (PDGFR) β (Chen et al. 1990; Backer et al. 1992; Prager et al. 1994; Wu et al. 2003a). NPxY motifs

interact with the phosphotyrosine-binding domains present in several monomeric adaptor proteins that, in turn, bind to the appendage domains of AP-2 and the amino-terminal domain of clathrin (Fig. 3) (Traub 2009).

Despite the importance of LL and NPxY motifs for internalization of IR and IGF-1R, the interaction of these receptors with AP-2 has not been shown. Such interaction was shown only for EGFR (Sorkin and Carpenter 1993). The YRAL motif of EGFR is responsible for binding to the μ2 subunit of AP-2 (Sorkin et al. 1996). However, mutation of this motif or inactivation of the tyrosine-binding interface of μ2 did not affect EGFR internalization (Sorkin et al. 1996; Nesterov et al. 1999). The LL motif in EGFR is important for phosphorylation of Tyr6 in the β2 subunit of AP-2, suggesting the role of this LL motif in EGFR:AP-2 interaction (Huang et al. 2003). Interestingly, tyrosine phosphorylation of β2 Tyr6 was predicted to facilitate LL motif interaction with AP-2 (Kelly et al. 2008). However, mutation of this motif also did not affect clathrin-dependent EGFR internalization. Overall, the data on the role of internalization motifs and AP-2 in RTK internalization are incomplete, and the contribution of RTK:AP-2 interactions in ligand-induced RTK endocytosis remains to be defined.

MECHANISMS AND ROLE OF RTK UBIQUITINATION IN CME: CBL AND GRB2

Growth factor binding to RTKs leads to activation of the tyrosine kinase and autophosphorylation of tyrosine residues in the cytoplasmic domain of RTKs, which serve as interaction sites for proteins mediating RTK ubiquitination, a posttranslational modification by the covalent attachment of the ubiquitin polypeptide to lysine residues (Fig. 2). Ubiquitination has historically been considered as the modification that targets proteins for proteosomal degradation, but more recently the role of ubiquitin as the sorting signal in membrane trafficking has emerged (Acconcia et al. 2009). PDGFR-β and EGFR were the first RTKs found to be ubiquitinated (Mori et al. 1992; Galcheva-Gargova et al. 1995). Later, ubiquitination of all

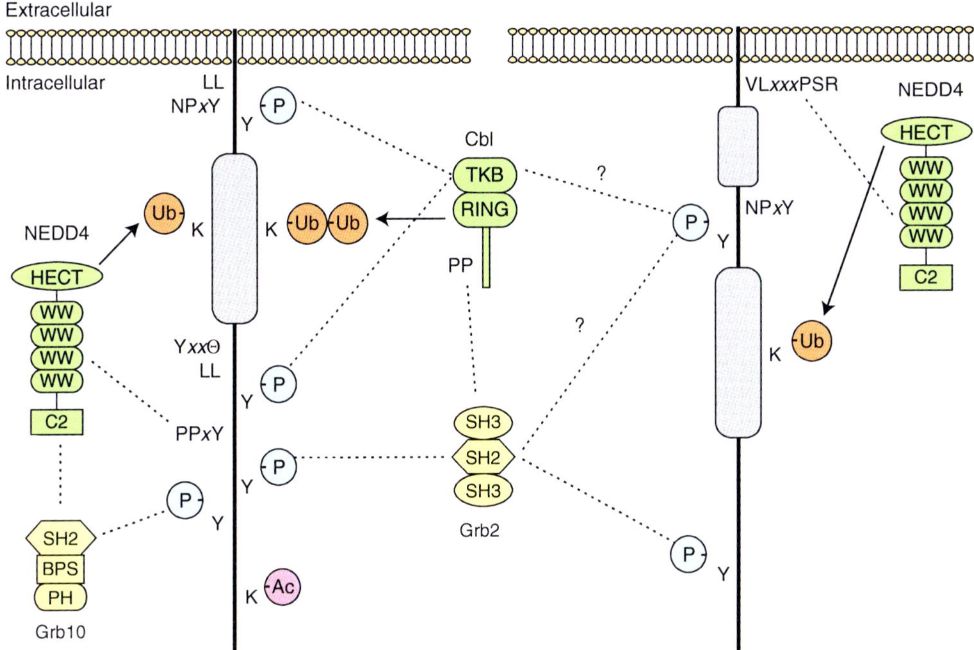

Figure 2. Sequence motifs and posttranslational modifications of RTKs involved in endocytosis. Examples of RTKs with a single kinase domain (EGFR, Met, IR/IGF-1R, Trk subtypes) and an insert-divided kinase domain (PDGFR, FGFR, kit, CSFR, VEGFR) are shown. Sequence motifs containing phosphorylated tyrosine residues (P-Y and Y-P) serve as binding sites for SH2 domains of Grb2, Grb10, and the tyrosine kinase binding (TKB) domain of c-Cbl, Cbl-b, and Cbl-3. Proline-rich motifs in Cbl-b and c-Cbl bind to SH3 domains in Grb2. WW domains of NEDD family E3 ligases bind to PPxY motifs in RTKs, and in case of FGFR, to an unconventional motif in the juxtamembrane domain of this receptor. C2 domain of NEDD4-1 binds to the Grb10 SH2 domain that also binds to phosphotyrosine-containing motifs. Lysine residues are conjugated to ubiquitin in the kinase domains by Cbl, NEDD4, and other E3 ligases. Acetylation of distal lysines in the carboxyl terminus of EGFR is shown. Internalization motifs NPxY, Yxx⊖ (⊖, bulky hydrophobic residue), and LL are located in juxtamembrane domains, kinase insert, and carboxy-terminal tails of receptors.

major subtypes of RTKs has been shown; thus ubiquitination is a characteristic modification of this receptor family.

The leading model of RTK endocytosis presumes that ubiquitinated RTKs are recruited into CCP by interacting with epsin, Eps15, and Eps15R, proteins that contain ubiquitin-binding domains (UBDs) and bind to clathrin or AP-2 (Fig. 3). However, direct evidence to support this model is still lacking for most RTKs. Ubiquitin-conjugation sites were mapped in EGFR and IGF-1R, and mutational analysis has been performed to assess the importance of ubiquitination sites for the internalization of EGFR, IGF-1R, and FGFR-2 (Huang et al. 2006; Haugsten et al. 2008; Mao et al. 2011). These studies

revealed that ubiquitination of EGFR and FGFR-2 is not essential for internalization. In contrast, mutation of lysines in the IGF-1R significantly decreased antibody-induced endosomal accumulation of this receptor (Mao et al. 2011).

The indirect evidence supporting the role of ubiquitination in RTK endocytosis is based on the effects of mutations or down-regulation of the components of the ubiquitination systems. Ubiquitination is mediated by the sequential action of the ubiquitin-activating (E1), ubiquitin-conjugating (E2), and ubiquitin ligase (E3) enzymes (Pickart 2004). E3 ligases determine the substrate specificity of ubiquitination. One large family of E3s that ubiquitinate RTKs is the RING finger containing ubiquitin ligases, such

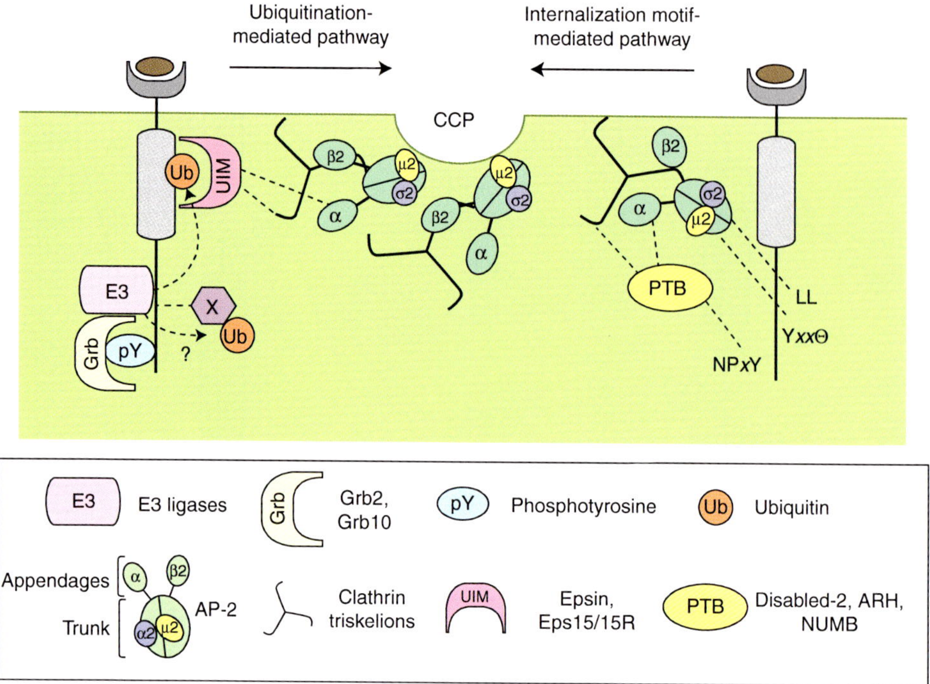

Figure 3. Hypothetic model of clathrin-mediated internalization of ligand-activated RTK. On ligand binding, RTKs are ubiquitinated by E3 ligases. Proteins, such as epsin, Eps15, and Eps15R, contain ubiquitin-interacting motifs (UIMs, a type of ubiquitin-binding domain [UBD]), and bind to ubiquitinated RTKs. UIM proteins bind to the appendage domains (mainly α) of AP-2, and epsin can also directly bind to the terminal domain of clathrin heavy chain (main components of clathrin triskelions). Interaction of RTKs with UIMs can occur before the recruitment of receptors into CCP and/or with UIMs present in assembled CCPs. The heterotetrameric AP-2 complex consists of four subunits: α, β2, μ2, and σ2. The trunk/core domain of AP-2 is formed by μ2, σ2, and the core domains of α and β2. YxxΘ and LL motifs directly interact with the μ2 and σ2/β2 subunits of AP-2, respectively. NPxY motifs interact with phosphotyrosine-binding domain (PTB) of adaptor proteins, such as Disabled-2, ARH, and NUMB, which themselves bind to the clathrin terminal domain and AP-2 appendage domains. The hinge region of β2 (between trunk and appendage) binds to clathrin heavy chain.

as Cbl. Cbl binds to and ubiquitinates EGFR, FGFR, PDGFR, VEGFR, CSF-1R, hepatocyte growth factor (c-Met), macrophage stimulating protein (Ron) and ephrin (EphR) receptors, as well as other RTKs (Levkowitz et al. 1999; Miyake et al. 1999; Wilhelmsen et al. 2002; Duval et al. 2003; Penengo et al. 2003; Marmor and Yarden 2004; Fasen et al. 2008). c-Cbl knockout, siRNA depletion of c-Cbl and Cbl-b, or overexpression of inactive Cbl mutants inhibit internalization of CSF-1R, EGFR, c-Kit, and several other RTKs (Lee et al. 1999; Levkowitz et al. 1999; Zeng et al. 2005; Huang et al. 2006). These studies implicate Cbl as a key mediator of RTK endocytosis. Besides Cbl, another

RING family E3, Nrdp1, ubiquitinates an EGFR homolog ErbB3 in a ligand (neuregulin)-independent manner, but this ubiquitination is augmented by neuregulin stimulation through the increased stability of Nrdp1 (Cao et al. 2007).

Cbl is a multidomain protein capable of interacting with various regulatory proteins. Activity of Cbl and its turnover are regulated by a range of mechanisms involving interaction with Sprouty2 and p85Cool-1, ubiquitination by NEDD4-1, and phosphorylation by Src (Wong et al. 2002; Bao et al. 2003; Wu et al. 2003b; Marmor and Yarden 2004; Thien and Langdon 2005). Therefore, ubiquitination of RTK can be indirectly regulated by these proteins. All three mammalian isoforms of Cbl (Cbl-b, Cbl-3, and c-Cbl) bind directly to specific phosphorylated tyrosines in RTKs, and long Cbl species (c-Cbl and Cbl-b) bind EGFR, c-Met, PDGFR-β, stem cell factor receptor (c-kit), and other RTKs indirectly through the SH2 adaptor protein Grb2 (Batzer et al. 1994; Peschard et al. 2001; Waterman et al. 2002; Jiang and Sorkin 2003; Li et al. 2007; Sun et al. 2007). The key role of Grb2 in ubiquitination and internalization of RTKs via Cbl has been shown using multiple experimental approaches in mammalian cells (Peschard 2001; Jiang 2003; Huang and Sorkin 2005; Johannessen 2006; Li 2007). In *Drosophila* and *Caenorhabditis elegans*, D-Cbl L binds to Grb2/Drk and Sli-1/Cbl binds to Grb2/SEM-5 to promote EGFR endocytosis (Yoon et al. 2000; Wang and Pai 2011). Finally, Grb2 and Cbl are colocalized with and remain bound to activated EGFR in CCPs (de Melker et al. 2001; Jiang et al. 2003; Johannessen et al. 2006). Collectively, these studies suggest that Grb2-Cbl complexes are required for the recruitment of RTKs into CCPs and, thereby, play an important role in receptor internalization.

The discrepancy between the requirement of Grb2 and Cbl for EGFR internalization and the lack of the role of EGFR ubiquitination sites might be explained by the existence of an unidentified receptor-associated protein that is (1) ubiquitinated by Cbl, and (2) mediates the internalization of the receptor (Fig. 3). It is important to note that the role of Grb2 and Cbl may be unrelated to ubiquitination. For instance, the SH2 domain of Grb2 binds to Tom1L protein that interacts with clathrin, thus linking EGFR to coated pits (Liu et al. 2009). It is unclear, however, how the SH2 domain of Grb2 binds simultaneously to both EGFR and Tom1L. Leucine repeat kinase Lrrk1 also binds to EGFR through Grb2 and promotes EGFR endocytosis (Hanafusa et al. 2011). On PDGF stimulation, Grb2 binds to PDGFR-β and the SH3 domain of Grb2 interacts with dynamin to facilitate internalization of the receptor (Kawada et al. 2009). A Cbl-interacting adaptor protein, CIN85, interacts with a plethora of partners through its five SH3 domains and is implicated in internalization of various RTKs

(Petrelli et al. 2002; Soubeyran et al. 2002; Szymkiewicz et al. 2002; Kowanetz et al. 2003; Schmidt et al. 2004; Buchse et al. 2011). However, the effects of knockdown or knockout of CIN85 on clathrin-mediated RTK internalization and the localization of CIN85 in CCPs have not been shown. It can be concluded that although the key role of Cbl and Grb2 in RTK endocytosis is well established, the role of RTK ubiquitination in this process remains unclear, and ubiquitination-independent mechanisms involving Grb2 and Cbl are not fully supported by functional and localization data.

Role of NEDD4 Family and Other E3 Ligases

The second large family of E3s, the HECT domain-containing NEDD4 ligases, directly interact with substrate PP*x*Y motifs through their WW domains (Fig. 2) (Rotin and Kumar 2009). NEDD4-2 was shown to ubiquitinate TrkA receptor by binding to its carboxyl terminus (Arevalo et al. 2006). Another NEDD4 family member, AIP4/Itch, binds to the PP*x*Y motif in ErbB4 (Sundvall et al. 2008). NEDD4-1 binds to an unconventional sequence of FGFR1 and ubiquitinates this receptor (Persaud et al. 2011). siRNA depletion of NEDD4 family or overexpression of their catalytically inactive forms inhibited endocytosis of TrkA, ErbB4, and FGFR1 (Arevalo et al. 2006; Sundvall et al. 2008; Persaud et al. 2011). IGF-1R and IR ubiquitination involves an indirect binding of NEDD4-1 through the Grb10 adaptor (Fig. 2) (Vecchione et al. 2003; Monami et al. 2008). The SH2 domain of Grb10 is capable of simultaneous binding to the C2 domain of NEDD4-1 and IGF-1R (Huang and Szebenyi 2010). Expression of Grb10 and NEDD4-1 mutants significantly reduced endocytosis of IGF-1R (Vecchione et al. 2003; Monami et al. 2008). Genetic knockout of NEDD4-1 caused reduction of IGF-1R on the cell surface of mouse fibroblasts, suggesting that the role of NEDD4-1 in this system is more complex (Cao et al. 2008). The latter study proposed that NEDD4-1 regulates Grb10 levels. In the absence of NEDD4-1, Grb10 is up-regulated, which results in increased endocytosis of IGF-1R. Although NEDD4-1/Grb10 complex appears to be involved in endocytosis of IR/IGF-1Rs, the localization of these proteins during early internalization events has not been determined.

The mechanisms by which IGF-1R is regulated by ubiquitination are convoluted because IGF-1R can also be ubiquitinated by the RING domain E3 ligase Mdm2, which is itself associated with IGF-1R through interaction with β-arrestins, and by c-Cbl (Girnita et al. 2005; Sehat et al. 2008). Indeed, siRNA knockdown of individual E3s implicated in IGF-1R ubiquitination did not affect IGF-1R endocytosis (Mao et al. 2011). The IGF-1R example of the potentially redundant role of multiple E3s in ubiquitination of a single RTK can be extended to EGFR (Cbl, Smurf2) (Levkowitz et al. 1999; Ray et al. 2011), ErbB4 (Itch, NEDD4, and WW1) (Sundvall et al. 2008; Li et al. 2009; Zeng et al. 2009), and TrkA (Cbl, NEDD4-2, and TRAF-6/UbcH7) (Geetha et al. 2005; Arevalo et al. 2006; Takahashi et al. 2011). Ubiquitination of a single RTK by different E3 ligases may result in different functional outcomes (Mao et al. 2011; Ray et al. 2011). Likely the contribution of different E3s is cell-type specific, depends on the experimental conditions, and can be indirect in some cases.

UBIQUITIN-BINDING PROTEINS IN CCPs

The model assuming that ubiquitin moieties target RTKs to CCPs implies that three UBD-containing proteins residing in CCPs (epsin-1/2, Eps15, and Eps15R) are involved in RTK internalization (Fig. 3). Binding of these proteins to ubiquitin moieties in vitro is well documented (Hawryluk et al. 2006), and coimmunoprecipitation of epsin-1 and Eps15 with activated EGFR has also been shown (Sigismund et al. 2005). There are, however, contrasting data regarding the importance of these proteins in EGFR internalization. siRNA knockdown of epsin-1 completely abolished EGF internalization in one study (Kazazic et al. 2009), whereas others report that individual or simultaneous depletion of epsin-1, Eps15, and Eps15R did not specifically block the CME of EGFR (Huang et al. 2004; Sigismund et al. 2005; Chen et al.

2009). Because epsin and Eps15/R contain multiple protein–protein interaction modules, overexpression of these proteins is not informative to their function (Sugiyama et al. 2005). Furthermore, epsin and Eps15/R may have distinct functions in cargo recruitment. Epsin is distributed throughout the entire CCP, whereas Eps15 is located at the edge of pits (Tebar et al. 1996; Kazazic et al. 2009). In general, direct interactions of epsins and Eps15/R with ubiquitinated RTKs are difficult to show, and the precise role of these three UBD-containing proteins in redundant or sequential recognition of ubiquitinated RTKs remains to be elucidated.

OTHER MECHANISMS OF RTK INTERNALIZATION

A tumor suppressor protein, RALT/Mig-6, was shown to promote EGFR endocytosis by a receptor-kinase-independent mechanism (Frosi et al. 2010). RALT/Mig6 binds to the EGFR kinase and inhibits its activity. RALT/Mig-6 is also capable of binding to AP-2 and intersectin, thus mediating internalization of EGFR through CCPs. Acetylation of three lysines at the carboxyl terminus of EGFR may also support internalization (Fig. 2). Mutation of these residues together with ubiquitination sites in the kinase domain of EGFR resulted in inefficient recruitment of EGFR into CCP, inability to phosphorylate the β2 subunit of AP-2, and significant inhibition of the EGFR CME (Goh et al. 2010). The precise role of EGFR acetylation in internalization is unknown, but could be related to the regulation of the receptor-kinase-substrate interactions. A unique Eph receptor system involves activation of Eph endocytosis by a membrane-anchored ligand (ephrin) (Pitulescu and Adams 2010). In neurons, endocytosis of ephrin-EphA4 receptor complex is mediated by the Rho family guanine exchange factor (GEF) Vav2 through an actin-dependent process (Cowan et al. 2005). Additionally, RIN1 (Ras binding and Rab5 GEF) is implicated in EphA4 endocytosis (Deininger et al. 2008). It is unclear, however, whether EphA4 endocytosis mediated by Vav2 or RIN1 is clathrin dependent.

POSTENDOCYTIC SORTING OF RTKs: MORPHOLOGY AND PATHWAYS

Endocytic carriers formed by CME and CIE deliver cargo to EEs located in the cell periphery. EEs have vesicular and tubular subcompartments; they can fuse to each other while moving along microtubules toward the centriole/Golgi area of the cell (Fig. 1). In the specialized and extreme example of neuronal cells, ligand-bound Trk receptors are internalized in the axonal terminal and transported for a long distance in EEs in a microtubule-dependent manner to the neuronal soma. The mechanisms and functions of the retrograde transport of endocytosed neurotrophin receptors were recently reviewed (Wu et al. 2009).

Movement toward the pericentriolar region is accompanied by the endosome maturation process that involves enlargement of the vesicular subcompartment, reduction in the amount of tubular extensions, and appearance of ILVs that are characteristic of MVBs. Changes in the biochemical composition of endosomes and an increasing luminal acidification also occur. Sorting of RTKs takes place in MVBs (Figs. 1 and 4). RTKs that are located in the limiting membrane of EEs and MVBs are capable of recycling back to the cell surface by moving to tubular extensions of these endosomes. RTKs that are packaged into ILVs are incapable of recycling and remain in MVBs that mature into late endosomes. Proteases and other lysosomal enzymes are delivered to MVB/late endosomes from the Golgi complex, leading to degradation of ILVs and RTKs.

MOLECULAR MECHANISMS OF RTK ENDOSOMAL SORTING

The current model of postendocytic sorting suggests that ubiquitinated RTK is recognized on the limiting membrane of a MVB by the heterodimeric complex of Hrs (hepatocyte growth factor regulated tyrosine-kinase substrate) and STAM1/2 (signal transducing adaptor molecule 1 and 2), which is the main component of ESCRT (endosomal sorting complexes required for transport)-0 (Fig. 4). UBDs in the Hrs and STAMs bind ubiquitinated RTKs and, possibly with the help of clathrin lattice associated with Hrs, retain RTKs from free diffusion in the limiting membrane and tubular extensions of MVBs, thus preventing RTK recycling. It has also been proposed that ubiquitinated EGFR interacts with GGA3, another UBD-containing adaptor, in EE before interacting with ESCRT-0 (Puertollano and Bonifacino 2004). Further, ESCRT-0 facilitates sequential recruitment of ESCRT-I, ESCRT-II, and ESCRT-III onto the MVB membrane. It is hypothesized that ubiquitinated cargo is passed along to ESCRT-I and -II, and ultimately packaged into ILV via a process of inward membrane invagination mediated by ESCRT-III (Henne et al. 2011).

The MVB sorting model is well supported by multiple lines of evidence. Degradation of EGFR, FGFR-2, and IGF-1R was highly sensitive to mutations of ubiquitin-conjugation and Cbl binding sites (Huang et al. 2006; Haugsten et al. 2008; Mao et al. 2011). Ubiquitin-deficient EGFR mutant was poorly incorporated into ILVs (Eden et al. 2011). The experimental manipulations with E3 ligases, their adaptors (Grb2 and Grb10), and other regulatory proteins cited above in the internalization section also have strong effects on the ligand-induced degradation of RTKs. For example, modulators of Cbl-mediated EGFR ubiquitination, such as Sprouty2, low-density lipoprotein receptor-related protein, Lrig1, and CIN85, also modulate the rate of lysosomal degradation of EGFR (Wong et al. 2002; Gur et al. 2004; Takayama et al. 2005; Dikic and Schmidt 2007; Running et al. 2011). Furthermore, the effects of overexpression of dominant−negative mutants of ESCRT components and their down-regulation by knockouts, knockdowns, and a microRNA on the lysosomal degradation of RTKs have been shown (Bache et al. 2003; Jekely and Rorth 2003; Bowers et al. 2006; Ewan et al. 2006; Wurdinger et al. 2008; Belleudi et al. 2009).

It should be emphasized that according to the ESCRT sorting model, RTK ubiquitination, and therefore, ligand binding, kinase activity, and Cbl association must be maintained until the formation of an ILV. To avoid the loss of ubiquitin, the cargo must also be deubiquiti-

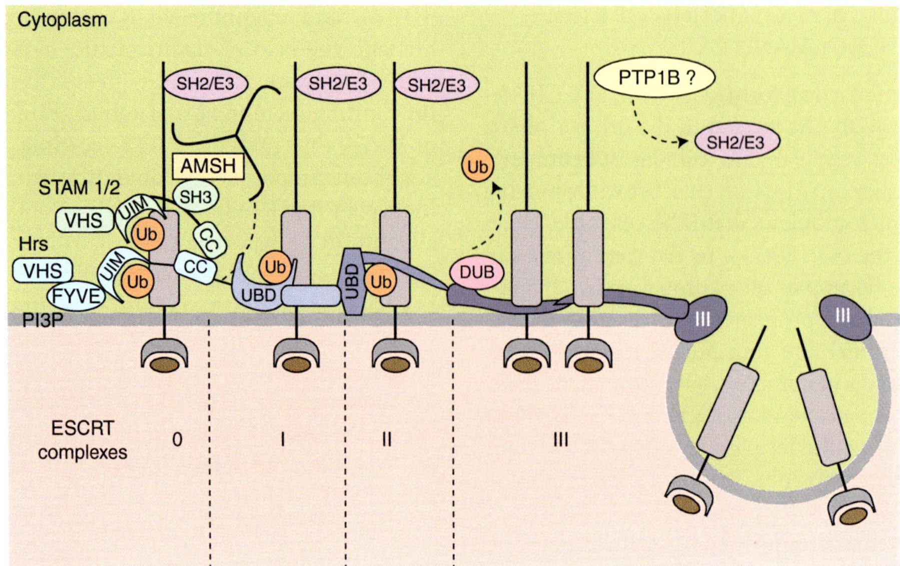

Figure 4. Hypothetic mechanism of RTK sorting into intraluminal vesicles of MVBs. Ubiquitinated RTK is recognized by UIMs of Hrs and STAM1/2 (ESCRT-0). Hrs is anchored to endosomal membrane through the interaction of its FYVE domain with phosphatidylinositol 3-phosphate (PI3P). Hrs also contains a coiled-coil domain that interacts with a similar domain of STAM, VHS domain, and a clathrin-binding motif. Binding of clathrin triskelions to Hrs nucleates an assembly of a flat clathrin lattice that further recruits additional Hrs molecules, leading to trapping of ubiquitinated cargo in the "Hrs microdomain." STAM contains the SH3 domain known to interact with AMSH. Accumulation of Hrs onto the endosomal membrane facilitates translocation of TSG101 and other components of ESCRT-I, and sequentially, components of ESCRT-II from cytosol to the MVB membrane. TSG101 and the ESCRT-II component EAP45/Vps36 have UBDs that may interact with ubiquitinated RTK. Ubiquitinated receptors appear to be transferred from ESCRT-0 to ESCRT-I and -II owing to increasing local concentrations of the latter complexes. ESCRT-III does not have ubiquitin-binding domains, and presumably traps receptors into forming ILVs by assembling into concentric hetero-oligomeric filaments and restricting diffusion of receptors. After formation of an ILV, ESCRT-III is disassembled by the Vps4 complex. The ESCRT model is reviewed by Teis et al. (2009). Before RTK entering the ILV, DUBs are proposed to remove ubiquitin from the receptor. Furthermore, receptor-associated proteins, such as SH2 adaptors and ubiquitin ligases, must also be removed before sequestration of receptors into ILV, possibly by means of receptor dephosphorylation by phosphotyrosine phosphatases like PTP1B (Eden et al. 2010). In addition, RTKs may facilitate ILV formation by phosphorylating annexin 1 (not shown) (White et al. 2006).

nated before entrance into ILVs. The requirement of such a tightly timed RTK deubiquitination for ILV incorporation, however, has not been directly shown. The deubiquitination enzyme (DUB) AMSH (associated molecule of SH3 domain of STAM) has been shown to associate with ESCRT-0 and also with the ESCRT-III component CHMP3 (Tanaka et al. 1999; Ma et al. 2007). AMSH directly deubiquitinates EGFR and PDGFR, and therefore decreases their degradation (McCullough et al. 2004, 2006; Bowers et al. 2006). Another DUB involved in cargo sorting is Usp8, which affects EGFR degradation by controlling the stability of STAM (Urbe et al. 2006) and Eps15 (Mizuno et al. 2006). However, there are contrasting effects of Usp8 knockdown on EGFR and c-Met degradation (Mizuno et al. 2005; Row et al. 2006). A conditional mouse knockout of Usp8 leads to decreased levels of EGFR, c-Met, and ErbB3 (Niendorf et al. 2007). Although it is clear that AMSH, Usp8, and possibly other DUBs play an important role in RTK sorting, further investigation is needed to distinguish between their function in deubiquitination of RTKs before ("sorting-negative" effect) or after

Cite this article as *Cold Spring Harb Perspect Biol* doi: 10.1101/cshperspect.a017459

ubiquitin-dependent RTK-ESCRT interactions ("sorting-positive" effect) and in deubiquitination of ESCRT proteins themselves.

The importance of a specific type of ubiquitination during RTK sorting in MVBs is another unresolved issue. Proteins can be modified by attachment of a single ubiquitin to a single lysine residue (monoubiquitination) or several lysines (multimonoubiquitination). Lysines in the ubiquitin itself can further serve as substrates for ubiquitin conjugation resulting in the formation of polyubiquitin chains. Mass spectrometry studies revealed that EGFR is ubiquitinated by Lys63-linked chains (Huang et al. 2006). Similarly, TrkA was proposed to be either monoubiquitinated or Lys63-chain polyubiquitinated (Geetha et al. 2005; Arevalo et al. 2006). In contrast, IGF-1R was found to be Lys48- and Lys29-polyubiquitinated during antibody-driven down-regulation (Mao et al. 2011). Unlike Lys48-linked chains that target proteins to proteasome, Lys63-linked polyubiquitination and monoubiquitination have nonproteosomal functions (Adhikari and Chen 2009). Importantly, Lys63-linked ubiquitin chains and monoubiquitin are recognized by UBDs in a similar manner (Varadan et al. 2004). Therefore, multimonoubiquitination and Lys63-polyubiquitination provide a platform for simultaneous RTK interactions with multiple UBDs in ESCRTs and, therefore, increase the avidity of weak ubiquitin-UBD interactions and the efficiency of sorting into ILVs (Ren and Hurley 2010). Considering the ample amount of evidence supporting the role of Lys63 chains in degradation of various ubiquitinated cargo, it can be proposed that Lys63-linked polyubiquitination is the major molecular signal involved in the endocytic sorting of mammalian RTKs. However, this hypothesis and its applicability to different subtypes of RTKs require further experimentation.

Finally, it should be noted that other mechanisms that do not appear to be related to ubiquitination have been implicated in RTK endosomal sorting. The LL motif in the carboxyl terminus of EGFR is important for EGFR degradation (Huang et al. 2003). Binding of RALT/MIG-6 associated with EGFR to the endosomal SNARE Syntaxin 8 also promotes EGFR degradation (Ying et al. 2010). The precise mechanisms underlying these regulations are poorly understood, and whether ESCRTs are involved is not determined.

RTK RECYCLING

As internalized activated RTKs are efficiently targeted for lysosomal degradation, it is generally assumed that recycling of these receptors is minimal. This notion, together with technical difficulties in recycling rate measurements, translates into a scarcity of information regarding RTK recycling mechanisms. Considerable recycling of ligand-receptor complexes has been shown for IR (Huecksteadt et al. 1986), EGFR (Sorkin et al. 1991), and TrkA (Zapf-Colby and Olefsky 1998). EGF–EGFR complexes are rapidly recycled back to the cell surface from peripheral EE and with a slower kinetics from pericentriolar endosomes (Fig. 1) (Sorkin et al. 1991). The fast recycling pathway or "retroendocytosis" is mediated by constitutive formation of recycling carriers from peripheral EE, whereas a slower pathway involves a sorting step in MVBs and late recycling compartments. Studies of the kinetics of EGF–EGFR sorting showed that the degradation pathway is saturable, and therefore, the size of the recycling pool of EGF–EGFR complexes is proportional to the concentration of internalized EGF–EGFR complexes (French et al. 1994). For instance, in cells expressing high levels of EGFR and stimulated with high EGF concentrations, up to 80% of internalized EGF-receptor complexes were recycled (Sorkin et al. 1991). In addition, unoccupied EGFRs that release EGF in endosomes were recycled back to the plasma membrane (Masui et al. 1993). It is estimated that ligand-occupied receptors recycle with a slower rate than unoccupied EGFR (Resat et al. 2003). Furthermore, a pool of unoccupied, and therefore, recycled EGFR in endosomes is significantly larger when EGFR is activated by ligands like transforming growth factor α that displays a higher pH sensitivity of binding and dissociates from the receptor in endosomes, as compared with EGF that remains mostly receptor bound

(French et al. 1995; Roepstorff et al. 2009). Also, alterations that are inhibitory to the kinase activity and ubiquitination, such as threonine phosphorylation in the juxtamembrane domain of EGFR by protein kinase C (PKC) or mutations of ubiquitination sites, elevate recycling of EGFR (Bao et al. 2000; Eden et al. 2011). Likewise, activation of PKC was necessary for recycling of PDGFR-β (Hellberg et al. 2009).

The kinetic analysis of EGFR trafficking is in line with the model whereby recycling is the default pathway of the cargo that is not trapped by ESCRT-0 in MVBs. Therefore, RTKs recycle through the pathways and compartments common with nutrient carrier receptors like the transferrin receptor. These recycling pathways are regulated by Rab4 in peripheral EE and Rab11 (and homologous Rabs, such as Rab22) in pericentriolar endosomes (Grant and Donaldson 2009). Specific "recycling" sequence motifs described for other classes of receptors have not been found in RTKs (Hanyaloglu and von Zastrow 2007). Recently, however, GGA3 was shown to bind c-Met via the Crk adaptor and promote c-Met recycling, suggesting that specific "recycling targeting" mechanisms may exist (Parachoniak et al. 2011).

CONCLUDING REMARKS

An emerging common theme in the RTK endocytosis literature is that RTK internalization is a highly robust process resistant to singular perturbations (e.g., mutations or altered expression of components of the endocytosis machinery). Possible explanations behind the robustness of the endocytic process relate to the existence of clathrin-dependent and -independent mechanisms. Furthermore, CME can itself be regulated by several redundant mechanisms. In the example of EGFR, simultaneous mutation of AP-2 binding motifs, multiple ubiquitin-conjugation and acetylation sites in EGFR is necessary to substantially suppress this receptor internalization via the CME pathway (Goh et al. 2010). Similarly, participation of AP-2 binding motifs and ubiquitination has been implicated in internalization of other RTKs. Moreover, ubiquitination of a single RTK can be mediated by several E3 ligases. In contrast, lysosomal targeting is mediated by a single major mechanism— ubiquitination—and, therefore, this step of trafficking is less robust and highly sensitive to the regulation of the ubiquitination and deubiquitination systems. In light of the redundancy of endocytosis mechanisms in mammalian cell culture in vitro, it is important to gear future research toward studying the mammalian RTK endocytosis in vivo so that the mechanistic analysis of endocytic pathways that are physiological and highly relevant to human disease can be revealed.

ACKNOWLEDGMENTS

We thank Kalen Dionne for his critical reading of the manuscript. This work is supported by National Cancer Institute grants CA089151 and CA112219.

REFERENCES

Acconcia F, Sigismund S, Polo S. 2009. Ubiquitin in trafficking: The network at work. *Exp Cell Res* **315:** 1610–1618.

Adhikari A, Chen ZJ. 2009. Diversity of polyubiquitin chains. *Dev Cell* **16:** 485–486.

Arevalo JC, Waite J, Rajagopal R, Beyna M, Chen ZY, Lee FS, Chao MV. 2006. Cell survival through Trk neurotrophin receptors is differentially regulated by ubiquitination. *Neuron* **50:** 549–559.

Bache KG, Raiborg C, Mehlum A, Stenmark H. 2003. STAM and Hrs are subunits of a multivalent ubiquitin-binding complex on early endosomes. *J Biol Chem* **278:** 12513–12521.

Backer JM, Shoelson SE, Haring E, White MF. 1991. Insulin receptors internalize by a rapid, saturable pathway requiring receptor autophosphorylation and an intact juxtamembrane region. *J Cell Biol* **115:** 1535–1545.

Backer JM, Shoelson SE, Weiss MA, Hua QX, Cheatham RB, Haring E, Cahill DC, White MF. 1992. The insulin receptor juxtamembrane region contains two independent tyrosine/β-turn internalization signals. *J Cell Biol* **118:** 831–839.

Bao J, Alroy I, Waterman H, Schejter ED, Brodie C, Gruenberg J, Yarden Y. 2000. Threonine phosphorylation diverts internalized epidermal growth factor receptors from a degradative pathway to the recycling endosome. *J Biol Chem* **275:** 26178–26186.

Bao J, Gur G, Yarden Y. 2003. Src promotes destruction of c-Cbl: Implications for oncogenic synergy between Src and growth factor receptors. *Proc Natl Acad Sci* **100:** 2438–2443.

Barbieri MA, Roberts RL, Gumusboga A, Highfield H, Alvarez-Dominguez C, Wells A, Stahl PD. 2000. Epidermal growth factor and membrane trafficking. EGF receptor activation of endocytosis requires Rab5a. *J Cell Biol* **151**: 539–550.

Batzer AG, Rotin D, Urena JM, Skolnik EY, Schlessinger J. 1994. Hierarchy of binding sites for Grb2 and Shc on the epidermal growth factor receptor. *Mol Cell Biol* **14**: 5192–5201.

Beattie EC, Howe CL, Wilde A, Brodsky FM, Mobley WC. 2000. NGF signals through TrkA to increase clathrin at the plasma membrane and enhance clathrin-mediated membrane trafficking. *J Neurosci* **20**: 7325–7333.

Beguinot L, Lyall RM, Willingham MC, Pastan I. 1984. Down-regulation of the epidermal growth factor receptor in KB cells is due to receptor internalization and subsequent degradation in lysosomes. *Proc Natl Acad Sci* **81**: 2384–2388.

Belleudi F, Leone L, Maggio M, Torrisi MR. 2009. Hrs regulates the endocytic sorting of the fibroblast growth factor receptor 2b. *Exp Cell Res* **315**: 2181–2191.

Bogdanovic E, Coombs N, Dumont DJ. 2009. Oligomerized Tie2 localizes to clathrin-coated pits in response to angiopoietin-1. *Histochem Cell Biol* **132**: 225–237.

Borg JP, Marchetto S, Le Bivic A, Ollendorff V, Jaulin-Bastard F, Saito H, Fournier E, Adelaide J, Margolis B, Birnbaum D. 2000. ERBIN: A basolateral PDZ protein that interacts with the mammalian ERBB2/HER2 receptor. *Nat Cell Biol* **2**: 407–414.

Bowers K, Piper SC, Edeling MA, Gray SR, Owen DJ, Lehner PJ, Luzio JP. 2006. Degradation of endocytosed epidermal growth factor and virally ubiquitinated major histocompatibility complex class I is independent of mammalian ESCRTII. *J Biol Chem* **281**: 5094–5105.

Buchse T, Horras N, Lenfert E, Krystal G, Korbel S, Schumann M, Krause E, Mikkat S, Tiedge M. 2011. CIN85 interacting proteins in B cells-specific role for SHIP-1. *Mol Cell Proteomics* **10**: M110.006239.

Burke PM, Wiley HS. 1999. Human mammary epithelial cells rapidly exchange empty EGFR between surface and intracellular pools. *J Cell Physiol* **180**: 448–460.

Cao Z, Wu X, Yen L, Sweeney C, Carraway KL 3rd. 2007. Neuregulin-induced ErbB3 downregulation is mediated by a protein stability cascade involving the E3 ubiquitin ligase Nrdp1. *Mol Cell Biol* **27**: 2180–2188.

Cao XR, Lill NL, Boase N, Shi PP, Croucher DR, Shan H, Qu J, Sweezer EM, Place T, Kirby PA, et al. 2008. Nedd4 controls animal growth by regulating IGF-1 signaling. *Sci Signal* **1**: ra5.

Carpentier JL, Gorden P, Anderson RG, Goldstein JL, Brown MS, Cohen S, Orci L. 1982. Co-localization of 125I-epidermal growth factor and ferritin-low density lipoprotein in coated pits: A quantitative electron microscopic study in normal and mutant human fibroblasts. *J Cell Biol* **95**: 73–77.

Chen WJ, Goldstein JL, Brown MS. 1990. NPXY, a sequence often found in cytoplasmic tails, is required for coated pit-mediated internalization of the low density lipoprotein receptor. *J Biol Chem* **265**: 3116–3123.

Chen H, Ko G, Zatti A, Di Giacomo G, Liu L, Raiteri E, Perucco E, Collesi C, Min W, Zeiss C, et al. 2009. Embryonic arrest at midgestation and disruption of Notch signaling produced by the absence of both epsin 1 and epsin 2 in mice. *Proc Natl Acad Sci* **106**: 13838–13843.

Cowan CW, Shao YR, Sahin M, Shamah SM, Lin MZ, Greer PL, Gao S, Griffith EC, Brugge JS, Greenberg ME. 2005. Vav family GEFs link activated Ephs to endocytosis and axon guidance. *Neuron* **46**: 205–217.

Deininger K, Eder M, Kramer ER, Zieglgansberger W, Dodt HU, Dornmair K, Colicelli J, Klein R. 2008. The Rab5 guanylate exchange factor Rin1 regulates endocytosis of the EphA4 receptor in mature excitatory neurons. *Proc Natl Acad Sci* **105**: 12539–12544.

de Melker AA, van der Horst G, Calafat J, Jansen H, Borst J. 2001. c-Cbl ubiquitinates the EGF receptor at the plasma membrane and remains receptor associated throughout the endocytic route. *J Cell Sci* **114**: 2167–2178.

Dikic I, Schmidt MH. 2007. Malfunctions within the Cbl interactome uncouple receptor tyrosine kinases from destructive transport. *Eur J Cell Biol* **86**: 505–512.

Duval M, Bedard-Goulet S, Delisle C, Gratton JP. 2003. Vascular endothelial growth factor-dependent down-regulation of Flk-1/KDR involves Cbl-mediated ubiquitination. Consequences on nitric oxide production from endothelial cells. *J Biol Chem* **278**: 20091–20097.

Eden ER, White IJ, Tsapara A, Futter CE. 2010. Membrane contacts between endosomes and ER provide sites for PTP1B-epidermal growth factor receptor interaction. *Nat Cell Biol* **12**: 267–272.

Eden ER, Huang F, Sorkin A, Futter CE. 2011. The role of EGF receptor ubiquitination in regulating its intracellular traffic. *Traffic* **13**: 329–337.

Ewan LC, Jopling HM, Jia H, Mittar S, Bagherzadeh A, Howell GJ, Walker JH, Zachary IC, Ponnambalam S. 2006. Intrinsic tyrosine kinase activity is required for vascular endothelial growth factor receptor 2 ubiquitination, sorting and degradation in endothelial cells. *Traffic* **7**: 1270–1282.

Fasen K, Cerretti DP, Huynh-Do U. 2008. Ligand binding induces Cbl-dependent EphB1 receptor degradation through the lysosomal pathway. *Traffic* **9**: 251–266.

Foti M, Moukil MA, Dudognon P, Carpentier JL. 2004. Insulin and IGF-1 receptor trafficking and signalling. *Novartis Found Symp* **262**: 265–268.

French AR, Sudlow GP, Wiley HS, Lauffenburger DA. 1994. Postendocytic trafficking of epidermal growth factor-receptor complexes is mediated through saturable and specific endosomal interactions. *J Biol Chem* **269**: 15749–15755.

French AR, Tadaki DK, Niyogi SK, Lauffenburger DA. 1995. Intracellular trafficking of epidermal growth factor family ligands is directly influenced by the pH sensitivity of the receptor/ligand interaction. *J Biol Chem* **270**: 4334–4340.

Frosi Y, Anastasi S, Ballaro C, Varsano G, Castellani L, Maspero E, Polo S, Alema S, Segatto O. 2010. A two-tiered mechanism of EGFR inhibition by RALT/MIG6 via kinase suppression and receptor degradation. *J Cell Biol* **189**: 557–571.

Galcheva-Gargova Z, Theroux SJ, Davis RJ. 1995. The epidermal growth factor receptor is covalently linked to ubiquitin. *Oncogene* **11**: 2649–2655.

Geetha T, Jiang J, Wooten MW. 2005. Lysine 63 polyubiquitination of the nerve growth factor receptor TrkA directs internalization and signaling. *Mol Cell* **20**: 301–312.

Girnita L, Shenoy SK, Sehat B, Vasilcanu R, Girnita A, Lefkowitz RJ, Larsson O. 2005. β-Arrestin is crucial for ubiquitination and down-regulation of the insulin-like growth factor-1 receptor by acting as adaptor for the MDM2 E3 ligase. *J Biol Chem* **280:** 24412–24419.

Goh LK, Huang F, Kim W, Gygi S, Sorkin A. 2010. Multiple mechanisms collectively regulate clathrin-mediated endocytosis of the epidermal growth factor receptor. *J Cell Biol* **189:** 871–883.

Gorden P, Carpentier JL, Cohen S, Orci L. 1978. Epidermal growth factor: Morphological demonstration of binding, internalization, and lysosomal association in human fibroblasts. *Proc Natl Acad Sci* **75:** 5025–5029.

Grant BD, Donaldson JG. 2009. Pathways and mechanisms of endocytic recycling. *Nat Rev Mol Cell Biol* **10:** 597–608.

Gur G, Rubin C, Katz M, Amit I, Citri A, Nilsson J, Amariglio N, Henriksson R, Rechavi G, Hedman H, et al. 2004. LRIG1 restricts growth factor signaling by enhancing receptor ubiquitylation and degradation. *EMBO J* **23:** 3270–3281.

Haft CR, Klausner RD, Taylor SI. 1994. Involvement of di-leucine motifs in the internalization and degradation of the insulin receptor. *J Biol Chem* **269:** 26286–26294.

Haigler HT, McKanna JA, Cohen S. 1979. Rapid stimulation of pinocytosis in human carcinoma cells A-431 by epidermal growth factor. *J Cell Biol* **83:** 82–90.

Hamer I, Haft CR, Paccaud JP, Maeder C, Taylor S, Carpentier JL. 1997. Dual role of a dileucine motif in insulin receptor endocytosis. *J Biol Chem* **272:** 21685–21691.

Hanafusa H, Ishikawa K, Kedashiro S, Saigo T, Iemura S, Natsume T, Komada M, Shibuya H, Nara A, Matsumoto K. 2011. Leucine-rich repeat kinase LRRK1 regulates endosomal trafficking of the EGF receptor. *Nat Commun* **2:** 158.

Hanyaloglu AC, von Zastrow M. 2007. A novel sorting sequence in the β2-adrenergic receptor switches recycling from default to the Hrs-dependent mechanism. *J Biol Chem* **282:** 3095–3104.

Haugsten EM, Malecki J, Bjorklund SM, Olsnes S, Wesche J. 2008. Ubiquitination of fibroblast growth factor receptor 1 is required for its intracellular sorting but not for its endocytosis. *Mol Biol Cell* **19:** 3390–3403.

Haugsten EM, Zakrzewska M, Brech A, Pust S, Olsnes S, Sandvig K, Wesche J. 2011. Clathrin- and dynamin-independent endocytosis of FGFR3—Implications for signalling. *PloS ONE* **6:** e21708.

Hawryluk MJ, Keyel PA, Mishra SK, Watkins SC, Heuser JE, Traub LM. 2006. Epsin 1 is a polyubiquitin-selective clathrin-associated sorting protein. *Traffic* **7:** 262–281.

Hellberg C, Schmees C, Karlsson S, Ahgren A, Heldin CH. 2009. Activation of protein kinase C α is necessary for sorting the PDGF β-receptor to Rab4a-dependent recycling. *Mol Biol Cell* **20:** 2856–2863.

Henne WM, Buchkovich NJ, Emr SD. 2011. The ESCRT pathway. *Dev Cell* **21:** 77–91.

Hommelgaard AM, Lerdrup M, van Deurs B. 2004. Association with membrane protrusions makes ErbB2 an internalization-resistant receptor. *Mol Biol Cell* **15:** 1557–1567.

Hopkins CR, Gibson A, Shipman M, Miller K. 1990. Movement of internalized ligand-receptor complexes along a continuous endosomal reticulum. *Nature* **346:** 335–339.

Huang F, Sorkin A. 2005. Grb2-mediated recruitment of the Cbl RING domain to the EGF receptor is essential and sufficient to support receptor endocytosis. *Mol Biol Cell* **16:** 1268–1281.

Huang Q, Szebenyi DM. 2010. Structural basis for the interaction between the growth factor-binding protein GRB10 and the E3 ubiquitin ligase NEDD4. *J Biol Chem* **285:** 42130–42139.

Huang F, Jiang X, Sorkin A. 2003. Tyrosine phosphorylation of the β2 subunit of clathrin adaptor complex AP-2 reveals the role of a di-leucine motif in the epidermal growth factor receptor trafficking. *J Biol Chem* **278:** 43411–43417.

Huang F, Khvorova A, Marshall W, Sorkin A. 2004. Analysis of clathrin-mediated endocytosis of epidermal growth factor receptor by RNA interference. *J Biol Chem* **279:** 16657–16661.

Huang F, Kirkpatrick D, Jiang X, Gygi S, Sorkin A. 2006. Differential regulation of EGF receptor internalization and degradation by multiubiquitination within the kinase domain. *Mol Cell* **21:** 737–748.

Huecksteadt T, Olefsky JM, Brandenberg D, Heidenreich KA. 1986. Recycling of photoaffinity-labeled insulin receptors in rat adipocytes. Dissociation of insulin-receptor complexes is not required for receptor recycling. *J Biol Chem* **261:** 8655–8659.

Jekely G, Rorth P. 2003. Hrs mediates downregulation of multiple signalling receptors in *Drosophila*. *EMBO Rep* **4:** 1163–1168.

Jiang X, Sorkin A. 2003. Epidermal growth factor receptor internalization through clathrin-coated pits requires Cbl RING finger and proline-rich domains but not receptor polyubiquitylation. *Traffic* **4:** 529–543.

Jiang X, Huang F, Marusyk A, Sorkin A. 2003. Grb2 regulates internalization of EGF receptors through clathrin-coated pits. *Mol Biol Cell* **14:** 858–870.

Johannessen LE, Pedersen NM, Pedersen KW, Madshus IH, Stang E. 2006. Activation of the epidermal growth factor (EGF) receptor induces formation of EGF receptor- and Grb2-containing clathrin-coated pits. *Mol Cell Biol* **26:** 389–401.

Jopling HM, Howell GJ, Gamper N, Ponnambalam S. 2011. The VEGFR2 receptor tyrosine kinase undergoes constitutive endosome-to-plasma membrane recycling. *Biochem Biophys Res Commun* **410:** 170–176.

Kawada K, Upadhyay G, Ferandon S, Janarthanan S, Hall M, Vilardaga JP, Yajnik V. 2009. Cell migration is regulated by platelet-derived growth factor receptor endocytosis. *Mol Cell Biol* **29:** 4508–4518.

Kazazic M, Bertelsen V, Pedersen KW, Vuong TT, Grandal MV, Rodland MS, Traub LM, Stang E, Madshus IH. 2009. Epsin 1 is involved in recruitment of ubiquitinated EGF receptors into clathrin-coated pits. *Traffic* **10:** 235–245.

Kelly BT, McCoy AJ, Spate K, Miller SE, Evans PR, Honing S, Owen DJ. 2008. A structural explanation for the binding of endocytic dileucine motifs by the AP2 complex. *Nature* **456:** 976–979.

Kowanetz K, Szymkiewicz I, Haglund K, Kowanetz M, Husnjak K, Taylor JD, Soubeyran P, Engstrom U, Ladbury JE, Dikic I. 2003. Identification of a novel proline-arginine motif involved in CIN85-dependent clustering of Cbl and down-regulation of epidermal growth factor receptors. *J Biol Chem* **278:** 39735–39746.

Lampugnani MG, Orsenigo F, Gagliani MC, Tacchetti C, Dejana E. 2006. Vascular endothelial cadherin controls VEGFR-2 internalization and signaling from intracellular compartments. *J Cell Biol* **174:** 593–604.

Lanahan AA, Hermans K, Claes F, Kerley-Hamilton JS, Zhuang ZW, Giordano FJ, Carmeliet P, Simons M. 2010. VEGF receptor 2 endocytic trafficking regulates arterial morphogenesis. *Dev Cell* **18:** 713–724.

Lazar CS, Cresson CM, Lauffenburger DA, Gill GN. 2004. The Na$^+$/H$^+$ exchanger regulatory factor stabilizes epidermal growth factor receptors at the cell surface. *Mol Biol Cell* **15:** 5470–5480.

Lee PS, Wang Y, Dominguez MG, Yeung YG, Murphy MA, Bowtell DD, Stanley ER. 1999. The Cbl protooncoprotein stimulates CSF-1 receptor multiubiquitination and endocytosis, and attenuates macrophage proliferation. *EMBO J* **18:** 3616–3628.

Levkowitz G, Waterman H, Ettenberg SA, Katz M, Tsygankov AY, Alroy I, Lavi S, Iwai K, Reiss Y, Ciechanover A, et al. 1999. Ubiquitin ligase activity and tyrosine phosphorylation underlie suppression of growth factor signaling by c-Cbl/Sli-1. *Mol Cell* **4:** 1029–1040.

Li N, Lorinczi M, Ireton K, Elferink LA. 2007. Specific Grb2-mediated interactions regulate clathrin-dependent endocytosis of the cMet-tyrosine kinase. *J Biol Chem* **282:** 16764–16775.

Li Y, Zhou Z, Alimandi M, Chen C. 2009. WW domain containing E3 ubiquitin protein ligase 1 targets the full-length ErbB4 for ubiquitin-mediated degradation in breast cancer. *Oncogene* **28:** 2948–2958.

Liu NS, Loo LS, Loh E, Seet LF, Hong W. 2009. Participation of Tom1L1 in EGF-stimulated endocytosis of EGF receptor. *EMBO J* **28:** 3485–3499.

Ma YM, Boucrot E, Villen J, Affar el B, Gygi SP, Gottlinger HG, Kirchhausen T. 2007. Targeting of AMSH to endosomes is required for epidermal growth factor receptor degradation. *J Biol Chem* **282:** 9805–9812.

Mao Y, Shang Y, Pham VC, Ernst JA, Lill JR, Scales SJ, Zha J. 2011. Poly-ubiquitination of the insulin-like growth factor I receptor (IGF-IR) activation loop promotes antibody-induced receptor internalization and down-regulation. *J Biol Chem* **286:** 41852–41861.

Marmor MD, Yarden Y. 2004. Role of protein ubiquitylation in regulating endocytosis of receptor tyrosine kinases. *Oncogene* **23:** 2057–2070.

Masui H, Castro L, Mendelsohn J. 1993. Consumption of EGF by A431 cells: Evidence for receptor recycling. *J Cell Biol* **120:** 85–93.

McClain DA. 1992. Mechanism and role of insulin receptor endocytosis. *Am J Med Sci* **304:** 192–201.

McCullough J, Clague MJ, Urbe S. 2004. AMSH is an endosome-associated ubiquitin isopeptidase. *J Cell Biol* **166:** 487–492.

McCullough J, Row PE, Lorenzo O, Doherty M, Beynon R, Clague MJ, Urbe S. 2006. Activation of the endosome-associated ubiquitin isopeptidase AMSH by STAM, a component of the multivesicular body-sorting machinery. *Curr Biol* **16:** 160–165.

Miller K, Beardmore J, Kanety H, Schlessinger J, Hopkins CR. 1986. Localization of the epidermal growth factor (EGF) receptor within the endosome of EGF-stimulated epidermoid carcinoma (A431) cells. *J Cell Biol* **102:** 500–509.

Mineo C, Gill GN, Anderson RG. 1999. Regulated migration of epidermal growth factor receptor from caveolae. *J Biol Chem* **274:** 30636–30643.

Miyake S, Mullane-Robinson KP, Lill NL, Douillard P, Band H. 1999. Cbl-mediated negative regulation of platelet-derived growth factor receptor-dependent cell proliferation. A critical role for Cbl tyrosine kinase-binding domain. *J Biol Chem* **274:** 16619–16628.

Mizuno E, Iura T, Mukai A, Yoshimori T, Kitamura N, Komada M. 2005. Regulation of epidermal growth factor receptor down-regulation by UBPY-mediated deubiquitination at endosomes. *Mol Biol Cell* **16:** 5163–5174.

Mizuno E, Kobayashi K, Yamamoto A, Kitamura N, Komada M. 2006. A deubiquitinating enzyme UBPY regulates the level of protein ubiquitination on endosomes. *Traffic* **7:** 1017–1031.

Monami G, Emiliozzi V, Morrione A. 2008. Grb10/Nedd4-mediated multiubiquitination of the insulin-like growth factor receptor regulates receptor internalization. *J Cell Physiol* **216:** 426–437.

Mori S, Heldin CH, Claesson-Welsh L. 1992. Ligand-induced polyubiquitination of the platelet-derived growth factor β-receptor. *J Biol Chem* **267:** 6429–6434.

Morrison P, Chung KC, Rosner MR. 1996. Mutation of Di-leucine residues in the juxtamembrane region alters EGF receptor expression. *Biochemistry* **35:** 14618–14624.

Nesterov A, Carter RE, Sorkina T, Gill GN, Sorkin A. 1999. Inhibition of the receptor-binding function of clathrin adaptor protein AP-2 by dominant-negative mutant μ2 subunit and its effects on endocytosis. *EMBO J* **18:** 2489–2499.

Niendorf S, Oksche A, Kisser A, Lohler J, Prinz M, Schorle H, Feller S, Lewitzky M, Horak I, Knobeloch KP. 2007. Essential role of ubiquitin-specific protease 8 for receptor tyrosine kinase stability and endocytic trafficking in vivo. *Mol Cell Biol* **27:** 5029–5039.

Orth JD, Krueger EW, Weller SG, McNiven MA. 2006. A novel endocytic mechanism of epidermal growth factor receptor sequestration and internalization. *Cancer Res* **66:** 3603–3610.

Parachoniak CA, Luo Y, Abella JV, Keen JH, Park M. 2011. GGA3 functions as a switch to promote Met receptor recycling, essential for sustained ERK and cell migration. *Dev Cell* **20:** 751–763.

Penengo L, Rubin C, Yarden Y, Gaudino G. 2003. c-Cbl is a critical modulator of the Ron tyrosine kinase receptor. *Oncogene* **22:** 3669–3679.

Persaud A, Alberts P, Hayes M, Guettler S, Clarke I, Sicheri F, Dirks P, Ciruna B, Rotin D. 2011. Nedd4-1 binds and ubiquitylates activated FGFR1 to control its endocytosis and function. *EMBO J* **30:** 3259–3273.

Peschard P, Fournier TM, Lamorte L, Naujokas MA, Band H, Langdon WY, Park M. 2001. Mutation of the c-Cbl TKB domain binding site on the Met receptor tyrosine kinase converts it into a transforming protein. *Mol Cell* **8:** 995–1004.

Petrelli A, Gilestro GF, Lanzardo S, Comoglio PM, Migone N, Giordano S. 2002. The endophilin-CIN85-Cbl complex mediates ligand-dependent downregulation of c-Met. *Nature* **416:** 187–190.

Pickart CM. 2004. Back to the future with ubiquitin. *Cell* **116:** 181–190.

Pitulescu ME, Adams RH. 2010. Eph/ephrin molecules—A hub for signaling and endocytosis. *Genes Dev* **24:** 2480–2492.

Prager D, Li HL, Yamasaki H, Melmed S. 1994. Human insulin-like growth factor I receptor internalization. Role of the juxtamembrane domain. *J Biol Chem* **269:** 11934–11937.

Puertollano R, Bonifacino JS. 2004. Interactions of GGA3 with the ubiquitin sorting machinery. *Nat Cell Biol* **6:** 244–251.

Ray D, Ahsan A, Helman A, Chen G, Hegde A, Gurjar SR, Zhao L, Kiyokawa H, Beer DG, Lawrence TS, et al. 2011. Regulation of EGFR protein stability by the HECT-type ubiquitin ligase SMURF2. *Neoplasia* **13:** 570–578.

Ren X, Hurley JH. 2010. VHS domains of ESCRT-0 cooperate in high-avidity binding to polyubiquitinated cargo. *EMBO J* **29:** 1045–1054.

Resat H, Ewald JA, Dixon DA, Wiley HS. 2003. An integrated model of epidermal growth factor receptor trafficking and signal transduction. *Biophys J* **85:** 730–743.

Roepstorff K, Grandal MV, Henriksen L, Knudsen SL, Lerdrup M, Grovdal L, Willumsen BM, van Deurs B. 2009. Differential effects of EGFR ligands on endocytic sorting of the receptor. *Traffic* **10:** 1115–1127.

Ronning SB, Pedersen NM, Madshus IH, Stang E. 2011. CIN85 regulates ubiquitination and degradative endosomal sorting of the EGF receptor. *Exp Cell Res* **317:** 1804–1816.

Rotin D, Kumar S. 2009. Physiological functions of the HECT family of ubiquitin ligases. *Nat Rev Mol Cell Biol* **10:** 398–409.

Row PE, Prior IA, McCullough J, Clague MJ, Urbe S. 2006. The ubiquitin isopeptidase UBPY regulates endosomal ubiquitin dynamics and is essential for receptor downregulation. *J Biol Chem* **281:** 12618–12624.

Salani B, Passalacqua M, Maffioli S, Briatore L, Hamoudane M, Contini P, Cordera R, Maggi D. 2010. IGF-IR internalizes with Caveolin-1 and PTRF/Cavin in HaCat cells. *PloS ONE* **5:** e14157.

Schmidt MH, Hoeller D, Yu J, Furnari FB, Cavenee WK, Dikic I, Bogler O. 2004. Alix/AIP1 antagonizes epidermal growth factor receptor downregulation by the Cbl-SETA/CIN85 complex. *Mol Cell Biol* **24:** 8981–8993.

Sehat B, Andersson S, Girnita L, Larsson O. 2008. Identification of c-Cbl as a new ligase for insulin-like growth factor-I receptor with distinct roles from Mdm2 in receptor ubiquitination and endocytosis. *Cancer Res* **68:** 5669–5677.

Sigismund S, Woelk T, Puri C, Maspero E, Tacchetti C, Transidico P, Di Fiore PP, Polo S. 2005. Clathrin-independent endocytosis of ubiquitinated cargos. *Proc Natl Acad Sci* **102:** 2760–2765.

Sorkin A, Carpenter G. 1993. Interaction of activated EGF receptors with coated pit adaptins. *Science* **261:** 612–615.

Sorkin A, Krolenko S, Kudrjavtceva N, Lazebnik J, Teslenko L, Soderquist AM, Nikolsky N. 1991. Recycling of epidermal growth factor-receptor complexes in A431 cells: Identification of dual pathways. *J Cell Biol* **112:** 55–63.

Sorkin A, Mazzotti M, Sorkina T, Scotto L, Beguinot L. 1996. Epidermal growth factor receptor interaction with clathrin adaptors is mediated by the Tyr974-containing internalization motif. *J Biol Chem* **271:** 13377–13384.

Soubeyran P, Kowanetz K, Szymkiewicz I, Langdon WY, Dikic I. 2002. Cbl-CIN85-endophilin complex mediates ligand-induced downregulation of EGF receptors. *Nature* **416:** 183–187.

Stoscheck CM, Carpenter G. 1984. Characterization of the metabolic turnover of epidermal growth factor receptor protein in A-431 cells. *J Cell Physiol* **120:** 296–302.

Sugiyama S, Kishida S, Chayama K, Koyama S, Kikuchi A. 2005. Ubiquitin-interacting motifs of Epsin are involved in the regulation of insulin-dependent endocytosis. *J Biochem* **137:** 355–364.

Sun J, Pedersen M, Bengtsson S, Ronnstrand L. 2007. Grb2 mediates negative regulation of stem cell factor receptor/c-Kit signaling by recruitment of Cbl. *Exp Cell Res* **313:** 3935–3942.

Sundvall M, Korhonen A, Paatero I, Gaudio E, Melino G, Croce CM, Aqeilan RI, Elenius K. 2008. Isoform-specific monoubiquitination, endocytosis, and degradation of alternatively spliced ErbB4 isoforms. *Proc Natl Acad Sci* **105:** 4162–4167.

Szymkiewicz I, Kowanetz K, Soubeyran P, Dinarina A, Lipkowitz S, Dikic I. 2002. CIN85 participates in Cbl-b-mediated down-regulation of receptor tyrosine kinases. *J Biol Chem* **277:** 39666–39672.

Takahashi Y, Shimokawa N, Esmaeili-Mahani S, Morita A, Masuda H, Iwasaki T, Tamura J, Haglund K, Koibuchi N. 2011. Ligand-induced downregulation of TrkA is partly regulated through ubiquitination by Cbl. *FEBS Lett* **585:** 1741–1747.

Takayama Y, May P, Anderson RG, Herz J. 2005. Low density lipoprotein receptor-related protein 1 (LRP1) controls endocytosis and c-CBL-mediated ubiquitination of the platelet-derived growth factor receptor β (PDGFR β). *J Biol Chem* **280:** 18504–18510.

Tanaka N, Kaneko K, Asao H, Kasai H, Endo Y, Fujita T, Takeshita T, Sugamura K. 1999. Possible involvement of a novel STAM-associated molecule "AMSH" in intracellular signal transduction mediated by cytokines. *J Biol Chem* **274:** 19129–19135.

Tebar F, Sorkina T, Sorkin A, Ericsson M, Kirchhausen T. 1996. Eps15 is a component of clathrin-coated pits and vesicles and is located at the rim of coated pits. *J Biol Chem* **271:** 28727–28730.

Teis D, Saksena S, Emr SD. 2009. SnapShot: The ESCRT machinery. *Cell* **137:** 182-182 e181.

Thien CB, Langdon WY. 2005. Negative regulation of PTK signalling by Cbl proteins. *Growth Factors* **23:** 161–167.

Traub LM. 2009. Tickets to ride: Selecting cargo for clathrin-regulated internalization. *Nat Rev Mol Cell Biol* **10:** 583–596.

Urbe S, McCullough J, Row P, Prior IA, Welchman R, Clague MJ. 2006. Control of growth factor receptor dynamics by reversible ubiquitination. *Biochem Soc Trans* **34:** 754–756.

Valdez G, Philippidou P, Rosenbaum J, Akmentin W, Shao Y, Halegoua S. 2007. Trk-signaling endosomes are generated by Rac-dependent macroendocytosis. *Proc Natl Acad Sci* **104:** 12270–12275.

Varadan R, Assfalg M, Haririnia A, Raasi S, Pickart C, Fushman D. 2004. Solution conformation of Lys63-linked di-ubiquitin chain provides clues to functional diversity of polyubiquitin signaling. *J Biol Chem* **279:** 7055–7063.

Vecchione A, Marchese A, Henry P, Rotin D, Morrione A. 2003. The Grb10/Nedd4 complex regulates ligand-induced ubiquitination and stability of the insulin-like growth factor I receptor. *Mol Cell Biol* **23:** 3363–3372.

von Kleist L, Stahlschmidt W, Bulut H, Gromova K, Puchkov D, Robertson MJ, MacGregor KA, Tomilin N, Pechstein A, Chau N, et al. 2011. Role of the clathrin terminal domain in regulating coated pit dynamics revealed by small molecule inhibition. *Cell* **146:** 471–484.

Wang PY, Pai LM. 2011. D-Cbl binding to Drk leads to dose-dependent down-regulation of EGFR signaling and increases receptor-ligand endocytosis. *PloS ONE* **6:** e17097.

Waterman H, Katz M, Rubin C, Shtiegman K, Lavi S, Elson A, Jovin T, Yarden Y. 2002. A mutant EGF-receptor defective in ubiquitylation and endocytosis unveils a role for Grb2 in negative signaling. *EMBO J* **21:** 303–313.

White IJ, Bailey LM, Aghakhani MR, Moss SE, Futter CE. 2006. EGF stimulates annexin 1-dependent inward vesiculation in a multivesicular endosome subpopulation. *EMBO J* **25:** 1–12.

Wiedlocha A, Sorensen V. 2004. Signaling, internalization, and intracellular activity of fibroblast growth factor. *Curr Top Microbiol Immunol* **286:** 45–79.

Wiley HS. 1988. Anomalous binding of epidermal growth factor to A431 cells is due to the effect of high receptor densities and a saturable endocytic system. *J Cell Biol* **107:** 801–810.

Wilhelmsen K, Burkhalter S, van der Geer P. 2002. C-Cbl binds the CSF 1 receptor at tyrosine 973, a novel phosphorylation site in the receptor's carboxy-terminus. *Oncogene* **21:** 1079–1089.

Wong ES, Fong CW, Lim J, Yusoff P, Low BC, Langdon WY, Guy GR. 2002. Sprouty2 attenuates epidermal growth factor receptor ubiquitylation and endocytosis, and consequently enhances Ras/ERK signalling. *EMBO J* **21:** 4796–4808.

Wu H, Windmiller DA, Wang L, Backer JM. 2003a. YXXM motifs in the PDGF-β receptor serve dual roles as phosphoinositide 3-kinase binding motifs and tyrosine-based endocytic sorting signals. *J Biol Chem* **278:** 40425–40428.

Wu WJ, Tu S, Cerione RA. 2003b. Activated Cdc42 sequesters c-Cbl and prevents EGF receptor degradation. *Cell* **114:** 715–725.

Wu C, Cui B, He L, Chen L, Mobley WC. 2009. The coming of age of axonal neurotrophin signaling endosomes. *J Proteomics* **72:** 46–55.

Wurdinger T, Tannous BA, Saydam O, Skog J, Grau S, Soutschek J, Weissleder R, Breakefield XO, Krichevsky AM. 2008. miR-296 regulates growth factor receptor overexpression in angiogenic endothelial cells. *Cancer Cell* **14:** 382–393.

Yamazaki T, Zaal K, Hailey D, Presley J, Lippincott-Schwartz J, Samelson LE. 2002. Role of Grb2 in EGF-stimulated EGFR internalization. *J Cell Sci* **115:** 1791–1802.

Yarden Y, Sliwkowski MX. 2001. Untangling the ErbB signalling network. *Nat Rev Mol Cell Biol* **2:** 127–137.

Ying H, Zheng H, Scott K, Wiedemeyer R, Yan H, Lim C, Huang J, Dhakal S, Ivanova E, Xiao Y, et al. 2010. Mig-6 controls EGFR trafficking and suppresses gliomagenesis. *Proc Natl Acad Sci* **107:** 6912–6917.

Yoon CH, Chang C, Hopper NA, Lesa GM, Sternberg PW. 2000. Requirements of multiple domains of SLI-1, a *Caenorhabditis elegans* homologue of c-Cbl, and an inhibitory tyrosine in LET-23 in regulating vulval differentiation. *Mol Biol Cell* **11:** 4019–4031.

Zapf-Colby A, Olefsky JM. 1998. Nerve growth factor processing and trafficking events following TrkA-mediated endocytosis. *Endocrinology* **139:** 3232–3240.

Zeng S, Xu Z, Lipkowitz S, Longley JB. 2005. Regulation of stem cell factor receptor signaling by Cbl family proteins (Cbl-b/c-Cbl). *Blood* **105:** 226–232.

Zeng F, Xu J, Harris RC. 2009. Nedd4 mediates ErbB4 JM-a/CYT-1 ICD ubiquitination and degradation in MDCK II cells. *FASEB J* **23:** 1935–1945.

Zheng J, Shen WH, Lu TJ, Zhou Y, Chen Q, Wang Z, Xiang T, Zhu YC, Zhang C, Duan S, et al. 2008. Clathrin-dependent endocytosis is required for TrkB-dependent Akt-mediated neuronal protection and dendritic growth. *J Biol Chem* **283:** 13280–13288.

Effects of Membrane Trafficking on Signaling by Receptor Tyrosine Kinases

Marta Miaczynska

International Institute of Molecular and Cell Biology, Laboratory of Cell Biology, 02-109 Warsaw, Poland
Correspondence: miaczynska@iimcb.gov.pl

The intracellular trafficking machinery contributes to the spatial and temporal control of signaling by receptor tyrosine kinases (RTKs). The primary role in this process is played by endocytic trafficking, which regulates the localization of RTKs and their downstream effectors, as well as the duration and the extent of their activity. The key regulatory points along the endocytic pathway are internalization of RTKs from the plasma membrane, their sorting to degradation or recycling, and their residence in various endosomal compartments. Here I will review factors and mechanisms that modulate RTK signaling by (1) affecting receptor internalization, (2) regulating the balance between degradation and recycling of RTK, and (3) compartmentalization of signals in endosomes and other organelles. Cumulatively, these mechanisms illustrate a multilayered control of RTK signaling exerted by the trafficking machinery.

At the cellular level, receptor tyrosine kinases (RTKs) need to be properly localized to function as signal-receiving and signal-transmitting devices (Lemmon and Schlessinger 2010). To receive signals (i.e., to bind extracellular ligands), RTKs have to be exposed at the surface of the plasma membrane. To transmit signals after ligand binding by RTKs, appropriate signaling components have to be available within intracellular compartments: in the cytoplasm, in association with membrane-bound organelles and in the cell nucleus. Importantly, the intracellular distribution of RTKs and their associated partners is not static but undergoes dynamic changes in different phases of signaling, as reflected for example by endocytic internalization of activated RTKs (Scita and Di Fiore 2010). Therefore, to function properly, the whole RTK signaling machinery within the cell has to be organized and tightly controlled both in space and in time. This organization and control are ensured by intracellular trafficking machineries, mainly by membrane transport systems such as endocytosis and secretion but also by other distribution systems (e.g., responsible for nucleocytoplasmic shuttling of proteins).

Recent years have brought increasing evidence that intracellular membrane trafficking, in particular endocytic internalization, degradation, and recycling, can profoundly affect the signaling properties of RTKs (Mukherjee et al. 2006; Abella and Park 2009; Lemmon and Schlessinger 2010; Scita and Di Fiore 2010; Grecco et al. 2011; Sigismund et al. 2012). The changes in the amounts of RTKs at the cell surface can alter the cellular responses when ligands are abundant (Grecco et al. 2011). In turn, the presence of a given RTK at the plasma membrane is determined by the rates of three trafficking processes: delivery of newly synthesized molecules by the secretory pathway, their internalization (occurring for both ligand-bound and ligand-free molecules), and endocytic recycling. Although the molecular details concerning the regulation of RTK delivery to the plasma membrane are not well known, numerous studies document various mechanisms by which internalization and recycling of RTKs can be modulated, thus affecting the signaling outputs (Le Roy and Wrana 2005). In addition to the regulation of RTKs at the cell surface, trafficking processes control the intracellular fate of endocytosed RTKs. Following internalization, RTKs can be either targeted for lysosomal degradation, or recycled back to the plasma membrane (Mukherjee et al. 2006; Abella and Park 2009; Scita and Di Fiore 2010). The first route results in the termination of signaling, whereas the second allows for sustained signaling if the ligand is available. Usually degradation and recycling of a given RTK can occur simultaneously but the balance between them is crucial to determine the net signaling output. Again, the molecular mechanisms that can shift the fate of internalized RTKs between degradation and recycling, thus changing RTK signaling, have begun to emerge in recent years (Polo and Di Fiore 2006; von Zastrow and Sorkin 2007; Sorkin and von Zastrow 2009; Sigismund et al. 2012). Finally, in contrast to an early view that only RTKs present at the plasma membrane are signaling competent, it is now accepted that in many cases activated RTKs can emit signals also after internalization into intracellular compartments (Miaczynska et al. 2004b; Miaczynska and Bar-Sagi 2010; Platta and Stenmark 2011). In some cell types (e.g., in neurons), such "signaling endosomes" are crucial for signal propagation within the cell and for the final cellular response. Moreover, endosomes can serve as platforms for amplification and compartmentalization of signals emitted by RTKs (Sadowski et al. 2009; Platta and Stenmark 2011).

In this article, I will review factors and mechanisms that modulate RTK signaling by (1) af-

fecting receptor internalization, (2) regulating the balance between degradation and recycling of RTK, and (3) compartmentalization of signals in endosomes and other organelles. As the membrane trafficking system of a cell is highly interconnected and can be considered a global dynamic continuum, it is important to note that often one primary alteration at a given stage of RTK trafficking may affect other transport steps or compartments, thus causing generalized changes in the intracellular routing and signaling of RTKs.

MODULATION OF RTK SIGNALING BY CHANGES IN ENDOCYTIC INTERNALIZATION

RTKs can undergo constitutive or ligand-induced internalization from the plasma membrane, the latter being responsible for downregulation of receptors from the cell surface following activation. Recent progress in the field of endocytosis revealed that internalization can occur via multiple structures formed from the plasma membrane. Among them are clathrin-coated vesicles (CCVs), caveolae, macropinosomes, clathrin- and dynamin-independent carriers (CLICs), and other less-characterized endocytic structures (Conner and Schmid 2003; Mayor and Pagano 2007; Howes et al. 2010; McMahon and Boucrot 2011). These different entry routes are often classified based on the involvement of a large GTPase dynamin as dynamin dependent (CCVs, caveolae) or dynamin independent (macropinocytosis, the CLIC pathway). As GTPase activity of dynamin can be inhibited by mutations or pharmacologically, these tools are often used to determine the impact of dynamin-mediated internalization on RTK signaling.

Historically, the use of a GTPase-deficient dynamin K44A mutant underlies one of the first reports that impaired endocytosis of epidermal growth factor receptor (EGFR) may affect a number of its downstream signaling molecules, with some being hyperactivated (phospholipase-Cγ and adaptor protein SHC) and others less active (phosphatidylinositol 3-kinase [PI3K] and mitogen-activated protein kinases

 Cite this article as *Cold Spring Harb Perspect Biol* doi: 10.1101/cshperspect.a009035

[MAPKs]) (Vieira et al. 1996). However, these early observations have been recently challenged by the study in which conditional depletion of dynamin in mouse fibroblasts inhibited EGFR endocytosis but did not affect the activation of MAPK or Akt on EGF stimulation (Sousa et al. 2012). Therefore, these data argued that the MAPK/Akt responses are elicited by EGFR at the plasma membrane and not by EGFR internalized into endosomes. Similarly, another recent study reported that the global transcriptional response to EGF is initiated primarily by the ligand-bound receptors at the plasma membrane (Brankatschk et al. 2012). These controversies indicate that signaling of activated RTKs after internalization is not an absolute requirement for their biological function. However, it cannot be excluded that in some cell types or under certain conditions signaling from RTKs localized in endosomes may contribute to the modulation of cellular responses.

An important recent concept is the notion that some factors, such as ligand concentration, can influence the preferred internalization route of a receptor, what determines its further sorting either toward degradation or recycling. Moreover, internalization of RTKs can be modulated by stress factors or stimuli other than their cognate ligands, affecting the signaling outcome. In addition, it appears that endocytic internalization and recycling of RTKs can occur in specialized plasma membrane microdomains, contributing to the determination of cell polarity. These aspects are discussed in detail below.

Different Entry Routes and Intracellular Fates of RTKs Dependent on Ligand Concentration

It appears that multiple cargo uptake mechanisms operate simultaneously in most cells and different entry routes can, to some extent, compensate for each other if one is blocked (Pelkmans et al. 2005). Different RTKs were shown to use various pathways, both dynamin dependent and independent, but the molecular basis for their selective use is still far from being understood. Ligand concentration is a postulated regulatory factor proposed for EGFR, which is the best-characterized RTK with respect to endo-

cytic trafficking (Fig. 1A). It was reported that at low ligand concentrations (~1 ng/mL EGF) EGFR is primarily internalized via clathrin-mediated endocytosis (Jiang and Sorkin 2003; Sigismund et al. 2005), whereas at higher concentrations (>20 ng/mL) EGFR enters cells both via clathrin-dependent and -independent routes (Sigismund et al. 2005). Both levels of ligand abundance are physiological, as the range of EGF concentration in different body fluids is rather broad (1–100 ng/mL). It was further proposed that clathrin-mediated internalization leads to increased recycling of EGFR and is required for the induction of DNA synthesis in a mitogenic response. In turn, lysosomal degradation is the main outcome of clathrin-independent endocytosis of EGFR (Fig. 1A) (Sigismund et al. 2008). In this way, low ligand concentrations would favor sustained and prolonged signaling via the continuous redelivery of EGFR to the plasma membrane, whereas high amounts of ligand would reduce the levels of the receptor at the cell surface, thus preventing cell overstimulation and eventually terminating signaling. It is, however, important to note that such dependence may not be universal but rather cell-type dependent, as others reported less degradation and higher recycling rates of EGFR on stimulation with high EGF concentrations (French et al. 1994) or no observable clathrin-independent endocytosis under these conditions (Kazazic et al. 2006).

With respect to other RTKs, low (1 ng/mL) or high (>5 ng/mL) concentrations of platelet-derived growth factor (PDGF) cause preferential activation of different signaling effectors and eventually result in PDGFR signaling toward cell migration or toward proliferation, respectively (De Donatis et al. 2008). By analogy to EGF, these differences in signaling were proposed to result from the selective use of distinct internalization pathways by PDGFR stimulated with various amounts of the ligand. Recently, PDGF was shown to undergo dynamin-dependent and -independent internalization but the activity of dynamin and high ligand concentration were specifically required for the mitogenic signaling via signal transducer and activator of transcription 3 (Stat3) and Myc (Sadowski

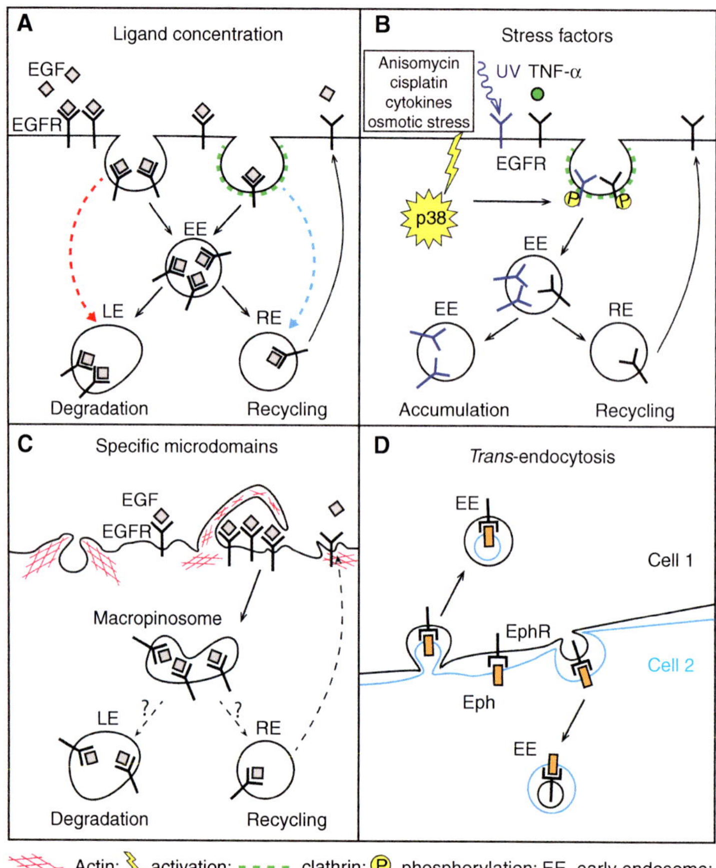

Figure 1. Different factors and mechanisms affecting RTK internalization from the plasma membrane. (*A*) Ligand concentration influences clathrin-dependent or -independent internalization of EGFR and its further trafficking toward degradation (high EGF concentration, red arrow) or recycling (low concentration, blue arrow). (*B*) Stress factors activating p38 MAPK stimulate ligand-independent internalization of EGFR. UV irradiation induces prolonged accumulation of EGFR in endosomes (marked in blue), whereas TNF-α increases receptor recycling (marked in black). (*C*) Dorsal ruffles represent plasma membrane microdomains for preferential localized internalization of EGFR into macropinosomes. Its further trafficking within the cell is not entirely clear. (*D*) *Trans*-endocytosis enables internalization of ephrin–Eph receptor complexes (Eph–EphR) into two neighboring cells in forward and reverse directions.

et al. 2013). Moreover, macropinocytosis of PDGFR-β (induced, for example, by an oncogenic mutant H-RasG12V) increased its signaling activity and anchorage-independent cell proliferation (Schmees et al. 2012).

The general concept of different signaling outcomes being dependent on a particular internalization route is supported by the data from the receptor systems other than RTKs. A serine/threonine kinase receptor for transform-

ing growth factor-β (TGF-βR) can be internalized via clathrin vesicles or caveolae. The former route activates downstream signaling effectors but the latter one leads to the rapid degradation of the receptor, thus terminating the signaling (Di Guglielmo et al. 2003). The opposite relationship was reported for the Wnt coreceptor LRP6. Wnt3a-stimulated uptake of LRP6 occurs via caveolae, which activates signaling via stabilization of β-catenin. In contrast, Wnt antagonist Dickkopf induces LRP6 internalization via the clathrin pathway, which prevents β-catenin signaling (Yamamoto et al. 2008). Cumulatively, these data argue that different internalization routes can be associated with the particular signaling outcomes but the exact functional relationships may vary for different ligand–receptor pairs and in different cell types.

Modulation of RTK Internalization by Stress Factors or Stimuli Other than the Cognate Ligands

As for any plasma membrane proteins, RTKs undergo constitutive turnover, including internalization, recycling, and degradation at basal cell type-specific rates. Ligand binding and receptor activation can significantly stimulate internalization of ligand–receptor complexes, as shown early on for insulin receptor, EGFR, PDGFR, hepatocyte growth factor (HGF) receptor Met, and others (Schlessinger et al. 1978; Heldin et al. 1982; Wiley et al. 1991; Naka et al. 1993). The molecular mechanism by which a ligand can regulate the rate of its own internalization in complex with a cognate RTK was proposed for EGFR. In this case EGF binding induces Ras-mediated activation of Rin1. This nucleotide exchange factor (GEF) activates in turn the small GTPase Rab5, which is a rate-limiting component for endocytic internalization (Tall et al. 2001). However, internalization rates of RTKs can be modulated also by factors other than the binding of cognate ligands, such as various types of stress, other signaling pathways, hypoxia, or molecules regulating cell adhesion (discussed below). These mechanisms provide an explanation for the often-reported cross talk between different signaling pathways. In the cases of such transmodulation, a heterologous stimulus can alter the levels of an RTK at the plasma membrane, thus regulating the overall cell responsiveness to a given ligand and a final signaling output of the receptor.

Stress Factors and Signaling Kinases

Several stress factors activating p38 MAPK, such as inflammatory cytokines (e.g., tumor

necrosis factor α [TNF-α]), UV irradiation, osmotic stress, or certain drugs (anisomycin, cisplatin), can stimulate internalization of ligand-free EGFR (Fig. 1B) (Vergarajauregui et al. 2006; Winograd-Katz and Levitzki 2006; Zwang and Yarden 2006; Singhirunnusorn et al. 2007). p38 MAPK stimulates clathrin-mediated internalization of EGFR by phosphorylating both EGFR and the components of the endocytic machinery, such as Rab GDP dissociation inhibitor (GDI) and Rab5 effectors EEA1 and Rabenosyn5 (Cavalli et al. 2001; Mace et al. 2005; Zwang and Yarden 2006). Interestingly, the fate of internalized EGFR can differ with various types of stimuli. For example, TNF-α treatment results in increased recycling of EGFR to the plasma membrane, whereas UV irradiation causes persistent intracellular accumulation of EGFR in endosomes, thus making the cells irresponsive to EGFR ligands (Fig. 1B) (Zwang and Yarden 2006). In addition to p38 MAPK-regulated trafficking, internalization of ligand-free EGFR can be induced on activation of protein kinase C (PKC) or inhibition of protein kinase A (PKA) (Beguinot et al. 1985; Lin et al. 1986; Salazar and Gonzalez 2002; Norambuena et al. 2010). In the case of VEGF, the activity of atypical PKC (aPKC) was shown to control vascular endothelial growth factor receptor (VEGFR) internalization and trafficking, thus regulating angiogenesis (Nakayama et al. 2013). Low activity of aPKC in endothelial sprouts of growing vessels contributes to high internalization and turnover of VEGFR, concomitant with increased signaling. In turn, high aPKC activity in mature vessels suppresses VEGFR endocytosis and signaling.

Hypoxia

Another mechanism modulating the internalization rates of RTKs is hypoxia. Under hypoxic conditions or on loss of a negative regulator of hypoxia, von Hippel–Lindau (VHL) protein, EGFR endocytosis is slowed down and the receptor half-life increased, leading to enhanced EGFR signaling (Wang et al. 2009). This is owing to the inhibition of Rab5-dependent fusion of early endosomes, resulting from the hypoxia-induced transcriptional down-regulation of

Rab5 effector Rabaptin-5. Moreover, the levels of Vps4B ATPase that regulate endosomal trafficking are down-regulated on hypoxia, which leads to the accumulation of EGFR and its increased signaling (Lin et al. 2012). Similarly, loss of VHL impairs internalization of fibroblast growth factor receptor 1 (FGFR1), causing its accumulation on the cell surface and enhanced signaling leading to increased cell motility on FGF stimulation (Hsu et al. 2006; Champion et al. 2008).

Regulators of Cell Adhesion

Molecules regulating cell adhesion or components of the extracellular matrix can affect the internalization rates of RTKs, both ligand-bound and ligand-free. Several types of cadherins were shown to interact with various RTKs, modulating their internalization and activation in complex, sometimes opposite, ways (Mukherjee et al. 2006; Orian-Rousseau and Ponta 2008; Delva and Kowalczyk 2009). For example, N-cadherin was reported to reduce internalization of the FGF2–FGFR1 complex, thereby causing sustained MAPK activation, increased transcription of MMP-9, and enhanced cell invasion (Suyama et al. 2002). Mutant E-cadherin associated with gastric cancer inhibited ligand-stimulated internalization of EGFR and enhanced its activation (Bremm et al. 2008). In contrast, wild-type E-cadherin was shown to block internalization of FGFR1 after stimulation with FGF1 or FGF2 but in this case MAPK signaling and FGFR1 translocation to the nucleus were inhibited, arguing that under these experimental conditions receptor endocytosis was required for sustained downstream signaling (Bryant et al. 2005). Similarly, as vascular endothelial growth factor receptor 2 (VEGFR2) was proposed to signal mainly from the intracellular compartments, inhibition of its internalization by VE-cadherin was therefore inhibitory for VEGFR2-induced cell proliferation (Lampugnani et al. 2006). Importantly, in many cases cadherins can be cointernalized along with the RTKs. This may provide a mechanism for the coordination of changes in cell adhesion with RTK-specific signaling. Finally, certain extracel-

lular molecules, particularly containing leucine-rich repeats such as decorin and LRIG1, were reported to affect internalization of several RTKs, such as ErbB receptors, Met, or insulin-like growth factor receptor (IGF-IR), affecting their signaling outputs (Gur et al. 2004; Laederich et al. 2004; Zhu et al. 2005; Shattuck et al. 2007; Goldoni and Iozzo 2008; Goldoni et al. 2009). Similarly, heparan sulfate proteoglycans, such as syndecans, can modulate endocytosis and signaling of RTKs, as shown recently for syndecan 4, which promotes macropinocytosis of FGFR1 and regulates its downstream signaling via MAPK (Elfenbein et al. 2012).

Spatial Regulation of RTK Internalization

Locally restricted internalization of RTKs emerges as an important although still poorly investigated aspect regulating signaling. It is now well accepted that the plasma membrane consists of various microdomains differing in composition and function, both in cells permanently polarized such as neurons or epithelia but also in cells, reversibly establishing short-term polarity like migrating fibroblasts (Mellman and Nelson 2008; Winckler and Mellman 2010; Schiefermeier et al. 2011). Therefore, it is reasonable to expect that the rates of endocytic uptake and the types of internalized cargo may differ in various microdomains of the plasma membrane (Disanza et al. 2009). For example, polarized endocytosis of FGFR2 on ligand stimulation occurs in the leading edge of migrating keratinocytes (Belleudi et al. 2011). Moreover, growth factor ligands of RTKs such as EGF, PDGF, or HGF change cell morphology by inducing plasma membrane ruffling in the form of peripheral (planar) or dorsal (circular) ruffles. These actin-rich protrusions can be used for macropinocytosis when ruffles fuse with each other enclosing large volumes of extracellular fluid, thus forming macropinosomes (Kerr and Teasdale 2009). Circular ruffles, which form transiently after growth factor stimulation at the dorsal side of a cell, were proposed to act as microdomains for preferential RTK endocytosis, as 50% of ligand-bound EGFR was shown to be internalized via these structures (Fig. 1C)

 Cite this article as *Cold Spring Harb Perspect Biol* doi: 10.1101/cshperspect.a009035

(Orth et al. 2006). In parallel, dorsal ruffles function also as sites for preferential RTK signaling required for biological responses, as evidenced for Met, which needs to signal from the dorsal ruffles to induce epithelial migration and morphogenesis (Abella et al. 2010a,b).

A special case of regulated internalization is represented by Eph receptors, which interact with membrane-bound ephrin ligands and signal in a cell-contact-dependent manner to control adhesion, migration, and various aspects of tissue morphogenesis (Pasquale 2008). Eph–ephrin complexes, in which ligand and receptor are embedded in the plasma membrane of two neighboring cells, are internalized locally by *trans*-endocytosis (Fig. 1D). In this poorly characterized process a piece of the plasma membrane of one cell is taken up into the other cell. It is postulated that *trans*-endocytosis (which can occur in forward and reverse directions) has profound consequences for the signaling outcomes depending on the cell type and other molecules cointernalized together with Eph–ephrin complexes (reviewed in Pitulescu and Adams 2010).

A key role of endocytic internalization and trafficking in spatial restriction of RTK signaling in vivo was shown in studies of collective migration of border cells during oogenesis in *Drosophila* (Jeffers et al. 1997; Jekely et al. 2005; Assaker et al. 2010). In this process, two RTKs, EGFR and PVR (a fly homolog of PDGFR and VEGFR), are localized specifically to the leading edge of the moving cells, maintained there by spatially restricted cycles of internalization and recycling. Mutations in several components of the endocytic machinery result in severe migration defects caused by delocalized activity of these RTKs, indicating a crucial role of membrane trafficking in establishing cell polarity.

MECHANISMS REGULATING THE BALANCE BETWEEN DEGRADATION AND RECYCLING OF RTKs AND THEIR IMPACT ON SIGNALING

As already mentioned, the balance between degradation and recycling for a given RTK may have important functional implications for its sig-naling properties, including the signal quality, duration, and magnitude (Lemmon and Schlessinger 2010; Grecco et al. 2011). The mechanisms specifying the intracellular routing of RTKs can be diverse. It appears that various ligands of the same RTK can elicit different trafficking, whereas the same ligand binding to distinct receptors can also be routed differently. This argues that the molecular determinants specifying the endocytic trafficking can be related to both ligands and receptors. Moreover, non-RTK coreceptors cooperating with the RTKs can also alter the intracellular trafficking of the ligand–receptor complexes and thus modulate the signaling output. Finally, the molecules and mechanisms that can switch the routing of a given RTK, thus regulating its signaling, are beginning to emerge.

Ligand-Dependent Sorting of a Given RTK

The best-characterized example of an RTK binding several ligands is EGFR, which associates with at least six other ligands in addition to EGF, all inducing EGFR internalization (Fig. 2A) (Harris et al. 2003). The trafficking of EGF and transforming growth factor α (TGF-α) was studied early on, revealing the opposite actions of the two ligands: EGF causing EGFR degradation and TGF-α stimulating receptor recycling (Decker 1990). This is caused by different pH sensitivities of ligand–receptor interactions: In acidic endosomal pH, the EGF–EGFR complex remains stable, whereas TGF-α dissociates from EGFR (Ebner and Derynck 1991; French et al. 1995). Recycling of TGF-α matches well its observed higher potency in evoking mitogenic signaling in comparison to EGF (Ebner and Derynck 1991; Waterman et al. 1998). Only very recently was the trafficking of other EGFR ligands studied in more detail (Stern et al. 2008; Baldys et al. 2009; Roepstorff et al. 2009). It was determined that similarly to EGF, heparin-binding EGF-like growth factor (HB-EGF) and β-cellulin induce EGFR degradation, whereas epiregulin and amphiregulin evoke EGFR recycling like TGF-α (Fig. 2A). At the mechanistic level, such differences can be correlated with the degree and duration of EGFR phosphorylation

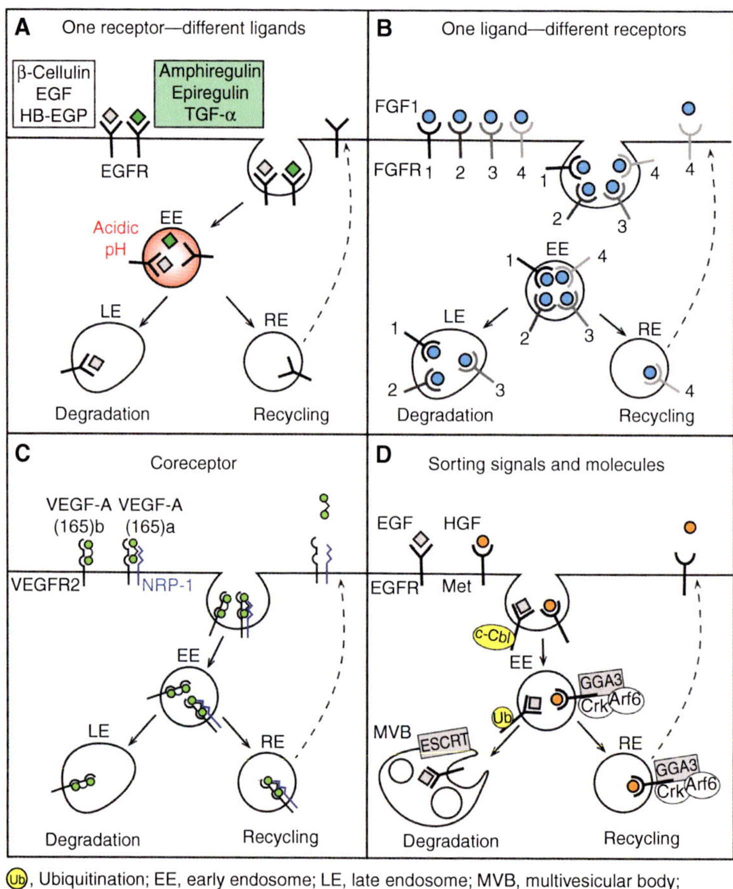

Figure 2. Selected mechanisms regulating the balance between degradation and recycling of RTKs. (*A*) Different ligands of EGFR induce receptor degradation or recycling. Ligands sorted to degradation (listed in the white box) are stably bound to EGFR in acidic pH of endosomes, while recycled ligands (green box) are dissociated under these conditions. (*B*) Different FGF receptors are trafficked either to degradation (FGFR1-3) or recycling (FGFR4) on binding of FGF1. (*C*) Coreceptor engagement directs trafficking of two isoforms of VEGF-A. Neuropilin-1 (NRP-1) coreceptor is involved in binding VEGF-A(165)a and causes its recycling. In contrast, it does not bind VEGF-A(165)b, which is degraded. (*D*) Selected molecules involved in RTK sorting. Many RTKs (here exemplified by EGFR) are ubiquitinated by c-Cbl ubiquitin ligase and sorted to degradation via the ESCRT complexes. RTKs can also be recycled and in the case of Met receptor, this process is mediated by a complex of GGA3, Crk, and Arf6.

and its Cbl-mediated ubiquitination induced by various ligands (Stern et al. 2008; Baldys et al. 2009; Roepstorff et al. 2009). Similar effects were reported for FGFR2, which is degraded on action of FGF7 and recycled on stimulation with FGF10, with the latter ligand showing higher mitogenic potential (Belleudi et al. 2007). In the case of FGFR1, neural cell adhesion molecule (NCAM) was proposed to act as an uncon-

ventional ligand that induces sustained FGFR1 recycling and promotes cell migration, in contrast to the natural ligand FGF2 targeting receptor for degradation (Francavilla et al. 2009).

Receptor-Mediated Sorting of a Given Ligand

It is known that one ligand can associate with several usually related RTKs. It was reported that

Cite this article as *Cold Spring Harb Perspect Biol* doi: 10.1101/cshperspect.a009035

FGF1, which can bind various FGFR family members, induces lysosomal degradation of FGFR1, FGFR2, and FGFR3, albeit with different efficiency (Fig. 2B). In contrast, FGFR4 is recycled on FGF1 binding, correlating with a lower degree of its ubiquitination in comparison to the other FGFRs (Haugsten et al. 2005). This may possibly cause sustained signaling downstream from FGFR4, which has been linked to poor prognosis of aggressive thyroid cancers (St Bernard et al. 2005). Moreover, some RTK types can heterodimerize with related family members and such complexes may be routed differently than the homodimers. On EGF binding, the heterodimers of EGFR-ErbB2 are preferentially recycled and thus have higher mitogenic potency than the homodimers of EGFR, which are targeted largely for degradation (Lenferink et al. 1998; Worthylake et al. 1999).

Coreceptor-Modulated Sorting of Ligand–RTK Complexes

Some ligands require an additional non-RTK coreceptor to act along with the RTK for proper signaling. A striking example of how a coreceptor can modulate the RTK trafficking and signaling was reported for VEGFR2 and its coreceptor neuropilin-1 (NRP-1) (Ballmer-Hofer et al. 2011). Two splicing variants of the VEGF-A ligand have opposite effects on the VEGFR2 trafficking and signaling: Proangiogenic VEGF-A(165)a leads to receptor recycling, whereas antiangiogenic VEGF-A(165)b causes its degradation (Fig. 2C). The decisive factor determining their endocytic routing is the NRP-1 coreceptor, which is involved in binding of VEGF-A(165)a but not VEGF-A(165)b. Engagement of the coreceptor is able to direct VEGFR2 recycling through the Rab11-positive recycling compartment. This in turn enhances p38 MAPK activation, which is required for vascular sprouting and which occurs only at a low level when NRP-1 is not engaged. There is a significant number of coreceptors for RTKs (e.g., p75^{NTR} neurotrophin receptor cooperating with the Trk receptors [Hempstead et al. 1991; Kaplan et al. 1991; Klein et al. 1991], GFRα coreceptor cooperating with Ret [Jing et al. 1996; Treanor

et al. 1996], and Agrin coreceptor Lrp4 cooperating with MuSK [Kim et al. 2008; Zhang et al. 2008]). Moreover, RTKs themselves play the role of coreceptors for other receptor types (e.g., Ror2 or PTK7/Otk in Wnt signaling) (Green et al. 2008; Peradziryi et al. 2011). It is thus conceivable that the impact of coreceptors on RTK trafficking and thus signaling may be broader than currently realized.

Mechanisms Determining Degradation or Recycling of RTKs and Their Impact on Signaling

Although a given receptor may be preferentially sorted either for degradation or recycling on ligand binding, there are certain molecular mechanisms—either physiological or pathological—that may shift the balance between these endocytic routes and thus affect signaling outputs. These mechanisms involve certain sorting signals in receptors, including ubiquitination, as well as specific trafficking proteins interacting with activated receptors.

RTK-Based Sorting Signals

Regulation of RTK trafficking via receptor ubiquitination is multilayered and occurs at different transport steps. A key E3 ubiquitin ligase modifying several RTKs is c-Cbl. Although the large body of literature on this topic is reviewed elsewhere (Swaminathan and Tsygankov 2006; Acconcia et al. 2009; Eden et al. 2011), for the purpose of this review it is important to stress the role of receptor ubiquitination in sorting to multivesicular endosomes, which precedes lysosomal degradation (Fig. 2D). This process is mediated by the ESCRT (endosomal sorting complex required for transport) complexes recognizing ubiquitinated receptors and sorting them into intraluminal vesicles of multivesicular endosomes (also termed multivesicular bodies) (Falguieres et al. 2009; Raiborg and Stenmark 2009; Babst 2011; Henne et al. 2011). At this stage, the receptors become separated from the bulk of cytoplasm, and therefore are no longer capable of active signaling. It was shown that RTK mutants unable to interact with c-Cbl

(EGFR Y1045, Met Y1003F), thus not properly sorted for degradation, elicit enhanced mitogenic signaling (Waterman et al. 2002; Abella et al. 2005).

In addition to ubiquitination, other factors can determine the balance between degradation and recycling of RTKs. For instance, different sorting signals localized in the juxtamembrane region of two related neurotrophin receptors TrkA and TrkB direct their trafficking to recycling or degradation, respectively. In consequence, activation of TrkA induced by nerve growth factor (NGF) causes enhanced trophic response of neurons in comparison to TrkB stimulation with brain-derived neurotrophic factor (BDNF) (Chen et al. 2005). In the case of PDGF receptors, increased phosphorylation of PDGFR-β caused by loss of T-cell protein tyrosine phosphatase (TC-PTP) diverts the receptor from degradation toward rapid recycling, while such an effect is not observed for the related PDGFR-α (Karlsson et al. 2006). In the case of Ret, alternative splicing creates two protein variants differing in their carboxyl termini (Ret9 and Ret51), which show distinct signaling and developmental roles. These differences were attributed to the unique trafficking properties of the two isoforms, both in terms of their de novo delivery to the plasma membrane and their internalization/recycling kinetics (Richardson et al. 2012).

Even more striking, mutations in a given RTK can change its intracellular trafficking, as shown for two activating Met mutants M1268T and D1246N found in human papillary renal carcinomas (Jeffers et al. 1997; Schmidt et al. 1997). Apart from being constitutively active, these mutants are degradation impaired and show increased rates of internalization and recycling both in the presence or absence of a ligand (Joffre et al. 2011). This leads to enhanced signaling and cell transformation manifested in vitro and in vivo. Importantly, blocking dynamin-dependent endocytosis inhibits migration and anchorage-independent growth of cells expressing either of the mutants as well as the ability of these cells to form tumors in nude mice (Joffre et al. 2011). These data argue that tumorigenic Met signaling is emitted mainly from intracellular compartments rather than from the plasma membrane. Moreover, modulating endocytic trafficking could be considered as a potential therapeutic strategy in some tumor types.

Proteins Sorting RTKs

With respect to sorting proteins directing the intracellular traffic of RTKs, recent work identified GGA3 (Golgi-localized γ-ear-containing Arf-binding protein 3) as a key factor mediating recycling of Met independently of its ubiquitination status (Fig. 2D) (Parachoniak et al. 2011). GGA3 along with Arf6-GTP and the Crk adaptor form a complex with activated Met, which sorts the receptor to Rab4-positive recycling endosomes. GGA3 depletion increases Met degradation on HGF stimulation, which impairs MAPK activation and cell migration (Parachoniak et al. 2011). However, this role of GGA3 is not universal, as another study showed that GGA3 depletion increased endosomal accumulation of EGFR and impaired its degradation (Puertollano and Bonifacino 2004). This was attributed to improper ESCRT- and ubiquitin-dependent sorting of EGFR to multivesicular endosomes in the absence of GGA3, which normally interacts with both ubiquitin and the ESCRT-I component Tsg101. These data argue that individual RTKs may use the same sorting adaptors for various purposes. This could be possibly achieved by RTK-dependent modification of such adaptors and indeed phosphorylation of GGA3 was reported to occur on EGFR activation and to regulate its association with membranes of intracellular organelles (Kametaka et al. 2005).

Similarly to the transmodulation of RTK internalization by heterologous stimuli, the balance between RTK degradation and recycling can be altered by other signaling pathways and regulatory proteins. For example, activation of PKCα mediates sorting of PDGFR-β to recycling (Hellberg et al. 2009), and PKC-dependent phosphorylation of EGFR at Thr654 diverts the EGF–EGFR complexes from degradation to recycling (Bao et al. 2000). Similarly, association of EGFR with α5β1 integrin via Rab coupling protein (RCP) can induce EGFR recycling, which increases Akt activation and promotes cell migration (Caswell et al. 2008). Moreover, the activation of Src kinase can modulate trafficking of RTKs, such as EGFR or FGFR (Sandilands et al. 2007; Medts et al. 2010).

THE ROLE OF ENDOSOMES AND OTHER INTRACELLULAR COMPARTMENTS IN RTK SIGNALING

Signaling Endosomes

A possible function of endosomal compartments in the propagation of RTK signaling was proposed in the mid-1990s when activated RTKs and their associated signaling molecules were detected in isolated fractions of endosomes (Di Guglielmo et al. 1994; Grimes et al. 1996). These studies argued that RTKs can remain signaling competent postinternalization, thus their active signaling may not be limited to their residence at the plasma membrane but can also occur from the endosomes. Since then, a number of reports confirmed and extended these observations for different RTKs and also for other receptor types (reviewed in Sadowski et al. 2009; Sorkin and von Zastrow 2009; Miaczynska and Bar-Sagi 2010; Platta and Stenmark 2011). In a current view, endosomal signaling may contribute to an overall signaling output, although this contribution is not obligatory and its exact role may differ for various ligand–receptor systems. Below I will review examples of how endosomes can participate in the delivery, amplification, or compartmentalization of signals emitted by RTKs.

Intracellular Signal Delivery via Endosomes

The concept of signaling endosomes originates from neurons (Grimes et al. 1996) and these cells provide the most striking example of endosomes serving as intracellular vehicles for signal delivery. Cell survival of NGF-responsive neurons depends on signaling initiated in axon terminals by NGF binding to its receptor TrkA. Following internalization, endosomes carrying activated TrkA and associated signaling effectors undergo long-distance, cytoskeleton-mediated transport to deliver the signal in a retrograde manner to the cell body (Fig. 3A) (Howe and Mobley 2005;

Cosker et al. 2008). In this case, passive cytoplasmic diffusion along the axon would not be sufficient to propagate the signal rapidly enough and therefore active endosome-mediated transport is required (Howe 2005). Over the years, several components of NGF signaling endosomes were identified, most of which represent general endocytic machinery operating also in other cell types (Howe et al. 2001; Delcroix et al. 2003; Deinhardt et al. 2006; Lin et al. 2006; Varsano et al. 2006; Valdez et al. 2007; Wu et al. 2007; McCaffrey et al. 2009; Philippidou et al. 2011). However, there is still an ongoing debate about whether the NGF signaling endosomes represent early endosomes, macropinosomes (termed macroendosomes), or multivesicular bodies (Wu et al. 2009). Intriguingly, neurotrophin NT3 can also bind TrkA; however, in contrast to NGF it cannot elicit retrograde endosomal transport and prosurvival signaling. This phenomenon was recently explained by different sensitivities of NGF–TrkA and NT3–TrkA complexes to an acidic pH of endosomes (Harrington et al. 2011). Only acid-stable NGF–TrkA complexes, but not acid-labile NT3–TrkA, can initiate actin depolymerization around endosomes, which is necessary to launch their retrograde transport. These data illustrate how certain properties of ligand–RTK complexes, which primarily determine their behavior within the endocytic system, can profoundly affect the functional state of endosomes and in consequence, the overall cellular response to a given ligand.

Amplification of Signals via Endosomes

Endosomes represent a confined environment enclosing ligand–receptor complexes in a tight volume when compared with their distribution over plasma membrane and, therefore, multiple rounds of receptor activation may be favored in endosomes (Grecco et al. 2011). In addition, the components of NADPH oxidase complex on endosomes may be involved in the localized generation of reactive oxygen species (ROS), among which hydrogen peroxide can reversibly inactivate tyrosine phosphatases, thus prolonging RTK signaling (Janssen-Heininger et al.

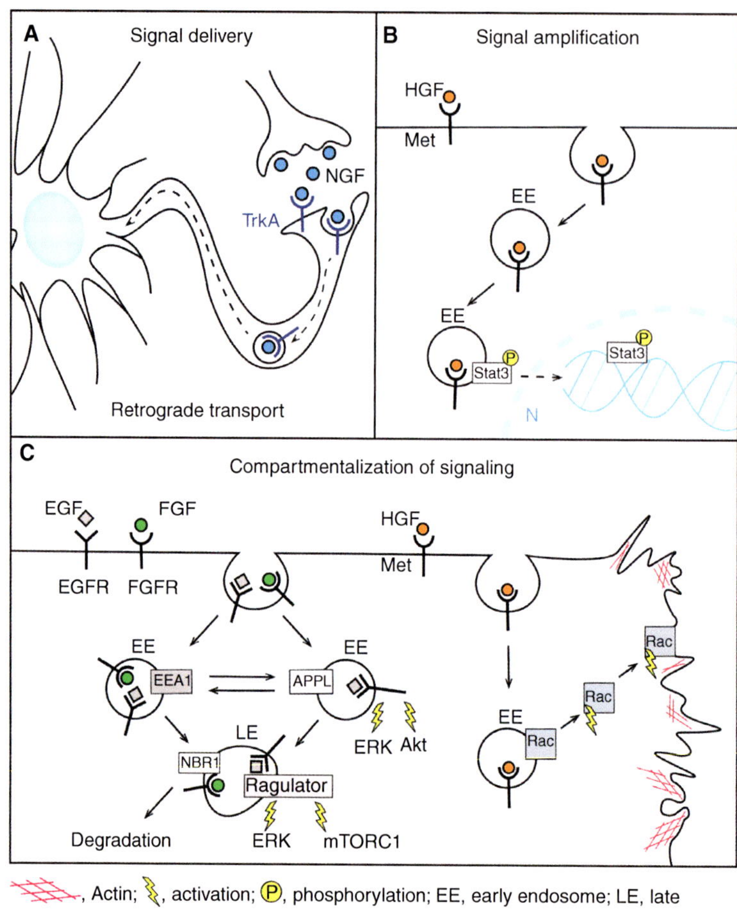

Figure 3. Mechanisms by which endosomes participate in signaling. (*A*) Endosomes in neurons deliver NGF-TrkA signaling complexes from axon terminals to the cell body. (*B*) On HGF stimulation, phosphorylation and nuclear translocation of Stat3 can occur efficiently only if activated Met receptors are transported via endosomes to the perinuclear region of the cell, thus amplifying weak signals. (*C*) Different populations of endosomes serve as platforms for activation of various signaling molecules. APPL endosomes participate in ERK and Akt signaling. Late endosomes carry NBR1 protein involved in FGF signaling and the Ragulator complex mediating mTORC1 and ERK signaling. On HGF stimulation Rac is activated on endosomes and subsequently recycled to the plasma membrane to stimulate the formation of migratory protrusions. See text for details.

2008; Oakley et al. 2009). Another mode of signal amplification involves endosome-based delivery of activated RTKs into the perinuclear region of the cell. This should enhance weak phosphorylation of transcription factors that would otherwise be destroyed by phosphatases during long-distance cytoplasmic diffusion (Fig. 3B). Such a mechanism was proposed for Stat3, weakly activated by HGF, for which phosphorylation and subsequent nuclear transloca-tion was observed only if activated Met receptor was localized on perinuclear endosomes (Kermorgant and Parker 2008). In contrast, robust Stat3 activation by the cytokine oncostatin-M did not require endosomal transport of its receptor. This argues that the localized endosome-based amplification is required only for weak signals. Finally, although not yet shown for RTKs, it is conceivable that multivesicular endosomes could contribute to signal amplification by sequestering in their intraluminal vesicles some negative cytoplasmic regulators of RTK signaling. Such a precedent was reported for the Wnt signaling, in which the inhibitory enzyme GSK3 is removed from the cytoplasm and entrapped within endosomes on Wnt stimulation, thus contributing to enhanced signaling (Taelman et al. 2010).

Compartmentalization of Signaling in Endosomes

Recent progress in the field of endocytosis indicates that the common classification of endosomes into early, late, and recycling is oversimplified, as cells harbor many subpopulations of endosomes, some of which may be cell-type-specific (Perret et al. 2005; Bokel et al. 2006; Zoncu et al. 2009). These endosomal compartments differ in their protein and lipid composition, dynamics, and function, therefore, cargo transported between them encounters different environments at subsequent trafficking steps. Several signaling components are localized on specific endosomes, thus enabling active signal propagation only when the activated receptor reaches a particular compartment (Sadowski et al. 2009; Platta and Stenmark 2011). For RTK signaling, one of the earliest examples of endosome-specific components was the scaffolding complex composed of MP1, p14 and p18, which tethers MEK1 to late endosomes and is required for full ERK activation on EGF stimulation (Teis et al. 2002, 2006; Nada et al. 2009). More recently, the same complex (now named Ragulator) was reported to activate mTORC1 (mammalian target of rapamycin complex 1) by recruiting it to lysosomes in response to amino acids (Fig. 3C) (Sancak et al. 2010). These findings open the possibility for direct cross talk between RTK-dependent ERK activation and mTOR signaling occurring on the late endosome/lysosome membrane. Another late endosomal protein NBR1 was shown to interact with an inhibitor of FGF signaling Spred2, thus participating in down-regulation of FGFR1 (Fig. 3C) (Mardakheh et al. 2009).

Various types of early endosomes can also serve as platforms for assembly of RTK signaling complexes, partly different from those formed at the plasma membrane (Burke et al. 2001). APPL endosomes, a subpopulation of early endosomes, were shown to participate in transport and signaling of EGF and NGF (Fig. 3C) (Miaczynska et al. 2004a; Lin et al. 2006; Varsano et al. 2006; Zoncu et al. 2009). Prolonged residence of EGFR in APPL endosomes increased activation of ERK and Akt (Zoncu et al. 2009), whereas depletion of APPL1 prevented their activation on NGF stimulation, thus inhibiting neurite outgrowth (Lin et al. 2006; Varsano et al. 2006). In zebrafish development, APPL endosomes transmit prosurvival signals by controlling the activity of Akt and its substrate specificity toward GSK3β but not TSC2 (Schenck et al. 2008).

Finally, endosomes enable compartmentalized signaling to direct cell migration, as shown by activation of Rac on early endosomes on HGF stimulation (Palamidessi et al. 2008). Activated Rac is subsequently recycled to specific microdomains on the plasma membrane to stimulate localized formation of migratory protrusions (Fig. 3C).

The Role of Endoplasmic Reticulum in Modulating RTK Signaling

In addition to the endosomal compartments, the organelles of the secretory pathway, in particular endoplasmic reticulum (ER), can also contribute to the modulation of RTK signaling by various mechanisms. First, ER-based ubiquitination and degradation mechanisms can regulate the levels of nascent RTKs eventually targeted to the plasma membrane, as recently shown for ErbB3 levels controlled by the E3 ubiquitin ligase Nrdp1 at the ER (Fry et al. 2011). Second, ER-localized protein tyrosine phosphatase PTP1B plays multiple roles in inactivating ligand-bound RTKs after internalization and in modulating their endocytic trafficking (Stuible and Tremblay 2010; Sangwan et al. 2011). A recent identification of direct contact sites between endosomes and ER provides a conceptual framework of how an ER-tethered enzyme can affect the function of RTKs on the endosomal membranes (Fig. 4A) (Eden et al. 2010). Third, there are cases of ER-specific ab-

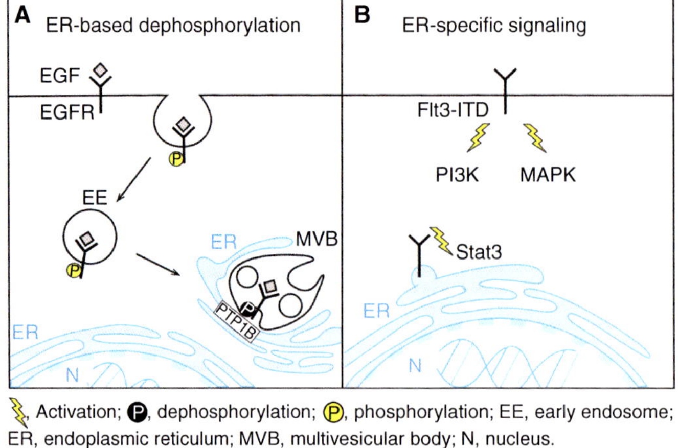

, Activation; , dephosphorylation; , phosphorylation; EE, early endosome; ER, endoplasmic reticulum; MVB, multivesicular body; N, nucleus.

Figure 4. The role of endoplasmic reticulum (ER) in modulating RTK signaling. (*A*) Direct contact sites between endosomes and the ER enable dephosphorylation of internalized RTKs by protein tyrosine phosphatase PTP1B localized on the ER membrane. (*B*) Constitutively active oncogenic mutant of Flt3 receptor (Flt3-ITD) activates different downstream effectors when localized in the ER or at the plasma membrane.

errant signaling elicited by oncogenic RTKs, as reported for the Flt3 mutant containing internal tandem duplication (Flt3-ITD), which causes its ligand-independent activation and predominant retention in the ER (Schmidt-Arras et al. 2005). Flt3-ITD is not only constitutively active in the ER but the pattern of tyrosine phosphorylation of the receptor and the set of activated downstream effectors are different from those observed for the wild-type Flt3 at the plasma membrane (Fig. 4B) (Choudhary et al. 2009). This argues that, at least under pathological conditions, ER can contribute to the overall signaling of a given RTK in a compartment-specific manner. It is further possible that, like ER, the Golgi apparatus may also play similar roles in the regulation of RTK signaling and trafficking, as reported for VEGFR1 and VEGFR2 (Mittar et al. 2009; Manickam et al. 2011).

Microvesicles for Extracellular Propagation of RTK Signals

Accumulating studies report that cells, in particular tumor cells, can release several types of microvesicles into the environment (reviewed in Lee et al. 2011). These vesicles are either of endosomal origins (produced on fusion of multivesicular endosomes with the plasma membrane as so-called exosomes [Lakkaraju and Rodriguez-Boulan 2008; Simons and Raposo 2009]) or can be produced by other mechanisms. They function as a means of communication among cells by exchanging proteins, lipids, and nucleic acids, as well as contributing to pathogen spread and immune surveillance. Microvesicles were shown to contain, among others, RTKs and their ligands (Lee et al. 2011). In human glioma cells, microvesicles termed oncosomes carried an oncogenic form of EGFR (EGFRvIII), which was incorporated into the membrane of recipient cells, contributing to their transformation and tumor spread (Al-Nedawi et al. 2008; Skog et al. 2008). Still, the exact mechanisms and the prevalence of this phenomenon need to be further investigated.

CONCLUDING REMARKS

As evident from this overview, the control of RTK signaling by intracellular trafficking is complex and occurs at multiple levels. Abundant evidence indicates that this regulation is of physiological importance in vivo, both during development and in adult tissues. Importantly, trafficking of RTKs can be modulated by non-RTK signaling pathways, providing a mechanism for pathway cross talk and coordinated regulation of cellular responses to various stimuli. Moreover, aberrant RTK trafficking can be oncogenic (Lanzetti and Di Fiore 2008; Mosesson et al. 2008; Abella and Park 2009), indicating that future therapeutic approaches could be directed to correct such defects.

Finally, as argued here, regulation of signaling by trafficking is important but the reverse is also true. RTK signaling affects the trafficking machinery, for example, to drive ligand-induced endocytosis and then intracellular sorting of RTKs, although the underlying molecular mechanisms are still poorly known. Much remains to be discovered on how activated RTKs may modify endocytic proteins, changing their activity or localization. Recent data indicate that different signaling pathways exert feedback control over the morphology and function of endosomal compartments (Collinet et al. 2010) but the knowledge of detailed mechanisms is missing. As currently the signaling and trafficking fields are integrated more than ever before, we are bound to face exciting progress and discover further principles of coordinated regulation between signal transduction and membrane transport of RTKs and other receptor types.

ACKNOWLEDGMENTS

I thank Dr. Anna Hupalowska for preparing the figures and the members of my group for a critical reading of the manuscript. Research in my laboratory is supported by the National Science Centre (MAESTRO grant 2011/02/A/NZ3/00149 and grant No. N301 296437), by a grant from Switzerland through the Swiss contribution to the enlarged European Union (Polish-Swiss Research Programme project PSPB-094/2010), and by the European Union FP7 grant FishMed GA No. 316125.

REFERENCES

Abella JV, Park M. 2009. Breakdown of endocytosis in the oncogenic activation of receptor tyrosine kinases. *Am J Physiol Endocrinol Metab* **296:** E973–E984.

Abella JV, Peschard P, Naujokas MA, Lin T, Saucier C, Urbe S, Park M. 2005. Met/Hepatocyte growth factor receptor ubiquitination suppresses transformation and is required for Hrs phosphorylation. *Mol Cell Biol* **25:** 9632–9645.

Abella JV, Parachoniak CA, Sangwan V, Park M. 2010a. Dorsal ruffle microdomains potentiate Met receptor tyrosine kinase signaling and down-regulation. *J Biol Chem* **285:** 24956–24967.

Abella JV, Vaillancourt R, Frigault MM, Ponzo MG, Zuo D, Sangwan V, Larose L, Park M. 2010b. The Gab1 scaffold regulates RTK-dependent dorsal ruffle formation through the adaptor Nck. *J Cell Sci* **123:** 1306–1319.

Acconcia F, Sigismund S, Polo S. 2009. Ubiquitin in trafficking: The network at work. *Exp Cell Res* **315:** 1610–1618.

Al-Nedawi K, Meehan B, Micallef J, Lhotak V, May L, Guha A, Rak J. 2008. Intercellular transfer of the oncogenic receptor EGFRvIII by microvesicles derived from tumour cells. *Nat Cell Biol* **10:** 619–624.

Assaker G, Ramel D, Wculek SK, Gonzalez-Gaitan M, Emery G. 2010. Spatial restriction of receptor tyrosine kinase activity through a polarized endocytic cycle controls border cell migration. *Proc Natl Acad Sci* **107:** 22558–22563.

Babst M. 2011. MVB vesicle formation: ESCRT-dependent, ESCRT-independent and everything in between. *Curr Opin Cell Biol* **23:** 452–457.

Baldys A, Gooz M, Morinelli TA, Lee MH, Raymond JR Jr, Luttrell LM, Raymond JR Sr. 2009. Essential role of c-Cbl in amphiregulin-induced recycling and signaling of the endogenous epidermal growth factor receptor. *Biochemistry* **48:** 1462–1473.

Ballmer-Hofer K, Andersson AE, Ratcliffe LE, Berger P. 2011. Neuropilin-1 promotes VEGFR-2 trafficking through Rab11 vesicles thereby specifying signal output. *Blood* **118:** 816–826.

Bao J, Alroy I, Waterman H, Schejter ED, Brodie C, Gruenberg J, Yarden Y. 2000. Threonine phosphorylation diverts internalized epidermal growth factor receptors from a degradative pathway to the recycling endosome. *J Biol Chem* **275:** 26178–26186.

Beguinot L, Hanover JA, Ito S, Richert ND, Willingham MC, Pastan I. 1985. Phorbol esters induce transient internalization without degradation of unoccupied epidermal growth factor receptors. *Proc Natl Acad Sci* **82:** 2774–2778.

Belleudi F, Leone L, Nobili V, Raffa S, Francescangeli F, Maggio M, Morrone S, Marchese C, Torrisi MR. 2007. Keratinocyte growth factor receptor ligands target the receptor to different intracellular pathways. *Traffic* **8:** 1854–1872.

Belleudi F, Scrofani C, Torrisi MR, Mancini P. 2011. Polarized endocytosis of the keratinocyte growth factor receptor in migrating cells: Role of SRC-signaling and cortactin. *PLoS ONE* **6:** e29159.

Bokel C, Schwabedissen A, Entchev E, Renaud O, Gonzalez-Gaitan M. 2006. Sara endosomes and the maintenance of Dpp signaling levels across mitosis. *Science* **314:** 1135–1139.

Brankatschk B, Wichert SP, Johnson SD, Schaad O, Rossner MJ, Gruenberg J. 2012. Regulation of the EGF transcriptional response by endocytic sorting. *Sci Signal* **5:** ra21.

Bremm A, Walch A, Fuchs M, Mages J, Duyster J, Keller G, Hermannstadter C, Becker KF, Rauser S, Langer R, et al. 2008. Enhanced activation of epidermal growth factor receptor caused by tumor-derived E-cadherin mutations. *Cancer Res* **68:** 707–714.

Bryant DM, Wylie FG, Stow JL. 2005. Regulation of endocytosis, nuclear translocation, and signaling of fibroblast growth factor receptor 1 by E-cadherin. *Mol Biol Cell* **16:** 14–23.

Burke P, Schooler K, Wiley HS. 2001. Regulation of epidermal growth factor receptor signaling by endocytosis and intracellular trafficking. *Mol Biol Cell* **12:** 1897–1910.

Caswell PT, Chan M, Lindsay AJ, McCaffrey MW, Boettiger D, Norman JC. 2008. Rab-coupling protein coordinates recycling of α5β1 integrin and EGFR1 to promote cell migration in 3D microenvironments. *J Cell Biol* **183:** 143–155.

Cavalli V, Vilbois F, Corti M, Marcote MJ, Tamura K, Karin M, Arkinstall S, Gruenberg J. 2001. The stress-induced MAP kinase 38 regulates endocytic trafficking via the GDI:Rab5 complex. *Mol Cell* **7:** 421–432.

Champion KJ, Guinea M, Dammai V, Hsu T. 2008. Endothelial function of von Hippel-Lindau tumor suppressor gene: Control of fibroblast growth factor receptor signaling. *Cancer Res* **68:** 4649–4657.

Chen ZY, Ieraci A, Tanowitz M, Lee FS. 2005. A novel endocytic recycling signal distinguishes biological responses of Trk neurotrophin receptors. *Mol Biol Cell* **16:** 5761–5772.

Choudhary C, Olsen JV, Brandts C, Cox J, Reddy PN, Bohmer FD, Gerke V, Schmidt-Arras DE, Berdel WE, Muller-Tidow C, et al. 2009. Mislocalized activation of oncogenic RTKs switches downstream signaling outcomes. *Mol Cell* **36:** 326–339.

Collinet C, Stoter M, Bradshaw CR, Samusik N, Rink JC, Kenski D, Habermann B, Buchholz F, Henschel R, Mueller MS, et al. 2010. Systems survey of endocytosis by multiparametric image analysis. *Nature* **464:** 243–249.

Conner SD, Schmid SL. 2003. Regulated portals of entry into the cell. *Nature* **422:** 37–44.

Cosker KE, Courchesne SL, Segal RA. 2008. Action in the axon: Generation and transport of signaling endosomes. *Curr Opin Neurobiol* **18:** 270–275.

Decker SJ. 1990. Epidermal growth factor and transforming growth factor-α induce differential processing of the epidermal growth factor receptor. *Biochem Biophys Res Commun* **166:** 615–621.

De Donatis A, Comito G, Buricchi F, Vinci MC, Parenti A, Caselli A, Camici G, Manao G, Ramponi G, Cirri P. 2008. Proliferation versus migration in platelet-derived growth factor signaling: The key role of endocytosis. *J Biol Chem* **283:** 19948–19956.

Deinhardt K, Salinas S, Verastegui C, Watson R, Worth D, Hanrahan S, Bucci C, Schiavo G. 2006. Rab5 and Rab7 control endocytic sorting along the axonal retrograde transport pathway. *Neuron* **52:** 293–305.

Delcroix JD, Valletta JS, Wu C, Hunt SJ, Kowal AS, Mobley WC. 2003. NGF signaling in sensory neurons: Evidence that early endosomes carry NGF retrograde signals. *Neuron* **39:** 69–84.

Delva E, Kowalczyk AP. 2009. Regulation of cadherin trafficking. *Traffic* **10:** 259–267.

Di Guglielmo GM, Baass PC, Ou WJ, Posner BI, Bergeron JJ. 1994. Compartmentalization of SHC, GRB2 and mSOS, and hyperphosphorylation of Raf-1 by EGF but not insulin in liver parenchyma. *EMBO J* **13:** 4269–4277.

Di Guglielmo GM, Le Roy C, Goodfellow AF, Wrana JL. 2003. Distinct endocytic pathways regulate TGF-β receptor signalling and turnover. *Nat Cell Biol* **5:** 410–421.

Disanza A, Frittoli E, Palamidessi A, Scita G. 2009. Endocytosis and spatial restriction of cell signaling. *Mol Oncol* **3:** 280–296.

Ebner R, Derynck R. 1991. Epidermal growth factor and transforming growth factor-α: Differential intracellular routing and processing of ligand-receptor complexes. *Cell Regul* **2:** 599–612.

Eden ER, White IJ, Tsapara A, Futter CE. 2010. Membrane contacts between endosomes and ER provide sites for PTP1B-epidermal growth factor receptor interaction. *Nat Cell Biol* **12:** 267–272.

Eden ER, Huang F, Sorkin A, Futter CE. 2012. The role of EGF receptor ubiquitination in regulating its intracellular traffic. *Traffic* **13:** 329–337.

Elfenbein A, Lanahan A, Zhou TX, Yamasaki A, Tkachenko E, Matsuda M, Simons M. 2012. Syndecan 4 regulates FGFR1 signaling in endothelial cells by directing macropinocytosis. *Sci Signal* **5:** ra36.

Falguieres T, Luyet PP, Gruenberg J. 2009. Molecular assemblies and membrane domains in multivesicular endosome dynamics. *Exp Cell Res* **315:** 1567–1573.

Francavilla C, Cattaneo P, Berezin V, Bock E, Ami D, de Marco A, Christofori G, Cavallaro U. 2009. The binding of NCAM to FGFR1 induces a specific cellular response mediated by receptor trafficking. *J Cell Biol* **187:** 1101–1116.

French AR, Sudlow GP, Wiley HS, Lauffenburger DA. 1994. Postendocytic trafficking of epidermal growth factor-receptor complexes is mediated through saturable and specific endosomal interactions. *J Biol Chem* **269:** 15749–15755.

French AR, Tadaki DK, Niyogi SK, Lauffenburger DA. 1995. Intracellular trafficking of epidermal growth factor family ligands is directly influenced by the pH sensitivity of the receptor/ligand interaction. *J Biol Chem* **270:** 4334–4340.

Fry WH, Simion C, Sweeney C, Carraway KL 3rd. 2011. Quantity control of the ErbB3 receptor tyrosine kinase at the endoplasmic reticulum. *Mol Cell Biol* **31:** 3009–3018.

Goldoni S, Iozzo RV. 2008. Tumor microenvironment: Modulation by decorin and related molecules harboring leucine-rich tandem motifs. *Int J Cancer* **123:** 2473–2479.

Goldoni S, Humphries A, Nystrom A, Sattar S, Owens RT, McQuillan DJ, Ireton K, Iozzo RV. 2009. Decorin is a novel antagonistic ligand of the Met receptor. *J Cell Biol* **185:** 743–754.

Grecco HE, Schmick M, Bastiaens PI. 2011. Signaling from the living plasma membrane. *Cell* **144:** 897–909.

Green JL, Kuntz SG, Sternberg PW. 2008. Ror receptor tyrosine kinases: Orphans no more. *Trends Cell Biol* **18:** 536–544.

Grimes ML, Zhou J, Beattie EC, Yuen EC, Hall DE, Valletta JS, Topp KS, LaVail JH, Bunnett NW, Mobley WC. 1996. Endocytosis of activated TrkA: Evidence that nerve growth factor induces formation of signaling endosomes. *J Neurosci* **16:** 7950–7964.

Cite this article as *Cold Spring Harb Perspect Biol* doi: 10.1101/cshperspect.a009035

Gur G, Rubin C, Katz M, Amit I, Citri A, Nilsson J, Amariglio N, Henriksson R, Rechavi G, Hedman H, et al. LRIG1 restricts growth factor signaling by enhancing receptor ubiquitylation and degradation. *EMBO J* **23:** 3270–3281.

Harrington AW, St Hillaire C, Zweifel LS, Glebova NO, Philippidou P, Halegoua S, Ginty DD. 2011. Recruitment of actin modifiers to TrkA endosomes governs retrograde NGF signaling and survival. *Cell* **146:** 421–434.

Harris RC, Chung E, Coffey RJ. 2003. EGF receptor ligands. *Exp Cell Res* **284:** 2–13.

Haugsten EM, Sorensen V, Brech A, Olsnes S, Wesche J. 2005. Different intracellular trafficking of FGF1 endocytosed by the four homologous FGF receptors. *J Cell Sci* **118:** 3869–3881.

Heldin CH, Wasteson A, Westermark B. 1982. Interaction of platelet-derived growth factor with its fibroblast receptor. Demonstration of ligand degradation and receptor modulation. *J Biol Chem* **257:** 4216–4221.

Hellberg C, Schmees C, Karlsson S, Ahgren A, Heldin CH. 2009. Activation of protein kinase C α is necessary for sorting the PDGF β-receptor to Rab4a-dependent recycling. *Mol Biol Cell* **20:** 2856–2863.

Hempstead BL, Martin-Zanca D, Kaplan DR, Parada LF, Chao MV. 1991. High-affinity NGF binding requires co-expression of the trk proto-oncogene and the low-affinity NGF receptor. *Nature* **350:** 678–683.

Henne WM, Buchkovich NJ, Emr SD. 2011. The ESCRT pathway. *Dev Cell* **21:** 77–91.

Howe CL. 2005. Modeling the signaling endosome hypothesis: Why a drive to the nucleus is better than a (random) walk. *Theor Biol Med Model* **2:** 43.

Howe CL, Mobley WC. 2005. Long-distance retrograde neurotrophic signaling. *Curr Opin Neurobiol* **15:** 40–48.

Howe CL, Valletta JS, Rusnak AS, Mobley WC. 2001. NGF signaling from clathrin-coated vesicles: Evidence that signaling endosomes serve as a platform for the Ras-MAPK pathway. *Neuron* **32:** 801–814.

Howes MT, Mayor S, Parton RG. 2010. Molecules, mechanisms, and cellular roles of clathrin-independent endocytosis. *Curr Opin Cell Biol* **22:** 519–527.

Hsu T, Adereth Y, Kose N, Dammai V. 2006. Endocytic function of von Hippel-Lindau tumor suppressor protein regulates surface localization of fibroblast growth factor receptor 1 and cell motility. *J Biol Chem* **281:** 12069–12080.

Janssen-Heininger YM, Mossman BT, Heintz NH, Forman HJ, Kalyanaraman B, Finkel T, Stamler JS, Rhee SG, van der Vliet A. 2008. Redox-based regulation of signal transduction: Principles, pitfalls, and promises. *Free Radic Biol Med* **45:** 1–17.

Jeffers M, Schmidt L, Nakaigawa N, Webb CP, Weirich G, Kishida T, Zbar B, Vande Woude GF. 1997. Activating mutations for the met tyrosine kinase receptor in human cancer. *Proc Natl Acad Sci* **94:** 11445–11450.

Jekely G, Sung HH, Luque CM, Rorth P. 2005. Regulators of endocytosis maintain localized receptor tyrosine kinase signaling in guided migration. *Dev Cell* **9:** 197–207.

Jiang X, Sorkin A. 2003. Epidermal growth factor receptor internalization through clathrin-coated pits requires Cbl RING finger and proline-rich domains but not receptor polyubiquitylation. *Traffic* **4:** 529–543.

Jing S, Wen D, Yu Y, Holst PL, Luo Y, Fang M, Tamir R, Antonio L, Hu Z, Cupples R, et al. 1996. GDNF-induced activation of the ret protein tyrosine kinase is mediated by GDNFR-α, a novel receptor for GDNF. *Cell* **85:** 1113–1124.

Joffre C, Barrow R, Menard L, Calleja V, Hart IR, Kermorgant S. 2011. A direct role for Met endocytosis in tumorigenesis. *Nat Cell Biol* **13:** 827–837.

Kametaka S, Mattera R, Bonifacino JS. 2005. Epidermal growth factor-dependent phosphorylation of the GGA3 adaptor protein regulates its recruitment to membranes. *Mol Cell Biol* **25:** 7988–8000.

Kaplan DR, Hempstead BL, Martin-Zanca D, Chao MV, Parada LF. 1991. The trk proto-oncogene product: A signal transducing receptor for nerve growth factor. *Science* **252:** 554–558.

Karlsson S, Kowanetz K, Sandin A, Persson C, Ostman A, Heldin CH, Hellberg C. 2006. Loss of T-cell protein tyrosine phosphatase induces recycling of the platelet-derived growth factor (PDGF) β-receptor but not the PDGF α-receptor. *Mol Biol Cell* **17:** 4846–4855.

Kazazic M, Roepstorff K, Johannessen LE, Pedersen NM, van Deurs B, Stang E, Madshus IH. 2006. EGF-induced activation of the EGF receptor does not trigger mobilization of caveolae. *Traffic* **7:** 1518–1527.

Kermorgant S, Parker PJ. 2008. Receptor trafficking controls weak signal delivery: A strategy used by c-Met for STAT3 nuclear accumulation. *J Cell Biol* **182:** 855–863.

Kerr MC, Teasdale RD. 2009. Defining macropinocytosis. *Traffic* **10:** 364–371.

Kim N, Stiegler AL, Cameron TO, Hallock PT, Gomez AM, Huang JH, Hubbard SR, Dustin ML, Burden SJ. 2008. Lrp4 is a receptor for Agrin and forms a complex with MuSK. *Cell* **135:** 334–342.

Klein R, Jing SQ, Nanduri V, O'Rourke E, Barbacid M. 1991. The trk proto-oncogene encodes a receptor for nerve growth factor. *Cell* **65:** 189–197.

Laederich MB, Funes-Duran M, Yen L, Ingalla E, Wu X, Carraway KL 3rd, Sweeney C. 2004. The leucine-rich repeat protein LRIG1 is a negative regulator of ErbB family receptor tyrosine kinases. *J Biol Chem* **279:** 47050–47056.

Lakkaraju A, Rodriguez-Boulan E. 2008. Itinerant exosomes: Emerging roles in cell and tissue polarity. *Trends Cell Biol* **18:** 199–209.

Lampugnani MG, Orsenigo F, Gagliani MC, Tacchetti C, Dejana E. 2006. Vascular endothelial cadherin controls VEGFR-2 internalization and signaling from intracellular compartments. *J Cell Biol* **174:** 593–604.

Lanzetti L, Di Fiore PP. 2008. Endocytosis and cancer: An "insider" network with dangerous liaisons. *Traffic* **9:** 2011–2021.

Lee TH, D'Asti E, Magnus N, Al-Nedawi K, Meehan B, Rak J. 2011. Microvesicles as mediators of intercellular communication in cancer—The emerging science of cellular "debris." *Semin Immunopathol* **33:** 455–467.

Lemmon MA, Schlessinger J. 2010. Cell signaling by receptor tyrosine kinases. *Cell* **141:** 1117–1134.

Lenferink AE, Pinkas-Kramarski R, van de Poll ML, van Vugt MJ, Klapper LN, Tzahar E, Waterman H, Sela M, van Zoelen EJ, Yarden Y. 1998. Differential endocytic routing of homo- and hetero-dimeric ErbB tyrosine kinases confers signaling superiority to receptor heterodimers. *EMBO J* **17**: 3385–3397.

Le Roy C, Wrana JL. 2005. Clathrin- and non-clathrin-mediated endocytic regulation of cell signalling. *Nat Rev Mol Cell Biol* **6**: 112–126.

Lin CR, Chen WS, Lazar CS, Carpenter CD, Gill GN, Evans RM, Rosenfeld MG. 1986. Protein kinase C phosphorylation at Thr 654 of the unoccupied EGF receptor and EGF binding regulate functional receptor loss by independent mechanisms. *Cell* **44**: 839–848.

Lin DC, Quevedo C, Brewer NE, Bell A, Testa JR, Grimes ML, Miller FD, Kaplan DR. 2006. APPL1 associates with TrkA and GIPC1 and is required for nerve growth factor-mediated signal transduction. *Mol Cell Biol* **26**: 8928–8941.

Lin HH, Li X, Chen JL, Sun X, Cooper FN, Chen YR, Zhang W, Chung Y, Li A, Cheng CT, et al. 2012. Identification of an AAA ATPase VPS4B-dependent pathway that modulates epidermal growth factor receptor abundance and signaling during hypoxia. *Mol Cell Biol* **32**: 1124–1138.

Mace G, Miaczynska M, Zerial M, Nebreda AR. 2005. Phosphorylation of EEA1 by 38 MAP kinase regulates mu opioid receptor endocytosis. *EMBO J* **24**: 3235–3246.

Manickam V, Tiwari A, Jung JJ, Bhattacharya R, Goel A, Mukhopadhyay D, Choudhury A. 2011. Regulation of vascular endothelial growth factor receptor 2 trafficking and angiogenesis by Golgi localized t-SNARE syntaxin 6. *Blood* **117**: 1425–1435.

Mardakheh FK, Yekezare M, Machesky LM, Heath JK. 2009. Spred2 interaction with the late endosomal protein NBR1 down-regulates fibroblast growth factor receptor signaling. *J Cell Biol* **187**: 265–277.

Mayor S, Pagano RE. 2007. Pathways of clathrin-independent endocytosis. *Nat Rev Mol Cell Biol* **8**: 603–612.

McCaffrey G, Welker J, Scott J, der Salm L, Grimes ML. 2009. High-resolution fractionation of signaling endosomes containing different receptors. *Traffic* **10**: 938–950.

McMahon HT, Boucrot E. 2011. Molecular mechanism and physiological functions of clathrin-mediated endocytosis. *Nat Rev Mol Cell Biol* **12**: 517–533.

Medts T, de Diesbach P, Cominelli A, N'Kuli F, Tyteca D, Courtoy PJ. 2010. Acute ligand-independent Src activation mimics low EGF-induced EGFR surface signalling and redistribution into recycling endosomes. *Exp Cell Res* **316**: 3239–3253.

Mellman I, Nelson WJ. 2008. Coordinated protein sorting, targeting and distribution in polarized cells. *Nat Rev Mol Cell Biol* **9**: 833–845.

Miaczynska M, Bar-Sagi D. 2010. Signaling endosomes: Seeing is believing. *Curr Opin Cell Biol* **22**: 535–540.

Miaczynska M, Christoforidis S, Giner A, Shevchenko A, Uttenweiler-Joseph S, Habermann B, Wilm M, Parton RG, Zerial M. 2004a. APPL proteins link Rab5 to nuclear signal transduction via an endosomal compartment. *Cell* **116**: 445–456.

Miaczynska M, Pelkmans L, Zerial M. 2004b. Not just a sink: Endosomes in control of signal transduction. *Curr Opin Cell Biol* **16**: 400–406.

Mittar S, Ulyatt C, Howell GJ, Bruns AF, Zachary I, Walker JH, Ponnambalam S. 2009. VEGFR1 receptor tyrosine kinase localization to the Golgi apparatus is calcium-dependent. *Exp Cell Res* **315**: 877–889.

Mosesson Y, Mills GB, Yarden Y. 2008. Derailed endocytosis: An emerging feature of cancer. *Nat Rev Cancer* **8**: 835–850.

Mukherjee S, Tessema M, Wandinger-Ness A. 2006. Vesicular trafficking of tyrosine kinase receptors and associated proteins in the regulation of signaling and vascular function. *Circ Res* **98**: 743–756.

Nada S, Hondo A, Kasai A, Koike M, Saito K, Uchiyama Y, Okada M. 2009. The novel lipid raft adaptor 18 controls endosome dynamics by anchoring the MEK-ERK pathway to late endosomes. *EMBO J* **28**: 477–489.

Naka D, Shimomura T, Yoshiyama Y, Sato M, Ishii T, Hara H. 1993. Internalization and degradation of hepatocyte growth factor in hepatocytes with down-regulation of the receptor/c-Met. *FEBS Lett* **329**: 147–152.

Nakayama M, Nakayama A, van Lessen M, Yamamoto H, Hoffmann S, Drexler HC, Itoh N, Hirose T, Breier G, Vestweber D, et al. 2013. Spatial regulation of VEGF receptor endocytosis in angiogenesis. *Nat Cell Biol* **15**: 249–260.

Norambuena A, Metz C, Jung JE, Silva A, Otero C, Cancino J, Retamal C, Valenzuela JC, Soza A, Gonzalez A. 2010. Phosphatidic acid induces ligand-independent epidermal growth factor receptor endocytic traffic through PDE4 activation. *Mol Biol Cell* **21**: 2916–2929.

Oakley FD, Abbott D, Li Q, Engelhardt JF. 2009. Signaling components of redox active endosomes: The redoxosomes. *Antioxid Redox Signal* **11**: 1313–1333.

Orian-Rousseau V, Ponta H. 2008. Adhesion proteins meet receptors: A common theme? *Adv Cancer Res* **101**: 63–92.

Orth JD, Krueger EW, Weller SG, McNiven MA. 2006. A novel endocytic mechanism of epidermal growth factor receptor sequestration and internalization. *Cancer Res* **66**: 3603–3610.

Palamidessi A, Frittoli E, Garre M, Faretta M, Mione M, Testa I, Diaspro A, Lanzetti L, Scita G, Di Fiore PP. 2008. Endocytic trafficking of Rac is required for the spatial restriction of signaling in cell migration. *Cell* **134**: 135–147.

Parachoniak CA, Luo Y, Abella JV, Keen JH, Park M. 2011. GGA3 functions as a switch to promote Met receptor recycling, essential for sustained ERK and cell migration. *Dev Cell* **20**: 751–763.

Pasquale EB. 2008. Eph-ephrin bidirectional signaling in physiology and disease. *Cell* **133**: 38–52.

Pelkmans L, Fava E, Grabner H, Hannus M, Habermann B, Krausz E, Zerial M. 2005. Genome-wide analysis of human kinases in clathrin- and caveolae/raft-mediated endocytosis. *Nature* **436**: 78–86.

Peradziryi H, Kaplan NA, Podleschny M, Liu X, Wehner P, Borchers A, Tolwinski NS. 2011. PTK7/Otk interacts with Wnts and inhibits canonical Wnt signalling. *EMBO J* **30**: 3729–3740.

Perret E, Lakkaraju A, Deborde S, Schreiner R, Rodriguez-Boulan E. 2005. Evolving endosomes: How many varieties and why? *Curr Opin Cell Biol* **17**: 423–434.

Philippidou P, Valdez G, Akmentin W, Bowers WJ, Federoff HJ, Halegoua S. 2011. Trk retrograde signaling requires persistent, Pincher-directed endosomes. *Proc Natl Acad Sci* **108**: 852–857.

Pitulescu ME, Adams RH. 2010. Eph/ephrin molecules— A hub for signaling and endocytosis. *Genes Dev* **24**: 2480–2492.

Platta HW, Stenmark H. 2011. Endocytosis and signaling. *Curr Opin Cell Biol* **23**: 393–403.

Polo S, Di Fiore PP. 2006. Endocytosis conducts the cell signaling orchestra. *Cell* **124**: 897–900.

Puertollano R, Bonifacino JS. 2004. Interactions of GGA3 with the ubiquitin sorting machinery. *Nat Cell Biol* **6**: 244–251.

Raiborg C, Stenmark H. 2009. The ESCRT machinery in endosomal sorting of ubiquitylated membrane proteins. *Nature* **458**: 445–452.

Richardson DS, Rodrigues DM, Hyndman BD, Crupi MJ, Nicolescu AC, Mulligan LM. 2012. Alternative splicing results in RET isoforms with distinct trafficking properties. *Mol Biol Cell* **23**: 3838–3850.

Roepstorff K, Grandal MV, Henriksen L, Knudsen SL, Lerdrup M, Grovdal L, Willumsen BM, van Deurs B. 2009. Differential effects of EGFR ligands on endocytic sorting of the receptor. *Traffic* **10**: 1115–1127.

Sadowski L, Pilecka I, Miaczynska M. 2009. Signaling from endosomes: Location makes a difference. *Exp Cell Res* **315**: 1601–1609.

Sadowski L, Jastrzebski K, Kalaidzidis Y, Heldin CH, Hellberg C, Miaczynska M. 2013. Dynamin inhibitors impair endocytosis and mitogenic signaling of PDGF. *Traffic* **14**: 725–736.

Salazar G, Gonzalez A. 2002. Novel mechanism for regulation of epidermal growth factor receptor endocytosis revealed by protein kinase A inhibition. *Mol Biol Cell* **13**: 1677–1693.

Sancak Y, Bar-Peled L, Zoncu R, Markhard AL, Nada S, Sabatini DM. 2010. Ragulator-Rag complex targets mTORC1 to the lysosomal surface and is necessary for its activation by amino acids. *Cell* **141**: 290–303.

Sandilands E, Akbarzadeh S, Vecchione A, McEwan DG, Frame MC, Heath JK. 2007. Src kinase modulates the activation, transport and signalling dynamics of fibroblast growth factor receptors. *EMBO Rep* **8**: 1162–1169.

Sangwan V, Abella J, Lai A, Bertos N, Stuible M, Tremblay ML, Park M. 2011. Protein-tyrosine phosphatase 1B modulates early endosome fusion and trafficking of Met and epidermal growth factor receptors. *J Biol Chem* **286**: 45000–45013.

Schenck A, Goto-Silva L, Collinet C, Rhinn M, Giner A, Habermann B, Brand M, Zerial M. 2008. The endosomal protein Appl1 mediates Akt substrate specificity and cell survival in vertebrate development. *Cell* **133**: 486–497.

Schiefermeier N, Teis D, Huber LA. 2011. Endosomal signaling and cell migration. *Curr Opin Cell Biol* **23**: 615–620.

Schlessinger J, Shechter Y, Willingham MC, Pastan I. 1978. Direct visualization of binding, aggregation, and internalization of insulin and epidermal growth factor on living fibroblastic cells. *Proc Natl Acad Sci* **75**: 2659–2663.

Schmees C, Villasenor R, Zheng W, Ma H, Zerial M, Heldin CH, Hellberg C. 2012. Macropinocytosis of the PDGF β-receptor promotes fibroblast transformation by H-RasG12V. *Mol Biol Cell* **23**: 2571–2582.

Schmidt L, Duh FM, Chen F, Kishida T, Glenn G, Choyke P, Scherer SW, Zhuang Z, Lubensky I, Dean M, et al. 1997. Germline and somatic mutations in the tyrosine kinase domain of the MET proto-oncogene in papillary renal carcinomas. *Nat Genet* **16**: 68–73.

Schmidt-Arras DE, Bohmer A, Markova B, Choudhary C, Serve H, Bohmer FD. 2005. Tyrosine phosphorylation regulates maturation of receptor tyrosine kinases. *Mol Cell Biol* **25**: 3690–3703.

Scita G, Di Fiore PP. 2010. The endocytic matrix. *Nature* **463**: 464–473.

Shattuck DL, Miller JK, Laederich M, Funes M, Petersen H, Carraway KL III, Sweeney C. 2007. LRIG1 is a novel negative regulator of the Met receptor and opposes Met and Her2 synergy. *Mol Cell Biol* **27**: 1934–1946.

Sigismund S, Woelk T, Puri C, Maspero E, Tacchetti C, Transidico P, Di Fiore PP, Polo S. 2005. Clathrin-independent endocytosis of ubiquitinated cargos. *Proc Natl Acad Sci* **102**: 2760–2765.

Sigismund S, Argenzio E, Tosoni D, Cavallaro E, Polo S, Di Fiore PP. 2008. Clathrin-mediated internalization is essential for sustained EGFR signaling but dispensable for degradation. *Dev Cell* **15**: 209–219.

Sigismund S, Confalonieri S, Ciliberto A, Polo S, Scita G, Di Fiore PP. 2012. Endocytosis and signaling: Cell logistics shape the eukaryotic cell plan. *Physiol Rev* **92**: 273–366.

Simons M, Raposo G. 2009. Exosomes—Vesicular carriers for intercellular communication. *Curr Opin Cell Biol* **21**: 575–581.

Singhirunnusorn P, Ueno Y, Matsuo M, Suzuki S, Saiki I, Sakurai H. 2007. Transient suppression of ligand-mediated activation of epidermal growth factor receptor by tumor necrosis factor-α through the TAK1-p38 signaling pathway. *J Biol Chem* **282**: 12698–12706.

Skog J, Wurdinger T, van Rijn S, Meijer DH, Gainche L, Sena-Esteves M, Curry WT Jr, Carter BS, Krichevsky AM, Breakefield XO. 2008. Glioblastoma microvesicles transport RNA and proteins that promote tumour growth and provide diagnostic biomarkers. *Nat Cell Biol* **10**: 1470–1476.

Sorkin A, von Zastrow M. 2009. Endocytosis and signalling: Intertwining molecular networks. *Nat Rev Mol Cell Biol* **10**: 609–622.

Sousa LP, Lax I, Shen H, Ferguson SM, De Camilli P, Schlessinger J. 2012. Suppression of EGFR endocytosis by dynamin depletion reveals that EGFR signaling occurs primarily at the plasma membrane. *Proc Natl Acad Sci* **109**: 4419–4424.

St Bernard R, Zheng L, Liu W, Winer D, Asa SL, Ezzat S. 2005. Fibroblast growth factor receptors as molecular targets in thyroid carcinoma. *Endocrinology* **146**: 1145–1153.

Stern KA, Place TL, Lill NL. 2008. EGF and amphiregulin differentially regulate Cbl recruitment to endosomes and EGF receptor fate. *Biochem J* **410**: 585–594.

Stuible M, Tremblay ML. 2010. In control at the ER: PTP1B and the down-regulation of RTKs by dephosphorylation and endocytosis. *Trends Cell Biol* **20**: 672–679.

Suyama K, Shapiro I, Guttman M, Hazan RB. 2002. A signaling pathway leading to metastasis is controlled by N-cadherin and the FGF receptor. *Cancer Cell* **2**: 301–314.

Swaminathan G, Tsygankov AY. 2006. The Cbl family proteins: Ring leaders in regulation of cell signaling. *J Cell Physiol* **209**: 21–43.

Taelman VF, Dobrowolski R, Plouhinec JL, Fuentealba LC, Vorwald PP, Gumper I, Sabatini DD, De Robertis EM. 2010. Wnt signaling requires sequestration of glycogen synthase kinase 3 inside multivesicular endosomes. *Cell* **143**: 1136–1148.

Tall GG, Barbieri MA, Stahl PD, Horazdovsky BF. 2001. Ras-activated endocytosis is mediated by the Rab5 guanine nucleotide exchange activity of RIN1. *Dev Cell* **1**: 73–82.

Teis D, Wunderlich W, Huber LA. 2002. Localization of the MP1-MAPK scaffold complex to endosomes is mediated by 14 and required for signal transduction. *Dev Cell* **3**: 803–814.

Teis D, Taub N, Kurzbauer R, Hilber D, de Araujo ME, Erlacher M, Offterdinger M, Villunger A, Geley S, Bohn G, et al. 2006. p14–MP1-MEK1 signaling regulates endosomal traffic and cellular proliferation during tissue homeostasis. *J Cell Biol* **175**: 861–868.

Treanor JJ, Goodman L, de Sauvage F, Stone DM, Poulsen KT, Beck CD, Gray C, Armanini MP, Pollock RA, Hefti F, et al. 1996. Characterization of a multicomponent receptor for GDNF. *Nature* **382**: 80–83.

Valdez G, Philippidou P, Rosenbaum J, Akmentin W, Shao Y, Halegoua S. 2007. Trk-signaling endosomes are generated by Rac-dependent macroendocytosis. *Proc Natl Acad Sci* **104**: 12270–12275.

Varsano T, Dong MQ, Niesman I, Gacula H, Lou X, Ma T, Testa JR, Yates JR 3rd, Farquhar MG. 2006. GIPC is recruited by APPL to peripheral TrkA endosomes and regulates TrkA trafficking and signaling. *Mol Cell Biol* **26**: 8942–8952.

Vergarajauregui S, San Miguel A, Puertollano R. 2006. Activation of 38 mitogen-activated protein kinase promotes epidermal growth factor receptor internalization. *Traffic* **7**: 686–698.

Vieira AV, Lamaze C, Schmid SL. 1996. Control of EGF receptor signaling by clathrin-mediated endocytosis. *Science* **274**: 2086–2089.

von Zastrow M, Sorkin A. 2007. Signaling on the endocytic pathway. *Curr Opin Cell Biol* **19**: 436–445.

Wang Y, Roche O, Yan MS, Finak G, Evans AJ, Metcalf JL, Hast BE, Hanna SC, Wondergem B, Furge KA, et al. 2009. Regulation of endocytosis via the oxygen-sensing pathway. *Nat Med* **15**: 319–324.

Waterman H, Sabanai I, Geiger B, Yarden Y. 1998. Alternative intracellular routing of ErbB receptors may determine signaling potency. *J Biol Chem* **273**: 13819–13827.

Waterman H, Katz M, Rubin C, Shtiegman K, Lavi S, Elson A, Jovin T, Yarden Y. 2002. A mutant EGF-receptor defective in ubiquitylation and endocytosis unveils a role for Grb2 in negative signaling. *EMBO J* **21**: 303–313.

Wiley HS, Herbst JJ, Walsh BJ, Lauffenburger DA, Rosenfeld MG, Gill GN. 1991. The role of tyrosine kinase activity in endocytosis, compartmentation, and down-regulation of the epidermal growth factor receptor. *J Biol Chem* **266**: 11083–11094.

Winckler B, Mellman I. 2010. Trafficking guidance receptors. *Cold Spring Harb Perspect Biol* **2**: a001826.

Winograd-Katz SE, Levitzki A. 2006. Cisplatin induces PKB/Akt activation and p38(MAPK) phosphorylation of the EGF receptor. *Oncogene* **25**: 7381–7390.

Worthylake R, Opresko LK, Wiley HS. 1999. ErbB-2 amplification inhibits down-regulation and induces constitutive activation of both ErbB-2 and epidermal growth factor receptors. *J Biol Chem* **274**: 8865–8874.

Wu C, Ramirez A, Cui B, Ding J, Delcroix JD, Valletta JS, Liu JJ, Yang Y, Chu S, Mobley WC. 2007. A functional dynein-microtubule network is required for NGF signaling through the Rap1/MAPK pathway. *Traffic* **8**: 1503–1520.

Wu C, Cui B, He L, Chen L, Mobley WC. 2009. The coming of age of axonal neurotrophin signaling endosomes. *J Proteomics* **72**: 46–55.

Yamamoto H, Sakane H, Yamamoto H, Michiue T, Kikuchi A. 2008. Wnt3a and Dkk1 regulate distinct internalization pathways of LRP6 to tune the activation of β-catenin signaling. *Dev Cell* **15**: 37–48.

Zhang B, Luo S, Wang Q, Suzuki T, Xiong WC, Mei L. 2008. LRP4 serves as a coreceptor of agrin. *Neuron* **60**: 285–297.

Zhu JX, Goldoni S, Bix G, Owens RT, McQuillan DJ, Reed CC, Iozzo RV. 2005. Decorin evokes protracted internalization and degradation of the epidermal growth factor receptor via caveolar endocytosis. *J Biol Chem* **280**: 32468–32479.

Zoncu R, Perera RM, Balkin DM, Pirruccello M, Toomre D, De Camilli P. 2009. A phosphoinositide switch controls the maturation and signaling properties of APPL endosomes. *Cell* **136**: 1110–1121.

Zwang Y, Yarden Y. 2006. p38 MAP kinase mediates stress-induced internalization of EGFR: Implications for cancer chemotherapy. *EMBO J* **25**: 4195–4206.

Regulation of Receptor Tyrosine Kinase Ligand Processing

Colin Adrain[1] and Matthew Freeman[2]

MRC Laboratory of Molecular Biology, Francis Crick Avenue, Cambridge Biomedical Campus, Cambridge CB2 0QH, United Kingdom

Correspondence: matthew.freeman@path.ox.ac.uk

A primary mode of regulating receptor tyrosine kinase (RTK) signaling is to control access of ligand to its receptor. Many RTK ligands are synthesized as transmembrane proteins. Frequently, the active ligand must be released from the membrane by proteolysis before signaling can occur. Here, we discuss RTK ligand shedding and describe the proteases that catalyze it in flies and mammals. We focus principally on the control of EGF receptor ligand shedding, but also refer to ligands of other RTKs. Two prominent themes emerge. First, control by regulated trafficking and cellular compartmentalization of the proteases and their ligand substrates plays a key role in shedding. Second, many external signals converge on the shedding proteases and their control machinery. Proteases therefore act as regulatory hubs that integrate information that the cell receives and translate it into precise outgoing signals. The activation of signaling by proteases is therefore an essential element of the cellular communication machinery.

Cells must talk to one another. This principle applies throughout the tree of life: from unicellular bacteria, to the trillions of cells that coordinate to make a mammal. Communication between cells requires dedicated machinery, capable of relaying information across membranes. Transmembrane proteins are therefore essential for signaling. Understanding how this is regulated is paramount. In mammals, receptor tyrosine kinases (RTKs) and their ligands are important examples of such machinery (Schlessinger 2000), controlling many biological processes including development, immunity, tissue repair, and metabolic homeostasis (Ull-rich and Schlessinger 1990). They are transmembrane proteins with an extracellular ligand-binding motif and an intracellular kinase domain. As discussed in other chapters, a common mode of RTK activation involves receptor dimerization induced by ligand binding (Lemmon and Schlessinger 2010).

Regulated access of ligand to receptor, over distance and time, is key to controlling signaling. Ligands are frequently synthesized as transmembrane forms; when they remain membrane-tethered and cannot diffuse, the range over which they can operate is limited to adjacent cells (Massague and Pandiella 1993; Singh and

---

[1]Present address: Instituto Gulbenkian de Ciência, Oeiras, Portugal.

[2]Present address: Dunn School of Pathology, University of Oxford, South Parks Road, Oxford OX1 3RE, United Kingdom.

Cite this article as *Cold Spring Harb Perspect Biol* doi: 10.1101/cshperspect.a008995

Harris 2005). Other ligands are soluble secretory proteins. This enables paracrine and endocrine signaling—communication between nonadjacent cells. A more complex mode of signaling exploits the characteristics of both of the above. Ligand is synthesized as a transmembrane precursor, which is then shed from the cell surface by proteolysis. This adds an additional and stringent regulatory step to a signaling network (Massague and Pandiella 1993).

This chapter will focus on RTK ligand cleavage and its regulation. We shall highlight how shedding is often critical for signaling, and describe the protease families that catalyze ligand release in flies and mammals. An emergent theme is that regulated trafficking and compartmentalization of ligand and protease modulate signaling. Another theme will be the range of stimuli that impinge on shedding.

The epidermal growth factor receptor (EGFR) is an excellent model RTK to illustrate the regulation of ligand proteolysis because the requirement for ligand cleavage in signaling is well established, and the major physiological sheddases have been identified (Blobel 2005). Where warranted, physiological evidence for the role of ligand shedding in the regulation of other RTKs will also be discussed. Whereas we shall deal mostly with ADAM proteases ("a disintegrin and metalloprotease"), which represent the canonical mammalian RTK ligand shedding machinery, the rhomboid family of intramembrane proteases will also be discussed.

THE EGFR—A PARADIGM FOR RTK REGULATION BY LIGAND CLEAVAGE

EGFR signaling has many developmental and physiological roles in flies and mammals (Shilo 2003; Sibilia et al. 2007). EGFR ligands in both species have the same domain organization and topology (Fig. 1) (Schneider and Wolf 2009). They are usually type I transmembrane proteins with an amino-terminal extracellular domain (ectodomain); therein lies a conserved motif called the EGF domain that is responsible for receptor binding. The EGF domain occurs in all EGFR ligands, but also in other contexts (Davis 1990). For cleavage to occur, the trans-

membrane ligand precursor must be trafficked into the same membranous compartment as its shedding protease, allowing proteolytic activation into a secreted ligand. We will now compare and contrast ligand shedding in flies and mammals.

REGULATION OF EGFR SIGNALING IN *DROSOPHILA* BY LIGAND PROTEOLYSIS

Flies have a relatively simple EGFR pathway: they have a single receptor and only four ligands (Shilo, 2003). As in mammals, most of the ligands (Spitz, Gurken, and Keren) are synthesized as transmembrane precursors, whereas Vein, which resembles neuregulins, is soluble (Freeman 1998). Spitz, a TGF-α-like molecule, is the most important EGFR-activating ligand; the others play subsidiary or tissue-specific roles (for example, Brown et al. 2007). For all three membrane-tethered ligands, their activation requires cleavage by rhomboid proteases (Urban et al. 2002).

Rhomboids are integral membrane proteins that contain six or seven transmembrane domains (Fig. 2) (Freeman 2009). First identified in flies, genetic analysis showed that rhomboid-1 was involved in EGFR signaling but its precise role was obscure (Ruohola-Baker et al. 1993; Sturtevant et al. 1993; Freeman 1994). Genetic mosaic experiments indicated that it acted in the signal-emitting cell, rather than the EGFR-expressing cell, implying a role in signal generation (Wasserman et al. 2000). This enigma was resolved when rhomboids were shown to be serine proteases (Urban et al. 2001) controlled by a catalytic dyad (Lemberg et al. 2005). Rhomboids cleave their substrates within or close to the upper part of the transmembrane domain (Urban et al. 2003; Strisovsky et al. 2009), but this raises the question of water accessibility for a proteolytic reaction in a membrane environment. This was resolved by high-resolution crystal structures of bacterial rhomboids. The catalytic site lies in a hydrophilic depression, just inside the lipid bilayer (Wang et al. 2006; reviewed in Lieberman and Wolfe 2007). In addition to regulating EGFR signaling in flies, rhomboids exist in all kingdoms of life. They

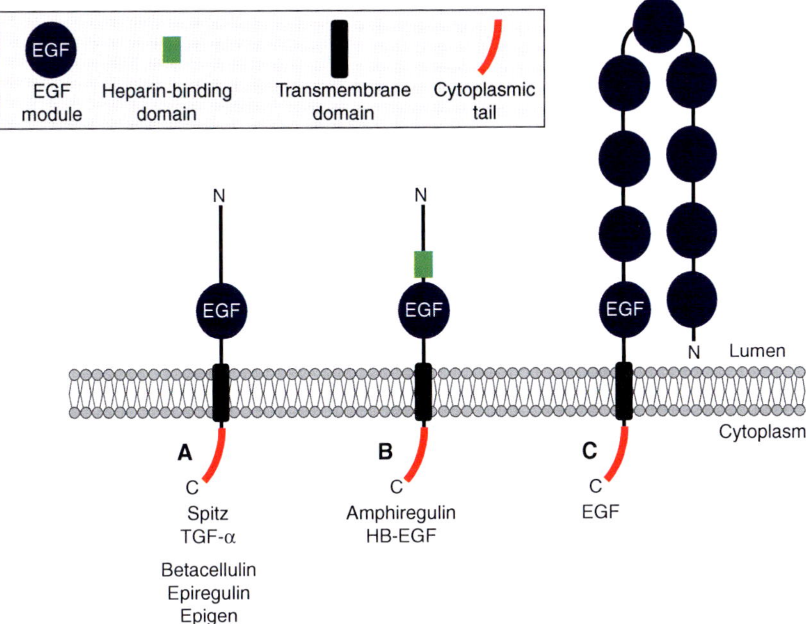

Figure 1. Topology of epidermal growth factor receptor (EGFR) ligands. EGFR ligands are type I transmembrane proteins with an extracellular (luminal) amino-terminus and a cytoplasmic carboxyl terminus. The domain structure of various EGFR ligands is indicated. (*A*) *Drosophila* Spitz, and the mammalian EGFR ligands TGF-α, Betacellulin, Epiregulin, and Epigen have a basic structure containing an amino-terminal prodomain and a bioactive EGF domain (indicated in blue). (*B*) Amphiregulin and HB-EGF contain a heparin binding motif amino terminal to the EGF domain (indicated in green); this facilitates binding to extracellular proteoglycans. Proteolytic cleavage occurs within the juxtamembrane domain between the EGF domain and the TMD; proteolytic removal of the amino-terminal prodomain also occurs (*A,B*). (*C*) Epidermal growth factor (EGF) contains additional EGF domains. The EGF domain closest to the membrane can activate the EGFR, whereas the remaining eight EGF domains cannot. The role of these is unclear, although they may play a role in regulating cell–cell adhesion. Cleavage liberates the bioactive EGF domain from the transmembrane precursor; depending on the tissue/context, the other EGF modules may either remain on the soluble molecule, or are cleaved off.

regulate processes as diverse as quorum sensing in bacteria, host invasion in *Plasmodium*, and mitochondrial homeostasis in eukaryotes (Urban and Dickey 2011), but their roles in mammals are less well understood.

Regulation of *Drosophila* EGF Ligand Cleavage by Trafficking

Transcription is a primary regulator of rhomboid in flies (Bier et al. 1990). Superimposed on

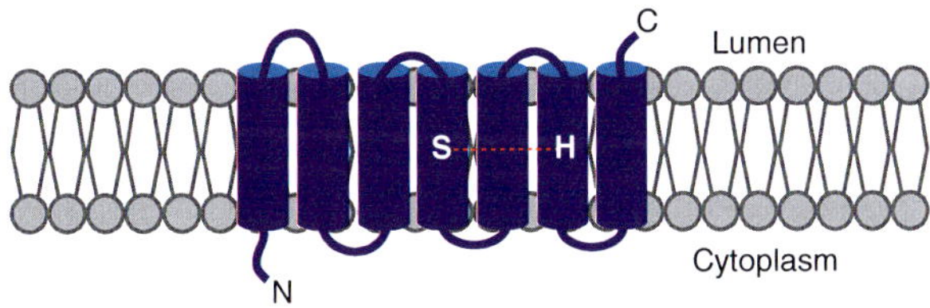

Figure 2. Domain structure of a rhomboid protease. Secretase rhomboids are polytopic transmembrane proteins with a cytoplasmic amino terminus and six or seven transmembrane domains. The catalytic serine and histidine residues are positioned within the upper third of transmembrane helices 4 and 6, respectively.

this, ligand cleavage in *Drosophila* is controlled by the compartmentalization of the growth factor and rhomboid (Fig. 3) (Lee et al. 2001). As for all proteins that enter the secretory pathway, *Drosophila* Spitz is synthesized in the endoplasmic reticulum (ER). Spitz is retained in the ER by a mechanism that depends on phospholipase C-γ (Schlesinger et al. 2004). Thus, Spitz is spatially separated from its shedding protease, rhomboid-1, which is Golgi-resident. This segregation is overcome by a trafficking partner called Star, whose role is to escort Spitz to the Golgi, where Spitz is cleaved by rhomboid-1 (Fig. 3) (Lee et al. 2001).

Recent evidence suggests the potential for even more intricate trafficking regulation. Another rhomboid, rhomboid-3, can cleave Spitz in both the ER and the later secretory pathway (Yogev et al. 2008). Spitz cleavage in different compartments exerts a radically different outcome on signaling. In the late secretory pathway, rhomboid-3 cleavage of Spitz is analogous to rhomboid-1: it is Star-dependent and leads to EGFR activation (Yogev et al. 2008). In contrast, ER cleavage of Spitz is an inactivating step because it leads to ER retention. In addition, it has been reported that proteolysis of Star by rhomboid-3 inhibits ER exit of Spitz, thereby attenuating EGFR activation (Tsruya et al. 2007).

In summary, the basis for maintaining control over EGFR activation in the fly involves keeping growth factor and enzyme apart, until signaling is required.

REGULATION OF EGFR SIGNALING IN MAMMALS

Mammalian EGFR Ligands

Although the mechanisms of regulation are different in mammals, the logic of regulated trafficking and compartmentalization is conserved. Regulation of shedding is more complex in mammals and includes posttranslational regulation of the shedding protease (Blobel 2005; Murphy 2008). Mammals have four members of the EGF receptor family: ErbB1–B4. Of these, only ErbB1 and ErbB4 are truly analogous to fly EGFR in that they bind ligand and are active RTKs; ErbB2 cannot bind ligand, whereas ErbB3 lacks kinase activity (Yarden and Sliwkowski 2001). Although ErbB2 and ErbB3 cannot therefore signal autonomously, they can form productive heterodimers with ErbB1 and ErbB4, thereby diversifying signaling properties (Citri et al. 2003).

For convenience, mammalian EGFR ligands can be separated into two classes, based on receptor-binding preferences (Harris et al. 2003). The first class, comprising ligands that bind to ErbB1, are amphiregulin (AREG), betacellulin (BTC), epidermal growth factor (EGF), epigen,

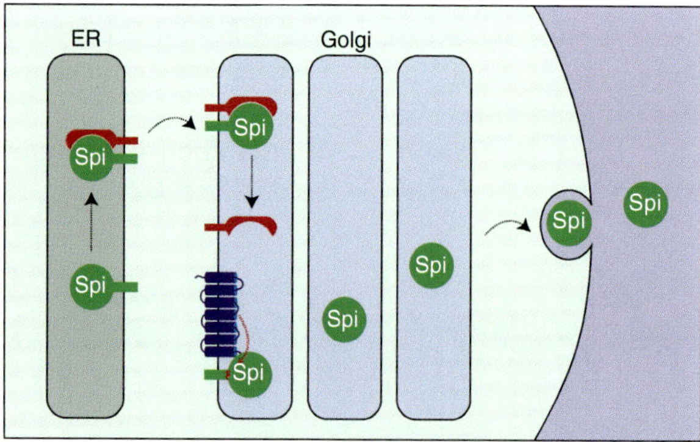

Figure 3. Regulated Spitz trafficking controls EGFR activation in *Drosophila*. Spitz is synthesized in the endoplasmic reticulum (ER) as a transmembrane precursor. Exit of Spitz from the ER to the Golgi requires the chaperone protein, Star (illustrated in red). On entry to the Golgi, Spitz encounters rhomboid (illustrated in blue) and undergoes proteolysis within the transmembrane domain. Spitz can now be secreted, thereby facilitating EGFR activation on a nearby cell.

epiregulin (EPR), heparin-binding EGF (HB-EGF), and transforming growth factor α (TGF-α) (Massague and Pandiella 1993; Schneider and Wolf 2009). Members of the neuregulin family (Nrg1-4) form the second group of ligands. These can bind to ErbB3/ErbB4 and include the multiple and complex splice variants of Nrg1 (Falls 2003). In addition, some members of the first class also show activity on ErbB4: BTC, HB-EGF, and EPR (Harris et al. 2003).

As shown in Figure 1, with the exception of EGF, most ligands have a domain structure resembling *Drosophila* Spitz. They are type I transmembrane proteins (their amino termini are on the luminal side of the membrane) and contain an amino-terminal prodomain, followed by a single EGF domain, a transmembrane domain, and cytoplasmic tail. Amphiregulin and HB-EGF have a heparin-binding motif located amino terminal to the EGF domain (Fig. 1) (Cook et al. 1991; Higashiyama et al. 1991). This allows binding to proteoglycans in the ex-

tracellular matrix and provides an extra mechanism to control ligand diffusion after cleavage (Piepkorn et al. 1998). EGF itself is somewhat unusual, because it contains eight EGF repeats amino terminal to the actual bioactive EGF domain (the module closest to the TMD) (Fig. 1).

ADAM PROTEASES

ADAM Protease Domain Structure

Mammalian EGFR ligand cleavage requires ADAM proteases instead of rhomboids. These are single-pass transmembrane proteins and their active sites are located on the extracellular side of the membrane, not within it (Fig. 4). Substrate cleavage typically occurs on the cell surface or within the late secretory pathway (Murphy 2008). Mammalian genomes contain multiple ADAM genes, a total of 21 in humans, of which 13 are predicted to be catalytically active (Blobel 2005). Rodent genomes have addi-

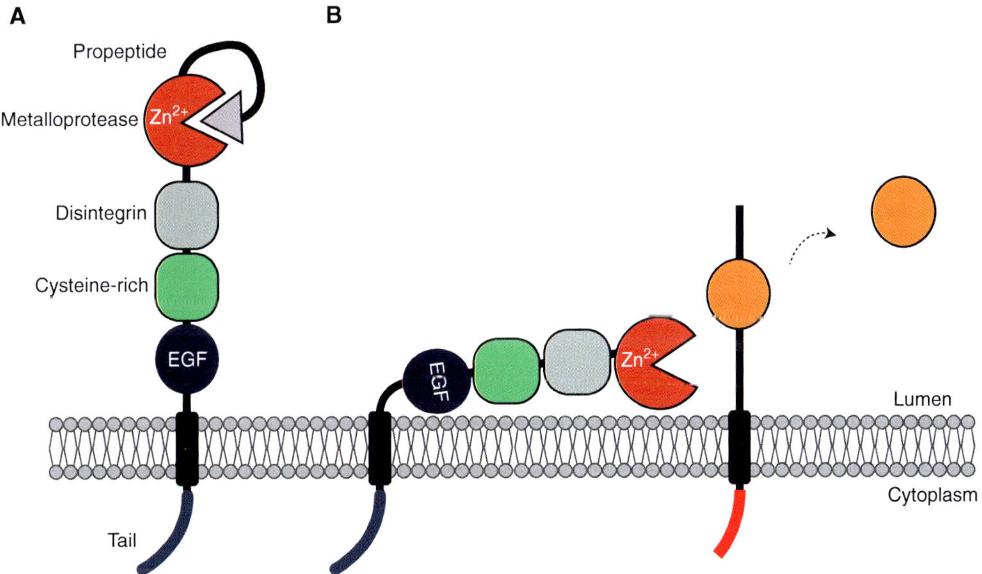

Figure 4. Domain structure and activation of ADAM metalloproteases. ADAMs are type I transmembrane proteins containing extracellular (luminal) amino termini and cytoplasmic carboxyl termini. (*A*) ADAMs are synthesized as a zymogen proform that lacks proteolytic activity, because the prodomain binds within the active site cleft. Removal of the prodomain by autocatalysis or by the proportein convertase, furin, is required before the enzyme can be active. (*B*) After processing, the prodomain may remain bound to the active site, and may require displacement before the ADAM can be active. When activated by signals, ADAMs cleave their substrates with a region just outside the membrane (within the juxtamembrane region).

tional ADAMs, many of which are testes-specific (Puente and Lopez-Otin 2004).

ADAMs contain a complicated domain structure (Fig. 4). Beyond the amino-terminal signal peptide, they contain a bifunctional prodomain. During biogenesis, it acts as a chaperone (Leonard et al. 2005) and subsequently maintains the enzyme in a zymogen (inactive) form during transit through the secretory pathway (Gonzales et al. 2008). Before ADAMs reach the cell surface, the inhibitory prodomain must be removed. Some ADAMs remove their prodomain autocatalytically (Murphy 2009), whereas for others (including ADAMs 10 and 17), it is removed in the late Golgi by the proprotein convertase furin (Peiretti et al. 2003a).

Adjacent to the prodomain is the catalytic heart of the enzyme, the metalloprotease domain, which contains the zinc-coordinating HEXHH catalytic motif (Blobel 2005). Next to this lies a disintegrin domain that derives its name from a similarity to an integrin-binding protein secreted in snake venom (Blobel and White 1992). In some ADAMs, extracellular matrix interactions with this domain may influence cell–cell adhesion (White 2003). The cysteine-rich region, which may influence substrate recognition (Smith et al. 2002), is located further toward the carboxyl terminus, followed by an EGF domain and transmembrane domain. On the other side of the membrane lies the cytoplasmic tail, which may be important for sensing cytoplasmic signals that regulate trafficking and control enzyme activity. Regulation of ADAM trafficking and activity is discussed later.

Physiological Importance of ADAMs in EGFR Ligand Shedding

It has been proposed that the membrane-tethered versions of some ligands can initiate juxtamembrane activation of the EGFR (Singh and Harris 2005). However, as in *Drosophila*, the importance of ligand shedding is very clear. First, as described below, ADAM mutant mice essentially phenocopy the EGFR ligand knockouts; second, pharmacological inhibition of metalloproteases blocks EGFR activation (Dong et al. 1999).

Although several ADAMs can cleave EGF ligands, physiological evidence suggests that TACE (tumor necrosis factor α-converting enzyme) also known as ADAM17, and ADAM10 are the major mammalian EGFR ligand sheddases. ADAM10 and TACE orthologs also exist in *Drosophila*, although they do not appear to be involved in EGFR ligand processing (Delwig and Rand 2008). In mammals, robust genetic and biochemical data suggest that TACE is most important: TACE knockout mice phenocopy many aspects of individual EGFR ligand knockouts, particularly TGF-α and HB-EGF knockouts. This includes defects in epithelial maturation, premature eyelid opening, hair follicle defects, lung branching morphogenesis defects, and heart valve malformations (Luetteke et al. 1993; Mann et al. 1993; Peschon et al. 1998). The importance of shedding is further highlighted by experiments in mice expressing a noncleavable version of HB-EGF. These animals show severe heart failure and enlarged heart valves: a phenocopy of the HB-EGF knockouts (Iwamoto et al. 2003; Yamazaki et al. 2003). Experiments using primary and immortalized cells (embryonic fibroblasts and keratinocytes) from TACE-null mice confirm that it sheds TGF-α, HB-EGF, amphiregulin, and epiregulin (Merlos-Suarez et al. 2001; Sunnarborg et al. 2002; Sahin et al. 2004).

ADAM10 knockout mice die in utero, because of Notch-related defects (on ligand binding, the receptor Notch is cleaved sequentially by ADAM10, then γ secretase; this triggers transcription of Notch target genes) (Hartmann et al. 2002). Because of this lethality, ADAM10's contribution to ligand shedding in individual tissues has not been examined. However, experiments using immortalized embryonic fibroblasts show that it is required for EGF and BTC cleavage (Sahin et al. 2004).

A number of other ADAMs can also cleave EGFR ligands (Edwards et al. 2008). For example, ADAMs 8, 9, 12, and 15 can process HB-EGF; EGF can be shed by ADAMs 8, 9, 12, and 19 and betacellulin by ADAMs 8, 12 and 19 (Izumi et al. 1998; Asakura et al. 2002; Schafer et al. 2004). Although these ADAMs are not major physiological EGFR ligand sheddases, there may be pathological contexts in which their deregulated expression or activity contributes to cleavage (Sahin et al. 2004; Horiuchi et al. 2007a). Related to this, ADAM10 can cleave TACE substrates when the activity of the latter is inhibited (Le Gall et al. 2009).

REGULATION OF ADAMS

EGFR ligand cleavage is not the only biological role of ADAM proteases. ADAM10 and TACE also regulate processes as diverse as inflammation, Notch signaling, cell adhesion, amyloid precursor protein processing, neuronal migration, and angiogenesis (Blobel 2005; Edwards et al. 2008). TACE itself was identified as the physiological sheddase for TNF (tumor necrosis factor) (Black et al. 1997); when secreted, this cytokine is the primary regulator of inflammatory responses (Palladino et al. 2003). For TACE alone, more than 50 substrates have been identified (Edwards et al. 2008). This raises an important question: How can a seemingly promiscuous sheddase control a given signaling pathway with the necessary precision?

Stimuli that Trigger ADAM Shedding

ADAM10 and TACE are controlled by distinct signaling circuits. TACE can be activated rapidly and powerfully by multiple signals including pharmacological stimuli like phorbol esters, which operate via protein kinase C (Fig. 5) (Rovida et al. 2001; Doedens et al. 2003). TACE is also activated by a more physiological but complex process known as transactivation (Fig. 5) (Gschwind et al. 2001; discussed by Ullrich and colleagues). Transactivation occurs when G-protein coupled receptor (GPCR) stimulation indirectly triggers activation of the EGFR. In outline, GPCR agonists (e.g., bombesin, angiotensin II, serotonin, hormones) trigger a rather poorly defined signaling cascade that activates TACE, resulting in ligand processing (principally HB-EGF) and consequent EGFR activation (Fig. 5). It is now clear that EGFR transactivation plays critical physiological roles in angiogenesis, heart development, and neurogenesis (Gschwind et al. 2001). It is also implicated extensively in tumor growth, invasion and metastasis (Lappano and Maggiolini 2011). TACE activity can also be induced by many other pathways including toll receptors, multiple kinases, and even cleaved TACE substrates (Diaz-Rodriguez et al. 2002; Soond et al. 2005). In summary, TACE lies at the heart of a complex regulatory network, responsible for multiple potent signaling events.

ADAM10 is also regulated by a different complement of signals. It is not affected by phor-

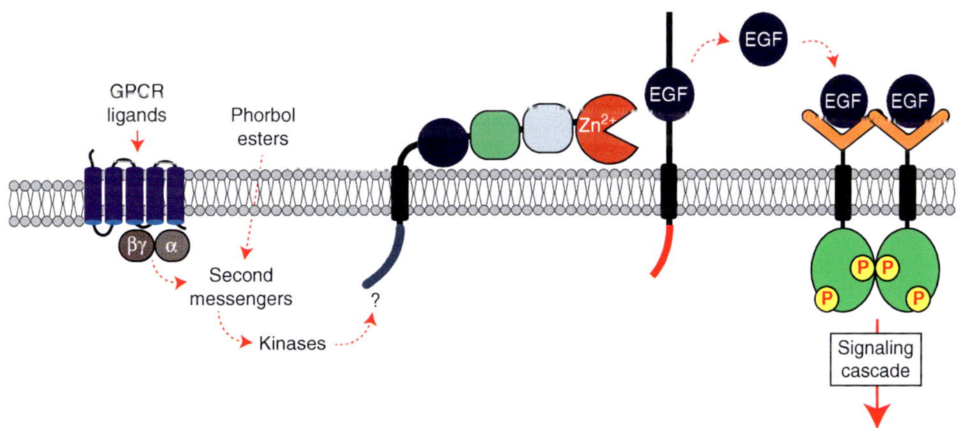

Figure 5. Transctivation of the EGFR by G protein-coupled receptors (GPCRs). Activation of GPCRs by agonists triggers a signaling cascade involving second messengers including Ca^{2+} and protein kinase C (PKC). This induces TACE cleavage of EGFR ligands (including HB-EGF), culminating in EGFR activation. How these signals trigger TACE activation remains unclear (see text). Phorbol esters such as PMA can also trigger TACE via a mechanism that involves PKC.

bol esters, but can be activated by calcium ionophores, calmodulin inhibition, and the metalloprotease-activating drug p-aminophenylmercuric acetate (Sanderson et al. 2005; Horiuchi et al. 2007a). More physiologically, activation of purinergic receptors can also trigger ADAM 10 activity (Le Gall et al. 2009), as can GPCRs (Lemjabbar and Basbaum 2002). For both TACE and ADAM10, the cytoplasmic tail may be required for receiving input stimuli, although the mechanism remains unclear (Horiuchi et al. 2007a).

How do these diverse stimuli activate TACE? Despite extensive study, this is still disappointingly unclear. Proposed mechanistic effects on TACE can be divided into three classes: regulation of enzyme activity on the plasma membrane, regulated access of enzyme to substrate, and control of TACE trafficking. Given the diversity of signals that can control TACE, it is quite possible that different mechanisms are used in different contexts.

Impact of Stimuli on TACE Activity on the Plasma Membrane

Signals may impinge directly on the activity of mature TACE on the plasma membrane. A recent study from Blobel and colleagues has suggested that phorbol esters can trigger a conformational change in the TACE active site, thereby enhancing proteolytic activity (Le Gall et al. 2010). This suggests a model whereby prodomain removal from TACE is not sufficient to render it fully active—an additional conformational rearrangement of the active site architecture is also required. It will be interesting to establish how signals trigger these potential active site conformational changes, whether this impacts directly on TACE or on a cofactor, and how this class of mechanism integrates with other forms of control.

The Timps (tissue inhibitors of metalloproteases) are another class of ADAM regulators that impact directly on enzymatic activity. These soluble proteins inhibit matrix metalloproteases and ADAMs by inserting their aminoterminal wedge-shaped cleft into the active site of the enzyme (Wisniewska et al. 2008). There

are four Timps in mammals; of these, ADAM10 can be inhibited by Timp1 and Timp3, whereas TACE is inhibited only by Timp3 (Murphy 2011). The impact of Timp on TACE shedding of EGF ligands is unclear; although some evidence suggests that loss of Timp3 affects EGFR ligand shedding in vivo (Murthy et al. 2010), a recent study suggests that loss of Timp3 in mouse embryonic fibroblasts has little impact on phorbol ester-triggered TACE shedding (Le Gall et al. 2010). Interestingly, the biggest physiological impact of loss of Timp3 is its effect on TNF shedding. As described above, as well as regulating EGFR signaling, TACE controls inflammation and apoptosis via cleavage of TNF. Reflecting this, deletion of Timp3 in mice causes uncontrolled TNF shedding by TACE, resulting in excessive inflammation caused by TNF perturbation (Mohammed et al. 2004; Guinea-Viniegra et al. 2009). Perhaps control by Timps of TACE activity is a means of regulating specificity in the face of multiple stimuli.

Tetraspanins: Spatial Control of ADAM Shedding

Another way of controlling TACE activity on the plasma membrane is by modulating access to substrates or regulators. This can be achieved by tetraspanins, which form a large family of proteins with four transmembrane domains in mammals (Yanez-Mo et al. 2011). The function of tetraspanins is diverse and not well understood in many cases but a common theme is the organization of membrane proteins into defined microdomains (Hemler 2005). They have been found in association not only with ADAMs (particularly ADAM10), but also EGF ligands and the EGFR itself, suggesting that they may have complex influences on signaling (Imhof et al. 2008; Murayama et al. 2008).

Tetraspanins can inhibit or promote ligand proteolysis, depending on the context. For instance, binding of tetraspanins, including CD9, CD81, and CD82, to ADAM10 promotes ligand shedding (Arduise et al. 2008). How this promotes cleavage is unclear, although an obvious possibility is regulating access of enzyme to substrate. In the case of tetraspanin-12, other possibilities include enhanced prodomain removal, or stabilizing the active enzyme (Xu et al. 2009).

In contrast, the tetraspanin CD9, which can bind to TACE and also to EGFR ligands, reduces shedding (Higashiyama et al. 1995; Imhof et al. 2008). By inhibiting ligand cleavage, tetraspanins may regulate switching from shedding to juxtamembrane signaling. In addition to corralling membrane proteins on the cell surface, CD9 may also influence ligand biogenesis/trafficking through the secretory pathway (Berditchevski and Odintsova 2007). Overall, it appears that rather than behaving as specific regulators of shedding, tetraspanins may represent a spatial anchoring network for many membrane proteins. They may control access of enzyme to substrate within membrane microdomains, or by coordinating access of signaling proteins that regulate ADAM activity.

Impact of Stimuli on TACE Trafficking

Clearly, before TACE can reach the cell surface as an active protease, it must first transit through the secretory pathway. Significant fractions of endogenous TACE are found within intracellular compartments (Schlondorff et al. 2000; Soond et al. 2005). This suggests that TACE biogenesis or trafficking is rate-limiting and may require chaperones or cofactors. Indeed, whereas many glycoproteins transit through the secretory pathway quite rapidly (within minutes to hours) (Ward and Kopito 1994; Janeens et al 2002), TACE trafficking is considerably slower (Schlondorff et al. 2000). Consistent with this, enhanced trafficking and furin cleavage has been suggested as a mechanism whereby signals activate TACE (Soond et al. 2005).

TACE Trafficking Regulators

A controversy concerns the role of the TACE cytoplasmic tail. In response to signals, the TACE cytoplasmic tail is phosphorylated by kinases, including extracellular-signal-regulated kinase (ERK), and this has been suggested to enhance TACE trafficking (Fan et al. 2003; Soond et al. 2005; Xu and Derynck 2010). Some studies,

however, have reported that the TACE tail is dispensable for signaling (Horiuchi et al. 2007a). In other approaches, it has been shown that the tails of many ADAMs possess SH3-binding sites; this has encouraged the search for binding partners (Seals and Courtneidge 2003). Yeast two-hybrid screens have identified TACE tail-interacting proteins, several of which contain SH3 domains and/or PDZ domains (Zheng et al. 2002; Peiretti et al. 2003b; Tanaka et al. 2004). Disappointingly, however, the physiological importance of these interactions remains to be shown. Overall, there is probably insufficient evidence to make a definitive judgment about the significance of the cytoplasmic tail of TACE; perhaps its role is different in different cellular contexts.

MAMMALIAN RHOMBOIDS AND RTKS

iRhoms: Pseudoprotease Regulation of RTK Ligand Shedding

As mentioned above, although TACE activity and trafficking is highly controlled, physiologically important regulators are lacking. However, we and others have recently shown that a protein called iRhom is essential for TACE trafficking (Adrain et al. 2012; McIlwain et al. 2012; Siggs et al. 2012). iRhoms are a metazoan subfamily of rhomboid-like proteins that lack key catalytic residues, rendering them proteolytically inactive (Fig. 6) (Adrain and Freeman 2012).

There are two mammalian iRhoms (also called RHBDF1 and RHBDF2). iRhom1/RHB DF1 has been broadly implicated in growth control of cancer cells and EGFR signaling, although the mechanistic basis for this is unclear (Nakagawa et al. 2005; Yan et al. 2008). Recently, however, the cellular function of mammalian iRhoms has been revealed: iRhom2 is essential for the export of TACE from the ER, and thereby for the shedding of TACE substrates (Adrain et al. 2012; McIlwain et al. 2012). In macrophages the primary role of TACE is TNF release, and accordingly iRhom2 mutant mice do not secrete TNF in response to immune challenge. Processing of the TACE substrate HB-EGF, is also defective in iRhom2-null cells, suggesting that mam-

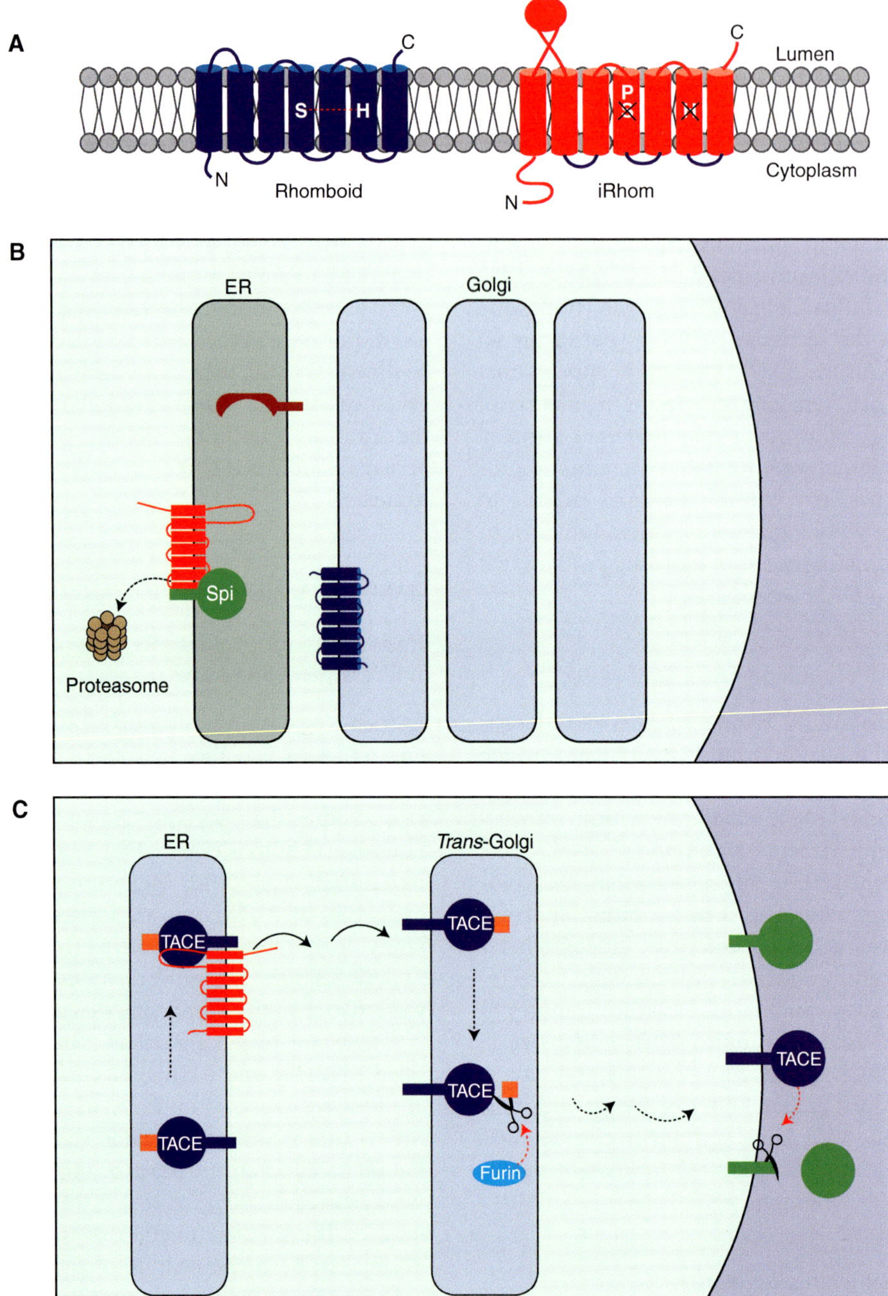

Figure 6. Regulation of RTK signaling by iRhoms. (*A*) Comparison of an active rhomboid and an iRhom. In comparison with an active rhomboid (*left*), iRhoms (*right*) contain an extended cytoplasmic amino terminus and a globular cysteine-rich domain called the iRhom homology domain (iRHD) within the lumen of the ER. All iRhoms have a conserved proline residue immediately amino terminal to the serine in TMD 4; this renders iRhoms proteolytically inactive. (*B*) *Drosophila* iRhom regulates ER-associated degradation of EGFR ligands. Spitz is normally trafficked out of the ER by Star and encounters rhomboid in the Golgi (Fig. 3). (*Legend continues on following page.*)

Cite this article as *Cold Spring Harb Perspect Biol* doi: 10.1101/cshperspect.a008995

malian iRhom can act as a positive regulator of the EGFR. Therefore iRhom is an important trafficking regulator required for the ER exit of TACE. How signals impact on the ability of iRhom to control TACE activation remains to be established.

In *Drosophila*, iRhom is also implicated in trafficking regulation, but in a different manner. It controls EGFR signaling, but unlike its effects on TACE, it is an inhibitor rather than an activator of trafficking. Flies null for iRhom show a severe activity phenotype (Zettl et al. 2011), which phenocopies the effect of rhomboid overexpression in the central nervous system (Foltenyi et al. 2007). At the cellular level, iRhom binds to EGFR ligands in the ER, and directs them into ERAD (ER-associated degradation); this blocks their onward trafficking and prevents them from activating EGFR signaling (Zettl et al. 2011). ERAD is a fundamental cellular quality control process, whereby misfolded proteins in the ER are retrotranslocated into the cytoplasm and degraded by the proteasome (Smith et al. 2011). In *Drosophila*, iRhoms exploit this mechanism to regulate signaling.

Rather confusingly, therefore, the *Drosophila* and mammalian results imply that in different contexts, iRhoms can either inhibit or promote signaling. The mechanism underlying this dual function remains to be determined, but it is clear that iRhoms are ER-localized trafficking regulators that impact on, among other pathways, EGFR signaling (Fig. 6). More generally, they add another example of compartmentalization as a way of controlling of RTK signaling.

Active Mammalian Rhomboids

We have highlighted that ADAMs are the principal mammalian EGFR ligand sheddases. But whether they are the only sheddases remains to be proven. There are suggestions of other ligand-shedding activities, including serine proteases (Pandiella et al. 1992; Le Gall et al. 2004). Tissue-specific heterogeneity in the molecular weight forms of cleaved ligands also suggests that different sheddases exist (Dempsey et al. 1997). Furthermore, given that many signaling components are conserved between flies and mammals, a role for rhomboids in mammalian EGFR ligand cleavage is also possible.

Four mammalian rhomboid proteases localize to the secretory pathway, although physiological substrates have yet to be identified for most (Lohi et al. 2004). Of these, RHBDL2 is the best characterized. RHBDL2 shows a restricted expression pattern in the mouse including intestine, stomach, prostate, bladder, and skin. These are tissues where ErbB1 and one of its ligands, EGF, are expressed. Consistent with this, EGF is an efficient substrate of RHBDL2 (but not other rhomboids) in cell culture assays (Adrain et al. 2011). Although the physiological significance of RHBDL2 cleavage of EGF is unknown, RHBDL2 contributes to the shedding of EGF in some tumor cells (Adrain et al. 2011). In these cases, ADAM inhibition alone cannot block shedding, suggesting that there may be some therapeutic contexts where blocking ADAMs may not necessarily inhibit EGF cleavage. Despite the possible relationship between RHBDL2 and EGF, most mammalian EGF family ligands appear not to be cleaved by rhomboids, implying other, as yet unknown functions for these intramembrane proteases. Notably, at least two other substrates of RHBDL2 have been reported, although, again, their biological relevance remains obscure (Lohi et al. 2004; Pascall and Brown 2004).

Figure 6. (*Continued*) However in the presence of iRhom, Spitz is retained in the ER and instead, shunted into the ER-associated degradation (ERAD). This results in its dislocation from the ER membrane and degradation by the proteasome. As a result, no Spitz enters the Golgi for cleavage and EGFR signaling is attenuated. (*C*) Regulation of TACE trafficking by mammalian iRhom2. TACE is synthesized in the ER as an inactive zymogen containing the prodomain (the TACE prodomain is indicated in orange). iRhom is required for trafficking of TACE into the Golgi, where it undergoes prodomain cleavage by furin. Active TACE can then cleave its substrates in the late Golgi or on the cell surface.

OTHER RTKS REGULATED BY LIGAND PROTEOLYSIS

Cleavage of Ephrins

Although the importance for ligand cleavage is most studied in the EGFR, shedding also regulates other RTKs. The regulation of Eph signaling is particularly interesting because it is atypical. Rather than being an activating step, ADAM cleavage is required to switch off signaling. Regulation of axonal guidance requires cell–cell repulsion, driven by the interaction between Eph receptors on one cell and membrane-tethered ephrin ligand on an adjacent cell (Pitulescu and Adams 2010). As the Eph:ephrin interaction is multivalent and high affinity, receptor: ligand complexes favor cell–cell adhesion. But successful repulsion requires signal termination, which necessitates breaking Eph:ephrin complexes. One way in which this is achieved is via proteolysis.

ADAM10 is the ephrin protease (Hattori et al. 2000; Janes et al. 2005). To terminate signaling, ADAM10 recognizes assembled Eph:ephrin complexes and severs them by cleaving the ligand in *trans* on the adjacent cell; unbound ligands are ignored (Davis et al. 1994). Failure to cleave ephrins delays axon withdrawal (Hattori et al. 2000). The role of ADAM proteolysis in Eph signaling may be conserved in flies, as mutants in the ADAM10 homolog, Kuzbanian, show an axonal extension defect (Fambrough et al. 1996). Although the physiological context is yet unclear, the mammalian rhomboid protease RHBDL2, can also cleave ephrinB3, leading to the possibility of analogous rhomboid regulation of Eph signaling (Pascall and Brown 2004).

In some contexts, Eph:ephrin signaling is bidirectional in nature. As well as triggering RTK activity in the receptor-bearing cells, a signal is triggered in the ligand-expressing cell (Georgakopoulos et al. 2006). This is important for the role of ephrinB2 in cardiac valve maturation and axon path finding (Cowan et al. 2004). On ADAM10 cleavage of receptor-bound ephrin, the transmembrane stub is further cleaved by γ-secretase and the carboxy-terminal fragment thus generated drives an Src-associated signaling cascade (Georgakopoulos et al. 2006).

Shedding of Ligands of the PDGF Family

Some members of the PDGF (platelet-derived growth factor) receptor subfamily are regulated by ligand shedding, including cKit/SCFR (Huang et al. 1992), FLT3/Flk2 (Horiuchi et al. 2009), and CSF1R/Ems (Horiuchi et al. 2007b). These RTK pathways play important immune roles including regulating haematopoiesis (cKit and FLT3) (Ashman 1999; Naoe and Kiyoi 2004) or macrophage and osteoclast development (CSF1R) (Cecchini et al. 1997). The ligand for cKit is expressed in two alternative splice forms: Kit ligand 1 (KL-1) contains an extra exon upstream of the juxtamembrane region that renders it readily susceptible to proteolytic shedding, whereas KL-2 lacks this exon and is cleaved less efficiently (Huang et al. 1992). As for many other RTK ligands, the principal Kit ligand sheddase is unknown, and several enzymes including ADAMs have been suggested (Huang et al. 1992). At least in mouse embryonic fibroblasts, TACE is essential for phorbol ester-induced shedding of KL-1 and KL-2, as well as for the constitutive shedding of KL-2 (Kawaguchi et al. 2007). However, constitutive shedding of KL-1 is not impaired in TACE null fibroblasts and cannot be inhibited by the metalloprotease inhibitor BB94, implicating another sheddase. Despite the importance of ligand shedding to activate Kit signaling, it is important to note that expression of soluble Kit ligand in mice cannot fully rescue the KitL knockout phenotype, illustrating that juxtamembrane signaling is also important (Brannan et al. 1991).

DEREGULATED LIGAND SHEDDING AND CANCER

A common theme for all RTK families is that once activated, the receptors can drive tumor growth, promote cell survival, migration, and resistance to chemotherapy (Lemmon and Schlessinger 2010; Sastry and Elferink 2011). Because proteolysis is irreversible, shedding can be the commitment point for unleashing potent signals. Amplification or mutation of RTKs can render them more sensitive to growth factors. When combined with excessive ligand shedding, this establishes an autocrine loop that enables cells to proliferate independently of the requirement for external cues (Di Marco et al. 1989). This is one route to malignant transformation (Sporn and Roberts 1985). EGFR amplification combined with increased growth factor shedding has been observed in diverse human cancers including gliomas and malignancies of the lung, bladder, gastric tract, esophagus, breast, ovaries, and head and neck tumors (Gullick 1991; Salomon et al. 1995). Other deregulated feedback loops can also drive cancer. Activation of the EGFR drives Ras activity, which can promote increased ligand transcription (Baselga et al. 1996), and kinases downstream of Ras can regulate TACE; this in turn amplifies EGFR signaling (Diaz-Rodriguez et al. 2002; Soond et al. 2005).

Increased expression of sheddases, including TACE, occurs in many tumors, contributing to metastasis and correlating with reduced patient survival rates (Murphy 2008). Increased crosstalk between many GPCRs and the EGFR also plays a significant role in cancer development and metastasis (Lappano and Maggiolini 2011). Another important factor is the interplay in signals between tumor and stromal cells. Recruitment of inflammatory cells to the tumor microenvironment is known to result in secretion of RTK ligands that enhance tumor growth (Wyckoff et al. 2004). These examples all highlight the medical importance of the control mechanisms discussed in this chapter and imply the therapeutic rationale of targeting the proteases that lie at the heart of much RTK activation. The ADAM proteases, for example, with their central position in both growth factor and cytokine activation, have been the focus of much pharmaceutical interest (Saftig and Reiss 2011). Attempts to inhibit them, however, have so far been disappointing, possibly because metalloproteases share a common active site architecture, resulting in off-target effects and toxicity (Coussens et al. 2002; DasGupta et al. 2009).

CONCLUDING REMARKS

It is clear that ligand shedding is a pivotal step for RTK activation in many contexts, particularly for the EGFR, but also for other RTKs. Whereas there has been progress in identifying the major physiological ligand proteases, much remains to be understood about the diverse mechanisms that trigger and regulate shedding. The primary mechanistic theme that we have emphasized is the role of regulated trafficking and compartmentalization of ligands and shedding proteases as a powerful and versatile mode of control. This regulatory logic, which exploits the exquisite regulation of cellular membrane trafficking systems, occurs repeatedly, even when specific components differ. Particular challenges for the future include understanding how the multiple overlapping signals and their regulators are integrated in vivo, and how pathological stimuli, including those emanating from GPCRs, control protease activity. Clearly sheddases are important drug targets; however, the ability to inhibit them specifically and potently still eludes us. Understanding more about control of proteolysis and the requirement for trafficking regulators and cofactors will no doubt identify new drugs targets for the many important pathways they control.

ACKNOWLEDGMENTS

We thank Viorica Lastun and Kvido Strisovsky for helpful comments on the manuscript. C.A. was supported by a long-term fellowship from The International Human Frontier Science Program Organization and was the recipient of an EMBO Long-Term Fellowship. The Freeman group is supported by the Medical Research Council Programme number U105178780.

REFERENCES

Adrain C, Freeman M. 2012. New lives for old: Evolution of pseudoenzyme function illustrated by iRhoms. *Nat Rev Mol Cell Biol* **13:** 489–498.

Adrain C, Strisovsky K, Zettl M, Hu L, Lemberg MK, Freeman M. 2011. Mammalian EGF receptor activation by the rhomboid protease RHBDL2. *EMBO Rep* **12:** 421–427.

Adrain C, Zettl M, Christova Y, Taylor N, Freeman M. 2012. Tumor necrosis factor signaling requires iRhom2 to promote trafficking and activation of TACE. *Science* **335:** 225–228.

Arduise C, Abache T, Li L, Billard M, Chabanon A, Ludwig A, Mauduit P, Boucheix C, Rubinstein E, Le Naour F. 2008. Tetraspanins regulate ADAM10-mediated cleavage of TNF-α and epidermal growth factor. *J Immunol* **181:** 7002–7013.

Asakura M, Kitakaze M, Takashima S, Liao Y, Ishikura F, Yoshinaka T, Ohmoto H, Node K, Yoshino K, Ishiguro H, et al. 2002. Cardiac hypertrophy is inhibited by antagonism of ADAM12 processing of HB-EGF: Metalloproteinase inhibitors as a new therapy. *Nat Med* **8:** 35–40.

Ashman LK. 1999. The biology of stem cell factor and its receptor C-kit. *Int J Biochem Cell Biol* **31:** 1037–1051.

Baselga J, Mendelsohn J, Kim YM, Pandiella A. 1996. Autocrine regulation of membrane transforming growth factor-α cleavage. *J Biol Chem* **271:** 3279–3284.

Berditchevski F, Odintsova E. 2007. Tetraspanins as regulators of protein trafficking. *Traffic* **8:** 89–96.

Bier E, Jan LY, Jan YN. 1990. Rhomboid, a gene required for dorsoventral axis establishment and peripheral nervous system development in *Drosophila* melanogaster. *Genes Dev* **4:** 190–203.

Black RA, Rauch CT, Kozlosky CJ, Peschon JJ, Slack JL, Wolfson MF, Castner BJ, Stocking KL, Reddy P, Srinivasan S, et al. 1997. A metalloproteinase disintegrin that releases tumour-necrosis factor-α from cells. *Nature* **385:** 729–733.

Blobel CP. 2005. ADAMs: Key components in EGFR signalling and development. *Nat Rev Mol Cell Biol* **6:** 32–43.

Blobel CP, White JM. 1992. Structure, function and evolutionary relationship of proteins containing a disintegrin domain. *Curr Opin Cell Biol* **4:** 760–765.

Brannan CI, Lyman SD, Williams DE, Eisenman J, Anderson DM, Cosman D, Bedell MA, Jenkins NA, Copeland NG. 1991. Steel-Dickie mutation encodes a c-kit ligand lacking transmembrane and cytoplasmic domains. *Proc Natl Acad Sci* **88:** 4671–4674.

Brown KE, Kerr M, Freeman M. 2007. The EGFR ligands Spitz and Keren act cooperatively in the *Drosophila* eye. *Dev Biol* **307:** 105–113.

Cecchini MG, Hofstetter W, Halasy J, Wetterwald A, Felix R. 1997. Role of CSF-1 in bone and bone marrow development. *Mol Reprod Dev* **46:** 75–83; discussion 83–4.

Citri A, Skaria KB, Yarden Y. 2003. The deaf and the dumb: The biology of ErbB-2 and ErbB-3. *Exp Cell Res* **284:** 54–65.

Cook PW, Mattox PA, Keeble WW, Pittelkow MR, Plowman GD, Shoyab M, Adelman JP, Shipley GD. 1991. A heparin sulfate-regulated human keratinocyte autocrine factor is similar or identical to amphiregulin. *Mol Cell Biol* **11:** 2547–2557.

Coussens LM, Fingleton B, Matrisian LM. 2002. Matrix metalloproteinase inhibitors and cancer: Trials and tribulations. *Science* **295:** 2387–2392.

Cowan CA, Yokoyama N, Saxena A, Chumley MJ, Silvany RE, Baker LA, Srivastava D, Henkemeyer M. 2004. Ephrin-B2 reverse signaling is required for axon pathfinding and cardiac valve formation but not early vascular development. *Dev Biol* **271:** 263–271.

DasGupta S, Murumkar PR, Giridhar R, Yadav MR. 2009. Current perspective of TACE inhibitors: A review. *Bioorg Med Chem* **17:** 444–459.

Davis CG. 1990. The many faces of epidermal growth factor repeats. *New Biol* **2:** 410–419.

Davis S, Gale NW, Aldrich TH, Maisonpierre PC, Lhotak V, Pawson T, Goldfarb M, Yancopoulos GD. 1994. Ligands for EPH-related receptor tyrosine kinases that require membrane attachment or clustering for activity. *Science* **266:** 816–819.

Delwig A, Rand MD. 2008. Kuz and TACE can activate Notch independent of ligand. *Cell Mol Life Sci* **65:** 2232–2243.

Dempsey PJ, Meise KS, Yoshitake Y, Nishikawa K, Coffey RJ. 1997. Apical enrichment of human EGF precursor in Madin-Darby canine kidney cells involves preferential basolateral ectodomain cleavage sensitive to a metalloprotease inhibitor. *J Cell Biol* **138:** 747–758.

Diaz-Rodriguez E, Montero JC, Esparis-Ogando A, Yuste L, Pandiella A. 2002. Extracellular signal-regulated kinase phosphorylates tumor necrosis factor α-converting enzyme at threonine 735: A potential role in regulated shedding. *Mol Biol Cell* **13:** 2031–2044.

Di Marco E, Pierce JH, Fleming TP, Kraus MH, Molloy CJ, Aaronson SA, Di Fiore PP. 1989. Autocrine interaction between TGF-α and the EGF-receptor: Quantitative requirements for induction of the malignant phenotype. *Oncogene* **4:** 831–838.

Doedens JR, Mahimkar RM, Black RA. 2003. TACE/ADAM-17 enzymatic activity is increased in response to cellular stimulation. *Biochem Biophys Res Commun* **308:** 331–338.

Dong J, Opresko LK, Dempsey PJ, Lauffenburger DA, Coffey RJ, Wiley HS. 1999. Metalloprotease-mediated ligand release regulates autocrine signaling through the epidermal growth factor receptor. *Proc Natl Acad Sci* **96:** 6235–6240.

Edwards DR, Handsley MM, Pennington CJ. 2008. The ADAM metalloproteinases. *Mol Aspects Med* **29:** 258–289.

Falls DL. 2003. Neuregulins: functions, forms, and signaling strategies. *Exp Cell Res* **284:** 14–30.

Fambrough D, Pan D, Rubin GM, Goodman CS. 1996. The cell surface metalloprotease/disintegrin Kuzbanian is required for axonal extension in *Drosophila*. *Proc Natl Acad Sci* **93:** 13233–13238.

Fan H, Turck CW, Derynck R. 2003. Characterization of growth factor-induced serine phosphorylation of tumor necrosis factor-α converting enzyme and of an alternatively translated polypeptide. *J Biol Chem* **278:** 18617–18627.

Foltenyi K, Greenspan RJ, Newport JW. 2007. Activation of EGFR and ERK by rhomboid signaling regulates the consolidation and maintenance of sleep in *Drosophila*. *Nat Neurosci* **10:** 1160–1167.

Freeman M. 1994. The Spitz gene is required for photoreceptor determination in the *Drosophila* eye where it interacts with the EGF receptor. *Mech Dev* **48:** 25–33.

Freeman M. 1998. Complexity of EGF receptor signalling revealed in *Drosophila*. *Curr Opin Genet Dev* **8:** 407–411.

Freeman M. 2009. Rhomboids: 7 years of a new protease family. *Semin Cell Dev Biol* **20:** 231–239.

Georgakopoulos A, Litterst C, Ghersi E, Baki L, Xu C, Serban G, Robakis NK. 2006. Metalloproteinase/Presenilin1 processing of ephrinB regulates EphB-induced Src phosphorylation and signaling. *EMBO J* **25:** 1242–1252.

Gonzales PE, Galli JD, Milla ME. 2008. Identification of key sequence determinants for the inhibitory function of the prodomain of TACE. *Biochemistry* **47:** 9911–9919.

Gschwind A, Zwick E, Prenzel N, Leserer M, Ullrich A. 2001. Cell communication networks: Epidermal growth factor receptor transactivation as the paradigm for interreceptor signal transmission. *Oncogene* **20:** 1594–1600.

Guinea-Viniegra J, Zenz R, Scheuch H, Hnisz D, Holcmann M, Bakiri L, Schonthaler HB, Sibilia M, Wagner EF. 2009. TNFα shedding and epidermal inflammation are controlled by Jun proteins. *Genes Dev* **23:** 2663–2674.

Gullick WJ. 1991. Prevalence of aberrant expression of the epidermal growth factor receptor in human cancers. *Br Med Bull* **47:** 87–98.

Harris RC, Chung E, Coffey RJ. 2003. EGF receptor ligands. *Exp Cell Res* **284:** 2–13.

Hartmann D, de Strooper B, Serneels L, Craessaerts K, Herreman A, Annaert W, Umans L, Lubke T, Lena Illert A, von Figura K, et al. 2002. The disintegrin/metalloprotease ADAM 10 is essential for Notch signalling but not for α-secretase activity in fibroblasts. *Hum Mol Genet* **11:** 2615–2624.

Hattori M, Osterfield M, Flanagan JG. 2000. Regulated cleavage of a contact-mediated axon repellent. *Science* **289:** 1360–1365.

Hemler ME. 2005. Tetraspanin functions and associated microdomains. *Nat Rev Mol Cell Biol* **6:** 801–811.

Higashiyama S, Abraham JA, Miller J, Fiddes JC, Klagsbrun M. 1991. A heparin-binding growth factor secreted by macrophage-like cells that is related to EGF. *Science* **251:** 936–939.

Higashiyama S, Iwamoto R, Goishi K, Raab G, Taniguchi N, Klagsbrun M, Mekada E. 1995. The membrane protein CD9/DRAP 27 potentiates the juxtacrine growth factor activity of the membrane-anchored heparin-binding EGF-like growth factor. *J Cell Biol* **128:** 929–938.

Horiuchi K, Le Gall S, Schulte M, Yamaguchi T, Reiss K, Murphy G, Toyama Y, Hartmann D, Saftig P, Blobel CP. 2007a. Substrate selectivity of epidermal growth factor-receptor ligand sheddases and their regulation by phorbol esters and calcium influx. *Mol Biol Cell* **18:** 176–188.

Horiuchi K, Miyamoto T, Takaishi H, Hakozaki A, Kosaki N, Miyauchi Y, Furukawa M, Takito J, Kaneko H, Matsuzaki K, et al. 2007b. Cell surface colony-stimulating factor 1 can be cleaved by TNF-α converting enzyme or endocytosed in a clathrin-dependent manner. *J Immunol* **179:** 6715–6724.

Horiuchi K, Morioka H, Takaishi H, Akiyama H, Blobel CP, Toyama Y. 2009. Ectodomain shedding of FLT3 ligand is mediated by TNF-α converting enzyme. *J Immunol* **182:** 7408–7414.

Huang EJ, Nocka KH, Buck J, Besmer P. 1992. Differential expression and processing of two cell associaent forms of the kit-ligand: KL-1 and KL-2. *Mol Biol Cell* **3:** 349–362.

Imhof I, Gasper WJ, Derynck R. 2008. Association of tetraspanin CD9 with transmembrane TGF-α confers alterations in cell-surface presentation of TGF-α and cytoskeletal organization. *J Cell Sci* **121:** 2265–2274.

Iwamoto R, Yamazaki S, Asakura M, Takashima S, Hasuwa H, Miyado K, Adachi S, Kitakaze M, Hashimoto K, Raab G, et al. 2003. Heparin-binding EGF-like growth factor and ErbB signaling is essential for heart function. *Proc Natl Acad Sci* **100:** 3221–3226.

Izumi Y, Hirata M, Hasuwa H, Iwamoto R, Umata T, Miyado K, Tamai Y, Kurisaki T, Sehara-Fujisawa A, Ohno S, et al. 1998. A metalloprotease-disintegrin, MDC9/meltrin-γ/

ADAM9 and PKCδ are involved in TPA-induced ectodomain shedding of membrane-anchored heparin-binding EGF-like growth factor. *EMBO J* **17:** 7260–7272.

Janes PW, Saha N, Barton WA, Kolev MV, Wimmer-Kleikamp SH, Nievergall E, Blobel CP, Himanen JP, Lackmann M, Nikolov DB. 2005. Adam meets Eph: An ADAM substrate recognition module acts as a molecular switch for ephrin cleavage in *trans*. *Cell* **123:** 291–304.

Jansens A, van Duijn E, Braakman I. 2002. Coordinated nonvectorial folding in a newly synthesized multidomain protein. *Science* **298:** 2401–2403.

Kawaguchi N, Horiuchi K, Becherer JD, Toyama Y, Besmer P, Blobel CP. 2007. Different ADAMs have distinct influences on Kit ligand processing: Phorbol-ester-stimulated ectodomain shedding of Kitl1 by ADAM17 is reduced by ADAM19. *J Cell Sci* **120:** 943–952.

Lappano R, Maggiolini M. 2011. G protein-coupled receptors: Novel targets for drug discovery in cancer. *Nat Rev Drug Discov* **10:** 47–60.

Lee JR, Urban S, Garvey CF, Freeman M. 2001. Regulated intracellular ligand transport and proteolysis control EGF signal activation in *Drosophila*. *Cell* **107:** 161–171.

Le Gall SM, Meneton P, Mauduit P, Dreux C. 2004. The sequential cleavage of membrane anchored pro-EGF requires a membrane serine protease other than kallikrein in rat kidney. *Regul Pept* **122:** 119–129.

Le Gall SM, Bobe P, Reiss K, Horiuchi K, Niu XD, Lundell D, Gibb DR, Conrad D, Saftig P, Blobel CP. 2009. ADAMs 10 and 17 represent differentially regulated components of a general shedding machinery for membrane proteins such as transforming growth factor α, L-selectin, and tumor necrosis factor α. *Mol Biol Cell* **20:** 1785–1794.

Le Gall SM, Maretzky T, Issuree PD, Niu XD, Reiss K, Saftig P, Khokha R, Lundell D, Blobel CP. 2010. ADAM17 is regulated by a rapid and reversible mechanism that controls access to its catalytic site. *J Cell Sci* **123:** 3913–3922.

Lemberg MK, Menendez J, Misik A, Garcia M, Koth CM, Freeman M. 2005. Mechanism of intramembrane proteolysis investigated with purified rhomboid proteases. *EMBO J* **24:** 464–472.

Lemjabbar H, Basbaum C. 2002. Platelet-activating factor receptor and ADAM10 mediate responses to *Staphylococcus aureus* in epithelial cells. *Nat Med* **8:** 41–46.

Lemmon MA, Schlessinger J. 2010. Cell signaling by receptor tyrosine kinases. *Cell* **141:** 1117–1134.

Leonard JD, Lin F, Milla ME. 2005. Chaperone-like properties of the prodomain of TNFα-converting enzyme (TACE) and the functional role of its cysteine switch. *Biochem J* **387:** 797–805.

Lieberman RL, Wolfe MS. 2007. From rhomboid function to structure and back again. *Proc Natl Acad Sci* **104:** 8199–8200.

Lohi O, Urban S, Freeman M. 2004. Diverse substrate recognition mechanisms for rhomboids; thrombomodulin is cleaved by Mammalian rhomboids. *Curr Biol* **14:** 236–241.

Luetteke NC, Qiu TH, Peiffer RL, Oliver P, Smithies O, Lee DC. 1993. TGF-α deficiency results in hair follicle and eye abnormalities in targeted and waved-1 mice. *Cell* **73:** 263–278.

Mann GB, Fowler KJ, Gabriel A, Nice EC, Williams RL, Dunn AR. 1993. Mice with a null mutation of the TGF-α gene have abnormal skin architecture, wavy hair, and curly whiskers and often develop corneal inflammation. *Cell* **73**: 249–261.

Massague J, Pandiella A. 1993. Membrane-anchored growth factors. *Annu Rev Biochem* **62**: 515–541.

McIlwain DR, Lang PA, Maretzky T, Hamada K, Kazuhito O, Kumar Maney S, Berger T, Murthy A, Duncan G, Xu HC, et al. 2012. iRhom2 regulates innate immunity via TACE/ADAM17. *Science* **335**: 229–232.

Merlos-Suarez A, Ruiz-Paz S, Baselga J, Arribas J. 2001. Metalloprotease-dependent protransforming growth factor-α ectodomain shedding in the absence of tumor necrosis factor-α-converting enzyme. *J Biol Chem* **276**: 48510–48517.

Mohammed FF, Smookler DS, Taylor SE, Fingleton B, Kassiri Z, Sanchez OH, English JL, Matrisian LM, Au B, Yeh WC, et al. 2004. Abnormal TNF activity in Timp3$^{-/-}$ mice leads to chronic hepatic inflammation and failure of liver regeneration. *Nat Genet* **36**: 969–977.

Murayama Y, Shinomura Y, Oritani K, Miyagawa J, Yoshida H, Nishida M, Katsube F, Shiraga M, Miyazaki T, Nakamoto T, et al. 2008. The tetraspanin CD9 modulates epidermal growth factor receptor signaling in cancer cells. *J Cell Physiol* **216**: 135–143.

Murphy G. 2008. The ADAMs: Signalling scissors in the tumour microenvironment. *Nat Rev Cancer* **8**: 929–941.

Murphy G. 2009. Regulation of the proteolytic disintegrin metalloproteinases, the "Sheddases." *Semin Cell Dev Biol* **20**: 138–145.

Murphy G. 2011. Tissue inhibitors of metalloproteinases. *Genome Biol* **12**: 233.

Murthy A, Defamie V, Smookler DS, Di Grappa MA, Horiuchi K, Federici M, Sibilia M, Blobel CP, Khokha R. 2010. Ectodomain shedding of EGFR ligands and TNFR1 dictates hepatocyte apoptosis during fulminant hepatitis in mice. *J Clin Invest* **120**: 2731–2744.

Nakagawa T, Guichard A, Castro CP, Xiao Y, Rizen M, Zhang HZ, Hu D, Bang A, Helms J, Bier E, et al. 2005. Characterization of a human rhomboid homolog, p100hRho/RHBDF1, which interacts with TGF-α family ligands. *Dev Dyn* **233**: 1315–1331.

Naoe T, Kiyoi H. 2004. Normal and oncogenic FLT3. *Cell Mol Life Sci* **61**: 2932–2938.

Palladino MA, Bahjat FR, Theodorakis EA, Moldawer LL. 2003. Anti-TNF-α therapies: The next generation. *Nat Rev Drug Discov* **2**: 736–746.

Pandiella A, Bosenberg MW, Huang EJ, Besmer P, Massague J. 1992. Cleavage of membrane-anchored growth factors involves distinct protease activities regulated through common mechanisms. *J Biol Chem* **267**: 24028–24033.

Pascall JC, Brown KD. 2004. Intramembrane cleavage of ephrinB3 by the human rhomboid family protease, RHBDL2. *Biochem Biophys Res Commun* **317**: 244–252.

Peiretti F, Canault M, Deprez-Beauclair P, Berthet V, Bonardo B, Juhan-Vague I, Nalbone G. 2003a. Intracellular maturation and transport of tumor necrosis factor α converting enzyme. *Exp Cell Res* **285**: 278–285.

Peiretti F, Deprez-Beauclair P, Bonardo B, Aubert H, Juhan-Vague I, Nalbone G. 2003b. Identification of SAP97 as an intracellular binding partner of TACE. *J Cell Sci* **116**: 1949–1957.

Peschon JJ, Slack JL, Reddy P, Stocking KL, Sunnarborg SW, Lee DC, Russell WE, Castner BJ, Johnson RS, Fitzner JN, et al. 1998. An essential role for ectodomain shedding in mammalian development. *Science* **282**: 1281–1284.

Piepkorn M, Pittelkow MR, Cook PW. 1998. Autocrine regulation of keratinocytes: The emerging role of heparin-binding, epidermal growth factor-related growth factors. *J Invest Dermatol* **111**: 715–721.

Pitulescu ME, Adams RH. 2010. Eph/ephrin molecules—A hub for signaling and endocytosis. *Genes Dev* **24**: 2480–2492.

Puente XS, Lopez-Otin C. 2004. A genomic analysis of rat proteases and protease inhibitors. *Genome Res* **14**: 609–622.

Rovida E, Paccagnini A, Del Rosso M, Peschon J, Dello Sbarba P. 2001. TNF-α-converting enzyme cleaves the macrophage colony-stimulating factor receptor in macrophages undergoing activation. *J Immunol* **166**: 1583–1589.

Ruohola-Baker H, Grell E, Chou TB, Baker D, Jan LY, Jan YN. 1993. Spatially localized rhomboid is required for establishment of the dorsal-ventral axis in *Drosophila* oogenesis. *Cell* **73**: 953–965.

Saftig P, Reiss K. 2011. The "A disintegrin and metalloproteases" ADAM10 and ADAM17: Novel drug targets with therapeutic potential? *Eur J Cell Biol* **90**: 527–535.

Sahin U, Weskamp G, Kelly K, Zhou HM, Higashiyama S, Peschon J, Hartmann D, Saftig P, Blobel CP. 2004. Distinct roles for ADAM10 and ADAM17 in ectodomain shedding of six EGFR ligands. *J Cell Biol* **164**: 769–779.

Salomon DS, Brandt R, Ciardiello F, Normanno N. 1995. Epidermal growth factor-related peptides and their receptors in human malignancies. *Crit Rev Oncol Hematol* **19**: 183–232.

Sanderson MP, Erickson SN, Gough PJ, Garton KJ, Wille PT, Raines EW, Dunbar AJ, Dempsey PJ. 2005. ADAM10 mediates ectodomain shedding of the betacellulin precursor activated by *p*-aminophenylmercuric acetate and extracellular calcium influx. *J Biol Chem* **280**: 1826–1837.

Sastry SK, Elferink LA. 2011. Checks and balances: Interplay of RTKs and PTPs in cancer progression. *Biochem Pharmacol* **82**: 435–440.

Schafer B, Marg B, Gschwind A, Ullrich A. 2004. Distinct ADAM metalloproteinases regulate G protein-coupled receptor-induced cell proliferation and survival. *J Biol Chem* **279**: 47929–47938.

Schlesinger A, Kiger A, Perrimon N, Shilo BZ. 2004. Small wing PLCγ is required for ER retention of cleaved Spitz during eye development in *Drosophila*. *Dev Cell* **7**: 535–545.

Schlessinger J. 2000. Cell signaling by receptor tyrosine kinases. *Cell* **103**: 211–225.

Schlondorff J, Becherer JD, Blobel CP. 2000. Intracellular maturation and localization of the tumour necrosis factor α convertase (TACE). *Biochem J* **347**: 131–138.

Schneider MR, Wolf E. 2009. The epidermal growth factor receptor ligands at a glance. *J Cell Physiol* **218**: 460–466.

Seals DF, Courtneidge SA. 2003. The ADAMs family of metalloproteases: Multidomain proteins with multiple functions. *Genes Dev* **17**: 7–30.

Shilo BZ. 2003. Signaling by the *Drosophila* epidermal growth factor receptor pathway during development. *Exp Cell Res* **284:** 140–149.

Sibilia M, Kroismayr R, Lichtenberger BM, Natarajan A, Hecking M, Holcmann M. 2007. The epidermal growth factor receptor: From development to tumorigenesis. *Differentiation* **75:** 770–787.

Siggs OM, Xiao N, Wang Y, Shi H, Tomisato W, Li X, Xia Y, Beutler B. 2012. iRhom2 is required for the secretion of mouse TNF α. *Blood* **119:** 5769–5771.

Singh AB, Harris RC. 2005. Autocrine, paracrine and juxtacrine signaling by EGFR ligands. *Cell Signal* **17:** 1183–1193.

Smith KM, Gaultier A, Cousin H, Alfandari D, White JM, DeSimone DW. 2002. The cysteine-rich domain regulates ADAM protease function in vivo. *J Cell Biol* **159:** 893–902.

Smith MH, Ploegh HL, Weissman JS. 2011. Road to ruin: Targeting proteins for degradation in the endoplasmic reticulum. *Science* **334:** 1086–1090.

Soond SM, Everson B, Riches DW, Murphy G. 2005. ERK-mediated phosphorylation of Thr735 in TNFα-converting enzyme and its potential role in TACE protein trafficking. *J Cell Sci* **118:** 2371–2380.

Sporn MB, Roberts AB. 1985. Autocrine growth factors and cancer. *Nature* **313:** 745–747.

Strisovsky K, Sharpe HJ, Freeman M. 2009. Sequence-specific intramembrane proteolysis: Identification of a recognition motif in rhomboid substrates. *Mol Cell* **36:** 1048–1059.

Sturtevant MA, Roark M, Bier E. 1993. The *Drosophila* rhomboid gene mediates the localized formation of wing veins and interacts genetically with components of the EGF-R signaling pathway. *Genes Dev* **7:** 961–973.

Sunnarborg SW, Hinkle CL, Stevenson M, Russell WE, Raska CS, Peschon JJ, Castner BJ, Gerhart MJ, Paxton RJ, Black RA, et al. 2002. Tumor necrosis factor-α converting enzyme (TACE) regulates epidermal growth factor receptor ligand availability. *J Biol Chem* **277:** 12838–12845.

Tanaka M, Nanba D, Mori S, Shiba F, Ishiguro H, Yoshino K, Matsuura N, Higashiyama S. 2004. ADAM binding protein Eve-1 is required for ectodomain shedding of epidermal growth factor receptor ligands. *J Biol Chem* **279:** 41950–41959.

Tsruya R, Wojtalla A, Carmon S, Yogev S, Reich A, Bibi E, Merdes G, Schejter E, Shilo BZ. 2007. Rhomboid cleaves Star to regulate the levels of secreted Spitz. *EMBO J* **26:** 1211–1220.

Ullrich A, Schlessinger J. 1990. Signal transduction by receptors with tyrosine kinase activity. *Cell* **61:** 203–212.

Urban S, Dickey SW. 2011. The rhomboid protease family: A decade of progress on function and mechanism. *Genome Biol* **12:** 231.

Urban S, Freeman M. 2003. Substrate specificity of rhomboid intramembrane proteases is governed by helix-breaking residues in the substrate transmembrane domain. *Mol Cell* **11:** 1425–1434.

Urban S, Lee JR, Freeman M. 2001. *Drosophila* rhomboid-1 defines a family of putative intramembrane serine proteases. *Cell* **107:** 173–182.

Urban S, Lee JR, Freeman M. 2002. A family of rhomboid intramembrane proteases activates all *Drosophila* membrane-tethered EGF ligands. *EMBO J* **21:** 4277–4286.

Wang Y, Zhang Y, Ha Y. 2006. Crystal structure of a rhomboid family intramembrane protease. *Nature* **444:** 179–180.

Ward CL, Kopito RR. 1994. Intracellular turnover of cystic fibrosis transmembrane conductance regulator. Inefficient processing and rapid degradation of wild-type and mutant proteins. *J Biol Chem* **269:** 25710–25718.

Wasserman JD, Urban S, Freeman M. 2000. A family of rhomboid-like genes: *Drosophila* rhomboid-1 and roughoid/rhomboid-3 cooperate to activate EGF receptor signaling. *Genes Dev* **14:** 1651–1663.

White JM. 2003. ADAMs: Modulators of cell–cell and cell–matrix interactions. *Curr Opin Cell Biol* **15:** 598–606.

Wisniewska M, Goettig P, Maskos K, Belouski E, Winters D, Hecht R, Black R, Bode W. 2008. Structural determinants of the ADAM inhibition by TIMP-3: Crystal structure of the TACE-N-TIMP-3 complex. *J Mol Biol* **381:** 1307–1319.

Wyckoff J, Wang W, Lin EY, Wang Y, Pixley F, Stanley ER, Graf T, Pollard JW, Segall J, Condeelis J. 2004. A paracrine loop between tumor cells and macrophages is required for tumor cell migration in mammary tumors. *Cancer Res* **64:** 7022–7029.

Xu P, Derynck R. 2010. Direct activation of TACE-mediated ectodomain shedding by p38 MAP kinase regulates EGF receptor-dependent cell proliferation. *Mol Cell* **37:** 551–566.

Xu D, Sharma C, Hemler ME. 2009. Tetraspanin12 regulates ADAM10-dependent cleavage of amyloid precursor protein. *FASEB J* **23:** 3674–3681.

Yamazaki S, Iwamoto R, Saeki K, Asakura M, Takashima S, Yamazaki A, Kimura R, Mizushima H, Moribe H, Higashiyama S, et al. 2003. Mice with defects in HB-EGF ectodomain shedding show severe developmental abnormalities. *J Cell Biol* **163:** 469–475.

Yan Z, Zou H, Tian F, Grandis JR, Mixson AJ, Lu PY, Li LY. 2008. Human rhomboid family-1 gene silencing causes apoptosis or autophagy to epithelial cancer cells and inhibits xenograft tumor growth. *Mol Cancer Ther* **7:** 1355–1364.

Yanez-Mo M, Gutierrez-Lopez MD, Cabanas C. 2011. Functional interplay between tetraspanins and proteases. *Cell Mol Life Sci* **68:** 3323–3335.

Yarden Y, Sliwkowski MX. 2001. Untangling the ErbB signalling network. *Nat Rev Mol Cell Biol* **2:** 127–137.

Yogev S, Schejter ED, Shilo BZ. 2008. *Drosophila* EGFR signalling is modulated by differential compartmentalization of rhomboid intramembrane proteases. *EMBO J* **27:** 1219–1230.

Zettl M, Adrain C, Strisovsky K, Lastun V, Freeman M. 2011. Rhomboid family pseudoproteases use the ER quality control machinery to regulate intercellular signaling. *Cell* **145:** 79–91.

Zheng Y, Schlondorff J, Blobel CP. 2002. Evidence for regulation of the tumor necrosis factor α-convertase (TACE) by protein-tyrosine phosphatase PTPH1. *J Biol Chem* **277:** 42463–42470.

Complexity of Receptor Tyrosine Kinase Signal Processing

Natalia Volinsky and Boris N. Kholodenko

Systems Biology Ireland, University College Dublin, Belfield, Dublin 4, Ireland

Correspondence: boris.kholodenko@ucd.ie

Our knowledge of molecular mechanisms of receptor tyrosine kinase (RTK) signaling advances with ever-increasing pace. Yet our understanding of how the spatiotemporal dynamics of RTK signaling control specific cellular outcomes has lagged behind. Systems-centered experimental and computational approaches can help reveal how overlapping networks of signal transducers downstream of RTKs orchestrate specific cell-fate decisions. We discuss how RTK network regulatory structures, which involve the immediate posttranslational and delayed transcriptional controls by multiple feed forward and feedback loops together with pathway cross talk, adapt cells to the combinatorial variety of external cues and conditions. This intricate network circuitry endows cells with emerging capabilities for RTK signal processing and decoding. We illustrate how mathematical modeling facilitates our understanding of RTK network behaviors by unraveling specific systems properties, including bistability, oscillations, excitable responses, and generation of intricate landscapes of signaling activities.

Since the first cloning of the cDNA encoding the epidermal growth factor (EGF) receptor (EGFR), signaling by receptor tyrosine kinases (RTKs) has been in the limelight of scientific interest owing to their central role in the regulation of development, cell motility, proliferation, differentiation, glucose metabolism, and apoptosis (Hunter 2000; Schlessinger 2000; Lemmon and Schlessinger 2010). The RTK family comprises more than 50 cell-surface receptors with intrinsic tyrosine kinase activity. All RTKs consist of three major domains: an extracellular domain for ligand binding, a membrane-spanning segment, and a cytoplasmic domain, which possesses tyrosine kinase activity and contains phosphorylation sites with tyrosine, serine, and threonine residues. Following

ligand binding, RTKs undergo dimerization (e.g., EGFR) or allosteric transitions (e.g., insulin receptor [IR] and insulin-like growth factor-1 receptor [IGF-1R] that are associated into oligomers before the ligand binding), resulting in receptor activation. Auto- and/or *trans*-phosphorylation of RTKs transmit biochemical signals to cytoplasmic adaptor proteins and enzymes, which contain characteristic protein domains, such as Src homology (SH2 and SH3), phosphotyrosine binding (PTB), and pleckstrin homology (PH) domains. These domains act as docking sites for phosphotyrosines or phospholipids thereby triggering the mobilization of these proteins to the cell surface (Pawson and Nash 2003). Subsequently, signals propagate through a tangled network of inter-

---

Cite this article as *Cold Spring Harb Perspect Biol* doi: 10.1101/cshperspect.a009043

connecting proteins and signaling cascades to the nucleus, inducing transcriptional responses of immediate early genes (IEG) and delayed early genes (DEG) (Avraham and Yarden 2011).

The idea of isolated linear pathways that relate signals and receptors to specific genes has given way to the concept of signaling networks, which allow a limited number of RTKs to generate an exponentially larger number of functional outcomes as a result of combinatorial interactions. Although early genetic experiments seem to support a linear pathway view, the current biochemical and imaging data suggest that any given RTK, downstream adaptors, small G-proteins (small GTPases, such as Ras, Rac, Rho, or Cdc42), and activated kinases interact with a large variety of signaling molecules resulting in highly interconnected networks (von Kriegsheim et al. 2009). These signaling networks not only transmit but also process and integrate signals. For instance, signals from different RTKs are integrated through common adaptor proteins and other points of pathway cross talk that include small GTPases and cytoplasmic kinases, such as the Src family kinases, phosphatidylinositol 3-kinase (PI3K), and mitogen activated protein kinases (MAPK) (Kholodenko et al. 2010).

Given that the protein complements of adaptor proteins, small GTPases, and kinases, which mediate RTK-induced signal transduction, overlap for all known RTKs (Lemmon and Schlessinger 2010), questions arise as to how cells can maintain specificity when activated by multiple cues? Also how can coherent cellular decisions, including whether to undergo proliferation, differentiation, or die, be made? The concept is emerging that for any RTK pathway, there is no single protein or gene responsible for signaling specificity. Rather, specificity is determined by the spatiotemporal dynamics of activation of signaling proteins, IEGs, and DEGs, downstream of RTKs (Murphy et al. 2004; Kholodenko 2006; Nakakuki et al. 2010). Yet many signaling events activated by RTKs are partially redundant, and different RTKs can compensate for each other in many cellular functions (Xu and Huang 2010).

A hallmark of complex signaling and early gene networks downstream of RTKs are multiple feed forward and feedback loops, both negative and positive. These regulations operate on different timescales, precisely tuning the signaling outcome and often convert analog input signals into digital outputs (Kholodenko et al. 2010). Immediate feedback or feedforward regulations occur through interactions between two proteins or a protein and a lipid or through posttranslational modifications, such as phosphorylation that alters protein activity. For instance, active extracellular signal regulated kinase (ERK) phosphorylates and inactivates the kinase Raf-1, which is upstream of ERK in the Ras/ERK signaling cascade (Dougherty et al. 2005). These immediate feedforward and feedback loops operate on the timescale of seconds to minutes, which are the characteristic times of the corresponding interactions or catalytic reactions. Another large group of feedback regulators, such as Sprouty, Spred, and Mitogen-inducible gene-6 (Mig-6)/receptor-associated late transducer (RALT) (Gotoh 2009; Murphy et al. 2010; Segatto et al. 2011) requires RTK-induced gene transcription and translation. These delayed feedbacks adapt cells to a more permanent external stimulation, persisting on the timescale of tens of minutes and hours. For instance, transcriptionally induced IEGs, such as dual specificity phosphatases (DUSP), attenuate RTK-induced MAPK signaling. Owing to these multiple feedback and feedforward loops, the understanding of the spatiotemporal dynamics of RTK networks requires more than knowledge of binding partners, their structures, and interactions (Kholodenko 2009).

In this review, we focus on a systems-centered view of the biology of RTK signaling networks. We discuss cross talk between different RTKs and pathways and illustrate how this cross talk may alter the functional outcome of signaling by individual RTKs. We analyze feedforward and feedback loops and show how this regulatory circuitry can bring about the intricate dynamic behavior of RTK networks, including bistability, oscillations, and excitable overshoot transitions. We illustrate how computational modeling and systems analysis can facilitate our understanding of the complex spatiotemporal behavior of RTK pathways based on the knowledge of molecular mechanisms of signal transduction. Finally, we discuss the challenges ahead that will stimulate further research.

RTK PATHWAY CROSS TALK: FINE BALANCES AT MULTIPLE LEVELS

Cells in situ are exposed to a plethora of signaling cues that activate multiple RTKs. Although individual RTKs have been extensively studied, how signaling networks integrate multiple cues is less understood. How do cells integrate and process external information by exploiting pathway cross talk? Cross talk between RTK-stimulated pathways exists at different signaling levels, which include the level of receptors, adaptor proteins, scaffolds, small GTPases, downstream kinases, and transcriptional responses (Pawson et al. 2001).

Cross talk at the receptor level can be mediated by several mechanisms. An RTK can induce activation of structurally unrelated RTKs, for instance, stimulation of IGF-1R can lead to EGFR (Ahmad et al. 2004) or ErbB2 activity (Balana et al. 2001). At least two mechanisms of cross talk were suggested: direct dimerization between structurally independent RTKs, such as IGF-1R or c-Met with ErbB family receptors (Balana et al. 2001; Ahmad et al. 2004; Tanizaki et al. 2011) and transactivation mediated by a cytoplasmic tyrosine kinase, such as Src. In the latter case, once activated by IGF-1R, Src binds and phosphorylates EGFR, thus promoting EGFR catalytic activity (Jones et al. 2006). Using high-throughput Western blotting and Bayesian interference techniques, different cross talk mechanisms were recently described (Ciaccio et al. 2010). Whereas EGF induced rapid phosphorylation of multiple sites on EGFR, phosphorylation of other RTKs, as well as some EGFR tyrosines, was detected at later time points. In particular, phosphorylation of the c-Met receptor was only detected after about 15 min of the EGFR phosphorylation. These delayed responses suggest the involvement of downstream signaling cascades in mediating RTK cross activation (Ciaccio et al. 2010).

Cross talk between distinct RTKs was shown in non-small cell lung cancer-derived cell lines. Using specific inhibitors of EGFR and c-Met, it was shown that these two RTKs positively influence each other, and that inhibition of one receptor negatively regulates the other (Guo et al. 2008). Another example is a direct interaction between c-Met and insulin receptors in hepatocytes, which facilitate optimal activation of downstream signaling pathways and glucose metabolism (Fafalios et al. 2011).

Different cross talk mechanisms, in which inhibition of one RTK leads to activation of the other RTK have also been observed and suggested to be one of the causes of chemoresistance in cancer patients (Jones et al. 2006; Xu and Huang 2010). In fact, a cocktail of inhibitors targeting several RTKs, such as EGFR, c-Met, and PDGF has been shown to be more effective than a single drug treatment (Stommel et al. 2007). RTK cross talk can also be mediated solely by downstream signaling interactions without affecting the receptors themselves. In this case, costimulation by two ligands, which often occurs in vivo, can lead to synergic or mutually inhibiting responses (Borisov et al. 2009; Martin et al. 2009).

When signals propagate through different branches converging at a common target, both branches can add to the overall response of the target (Kholodenko et al. 1997). Feedforward and feedback loops embracing interacting RTK pathways make the input–output response relationships of one pathway dependent on the activity of the other pathway, thereby creating context-dependent signaling output. A recent study explored how a concordant interplay between RTK pathways stimulated by EGF and insulin can potentiate signaling by the ERK/MAPK at physiological, low EGF levels in HEK293 cells (Borisov et al. 2009). Although the EGFR and IR networks share many downstream components, their responses to cognate stimuli are different. In HEK293 cells, EGF causes strong ERK activation, whereas insulin poorly activates ERK. The main insulin function is metabolic, including the control of glucose metabolism, and stimulation of protein and lipid syntheses. Using a combined experimental and computational analysis, it was shown that at low physiological concentration of EGF, cross talk between the EGFR and IR networks converts the insulin-induced increase in the PIP_3 concentration into enhanced ERK activation, whereas at saturating EGF levels the insulin effect becomes insignificant (Borisov et al. 2009).

The data and computational modeling results suggest that major cross talk mechanisms that amplify ERK signaling by insulin are localized upstream of Ras and at the Ras/Raf level (Borisov et al. 2009). Among the variety of cross talk interactions affecting multiple Ras activation and inactivation routes, five key network nodes are shown to be crucial for the EGF-insulin synergy. These nodes involve the adaptor proteins, Grb2-associated binder-1 (GAB1) and insulin receptor substrates (IRS), $PI3K/PIP_3$ node, soluble tyrosine kinase Src, and the SH2-domain containing protein tyrosine phosphatase-2 (SHP2) that is located upstream of the ERK kinase MEK. The computational model reveals key features of the EGFR and IR network signal processing brought about by: (1) coincidence detection of EGF and insulin stimuli through GAB1 phosphorylation response, (2) coherent feedforward loop from EGFR to Raf via Src (this is a network motif in which an initial signal induces an intermediate input, and both the initial and intermediate inputs are needed to generate the final output [Mangan et al. 2003], see also below), and (3) multiple positive ($PI3K \rightarrow PIP_3 \rightarrow GAB1 \rightarrow PI3K$) and negative (ERK ⊣ GAB1, ERK ⊣ Son of Sevenless (SOS), mammalian target of rapamycin (mTOR) ⊣ IRS) feedback loops. The simplified scheme, shown in Figure 1, illustrates how insulin enhances EGF-induced mitogenesis through two partially redundant and compensating signaling branches via IRS and GAB1.

Receptor cross activation can also be mediated via transcriptional induction of an RTK (or its ligands) by another RTK (Jones et al. 2006; Esposito et al. 2008; Gujral et al. 2012; Velpula et al. 2012). For instance, in neuroblastoma-derived cell lines retinoic acid induces activation of the RTK Ret, leading to cell differentiation, as indicated by morphological changes and expression of several differentiation markers. Several studies suggested that TrkB (another RTK and indicator of poor prognosis for neuroblastoma tumors) can link retinoic acid stimulation and the phenotype change. Ret activation induces expression of both TrkB and its primary ligand brain-derived neurotrophic factor (BDNF) (Kaplan et al. 1993; Esposito et al. 2008). Moreover, TrkB knockdown prevents cell differentiation (Esposito et al. 2008). "Reverse" cross talk, in which TrkA and TrkB receptors induce activation of Ret, was also observed (Tsui-Pierchala et al. 2002; Esposito et al. 2008). Although in this case positive regulation of Ret

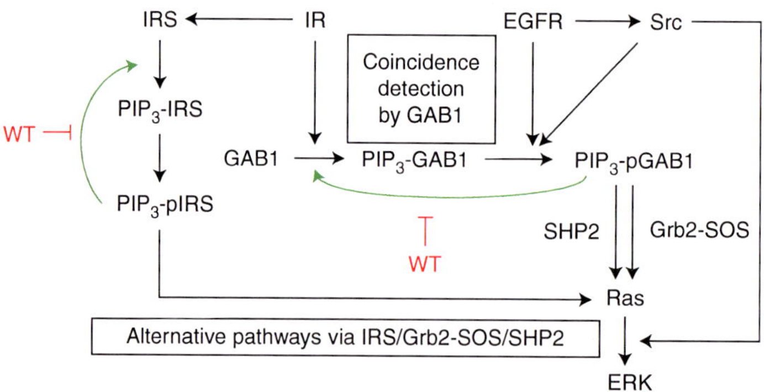

Figure 1. Mechanisms of insulin-EGF signal integration. Synergistic ERK activation arises from coincidence detection of insulin and EGF stimuli at the level of GAB1 adaptor protein. GAB1 is massively recruited to the membrane by IR signaling ($GAB1 \rightarrow PIP_3$-GAB1) and subsequently phosphorylated by EGFR and activated Src (PIP_3-GAB1$\rightarrow PIP_3$-pGAB1). (Modified from Borisov et al. 2009; reprinted, with permission, from the authors.)

Cite this article as *Cold Spring Harb Perspect Biol* doi: 10.1101/cshperspect.a009043

does not involve transcriptional regulation (Esposito et al. 2008), TrkB→Ret→TrkB positive feedback may lead to emerging systems properties, such as bistability (see section below "Intricate RTK Network Dynamics Are Brought about by Feedback Loops").

In a recent study (Gujral et al. 2012), breast cancer samples were subjected to "reverse-phase" protein microarrays (Sevecka and MacBeath 2006) and phosphorylation or total protein levels of multiple signaling molecules were tested. The resulted data were analyzed by unsupervised hierarchical clustering (Herrero et al. 2001), revealing 12 major clusters of correlating proteins. Interestingly, c-Met RTK phosphorylated on Tyr1349 was clustered together with the total protein level of Axl, which is a structurally independent RTK. This finding was further experimentally validated; siRNA-mediated c-Met targeting resulted in a significant decrease of Axl mRNA levels. These receptors were also shown to interact on a protein level, c-Met stimulation with hepatocyte growth factor-induced activation of Axl. Moreover, both receptors physically interacted with each other, at least when ectopically expressed. Surprisingly, Axl down-regulation or stimulation did not affect c-Met function on any level, suggesting complex interplay of these two RTKs (Gujral et al. 2012).

Additional levels of complexity, which we do not discuss here, involve cross talk between RTKs and other receptors, such as G-protein-coupled receptors, adhesion molecules, and nuclear receptors.

RTK NETWORKS ARE TIGHTLY CONTROLLED BY FEEDBACK AND FEEDFORWARD LOOPS

Coherent and Incoherent Feedforward Loops

A useful approach toward understanding of signaling and transcriptional networks is the analysis of so-called network motifs (Milo et al. 2002). In addition to the simplest linear signaling design in which the initial signal A regulates the intermediate output B, and B regulates the output C, signaling networks contain more complicated motifs including feedforward loops (FFL) where A regulates B, while A and B jointly regulate C (Fig. 2) (Mangan and Alon

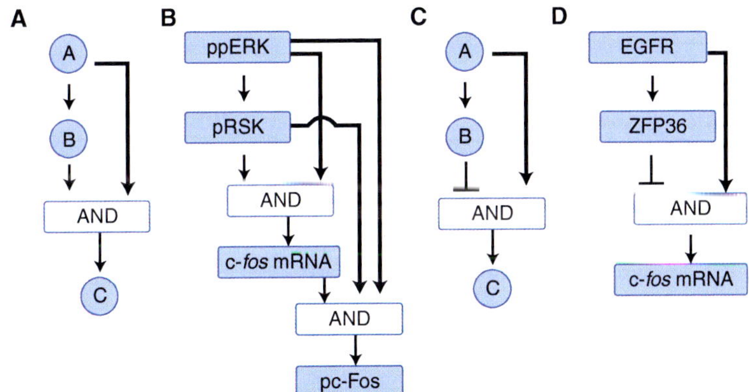

Figure 2. Coherent and incoherent feedforward motifs. (*A*) Basic structure of coherent feed forward loop (coherent FFL [Mangan and Alon 2003]). (*B*) Schematic representation of ERK-induced c-Fos expression and activation that includes a cascade of coherent FFLs. Active ERK (ppERK) and RSK (pRSK) activate transcription factors required for c-Fos expression and therefore, *c-fos* mRNA expression. ERK and RSK stabilize and activate the nascent c-Fos protein by phosphorylation making an additional AND gate (based on data from Nakakuki et al. 2010). (*C*) Basic structure of incoherent feed forward loop of type I (incoherent FFL [Mangan and Alon 2003]). (*D*) Schematic representation of EGFR regulated *c-fos* mRNA availability in terms of incoherent FFL. On stimulation, EGFR induces expression of *c-fos* (*B*). ZFP36 is also induced by EGFR and mediates *c-fos* RNA degradation. (Based on data from Avraham and Yarden 2011.)

2003; Shoval and Alon 2010). In physiological systems, FFL motifs may be combined into larger integrated structures, leading to complex signaling and transcriptional circuits (Shoval and Alon 2010).

In coherent FFLs, the initial input A activates the intermediate output B, while A and B form the logical gates "AND" or "OR," thereby providing different regulation of the outcome C. This final outcome can be a downstream effector or process, such as gene promoter or protein activation, that is responsive to two inputs, in which only one or both inputs are required in OR or AND gates, respectively. For an AND gate, increasing time delays related to accumulation of B may be required to activate C, and consequently, an AND motif shows delayed "ON" and immediate "OFF" responses (Mangan and Alon 2003). An OR gate motif is characterized by immediate ON and delayed OFF responses (Shoval and Alon 2010). Interestingly, an AND coherent FFL motif distinguishes between transient and sustained signals. This regulatory motif is found in the networks stimulated by two different RTK ligands, EGF and platelet-derived growth factor (PDGF), which induce transient and sustained ERK activation, respectively (Murphy et al. 2002, 2004). Expression of several IEGs, including *c-fos*, is induced by active ERK and its downstream effector, the p90 ribosomal S6 kinase (RSK). However, the nascent c-Fos protein is unstable, and phosphorylation by ERK and RSK is required for its stabilization and activation. If ERK and RSK signals are transient, there will be no appreciable kinase activity at the time when the c-Fos protein is de novo synthesized. Thus, c-Fos will not be phosphorylated, and will be rapidly degraded. Therefore, only sustained ERK and RSK signaling can allow for a strong c-Fos induction (Fig. 2B) (Murphy et al. 2002, 2004; Nakakuki et al. 2010). This mechanism of discrimination between transient and sustained ERK activity at the IEG level is common for different RTKs and cell types and may lead to distinct cell-fate decisions, such as proliferation or tumorigenicity (Nakakuki et al. 2010).

Incoherent FFLs where A activates C, but also activates B, which is the repressor of C, is another common motif in network regulation (Fig. 2C). In case of an AND gate, both the presence of an activator A and the lack of a repressor B are required for obtaining the outcome C. For instance, if two independent elements on a promoter bind an activator A and a repressor B, appreciable transcriptional response will only be observed when A is bound and B is absent (Mangan and Alon 2003). Depending on parameters, this motif can endow the system with noise filtering and adaptation (Goentoro et al. 2009; Ma et al. 2009). The repressor B serves as a "memory" element keeping track of the previous levels of the activator A, and the output C detects fold-changes of A rather than the absolute A values (Goentoro et al. 2009; Ma et al. 2009). For instance, a sustained input will result in peak-like transient response, followed by full or partial adaptation, because of accumulating levels of the repressor (Ma et al. 2009).

An incoherent FFL in EGFR-mediated transcriptional circuits was recently described (Amit et al. 2007). EGFR induces expression of the Zink Finger Protein 36 (ZFP36), which binds to AU-rich elements, predominantly found in the 3'-untranslated regions (3'-UTR) of unstable mRNA molecules, and promotes their deadenylation and subsequent degradation (Chen et al. 2001). A systematic analysis of EGF-induced genes revealed that many of these genes contain such 3'-UTR elements, thus suggesting a widespread role of ZFP36 in IEG and DEG regulation (Amit et al. 2007). Although ZFP36 does not affect gene transcription directly, it regulates the availability of multiple mRNA molecules, thereby forming an incoherent FFL (Fig. 2D) (Amit et al. 2007; Avraham and Yarden 2011).

Positive and Negative Feedback Regulation

The concept of feedback control goes back to the ancient Greeks who used float regulators to keep a constant level of water in a tank or oil in a lamp (Sauro and Kholodenko 2004). In biological control systems, negative and positive feedback loops are the most fundamental features. Positive feedback amplifies the signal, whereas negative feedback attenuates it. RTK network topologies involve comprehensive and intricate

     Cite this article as *Cold Spring Harb Perspect Biol* doi: 10.1101/cshperspect.a009043

regulatory structures of immediate and delayed feedback loops (Fig. 3) (Kiyatkin et al. 2006; Amit et al. 2007; Gotoh 2009; Lemmon and Schlessinger 2010; Sturm et al. 2010; Avraham and Yarden 2011; Segatto et al. 2011).

Positive and Negative Feedbacks at the Receptor Level

In RTK pathways, both the receptor abundance and ligand availability are tightly controlled by positive and negative feedback loops. For instance, RTK ubiquitination by the E3 ubiquitin ligase Cbl and subsequent degradation of the receptor in lysosomes creates negative feedback at the receptor level. Cbl is recruited to the phosphorylated receptor either directly or via the Grb2 adaptor and is further phosphorylated by RTK or c-Src to become activated (Zwang and Yarden 2009). Positive feedback can be mediated by reactive oxygen species, which are produced in response to RTK activation and inhibit

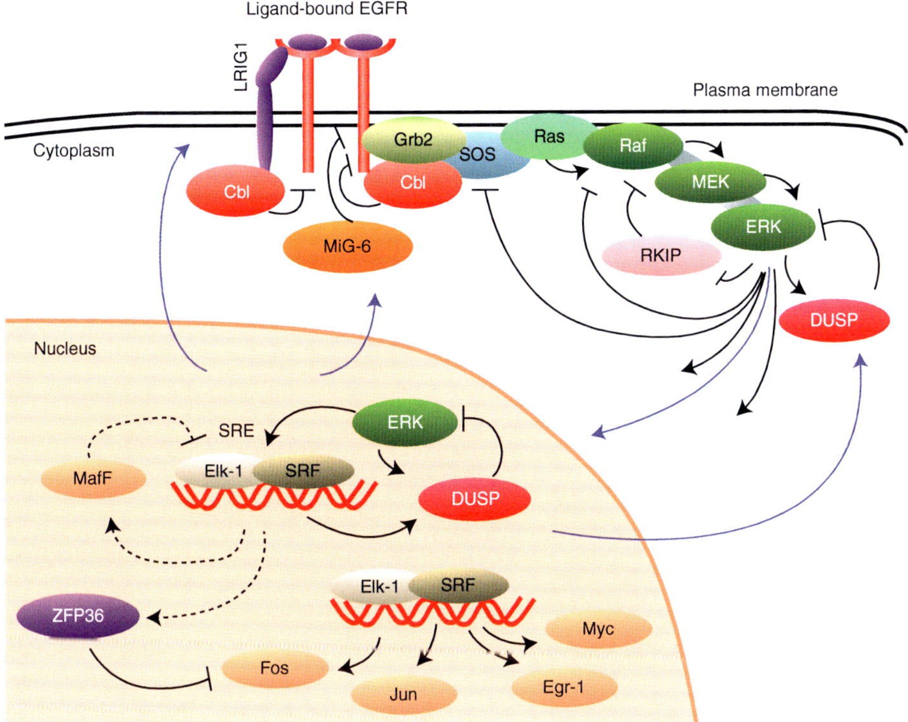

Figure 3. Simplified scheme presenting multiple feedback loops in the EGFR pathway. Upon EGFR ligation and activation, protein complexes nucleated by scaffolds are formed at the plasma membrane. Activated Ras induces activation of the Raf-MEK-ERK signaling cascade. Active ERK phosphorylates several upstream signaling regulators, including Raf and SOS, forming negative feedback loops. However, ERK also phosphorylates and inhibits RKIP, which is a negative regulator of Raf, thus creating positive feedback. Active ERK translocates to the nucleus where it phosphorylates several transcription factors, inducing transcription of several IEGs that are negative regulators of EGFR signaling. DUSPs are rapidly induced upon ERK activation, and some DUSPs also require phosphorylation by ERK to become fully active in order to dephosphorylate ERK. Another inducible inhibitor of EGFR, Mig-6, inhibits EGFR by blocking its kinase activity and mediating ubiquitin-independent degradation of EGFR. Several transcription factors induced by EGFR, such as MafF, inhibit transcription of genes that contain SRE in the promoter. Additional mechanism of negative regulation is mediated by ZFP36 that binds to the AU-rich 3′-UTR of mRNA molecules, such as *c-fos* mRNA and other IEG, targeting them for degradation. Dashed arrows represent indirect or unknown regulation; blue arrows represent mechanisms involving transports between the cytoplasmic and nuclear compartments.

protein tyrosine phosphatases (PTPs) that inactivate RTKs. Internalization of active RTKs into endosomes also creates feedback loops, which serve as both positive and negative regulators (Kholodenko 2002; Wiley 2003; Polo and Di Fiore 2006; Sorkin and von Zastrow 2009).

One of downstream targets of RTKs is the ADAM (a disintegrin and metalloproteinase) family of proteases responsible for shedding and release of growth factors such as heparin-binding EGF (Hynes and Schlange 2006; Zhou et al. 2006; Mendelson et al. 2010; Maretzky et al. 2011a,b; Rao et al. 2011). A recent study of head and neck squamous cell carcinoma cell lines has shown that ErbB2 and EGFR mediate ADAM12 protease expression in a PI3K and mTORC1 dependent manner (Rao et al. 2011). In turn, ADAM12 positively regulates ErbB2 expression, thus forming a positive feedback loop. Activated ADAM proteins are also associated with receptor cross talk. Ligands such as vascular endothelial growth factor (VEGF) and fibroblast growth factor (FGF) induce ERK1/2 activation and cell migration as a result of EGFR stimulation by heparin-binding EGF, which is released in ADAM-dependent manner (Maretzky et al. 2011a).

Positive and Negative Feedbacks at the Signaling Level

ERK, which is the terminal kinase in the three-layered MAPK/ERK cascade, phosphorylates multiple signaling molecules and transcription factors (Yoon and Seger 2006; von Kriegsheim et al. 2009; Yang et al. 2013). An example of ERK-mediated negative feedback is the phosphorylation of the guanine nucleotide exchange factor SOS. Upon RTK activation, the SOS-Grb2 complex is recruited to the plasma membrane where SOS mediates Ras activation. SOS phosphorylation by ERK or by its downstream kinase RSK-2 leads to dissociation of the SOS–Grb2 complex, thereby terminating Ras activation (Langlois et al. 1995; Dong et al. 1996; Douville and Downward 1997).

Raf kinase inhibitor protein (RKIP) is an endogenous inhibitor of the ERK pathway. By binding both Raf-1 and MEK, RKIP prevents

their physical interaction and MEK phosphorylation (Yeung et al. 1999, 2000). Activated ERK forms a positive feedback loop by phosphorylating RKIP and inhibiting its interaction with Raf-1. This positive feedback loop is imbedded into a long negative feedback loop from ERK to SOS and contributes to oscillations in ERK activity (Shin et al. 2009).

Disruption of RTK-mediated signaling complexes assembled on scaffolds is a common mechanism of negative feedback. For instance, following IR or IGF-1R-mediated phosphorylation of numerous tyrosine residues on the IRS scaffold proteins, these scaffolds become capable of recruiting multiple signaling complexes, including PI3K (Schmitz-Peiffer and Whitehead 2003; Boura-Halfon and Zick 2009). This facilitates PI3K activation, production of PIP_3 and activation of the PI3K/AKT/mTOR cascade. Subsequent phosphorylation of IRS on serine residues by downstream kinases, such as mTOR, its effector S6K1, and MAPKs, disrupts the signaling complexes assembled on IRS and down-regulates the PI3K pathway (Gual et al. 2005; Boura-Halfon and Zick 2009).

Feedback Loops via Transcription

RTK signaling induces transcriptional activation of multiple IEGs and DEGs, which, in turn, creates multiple positive and negative feedback loops (Avraham and Yarden 2011). Several induced proteins, including Mig-6/RALT and the leucine-rich repeats and immunoglobulin-like domains protein 1 (LRIG1), generate feedback by acting on the receptor. The transmembrane protein LRIG1 is expressed several hours after the onset of EGFR stimulation. While LRIG1 interacts with EGFR via the extracellular domain, the intracellular domain of LRIG1 recruits Cbl, thereby providing additional mechanism of Cbl-mediated EGFR degradation (Segatto et al. 2011). Whereas Cbl activation depends on EGFR activity, LRIG1 reduces the number of surface EGFR molecules both in basal conditions and upon stimulation, suggesting additional mechanisms of EGFR down-regulation (Segatto et al. 2011). Mig-6/RALT is expressed within an hour after EGFR activation

Cite this article as *Cold Spring Harb Perspect Biol* doi: 10.1101/cshperspect.a009043

and binds the kinase domain of all ErbB receptors in a ligand-dependent manner, suppressing their catalytic activities (Gotoh 2009; Segatto et al. 2011). Mig-6 targets EGFR to lysosomes and mediates EGFR degradation independently of receptor phosphorylation or ubiquitination, thus showing the second mechanism of negative regulation (Segatto et al. 2011). The Mig-6 protein also contains a Cdc42/Rac interaction and binding (CRIB) domain that selectively binds an active form of the small G protein Cdc42, which is an important regulator of cell migration. Indeed, by sequestering Cdc42, Mig-6 inhibits hepatocyte growth factor (HGF)-induced cell migration (Gotoh 2009).

DUSP proteins dephosphorylate and inactivate MAPKs. Different DUSPs have different preferred MAPK substrates and distinct intracellular localization. Also, often MAPK binding selectively facilitates DUSP activation. For instance, binding of ERK but not JNK or p38 MAPKs is required for catalytic activation of the cytoplasmic DUSP6 (Camps et al. 1998). The class I DUSPs (1/2/4/5) are inducible nuclear phosphatases, whose expression can be detected within an hour after activation of the ErbB receptors (Owens and Keyse 2007; Nakakuki et al. 2010). A recent study found transient ERK activity in the nucleus, whereas ERK activity in the cytoplasm was sustained. These different ERK kinetics in the cytoplasm and the nucleus were explained by inducible expression of several nuclear DUSPs since simultaneous knockdown of these phosphatases substantially prolonged ERK activation in the nucleus (Nakakuki et al. 2010).

Sprouty, whose expression is also induced by RTKs, is another regulator of upstream signaling. De novo synthesized Sprouty translocates to the plasma membrane where it is phosphorylated by the Src family kinases (Mason et al. 2006; Edwin et al. 2009). This phosphorylation (at Tyr55 for the Sprouty2 sequence) induces a conformational change that allows Sprouty to bind Grb2, disrupting the Grb2–SOS complex and leading to inhibition of Ras activation. The same phosphorylation at Tyr55 triggers Sprouty–Cbl interaction that sequesters Cbl from RTK, thereby preventing RTK degra-

dation. Because of this dual role in RTK regulation, Sprouty acts as an RTK signaling modulator rather than inhibitor (Edwin et al. 2009).

Additional example of positive feedback regulation was observed in breast tumor initiating cells (Aceto et al. 2012). Src-homology 2 domain-containing phosphatase 2 (SHP2) can act as tumor promoter by facilitating RTK-induced mitogenic signaling (Ostman et al. 2006). In a recent study SHP2 was shown to activate transcription factors ZEB1 and c-Myc. Whereas ZEB1 drives epithelial-to-mesenchymal transition (EMT) (Schmalhofer et al. 2009), c-myc induces expression of lin-28 homolog B, a repressor of microRNA biogenesis (Chang et al. 2009). This leads to let-7 microRNA repression and, therefore, results in overexpression of let-7 targets, including Ras and c-Myc itself (Aceto et al. 2012). Thus, SHP2-mediated positive feedback loop is required for maintenance and invasiveness of breast tumors (Aceto et al. 2012). Importantly, this is only one example of many in which microRNA molecules take part in signaling and transcriptional regulation downstream to RTKs (Avraham and Yarden 2012).

Feedback Loops Operating within Transcriptional Circuits

Additional mechanisms of positive or negative regulation involve de novo expression of proteins that affect mainly transcriptional events rather than upstream signaling. The first wave of the EGFR-induced transcriptional response (IEGs; immediate early genes) contains predominantly transcriptional activators; the second, delayed response wave (DEGs), involves multiple inhibitors of gene transcription that act at different levels (Amit et al. 2007). For instance, the proteins MafF and Kruppel-like factor 2 (KLF2) induced by RTKs (Amit et al. 2007; Dijkmans et al. 2009; Dutta et al. 2011) are negative regulators of several promoter elements, including the serum response element (SRE) (Amit et al. 2007). SRE is found within promoters of multiple IEG, such as transcription factors *c-fos* and *egr-1*, and a negative signaling regulator DUSP6 (Christy and Nathans 1989; Rivera et al.

1990; Treisman 1995; Bluthgen et al. 2009). Although the molecular mechanism of MafF- and KLF2-mediated transcriptional inhibition is unknown, the diversity of SRE containing genes clearly indicates that this regulation might have both positive and negative impact on signaling and transcriptional events.

Feedback mediated by posttranslational modifications develops on the scale of minutes, whereas feedback generated by de novo expressed proteins is much slower and operates on a longer timescale ranging from 30 min to several hours and more. The delayed negative feedback mechanisms may prevent permanent activation by continuous or recurrent signals, if de novo expressed proteins are present in the system for a prolonged time. In this case, a refractory period when cells are not responsive to stimuli is often observed. The refractory mechanism might be necessary for preventing continuous proliferation of cells exposed to multiple pulses of growth factors.

INTRICATE RTK NETWORK DYNAMICS ARE BROUGHT ABOUT BY FEEDBACK LOOPS: BISTABILITY, OSCILLATIONS, AND EXCITABILITY

Immediate and delayed feedback loops not only change steady-state input–output responses, but they also bring about dynamic instabilities. When a steady state becomes unstable, a system can jump to another stable state, start to oscillate, or show chaotic behavior. For instance, signals downstream of RTKs are often amplified by positive feedback, but if the feedback is sufficiently strong and a part of the pathway within the feedback shows ultrasensitive behavior with respect to the input signal, this system can become *bistable* (Ferrell 1999). A bistable system can switch between two distinct stable steady states, but cannot rest in intermediate states. Because of the coexistence of two alternative states, a bistable system displays *hysteresis*, meaning that the stimulus needs to exceed a threshold value to flip the system to another steady state at which it may remain when the stimulus returns to its initial value. In fact, hysteresis and bistability in MAPK cascades brought about by positive

feedback loops have been suggested to contribute to cell differentiation and long-term synaptic potentiation (Santos et al. 2007; Smolen et al. 2008). In addition, positive feedback either alone or in combination with negative feedback can trigger oscillations. Such positive–negative feedback oscillations generally do not have sinusoidal shapes and operate in a pulsatory manner: A part of a dynamic system is bistable, and there is a slow process induced by negative feedback that periodically forces the system to jump between "Off" and "On" states, generating oscillations (Pomerening et al. 2003; Sha et al. 2003).

Negative feedback in the MAPK/ERK cascade endows robustness to variations of parameters within the feedback loop and stabilizes the active ERK concentration when demand for active ERK fluctuates (Sauro and Kholodenko 2004). Using a combination of modeling and validating experiments, it was shown that the integration of a three-tier MAPK cascade structure with negative feedback generates emergent systems properties that resemble the features of a negative feedback amplifier (NFA), known from engineering (Sturm et al. 2010). These properties have decisive effects on the MAPK cascade drug sensitivity and adaptation to perturbations. In particular, the active ERK concentration was shown to be stabilized in response to drug-induced perturbations of the upstream ERK kinase, MEK (Sturm et al. 2010). Yet, above a certain threshold strength, negative feedback can induce damped or sustained oscillations in the system. These oscillations are caused by the time delay within the negative feedback loop and also require some degree of ultrasensitivity within the cascade (Kholodenko 2000). In fact, oscillations in the MAPK/ERK pathway were recently discovered experimentally, confirming previous theoretical predictions (Nakayama et al. 2008; Shankaran et al. 2009; Shin et al. 2009; Hu et al. 2013). An intriguing systems property of protein modification cascades and even simpler signaling systems is excitable and overshoot signaling responses to transient stimuli (Kaimachnikov and Kholodenko 2009). In this case, the entire RTK pathway behaves as an excitable device with a built-in excitability

   Cite this article as *Cold Spring Harb Perspect Biol* doi: 10.1101/cshperspect.a009043

threshold. Depending on the magnitude and duration of a transient stimulus, activation responses of key kinases and/or small GTPases fit into one of two distinct classes of either low or high amplitude responses, whereas there are no intermediate responses, merely proportional to the stimulus. For instance, the excitable behavior of small GTPases such as Rac or Rho, which does not respond to under-threshold stimuli, but yields high-activity pulses in response to over-threshold stimuli, can precisely control cell movement (Ridley 2001; Tsyganov et al. 2012).

A fundamental question is how cell-to-cell variability affects the RTK network function and dynamics. In any cell, transcription and translation are inherently stochastic and give rise to large cell-to-cell variability in protein levels observed even in genetically identical cells, such as clonal cell populations (McAdams and Arkin 1997; Spencer et al. 2009). Theoretical findings suggest that nonlinear signal processing by signaling networks *always* broadens the distribution of cellular responses to any signal, perturbation, or drug. Although bimodal distributions of cellular responses to input signal are generally thought to manifest bistable or ultrasensitive behavior in single cells, it has been shown recently that such bimodal responses can occur as a result of the protein abundance variability (Birtwistle et al. 2012; Kim and Sauro 2012). In particular, experiments and computer simulations show that a simple MAPK/ERK-cascade model with negative feedback that displays graded, analog ERK responses at a single cell level may show bimodal responses to EGF at the cell population level (Birtwistle et al. 2012). These results show that bimodal signaling response distributions do not necessarily imply digital (ultrasensitive or bistable) single cell signaling. The interplay between protein expression noise and network topologies can bring about digital population responses from analog single-cell dose responses. Thus, cells can retain the benefits of robustness arising from negative feedback, while simultaneously generating population-level on/off responses that are thought to be critical for regulating cell-fate decisions.

The extremely rich repertoire of different dynamic behaviors allow RTK signaling pathways to serve as analog-to-digital converters, transforming a transient or sustained growth factor activation into distinct responses, such as all-or-none, oscillatory, and pulsatory responses. Subsequent deconvolution of these different responses by the transcriptional machinery can be translated into different cell-fate decisions (Murphy et al. 2004; Kholodenko et al. 2010; Nakakuki et al. 2010).

SPATIAL SIGNAL PROPAGATION

Imaging data and computational models suggest that the propagation of RTK signaling from the plasma membrane to targets in the nucleus is tightly controlled by a variety of regulatory mechanisms (Kholodenko et al. 2010; Vartak and Bastiaens 2010; Grecco et al. 2011; Alam-Nazki and Krishnan 2012). RTK signaling pathways are highly spatially organized within cells. Activator enzymes, such as kinases or guanine nucleotide exchange factors (GEFs), and inactivating enzymes (e.g., phosphatases and GTPase-activating proteins [GAPs]) often localize to different cellular locales (Kholodenko 2009). For a protein phosphorylated by a membrane-bound kinase and dephosphorylated by a cytosolic phosphatase, it was predicted that there can be a gradient of the phosphorylated form, which is high close to the membrane and low within the cell (Brown and Kholodenko 1999). Instructively, the shape of the gradient depends mainly on the phosphatase activity. If the phosphatase is not saturated, the concentration profile of the active, phosphorylated form of the target protein decays almost exponentially with the distance from the membrane. Such exponential decrease in the activity of G proteins can also be observed if GEF activity is associated with a cellular structure, such as chromatin, and when GAP activity is excluded from this structure (Kholodenko 2006). Spatial gradients of protein activities organize signaling around cellular structures, such as membranes, chromosomes and scaffolds, and provide positional cues for key processes, including cell division. Such intracellular gradients of protein activities have been detected in live cells using imaging technologies based on fluorescence resonance energy (Maeder et al. 2007).

Raf-1, the initial kinase in the MAPK/ERK cascade is activated near the plasma membrane where activated downstream of RTKs, the small GTPase Ras resides. How can the phosphorylation signal that was initiated at the plasma membrane propagate through the cytoplasm where it is terminated by phosphatases? Central mechanisms of spatial signal propagation have been suggested. First, a kinase cascade can be assembled on a scaffold protein that protects transmission of phosphorylation against phosphatase activity. Second, even for a soluble cascade of (de)phosphorylation cycles, the phosphorylation signal reaches further into the cell interior, when the cascade has more levels, and this might be one of the reasons that cascades exist (Munoz-Garcia et al. 2009). Third, kinesin motor-mediated movement of the endosomes and kinase complexes along microtubules can transfer phosphorylation signals, guarding against dephosphorylation (Kholodenko 2002; Perlson et al. 2005, 2006). Finally, and most intriguingly, it was suggested that RTK signals can propagate as nonlinear traveling waves that create global spatial switches or pulses of kinase and GTPase (in)activation (Munoz-Garcia and Kholodenko 2010). These mechanisms facilitate signal propagation from activated RTKs across single cells.

CONCLUDING REMARKS AND FUTURE DIRECTIONS

Versatility and diversity of RTKs and their malfunctioning in major pathological conditions have made RTK signaling a focus point of extensive genetic and biochemical studies. Recent changes in our perception of RTK signaling pathways from linear pipelines to combinatorial complex interaction networks have highlighted the need for novel conceptual and technological approaches to move forward our comprehension of cell signaling by RTKs. An ever-growing contribution of mathematical models helps us understand RTK signaling in a cell as an integrated system rather than a list of proteins and genes.

Signaling from different RTKs leads to distinct cellular outcomes, yet these signals are transmitted through the overlapping cascades of common signal transducers. As described in this review, we have begun to realize that signal specificity is generated by a multiplicity of feedforward and feedback loops, combinatorial protein assemblies and their spatiotemporal dynamics rather than by a large number of genes with specific functions. However, many questions remain open. For instance, how is specificity maintained, if activation of one RTK results in cross talks with many other RTKs? We can speculate that whereas activation of a "primary" RTK occurs almost immediately following the ligand binding, the activation of "secondary" receptors is delayed. This time delay, combined with multiple negative feedbacks that are initiated by "primary" receptor, is expected to have a negative impact on the amplitude and duration of signals induced by the "secondary" receptor. Therefore, in a physiologic system that contains intact feedbacks, secondary activation is expected to serve only for fine-tuning of the responses.

Additional complexity arises from multiple protein isoforms that can contribute to both signaling specificity and redundancy. For instance, the growing body of evidence suggests that ERK1 and ERK2 might also have isoform specific, nonredundant functions in proliferation, differentiation, and EMT (Li and Johnson 2006; Vantaggiato et al. 2006; Shin et al. 2010). Future research will elaborate how different protein isoforms, such as the phosphatase SHP1 versus SHP2 or distinct isoforms of the scaffolds IRS and GAB, contribute to specific signaling responses and, thereby, to specific cellular phenotypes.

A feature of RTK networks highlighted here is the coexistence of multiple mechanisms that often play similar regulatory roles. For instance, negative feedback loops from ERK to SOS and from ERK to Raf-1 and EGFR degradation are redundant because each of these processes is sufficient for signal termination (Orton et al. 2008). In fact, mathematical models show that the system control is always distributed over multiple processes (von Kriegsheim et al. 2009).

 Cite this article as *Cold Spring Harb Perspect Biol* doi: 10.1101/cshperspect.a009043

This high degree of redundancy is expected to make the system robust toward a multitude of perturbations. Although system robustness is important for cell survival, it also makes cancer cells resistant to multiple drug treatments. This is the place where systems approach can be helpful in predicting the fragility points of pathological RTK networks to be exploited by combinatorial drug therapies.

ACKNOWLEDGMENTS

We thank David Croucher and Walter Kolch for discussions and critical reading of the manuscript. This work is supported by Science Foundation Ireland under grant No. 06/CE/B1129. We apologize that we could not cite many pertinent contributions to the field because of space limitations.

REFERENCES

Aceto N, Sausgruber N, Brinkhaus H, Gaidatzis D, Martiny-Baron G, Mazzarol G, Confalonieri S, Quarto M, Hu G, Balwierz PJ, et al. 2012. Tyrosine phosphatase SHP2 promotes breast cancer progression and maintains tumor-initiating cells via activation of key transcription factors and a positive feedback signaling loop. *Nat Med* **18:** 529–537.

Ahmad T, Farnie G, Bundred NJ, Anderson NG. 2004. The mitogenic action of insulin-like growth factor I in normal human mammary epithelial cells requires the epidermal growth factor receptor tyrosine kinase. *J Biol Chem* **279:** 1713–1719.

Alam-Nazki A, Krishnan J. 2012. An investigation of spatial signal transduction in cellular networks. *BMC Syst Biol* **6:** 83.

Amit I, Citri A, Shay T, Lu Y, Katz M, Zhang F, Tarcic G, Siwak D, Lahad J, Jacob-Hirsch J, et al. 2007. A module of negative feedback regulators defines growth factor signaling. *Nat Genet* **39:** 503–512.

Avraham R, Yarden Y. 2011. Feedback regulation of EGFR signalling: Decision making by early and delayed loops. *Nat Rev Mol Cell Biol* **12:** 104–117.

Avraham R, Yarden Y. 2012. Regulation of signalling by microRNAs. *Biochem Soc Trans* **40:** 26–30.

Balana ME, Labriola L, Salatino M, Movsichoff F, Peters G, Charreau EH, Elizalde PV. 2001. Activation of ErbB-2 via a hierarchical interaction between ErbB-2 and type I insulin-like growth factor receptor in mammary tumor cells. *Oncogene* **20:** 34–47.

Birtwistle MR, Rauch J, Kiyatkin A, Aksamitiene E, Dobrzynski M, Hoek JB, Kolch W, Ogunnaike BA, Kholodenko BN. 2012. Emergence of bimodal cell population responses from the interplay between analog single-cell signaling and protein expression noise. *BMC Syst Biol* **6:** 109.

Bluthgen N, Legewie S, Kielbasa SM, Schramme A, Tchernitsa O, Keil J, Solf A, Vingron M, Schafer R, Herzel H, et al. 2009. A systems biological approach suggests that transcriptional feedback regulation by dual-specificity phosphatase 6 shapes extracellular signal-related kinase activity in RAS-transformed fibroblasts. *FEBS J* **276:** 1024–1035.

Borisov N, Aksamitiene E, Kiyatkin A, Legewie S, Berkhout J, Maiwald T, Kaimachnikov NP, Timmer J, Hoek JB, Kholodenko BN. 2009. Systems-level interactions between insulin-EGF networks amplify mitogenic signaling. *Mol Syst Biol* **5:** 256.

Boura-Halfon S, Zick Y. 2009. Phosphorylation of IRS proteins, insulin action, and insulin resistance. *Am J Physiol Endocrinol Metab* **296:** E581–E591.

Brown GC, Kholodenko BN. 1999. Spatial gradients of cellular phospho-proteins. *FEBS Lett* **457:** 452–454.

Camps M, Nichols A, Gillieron C, Antonsson B, Muda M, Chabert C, Boschert U, Arkinstall S. 1998. Catalytic activation of the phosphatase MKP-3 by ERK2 mitogen-activated protein kinase. *Science* **280:** 1262–1265.

Chang TC, Zeitels LR, Hwang HW, Chivukula RR, Wentzel EA, Dews M, Jung J, Gao P, Dang CV, Beer MA, et al. 2009. Lin-28B transactivation is necessary for Myc-mediated let-7 repression and proliferation. *Proc Natl Acad Sci* **106:** 3384–3389.

Chen CY, Gherzi R, Ong SE, Chan EL, Raijmakers R, Pruijn GJ, Stoecklin G, Moroni C, Mann M, Karin M. 2001. AU binding proteins recruit the exosome to degrade ARE-containing mRNAs. *Cell* **107:** 451–464.

Christy B, Nathans D. 1989. Functional serum response elements upstream of the growth factor-inducible gene zif268. *Mol Cell Biol* **9:** 4889–4895.

Ciaccio MF, Wagner JP, Chuu CP, Lauffenburger DA, Jones RB. 2010. Systems analysis of EGF receptor signaling dynamics with microwestern arrays. *Nat Methods* **7:** 148–155.

Dijkmans TF, van Hooijdonk LW, Schouten TG, Kamphorst JT, Fitzsimons CP, Vreugdenhil E. 2009. Identification of new nerve growth factor-responsive immediate-early genes. *Brain Res* **1249:** 19–33.

Dong C, Waters SB, Holt KH, Pessin JE. 1996. SOS phosphorylation and disassociation of the Grb2–SOS complex by the ERK and JNK signaling pathways. *J Biol Chem* **271:** 6328–6332.

Dougherty MK, Muller J, Ritt DA, Zhou M, Zhou XZ, Copeland TD, Conrads TP, Veenstra TD, Lu KP, Morrison DK. 2005. Regulation of Raf-1 by direct feedback phosphorylation. *Mol Cell* **17:** 215–224.

Douville E, Downward J. 1997. EGF induced SOS phosphorylation in PC12 cells involves P90 RSK-2. *Oncogene* **15:** 373–383.

Dutta P, Koch A, Breyer B, Schneider H, Dittrich-Breiholz O, Kracht M, Tamura T. 2011. Identification of novel target genes of nerve growth factor (NGF) in human mastocytoma cell line (HMC-1 [V560G c-Kit]) by transcriptome analysis. *BMC Genomics* **12:** 196.

Edwin F, Anderson K, Ying C, Patel TB. 2009. Intermolecular interactions of Sprouty proteins and their implications in development and disease. *Mol Pharmacol* **76:** 679–691.

Esposito CL, D'Alessio A, de Franciscis V, Cerchia L. 2008. A cross talk between TrkB and Ret tyrosine kinases receptors mediates neuroblastoma cells differentiation. *PLoS ONE* **3:** e1643.

Fafalios A, Ma J, Tan X, Stoops J, Luo J, Defrances MC, Zarnegar R. 2011. A hepatocyte growth factor receptor (Met)-insulin receptor hybrid governs hepatic glucose metabolism. *Nat Med* **17:** 1577–1584.

Ferrell JE Jr. 1999. Building a cellular switch: More lessons from a good egg. *Bioessays* **21:** 866–870.

Goentoro L, Shoval O, Kirschner MW, Alon U. 2009. The incoherent feedforward loop can provide fold-change detection in gene regulation. *Mol Cell* **36:** 894–899.

Gotoh N. 2009. Feedback inhibitors of the epidermal growth factor receptor signaling pathways. *Int J Biochem Cell Biol* **41:** 511–515.

Grecco HE, Schmick M, Bastiaens PI. 2011. Signaling from the living plasma membrane. *Cell* **144:** 897–909.

Gual P, Le Marchand-Brustel Y, Tanti JF. 2005. Positive and negative regulation of insulin signaling through IRS-1 phosphorylation. *Biochimie* **87:** 99–109.

Gujral TS, Karp RL, Finski A, Chan M, Schwartz PE, Macbeath G, Sorger P. 2012. Profiling phospho-signaling networks in breast cancer using reverse-phase protein arrays. *Oncogene* doi: 10.1038/onc.2012.378.

Guo A, Villen J, Kornhauser J, Lee KA, Stokes MP, Rikova K, Possemato A, Nardone J, Innocenti G, Wetzel R, et al. 2008. Signaling networks assembled by oncogenic EGFR and c-Met. *Proc Natl Acad Sci* **105:** 692–697.

Herrero J, Valencia A, Dopazo J. 2001. A hierarchical unsupervised growing neural network for clustering gene expression patterns. *Bioinformatics* **17:** 126–136.

Hu H, Goltsov A, Bown JL, Sims AH, Langdon SP, Harrison DJ, Faratian D. 2013. Feedforward and feedback regulation of the MAPK and PI3K oscillatory circuit in breast cancer. *Cell Signal* **25:** 26–32.

Hunter T. 2000. Signaling—2000 and beyond. *Cell* **100:** 113–127.

Hynes NE, Schlange T. 2006. Targeting ADAMS and ERBBs in lung cancer. *Cancer Cell* **10:** 7–11.

Jones HE, Gee JM, Hutcheson IR, Knowlden JM, Barrow D, Nicholson RI. 2006. Growth factor receptor interplay and resistance in cancer. *Endocr Relat Cancer* **13:** S45–S51.

Kaimachnikov NP, Kholodenko BN. 2009. Toggle switches, pulses and oscillations are intrinsic properties of the Src activation/deactivation cycle. *FEBS J* **276:** 4102–4118.

Kaplan DR, Matsumoto K, Lucarelli E, Thiele CJ. 1993. Induction of TrkB by retinoic acid mediates biologic responsiveness to BDNF and differentiation of human neuroblastoma cells. Eukaryotic Signal Transduction Group. *Neuron* **11:** 321–331.

Kholodenko BN. 2000. Negative feedback and ultrasensitivity can bring about oscillations in the mitogen-activated protein kinase cascades. *Eur J Biochem* **267:** 1583–1588.

Kholodenko BN. 2002. MAP kinase cascade signaling and endocytic trafficking: A marriage of convenience? *Trends Cell Biol* **12:** 173–177.

Kholodenko BN. 2006. Cell-signalling dynamics in time and space. *Nat Rev Mol Cell Biol* **7:** 165–176.

Kholodenko BN. 2009. Spatially distributed cell signalling. *FEBS Lett* **583:** 4006–4012.

Kholodenko BN, Hoek JB, Westerhoff HV, Brown GC. 1997. Quantification of information transfer via cellular signal transduction pathways. *FEBS Lett* **414:** 430–434; erratum, **419:** 150.

Kholodenko BN, Hancock JF, Kolch W. 2010. Signalling ballet in space and time. *Nat Rev Mol Cell Biol* **11:** 414–426.

Kim KH, Sauro HM. 2012. In search of noise-induced bimodality. *BMC Biol* **10:** 89.

Kiyatkin A, Aksamitiene E, Markevich NI, Borisov NM, Hoek JB, Kholodenko BN. 2006. Scaffolding protein Grb2-associated binder 1 sustains epidermal growth factor-induced mitogenic and survival signaling by multiple positive feedback loops. *J Biol Chem* **281:** 19925–19938.

Langlois WJ, Sasaoka T, Saltiel AR, Olefsky JM. 1995. Negative feedback regulation and desensitization of insulin- and epidermal growth factor-stimulated p21ras activation. *J Biol Chem* **270:** 25320–25323.

Lemmon MA, Schlessinger J. 2010. Cell signaling by receptor tyrosine kinases. *Cell* **141:** 1117–1134.

Li J, Johnson SE. 2006. ERK2 is required for efficient terminal differentiation of skeletal myoblasts. *Biochem Biophys Res Commun* **345:** 1425–1433.

Ma W, Trusina A, El-Samad H, Lim WA, Tang C. 2009. Defining network topologies that can achieve biochemical adaptation. *Cell* **138:** 760–773.

Maeder CI, Hink MA, Kinkhabwala A, Mayr R, Bastiaens PI, Knop M. 2007. Spatial regulation of Fus3 MAP kinase activity through a reaction-diffusion mechanism in yeast pheromone signalling. *Nat Cell Biol* **9:** 1319–1326.

Mangan S, Alon U. 2003. Structure and function of the feedforward loop network motif. *Proc Natl Acad Sci* **100:** 11980–11985.

Mangan S, Zaslaver A, Alon U. 2003. The coherent feedforward loop serves as a sign-sensitive delay element in transcription networks. *J Mol Biol* **334:** 197–204.

Maretzky T, Evers A, Zhou W, Swendeman SL, Wong PM, Rafii S, Reiss K, Blobel CP. 2011a. Migration of growth factor-stimulated epithelial and endothelial cells depends on EGFR transactivation by ADAM17. *Nat Commun* **2:** 229.

Maretzky T, Zhou W, Huang XY, Blobel CP. 2011b. A transforming Src mutant increases the bioavailability of EGFR ligands via stimulation of the cell-surface metalloproteinase ADAM17. *Oncogene* **30:** 611–618.

Martin B, Brenneman R, Golden E, Walent T, Becker KG, Prabhu VV, Wood W 3rd, Ladenheim B, Cadet JL, Maudsley S. 2009. Growth factor signals in neural cells: Coherent patterns of interaction control multiple levels of molecular and phenotypic responses. *J Biol Chem* **284:** 2493–2511.

Mason JM, Morrison DJ, Basson MA, Licht JD. 2006. Sprouty proteins: Multifaceted negative-feedback regulators of receptor tyrosine kinase signaling. *Trends Cell Biol* **16:** 45–54.

McAdams HH, Arkin A. 1997. Stochastic mechanisms in gene expression. *Proc Natl Acad Sci* **94:** 814–819.

Mendelson K, Swendeman S, Saftig P, Blobel CP. 2010. Stimulation of platelet-derived growth factor receptor β (PDGFRβ) activates ADAM17 and promotes metalloproteinase-dependent cross-talk between the PDGFRβ and epidermal growth factor receptor (EGFR) signaling pathways. *J Biol Chem* **285:** 25024–25032.

Milo R, Shen-Orr S, Itzkovitz S, Kashtan N, Chklovskii D, Alon U. 2002. Network motifs: simple building blocks of complex networks. *Science* **298:** 824–827.

Munoz-Garcia J, Kholodenko BN. 2010. Signalling over a distance: Gradient patterns and phosphorylation waves within single cells. *Biochem Soc Trans* **38:** 1235–1241.

Munoz-Garcia J, Neufeld Z, Kholodenko BN. 2009. Positional information generated by spatially distributed signaling cascades. *PLoS Comput Biol* **5:** e1000330.

Murphy LO, Smith S, Chen RH, Fingar DC, Blenis J. 2002. Molecular interpretation of ERK signal duration by immediate early gene products. *Nat Cell Biol* **4:** 556–564.

Murphy LO, MacKeigan JP, Blenis J. 2004. A network of immediate early gene products propagates subtle differences in mitogen-activated protein kinase signal amplitude and duration. *Mol Cell Biol* **24:** 144–153.

Murphy T, Hori S, Sewell J, Gnanapragasam VJ. 2010. Expression and functional role of negative signalling regulators in tumour development and progression. *Int J Cancer* **127:** 2491–2499.

Nakakuki T, Birtwistle MR, Saeki Y, Yumoto N, Ide K, Nagashima T, Brusch L, Ogunnaike BA, Okada-Hatakeyama M, Kholodenko BN. 2010. Ligand-specific c-Fos expression emerges from the spatiotemporal control of ErbB network dynamics. *Cell* **141:** 884–896.

Nakayama K, Satoh T, Igari A, Kageyama R, Nishida E. 2008. FGF induces oscillations of Hes1 expression and Ras/ERK activation. *Curr Biol* **18:** R332–334.

Orton RJ, Sturm OE, Gormand A, Wolch W, Gilbert DR. 2008. Computational modelling reveals feedback redundancy within the epidermal growth factor receptor/extracellular-signal regulated kinase signalling pathway. *IET Syst Biol* **2:** 173–183.

Ostman A, Hellberg C, Bohmer FD. 2006. Protein-tyrosine phosphatases and cancer. *Nat Rev Cancer* **6:** 307–320.

Owens DM, Keyse SM. 2007. Differential regulation of MAP kinase signalling by dual-specificity protein phosphatases. *Oncogene* **26:** 3203–3213.

Pawson T, Nash P. 2003. Assembly of cell regulatory systems through protein interaction domains. *Science* **300:** 445–452.

Pawson T, Gish GD, Nash P. 2001. SH2 domains, interaction modules and cellular wiring. *Trends Cell Biol* **11:** 504–511.

Perlson E, Hanz S, Ben-Yaakov K, Segal-Ruder Y, Seger R, Fainzilber M. 2005. Vimentin-dependent spatial translocation of an activated MAP kinase in injured nerve. *Neuron* **45:** 715–726.

Perlson E, Michaelevski I, Kowalsman N, Ben-Yaakov K, Shaked M, Seger R, Eisenstein M, Fainzilber M. 2006. Vimentin binding to phosphorylated Erk sterically hinders enzymatic dephosphorylation of the kinase. *J Mol Biol* **364:** 938–944.

Polo S, Di Fiore PP. 2006. Endocytosis conducts the cell signaling orchestra. *Cell* **124:** 897–900.

Pomerening JR, Sontag ED, Ferrell JE. 2003. Building a cell cycle oscillator: Hysteresis and bistability in the activation of Cdc2. *Nat Cell Biol* **5:** 346–351.

Rao VH, Kandel A, Lynch D, Pena Z, Marwaha N, Deng C, Watson P, Hansen LA. 2011. A positive feedback loop between HER2 and ADAM12 in human head and neck cancer cells increases migration and invasion. *Oncogene* **31:** 2888–2898.

Ridley AJ. 2001. Rho GTPases and cell migration. *J Cell Sci* **114:** 2713–2722.

Rivera VM, Sheng M, Greenberg ME. 1990. The inner core of the serum response element mediates both the rapid induction and subsequent repression of c-fos transcription following serum stimulation. *Genes Dev* **4:** 255–268.

Santos SD, Verveer PJ, Bastiaens PI. 2007. Growth factor-induced MAPK network topology shapes Erk response determining PC-12 cell fate. *Nat Cell Biol* **9:** 324–330.

Sauro HM, Kholodenko BN. 2004. Quantitative analysis of signaling networks. *Prog Biophys Mol Biol* **86:** 5–43.

Schlessinger J. 2000. Cell signaling by receptor tyrosine kinases. *Cell* **103:** 211–225.

Schmalhofer O, Brabletz S, Brabletz T. 2009. E-cadherin, β-catenin, and ZEB1 in malignant progression of cancer. *Cancer Metastasis Rev* **28:** 151–166.

Schmitz-Peiffer C, Whitehead JP. 2003. IRS-1 regulation in health and disease. *IUBMB Life* **55:** 367–374.

Segatto O, Anastasi S, Alema S. 2011. Regulation of epidermal growth factor receptor signalling by inducible feedback inhibitors. *J Cell Sci* **124:** 1785–1793.

Sevecka M, MacBeath G. 2006. State-based discovery: A multidimensional screen for small-molecule modulators of EGF signaling. *Nat Methods* **3:** 825–831.

Sha W, Moore J, Chen K, Lassaletta AD, Yi CS, Tyson JJ, Sible JC. 2003. Hysteresis drives cell-cycle transitions in *Xenopus laevis* egg extracts. *Proc Natl Acad Sci* **100:** 975–980.

Shankaran H, Ippolito DL, Chrisler WB, Resat H, Bollinger N, Opresko LK, Wiley HS. 2009. Rapid and sustained nuclear-cytoplasmic ERK oscillations induced by epidermal growth factor. *Mol Syst Biol* **5:** 332.

Shin SY, Rath O, Choo SM, Fee F, McFerran B, Kolch W, Cho KH. 2009. Positive- and negative-feedback regulations coordinate the dynamic behavior of the Ras-Raf-MEK-ERK signal transduction pathway. *J Cell Sci* **122:** 425–435.

Shin S, Dimitri CA, Yoon SO, Dowdle W, Blenis J. 2010. ERK2 but not ERK1 induces epithelial-to-mesenchymal transformation via DEF motif-dependent signaling events. *Mol Cell* **38:** 114–127.

Shoval O, Alon U. 2010. SnapShot: Network motifs. *Cell* **143:** 326–e321.

Smolen P, Baxter DA, Byrne JH. 2008. Bistable MAP kinase activity: A plausible mechanism contributing to maintenance of late long-term potentiation. *Am J Physiol Cell Physiol* **294:** C503–C515.

Sorkin A, von Zastrow M. 2009. Endocytosis and signalling: Intertwining molecular networks. *Nat Rev Mol Cell Biol* **10:** 609–622.

Spencer SL, Gaudet S, Albeck JG, Burke JM, Sorger PK. 2009. Non-genetic origins of cell-to-cell variability in TRAIL-induced apoptosis. *Nature* **459:** 428–432.

Stommel JM, Kimmelman AC, Ying H, Nabioullin R, Ponugoti AH, Wiedemeyer R, Stegh AH, Bradner JE, Ligon KL, Brennan C, et al. 2007. Coactivation of receptor tyrosine kinases affects the response of tumor cells to targeted therapies. *Science* **318:** 287–290.

Sturm OE, Orton R, Grindlay J, Birtwistle M, Vyshemirsky V, Gilbert D, Calder M, Pitt A, Kholodenko B, Kolch W. 2010. The mammalian MAPK/ERK pathway exhibits properties of a negative feedback amplifier. *Sci Signal* **3:** ra90.

Tanizaki J, Okamoto I, Sakai K, Nakagawa K. 2011. Differential roles of trans-phosphorylated EGFR, HER2, HER3, and RET as heterodimerisation partners of MET in lung cancer with MET amplification. *Br J Cancer* **105:** 807–813.

Treisman R. 1995. Journey to the surface of the cell: Fos regulation and the SRE. *EMBO J* **14:** 4905–4913.

Tsui-Pierchala BA, Milbrandt J, Johnson EM Jr. 2002. NGF utilizes c-Ret via a novel GFL-independent, inter-RTK signaling mechanism to maintain the trophic status of mature sympathetic neurons. *Neuron* **33:** 261–273.

Tsyganov MA, Kolch W, Kholodenko BN. 2012. The topology design principles that determine the spatiotemporal dynamics of G-protein cascades. *Mol Biosyst* **8:** 730–743.

Vantaggiato C, Formentini I, Bondanza A, Bonini C, Naldini L, Brambilla R. 2006. ERK1 and ERK2 mitogen-activated protein kinases affect Ras-dependent cell signaling differentially. *J Biol* **5:** 14.

Vartak N, Bastiaens P. 2010. Spatial cycles in G-protein crowd control. *EMBO J* **29:** 2689–2699.

Velpula KK, Dasari VR, Asuthkar S, Gorantla B, Tsung AJ. 2012. EGFR and c-Met cross talk in glioblastoma and its regulation by human cord blood stem cells. *Transl Oncol* **5:** 379–392.

von Kriegsheim A, Baiocchi D, Birtwistle M, Sumpton D, Bienvenut W, Morrice N, Yamada K, Lamond A, Kalna G, Orton R, et al. 2009. Cell fate decisions are specified by the dynamic ERK interactome. *Nat Cell Biol* **11:** 1458–1464.

Wiley HS. 2003. Trafficking of the ErbB receptors and its influence on signaling. *Exp Cell Res* **284:** 78–88.

Xu AM, Huang PH. 2010. Receptor tyrosine kinase coactivation networks in cancer. *Cancer Res* **70:** 3857–3860.

Yang SH, Sharrocks AD, Whitmarsh AJ. 2013. MAP kinase signalling cascades and transcriptional regulation. *Gene* **513:** 1–13.

Yeung K, Seitz T, Li S, Janosch P, McFerran B, Kaiser C, Fee F, Katsanakis KD, Rose DW, Mischak H, et al. 1999. Suppression of Raf-1 kinase activity and MAP kinase signalling by RKIP. *Nature* **401:** 173–177.

Yeung K, Janosch P, McFerran B, Rose DW, Mischak H, Sedivy JM, Kolch W. 2000. Mechanism of suppression of the Raf/MEK/extracellular signal-regulated kinase pathway by the raf kinase inhibitor protein. *Mol Cell Biol* **20:** 3079–3085.

Yoon S, Seger R. 2006. The extracellular signal-regulated kinase: Multiple substrates regulate diverse cellular functions. *Growth Factors* **24:** 21–44.

Zhou BB, Peyton M, He B, Liu C, Girard L, Caudler E, Lo Y, Baribaud F, Mikami I, Reguart N, et al. 2006. Targeting ADAM-mediated ligand cleavage to inhibit HER3 and EGFR pathways in non-small cell lung cancer. *Cancer Cell* **10:** 39–50.

Zwang Y, Yarden Y. 2009. Systems biology of growth factor-induced receptor endocytosis. *Traffic* **10:** 349–363.

Cite this article as *Cold Spring Harb Perspect Biol* doi: 10.1101/cshperspect.a009043

The Insulin Receptor: Both a Prototypical and Atypical Receptor Tyrosine Kinase

Stevan R. Hubbard

Kimmel Center for Biology and Medicine of the Skirball Institute and Department of Biochemistry and Molecular Pharmacology, New York University School of Medicine, New York, New York 10016

Correspondence: stevan.hubbard@med.nyu.edu

Unlike prototypical receptor tyrosine kinases (RTKs), which are single-chain polypeptides, the insulin receptor (InsR) is a preformed, covalently linked tetramer with two extracellular α subunits and two membrane-spanning, tyrosine kinase-containing β subunits. A single molecule of insulin binds asymmetrically to the ectodomain, triggering a conformational change that is transmitted to the cytoplasmic kinase domains, which facilitates their *trans*-phosphorylation. As in prototypical RTKs, tyrosine phosphorylation in the juxtamembrane region of InsR creates recruitment sites for downstream signaling proteins (IRS [InsR substrate] proteins, Shc) containing a phosphotyrosine-binding (PTB) domain, and tyrosine phosphorylation in the kinase activation loop stimulates InsR's catalytic activity. For InsR, phosphorylation of the activation loop, which contains three tyrosine residues, also creates docking sites for adaptor proteins (Grb10/14, SH2B2) that possess specialized Src homology-2 (SH2) domains, which are dimeric and engage two phosphotyrosines in the activation loop.

Insulin is a highly potent anabolic hormone that is critical for tissue development and for glucose homeostasis (Taniguchi et al. 2006). Released from the β cells of the pancreas, insulin regulates glucose output from the liver and glucose uptake into (primarily) skeletal muscle and adipose tissue. In addition, insulin promotes the synthesis and storage of carbohydrates, lipids, and protein. Insulin's actions are mediated by the insulin receptor (InsR), a plasma membrane-resident glycoprotein and member of the receptor tyrosine kinase (RTK) family. Other members of the InsR subfamily of RTKs include the insulinlike growth factor-1 receptor (IGF1R) and insulin receptor-related receptor, the latter of which has no known ligand. As an RTK, InsR is ligand-activated through mechanisms that are both prototypical and atypical of RTKs. These mechanisms will be the focus of this article.

InsR ARCHITECTURE

InsR comprises two α and two β subunits, forming an $\alpha_2\beta_2$ heterotetramer (Fig. 1). The two α subunits are extracellular, whereas the two β subunits begin on the extracellular side of the membrane and then traverse the membrane into the cytoplasmic region. The β subunits contain the cytoplasmic tyrosine kinase domains. The α subunits are disulfide-bridged by at least two (possibly four) cysteine residues, and each α

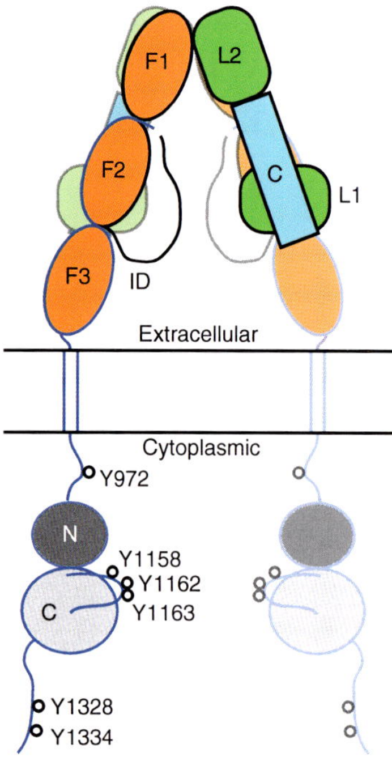

Figure 1. Structure-based schematic diagram of InsR. The α subunits (extracellular, denoted in black outline) comprise two leucine-rich domains (L1 and L2) with an intervening cysteine-rich domain (C), one intact fibronectin type III (FnIII) domain (F1), a partial FnIII domain (F2), and a long insert domain (ID) that contains the site of furin cleavage (L1-C-L2-F1-F2-ID). The β subunits (extracellular and cytoplasmic, denoted in blue outline) contain the completion of the second FnIII domain (F2), followed by an intact FnIII domain (F3), a transmembrane helix, and a cytoplasmic tyrosine kinase domain (N and C lobes). The second αβ half-receptor is shown semitransparent. The major sites of tyrosine autophosphorylation in the cytoplasmic domain are indicated (InsR-B isoform numbering). The molecular twofold axis is vertical, in the plane of the figure.

subunit is paired to one β subunit through a single disulfide bridge (Sparrow et al. 1997).

The α subunit consists of a leucine-rich domain followed by a cysteine-rich domain, a second leucine-rich domain, one complete fibronectin type III (FnIII) domain, a partial FnIII domain, and a long carboxy-terminal segment that, during maturation, is cleaved by furin to yield α and β subunits. The β subunit begins (after a short amino-terminal segment) with the completion of the second FnIII domain,

followed by a third FnIII domain, which leads into the transmembrane segment. The transmembrane segment is believed to adopt an α-helical conformation.

The cytoplasmic region of the β subunit consists of a juxtamembrane region (~30 residues; between the transmembrane helix and the tyrosine kinase domain), the tyrosine kinase domain (~300 residues), and a carboxy-terminal region (~70 residues). Tyrosine autophosphorylation sites have been mapped in all three regions (White et al. 1988; Kohanski 1993), with Y972 (+exon 11 numbering) in the juxtamembrane region and Y1158, Y1162, and Y1163 in the kinase domain having been shown to be functionally important. The juxtamembrane and carboxy-terminal regions are thought to be unstructured polypeptide segments, save for possible β turns in the juxtamembrane region near Y965 and Y972 (Backer et al. 1992).

InsR THREE-DIMENSIONAL STRUCTURE

Because InsR is a large, heavily glycosylated membrane-spanning protein, structural studies have been confined thus far to crystallographic studies of either the soluble ectodomain or the soluble cytoplasmic kinase domain. A crystal structure of the full (disulfide-linked) dimeric ectodomain was reported in 2006 (McKern et al. 2006) and later re-refined (Smith et al. 2010). The structure shows an antiparallel "inverted V" arrangement (Fig. 2, left). Previous photo-labeling cross-linking studies and alanine-scanning mutagenesis have implicated two regions on InsR in insulin binding (Whittaker et al. 2008). Site 1 is formed by the first leucine-rich domain in one α subunit and the carboxy-terminal segment of the other α subunit. Site 2 is thought to involve loops from the first and second FnIII domains of the other αβ half-receptor (Fig. 2, right).

Crystal structures of the InsR tyrosine kinase domain (IRK) have been determined in several different phosphorylation states and with bound substrates (Mg-ATP and substrate peptide) (Hubbard et al. 1994; Hubbard 1997). IRK possesses a canonical protein-kinase architecture, with an amino-terminal lobe (N lobe)

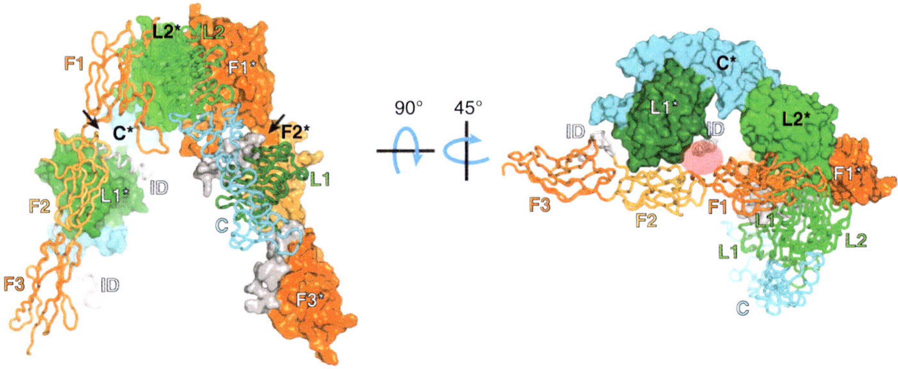

Figure 2. Structure of the ectodomain of InsR. In this disulfide-linked symmetric dimer (apo—no insulin) (McKern et al. 2006; Smith et al. 2010), one of the αβ half-receptors (in *front*) is shown as a Cα trace, and the other αβ half-receptor (in *back*) is shown with a molecular surface. The molecular twofold axis is vertical, in the plane of the figure. The two L domains are labeled L1 and L2, the cysteine-rich domain is labeled C, the FnIII domains are labeled F1–3, and the insert domain (end of the α subunit, beginning of the β subunit) is labeled ID. An asterisk in the label denotes a domain from the second half-receptor. The arrows point to the two (identical) insulin-binding sites (although binding of a single insulin molecule is sufficient for receptor activation). In the *right* panel, the structure has been rotated first by 90°, then by 45°, as indicated. The red oval shows where insulin is thought to bind (corresponds to the arrow on the left in the *left* panel).

comprising a five-stranded β sheet and a single α helix (αC), and a C lobe that is mainly helical and contains most of the catalytic residues, in the so-called catalytic and activation loops (Fig. 3). One deviation in the C-lobe architecture relative to other RTKs is the presence of an additional α helix (αJ) at the carboxy-terminal end. (The function of this extra helix has not been established, but in the structure of a complex between IRK and the protein tyrosine phosphatase PTP1B (Li et al. 2005), this helix is part of the phosphatase binding site.) Tyrosine kinase activity is regulated by the phosphorylation state of the activation loop in the C lobe, which begins with the kinase-conserved ^{1150}DFG motif and ends with a conserved proline (P1172). The IRK activation loop contains three sites of tyrosine autophosphorylation, Y1158, Y1162, and Y1163, which are phosphorylated in *trans* on insulin binding to the ectodomain.

InsR ACTIVATION

As described above, InsR is a preformed disulfide-linked dimer, and there is compelling evidence that *trans*-phosphorylation/activation occurs within the dimer rather than through

higher-order oligomerization (Frattali et al. 1992). Although InsR is a symmetric αβ homodimer (or $\alpha_2\beta_2$ heterotetramer), the active signaling complex is thought to be a single molecule of insulin (which circulates in the bloodstream in a hexameric form) bound asymmetrically to InsR with subnanomolar affinity, and thus the conformational change is induced in (or mainly in) only one αβ half-receptor. Higher concentrations of insulin display negative cooperativity for binding to the two equivalent sites on InsR (De Meyts 2008). Even though insulin is not present in the available crystal structure of the ectodomain (McKern et al. 2006), sites 1 and 2 are already closely apposed (Fig. 2, right), suggesting that the conformational changes induced by insulin binding will be relatively subtle (Renteria et al. 2008; Ward and Lawrence 2012).

In the basal (noninsulin) state, *trans*-phosphorylation of the kinase domains is limited by an unknown mechanism, and insulin binding to the ectodomain induces a structural transition within the receptor that facilitates *trans*-phosphorylation of the cytoplasmic kinase domains. Two scenarios can be envisioned for the basal state. In the first, the kinase domains are

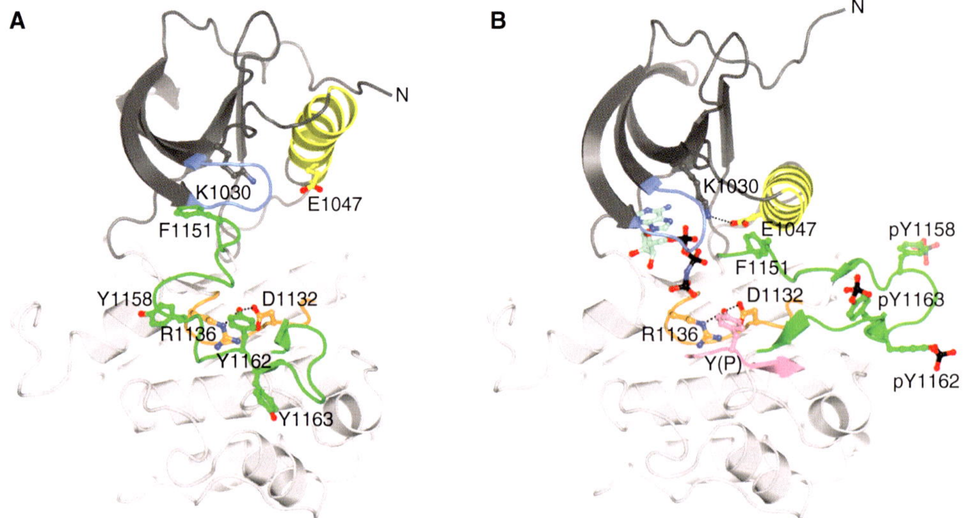

Figure 3. Structure of the tyrosine kinase domain of InsR. (*A*) Ribbon diagram of inactive, unphosphorylated IRK (Hubbard et al. 1994). The N lobe is colored dark gray except for αC and the nucleotide binding loop (blue), and the C lobe is colored light gray except for the catalytic loop (orange) and activation loop (green). Carbon atoms are colored according to their position in the structure, oxygen atoms are colored red, and nitrogen atoms are colored blue. Select side chains are shown in ball-and-stick representation. Hydrogen bonds between Y1162 (the "pseudosubstrate") in the activation loop and D1132 and R1136 in the catalytic loop are shown as black dashed lines. The amino terminus is labeled (N) (the carboxyl terminus is hidden behind the structure). (*B*) Ribbon diagram of active, tris-phosphorylated IRK in complex with an ATP analog (AMPPNP) and a peptide substrate (Hubbard 1997). Same coloring as in (*A*) and, in addition, phosphorus atoms are colored black, carbon atoms of AMPPNP are colored pale green, and the peptide substrate is colored pink. F1151, at the beginning of the activation loop, is part of the DFG motif. Hydrogen bonds between the substrate tyrosine (Y[P]) and D1132 and R1136 are shown by black dashed lines as is the salt bridge between conserved residues K1030 (β3) and E1047 (αC).

spatially separated and thus cannot undergo efficient *trans*-phosphorylation. In the second, the kinase domains are already within "striking distance," but are arranged in such a way as to prevent *trans*-phosphorylation (perhaps as an inhibitory dimer). A partially activating point mutation (Y984A) in the juxtamembrane region of InsR is more consistent with the latter mechanism (Li et al. 2003).

cis-INHIBITION AND *trans*-ACTIVATION IN THE InsR KINASE DOMAIN

As observed in the first crystal structure of IRK (Hubbard et al. 1994), the unphosphorylated (basal state) activation loop, in the absence of Mg-ATP, adopts an autoinhibitory configuration in which Y1162, the second of the three

activation-loop tyrosines, is bound in the kinase active site, hydrogen-bonded to D1132 and R1136 in the catalytic loop (Fig. 3A). In this configuration, although Y1162 is positioned for phosphoryl transfer, the amino-terminal end of the activation loop (^{1150}DFG motif) occludes the ATP-binding site, preventing phosphorylation of Y1162 in *cis*. That is, because of the finite length of the activation loop, it is not sterically possible to bind simultaneously ATP in its binding cleft and Y1162 in the active site in *cis*. This *cis*-inhibitory/pseudosubstrate (Y1162) conformation first viewed in IRK has now been observed in several other RTKs, including IGF1R (Munshi et al. 2002), TrkA-B (Artim et al. 2012; Bertrand et al. 2012), MuSK (Till et al. 2002), and Ror2 (Artim et al. 2012). The activation loops of all of these RTKs

     Cite this article as *Cold Spring Harb Perspect Biol* doi: 10.1101/cshperspect.a008946

have in common three (phosphorylatable) tyrosines in a YXXDYY motif (where X is any residue), in which the second tyrosine is bound in the active site as a pseudosubstrate. The activation loops of ALK (anaplastic lymphoma kinase) and Met also contain three tyrosines in the same positions, but lack the aspartic acid preceding the second tyrosine (a key residue in stabilizing the pseudosubstrate conformation of InsR [Till et al. 2001]), and their conformations in the unphosphorylated state differ (Wang et al. 2006; Lee et al. 2010). In the presence of cellular (millimolar) concentrations of Mg-ATP, the unphosphorylated activation loop probably adopts multiple conformations, including the pseudosubstrate conformation, as well as conformations in which Mg-ATP is bound and the active site is available for *trans*-phosphorylation of a tyrosine from the neighboring kinase domain.

It was shown many years ago that Y1162 was the first of the three tyrosines in the IRK activation loop to undergo autophosphorylation, followed by Y1158 and Y1163 (Wei et al. 1995). Several years ago, a crystal structure of the IGF1R kinase domain (IGF1RK) was determined in which the tyrosine equivalent to Y1162, Y1135, was bound in the active site of a symmetry-related kinase domain (in *trans*) (Wu et al. 2008a). This configuration was made possible through the tight binding of an ATP-competitive small-molecule inhibitor, which forced the activation loop out of its pseudosubstrate configuration. This structure provides a snapshot of the first *trans*-phosphorylation reaction in IGF1RK as well as in IRK (100% sequence conservation in the activation loop), and afforded rationalization of a mutation in the activation loop of InsR, R1164Q, which causes loss of InsR autophosphorylation (Formisano et al. 1993). The structure indicates that R1137 (R1164 in InsR) and E1132 (E1159 in InsR) in the IGF1R activation loop form a salt bridge that is probably important for positioning Y1135 (Y1162 in InsR) in the active site for *trans*-phosphorylation.

On *trans*-phosphorylation of the IRK activation loop on Y1158, Y1162, and Y1163, the loop is stabilized in a conformation that is optimized for binding of substrates (Mg-ATP, tyrosine) and for catalysis (Fig. 3B). Of the three phosphotyrosines, pY1163 is the most important one for loop stabilization, corresponding structurally to the single phosphotyrosine in the activation loop of Src-family tyrosine kinases (Yamaguchi and Hendrickson 1996). Although the unphosphorylated activation loops of RTKs display (in crystal structures) a number of inactive conformations, the active (usually phosphorylated) conformations are remarkably similar (Huse and Kuriyan 2002). Although the other two phosphotyrosines in the InsR activation loop, pY1158 and pY1162, may contribute to stabilization, they are also directly involved in binding of downstream signaling proteins (discussed below).

RECRUITMENT OF DOWNSTREAM SIGNALING PROTEINS TO THE ACTIVATED InsR

On insulin-stimulated activation/autophosphorylation of InsR, several proteins are recruited to the receptor for downstream signal propagation, including the IRS proteins, Shc, SH2B1 (formerly SH2-B), and SH2B2 (formerly APS). These proteins contain either a phosphotyrosine-binding (PTB) domain (IRS proteins, Shc) or an Src homology-2 (SH2) domain (SH2B1-2). The PTB domains of the IRS proteins and Shc bind to the juxtamembrane autophosphorylation site pY972, a canonical PTB-domain binding site (NPXpY). The SH2 domains of SH2B1−2 and of Grb10 and Grb14 (negative regulators described below) bind to the triply phosphorylated IRK activation loop.

One unusual feature of pY972 in the InsR juxtamembrane region is that it is a low-affinity binding site for the PTB domain of IRS1, arguably the most important substrate (along with IRS2) of InsR. A phosphopeptide representing the native pY972 sequence binds to the isolated IRS1 PTB domain with a dissociation constant (K_d) of 87 μM (Farooq et al. 1999), whereas most physiological PTB domain−phosphopeptide interactions are characterized by K_d values in the low micromolar range. The residue that precedes Y972 is E971, and an E971A substitution dramatically lowers the K_d to 2 μM (Farooq et al. 1999). The high K_d for binding to the native NPEpY sequence can be rationalized from the IRS1 PTB-domain structure (Eck et al. 1996). Owing to an extended carboxy-terminal helix, the pocket on the PTB domain for the residue preceding pY972 (E971) is shallow and suboptimal for glutamic acid. In fact, a phosphopeptide harboring E971A was used for cocrystallization.

At least one reason why glutamate precedes Y972 is that an acidic residue at the P-1 position (relative to the substrate tyrosine) decreases the substrate K_m (Shoelson et al. 1992). Indeed, substitution of E971 with alanine in InsR results in loss of autophosphorylation of Y972 (SR Hubbard, unpubl.). That is, E971 is critical for autophosphorylation of Y972, which resides in a nonoptimal substrate sequence for InsR. Interestingly, the PTB domain of Shc binds to the native pY972 phosphopeptide (NPEpY) with a K_d of 4 μM (Farooq et al. 1999) (i.e., within the typical range of PTB domain–phosphopeptide-binding affinities). IRS1–2 contain a pleckstrin homology (PH) domain upstream of the PTB domain (Dhe-Paganon et al. 1999), which plays a major role in recruitment of IRS1–2 to InsR (Yenush et al. 1996). Dimerization of the PH-PTB domains of IRS1 may also facilitate binding to the dimeric InsR (SR Hubbard, unpubl.).

IRS1–2 contain numerous tyrosine phosphorylation sites in their carboxy-terminal regions, which reside in a YΦXM motif (where Φ is hydrophobic). These sites, once phosphorylated by InsR, serve as recruitment sites for the SH2 domains of phosphatidylinositol 3-kinase (PI3K) (Myers et al. 1992), which, on activation, leads to activation of Akt. Other tyrosine phosphorylation sites in IRS1–2 recruit the adaptor protein Grb2 (which, through Sos, activates Ras) and the protein tyrosine phosphatase SHP2 (White 2002). IRS1–2 also possess numerous sites of serine/threonine phosphorylation that negatively regulate tyrosine phosphorylation, either in the course of normal negative feedback or in pathological insulin resistance (Pirola et al. 2004).

Previous yeast two-hybrid studies showed that a second InsR-interacting region (in addi-

tion to the PTB domain) existed in IRS2, but was lacking in IRS1, and this region was named the kinase regulatory-loop binding (KRLB) region (Sawka-Verhelle et al. 1996, 1997) or receptor binding domain-2 (RBD2) (He et al. 1996). These studies showed that the KRLB region binds to the kinase domain of InsR in a phosphorylation-dependent manner (He et al. 1996; Sawka-Verhelle et al. 1996). The KRLB region was coarsely mapped to residues 591–733, starting $\sim$300 residues carboxy terminal to the PTB domain. This region is predicted to lack secondary and tertiary structure (it contains a high proportion of glycine, serine, and proline residues), and mutagenesis studies identified two non-YΦXM tyrosines, Y624 and Y628, as critical residues in the KRLB–IRK interaction (Sawka-Verhelle et al. 1997).

The molecular basis for the interaction of the IRS2 KRLB region with InsR was elucidated through a cocrystal structure of a 15-residue peptide from the KRLB region (containing Y624 and Y628) bound to phosphorylated IRK (Wu et al. 2008b). The structure revealed that the KRLB region binds in the active site of IRK, with Y628 positioned for phosphorylation. Biochemical experiments showed that Y628 is phosphorylated by IRK, but with a K_m (ATP) that is unusually high compared to that of a YΦXM substrate (1.6 mM vs. 40 μM) (Wu et al. 2008b). In addition, the phosphorylated Y628 peptide retains significant binding affinity in the IRK active site, which results in poor substrate turnover. Consequently, this segment of IRS2 inhibits phosphorylation by InsR of other tyrosine sites in IRS2. Importantly, Y628 in IRS2 was mapped as an insulin-stimulated phosphorylation site in cells (Schmelzle et al. 2006). A possible functional role for the KRLB region is to suppress tyrosine phosphorylation of IRS2, setting a threshold such that only metabolic pathways via PI3K are stimulated by insulin and not mitogenic pathways via Grb2/Ras; there are 10 potential PI3K recruitment sites in IRS2 versus a single Grb2 site.

InsR-interacting proteins that contain an SH2 domain (SH2B2, Grb10/14) bind to the phosphorylated activation loop of InsR. The phosphorylated activation loop represents an

Cite this article as *Cold Spring Harb Perspect Biol* doi: 10.1101/cshperspect.a008946

atypical SH2-domain binding site, in that the loop is multiply phosphorylated and is stabilized in a turn-containing (rather than an extended) conformation. Indeed, these SH2 domains possess unique features that allow them to bind efficiently to the phosphorylated activation loop of InsR (Hu et al. 2003; Stein et al. 2003; Depetris et al. 2005). First, they are dimeric. Second, each protomer in the dimer binds to two phosphotyrosines in the activation loop: pY1158 canonically, through the invariant arginine (βB5) in the SH2 domain, and pY1163 noncanonically, through two lysine residues (βD1 and βD3). Third, they have evolved to bind poorly to canonical phosphotyrosine sequences (presumably, to reduce competition), those containing a pYXXΦ sequence, by disrupting the pY+3 (hydrophobic) binding pocket. Although other RTKs such as TrkA-C, MuSK, Met, Ror2, and ALK also contain three activation-loop tyrosines at the equivalent positions of Y1158, Y1162, and Y1163 in InsR, only in the case of Trks have SH2 domain-containing proteins (SH2B1 and SH2B2) been implicated in binding to the phosphorylated activation loop (Qian et al. 1998).

A crystal structure of the SH2 domain of SH2B2 bound to IRK (Hu et al. 2003) revealed that this SH2 domain possesses a noncanonical architecture, in which the carboxy-terminal half of the domain forms a long α helix rather than two β strands (βE and βF) and a shorter α helix (αB), as found in a typical SH2 domain. This structural rearrangement facilitates formation of a novel SH2-domain dimer, which interacts with the phosphorylated activation loops in the two kinase domains of InsR (Fig. 4A). Each protomer of the SH2-domain dimer coordinates two phosphotyrosines, the first (pY1158) in the canonical (arginine-containing) phosphate-binding pocket and the second (pY1162) by two lysines in βD. Binding of the SH2B2 SH2 domain does not perturb the conformation of the phosphorylated activation loop, but rather this mode of recruitment facilitates phosphorylation by InsR of a carboxy-terminal tyrosine (Y618) in SH2B2 (Liu et al. 2002). Of note, despite 79% sequence identity between the SH2 domains of SH2B1 and SH2B2, the

SH2B1 SH2 domain is monomeric, which switches its binding preference from the phosphorylated activation loop of IRK to pY813 of Jak2, a conventional SH2-domain ligand with a pY+3 hydrophobic residue (Hu and Hubbard 2006).

RECRUITMENT OF NEGATIVE REGULATORS TO THE ACTIVATED InsR

In addition to recruitment to InsR of positive regulators of insulin signaling, several proteins are recruited to the activated InsR to attenuate signaling, including the adaptor proteins Grb10 (Smith et al. 2007) and Grb14 (Cooney et al. 2004) and the protein tyrosine phosphatase PTP1B (Elchebly et al. 1999; Klaman et al. 2000). As mentioned above, Grb10/14 contains a carboxy-terminal SH2 domain that binds directly to the phosphorylated activation loop of InsR. In addition to the SH2 domain, Grb10/14 possess several other signaling modules, including a Ras-associating (RA) domain, a PH domain, and a ~45-residue region known as BPS (between PH and SH2) (He et al. 1998) or PIR (phosphorylated insulin receptor-interacting region) (Kasus-Jacobi et al. 1998), which is unique to this adaptor family.

Biochemical studies showed that the BPS region of Grb10/14 is capable of directly inhibiting the catalytic activity of InsR (Stein et al. 2001; Bereziat et al. 2002). A crystal structure of the Grb14 BPS region in complex with phosphorylated IRK revealed at least one mechanism by which Grb14 inhibits signaling by InsR (Depetris et al. 2005). The amino-terminal portion of the BPS region binds in the substrate-binding groove in the C lobe of IRK, with a leucine (L376) inserted into the active site instead of a tyrosine (Fig. 4B). Hydrophobic residues at the +1, +3, and +5 positions relative to L376 mimic binding of an optimal IRK substrate (YΦXMXΦ). Thus, this segment of the BPS region functions as a pseudosubstrate inhibitor of IRK.

Carboxy terminal to the pseudosubstrate segment, the Grb14 BPS region adopts a 16-residue α helix whose residues make interactions with the phosphorylated activation loop.

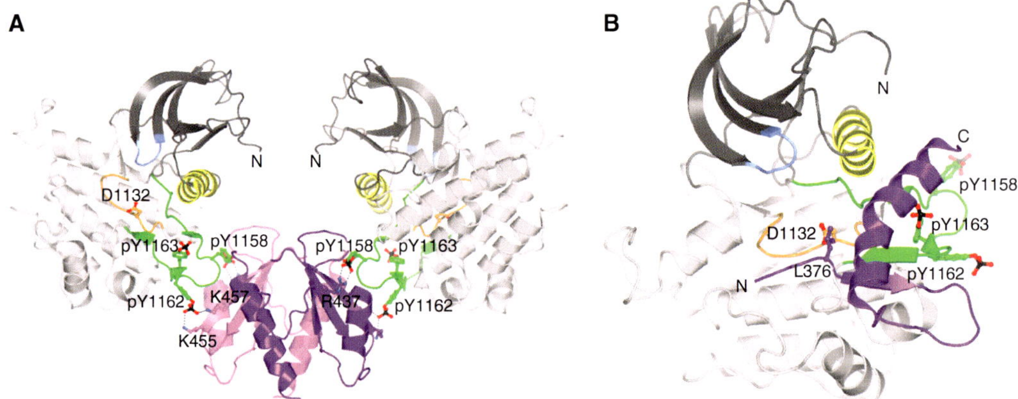

Figure 4. Structures of InsR-interacting proteins. (*A*) Ribbon diagram of the dimeric SH2 domain of SH2B2 bound to two IRK (InsR tyrosine kinase domain) molecules (Hu et al. 2003). The two SH2-domain protomers are colored pink and purple. Tris-phosphorylated IRK is colored as in Figure 3B. The molecular twofold axis is vertical. D1132 (catalytic loop), pY1158, pY1162, and pY1163 (activation loop) of IRK are shown in ball-and-stick representation, as are R437, K455, and K457 of SH2B2. Salt bridges are shown by black dashed lines. (*B*) Ribbon diagram of the BPS region of Grb14 bound to tris-phosphorylated IRK (Depetris et al. 2005). Grb14 BPS is colored purple and the pseudosubstrate residue L376 is shown in ball-and-stick representation. The amino and carboxyl termini of the BPS region are labeled N and C.

These interactions fortify the BPS–IRK interaction and provide specificity; the BPS region of Grb14/10 inhibits only InsR and the highly related IGF1R. The BPS α helix ends near the kinase N lobe, which positions the carboxy-terminal SH2 domain for interaction with pY1158 and pY1162 in the activation loop (Depetris et al. 2005). The SH2 domain of Grb14/10 is dimeric (Stein et al. 2003; Depetris et al. 2005), like the SH2B2 SH2 domain (Hu et al. 2003), but each protomer possesses a prototypical SH2-domain architecture. Binding of canonical phosphotyrosine sequences (with P+3 hydrophobic residues) are selected against by a valine in the BG loop (glycine in Src and Abl), which effectively seals off the P+3 binding pocket (Stein et al. 2003).

In recent studies, two groups have identified Grb10 as a downstream phosphorylation target of mTORC1 (Hsu et al. 2011; Yu et al. 2011). Grb10 provides the "missing link" in a negative-feedback system involving insulin/IGF1, PI3K/Akt, and mTORC1, in which activation of mTORC1 leads to inhibition of PI3K/Akt signaling. Grb10 negatively regulates PI3K/Akt signaling by binding to and inhibiting the cat-alytic activity of the insulin/IGF1 receptor (as described above for Grb14). Phosphorylation of Grb10 by mTORC1 was shown to increase the stability of Grb10 protein levels in cells (Hsu et al. 2011; Yu et al. 2011), which provides one mechanism by which mTORC1-mediated phosphorylation of Grb10 potentiates PI3K/Akt inhibition. Another possible mechanism is that phosphorylation of Grb10 increases the interaction between Grb10 and IRK/IGF1RK and thus the Grb10 inhibitory effect. The mTORC1-mediated serine/threonine sites in Grb10 have been mapped (Hsu et al. 2011; Yu et al. 2011), and several are located in the BPS region and the BPS-SH2 linker, regions that interact (BPS) or potentially interact (BPS-SH2 linker) directly with IRK/IGF1K. In particular, S428 is four residues upstream of the leucine pseudosubstrate in the Grb10 BPS region (see Fig. 4B), and because IRK/IGF1RK prefers tyrosine substrates with acidic residues in this region (Shoelson et al. 1992), it is plausible that pS428 (now acidic) increases Grb10 binding affinity.

PTP1B is capable of dephosphorylating numerous RTKs in vitro and in cells in which it is overexpressed, yet PTP1B-null mice are of normal size (and not tumor-prone) and show two major phenotypes: insulin hypersensitivity and resistance to high-fat-diet-induced weight gain (Elchebly et al. 1999; Klaman et al. 2000). The two main targets of PTP1B's action are thought to be InsR and the cytoplasmic tyrosine kinase Jak2 (Myers et al. 2001). A crystal structure of PTP1B with a bound phosphopeptide representing the activation loop of IRK indicated that specificity for InsR may be dictated in part by the twin phosphotyrosines pY1162/1163, in the context (E/D)pYpY(K/R) (Salmeen et al. 2000). Jak2 also contains twin phosphotyrosines (pY1007/1008) in this sequence motif in its activation loop, but other RTKs do as well, including IGF1R, fibroblast growth factor receptors, Trks, MuSK, and Mer. Therefore, additional determinants of PTP1B substrate specificity must exist.

A crystal structure of a complex between PTP1B and phosphorylated IRK was determined (Li et al. 2005), which revealed an unusual mode of interaction. Rather than binding to the phosphorylated activation loop, PTP1B was bound to the "backside" of IRK. This binding mode was clearly facilitated by high concentrations of ammonium sulfate, a common precipitating agent for protein crystallization. Although this binding mode between PTP1B and InsR remains to be corroborated in cell-based assays, the specificity of the interaction is intriguing: even the highly related IGF1R would not be predicted to engage PTP1B in this manner. Furthermore, two tyrosines in PTP1B (Y152, Y153) that are found in the PTP1B–IRK interface were shown (before the structure was solved) to be important in the interaction between PTP1B and InsR in cells (Bandyopadhyay et al. 1997; Dadke et al. 2000). It thus remains an intriguing hypothesis that the mode of binding observed in the PTP1B–IRK crystal structure represents a recruitment interaction to localize PTP1B to InsR. No crystal structure of the catalytic mode of interaction between PTP1B and phosphorylated IRK has been reported. Presumably, this structure would reveal additional specificity determinants not highlighted by the PTP1B-phosphopeptide

(IRK activation loop) structure (Salmeen et al. 2000).

Because Grb14 and Grb10 engage all three activation-loop phosphotyrosines of InsR via their SH2 domains and BPS regions (Depetris et al. 2005), these negative regulators would be predicted to antagonize PTP1B-catalyzed dephosphorylation of InsR. This indeed appears to be the case, at least for Grb14, based on both in vitro and cell-based experiments (Bereziat et al. 2002; Nouaille et al. 2006). It is conceivable that, in cells in which Grb14 is present (muscle and liver), Grb14 acts initially to attenuate InsR signaling at the plasma membrane and, on internalization of InsR, PTP1B (tethered to the endoplasmic reticulum) outcompetes Grb14 for binding and dephosphorylates InsR.

CONCLUDING REMARKS

Since the cloning of human InsR in the mid-1980s (Ebina et al. 1985; Ullrich et al. 1985), a vast amount of knowledge has been gained on the biochemical, structural, and functional properties of this crucial RTK in human biology. The key mechanistic questions that remain to be answered are (i) What prevents the kinase domains from *trans*-phosphorylating in the absence of insulin (i.e., how is the basal state maintained?) (ii) What is the insulin-induced conformational change in the ectodomain? (iii) How is this conformational change transduced to the kinase domains to facilitate *trans*-phosphorylation? (iv) On activation, do the kinase domains function in tandem or individually in the phosphorylation of downstream substrates? The prevalence of type II diabetes in developed countries is increasing at an alarming rate. One therapeutic strategy (out of many) for treating this disease is to design or screen for agents that directly potentiate the signaling output from InsR. Our current knowledge and that gained from answering these mechanistic questions should prove invaluable in the quest for such therapeutic agents. A recent crystal structure reveals how insulin binds to its primary site on InsR (Menting et al. 2013).

ACKNOWLEDGMENTS

Past support for InsR research is acknowledged from the National Institutes of Health (DK052916).

REFERENCES

Artim SC, Mendrola JM, Lemmon MA. 2012. Assessing the range of kinase autoinhibition mechanisms in the insulin receptor family. *Biochem J* **448:** 213–220.

Backer JM, Shoelson SE, Weiss MA, Hua QX, Cheatham RB, Haring E, Cahill DC, White MF. 1992. The insulin receptor juxtamembrane region contains two independent tyrosine/β-turn internalization signals. *J Cell Biol* **118:** 831–839.

Bandyopadhyay D, Kusari A, Kenner KA, Liu F, Chernoff J, Gustafson TA, Kusari J. 1997. Protein-tyrosine phosphatase 1B complexes with the insulin receptor in vivo and is tyrosine-phosphorylated in the presence of insulin. *J Biol Chem* **272:** 1639–1645.

Bereziat V, Kasus-Jacobi A, Perdereau D, Cariou B, Girard J, Burnol AF. 2002. Inhibition of insulin receptor catalytic activity by the molecular adapter Grb14. *J Biol Chem* **277:** 4845–4852.

Bertrand T, Kothe M, Liu J, Dupuy A, Rak A, Berne PF, Davis S, Gladysheva T, Valtre C, Crenne JY, et al. 2012. The crystal structures of TrkA and TrkB suggest key regions for achieving selective inhibition. *J Mol Biol* **423:** 439–453.

Cooney GJ, Lyons RJ, Crew AJ, Jensen TE, Molero JC, Mitchell CJ, Biden TJ, Ormandy CJ, James DE, Daly RJ. 2004. Improved glucose homeostasis and enhanced insulin signalling in Grb14-deficient mice. *EMBO J* **23:** 582–593.

Dadke S, Kusari J, Chernoff J. 2000. Down-regulation of insulin signaling by protein-tyrosine phosphatase 1B is mediated by an N-terminal binding region. *J Biol Chem* **275:** 23642–23647.

De Meyts P. 2008. The insulin receptor: A prototype for dimeric, allosteric membrane receptors? *Trends Biochem Sci* **33:** 376–384.

Depetris RS, Hu J, Gimpelevich I, Holt LJ, Hubbard SR. 2005. Structural basis for inhibition of the insulin receptor by the adaptor protein Grb14. *Mol Cell* **20:** 325–333.

Dhe-Paganon S, Ottinger EA, Nolte RT, Eck MJ, Shoelson SE. 1999. Crystal structure of the pleckstrin homology-phosphotyrosine binding (PH-PTB) targeting region of insulin receptor substrate 1. *Proc Natl Acad Sci* **96:** 8378–8383.

Ebina Y, Ellis L, Jarnagin K, Edery M, Graf L, Clauser E, Ou JH, Masiarz F, Kan YW, Goldfine ID, et al. 1985. The human insulin receptor cDNA: The structural basis for hormone-activated transmembrane signalling. *Cell* **40:** 747–758.

Eck MJ, Dhe-Paganon S, Trub T, Nolte RT, Shoelson SE. 1996. Structure of the IRS-1 PTB domain bound to the juxtamembrane region of the insulin receptor. *Cell* **85:** 695–705.

Elchebly M, Payette P, Michaliszyn E, Cromlish W, Collins S, Loy AL, Normandin D, Cheng A, Himms-Hagen J, Chan CC, et al. 1999. Increased insulin sensitivity and obesity resistance in mice lacking the protein tyrosine phosphatase-1B gene. *Science* **283:** 1544–1548.

Farooq A, Plotnikova O, Zeng L, Zhou MM. 1999. Phosphotyrosine binding domains of Shc and insulin receptor substrate 1 recognize the NPXpY motif in a thermodynamically distinct manner. *J Biol Chem* **274:** 6114–6121.

Formisano P, Sohn KJ, Miele C, Di Finizio B, Petruzziello A, Riccardi G, Beguinot L, Beguinot F. 1993. Mutation in a conserved motif next to the insulin receptor key autophosphorylation sites de-regulates kinase activity and impairs insulin action. *J Biol Chem* **268:** 5241–5248.

Frattali AL, Treadway JL, Pessin JE. 1992. Transmembrane signaling by the human insulin receptor kinase. *J Biol Chem* **267:** 19521–19528.

He W, Craparo A, Zhu Y, O'Neill TJ, Wang L-M, Pierce J, Gustafson TA. 1996. Interaction of insulin receptor substrate-2 (IRS-2) with the insulin and insulin-like growth factor I receptors. *J Biol Chem* **271:** 11641–11645.

He W, Rose DW, Olefsky JM, Gustafson TA. 1998. Grb10 interacts differentially with the insulin receptor, insulin-like growth factor I receptor, and epidermal growth factor receptor via the Grb10 Src homology 2 (SH2) domain and a second novel domain located between the pleckstrin homology and SH2 domains. *J Biol Chem* **273:** 6860–6867.

Hsu PP, Kang SA, Rameseder J, Zhang Y, Ottina KA, Lim D, Peterson TR, Choi Y, Gray NS, Yaffe MB, et al. 2011. The mTOR-regulated phosphoproteome reveals a mechanism of mTORC1-mediated inhibition of growth factor signaling. *Science* **332:** 1317–1322.

Hu J, Hubbard SR. 2006. Structural basis for phosphotyrosine recognition by the Src homology-2 domains of the adapter proteins SH2-B and APS. *J Mol Biol* **361:** 69–79.

Hu J, Liu J, Ghirlando R, Saltiel AR, Hubbard SR. 2003. Structural basis for recruitment of the adaptor protein APS to the activated insulin receptor. *Mol Cell* **12:** 1379–1389.

Hubbard SR. 1997. Crystal structure of the activated insulin receptor tyrosine kinase in complex with peptide substrate and ATP analog. *EMBO J* **16:** 5572–5581.

Hubbard SR, Wei L, Ellis L, Hendrickson WA. 1994. Crystal structure of the tyrosine kinase domain of the human insulin receptor. *Nature* **372:** 746–754.

Huse M, Kuriyan J. 2002. The conformational plasticity of protein kinases. *Cell* **109:** 275–282.

Kasus-Jacobi A, Perdereau D, Auzan C, Clauser E, Van Obberghen E, Mauvais-Jarvis F, Girard J, Burnol AF. 1998. Identification of the rat adapter Grb14 as an inhibitor of insulin actions. *J Biol Chem* **273:** 26026–26035.

Klaman LD, Boss O, Peroni OD, Kim JK, Martino JL, Zabolotny JM, Moghal N, Lubkin M, Kim YB, Sharpe AH, et al. 2000. Increased energy expenditure, decreased adiposity, and tissue-specific insulin sensitivity in protein-tyrosine phosphatase 1B-deficient mice. *Mol Cell Biol* **20:** 5479–5489.

Kohanski RA. 1993. Insulin receptor autophosphorylation. II. Determination of autophosphorylation sites by chemical sequence analysis and identification of the juxtamembrane sites. *Biochemistry* **32:** 5773–5780.

Lee CC, Jia Y, Li N, Sun X, Ng K, Ambing E, Gao MY, Hua S, Chen C, Kim S, et al. 2010. Crystal structure of the ALK (anaplastic lymphoma kinase) catalytic domain. *Biochem J* **430:** 425–437.

Li S, Covino ND, Stein EG, Till JH, Hubbard SR. 2003. Structural and biochemical evidence for an autoinhibitory role for tyrosine 984 in the juxtamembrane region of the insulin receptor. *J Biol Chem* **278:** 26007–26014.

Li S, Depetris RS, Barford D, Chernoff J, Hubbard SR. 2005. Crystal structure of a complex between protein tyrosine phosphatase 1B and the insulin receptor tyrosine kinase. *Structure* **13:** 1643–1651.

Liu J, Kimura A, Baumann CA, Saltiel AR. 2002. APS facilitates c-Cbl tyrosine phosphorylation and GLUT4 translocation in response to insulin in 3T3-L1 adipocytes. *Mol Cell Biol* **22:** 3599–3609.

McKern NM, Lawrence MC, Streltsov VA, Lou MZ, Adams TE, Lovrecz GO, Elleman TC, Richards KM, Bentley JD, Pilling PA, et al. 2006. Structure of the insulin receptor ectodomain reveals a folded-over conformation. *Nature* **443:** 218–221.

Menting JG, Whittaker J, Margetts MB, Whittaker LJ, Kong GK-W, Smith BJ, Watson CJ, Žáková L, Kletvíková E, Jiráček J, et al. 2013. How insulin engages its primary binding site on the insulin receptor. *Nature* **493:** 241–245.

Munshi S, Kornienko M, Hall DL, Reid JC, Waxman L, Stirdivant SM, Darke PL, Kuo LC. 2002. Crystal structure of the Apo, unactivated insulin-like growth factor-1 receptor kinase. Implication for inhibitor specificity. *J Biol Chem* **277:** 38797–38802.

Myers MG Jr, Backer JM, Sun XJ, Shoelson S, Hu P, Schlessinger J, Yoakim M, Schaffhausen B, White MF. 1992. IRS-1 activates phosphatidylinositol 3'-kinase by associating with src homology 2 domains of p85. *Proc Natl Acad Sci* **89:** 10350–10354.

Myers MP, Andersen JN, Cheng A, Tremblay ML, Horvath CM, Parisien JP, Salmeen A, Barford D, Tonks NK. 2001. TYK2 and JAK2 are substrates of protein-tyrosine phosphatase 1B. *J Biol Chem* **276:** 47771–47774.

Nouaille S, Blanquart C, Zilberfarb V, Boute N, Perdereau D, Burnol AF, Issad T. 2006. Interaction between the insulin receptor and Grb14: A dynamic study in living cells using BRET. *Biochem Pharmacol* **72:** 1355–1366.

Pirola L, Johnston AM, Van Obberghen E. 2004. Modulation of insulin action. *Diabetologia* **47:** 170–184

Qian X, Riccio A, Zhang Y, Ginty DD. 1998. Identification and characterization of novel substrates of Trk receptors in developing neurons. *Neuron* **21:** 1017–1029.

Renteria ME, Gandhi NS, Vinuesa P, Helmerhorst E, Mancera RL. 2008. A comparative structural bioinformatics analysis of the insulin receptor family ectodomain based on phylogenetic information. *PloS ONE* **3:** e3667.

Salmeen A, Andersen JN, Myers MP, Tonks NK, Barford D. 2000. Molecular basis for the dephosphorylation of the activation segment of the insulin receptor by protein tyrosine phosphatase 1B. *Mol Cell* **6:** 1401–1412.

Sawka-Verhelle D, Tartare-Deckert S, White MF, Van Obberghen E. 1996. Insulin receptor substrate-2 binds to the insulin receptor through its phosphotyrosine-binding domain and through a newly identified domain comprising amino acids 591–786. *J Biol Chem* **271:** 5980–5983.

Sawka-Verhelle D, Baron V, Mothe I, Filloux C, White MF, Van Obberghen E. 1997. Tyr624 and Tyr628 in insulin receptor substrate-2 mediate its association with the insulin receptor. *J Biol Chem* **272:** 16414–16420.

Schmelzle K, Kane S, Gridley S, Lienhard GE, White FM. 2006. Temporal dynamics of tyrosine phosphorylation in insulin signaling. *Diabetes* **55:** 2171–2179.

Shoelson SE, Chatterjee S, Chaudhuri M, White MF. 1992. YMXM motifs of IRS-1 define substrate specificity of the insulin receptor kinase. *Proc Natl Acad Sci* **89:** 2027–2031.

Smith FM, Holt LJ, Garfield AS, Charalambous M, Koumanov F, Perry M, Bazzani R, Sheardown SA, Hegarty BD, Lyons RJ, et al. 2007. Mice with a disruption of the imprinted Grb10 gene exhibit altered body composition, glucose homeostasis, and insulin signaling during postnatal life. *Mol Cell Biol* **27:** 5871–5886.

Smith BJ, Huang K, Kong G, Chan SJ, Nakagawa S, Menting JG, Hu SQ, Whittaker J, Steiner DF, Katsoyannis PG, et al. 2010. Structural resolution of a tandem hormone-binding element in the insulin receptor and its implications for design of peptide agonists. *Proc Natl Acad Sci* **107:** 6771–6776.

Sparrow LG, McKern NM, Gorman JJ, Strike PM, Robinson CP, Bentley JD, Ward CW. 1997. The disulfide bonds in the C-terminal domains of the human insulin receptor ectodomain. *J Biol Chem* **272:** 29460–29467.

Stein EG, Gustafson TA, Hubbard SR. 2001. The BPS domain of Grb10 inhibits the catalytic activity of the insulin and IGF1 receptors. *FEBS Lett* **493:** 106–111.

Stein EG, Ghirlando R, Hubbard SR. 2003. Structural basis for dimerization of the Grb10 Src homology 2 domain. Implications for ligand specificity. *J Biol Chem* **278:** 13257–13264.

Taniguchi CM, Emanuelli B, Kahn CR. 2006. Critical nodes in signalling pathways: Insights into insulin action. *Nat Rev Mol Cell Biol* **7:** 85–96.

Till JH, Ablooglu AJ, Frankel M, Bishop SM, Kohanski RA, Hubbard SR. 2001. Crystallographic and solution studies of an activation loop mutant of the insulin receptor tyrosine kinase: Insights into kinase mechanism. *J Biol Chem* **276:** 10049–10055.

Till JH, Becerra M, Watty W, Lu Y, Ma Y, Neubert TA, Burden SJ, Hubbard SR. 2002. Crystal structure of the MuSK tyrosine kinase: Insights into receptor autoregulation. *Structure* **10:** 1187–1196.

Ullrich A, Bell JR, Chen EY, Herrera R, Petruzzelli LM, Dull TJ, Gray A, Coussens L, Liao YC, Tsubokawa M, et al. 1985. Human insulin receptor and its relationship to the tyrosine kinase family of oncogenes. *Nature* **313:** 756–761.

Wang W, Marimuthu A, Tsai J, Kumar A, Krupka HI, Zhang C, Powell B, Suzuki Y, Nguyen H, Tabrizizad M, et al. 2006. Structural characterization of autoinhibited c-Met kinase produced by coexpression in bacteria with phosphatase. *Proc Natl Acad Sci* **103:** 3563–3568.

Ward CW, Lawrence MC. 2009. Ligand-induced activation of the insulin receptor: A multi-step process involving structural changes in both the ligand and the receptor. *BioEssays* **31:** 422–434.

Ward CW, Lawrence MC. 2012. Similar but different: Ligand-induced activation of the insulin and epidermal growth factor receptor families. *Curr Opin Struct Biol* **22:** 360–366.

Wei L, Hubbard SR, Hendrickson WA, Ellis L. 1995. Expression, characterization, and crystallization of the catalytic core of the human insulin receptor protein-tyrosine kinase domain. *J Biol Chem* **270:** 8122–8130.

White MF. 2002. IRS proteins and the common path to diabetes. *Am J Physiol Endocrinol Metab* **283:** E413–E422.

White MF, Shoelson SE, Keutmann H, Kahn CR. 1988. A cascade of tyrosine autophosphorylation in the β-subunit activates the phosphotransferase of the insulin receptor. *J Biol Chem* **263:** 2969–2980.

Whittaker L, Hao C, Fu W, Whittaker J. 2008. High-affinity insulin binding: Insulin interacts with two receptor ligand binding sites. *Biochemistry* **47:** 12900–12909.

Wu J, Li W, Craddock BP, Foreman KW, Mulvihill MJ, Ji QS, Miller WT, Hubbard SR. 2008a. Small-molecule inhibition and activation-loop trans-phosphorylation of the IGF1 receptor. *EMBO J* **27:** 1985–1994.

Wu J, Tseng YD, Xu CF, Neubert TA, White MF, Hubbard SR. 2008b. Structural and biochemical characterization of the KRLB region in insulin receptor substrate-2. *Nat Struct Mol Biol* **15:** 251–258.

Yamaguchi H, Hendrickson WA. 1996. Structural basis for activation of the human lymphocyte kinase Lck upon tyrosine phosphorylation. *Nature* **384:** 484–489.

Yenush L, Makati KJ, Smith-Hall J, Ishibashi O, Myers MG Jr, White MF. 1996. The pleckstrin homology domain is the principal link between the insulin receptor and IRS-1. *J Biol Chem* **271:** 24300–24306.

Yu Y, Yoon SO, Poulogiannis G, Yang Q, Ma XM, Villen J, Kubica N, Hoffman GR, Cantley LC, Gygi SP, et al. 2011. Phosphoproteomic analysis identifies Grb10 as an mTORC1 substrate that negatively regulates insulin signaling. *Science* **332:** 1322–1326.

The EGFR Family: Not So Prototypical Receptor Tyrosine Kinases

Mark A. Lemmon[1], Joseph Schlessinger[2], and Kathryn M. Ferguson[3]

[1]Department of Biochemistry and Biophysics, University of Pennsylvania Perelman School of Medicine, Philadelphia, Pennsylvania 19104

[2]Department of Pharmacology, Yale University School of Medicine, New Haven, Connecticut 06520

[3]Department of Physiology, University of Pennsylvania Perelman School of Medicine, Philadelphia, Pennsylvania 19104

Correspondence: mlemmon@mail.med.upenn.edu

The epidermal growth factor receptor (EGFR) was among the first receptor tyrosine kinases (RTKs) for which ligand binding was studied and for which the importance of ligand-induced dimerization was established. As a result, EGFR and its relatives have frequently been termed "prototypical" RTKs. Many years of mechanistic studies, however, have revealed that—far from being prototypical—the EGFR family is quite unique. As we discuss in this review, the EGFR family uses a distinctive "receptor-mediated" dimerization mechanism, with ligand binding inducing a dramatic conformational change that exposes a dimerization arm. Intracellular kinase domain regulation in this family is also unique, being driven by allosteric changes induced by asymmetric dimer formation rather than the more typical activation-loop phosphorylation. EGFR family members also distinguish themselves from other RTKs in having an intracellular juxtamembrane (JM) domain that activates (rather than autoinhibits) the receptor and a very large carboxy-terminal tail that contains autophosphorylation sites and serves an autoregulatory function. We discuss recent advances in mechanistic aspects of all of these components of EGFR family members, attempting to integrate them into a view of how RTKs in this important class are regulated at the cell surface.

The epidermal growth factor receptor (EGFR) is often considered the "prototypical" receptor tyrosine kinase (RTK) and has been intensively studied. It is one of a family of four RTKs in humans, the others being ErbB2/HER2, ErbB3/HER3, and ErbB4/HER4 (Fig. 1). EGFR and its relatives are known oncogenic drivers in cancers such as lung cancer (Mok 2011), breast cancer (Arteaga et al. 2011), and glioblastoma (Libermann et al. 1985; Lee et al. 2006a; Vivanco et al. 2012), and inhibitors of these receptors have been among the most successful examples of targeted cancer therapies to date (Arteaga 2003; Moasser 2007; Zhang et al. 2007), including antibody therapeutics (e.g., trastuzumab and cetuximab) and small-molecule tyrosine kinase inhibitors (e.g., erlotinib, gefitinib, lapatinib).

Far from being prototypical, however, it is now clear that regulation of EGFR family mem-

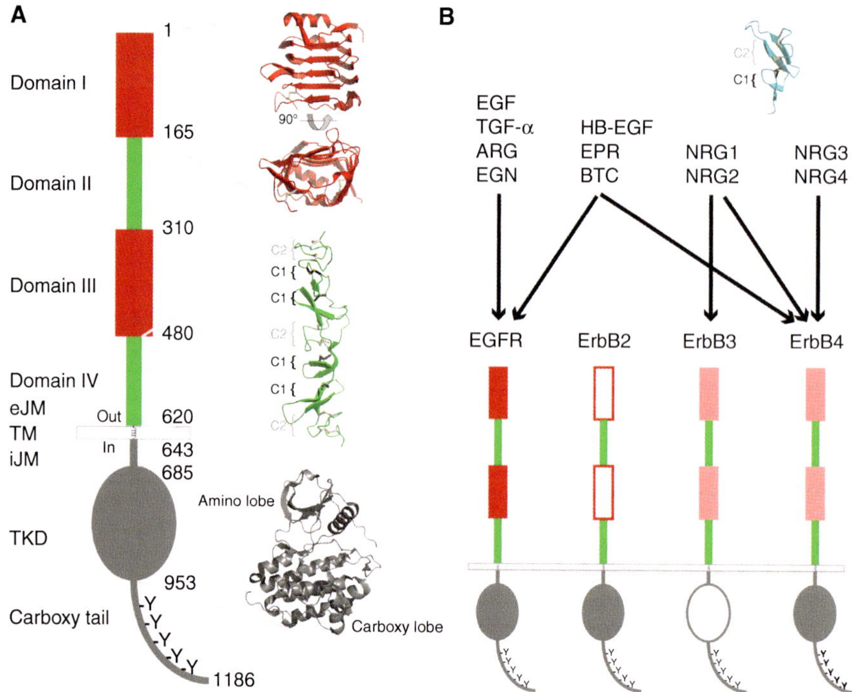

Figure 1. Schematic representation of EGFR/ErbB family receptors and their ligands. (*A*) The domain composition of human EGFR is shown. The extracellular region contains four domains: Domain I (amino acids 1–165), domain II (amino acids 165–310), domain III (amino acids 310–480), and domain IV (amino acids 480–620). Domains I and III are closely related in sequence, as are domains II and IV. Shown are representations of the structures of domains I and IV. Domain IV contains two types of disulfide-bonded module (C1 and C2). In C1 domains, a single disulfide constrains an intervening bow-like loop. In C2 modules, two disulfides link four successive cysteines in the patterns C1–C3 and C2–C4 to give a knot-like structure. A short extracellular juxtamembrane (eJM) region separates the extracellular region from the ∼23-amino-acid transmembrane (TM) domain. Within the cell, a short intracellular juxtamembrane (iJM) region separates the tyrosine kinase domain (TKD) from the membrane. A representative EGFR TKD structure is shown. The TKD is followed by a carboxy-terminal largely unstructured tail (amino acids 953–1186) that contains at least five tyrosine autophosphorylation sites. (*B*) EGFR is one of four members of the EGFR/ErbB family in humans. The other members are ErbB2/HER2, for which no soluble activating ligand is shown; ErbB3/HER3, which has a significantly impaired kinase domain (Jura et al. 2009b; Shi et al. 2010); and ErbB4/HER4. The primary active moiety of the ligands for these receptors is the EGF-like domain, shown as a cartoon structure (*top right*). EGFR is activated by the EGFR agonists: EGF itself, TGF-α (transforming growth factor α), ARG (amphiregulin), and EGN (epigen). The bispecific ligands regulate both EGFR and ErbB4: HB-EGF (heparin-binding EGF-like growth factor), EPR (epiregulin), and BTC (betacellulin). Neuregulins (NRGs) 1 and 2 regulate ErbB3 and ErbB4, whereas NRG3 and NRG4 appear to be specific for ErbB4 (Wilson et al. 2009).

bers is unique among RTKs (Ferguson 2008; Lemmon 2009; Lemmon and Schlessinger 2010). Structural studies have revealed how the ∼620-amino-acid isolated extracellular region is induced to dimerize after growth factor binding (Burgess et al. 2003) and how the isolated intracellular tyrosine kinase domain (TKD) becomes allosterically activated after forming an asymmetric dimer (Zhang et al. 2006; Jura et al. 2009a; Red Brewer et al. 2009). These findings have typically been interpreted in the context of a model in which EGF family receptors are regulated through ligand-induced receptor homodimerization or heterodimerization, with growth factor binding converting the receptor from an inactive monomeric configuration to

Cite this article as *Cold Spring Harb Perspect Biol* doi: 10.1101/cshperspect.a020768

an active dimeric conformation (Yarden and Schlessinger 1987; Schlessinger 1988, 2014; Ullrich and Schlessinger 1990). Although this original model has stood the test of time and initiated a whole field of studies of ligand-induced RTK dimerization, recent work has provided highly sophisticated views of a unique mode of allosteric regulation used by EGFR.

ErbB RECEPTORS AND THEIR LIGANDS

Although EGFR/ErbB family members are most frequently considered in the context of human cancer—where EGFR, ErbB2, and ErbB3 are validated therapeutic targets—the biology of these broadly expressed receptors is very complicated (Yarden and Sliwkowski 2001; Burgess 2008). The effect of knocking out the *EGFR* gene in mice ranges from embryonic lethality in one genetic background to death at birth in another, to postnatal death in yet another (Sibilia et al. 2007). Defects are seen in bone, brain, heart, and various epithelia—notably skin, hair, eyes, and lungs. Mouse knockouts of ErbB2, ErbB3, or ErbB4 are all embryonic lethal, also with neurodevelopmental and cardiac defects in each case (Sibilia et al. 2007; Burgess 2008). It is now clear that ErbB 2/3/4 signaling has a key role in both cardiac development and maintenance of cardiac function in the adult (Pentassuglia and Sawyer 2009).

EGFR is regulated by at least seven different activating ligands in humans (Harris et al. 2003; Schneider and Wolf 2009) listed in Figure 1: EGF itself, transforming growth factor α (TGF-α), betacellulin (BTC), heparin-binding EGF-like growth factor (HB-EGF), amphiregulin (ARG), epiregulin (EPR), and epigen (EGN). Each contains an EGF-like domain that is responsible for receptor binding and activation, with a characteristic pattern of six spatially conserved cysteines (that form three intramolecular disulfides). The EGFR ligands are all produced as membrane-bound precursor proteins (Harris et al. 2003) and are cleaved by cell-surface proteases to yield the active growth factor species as described by Adrain and Freeman (2014). Although defects in EGFR affect a wide range of processes, it remains unclear which ligands are

responsible in which context—with a few exceptions (Fiske et al. 2009). ErbB3 and ErbB4 are regulated by neuregulins (NRGs) (Falls 2003)—also called heregulins (HRGs), a family of ligands produced from four genes (*NRG1—NRG4*) in a wide variety of different isoforms that all contain an EGF-like domain. NRG1 and NRG2 bind both ErbB3 and ErbB4, whereas NRG3 and NRG4 appear to be ErbB4 specific (Wilson et al. 2009). The NRGs and their receptors have an important role in nervous system development (Birchmeier 2009). NRG1 and ErbB4 have been linked to schizophrenia (Rico and Marín 2011). Three of the EGFR ligands mentioned above (BTC, EPR, and HB-EGF) also bind and activate ErbB4 and are termed "bispecific" ligands (Riese and Stern 1998; Wilson et al. 2009). For ErbB2, no soluble ligand has been identified despite a great deal of investigation in the 1990s. This orphan receptor is generally assumed only to be regulated by heterodimerization with other ErbB family receptors, although its close structural resemblance to the *Drosophila melanogaster* EGFR suggests that membrane-associated ErbB2 ligands may remain to be discovered (Alvarado et al. 2009).

Each ErbB ligand is produced as a membrane-bound precursor that is processed in a ligand-specific manner (Buonanno and Fischbach 2001). Although the EGF-like domains of the NRGs and the EGFR ligands appear to be sufficient for much of their biological effect, it is clear that other parts of the full-length ligands *do* influence signaling—although in ways that are not yet fully understood. The discussion in this review is limited to receptor activation by the EGF-like domains. The receptors themselves (Fig. 1) all have a large extracellular region of ~620 amino acids that can be subdivided into four domains. Domains I and III are related to one another (and to similar domains in the insulin receptor) and serve as the primary ligand-binding regions. Domains II and IV are cysteine-rich domains that share similarities with laminin repeats (Ward et al. 1995) and contain a string of disulfide-bonded modules. A single transmembrane domain links the extracellular region to the ~540-amino-acid intracellular region of the receptor that contains a tyrosine kinase domain as well as a carboxy-terminal "tail" of ~230 amino acids. Binding of activating ligands to EGFR or ErbB4 promotes strong homodimerization of these receptors (Ferguson et al. 2000). In addition, it is thought that the four members of the family form various heterodimers (Yarden and Sliwkowski 2001). In particular, ErbB2 and ErbB3—which do not form homodimers (Ferguson et al. 2000; Cho et al. 2003; Berger et al. 2004)—signal only through heterodimerization (with one another and with other ErbB receptors). Following activation, a series of tyrosines in the carboxy-terminal tail become autophosphorylated in *trans* (Honegger et al. 1989) and serve as "docking" sites for phosphotyrosine-binding SH2 and PTB domains as discussed in Wagner et al. (2013). Different ErbB family members harbor different complements of tyrosine phosphorylation sites in their carboxy-terminal tails (Lemmon and Schlessinger 1994; Yarden and Sliwkowski 2001; Jones et al. 2006), thus eliciting distinct but overlapping sets of responses.

STRUCTURAL BASIS FOR LIGAND-INDUCED DIMERIZATION OF ErbB RECEPTORS

A series of crystal structures published in 2002 and 2003 provided a dramatic leap forward in our understanding of transmembrane signaling by ErbB receptors (Burgess et al. 2003) and laid the foundation for much of our current understanding of how these receptors function. Structures of the EGFR extracellular region with and without bound growth factor provided a satisfying, and unexpected, model for ligand-induced receptor dimerization. They showed that—in contrast with other RTKs—dimerization of the EGFR extracellular region is mediated entirely by receptor–receptor contacts. With the exception of the insulin receptor, this makes EGFR and its relatives unique among the RTKs discussed in this collection. In all other cases that are understood, the activating growth factor ligand binds at the dimer interface, effectively serving to cross-link the two receptors into a dimer (Lemmon and Schlessinger 2010)—typ-

ically with the aid of additional receptor–receptor contacts. As shown on the right side of Figure 2, the bound ligands in an EGFR dimer are as far as they possibly could be from the dimer interface (Garrett et al. 2002; Ogiso et al. 2002)—and, indeed, from one another. Each bound ligand is bivalent but simultaneously contacts two points (domains I and III) in the extracellular region of a single receptor molecule, rather than spanning two receptors. Domains I and III are rigid β-helix/solenoid structures that do not change conformation after ligand binding (Ferguson 2004). Dimerization is driven almost exclusively by a "dimerization arm" that projects from domain II, although inter-receptor domain IV contacts (Fig. 2A) may also contribute weakly (Dawson et al. 2005; Lu et al. 2010). Activation (by way of inducing dimerization) involves an unexpectedly large ligand-induced conformational change that exposes this domain II dimerization arm. In the absence of bound ligand, the EGFR, ErbB3, and ErbB4 extracellular regions form the "tethered" structure shown on the left side of Figure 2A, in which the domain II dimerization arm is buried by intramolecular interactions with domain IV (Cho and Leahy 2002; Ferguson et al. 2003; Bouyain et al. 2005). Ligand binding to domains I and III effectively pulls these domains together and "extends" the receptor's extracellular region to break the domain II/IV tether and expose the dimerization site—yielding a dimerization-competent ligand-bound EGFR extracellular region that self-associates strongly.

ACTIVATION OF THE KINASE DOMAIN BY FORMATION OF AN ASYMMETRIC DIMER

The first crystallographic view of the EGFR TKD (Stamos et al. 2002) confirmed previous reports that activation-loop phosphorylation is not required for it to adopt an active-like structure (Gotoh et al. 1992; Burgess et al. 2003) but provided little insight into how it is activated by receptor dimerization. The likely activation mechanism was revealed in 2006 through insightful analysis of additional crystal structures

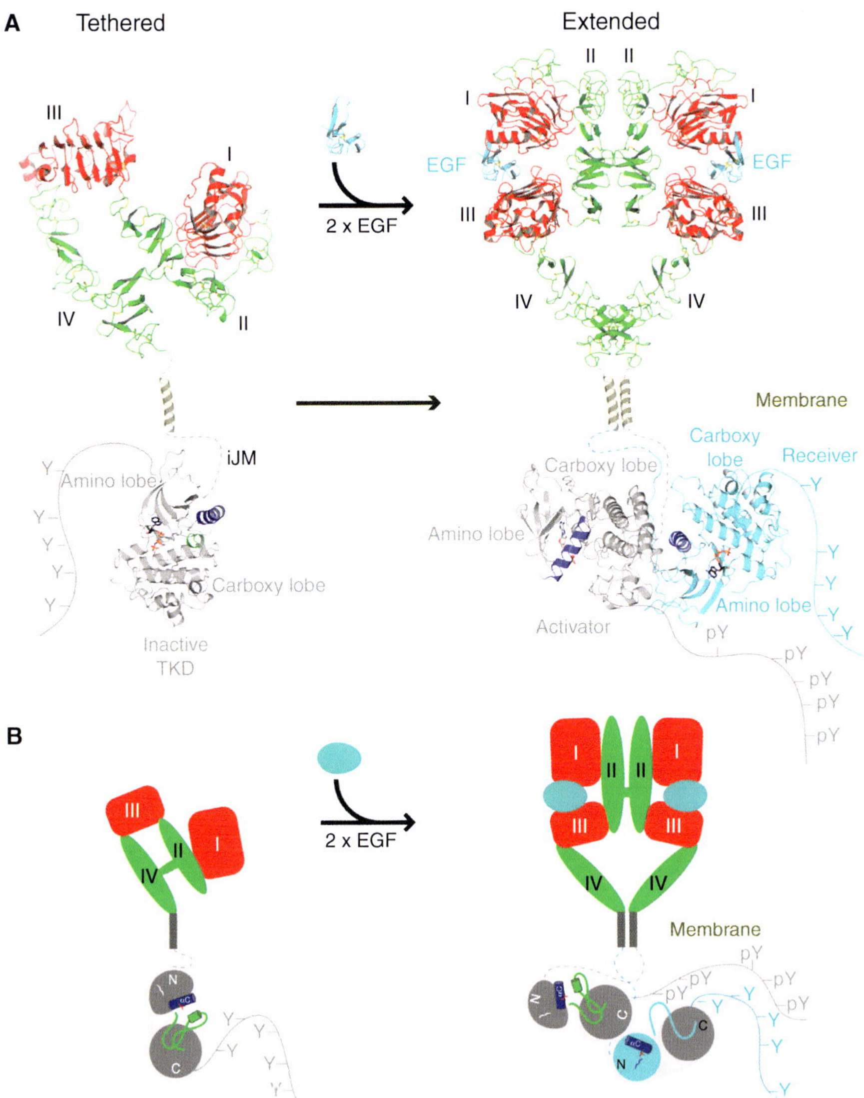

Figure 2. Basic model for EGF-induced dimerization and activation of EGFR. (*A*) In the absence of bound EGF, the human receptor is largely monomeric. The intracellular TKD is inactive, and the extracellular region adopts a "tethered" configuration in which a β-hairpin from domain II (the dimerization arm) forms intramolecular autoinhibitory interactions with domain IV. EGF binds to both domains I and III and induces a dramatic conformational change that "extends" the extracellular region and exposes the dimerization arm. With the domain II dimerization arm exposed, the EGFR extracellular region dimerizes (Burgess et al. 2003), bringing the intracellular TKDs into close proximity so that they can form the asymmetric dimer that leads to kinase activation (Zhang et al. 2006). In the asymmetric dimer, one TKD (gray) serves as the "activator," and the other (cyan) is the "receiver" that becomes allosterically activated and *trans*-phosphorylates tyrosines in the tail of the activator. (*B*) A cartoon representation of the structural changes shown in (*A*). (From Ferguson 2008; adapted, with permission, from the author.)

of the active EGFR TKD by the Kuriyan laboratory (Zhang et al. 2006). A characteristic asymmetric dimer was seen in the crystals used to solve each of these structures, in which the carboxy lobe of one TKD abuts the amino lobe of the other TKD, as depicted in the right-hand sides of Figure 2A and B. This relationship resembles that seen between cyclins and the amino lobe of the cyclin-dependent kinases (CDKs) that they activate, and the carboxy lobe of the "activator" kinase in an EGFR TKD dimer is thought to induce allosteric changes in the amino lobe of the "receiver" kinase and thus activate it. Analysis of a series of mutations in the asymmetric dimer interface confirmed its importance for activation of intact EGFR (Zhang et al. 2006), and subsequent studies showed that the NRG receptor ErbB4 is regulated in much the same way (Qiu et al. 2008).

These studies revealed that just as ErbB receptors are anything but prototypical RTKs in terms of their extracellular regulation, so are they unique in their intracellular regulation. Whereas activation-loop phosphorylation is a key step in regulating the TKDs of most RTKs (see Belov and Mohammadi 2013; Heldin and Lennartsson 2013; Hubbard 2013), it plays no part in the ErbB family (Gotoh et al. 1992; Burgess et al. 2003; Jura et al. 2011). Consistent with this distinction, the "inactive" conformation of ErbB family TKDs differs substantially from that seen in other RTKs (Wood et al. 2004; Zhang et al. 2006) and instead resembles inactive Src family kinases and CDKs. The model for allosteric activation of the EGFR TKD by its asymmetric dimerization following extracellular ligand binding provides a satisfying explanation for how this receptor can be regulated without activation-loop phosphorylation. In brief, the carboxy lobe of the activator kinase interacts with the amino lobe of the receiver and induces several conformational changes in the receiver amino lobe (Fig. 3A,B). In particular, helix αH of the activator interacts directly with helix αC of the receiver, causing the αC helix to rotate from its "out" position in the inactive EGFR TKD to the "in" position seen in active kinases. This movement disrupts interactions between the αC helix and the short helix in the activation loop—freeing the activation loop to adopt the conformation typically seen in active kinases. In addition, rotation of helix αC brings a key glutamate in this helix (E738 in mature receptor numbering) close to the side chain of lysine 721, with which it can form a salt bridge to promote ATP binding. Thus, interaction with the carboxy lobe of the activator promotes an inactive-to-active conformational transition in the receiver TKD (Zhang et al. 2006), and the EGFR kinase is activated allosterically rather than through activation-loop phosphorylation. The models in Figures 2 and 3B raise the question as to whether this is "vectorial," with only one EGFR becoming autophosphorylated in its carboxy-terminal tail. Presumably, however, the two molecules in the dimer dissociate and reassociate sufficiently rapidly that both can occupy the activator position (and thus be *trans*-autophosphorylated) for at least part of the time—although it should be noted that *cis*-autophosphorylation has not been formally excluded as a possibility.

EGFR TKD MUTATIONS IN LUNG CANCER

Importantly, the allosteric activation mechanism described above provides a satisfying explanation for the effects of oncogenic driver mutations in the EGFR TKD found in ~10% of non-small-cell lung cancer patients (Sharma et al. 2007). These mutations were identified in 2004 in lung cancer patients who responded dramatically to the EGFR tyrosine kinase inhibitors (TKIs) erlotinib and gefitinib (Lynch et al. 2004; Paez et al. 2004; Pao et al. 2004). Many of these mutations—notably those at L834 and L837 (Fig. 3A)—occur at residues that stabilize "autoinhibitory" interactions in the inactive EGFR TKD. In particular, mutations such as L834R and L837Q disrupt autoinhibitory interactions between an α-helix in the activation loop (green in Fig. 3A; seen only in the inactive TKD) and the αC helix (blue in Fig. 3A) that hold the αC helix away from the position that it adopts in the active kinase. When this autoinhibitory interaction is disrupted, the inactive TKD conformation is destabilized, leading to its constitutive activation and the acquisition of ligand-independent (oncogenic) signaling properties

          Cite this article as *Cold Spring Harb Perspect Biol* doi: 10.1101/cshperspect.a020768

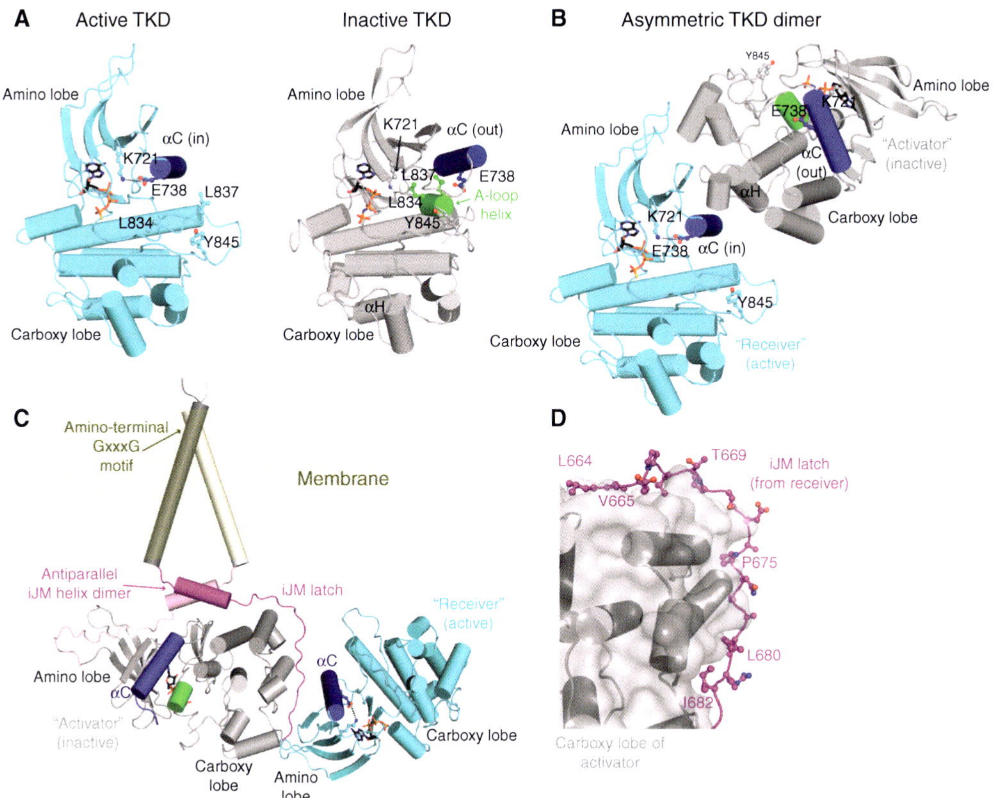

Figure 3. Intracellular EGFR activation. (*A*) EGFR TKD is shown in its active (*left:* PDB entry 2GS6) and inactive (*right:* PDB entry 2GS7) configurations (Zhang et al. 2006). The amino and carboxy lobes are marked, and the nucleotide moiety is shown in stick representation. The αC helix (dark blue) occupies the "in" position in the active TKD and the "out" position in the inactive TKD. As a result, the E738 side chain is brought sufficiently close to the K721 side chain to form a salt bridge only in the active TKD. An additional short α-helix forms in the activation loop only in the inactive TKD (green) and interacts with the αC helix to promote its displacement. Mutations at two residues in this short A-loop helix (L834 and L837 in mature EGFR, L858 and L861 in pro-EGFR) are among the most common seen in EGFR-driven non-small-cell lung cancer (Sharma et al. 2007). Y845 (labeled) in the A-loop is equivalent to the key site of activating autophosphorylation in other RTKs, but its phosphorylation is not required for EGFR activation. (*B*) A model of an EGFR TKD asymmetric dimer, based on crystal packing in the 2GS6 structure (Zhang et al. 2006) in which an inactive TKD (from 2GS7) has been modeled in the activator position. The activator (gray) remains inactive in this discrete asymmetric dimer, and the receiver (cyan) becomes activated. (*C*) The same asymmetric dimer shown in (*B*) is shown linked to the TM domain based on a composite view of the TM and iJM structure derived from crystallographic, NMR, and computational studies (Jura et al. 2009a; Red Brewer et al. 2009; Endres et al. 2013). The TM domain forms a symmetric dimer mediated by its more amino-terminal GxxxG motif, and this, in turn, is thought to drive formation of an antiparallel dimer between helices in the amino-terminal part of the iJM region (magenta). The remainder of the iJM region in the receiver "cradles" the carboxy lobe of the activator to form the iJM "latch" that stabilizes the activating asymmetric dimer. The structure of the iJM latch region in the activator is not defined. (*D*) Close-up view of the iJM latch cradling the carboxy lobe of the activator TKD. Mutations at L664, V665, P675, L680, I682, and other residues in the latch impair EGFR activation, and a V665M mutation is activating (Jura et al. 2009a; Red Brewer et al. 2009).

by the EGF receptor. Numerous structures of mutated EGFR TKD variants found in cancer patients have provided a rich understanding of how somatic mutations activate the receptor to become oncogenic drivers, as reviewed elsewhere (Eck and Yun 2010). Combined with biochemical analysis, these studies have also shown how secondary mutations can promote drug resistance and have provided important insight into the requirements for "next-generation" EGFR-targeted TKIs (Ohashi et al. 2013).

PIECING TOGETHER INTACT ErbB RECEPTORS THROUGH EXPERIMENT

The Transmembrane (TM) Domain

Although the models presented in Figures 2 and 3B likely capture many of the key interactions involved in EGFR regulation, they are clearly missing several important elements. For example, it has been known for some time that the single transmembrane (TM) domain of each ErbB family member self-associates in lipid membranes (Mendrola et al. 2002) through one or more GxxxG dimerization motifs (Lemmon et al. 1994). Cysteine cross-linking studies in cell membranes suggest that the two TM domains in an EGF-stimulated EGFR dimer contact one another in the more amino terminal of two GxxxG-motif regions (Lu et al. 2010). However, EGFR signaling appears to be highly resilient to mutations in the TM domain of the receptor when ligand-induced autophosphorylation is assessed in qualitative and semiquantitative studies (Kashles et al. 1988; Carpenter et al. 1991; Lu et al. 2010). Moreover, an EGFR variant with its TM domain replaced with polyleucine retains signaling activity in typical cell-based assays of this sort (JM Mendrola and MA Lemmon, unpubl.). Such observations led Springer and colleagues to argue that the intracellular and extracellular regions of EGFR are "loosely" linked (Lu et al. 2010)—and that no specific transmembrane (or extracellular juxtamembrane) interfaces are required for EGF-induced autophosphorylation of EGFR. However, more quantitative immunofluorescence-based

studies of EGFR autophosphorylation have recently indicated that the TM domain contributes to EGFR dimerization (and activation). An NMR-guided structural model suggested that the EGFR TM region dimerizes through the more amino terminal of its two GxxxG motif regions (Endres et al. 2013), as shown in Figure 3C. Interestingly, this region has the sequence TGMVGA, which contains two overlapping G/smallxxxG/small motifs as defined (Russ and Engelman 2000): TxxxG and GxxxA. Consistent with structural "plasticity" in this region, no effect on EGFR signaling was seen if either GxxxG motif was mutated individually. Replacing all four "small" residues with isoleucines, however, thus disrupting both motifs, resulted in some (~50%) reduction in EGF-induced EGFR autophosphorylation (Endres et al. 2013).

The Extracellular Juxtamembrane (eJM) Region

There is relatively little information on the properties of, and interactions mediated by, the extracellular juxtamembrane (eJM) region of EGFR family members. This is a short stretch of polypeptide, with just seven amino acids separating the most carboxy-terminal residue of domain IV seen in crystal structures of the EGFR extracellular region (Ferguson et al. 2003; Lu et al. 2010) and the hydrophobic region of the TM domain. Cysteine cross-linking studies suggest that the eJM regions of the two receptors in an EGF-induced dimer come into close proximity, but do not indicate a specific or well-defined mode of association (Lu et al. 2010). Intriguingly, however, inserting an additional flexible linker of 20–40 residues into this region appears to promote ligand-independent activation of EGFR (Sorokin 1995; Endres et al. 2013). This result might be taken to suggest that introducing flexibility between the extracellular and TM regions of EGFR leads to constitutive activation and/or that the spatial relationship between the extracellular region and the membrane is important. The eJM region of EGFR has also been suggested as the site of interaction with the ganglioside GM3, which inhibits

     Cite this article as *Cold Spring Harb Perspect Biol* doi: 10.1101/cshperspect.a020768

allosteric activation of the receptor (Coskun et al. 2011). Mutation of a lysine in the middle of the seven-residue JM linker abolishes GM3's inhibitory effect.

Differences in the eJM region can profoundly affect ErbB receptor function. ErbB4 exists as two isoforms (JM-a and JM-b) that differ in their eJM region (Elenius et al. 1997). The JM-a isoform is cleaved extracellularly following NRG binding, and is then cleaved again to generate an intracellular soluble species that is translocated to the nucleus (see Carpenter and Liao 2013; Song et al. 2013). In contrast, the JM-b isoform functions only as a membrane-bound receptor. Similar variation is not seen for the other ErbB family members.

The Intracellular Juxtamembrane (iJM) Region

Recent work has shed important light on the intracellular JM (iJM) region of EGFR and its relatives. As with other aspects, increased understanding of the iJM region has shown EGFR to be anything but a prototype for other RTKs. Whereas most studies of RTK iJM regions have shown them to have an autoinhibitory role (Hubbard 2004), it is now clear that the iJM of EGFR has a key positive role in receptor activation. This observation was first reported by Graham Carpenter's laboratory (Thiel and Carpenter 2007). They found that an intact iJM region is required for EGF-induced autophosphorylation of EGFR and is primarily required in the "receiver" kinase shown in Figures 2 and 3B. Subsequent alanine-scanning mutagenesis studies defined an activation domain in the iJM region (Red Brewer et al. 2009), extending from L664 to I682. Residues G672–I682 were already known to lie in the "amino-terminal extension" of the EGFR TKD (Zhang et al. 2006) and to play a key part in the asymmetric dimer interface that is responsible for EGFR activation. Mutating residues 675, 680, or 682 in this region disrupts the asymmetric dimer (Zhang et al. 2006). Crystallographic studies of an EGFR TKD variant containing the complete iJM showed that, in the asymmetric dimer, a region beyond the amino-terminal extension of

the receiver TKD "cradles" the carboxy lobe of the activator (Fig. 3C,D), apparently stabilizing asymmetric interactions between the two TKDs (Red Brewer et al. 2009). A similar arrangement was also reported in asymmetric ErbB4 TKD dimers (Wood et al. 2008). Based on these structures, residues 664–672 are considered to constitute a "juxtamembrane latch" (Jura et al. 2009a), enhancing EGFR activation by promoting formation of the asymmetric dimer. Interestingly, phosphorylation of a threonine (T669) in the juxtamembrane latch (by MAP kinase) modulates EGFR activation and downregulation (Morrison et al. 1993; Li et al. 2008), and mutation of V665 to methionine—which would be predicted to strengthen the latch (Fig. 3D)—appears to be an oncogenic driver mutation in non-small-cell lung cancer (Red Brewer et al. 2009).

The juxtamembrane latch constitutes about half of the iJM region of EGFR and is preceded by an α-helix that extends from T654, an inhibitory PKC site (Welsh et al. 1991), to L664 (Red Brewer et al. 2009). Deletion of this region—even when the juxtamembrane latch is intact—impairs EGFR activation (Thiel and Carpenter 2007; Jura et al. 2009a). Intriguingly, this region appears to form an antiparallel helical dimer in the NMR-guided structural model of the TM domain plus iJM helix in Figure 3C (Endres et al. 2013). A chemical biology approach using bipartite tetracysteine display has also provided evidence for formation of this antiparallel helical dimer in the intact receptor in cells (Scheck et al. 2012). Formation of such a dimer offers a way to break the artifactual "daisy chain" seen in crystals of the isolated EGFR TKD, in which the iJM region of each molecule cradles the carboxy lobe of its neighbor, and each TKD simultaneously acts as activator and receiver. As shown in the hypothetical model presented in Figure 3C, the TM-domain dimer appears to promote antiparallel association of the iJM helices of two TKDs. Formation of this dimer displaces the remainder of the iJM in the activator TKD to break the daisy chain seen in crystals while allowing the iJM of the receiver to function as the juxtamembrane latch shown in Figure 3C,D.

The models described above, and depicted in Figure 3, consider the TM and JM regions in only the activated EGFR. By analogy with the extracellular region and TKD, however, it is likely that these are subject to autoinhibitory interactions in the absence of ligand. Indeed, the second GxxxG motif in the TM domain may offer an alternative mode of TM packing in much-discussed preformed dimers of EGFR, which has been proposed to be autoinhibitory (Endres et al. 2013). Reorientation of the extracellular regions after ligand binding is presumed to disrupt these putative autoinhibitory TM-domain interactions. Turning attention to the iJM region, McLaughlin and colleagues (2005) made an intriguing proposal that EGFR autoinhibition involves binding of both a basic region within the iJM (including the iJM α-helix) and a basic patch on the TKD surface to anionic phospholipids in the plasma membrane inner leaflet. They showed that a peptide corresponding to the amino-terminal iJM region (residues 645–660) binds strongly to negatively charged membrane surfaces. Although they initially suggested that calmodulin binding to the iJM region might dissociate it from the membrane (and thus activate EGFR), more recent NMR studies have suggested that TM-domain dimerization or reorientation may be sufficient to dissociate the iJM region from the membrane, presumably by stabilizing the antiparallel α-helix dimer shown in Figure 3C (Matsushita et al. 2013). These possibilities clearly need further investigation in the context of the intact receptor.

The Carboxy-Terminal Tail

The ~230-amino-acid carboxy-terminal tail of EGFR (residues 956–1186) and its relatives accounts for ~20% of the receptor molecule. This region contains all of the known autophosphorylation sites in EGFR (except Y845 in the TKD) but is poorly characterized and is generally viewed as an unstructured region to which multiple SH2 domains bind (Gajiwala 2013). Fluorescence anisotropy and fluorescense resonance energy transfer (FRET) studies have clearly shown changes in the conformation and dynamics of the EGFR carboxy-terminal tail after autophosphorylation (Lee and Koland 2005; Lee et al. 2006b). Moreover, truncation mutants in this region have shown altered kinase activity toward cellular substrates, leading to the suggestion that some regions within the carboxy-terminal tail function as autoinhibitory domains (requiring tyrosine phosphorylation to release the inhibition) and others as positive modulators of activity (Walton et al. 1990; Alvarez et al. 1995). A regulatory function for the region of the carboxy-terminal tail closest to the kinase domain is also supported by crystallographic studies. A structure of EGFR TKD in its inactive conformation (Wood et al. 2004) showed two TKD-proximal regions of the carboxy-terminal tail packed against the amino lobe of the kinase in the manner expected if they were to play an autoinhibitory part (Fig. 4A). Residues 971–980 form a short α-helix that lies adjacent to the hinge region between the TKD amino and carboxy lobes. Residues 986–994 form another partly helical region that contacts elements of strands β3–β5 in the TKD amino lobe. Interestingly, Y992 (a known major autophosphorylation site in EGFR) and Y974 (at which EGF-induced phosphorylation promotes endocytosis) (Tong et al. 2009) are positioned to stabilize these potentially autoinhibitory intramolecular interactions. The short helix that contains Y992 was also seen in a very similar location and orientation in another crystal structure of inactive EGFR TKD (Zhang et al. 2006) but is also seen (only slightly shifted) in many active EGFR TKD structures (Fig. 4B). In a symmetric dimer of inactive EGFR TKD (Jura et al. 2009a), the intramolecularly associated 986–994 helix is replaced by a longer helix (with the opposite orientation) involving residues 967–978 of a neighboring molecule. Several other locations and orientations of these regions of the carboxy-terminal EGFR TKD tail have now been seen in other crystal structures (Gajiwala 2013). Although a self-consistent picture of the mechanism of autoinhibition by the carboxy-terminal tail (and how it might be reversed by autophosphorylation) has yet to emerge, it seems highly likely that these interactions involving the carboxy-terminal tail represent an additional important regulatory element.

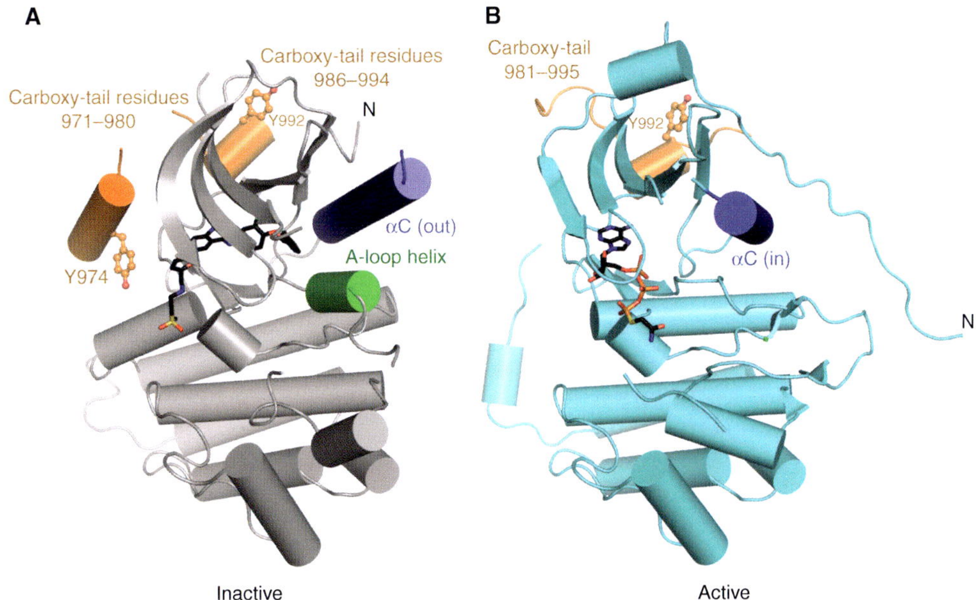

Figure 4. Location of the small section of the EGFR carboxy-terminal tail with known structure. (*A*) A structure of the inactive EGFR TKD (PDB entry 1XKK) shows that two TKD-proximal regions of the carboxy-terminal tail form short helices (colored orange) that pack against the amino lobe as described in the text. Tyrosines 974 and 992 are marked, phosphorylation of which could be involved in EGFR regulation. (*B*) Structures of the active EGFR TKD, including this one (PDB entry 2GS6), also show the 986–994 region, including Y992, in a similar location—although not the 971–980 helix.

TOWARD AN UNDERSTANDING OF THE INTACT EGF RECEPTOR

Negative Cooperativity

Although initially satisfying, the basic model presented in Figure 2 for EGFR activation fails to explain several long-standing observations about the receptor—despite the additional complexity added by studies of the JM, TM, and carboxy-terminal tail regions. In particular, as pointed out as the first structures were solved (Burgess et al. 2003), this model fails to capture key qualitative features of EGF binding to the cell surface. As described by Schlessinger (2014), EGFR and the insulin receptor were the first RTKs for which cell-surface ligand binding was analyzed in detail, in the 1970s. Concave-up Scatchard plots were seen in both cases, indicating either heterogeneity of sites or negative cooperativity. In the case of insulin, further experiments showed the binding to be negatively cooperative (De Meyts 2008). For EGFR, in

contrast, heterogeneity of sites was generally assumed, and a great deal of effort was put into understanding the difference(s) between the high- and low-affinity EGF-binding sites (Wofsy et al. 1992; Burgess et al. 2003; Klein et al. 2004)—with the former considered to be the site relevant for cell signaling. More recently, Pike and colleagues (Macdonald and Pike 2008) provided convincing evidence for negative cooperativity in EGF binding to its receptor from experiments in which they globally fit binding data obtained using cells expressing EGFR at a range of different levels. Negative cooperativity (and the characteristic concave-up Scatchards) is seen only for the intact receptor. Deletion of the cytoplasmic domain from EGFR appears to abolish negative cooperativity (Livneh et al. 1986), and further analysis has placed particular emphasis on interactions that involve the iJM region (Macdonald-Obermann and Pike 2009). As is the case with the insulin receptor (De Meyts 2008), negative cooperativity is not observed in studies of the soluble extracellular region of the human EGF receptor. The EGFR extracellular region forms a 1:1 complex with EGF ($K_D \sim$ 400 nm) that subsequently dimerizes (with $K_D \sim 3\ \mu$m) to form a 2:2 EGF:EGFR complex (Lemmon et al. 1997) with no significant cooperativity. Interactions involving the intracellular and iJM regions are clearly required for occupation of one binding site in an EGFR dimer to influence binding to the other in negative cooperativity.

Studies of the *D. melanogaster* EGFR (dEGFR) have provided a structural view of how negative cooperativity can arise in ligand binding to a member of the ErbB family (Alvarado et al. 2010). Unlike its human counterpart, dEGFR retains negative cooperativity in ligand binding when the soluble (overexpressed) extracellular region is studied. This protein can also be crystallized as an asymmetric singly ligated extracellular dimer that suggests a straightforward explanation for half-of-the-sites negative cooperativity (Levitzki et al. 1971), as summarized in Figure 5. Intriguingly, dEGFR crystallizes as a dimer even when not bound to ligand (Alvarado et al. 2009), providing a potential structural model for the preformed dimers of EGFR that have been reported in many studies (discussed in Valley et al. 2014 and Arndt-Jovin et al. 2014). This dimer also forms in solution (albeit weakly) and is mediated by the dimerization arm mentioned above. The scheme shown in Figure 5 is based on actual crystal structures. In the absence of ligand, all binding sites are equivalent (as monomers or in symmetric preformed dimers). Upon binding to the left-hand molecule in Figure 5, the Spitz ligand (cyan) "wedges" itself between domains I and III and forces them apart. As a result, domain II (which connects domains I and III) becomes bent and is forced up against domain II from the right-hand molecule, with which it makes a large number of new interactions. The extent of the dimer interface is thus increased substantially in the asymmetric dimer seen in the middle of Figure 5—consistent with the observed ligand-induced dimerization (Alvarado et al. 2010). Importantly, the new asymmetric dimer interface imposes restraints on the

right-hand binding site of the dimer. When a second ligand binds to the dimer, it cannot push domains I and III apart without compromising the asymmetric dimer interface. Indeed, asymmetry is retained in a 2:2 ligand:receptor complex (rightmost structure in Fig. 5), and the second ligand forms a compromised set of interactions consistent with its lower affinity. Thus, binding of the first ligand to a symmetric dimer imposes an asymmetry that reduces the binding affinity of the second ligand in an example of classic half-of-the-sites negative cooperativity (Levitzki et al. 1971; Alvarado et al. 2010).

Negative cooperativity provides a potential mechanism for graded activation of EGFR in response to a concentration gradient of activating ligand (as morphogen), known to be important in several contexts for Spitz and EGFR in *D. melanogaster* (Golembo et al. 1996). It is also likely to be relevant in mammals. Although a singly ligated asymmetric dimer has not been seen for a mammalian ErbB receptor extracellular region, detailed comparison of crystal structures of ligand-bound EGFR and ErbB4 extracellular dimers (Liu et al. 2012) provides compelling arguments that the model shown in Figure 5 is also relevant for human receptors. Moreover, by coexpressing mutated ErbB receptors with defects in ligand binding and kinase activity, respectively, the Leahy laboratory has provided evidence that singly ligated dimers of EGFR and ErbB4 are competent to signal (Liu et al. 2012).

Linkage between Extracellular and Intracellular Regions

As described above, there is now clear evidence for formation of asymmetric dimers in both the extracellular and intracellular regions of EGFR, raising the question of how the two are linked. It would seem reasonable to argue that asymmetry in the extracellular region might promote asymmetry in the intracellular region in a well-defined way, suggesting a view of intact ErbB receptors as allosterically regulated dimeric enzymes with a dimer interface that happens to span the membrane (Lemmon 2009). Experi-

 Cite this article as *Cold Spring Harb Perspect Biol* doi: 10.1101/cshperspect.a020768

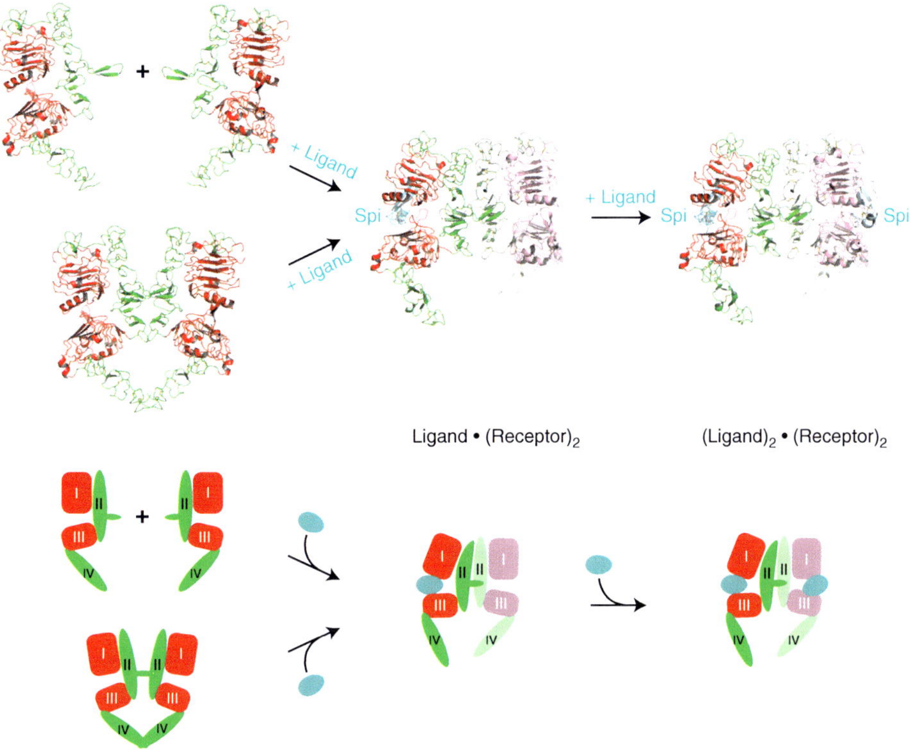

Figure 5. Structural basis for negative cooperativity in an EGF receptor. Negatively cooperative ligand binding is retained by the isolated extracellular region of the *D. melanogaster* EGFR, which binds to the EGF homolog Spitz (Spi), whereas it is lost in studies of the human receptor. Crystallographic studies have revealed the structural basis for this negative cooperativity. The *D. melanogaster* EGFR extracellular region does not form the tethered structure depicted in Figure 2A, remaining as an untethered monomer or forming a symmetric (crystallographic) dimer even in the absence of bound ligand (Alvarado et al. 2009). After binding of ligand to the left-hand molecule in the symmetric dimer, domains I and III are wedged apart as described in the text so that domain II of the left-hand molecule "collapses" onto its counterpart in the right-hand molecule. The result is a singly ligated asymmetric dimer. This transition restrains the second ligand-binding site so that the wedging apart of domains I and III is disfavored, reducing the affinity of ligand for the second site—resulting in negative cooperativity. The structures in the *upper* panel are actual crystal structures (PDB entries 3I2T, 3LTG, and 3LTF) and are redrawn as cartoons in the *lower* part of the figure for clarity (Alvarado et al. 2009, 2010).

mental studies with intact receptors, however, give quite different impressions. As mentioned above, cysteine cross-linking studies in the Springer laboratory (Lu et al. 2010) suggested that the extracellular and intracellular regions of EGFR in cell membranes are only loosely linked. Negative-stain electron microscopy studies of nearly intact EGFR provide a similar impression, suggesting that even with a single dimeric configuration of the EGFR extracellular region, the intracellular region of the same molecule is free to adopt a range of different confor-

mations (Mi et al. 2011). Moreover, in studying signaling by singly ligated EGFR and ErbB4 dimers, Liu et al. (2012) reported that the part of "activator" in the intracellular asymmetric TKD dimer could be played by either the ligand-bound or the ligand-free receptor—again suggesting flexibility or plasticity. In contrast, other studies using mutated receptors have suggested that the extracellular asymmetry in an EGFR/ErbB2 heterodimer is strictly maintained in the intracellular region (Macdonald-Obermann et al. 2012)—with the EGFR TKD obligately adopting the receiver position and being activated first.

A flexible linkage between the extra- and intracellular regions of EGFR family members would certainly be sufficient for a signaling mechanism involving straightforward ligand-induced receptor dimerization. It is difficult, however, to understand how preformed receptor dimers could be regulated by ligand binding (Chung et al. 2010) without rigid coupling of extra- and intracellular conformations. Moreover, it seems unlikely that intracellular mutations or deletions in EGFR (or T654 phosphorylation) would be able to influence the cooperativity of extracellular ligand binding (Macdonald-Obermann and Pike 2009) if intra/extracellular linkage is loose. In addition, is not clear how the different ligands that bind to a given ErbB family member could induce the distinct signaling responses that have been reported (Wilson et al. 2009) if the extra- and intracellular regions were flexibly linked— and there would be no basis for expecting distinct signaling by singly and doubly ligated receptor dimers. Interpreting studies of purified intact ErbB receptors is made difficult by the need to recapitulate the correct membrane environment—including acidic phospholipids and possibly gangliosides—and its asymmetry in order to be sure that the extracellular and intracellular (and, more importantly, the JM) regions are placed in physiologically relevant contexts. Cellular studies are inherently less detailed and quantitative, and—for negatively cooperative systems—it is difficult to be certain that a single receptor conformation is being studied (rather than a mixture of singly and doubly ligated dimers, for example). As discussed elsewhere, however, technologies for studying intact receptors in cells are improving greatly and will soon be at the stage where cellular and structural studies can be productively linked.

CONCLUDING REMARKS

Once considered prototypical RTKs, it is now clear that the EGFR and its relatives are actually quite unique in their properties. Our understanding of their regulation is now quite sophis-

ticated, with multiple structures of extracellular regions and TKDs, and a growing knowledge of the JM and TM regions. Mechanistic studies have provided clear explanations for how oncogenic mutations in the TKD activate EGFR in non-small-cell lung cancer and have aided development of next-generation kinase inhibitors to combat resistance to existing agents. Advances in our understanding of extracellular regulation also promise to explain EGFR mutations in glioblastoma in the future (Lee et al. 2006a; Vivanco et al. 2012). EGFR family members are the only RTKs with a TKD that are regulated without requiring activation-loop phosphorylation—instead using a unique mode of allosteric regulation through asymmetric dimerization. EGFR is also the only well-understood RTK in which bound ligand does not contribute directly to the dimer interface—instead controlling receptor-mediated dimerization. The EGFR family shares more similarities with the insulin receptor than with most other RTKs— including negative cooperativity in its ligand binding, which may reflect its function as a detector of ligand concentration in morphogen gradients. One of the next frontiers in understanding EGFR family signaling will be detailed studies of intact receptors—so that the JM regions can be studied in physiologically (and structurally) relevant environments. Another will be thorough investigation of how the flexible carboxy-terminal tail contributes to EGFR regulation and coupling to downstream signaling. Given the plethora of ErbB family ligands and the wide range of processes in which ErbB receptors participate, it will also be important to relate signaling outcomes to specific ligands (or concentrations of ligands) to unravel the complicated biology of these receptors.

REFERENCES

*Reference is also in this collection.

* Adrain C, Freeman M. 2014. Regulation of receptor tyrosine kinase ligand processing. *Cold Spring Harb Perspect Biol* doi: 10.1101/cshperspect.a008995.
Alvarado D, Klein DE, Lemmon MA. 2009. ErbB2 resembles an autoinhibited invertebrate epidermal growth factor receptor. *Nature* **461:** 287–291.

Alvarado D, Klein DE, Lemmon MA. 2010. Structural basis for negative cooperativity in growth factor binding to an EGF receptor. *Cell* **142:** 568–579.

Alvarez CV, Shon KJ, Miloso M, Beguinot L. 1995. Structural requirements of the epidermal growth factor receptor for tyrosine phosphorylation of eps8 and eps15, substrates lacking Src SH2 homology domains. *J Biol Chem* **270:** 16271–16276.

* Arndt-Jovin DJ, Botelho MG, Jovin TM. 2014. Structure–function relationships of ErbB RTKs in the plasma membranes of living cells. *Cold Spring Harb Perspect Biol* doi: 10.1101/cshperspect.a008961.

Arteaga CL. 2003. ErbB-targeted therapeutic approaches in human cancer. *Exp Cell Res* **284:** 122–130.

Arteaga CL, Sliwkowski MX, Osborne CK, Perez EA, Puglisi F, Gianni L. 2011. Treatment of HER2-positive breast cancer: Current status and future perspectives. *Nat Rev Clin Oncol* **9:** 16–32.

* Belov AA, Mohammadi M. 2013. Advances in the molecular mechanisms of FGF signaling in physiology and pathology. *Cold Spring Harb Perspect Biol* **5:** 015958.

Berger MB, Mendrola JM, Lemmon MA. 2004. ErbB3/HER3 does not homodimerize upon neuregulin binding at the cell surface. *FEBS Letts* **569:** 332–336.

Birchmeier C. 2009. ErbB receptors and the development of the nervous system. *Exp Cell Res* **315:** 611–618.

Bouyain S, Longo PA, Li S, Ferguson KM, Leahy DJ. 2005. The extracellular region of ErbB4 adopts a tethered conformation in the absence of ligand. *Proc Natl Acad Sci* **102:** 15024–15029.

Buonanno A, Fischbach GD. 2001. Neuregulin and ErbB receptor signaling pathways in the nervous system. *Curr Opin Neurobiol* **11:** 287–296.

Burgess AW. 2008. EGFR family: Structure physiology signalling and therapeutic targets. *Growth Factors* **26:** 263–274.

Burgess AW, Cho HS, Eigenbrot C, Ferguson KM, Garrett TP, Leahy DJ, Lemmon MA, Sliwkowski MX, Ward CW, Yokoyama S. 2003. An open-and-shut case? Recent insights into the activation of EGF/ErbB receptors. *Mol Cell* **12:** 541–552.

* Carpenter G, Liao H-J. 2013. Receptor tyrosine kinases in the nucleus. *Cold Spring Harb Perspect Biol* doi: 10.1101/cshperspect.a008979.

Carpenter CD, Ingraham HA, Cochet C, Walton GM, Lazar CS, Sowadski JM, Rosenfeld MG, Gill GN. 1991. Structural analysis of the transmembrane domain of the epidermal growth factor receptor. *J Biol Chem* **266:** 5750–5755.

Cho HS, Leahy DJ. 2002. Structure of the extracellular region of HER3 reveals an interdomain tether. *Science* **297:** 1330–1333.

Cho HS, Mason K, Ramyar KX, Stanley AM, Gabelli SB, Denney DW Jr, Leahy DJ. 2003. Structure of the extracellular region of HER2 alone and in complex with the Herceptin Fab. *Nature* **421:** 756–760.

Chung I, Akita R, Vandlen R, Toomre D, Schlessinger J, Mellman I. 2010. Spatial control of EGF receptor activation by reversible dimerization on living cells. *Nature* **464:** 783–787.

Coskun Ü, Grzybek M, Drechsel D, Simons K. 2011. Regulation of human EGF receptor by lipids. *Proc Natl Acad Sci* **108:** 9044–9048.

Dawson JP, Berger MB, Lin CC, Schlessinger J, Lemmon MA, Ferguson KM. 2005. Epidermal growth factor receptor dimerization and activation require ligand-induced conformational changes in the dimer interface. *Mol Cell Biol* **25:** 7734–7742.

De Meyts P. 2008. The insulin receptor: A prototype for dimeric, allosteric membrane receptors? *Trends Biochem Sci* **33:** 376–384.

Eck MJ, Yun CH. 2010. Structural and mechanistic underpinnings of the differential drug sensitivity of EGFR mutations in non-small cell lung cancer. *Biochim Biophys Acta* **1804:** 559–566.

Elenius K, Corfas G, Paul S, Choi CJ, Rio C, Plowman GD, Klagsbrun M. 1997. A novel juxtamembrane domain isoform of HER4/ErbB4. Isoform-specific tissue distribution and differential processing in response to phorbol ester. *J Biol Chem* **272:** 26761–26768.

Endres NF, Das R, Smith AW, Arkhipov A, Kovacs E, Huang Y, Pelton JG, Shan Y, Shaw DE, Wemmer DE, et al. 2013. Conformational coupling across the plasma membrane in activation of the EGF receptor. *Cell* **152:** 543–556.

Falls DL. 2003. Neuregulins: Functions, forms, and signaling strategies. *Exp Cell Res* **284:** 14–30.

Ferguson KM. 2004. Active and inactive conformations of the epidermal growth factor receptor. *Biochem Soc Trans* **32:** 742–745.

Ferguson KM. 2008. Structure-based view of epidermal growth factor receptor regulation. *Annu Rev Biophys* **37:** 353–373.

Ferguson KM, Darling PJ, Mohan MJ, Macatee TL, Lemmon MA. 2000. Extracellular domains drive homo- but not hetero-dimerization of erbB receptors. *EMBO J* **19:** 4632–4643.

Ferguson KM, Berger MB, Mendrola JM, Cho HS, Leahy DJ, Lemmon MA. 2003. EGF activates its receptor by removing interactions that autoinhibit ectodomain dimerization. *Mol Cell* **11:** 507–517.

Fiske WH, Threadgill D, Coffey RJ. 2009. ERBBs in the gastrointestinal tract: Recent progress and new perspectives. *Exp Cell Res* **315:** 583–601.

Gajiwala KS. 2013. EGFR: Tale of the C-terminal tail. *Prot Sci* **22:** 995–999.

Garrett TP, McKern NM, Lou M, Elleman TC, Adams TE, Lovrecz GO, Zhu HJ, Walker F, Frenkel MJ, Hoyne PA, et al. 2002. Crystal structure of a truncated epidermal growth factor receptor extracellular domain bound to transforming growth factor α. *Cell* **110:** 763–773.

Golembo M, Raz E, Shilo BZ. 1996. The *Drosophila* embryonic midline is the site of Spitz processing, and induces activation of the EGF receptor in the ventral ectoderm. *Development* **122:** 3363–3370.

Gotoh N, Tojo A, Hino M, Yazaki Y, Shibuya M. 1992. A highly conserved tyrosine residue at codon 845 within the kinase domain is not required for the transforming activity of human epidermal growth factor receptor. *Biochem Biophys Res Commun* **186:** 768–774.

Harris RC, Chung E, Coffey RJ. 2003. EGF receptor ligands. *Exp Cell Res* **284:** 2–13.

* Heldin C-H, Lennartsson J. 2013. Structural and functional properties of platelet-derived growth factor and stem cell factor receptors. *Cold Spring Harb Perspect Biol* **5:** a009100.

Honegger AM, Kris RM, Ullrich A, Schlessinger J. 1989. Evidence that autophosphorylation of solubilized receptors for epidermal growth factor is mediated by intermolecular cross-phosphorylation. *Proc Natl Acad Sci* **86:** 925–929.

Hubbard SR. 2004. Juxtamembrane autoinhibition in receptor tyrosine kinases. *Nat Rev Mol Cell Biol* **5:** 464–471.

* Hubbard SR. 2013. The insulin receptor: Both a prototypical and a typical receptor tyrosine kinase. *Cold Spring Harb Perspect Biol* **5:** a008946.

Jones RB, Gordus A, Krall JA, MacBeath G. 2006. A quantitative protein interaction network for the ErbB receptors using protein microarrays. *Nature* **439:** 168–174.

Jura N, Endres NF, Engel K, Deindl S, Das R, Lamers MH, Wemmer DE, Zhang X, Kuriyan J. 2009a. Mechanism for activation of the EGF receptor catalytic domain by the juxtamembrane segment. *Cell* **137:** 1293–1307.

Jura N, Shan Y, Cao X, Shaw DE, Kuriyan J. 2009b. Structural analysis of the catalytically inactive kinase domain of the human EGF receptor 3. *Proc Natl Acad Sci* **106:** 21608–21613.

Jura N, Zhang X, Endres NF, Seeliger MA, Schindler T, Kuriyan J. 2011. Catalytic control in the EGF receptor and its connection to general kinase regulatory mechanisms. *Mol Cell* **42:** 9–22.

Kashles O, Szapary D, Bellot F, Ullrich A, Schlessinger J, Schmidt A. 1988. Ligand-induced stimulation of epidermal growth factor receptor mutants with altered transmembrane regions. *Proc Natl Acad Sci* **85:** 9567–9571.

Klein P, Mattoon D, Lemmon MA, Schlessinger J. 2004. A structure-based model for ligand binding and dimerization of EGF receptors. *Proc Natl Acad Sci* **101:** 929–934.

Lee NY, Koland JG. 2005. Conformational changes accompany phosphorylation of the epidermal growth factor receptor C-terminal domain. *Prot Sci* **14:** 2793–2803.

Lee JC, Vivanco I, Beroukhim R, Huang JH, Feng WL, DeBiasi RM, Yoshimoto K, King JC, Nghiemphu P, Yuza Y, et al. 2006a. Epidermal growth factor receptor activation in glioblastoma through novel missense mutations in the extracellular domain. *PLoS Med* **3:** e485.

Lee NY, Hazlett TL, Koland JG. 2006b. Structure and dynamics of the epidermal growth factor receptor C-terminal phosphorylation domain. *Prot Sci* **15:** 1142–1152.

Lemmon MA. 2009. Ligand-induced ErbB receptor dimerization. *Exp Cell Res* **315:** 638–648.

Lemmon MA, Schlessinger J. 1994. Regulation of signal transduction and signal diversity by receptor oligomerization. *Trends Biochem Sci* **19:** 459–463.

Lemmon MA, Schlessinger J. 2010. Cell signaling by receptor tyrosine kinases. *Cell* **141:** 1117–1134.

Lemmon MA, Treutlein HR, Adams PD, Brünger AT, Engelman DM. 1994. A dimerization motif for transmembrane α-helices. *Nat Struct Biol* **1:** 157–163.

Lemmon MA, Bu Z, Ladbury JE, Zhou M, Pinchasi D, Lax I, Engelman DM, Schlessinger J. 1997. Two EGF molecules contribute additively to stabilization of the EGFR dimer. *EMBO J* **16:** 281–294.

Levitzki A, Stallcup WB, Koshland DE Jr. 1971. Half-of-the-sites reactivity and the conformational states of cytidine triphosphate synthetase. *Biochemistry* **10:** 3371–3378.

Li X, Huang Y, Jiang J, Frank SJ. 2008. ERK-dependent threonine phosphorylation of EGF receptor modulates receptor downregulation and signaling. *Cell Signal* **20:** 2145–2155.

Libermann TA, Nusbaum HR, Razon N, Kris R, Lax I, Soreq H, Whittle N, Waterfield MD, Ullrich A, Schlessinger J. 1985. Amplification, enhanced expression and possible rearrangement of EGF receptor gene in primary human brain tumours of glial origin. *Nature* **313:** 144–147.

Liu P, Cleveland TE 4th, Bouyain S, Byrne PO, Longo PA, Leahy DJ. 2012. A single ligand is sufficient to activate EGFR dimers. *Proc Natl Acad Sci* **109:** 10861–10866.

Livneh E, Prywes R, Kashles O, Reiss N, Sasson I, Mory Y, Ullrich A, Schlessinger J. 1986. Reconstitution of human epidermal growth factor receptors and its deletion mutants in cultured hamster cells. *J Biol Chem* **261:** 12490–12497.

Lu C, Mi LZ, Grey MJ, Zhu J, Graef E, Yokoyama S, Springer TA. 2010. Structural evidence for loose linkage between ligand binding and kinase activation in the epidermal growth factor receptor. *Mol Cell Biol* **30:** 5432–5443.

Lynch TJ, Bell DW, Sordella R, Gurubhagavatula S, Okimoto RA, Brannigan BW, Harris PL, Haserlat SM, Supko JG, Haluska FG, et al. 2004. Activating mutations in the epidermal growth factor receptor underlying responsiveness of non-small-cell lung cancer to gefitinib. *N Engl J Med* **350:** 2129–2139.

Macdonald JL, Pike LJ. 2008. Heterogeneity in EGF-binding affinities arises from negative cooperativity in an aggregating system. *Proc Natl Acad Sci* **105:** 112–117.

Macdonald-Obermann JL, Pike LJ. 2009. The intracellular juxtamembrane domain of the epidermal growth factor (EGF) receptor is responsible for the allosteric regulation of EGF binding. *J Biol Chem* **284:** 13570–13576.

Macdonald-Obermann JL, Piwnica-Worms D, Pike LJ. 2012. Mechanics of EGF receptor/ErbB2 kinase activation revealed by luciferase fragment complementation imaging. *Proc Natl Acad Sci* **109:** 137–142.

Matsushita C, Tamagaki H, Miyazawa Y, Aimoto S, Smith SO, Sato T. 2013. Transmembrane helix orientation influences membrane binding of the intracellular juxtamembrane domain in Neu receptor peptides. *Proc Natl Acad Sci* **110:** 1646–1651.

McLaughlin S, Smith SO, Hayman MJ, Murray D. 2005. An electrostatic engine model for autoinhibition and activation of the epidermal growth factor receptor (EGFR/ErbB) family. *J Gen Physiol* **126:** 41–53.

Mendrola JM, Berger MB, King MC, Lemmon MA. 2002. The single transmembrane domains of ErbB receptors self-associate in cell membranes. *J Biol Chem* **277:** 4704–4712.

Mi LZ, Lu C, Li Z, Nishida N, Walz T, Springer TA. 2011. Simultaneous visualization of the extracellular and cytoplasmic domains of the epidermal growth factor receptor. *Nat Struct Mol Biol* **18:** 984–989.

Moasser MM. 2007. Targeting the function of the *HER2* oncogene in human cancer therapeutics. *Oncogene* **26:** 6577–6592.

Mok TS. 2011. Personalized medicine in lung cancer: What we need to know. *Nat Rev Clin Oncol* **8:** 661–668.

Morrison P, Takishima K, Rosner MR. 1993. Role of threonine residues in regulation of the epidermal growth factor receptor by protein kinase C and mitogen-activated protein kinase. *J Biol Chem* **268:** 15536–15543.

Ogiso H, Ishitani R, Nureki O, Fukai S, Yamanaka M, Kim JH, Saito K, Sakamoto A, Inoue M, Shirouzu M, et al. 2002. Crystal structure of the complex of human epidermal growth factor and receptor extracellular domains. *Cell* **110:** 775–787.

Ohashi K, Maruvka YE, Michor F, Pao W. 2013. Epidermal growth factor receptor tyrosine kinase inhibitor-resistant disease. *J Clin Oncol* **31:** 1070–1080.

Paez JG, Jänne PA, Lee JC, Tracy S, Greulich H, Gabriel S, Herman P, Kaye FJ, Lindeman N, Boggon TJ, et al. 2004. EGFR mutations in lung cancer: Correlation with clinical response to gefitinib therapy. *Science* **304:** 1497–1500.

Pao W, Miller V, Zakowski M, Doherty J, Politi K, Sarkaria I, Singh B, Heelan R, Rusch V, Fulton L, et al. 2004. EGF receptor gene mutations are common in lung cancers from "never smokers" and are associated with sensitivity of tumors to gefitinib and erlotinib. *Proc Natl Acad Sci* **101:** 13306–13311.

Pentassuglia L, Sawyer DB. 2009. The role of Neuregulin-1β/ErbB signaling in the heart. *Exp Cell Res* **315:** 627–637.

Qiu C, Tarrant MK, Choi SH, Sathyamurthy A, Bose R, Banjade S, Pal A, Bornmann WG, Lemmon MA, Cole PA, et al. 2008. Mechanism of activation and inhibition of the HER4/ErbB4 kinase. *Structure* **16:** 460–467.

Red Brewer M, Choi SH, Alvarado D, Moravcevic K, Pozzi A, Lemmon MA, Carpenter G. 2009. The juxtamembrane region of the EGF receptor functions as an activation domain. *Mol Cell* **34:** 641–651.

Rico B, Marín O. 2011. Neuregulin signaling, cortical circuitry development and schizophrenia. *Curr Opin Genet Dev* **21:** 262–270.

Riese DJ II, Stern DF. 1998. Specificity within the EGF family/ErbB receptor family signaling network. *Bioessays* **20:** 41–48.

Russ WP, Engelman DM. 2000. The GxxxG motif: A framework for transmembrane helix–helix association. *J Mol Biol* **296:** 911–919.

Scheck RA, Lowder MA, Appelbaum JS, Schepartz A. 2012. Bipartite tetracysteine display reveals allosteric control of ligand-specific EGFR activation. *ACS Chem* **7:** 1367–1376.

Schlessinger J. 1988. Signal transduction by allosteric receptor oligomerization. *Trends Biochem Sci* **13:** 443–447.

* Schlessinger J. 2014. Receptor tyrosine kinases: Legacy of the first two decades. *Cold Spring Harb Perspect Biol* doi: 10.1101/cshperspect.a008912.

Schneider MR, Wolf E. 2009. The epidermal growth factor receptor ligands at a glance. *J Cell Physiol* **218:** 460–466.

Sharma SV, Bell DW, Settleman J, Haber DA. 2007. Epidermal growth factor receptor mutations in lung cancer. *Nat Rev Cancer* **7:** 169–181.

Shi F, Telesco SE, Liu Y, Radhakrishnan R, Lemmon MA. 2010. ErbB3/HER3 intracellular domain is competent to bind ATP and catalyze autophosphorylation. *Proc Natl Acad Sci* **107:** 7692–7697.

Sibilia M, Kroismayr R, Lichtenberger BM, Natarajan A, Hecking M, Holcmann M. 2007. The epidermal growth factor receptor: From development to tumorigenesis. *Differentiation* **75:** 770–787.

* Song S, Rosen KM, Corfas G. 2013. Biological function of nuclear receptor tyrosine kinase action. *Cold Spring Harb Perspect Biol* **5:** a009001.

Sorokin A. 1995. Activation of the EGF receptor by insertional mutations in its juxtamembrane regions. *Oncogene* **11:** 1531–1540.

Stamos J, Sliwkowski MX, Eigenbrot C. 2002. Structure of the epidermal growth factor receptor kinase domain alone and in complex with a 4-anilinoquinazoline inhibitor. *J Biol Chem* **277:** 46265–46272.

Thiel KW, Carpenter G. 2007. Epidermal growth factor receptor juxtamembrane region regulates allosteric tyrosine kinase activation. *Proc Natl Acad Sci* **104:** 19238–19243.

Tong J, Taylor P, Peterman SM, Prakash A, Moran MF. 2009. Epidermal growth factor receptor phosphorylation sites Ser991 and Tyr998 are implicated in the regulation of receptor endocytosis and phosphorylations at Ser1039 and Thr1041. *Mol Cell Proteomics* **8:** 2131–2144.

Ullrich A, Schlessinger J. 1990. Signal transduction by receptors with tyrosine kinase activity. *Cell* **61:** 203–212.

* Valley CC, Lidke KA, Lidke DS. 2014. The spatiotemporal organization of erbB receptors: Insights from microscopy. *Cold Spring Harb Perspect Biol* doi: 10.1101/cshperspect.a020735.

Vivanco I, Robins HI, Rohle D, Campos C, Grommes C, Nghiemphu PL, Kubek S, Oldrini B, Chheda MG, Yannuzzi N, et al. 2012. Differential sensitivity of glioma- versus lung cancer-specific EGFR mutations to EGFR kinase inhibitors. *Cancer Discov* **2:** 458–471.

* Wagner MJ, Stacey MM, Liu BA, Pawson T. 2013. Molecular mechanisms of SH2- and PTB-domain-containing proteins in receptor tyrosine kinase signaling. *Cold Spring Harb Perspect Biol* doi: 10.1101/cshperspect.a008987.

Walton GM, Chen WS, Rosenfeld MG, Gill GN. 1990. Analysis of deletions of the carboxyl terminus of the epidermal growth factor receptor reveals self-phosphorylation at tyrosine 992 and enhanced in vivo tyrosine phosphorylation of cell substrates. *J Biol Chem* **265:** 1750–1754.

Ward CW, Hoyne PA, Flegg RH. 1995. Insulin and epidermal growth factor receptors contain the cysteine repeat motif found in the tumor necrosis factor receptor. *Proteins* **22:** 141–153.

Welsh JB, Gill GN, Rosenfeld MG, Wells A. 1991. A negative feedback loop attenuates EGF-induced morphological changes. *J Cell Biol* **114:** 533–543.

Wilson KJ, Gilmore JL, Foley J, Lemmon MA, Riese DJ II. 2009. Functional selectivity of EGF family peptide growth factors: Implications for cancer. *Pharmacol Ther* **122:** 1–8.

Wofsy C, Goldstein B, Lund K, Wiley HS. 1992. Implications of epidermal growth factor (EGF) induced egf receptor aggregation. *Biophys J* **63:** 98–110.

Wood ER, Truesdale AT, McDonald OB, Yuan D, Hassell A, Dickerson SH, Ellis B, Pennisi C, Horne E, Lackey K, et al. 2004. A unique structure for epidermal growth factor receptor bound to GW572016 (Lapatinib): Relationships among protein conformation, inhibitor off-rate, and receptor activity in tumor cells. *Cancer Res* **64:** 6652–6659.

Wood ER, Shewchuk LM, Ellis B, Brignola P, Brashear RL, Caferro TR, Dickerson SH, Dickson HD, Donaldson KH, Gaul M, et al. 2008. 6-Ethynylthieno[3,2-d]- and 6-ethynylthieno[2,3-d]pyrimidin-4-anilines as tunable covalent modifiers of ErbB kinases. *Proc Natl Acad Sci* **105:** 2773–2778.

Yarden Y, Schlessinger J. 1987. Epidermal growth factor induces rapid, reversible aggregation of the purified epidermal growth factor receptor. *Biochemistry* **26:** 1443–1451.

Yarden Y, Sliwkowski MX. 2001. Untangling the ErbB signalling network. *Nat Rev Mol Cell Biol* **2:** 127–137.

Zhang X, Gureasko J, Shen K, Cole PA, Kuriyan J. 2006. An allosteric mechanism for activation of the kinase domain of epidermal growth factor receptor. *Cell* **125:** 1137–1149.

Zhang H, Berezov A, Wang Q, Zhang G, Drebin J, Murali R, Greene MI. 2007. ErbB receptors: From oncogenes to targeted cancer therapies. *J Clin Invest* **117:** 2051–2058.

Cite this article as *Cold Spring Harb Perspect Biol* doi: 10.1101/cshperspect.a020768

The Spatiotemporal Organization of ErbB Receptors: Insights from Microscopy

Christopher C. Valley[1], Keith A. Lidke[2], and Diane S. Lidke[1]

[1]Department of Pathology and the Cancer Research and Treatment Center, University of New Mexico, Albuquerque, New Mexico 87131

[2]Department of Physics and Astronomy, University of New Mexico, Albuquerque, New Mexico 87131

Correspondence: dlidke@salud.unm.edu

Signal transduction is regulated by protein–protein interactions. In the case of the ErbB family of receptor tyrosine kinases (RTKs), the precise nature of these interactions remains a topic of debate. In this review, we describe state-of-the-art imaging techniques that are providing new details into receptor dynamics, clustering, and interactions. We present the general principles of these techniques, their limitations, and the unique observations they provide about ErbB spatiotemporal organization.

Signal transduction is associated with dramatic spatial and temporal changes in membrane protein distribution. Although the biochemical events downstream of membrane receptor activation are often well characterized, the initiating events within the plasma membrane remain unclear. Many cell surface receptors have been shown to redistribute into clusters in response to ligand binding (Metzger 1992). Therefore, correlating membrane receptor activation with dynamics and aggregation state is essential to understanding cell signaling.

The role of receptor aggregation is of particular interest in the case of receptor tyrosine kinases (RTKs). It is generally accepted that ligand binding to the extracellular domain of RTKs induces dimerization, whether ligand- or receptor-mediated (Lemmon and Schlessinger 2010). However, there is evidence that some RTKs exist as oligomers in the absence of ligand, whereas others require higher-order oligomerization for activation (Lemmon and Schlessinger 2010). Understanding the fundamental interactions that regulate RTK signaling still remains an important focus in the field.

Over the past decade, imaging technologies and biological tools have developed to a point such that questions about protein dynamics, clustering, and interactions can now be addressed in living cells (Fig. 1). These techniques reveal information about protein behavior on a spatial and temporal scale that is not provided by traditional biochemical assays. In this review, we will discuss the application of these advanced imaging technologies to the study of the ErbB family of RTKs.

ErbB SIGNAL TRANSDUCTION

The ErbB family consists of four members: ErbB1 (the classical EGFR, HER1), ErbB2 (HER2), ErbB3 (HER3), and ErbB4 (HER4).

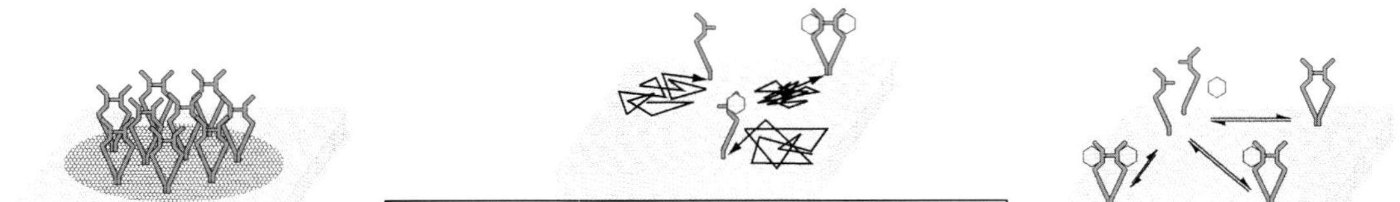

	Clustering				Dynamics		Interactions	
	Electron microscopy	NSOM	Localization microscopy	Number and brightness	SPIDA	SPT	Fluoresence complementation	FRET
Information	Protein proximity Protein distribution Coclustering	Cluster size Number of proteins per cluster Coclustering	Cluster size Coclustering	Cluster size Dynamics of cluster formation	Cluster size Protein concentration Dynamics of cluster formation	Diffusion Modes of motion Confinement zones Dimer lifetimes Viscosity mapping	Protein interactions on live cells	Protein proximity less than 10 nm
Spatial scale	1–10 nm	~70 nm	20–50 nm	~250 nm	~250 nm	~10–50 nm	Interaction distance: <10 nm	Interaction distance: <10 nm
Temporal resolution	Fixed cells	Fixed cells	Ultimate resolution requires fixed cells	Seconds	Milliseconds to seconds	Milliseconds	Seconds	Seconds
Limitations	Complicated sample preparation Large gold probes (nanoparticles) have reduced labeling efficiency	Limited to scanning of apical cell surface Technically challenging	Some require specialized buffers not compatible with live cells Long acquisition times Limited dye selection	Measures averaged protein behavior Analysis is model-dependent	Measures averaged protein behavior Analysis is model-dependent	Requires sparse labeling density Tracking is limited by probe photostability	Irreversible signal Complementary fluorescent proteins may affect binding and unbinding kinetics Luciferase is rapidly degraded in live cells	Stoichiometry of donor/acceptor is critical Analysis is model-dependent Results are highly influenced by fluorophore placement
Citations	Yang et al. 2007 Lidke et al. 2012	Betzig et al. 1986 Nagy et al. 1999	Hell 2007 Huang et al. 2009 Patterson et al. 2010	Digman et al. 2009 Nagy et al. 2010	Godin et al. 2011 Swift et al. 2011	Kusumi et al. 1993 Schmidt et al. 1996 Weiss 1999 Sako et al. 2000 Low-Nam et al. 2011	Luker et al. 2004 Kerppola 2008 Macdonald-Obermann et al. 2012	Gadella and Jovin 1995 Jares-Erijman and Jovin 2003 Lidke et al. 2003 Hofman et al. 2010

Figure 1. Summary of imaging techniques for quantifying receptor clustering, dynamics, and interactions.

In general, these proteins are single-pass transmembrane proteins with an extracellular ligand-binding domain and a cytoplasmic tail containing a tyrosine kinase domain (Yarden and Sliwkowski 2001; Lemmon and Schlessinger 2010). As exceptions, ErbB2 has no known ligand, and ErbB3 has a defective kinase domain. Ligand binding to these transmembrane proteins leads to conformational changes, receptor homo- and hetero-oligomerization, kinase activation, and the transphosphorylation of multiple cytoplasmic tail tyrosines (Schlessinger 2002). These phosphotyrosines, in turn, provide sites for the recruitment and activation of cytoplasmic adaptor proteins, initiating signaling cascades that control numerous cellular processes such as gene expression, cell migration, and cell division. There exists a rich literature of biochemical and structural data describing the mechanisms of ErbB activation. However, despite being a focus of study for more than 25 years, fundamental questions concerning the association states and activation of the ErbB family still remain.

In 2002, Ogiso et al. (2002) solved the X-ray crystal structure of the extracellular domains of the 2:2 epidermal growth factor (EGF):ErbB1 dimer, showing that dimerization of ErbB1 is achieved through a physical interaction between two receptors rather than through the ligand. Comparison with the unliganded or "tethered" structure suggests that on ligand binding, domain I undergoes a large ($\sim$130°) rotation to form a binding pocket for EGF with domain III, known as the "extended" structure. This conformational change subsequently rearranges domain II, exposing the "dimerization arm" to which another EGF-bound ErbB1 in the extended conformation can bind, forming a "back-to-back" dimer. Furthermore, it has also been suggested that ErbB2 is the preferential dimerization partner for ErbB1 because the crystal structure of the ErbB2 ectodomain displays the extended structure (Garrett et al. 2003).

Although the ectodomain crystal structures formed the basis for the model of receptor-driven dimerization, ideas about the size and ligand-occupancy of the ErbB1 signaling complex remain controversial (Sako et al. 2000;

Clayton et al. 2005; Lidke et al. 2005; Macdonald and Pike 2008; Chung et al. 2010; Hofman et al. 2010). For example, there have been reports that the 1:2 EGF:ErbB1 dimer should be the predominant signaling species (Macdonald and Pike 2008), whereas other groups have suggested that a transition from dimers to tetramers is the critical step (Clayton et al. 2005). Additionally, there is evidence for "predimers" that form in the absence of ligand (Hofman et al. 2010).

Recent studies have shown that the cytoplasmic kinase domain is also involved in ErbB1 regulation (Zhang et al. 2006; Choi et al. 2007). Crystal structures of the kinase domain (Zhang et al. 2006) have revealed an autoinhibitory conformation, similar to that of Src and cyclin-dependent kinases. The discovery of an allosteric mechanism for ErbB1 kinase activation suggests additional intracellular complexities in the regulation in ErbB signaling. More recently, structural studies of near full-length, purified ErbB1 have shown that activating mutations in the kinase domain can lead to asymmetric dimerization of the kinase domain without the need for ligand or extracellular domain contacts (Wang et al. 2011).

The insights provided by biochemical, biophysical, and structural studies have been invaluable to our understanding of ErbB signaling. However, these studies typically cannot be performed in the context of the intact plasma membrane. With cutting-edge imaging techniques (summarized in Fig. 1), quantitative information of receptor behavior in the cell membrane can be integrated with structural knowledge to gain a more complete mechanistic picture of ErbB function. Importantly, live cell imaging allows for detailed observation of homo- and heterointeractions with high spatiotemporal resolution, as well as the ability to investigate the role of membrane microdomains in regulating receptor function.

DETERMINING CHANGES IN RECEPTOR CLUSTER SIZE USING CONVENTIONAL MICROSCOPY

Receptor clustering and/or oligomerization occur on scales $<$100 nm, a distance too small to be resolved by conventional, diffraction-limited microscopy, such as confocal or total internal reflection fluorescence (TIRF) microscopy. Furthermore, even at low (physiological) expression levels, the receptors are present at a high enough concentration in the membrane that, on average, there is significantly more than one receptor per diffraction-limited area of the conventional microscope. This makes direct counting of oligomer size (e.g., by brightness) difficult. Nevertheless, several techniques have been developed that use the spatial or temporal information that is present in conventional microscope images or time series to infer the oligomerization state (Kolin and Wiseman 2007; Digman and Gratton 2011).

One method to measure concentration and stoichiometry of complexes is called number and brightness (N&B) (Digman et al. 2009). This imaging technique uses the temporal fluctuations in pixel intensity to determine the average concentration in a pixel as well as the brightness of the fluorophores. The brightness is then correlated with the stoichiometry by comparing to the known brightness of single labels. This works because stochastic fluctuations around the mean follow the relationship that variance/mean^2 is proportional to $1/N$, in which N is the average number of particles in the area imaged by one pixel. Nagy et al. (2010) used N&B to examine the aggregation of ErbB1 on the plasma membrane of living cells. They found that ErbB1 primarily exists as monomers in low-expressing cells. Yet, even in the absence of ligand, the fraction of ErbB1 homodimers increased with expression level. This result suggests that there is some affinity for the unliganded dimer and that mass action kinetics drives the system toward dimers as receptor concentration increases. In contrast, ErbB2 was found to be in larger clusters, even at low expression. EGF stimulation led to almost complete ErbB1 homodimerization and the subsequent formation of higher-order oligomers that were associated with clathrin structures (Nagy et al. 2010).

Another image analysis method called spatial intensity distribution analysis (SpIDA) has recently been developed (Godin et al. 2011). In this technique, a histogram is generated from the pixel intensities in a single image. The histogram is fit to a model that assumes a random (spatial Poisson) distribution of either monomers or a mixture of monomers and dimers. Because this analysis does not depend on temporal correlation, it can extract data from a single image. Swift et al. (2011) applied SpIDA to simultaneously monitor ErbB1 dimerization and internalization (concentration at the membrane) in response to transactivation via G-protein coupled receptors (GPCR). They found a differential response depending on the specific GPCR involved; all GPCR transactivation induced ErbB1 dimerization, yet not all of the GPCRs induced rapid receptor internalization.

These techniques are useful in their ability to extract clustering information from intact and living cells without the need for specialized instrumentation. Particularly useful is the ability to follow changes in live cells over time to provide dynamic information. However, it is important to note that these techniques critically rely on the assumption that the true dynamics and clustering behavior follow the model on which the parameter estimation is derived, and therefore the results are highly dependent on the models assumed. The model most often assumed was a mixture of oligomers (or just monomers and dimers) undergoing free Brownian motion on the cell membrane. Often, a range of alternative hypotheses was not tested. For example, in both N&B and SpIDA, it may be very difficult to distinguish between dimers and a collection of higher-order oligomers. More complicated behavior such as transient immobilization, corralled diffusion, or movement of proteins transiently confined to lipid rafts or other heterogeneous membrane structures render the interpretation of results more difficult. Finally, these techniques alone cannot distinguish protein proximity from functional protein–protein interactions.

HIGH-RESOLUTION IMAGING OF PROTEIN DISTRIBUTION

There are a number of techniques that can be used to directly map the distribution of receptors with high spatial resolution. These include electron microscopy (EM), near-field scanning optical microscopy (NSOM), and super-resolution (SR) imaging. Each of these can provide a nanometer-scale image of protein distribution.

NSOM is capable of mapping out the nanoscale organization of the membrane by scanning an optical fiber across the surface of the cell such that excitation and fluorescence collection only occurs near the tip (Betzig et al. 1986). Nagy et al. (1999) used NSOM to show the presence of ErbB2 clusters on the surface of SKBR3 cells and that the cluster size was associated with activation. Furthermore, they showed that ErbB2 cluster size increased with EGF stimulation of ErbB1. More recently, NSOM was used to show that ErbB1 exists in clusters on the surface of HeLa cells (Abulrob et al. 2010). In contrast to the changes seen for ErbB2, ErbB1 cluster size was not observed to change with ligand activation, although cluster size did decrease with ErbB1 kinase domain inhibitors. Using two-color imaging, Abulrob et al. also showed that ErbB1 could be found in three states: colocalized with caveolae, colocalized with lipid rafts, or not associated with either compartment.

Using EM, Wilson and colleagues have also shown that resting ErbB receptors are found in clusters on the cell surface (Fig. 2A,B) and these clusters did not increase in size on stimulation (Yang et al. 2007). However, they did note a large increase in ErbB3 cluster size in response to heregulin stimulation. By labeling with two sizes of gold particles, they found that pairs of family members (ErbB1/ErbB2, ErbB2/ErbB3, ErbB1/ErbB3) showed very little coclustering, even after ligand addition. This unexpected segregation of receptors suggests a possible mechanism through which membrane microdomains may regulate heterointeractions.

Emerging super-resolution techniques in fluorescence microscopy are allowing for imaging below the diffraction limit of the light microscope (Fig. 2C–H), providing 100 nm or better resolution with far-field microscopy (Leung and Chou 2011; Lidke and Lidke 2012). These techniques include photoactivation localization microscopy (PALM), stochastic optical reconstruction microscopy (STORM), stimulated

emission depletion (STED), and structured illumination microscopy (SIM) (Hell 2007; van de Linde et al. 2008; Huang et al. 2009; Patterson et al. 2010; Lidke and Lidke 2012). The single-molecule localization techniques, such as PALM and STORM, are powerful ways to map protein distribution, but have suffered from difficulties in quantifying protein numbers. Because localization microscopy typically relies on the sequential localization of individual fluorophores, repeated observation of the same fluorophore can lead to overcounting and apparent clustering of proteins. In addition, unlabeled, photobleached, or nonfunctional probes can lead to undercounting (Renz et al. 2012). Therefore, although information regarding protein distribution may be easily obtained, development of quantitative methods to identify protein clustering with precise protein numbers has remained a challenge within the field. There are many groups working to improve the quantitative analysis of superresolution images. Veatch and colleagues have used pair correlation analysis to determine cluster size from super-resolution images (Sengupta et al. 2011; Veatch et al. 2012). However, this method reports only the average cluster size assuming homogeneous clusters. Needham et al. (2013) have developed a super-resolution imaging method to identify receptor separation down to 10 nm using fluorophore localization and photobleaching. They find five discrete inter-ErbB1 receptor separations from 8–57 nm, in the absence of ligand. They postulate that this distribution is due to alignment of resting ErbB1 along cortical actin filaments, forming polymers of up to 10 receptors (Needham et al. 2013). We note that, as with many antibody labeling protocols, in this study the cells were labeled at 4°C before fixation.

It is important to emphasize that although these studies can reveal high-resolution information about protein proximity and distributions, they cannot distinguish clustering from functional protein interactions. Nevertheless, information on cluster distributions, receptor coclustering, and colocalization with membrane compartments has provided a unique view of membrane organization. Importantly, the observed clustering of receptors, particularly in

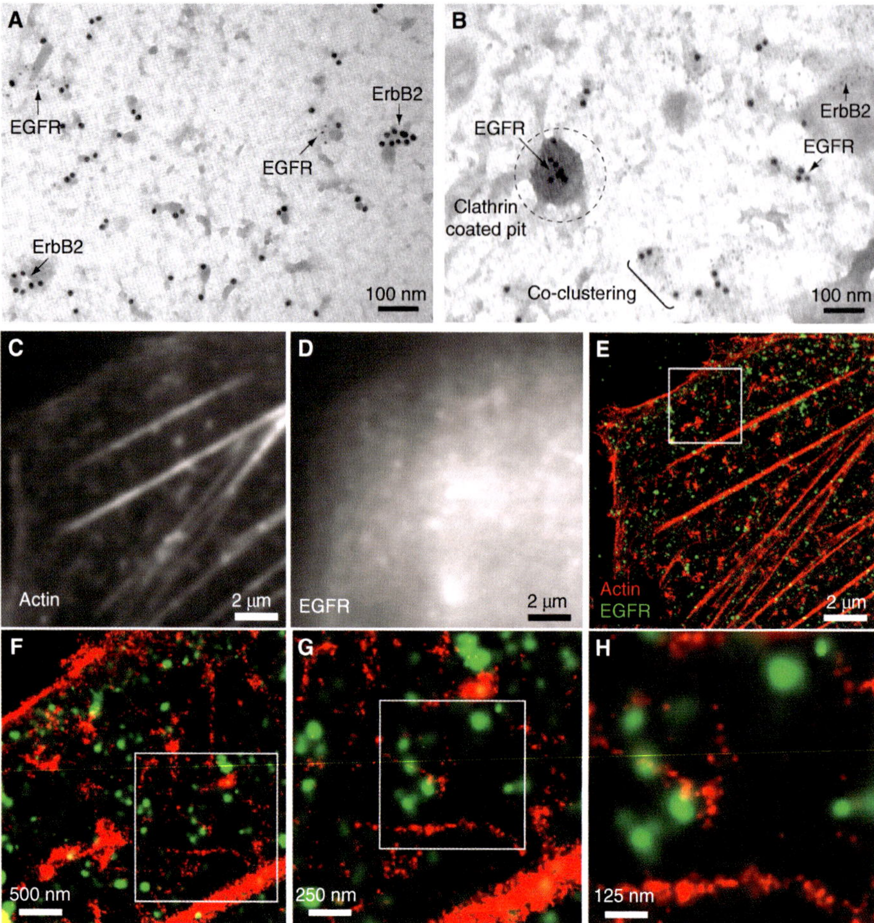

Figure 2. High-resolution imaging of receptor distribution. (*A,B*) Transmission electron microscopy images of ErbB1/ErbB2 distribution on SKBR3 membrane sheets. Dual labeling (10-nm and 5-nm gold) allows for mapping of ErbB1 and ErbB2 proximity. Note the localization of ErbB1 to a clathrin-coated pit in (*B*). Images courtesy of Bridget Wilson created from data based on Yang et al. (2007). (*C*)–(*H*) Two-color stochastic optical reconstruction microscopy (STORM) imaging of ErbB1 with respect to actin on Chinese hamster ovary cells stably expressing ErbB1. Diffraction-limited images of actin (*C*) and ErbB1 (*D*) as imaged on the basal cell surface using TIRF microscopy. Super-resolution reconstruction overlay (*E*) of actin (red, phalloidin-Alexa-Fluor647) and ErbB1 (green, anti-ErbB1 antibody conjugated to Cy3b) showing the subdiffraction resolution attained using STORM and the localization of ErbB1 within the actin cytoskeleton. The reconstruction overlay is shown at increasing zoom (white boxes) to highlight imaging resolution (*F,G,H*).

the absence of ligand, has reinforced the importance of microdomains in organizing receptors on the cell surface.

ENSEMBLE PROXIMITY ASSAYS TO DETERMINE PROTEIN INTERACTIONS

To study protein–protein interactions in the ErbB family, a number of groups have used För-

ster (or fluorescence) resonance energy transfer (FRET). FRET can determine the proximity of fluorophore-labeled proteins based on the level of excited state energy transferred from one fluorophore (donor) to a second fluorophore (acceptor), within a range of ~2–10 nm (Jares-Erijman and Jovin 2003). The classical FRET studies by Gadella and Jovin confirmed that ligand-mediated interaction of receptors within the plasma membrane is a key initiating step in ErbB signaling (Gadella and Jovin 1995). Energy transfer between proximal, ligand-bound receptors was determined by donor photobleaching FRET. From these measurements, it was concluded that the ligand-induced dimeric association of receptors occurs within minutes of incubation with EGF and that the high-affinity receptors are found as preformed dimers. In these experiments, they also revealed a dependence of dimer formation on temperature, with a shift from 13% dimers at 4°C to 69% at 37°C. This result emphasizes the importance of performing experiments at physiological temperatures.

Since those early experiments, other groups have used FRET to examine ErbB1 interactions that have actually led to variations on the model of ErbB1 activation. For example, Martin-Fernandez et al. (2002) showed that ligand binding causes a rapid and transient structural transition followed by oligomer "dissociation" on a longer timescale, possibly explained by negative cooperativity of EGF-ErbB1 binding/dimerization characterized by others based on Scatchard plots (Macdonald and Pike 2008). In contrast, Clayton and colleagues have used FRET in combination with image correlation spectroscopy (ICS) measurements to suggest that a dimer-to-tetramer transition is the key event in ErbB1 activation (Clayton et al. 2005).

Energy transfer between identical fluorophores, known as homo-FRET, has been used to quantify the degree of protein clustering based on the degree of emission polarization or anisotropy (Lidke et al. 2003; Bader et al. 2007, 2009, 2011; Hofman et al. 2010). Similar to conventional hetero-FRET, homo-FRET involves the transfer of energy from the excited state of one fluorophore molecule to a proximal fluorophore (or fluorophores), in which the fluorescence emission of the latter is essentially depolarized with respect to the original excitation dipole. Thus, energy transfer between identical fluorophores results in a decrease in the steady-state fluorescence anisotropy or time-resolved anisotropy decay and can be used to quantify cluster size of fluorescent molecules. Homo-FRET systems are ideal for studying the

dimerization and oligomerization events in like molecules without the need for distinct donor/acceptor labels in appropriate ratios.

Using microscopy-based fluorescence anisotropy measurements of ErbB1-green fluorescent protein (GFP), Hofman et al. (2010) found that ErbB1 forms clusters in the absence of ligand as measured by a decrease in fluorescence anisotropy. Converting fluorescence anisotropy values into cluster sizes (Bader et al. 2009), it was predicted that ~40% of ErbB1-GFP exists as predimers in the absence of ligand and the percentage is not dependent on kinase activity. The investigators concluded that predimerization is likely driven through protein interactions within the cytosolic domain, which is consistent with the putative autoinhibitory symmetric dimer as observed crystallographically from the kinase domain crystal structure (Jura et al. 2009). Higher receptor clustering was dependent on kinase activity, suggesting that high-order oligomers are stabilized through recruitment of cytoplasmic adaptor proteins. Furthermore, because anisotropy is calculated spatially, the investigators show that ErbB1 remains clustered within early endosomes, suggesting that signaling may continue through the process of endocytosis.

An alternative method for visualizing protein interactions involves conjugating nonfluorescent, complementary amino- and carboxy-terminal fragments of a fluorescent protein to multiple proteins of interest. Although neither fragment in isolation shows fluorescent properties, the complementary fragments form a complete, photoactive fluorophore when their respective conjugated proteins are in close proximity. This method, known as BiFC, bimolecular fluorescence complementation (Kerppola 2008), allows for visualization of protein interactions occurring at the nanometer scale and localization of those interactions within the diffraction limit. Conjugation and coexpression of complementary fluorescent protein fragments to ErbB1 showed the formation of ligand-independent interactions at the cell surface, which internalized on the addition of EGF. They found that the overall fluorescence intensity did not change in response to ligand, suggesting that most receptors exist in preformed dimers in the resting cell. It is important to note, however, that the favorable interaction between complementary fluorescent protein fragments may bias quantified results toward the receptor complexed state. BiFC was further used to show the presence of ErbB2 and ErbB4 homodimers at the cell surface, as well as ErbB3 homodimers within the nucleus. As with FRET, BiFC is particularly useful in studying the interaction of different protein species. As such, heterodimers within the ErbB family were also identified, as was their subcellular localization (plasma membrane, nucleus, or cytosolic). The investigators' conclusions that ligand-independent dimerization occurs primarily in the endoplasmic reticulum highlights the importance of microscopy-based methods for spatially characterizing protein interactions (Tao and Maruyama 2008).

Pike and colleagues have recently generated a complementation system using fragments of firefly luciferase (Luker et al. 2004; Macdonald-Obermann et al. 2012). With this clever system, they were able to determine not only the formation of ErbB1/ErbB2 homo- and heterodimers, but also the sequence of activation between the kinase domains. They showed that ErB1/ErbB2 heterodimerization is dependent on EGF binding to ErbB1, and that ErbB1 phosphorylates ErbB2 within the heterodimer. These results support the asymmetric kinase domain dimer structural model and introduce the idea that kinase domain interactions are directional.

The ability to express proteins of interest in a variety of cellular backgrounds, along with the ability to genetically encode fluorescent tags on proteins, has revolutionized the field of cellular imaging. However, with any heterologous expression system, it is important to consider that overexpression may cause improper post-translational modification (such as receptor glycosylation) and subsequent misfolding, which can lead to aggregation within intracellular compartments. Further complexities in protein folding and processing may occur on the addition of GFP (or GFP fragments); therefore, the assessment of protein folding, trafficking, and function of GFP-tagged proteins is critical in such experimental systems.

DIRECT OBSERVATION OF PROTEIN–PROTEIN INTERACTIONS: SINGLE-MOLECULE STUDIES

Evaluation of dimerization and higher-order clustering of ErbB1 using FRET and complementation assays are typically limited to time- and ensemble-averaged techniques. Moreover, the timescale with which FRET and BiFC data are collected greatly exceeds that of protein dynamics. Recent advances in single-molecule imaging allow for the localization of individual receptor molecules on the surface of living cells with high spatial and temporal resolution (Schmidt et al. 1996; Weiss 1999). These single-particle tracking (SPT) techniques involve the subdiffraction localization of a subset of membrane receptors at the millisecond timescale and have been most often been used to characterize diffusion of receptor molecules within the plasma membrane (Kusumi et al. 1993; Orr et al. 2005; Chung et al. 2010; Subach et al. 2010). SPT has recently proven to be a powerful technique for visualizing protein interactions by directly observing dimerization at the single-molecule level (Fig. 3A) (Sako et al. 2000; Low-Nam et al. 2011).

Using small organic dyes, Sako and colleagues provided the first microscopy data showing direct dimerization of EGF-bound ErbB1 within the plasma membrane (Sako et al. 2000). By labeling A431 cells with a low concentration of fluorescent EGF, they reported diffusion and dimerization of two ligand-bound ErbB1 monomers. Interestingly, in their one-color system, in which dimers are classified based on relative intensity, they more commonly observed a sudden increase in fluorescence of a single ligand-bound ErbB1 on the cell surface, followed by a two-step bleaching process. This suggests either ligand binding to predimerized receptors, or that ligand binding to one ErbB1 results in a dimerization event that further increases the affinity for another EGF molecule to bind. However, the latter model is in contrast to existing models of negative cooperativity based on Scatchard plot and structural data, in which binding of one EGF molecule causes lower affinity ligand binding for the second EGF molecule via structural constraints (Macdonald and Pike

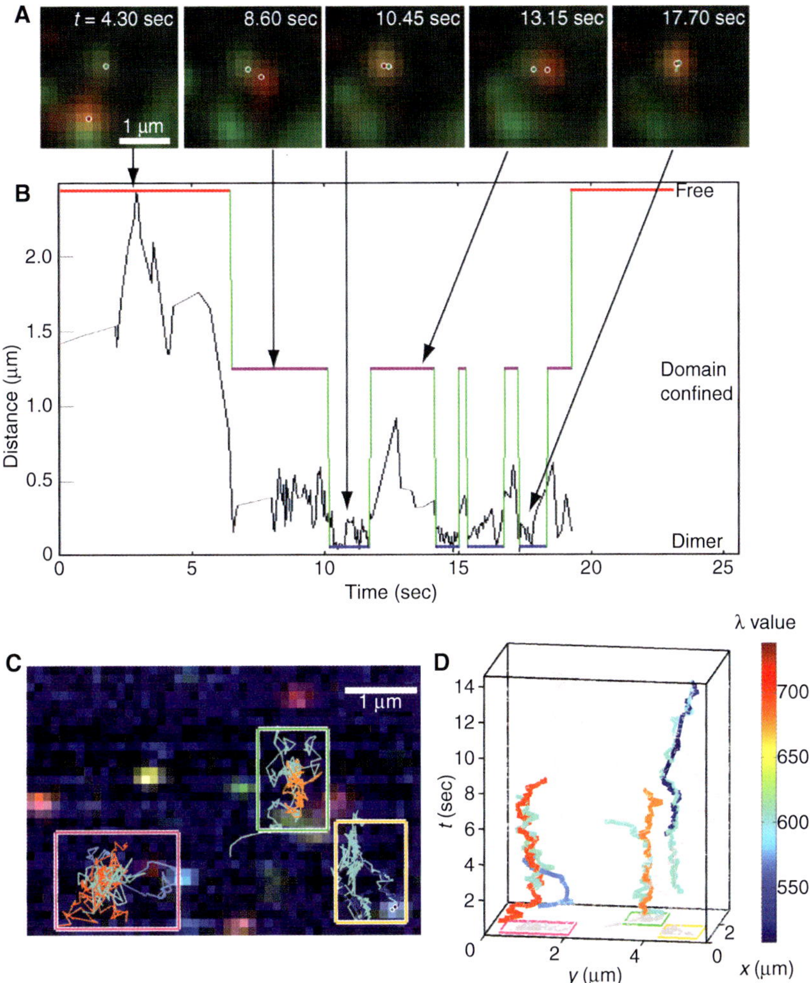

Figure 3. SPT captures dimerization of EGF-bound ErbB1. (*A*) Images from a time series in which EGF-QD585 (green) and EGF-QD655 (red) are simultaneously tracked. The pixelated image shows the raw data, and the circles indicate the subdiffraction localization of individual receptor molecules. (*B*) Plot showing the separation over time of the two receptors in (*A*). Particle trajectories are analyzed using a three-state hidden Markov model based on interparticle distance to distinguish between free (red line), confined (purple line), and dimeric (blue line) receptor populations. In this example, the EGF-bound receptors are seen to undergo repeated (four) dimerization events (see also Low-Nam et al. 2011). (*C*) Hyperspectral imaging of eight spectrally distinct QDs bound to EGF. Pseudocoloring of the data is generated based on each quantum dot's (QD) peak spectral wavelength (λ). (*D*) Three-dimensional particle trajectories (*x,y,t*) of the corresponding boxed regions in (*C*). Correlated motion of dimerized receptors can be observed. Color map (*right*) indicates QD emission peak. (Images from Cutler et al. 2013; reprinted, with permission, from the investigators.)

2008). Additionally, two-color tracking and single-pair FRET was performed using the Cy3- and Cy5-labeled EGF showing energy transfer in co-localized receptors, confirming dimerization at the nanometer scale. An important result from this work was that doubly liganded receptor di-

mers were also seen to label with an antibody that recognizes phosphorylated ErbB1, consistent with the hypothesis that dimerization is sufficient for activation.

The previous work (Sako et al. 2000) revealed many insights about ErbB1 interactions at the single-molecule level. However, they were limited in the duration of the measurements by photobleaching of the organic fluorophores used and only tracked the EGF-bound receptors. In our own work, we have used highly photostable quantum dot (QD) probes to monitor diffusion and dimerization of both liganded and unliganded receptors over longer timescales (Lidke et al. 2004, 2005; Low-Nam et al. 2011). Two-color single-QD tracking of EGF-bound QDs was first used to determine that dimerization was sufficient to initiate retrograde transport of ErbB1 along cellular filopodia (Lidke et al. 2005). Because retrograde transport was shown to be dependent on phosphorylation of ErbB1, the transport of dimers was consistent with the dimer being the minimal signaling unit of ErbB1. More recently, we have used single-QD tracking to quantify ErbB1 homodimerization kinetics at the single-molecule level on the apical surface of living cells (Low-Nam et al. 2011). Direct observation of dimerization using two-color QD tracking revealed a fundamental relationship between receptor ligand occupancy and dimerization. Using a three-state hidden Markov model (Fig. 3B), dimerization events were identified and the dimer off rates were calculated. Interactions between unliganded receptors were observed, but were transient, whereas two ligand-bound receptors formed long-lived dimers. In contrast to Sako et al. (2000), we often observed the coming together of independent receptors, rather than the spontaneous recruitment of the second ligand. The 2:2 EGF:ErbB1 dimers showed a marked reduction in mobility that was depended on kinase activity. Interestingly, a 1:2 EGF:ErbB1 dimer was found to have an intermediate dimer lifetime, but did not show reduced mobility, suggesting that these less stable interactions are not as efficient at initiating signaling events. This work also indicated that the actin cytoskeleton promotes dimer formation by coconfining receptors. Taken together, these results support the model suggested by the crystal structure that ligand binding to the extracellular domain leads to a conformational change that promotes dimerization and receptor activation.

Although evidence for functional ErbB1 homodimers has been provided from single-molecule studies, there is still the question of whether higher-order oligomers occur and what their significance may be. Recent advances in multicolor imaging are making it possible to explore these questions of higher-order oligomers (Arnspang et al. 2012; Cutler et al. 2013). In particular, a novel high-speed hyperspectral microscope developed in the laboratory of Keith Lidke allows for simultaneous single particle tracking of up to eight spectrally distinct QDs at video rate (Fig. 3C). Using this instrument, we have consistently observed the formation of ErbB1 homodimers by tracking QD-EGF (Fig. 3D) (Cutler et al. 2013). Notably, we have not yet observed higher-order interactions. This instrument also extends our ability to examine oligomerization with respect to the membrane landscape (Cutler et al. 2013). Previous work using simultaneous imaging of GFP-actin and single-QD tracking of the immunoreceptor FcεRI showed that membrane receptor motion can be restricted by the membrane-proximal actin bundles (Andrews et al. 2008). The ability to perform multicolor SPT with markers of membrane domains and cytoskeletal proteins will yield new information about receptor mobility and oligomerization within the context of the cellular and membrane architecture.

SPT provides information on the stochastic behavior of individual proteins that is lost in ensemble techniques. However, SPT is limited to sparse labeling densities such that only a subset of proteins is labeled at one time, making it difficult to determine the oligomer distribution accurately. Despite this limitation, SPT is a critical tool that provides direct observation of protein dynamics and interactions within the native cell membrane that cannot be achieved by any other method.

IMAGING RECEPTOR ACTIVATION AND DOWNSTREAM SIGNALING

Microscopy-based characterization of ErbB1 signaling has focused primarily on events in the membrane. Direct imaging of downstream signaling events in live cells is difficult, owing

 Cite this article as *Cold Spring Harb Perspect Biol* doi: 10.1101/cshperspect.a020735

largely to the challenge of delivering fluorophores to the cytosol. Antibody labeling of cytoplasmic proteins in live cells requires mild membrane permeabilization (Sako et al. 2000), and it is unknown whether such membrane perturbations affect ErbB mobility, interactions, or signal initiation. An alternative to antibody labeling is the use of FRET biosensors in which FRET via GFP spectral variants are used to detect intracellular signaling events such as protein recruitment, posttranslational modifications, and/or structural changes (Hahn and Toutchkine 2002). FRET-based methods have been used in several recent studies to address questions of intracellular events in ErbB signaling. Sorkin and colleagues characterized the recruitment of Grb2 to activated ErbB1 by coexpression of Erb1-CFP donor and Grb2-YFP acceptor molecules (Sorkin et al. 2000). FRET between three spectral variants of GFP (3-FRET) has been used to visualize the simultaneous recruitment of multiple downstream factors in live cells. Galperin et al. (2004) used 3-FRET to determine the corecruitment of Grb2 and Cbl to activated ErbB1 through direct and indirect interactions, respectively. BiFC has also been used to show STAT5 recruitment to ErbB1 (Chu et al. 2009). An alternative method has been developed to study protein phosphorylation using a FRET biosensor, in which intramolecular FRET is dependent on the protein phosphorylation state (Sato and Umezawa 2004). Coupling these signaling readouts to receptor dynamics and oligomerization will be an import future goal.

CONCLUDING REMARKS

Efficient signal transduction along the ErbB signaling pathways involves the formation of stable interactions between inherently dynamic proteins. Our current understanding of ErbB structure–function relationships has relied heavily on the detailed structural information of crystallography studies. Crystallography, however, cannot provide the dynamic information that is clearly important in regulating protein functions and requires isolating the protein from its physiological environment. Advanced techniques in high-resolution imaging are being used to complement the structural models by providing information on receptor dynamics, including the direct observation of receptor dimerization. The ability to measure protein behavior in living cells is making it possible to directly test the existence of structural predictions and their functional consequences.

Despite their advantages, these new technologies have not provided a definitive mechanism for ErbB activation. As shown in this review, there exist imaging results that are consistent with seemingly disparate ideas. For example, FRET and SPT have suggested both the existence of ligand-independent dimerization (Chung et al. 2010; Hofman et al. 2010) and ligand-dependent dimer stabilization (Gadella and Jovin 1995; Lidke et al. 2005; Low-Nam et al. 2011). Even the fundamental signaling unit of the receptor is debated, with some data supporting the concept that dimerization is key (Sako et al. 2000; Lidke et al. 2005; Low-Nam et al. 2011), and other results suggesting that a dimer-to-tetramer transition is the critical step (Clayton et al. 2005). This confusion may be partially due to variations in sample treatment (fixed vs. live cells, imaging temperature, steric hindrance due to labeling, protein overexpression, cell type, etc.) or the choice of models used to quantify and interpret data. Moreover, each technique has its own limitations that must be taken into consideration (Fig. 1). Or, perhaps, each of these mechanisms plays some role under the appropriate physiological conditions. A future challenge in the field will be to determine whether all these observations can be unified into one comprehensive model for ErbB activation. Ultimately, it will be the integration of structural, biochemical, and biophysical results that provide a complete picture of ErbB signal transduction.

ACKNOWLEDGMENTS

The work is supported by National Science Foundation (NSF) CAREER Award MCB-0845062 to D.S.L., NSF CAREER Award #0954836 to K.A.L., and the New Mexico Spatiotemporal Modeling Center (National Institutes of Health P50GM085273).

REFERENCES

Abulrob A, Lu Z, Baumann E, Vobornik D, Taylor R, Stanimirovic D, Johnston LJ. 2010. Nanoscale imaging of epidermal growth factor receptor clustering: Effects of inhibitors. *J Biol Chem* **285:** 3145–3156.

Andrews NL, Lidke KA, Pfeiffer JR, Burns AR, Wilson BS, Oliver JM, Lidke DS. 2008. Actin restricts FcepsilonRI diffusion and facilitates antigen-induced receptor immobilization. *Nat Cell Biol* **10:** 955–963.

Arnspang EC, Brewer JR, Lagerholm BC. 2012. Multi-color single particle tracking with quantum dots. *PLoS ONE* **7:** e48521.

Bader AN, Hofman EG, van Bergen En Henegouwen PM, Gerritsen HC. 2007. Imaging of protein cluster sizes by means of confocal time-gated fluorescence anisotropy microscopy. *Opt Express* **15:** 6934–6945.

Bader AN, Hofman EG, Voortman J, en Henegouwen PM, Gerritsen HC. 2009. Homo-FRET imaging enables quantification of protein cluster sizes with subcellular resolution. *Biophys J* **97:** 2613–2622.

Bader AN, Hoetzl S, Hofman EG, Voortman J, van Bergen en Henegouwen PM, van Meer G, Gerritsen HC. 2011. Homo-FRET imaging as a tool to quantify protein and lipid clustering. *Chemphyschem* **12:** 475–483.

Betzig E, Lewis A, Harootunian A, Isaacson M, Kratschmer E. 1986. Near field scanning optical microscopy (NSOM): Development and biophysical applications. *Biophys J* **49:** 269–279.

Choi SH, Mendrola JM, Lemmon MA. 2007. EGF-independent activation of cell-surface EGF receptors harboring mutations found in gefitinib-sensitive lung cancer. *Oncogene* **26:** 1567–1576.

Chu J, Zhang Z, Zheng Y, Yang J, Qin L, Lu J, Huang ZL, Zeng S, Luo Q. 2009. A novel far-red bimolecular fluorescence complementation system that allows for efficient visualization of protein interactions under physiological conditions. *Biosens Bioelectron* **25:** 234–239.

Chung I, Akita R, Vandlen R, Toomre D, Schlessinger J, Mellman I. 2010. Spatial control of EGF receptor activation by reversible dimerization on living cells. *Nature* **464:** 783–787.

Clayton AH, Walker F, Orchard SG, Henderson C, Fuchs D, Rothacker J, Nice EC, Burgess AW. 2005. Ligand-induced dimer-tetramer transition during the activation of the cell surface epidermal growth factor receptor-A multidimensional microscopy analysis. *J Biol Chem* **280:** 30392–30399.

Cutler PJ, Malik MD, Liu S, Byars JM, Lidke DS, Lidke KA. 2013. Multi-color quantum dot tracking using a high-speed hyperspectral line-scanning microscope. *PLoS ONE* **8:** e64320.

Digman MA, Gratton E. 2011. Lessons in fluctuation correlation spectroscopy. *Annu Rev Phys Chem* **62:** 645–668.

Digman MA, Wiseman PW, Choi C, Horwitz AR, Gratton E. 2009. Stoichiometry of molecular complexes at adhesions in living cells. *Proc Natl Acad Sci* **106:** 2170–2175.

Gadella TW Jr, Jovin TM. 1995. Oligomerization of epidermal growth factor receptors on A431 cells studied by time-resolved fluorescence imaging microscopy. A stereochemical model for tyrosine kinase receptor activation. *J Cell Biol* **129:** 1543–1558.

Galperin E, Verkhusha VV, Sorkin A. 2004. Three-chromophore FRET microscopy to analyze multiprotein interactions in living cells. *Nat Methods* **1:** 209–217.

Garrett TP, McKern NM, Lou M, Elleman TC, Adams TE, Lovrecz GO, Kofler M, Jorissen RN, Nice EC, Burgess AW, et al. 2003. The crystal structure of a truncated ErbB2 ectodomain reveals an active conformation, poised to interact with other ErbB receptors. *Mol Cell* **11:** 495–505.

Godin AG, Costantino S, Lorenzo LE, Swift JL, Sergeev M, Ribeiro-da-Silva A, De Koninck Y, Wiseman PW. 2011. Revealing protein oligomerization and densities in situ using spatial intensity distribution analysis. *Proc Natl Acad Sci* **108:** 7010–7015.

Hahn K, Toutchkine A. 2002. Live-cell fluorescent biosensors for activated signaling proteins. *Curr Opin Cell Biol* **14:** 167–172.

Hell SW. 2007. Far-field optical nanoscopy. *Science* **316:** 1153–1158.

Hofman EG, Bader AN, Voortman J, van den Heuvel DJ, Sigismund S, Verkleij AJ, Gerritsen HC, van Bergen en Henegouwen PM. 2010. Ligand-induced EGF receptor oligomerization is kinase-dependent and enhances internalization. *J Biol Chem* **285:** 39481–39489.

Huang B, Bates M, Zhuang X. 2009. Super-resolution fluorescence microscopy. *Annu Rev Biochem* **78:** 993–1016.

Jares-Erijman EA, Jovin TM. 2003. FRET imaging. *Nat Biotechnol* **21:** 1387–1395.

Jura N, Shan Y, Cao X, Shaw DE, Kuriyan J. 2009. Structural analysis of the catalytically inactive kinase domain of the human EGF receptor 3. *Proc Natl Acad Sci* **106:** 21608–21613.

Kerppola TK. 2008. Bimolecular fluorescence complementation (BiFC) analysis as a probe of protein interactions in living cells. *Annu Rev Biophys* **37:** 465–487.

Kolin DL, Wiseman PW. 2007. Advances in image correlation spectroscopy: Measuring number densities, aggregation states, and dynamics of fluorescently labeled macromolecules in cells. *Cell Biochem Biophys* **49:** 141–164.

Kusumi A, Sako Y, Yamamoto M. 1993. Confined lateral diffusion of membrane receptors as studied by single particle tracking (nanovid microscopy). Effects of calcium-induced differentiation in cultured epithelial cells. *Biophys J* **65:** 2021–2040.

Lemmon MA, Schlessinger J. 2010. Cell signaling by receptor tyrosine kinases. *Cell* **141:** 1117–1134.

Leung BO, Chou KC. 2011. Review of super-resolution fluorescence microscopy for biology. *Appl Spectrosc* **65:** 967–980.

Lidke DS, Lidke KA. 2012. Advances in high-resolution imaging—Techniques for three-dimensional imaging of cellular structures. *J Cell Sci* **125:** 2571–2580.

Lidke DS, Nagy P, Barisas BG, Heintzmann R, Post JN, Lidke KA, Clayton AH, Arndt-Jovin DJ, Jovin TM. 2003. Imaging molecular interactions in cells by dynamic and static fluorescence anisotropy (rFLIM and emFRET). *Biochem Soc Trans* **31:** 1020–1027.

Cite this article as *Cold Spring Harb Perspect Biol* doi: 10.1101/cshperspect.a020735

Lidke DS, Nagy P, Heintzmann R, Arndt-Jovin DJ, Post JN, Grecco HE, Jares-Erijman EA, Jovin TM. 2004. Quantum dot ligands provide new insights into erbB/HER receptor-mediated signal transduction. *Nat Biotechnol* **22:** 198–203.

Lidke DS, Lidke KA, Rieger B, Jovin TM, Arndt-Jovin DJ. 2005. Reaching out for signals: Filopodia sense EGF and respond by directed retrograde transport of activated receptors. *J Cell Biol* **170:** 619–626.

Low-Nam ST, Lidke KA, Cutler PJ, Roovers RC, van Bergen en Henegouwen PM, Wilson BS, Lidke DS. 2011. ErbB1 dimerization is promoted by domain co-confinement and stabilized by ligand binding. *Nat Struct Mol Biol* **18:** 1244–1249.

Luker KE, Smith MC, Luker GD, Gammon ST, Piwnica-Worms H, Piwnica-Worms D. 2004. Kinetics of regulated protein-protein interactions revealed with firefly luciferase complementation imaging in cells and living animals. *Proc Natl Acad Sci* **101:** 12288–12293.

Macdonald JL, Pike LJ. 2008. Heterogeneity in EGF-binding affinities arises from negative cooperativity in an aggregating system. *Proc Natl Acad Sci* **105:** 112–117.

Macdonald-Obermann JL, Piwnica-Worms D, Pike LJ. 2012. Mechanics of EGF receptor/ErbB2 kinase activation revealed by luciferase fragment complementation imaging. *Proc Natl Acad Sci* **109:** 137–142.

Martin-Fernandez M, Clarke DT, Tobin MJ, Jones SV, Jones GR. 2002. Preformed oligomeric epidermal growth factor receptors undergo an ectodomain structure change during signaling. *Biophys J* **82:** 2415–2427.

Metzger H. 1992. Transmembrane signaling: The joy of aggregation. *J Immunol* **149:** 1477–1487.

Nagy P, Jenei A, Kirsch AK, Szollosi J, Damjanovich S, Jovin TM. 1999. Activation-dependent clustering of the erbB2 receptor tyrosine kinase detected by scanning near-field optical microscopy. *J Cell Sci* **112:** 1733–1741.

Nagy P, Claus J, Jovin TM, Arndt-Jovin DJ. 2010. Distribution of resting and ligand-bound ErbB1 and ErbB2 receptor tyrosine kinases in living cells using number and brightness analysis. *Proc Natl Acad Sci* **107:** 16524–16529.

Needham SR, Hirsch M, Rolfe DJ, Clarke DT, Zanetti-Domingues LC, Wareham R, Martin Fernandez ML. 2013. Measuring EGFR separations on cells with approximately 10 nm resolution via fluorophore localization imaging with photobleaching. *PLoS ONE* **8:** e62331.

Ogiso H, Ishitani R, Nureki O, Fukai S, Yamanaka M, Kim JH, Saito K, Sakamoto A, Inoue M, Shirouzu M, et al. 2002. Crystal structure of the complex of human epidermal growth factor and receptor extracellular domains. *Cell* **110:** 775–787.

Orr G, Hu D, Ozcelik S, Opresko LK, Wiley HS, Colson SD. 2005. Cholesterol dictates the freedom of EGF receptors and HER2 in the plane of the membrane. *Biophys J* **89:** 1362–1373.

Patterson G, Davidson M, Manley S, Lippincott-Schwartz J. 2010. Superresolution imaging using single-molecule localization. *Ann Rev Phys Chem* **61:** 345–367.

Renz M, Daniels BR, Vamosi G, Arias IM, Lippincott-Schwartz J. 2012. Plasticity of the asialoglycoprotein receptor deciphered by ensemble FRET imaging and single-molecule counting PALM imaging. *Proc Natl Acad Sci* **109:** E2989–E2997.

Sako Y, Minoghchi S, Yanagida T. 2000. Single-molecule imaging of EGFR signalling on the surface of living cells. *Nat Cell Biol* **2:** 168–172.

Sato M, Umezawa Y. 2004. Imaging protein phosphorylation by fluorescence in single living cells. *Methods* **32:** 451–455.

Schlessinger J. 2002. Ligand-induced, receptor-mediated dimerization and activation of EGF receptor. *Cell* **110:** 669–672.

Schmidt T, Schutz GJ, Baumgartner W, Gruber HJ, Schindler H. 1996. Imaging of single molecule diffusion. *Proc Natl Acad Sci* **93:** 2926–2929.

Sengupta P, Jovanovic-Talisman T, Skoko D, Renz M, Veatch SL, Lippincott-Schwartz J. 2011. Probing protein heterogeneity in the plasma membrane using PALM and pair correlation analysis. *Nat Methods* **8:** 969–975.

Sorkin A, McClure M, Huang F, Carter R. 2000. Interaction of EGF receptor and grb2 in living cells visualized by fluorescence resonance energy transfer (FRET) microscopy. *Curr Biol* **10:** 1395–1398.

Subach FV, Patterson GH, Renz M, Lippincott-Schwartz J, Verkhusha VV. 2010. Bright monomeric photoactivatable red fluorescent protein for two-color super-resolution sptPALM of live cells. *J Am Chem Soc* **132:** 6481–6491.

Swift JL, Godin AG, Dore K, Freland L, Bouchard N, Nimmo C, Sergeev M, De Koninck Y, Wiseman PW, Beaulieu JM. 2011. Quantification of receptor tyrosine kinase transactivation through direct dimerization and surface density measurements in single cells. *Proc Natl Acad Sci* **108:** 7016–7021.

Tao RH, Maruyama IN. 2008. All EGF(ErbB) receptors have preformed homo- and heterodimeric structures in living cells. *J Cell Sci* **121:** 3207–3217.

van de Linde S, Kasper R, Heilemann M, Sauer M. 2008. Photoswitching microscopy with standard fluorophores. *Appl Phys B* **93:** 725–731.

Veatch SL, Machta BB, Shelby SA, Chiang EN, Holowka DA, Baird BA. 2012. Correlation functions quantify super-resolution images and estimate apparent clustering due to over-counting. *PLoS ONE* **7:** e31457.

Wang Z, Longo PA, Tarrant MK, Kim K, Head S, Leahy DJ, Cole PA. 2011. Mechanistic insights into the activation of oncogenic forms of EGF receptor. *Nat Struct Mol Biol* **18:** 1388–1393.

Weiss S. 1999. Fluorescence spectroscopy of single biomolecules. *Science* **283:** 1676–1683.

Yang S, Raymond-Stintz MA, Ying W, Zhang J, Lidke DS, Steinberg SL, Williams L, Oliver JM, Wilson BS. 2007. Mapping ErbB receptors on breast cancer cell membranes during signal transduction. *J Cell Sci* **120:** 2763–2773.

Yarden Y, Sliwkowski MX. 2001. Untangling the ErbB signalling network. *Nat Rev Mol Cell Biol* **2:** 127–137.

Zhang X, Gureasko J, Shen K, Cole PA, Kuriyan J. 2006. An allosteric mechanism for activation of the kinase domain of epidermal growth factor receptor. *Cell* **125:** 1137–1149.

Structure-Function Relationships of ErbB RTKs in the Plasma Membrane of Living Cells

Donna J. Arndt-Jovin, Michelle G. Botelho, and Thomas M. Jovin

Laboratory of Cellular Dynamics, Max Planck Institute for Biophysical Chemistry, 37077 Göttingen, Germany
Correspondence: djovin@gwdg.de

We review the states of the ErbB family of receptor tyrosine kinases (RTKs), primarily the EGF receptor (EGFR, ErbB1, HER1) and the orphan receptor ErbB2 as they exist in living mammalian cells, focusing on four main aspects: (1) aggregation state and distribution in the plasma membrane; (2) conformational features of the receptors situated in the plasma membrane, compared to the crystallographic structures of the isolated extracellular domains; (3) coupling of receptor disposition on filopodia with the transduction of signaling ligand gradients; and (4) ligand-independent receptor activation by application of a magnetic field.

This review deals exclusively with the disposition and function of the human ErbB (HER) family of receptor tyrosine kinases (RTKs) in the plasma membrane of living cells. We have divided the material into four main topics: (1) distribution and aggregation state of ErbB family members; (2) the 3D structure of the ErbB1 and ErbB2 receptors; (3) the role of receptors located on extensions (filopodia) of the cell body; and (4) the phenomenon and implications of nonligand-dependent activation of the receptors.

DISTRIBUTION OF ErbB FAMILY MEMBERS IN THE PLASMA MEMBRANE OF LIVING CELLS

The association state(s) and activities of the ErbB family members in intact living cells differ widely depending on the expression level (number per cell) and the distribution (density, state of homo- and heteroassociation) of each receptor. There are numerous, highly contradictory views about the aggregation state of the receptors in the literature. During the last few years more accurate assessments of receptor distribution have been conducted on living rather than fixed cells. In the latter case, numerous artifacts account for important quantitative discrepancies. One example is the failure of formaldehyde fixation to prevent receptor redistribution (Brock et al. 1999; Tanaka et al. 2010). A more fundamental consideration relates to the hierarchical organization of the plasma membrane, extensively investigated and most recently reviewed by Kusumi et al. (2011). These investigators propose the existence of (1) a mesoscale (40–300 nm) compartmentalization with actin cytoskeletal "fences" and transmembrane protein "pickets," (2) lipid raft domains (2–

20 nm), and (3) oligomeric complexes of membrane-associated proteins (3–10 nm). Signal transduction is mediated by cooperative interactions (varying in strength) of and in these domains. Such interactions involve thermal fluctuations, and molecular redistribution and conformational transitions induced by binding and enzymatic action. Temperature clearly plays a critical role in all of these phenomena. Eukaryotic cells exist primarily between 18°C and 37°C in their respective organisms, undergoing cell death outside this range. The membrane composition and domain structure are intimately connected to these optimum temperatures. From the biophysical perspective, the challenge is to assess the spatial-temporal organization of activation and deactivation processes and their functional consequences.

There exists a persistent controversy about the existence, nature, and extent of preformed, unliganded dimers and higher-order associated states of the EGF receptor (EGFR, ErbB1, HER1). Much of the published data was acquired using labeled ligands as probes (Gadella and Jovin 1995; Sako et al. 2000; Moriki et al. 2001; Martin-Fernandez et al. 2002; Tynan et al. 2012) and/or cells expressing high levels of receptor. In the following section we summarize the latest information derived from strictly live cell analyses at room temperature or 37°C.

Starved cells (HeLa or transfected Chinese hamster ovary [CHO] cells) with 50,000 or fewer ErbB1 receptors and in the absence of serum and autocrine expression of growth factors have no detectable level of activated ErbB1. Two studies using carboxy-terminal, fused green fluorescent protein (GFP) chimeric ErbB1 examined explicitly the levels of receptor monomers, dimers, and oligomers in live cells. Measurements of diffusion by fluorescence correlation spectroscopy (FCS) (Saffarian et al. 2007) and determinations of receptor distribution by number and brightness analysis (N&B) (Nagy et al. 2010), a correlation technique that computes the fluorescence intensities corresponding to single (and thus multiple) receptors (Digman et al. 2008), led to the conclusion that ErbB1 exists almost exclusively as a monomer in such cells and only forms detectable di-

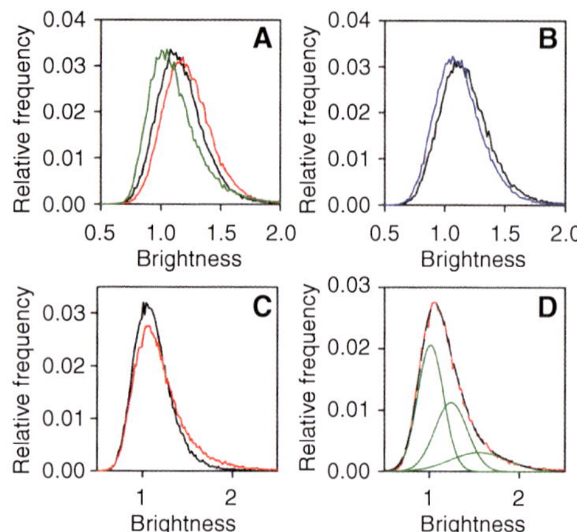

Figure 1. Clustering of ErbB1 in quiescent and EGF-stimulated cells measured by N&B. (*A*) Starved F1-4 cells expressing 5×10^5 receptors were measured before and after stimulation with EGF for 3 min. The brightness histograms are plotted for quiescent (black line) and the EGF-stimulated cells (red line). The brightness histogram of soluble monomeric eGFP is shown by the green line. (*B*) Starved F1-4 cells were photobleached until their fluorescence intensity reached ∼20% of the initial value and the brightness histograms of the control (black line) and the bleached (blue line) cells are shown. (*C*) Brightness histograms of F1-10 cells expressing 5×10^4 receptors before ligand (starved [black line]) and after ligand (EGF-stimulated [red line]) addition for 3 min. (*D*) The brightness histogram of EGF-stimulated F1-10 cells (red line) was fitted by three overlapping Gaussians (green lines). The sum of the three fitted Gaussian distributions perfectly overlapping the experimentally determined histogram is shown by the black, long dashed line. (From Nagy et al. 2010; reproduced, with permission, from the National Academy of Sciences © 2010.)

mers and some higher-order aggregates at expression levels of $>10^5$ receptors (Fig. 1; Table 1). In the N&B measurements, addition of EGF caused a rapid dimerization of the receptors, which then associated with clathrin-coated pits and internalized. In the case of cells harboring 5×10^5 receptors there was a concerted shift in the ErbB1 population to a dimer distribution (Fig. 1).

Two independent studies used single-particle tracking (SPT) to assess the coconfinement of ErbB1 receptors within the optical resolu-

Table 1. Molecular brightness values calculated from data as in Figure 1 for ErbB1-eGFP-expressing cells without ligand and after exposure to EGF

F1-4 cells		
Molecular brightness	Starved cells	0.158 ± 0.002 (1.3)
(mol/cluster)	Starved cells after photobleaching	0.128 ± 0.002 (1.0)
	Starved cells plus EGF for 3 min	0.217 ± 0.003 (1.8)
F1-10 cells		
Molecular brightness	Starved cells	0.122 ± 0.001 (1.0)
(mol/cluster)	Starved cells plus EGF for 3 min	0.175 ± 0.003 (1.5)

Molecular brightness values ($\pm$ s.e.m.) of quiescent and EGF-stimulated cells are shown in the table. The numbers in parentheses represent the average number of ErbB1-eGFP in a molecular cluster determined by dividing the molecular brightness values of cells with the brightness value of 0.118 ± 0.002 for soluble monomeric eGFP fluorescent protein. The displayed values were calculated from 5 to 10 cells. (From Nagy et al. 2010; modified, with permission.)

tion of the microscope (200–300 nm). The diffusion in and out of corrals was determined with Fab fragments (Chung et al. 2010) or nanobodies (Low-Nam et al. 2011) coupled to quantum dots (QDs). Chung et al. found that coconfined receptors had finite "dimer" lifetimes with an increased probability in the lamellipodia. These coconfined receptors did not show the slow diffusion of the activated receptors. However, there was a significant difference in the diffusion of receptors whose "dimers" had a defective dimerization arm in the extracellular domain 2 (revealed by crystallography [Ogiso et al. 2002]). Using two-color QDs Low-Nam et al. calculated off rates for coconfinement and "dimer" formation. These data were consistent with very short-lived unliganded dimers and the absence of correlated diffusion. However, these studies also showed a facilitated coconfinement and more probable dimer association of receptors in cells with high receptor densities such as A431 (2×10^6) compared to HeLa cells with a low density (5×10^4). The formation of dimers or higher-order aggregates was not observed without addition of EGF (i.e., a correlated motion of the two receptors was absent). Low-Nam et al. (2011) calculated a decrease in the dimer dissociation rate of only approximately eightfold after EGF binding, but this value may have been an underestimate because tracks were only analyzed for 50 sec. Using long-term tracking, we have demonstrated the continued association of activated dimers for >400 sec (MG Botelho et al., unpubl.) and conclude that SPT confirms the

biophysical data gained from the FCS and N&B analyses. An additional finding of functional significance was the peripheral enrichment in the actin-dependent cortical compartments (Kusumi et al. 2011), offering a means for polarized responses to growth factors (Chung et al. 2010) albeit with a reduced mobility of signaling-competent ErbB1 dimers (Low-Nam et al. 2011).

A number of experiments have been reported utilizing chimeric ErbB1 transgenes with a platelet-derived growth factor (PDGF)-binding domain (Tyson et al. 2003) or extracellular and intracellular domains exchanged with those of other ErbB family members (Wang et al. 1999; Berger et al. 2004). The aim has been to elucidate the mechanism of activation and the importance of the carboxy-terminal tail, the juxtamembrane domain, as well as the transmembrane segment. These issues are more thoroughly treated in other articles in this collection on structural analysis. Suffice it to say that all of the data are compatible with dimerization as a necessary step leading to transphosphorylation. We will deal later in this report with high local densities of ErbB1, achieved by physical means, that may signal without the intervention of a specific ligand.

In contrast to ErbB1, the orphan receptor ErbB2 is found predominantly in aggregates of three to eight molecules as well as in large clusters comprising many hundreds of receptors, as was shown in living cells by N&B measurements (Nagy et al. 2010), in Förster resonance energy transfer (FRET) studies using mAbs (Nagy et al. 1998; Szabó et al. 2008), and by high-resolution near-field scanning microscopy (Nagy et al. 1999) and transmission electron microscopy (TEM) (Yang et al. 2007) of fixed samples labeled by immunofluorescence or immunogold, respectively. The dispersion of the large ErbB2 clusters, preferentially associated with rafts, upon addition of cholera toxin (Nagy et al. 2002) or activation by EGF is attributable to preferential formation of ErbB2/ErbB3 heterodimers (Nagy et al. 1999) or ErbB1/ErbB2 heterodimers (Nagy et al. 2010); see Table 2, where data is modified from Nagy et al. (2010). In cells with high receptor density the ErbB2 clusters are by themselves capable of low-level self-activation in the absence of heterodimerization. Another study showed that cells with increased levels of ErbB2 expression exhibited prolonged signaling from ErbB1 from the cellular membrane, presumably because heterodimerization inhibits the normal endocytosis of ErbB1 homodimers through clathrin-coated pits (Offterdinger and Bastiaens 2008). This effect was reversed by addition of an antibody that blocks the dimerization arm of ErbB2.

ErbB3/ErbB2 heterodimers and neuregulin autocrine activation are the hallmarks of ovarian carcinomas and many breast cancers (Gilmour et al. 2002; Holbro et al. 2003). Both cellular proliferation and migration are promoted by this RTK heterodimer through strong upregulation of the PI3-kinase pathway (Smirnova et al. 2011). In fact, the activated ErbB3/ErbB2 heterodimer shows the highest proliferation index (Rubin and Yarden 2001; Citri et al. 2003; Holbro et al. 2003). However, using cross-linking aptamers, Park et al. (2008) showed that ErbB3 is not interspersed in ErbB2 domains but can form inactive homodimers even at levels of 25,000 or fewer receptors/cell, which reside in a different membrane environment than that of activated ErbB2/ErbB3 heterodimers.

Because various cell types express different numbers of the four ErbB family members it has been difficult to determine the comparative signaling strengths of homo- and heterodimer pairs. Zhang et al. (2009) created a clone library of human mammary epithelial cells expressing different levels of ErbB1, ErbB2, and ErbB3 and conducted quantitative analyses of the signal transduction pathway outputs in the presence of EGF or NRG. They reached relevant conclusions about the network structures and the relative activation strength of dimer pairs. The modeling of the network dynamics of the ErbB RTKs is discussed in detail in Volinsky and Kholodenko (2013).

The lipid environment as well as the presence and nature of proteoglycans on the cell membrane influence the distribution of the ErbB family members in different ways and can account for inhibition of activity in given cases as well as resistance to monoclonal antibody tumor therapy. Hakamori's group investigated the role of GM3 in EGFR kinase inhibition, and showed that modification of the EGFR with N-linked glycan terminated with GlcNAc

Table 2. Molecular brightness of ErbB2-mYFP and Atto565-labeled ACP-erbB2

	Soluble mYFP	**ErbB2-short-mYFP**	
Molecular brightness (mol/cluster)	0.063 ± 0.01	Inside macroclusters	0.89 ± 0.06 (14.1)
		Outside macroclusters	0.38 ± 0.05 (6.0)
	Soluble Atto565	**SfP-ErbB2**	
Molecular brightness (mol/cluster)	0.19 ± 0.01	Inside macroclusters	0.81 ± 0.04 (7.8)
		Outside macroclusters	0.44 ± 0.03 (4.2)

Cells expressing a 13-amino acid SfP-amino terminus (Garrett et al. 2002; Yin et al. 2006) tagged ErbB2 were labeled with Atto565-CoA using SfP transferase 2 days after transfection. Atto565-labeled and stably transfected ErbB2-mYFP-expressing cells were analyzed by the number and molecular brightness (N&B) method. The molecular brightness of ErbB2 analyzed inside and outside macroclusters (± s.e.m.) of ~10 cells are displayed as well as the brightness of the soluble monomeric fluorophores. To determine the average number of ErbB2 molecules in a molecular cluster (displayed in parentheses), cellular molecular brightness values were divided by that of the soluble fluorophore.

 Cite this article as *Cold Spring Harb Perspect Biol* doi: 10.1101/cshperspect.a008961

promotes a direct interaction with GM3 (Yoon et al. 2006). In vitro studies with liposomes confirmed the exclusive interaction of ganglioside GM3 with the extracellular juxtamembrane domain of EGFR and a corresponding reduction of transphosphorylation, suggesting that this interaction causes an allosteric, inhibitory effect on autophosphorylation (Coskun et al. 2011). In contrast, ErbB2 is enriched in lipid rafts (Nagy et al. 2002) and is activated by its interaction with hyaluronic acid and CD44, both of which are up-regulated in several breast cancer lines (Ghatak et al. 2005). Interactions with the ECM and production of mucins may also lead to resistance to herceptin therapy (Friedländer et al. 2005; Nagy et al. 2005).

WHAT IS THE STRUCTURE OF THE ErbB1 IN LIVING CELLS?

The X-ray crystallographic structures of the extracellular and kinase domains of the ErbB family of receptors are discussed in detail in other articles. They have provided the basis for understanding (1) the requirement for dimerization for kinase activation, and (2) the mechanism of transphosphorylation (Shibuya 2013; Song et al. 2013; Barton et al. 2014; Hunter 2014). The first crystal structures of the liganded ErbB1 receptor (Garrett et al. 2002; Ogiso et al. 2002) revealed an extended extracellular structure with ligand binding between domains 1 and 3 and dimerization through domain 2, confirming an earlier model of Lemmon et al. (1997) of two EGF:two ErbB1 receptors. The subsequent ErbB2 structure of Garrett et al. (2003) also showed this extended structure. However, the unliganded, unactivated ErbB1 ectodomain crystallized in a self-inhibitory structure that oriented domain 1 in close proximity to the cell membrane (Ferguson et al. 2003). This structure has been presumed to be the form adopted by the inactive receptor in the plasma membrane. A number of in vivo studies on mutated ErbB1 receptors (Ozcan et al. 2006), antibody binding (Schmitz and Ferguson 2009), and cocrystal structures of antibody with ErbB1 (Li et al. 2005) have confirmed the existence of a configuration in which the EGF-binding face

of domain 3 is exposed and ErbB1 is situated close to the cell membrane. Although ligand-binding transfers structural changes via the transmembrane domain in a manner causing reorganization of intracellular juxtamembrane and kinase domains into an active conformation (Yarden and Schlessinger 1987; Jura et al. 2009), it is unclear which extracellular configuration predominates in the presence of kinase inhibitors (e.g., quinazoline or PD168393) that stabilize the active state of the kinase domain (Stamos et al. 2002; Lu et al. 2012). A molecular dynamics (MD) simulation of EGFR in the presence of type I or type 1 1/2 kinase inhibitors concludes that better efficacy in tumor therapy is achieved by stabilizing the inactive kinase conformation and specificity may be increased by understanding the flexibility of the various loops and helices in the presence of the inhibitors (Songtawee et al. 2013). Mutational studies of the juxtamembrane domain have confirmed its importance in activation (Red Brewer et al. 2009). Other MD simulations starting from the X-ray crystallographic structures of the extracellular domains or the kinase domains (Arkhipov et al. 2013) have addressed the influences of the membrane environment on conformation and thus, functional state. The importance of testing hypotheses generated from MD calculations with mutational studies or activity assays has been shown by Endres et al. (2013) and is certain to bear more fruit in the future.

There have been few studies of the structural features of the complete native ErbB1 receptor in the plasma membrane of intact living cells in the absence of ligand. Kozer et al. (2011) constructed an amino-terminally fused YFP EGFR chimera and measured FRET to a rhodamine labeled lipid probe in the plasma membrane. The investigators saw little or no change in the FRET efficiency after addition of EGF and concluded that the unliganded receptor is present in the plasma membrane primarily in the extended conformation. This transgene, however, shows spontaneous activation of the receptor even without addition of EGF, suggesting it may dimerize spontaneously via the YFP moiety and not adopt the normal configuration of the wild-type membrane-associated receptor. It is well known that YFP lacking the A206K mutation has a strong tendency to dimerize (Zacharias et al. 2002). In addition, these investigators used 20 μm phenyl arsine oxide (PAO) to inhibit endocytosis of the EGFR. PAO is a potent tyrosine phosphatase inhibitor (Garcia-Morales et al. 1990), and also specifically inhibits protein kinase IIa, which is responsible for maintaining the balance of PIP2 in the plasma membrane (Santos et al. 2013). PIP2 has been shown to influence the activity of the ErbB1 by binding to the intracellular juxtamembrane domain (Michailidis et al. 2011). The pleiotropic effects of this cellular effector, and the general effects of phosphatase inhibition, therefore, complicate the interpretation of these data.

We recently reported a biophysical determination of the orientation(s) of the ectodomain 1 of unliganded, nonactivated ErbB1 in living cells using FRET. The states before and after addition of EGF were assessed. An ErbB1-specific kinase inhibitor (PD153035) was used that prevents endocytosis but does not inhibit EGF binding (Ziomkiewicz et al. 2013). For these studies, we constructed a transgene ErbB1 with an acyl carrier protein sequence (Vivero-Pol et al. 2005; George 2006; Yin et al. 2006) between the signal peptide and the ErbB1 mature protein sequence. The mutated protein has unchanged functional properties, compared to wild-type ErbB1, and can be covalently labeled with a fluorophore at a specific serine in the acyl carrier tag. By measuring FRET between the acyl carrier tag and a novel lipid probe that localizes exclusively to the outer leaflet of the cell membrane (Kucherak et al. 2010), we determined the apparent relative distance to the membrane of the ACP tag under nonactivation and activation conditions. The fluorescence lifetime imaging microscopy (FLIM) experiments showed that the unliganded receptor is situated with its amino terminus significantly closer to the membrane than after EGF addition, thereby supporting the model in which most of the inactive receptor in the plasma membane of quiescent cells is in the self-inhibited, tethered configuration (Ziomkiewicz et al. 2013). Figure 2 shows representative FLIM images for the quiescent and activated conditions.

SIGNALING FROM FILOPODIA

RTKs expressed on nononcogenic cells under physiological conditions are required to sense directional gradients of growth factors for initiating rapid cellular division and directed migration in response to injury to the epithelium or for guidance during differentiation and development. Most experiments in the laboratory, however, involve isotropic exposure of tissue culture cells to growth factors in the working solution and monitoring the resulting cellular responses. The cells used are usually not polarized, and nor are they situated in a tissue context. It is known that very high levels of EGF can be antagonistic, causing cell death rather than activation or migration. Several years ago, we investigated the response of living cells to very low levels of EGF (5–50 pM) by tracking single receptors binding EGF coupled to QDs (Lidke et al. 2004, 2005). This technique allowed the observation of all stages of the receptor trafficking. The data also revealed a new paradigm according to which receptors situated far from the cell body on filopodia bind EGF, dimerize, and then attach to the actin filaments constituting the filapodial core. Actin treadmilling initiates retrograde transport of the anchored receptors toward the cell body, where they undergo endocytosis via clathrin-coated pits. The scheme depicting this mechanism is shown in Figure 3, and online Movie 1 shows the phenomenon (see online Movie 1 at http://cshperspectives.cshlp.org) for the online version. Our ongoing study of retrograde transport has revealed that Shc1 is essential for the coupling of the activated ErbB1 to the actin filaments (MG Botelho et al., in prep.). Several of the other downstream effectors of the Erb1-associated signaling cascade are not membrane associated and interact with the receptor presenting on the surface of endocytic vesicles. We conclude that the cell may use receptors situated on the filopodia as sensors of growth factor gradients, serving to transduce the responses only if defined threshold levels are exceeded. That is, owing to the inherently nonlinear property of receptor activation, the filopodial mechanism of retrograde transport provides the cell with the capacity for selec-

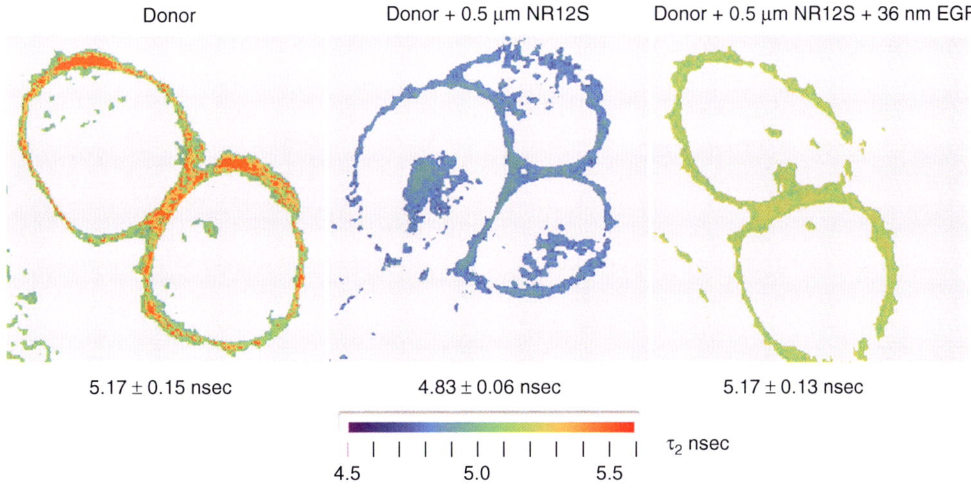

Figure 2. Conformational change of membrane-bound ErbB1 upon EGF binding as determined from FRET data. Stable Chinese hamster ovary (CHO) cell line transfected with the acyl carrier protein tag fused to the amino terminus of the ErbB1 ectodomain. The target serine residue was covalently labeled by phosphopantetheine transferase with Atto390-CoA. Atto390 emits at 480 nm and served as the FRET donor. Living cells were examined in an Olympus-PicoQuant FLIM microscope system with two-photon, 80-MHz excitation at 740 nm. Lifetime images for representative cells under three conditions are shown: (*left*) donor only; (*middle*) donor plus Nile Red12S (NR12S) membrane dye (applied at 0.5 µm) that localizes in the outer leaflet and acted as the FRET acceptor (Kucherak et al. 2010). NR12S absorbs broadly from 450 to 550 with a Förster R_0 of 5.4 nm and emits at >590 nm; and (*right*) donor and acceptor labeled cells bathed in 1 µm ErbB1 kinase inhibitor PD153035 and exposed to 32 nm EGF (room temperature, 5 min). Image acquisition was performed until the mean value for the brightest pixels reached 1000 counts using the PicoQuant SymphoTime TCSCP software. Pixel data (photon arrival times) were entered into a laboratory-generated *Mathematica* program for (i) selecting an analysis region of interest (ROI), (ii) filtering (masking and intensity ordering), and (iii) computing lifetimes from the intensity binned pixel data, assuming a two-component decay and using an analytical function to represent the experimental impulse response function (IRF) and the corresponding fluorescence time course (see Ziomkiewicz et al. 2013). The mean ± s.e. lifetime values are given below each panel. There was an ~7% drop in mean lifetime upon addition of the NR12S acceptor in the absence of EGF; this effect was substantially reversed upon addition of EGF (in the presence of kinase inhibitor) although the peak values originally observed in the absence of acceptor were not achieved. Data derived from experiments similar to those in Ziomkiewicz et al. (2013). We conclude that EGF binding, even in the absence of (auto)phosphorylation, leads to a conformational change, extending the ectodomain and thereby increasing the distance of the amino terminus from the plasma membrane in accordance with the model proposed from crystallographic analysis of isolated ErbB1 and its fragments. Larger separation correlates with a lower FLIM-detected FRET efficiency.

tively attenuating the response to low, physiologically nonrelevant levels of ligand, suppressing cellular migration and division under these conditions.

Epithelial cells have extensive filopodia and are covered with microvilli. These structures are enriched in growth factor and regulatory receptors and may regulate communication and signaling. Particularly long filopodia have been observed on the trophoectoderm in developing embryos which may promote communicating with the inner cell mass (Salas-Vidal and Lomeli 2004). Very short exposure to subsaturating levels of ligand results in predominance of activated receptor on microvilli and filopodia on A431 cells, as shown in Figure 4. Chung et al. (2010) calculated from single-particle analysis of mAb (Fab) binding that there was a predominance of receptors at the periphery of the cell. Surface plasmon resonance measurements based on gold immunolabeling reported by Wang et al. (2011) indicate that the density of ErbB1 in A431 cells is higher on the filopodia than on the cell body.

SIGNALING POTENTIATION AND SIGNALING IN THE ABSENCE OF NATIVE LIGAND

The extent of protein-tyrosine phosphorylation in a cell reflects an equilibrium between the actions of protein-tyrosine kinases (PTKs) and protein-tyrosine phosphatases (PTPs), as has been shown in living cells by the Bastiaens group using FLIM to measure FRET between tagged phosphotyrosine antibodies and tagged ErbB proteins (Reynolds et al. 2003; Offterdinger et al. 2004; Grecco et al. 2010). Rhee and coworkers have studied the effects of cysteine oxidation on the inhibition of specific phosphotyrosine phosphatases as well as mitogen-activated protein kinase phosphatases (reviewed in Lee et al. 1998; Rhee et al. 2000; Janssen-Heininger et al. 2008) and the enhancement of RTK activity by localized hydrogen peroxide

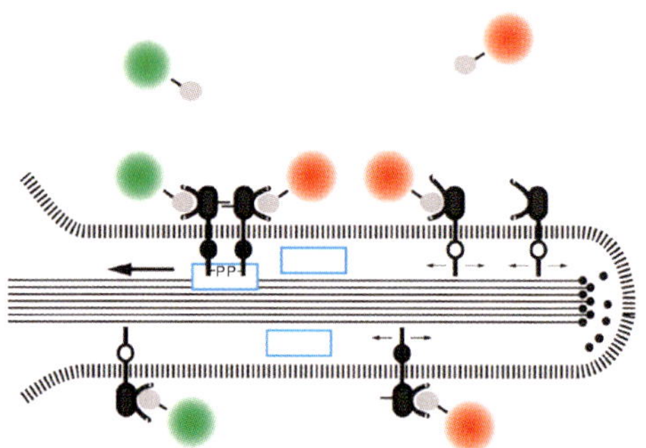

Figure 3. Filopodial retrograde transport of ErbB1. Recent studies identify Shc1 as a component (blue box) linking the activated receptor to the actin filaments constituting the core of the filopodium (see text). (From Lidke et al. 2005; adapted, with permission, from Rockefeller University Press © 2005.)

production. The cell type and state-dependent regulation of the balanced PTK-PTP activity presumably accounts for apparent discrepancies in studies reporting the confinement or, alternatively, propagation of locally applied (microbead-mediated) activation of the ErbB1 (Ver-

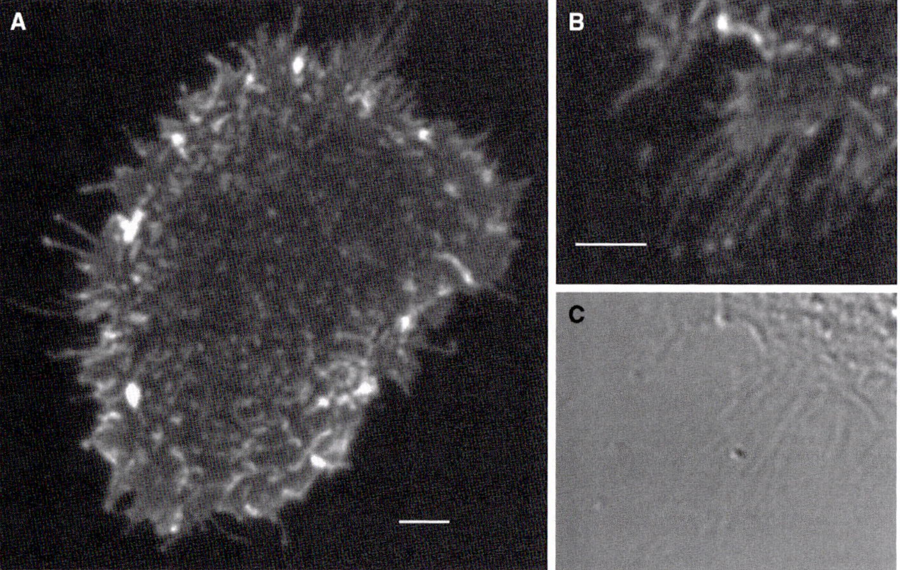

Figure 4. Preferential activation of ErbB1 on filopodia and microvilli with short exposure to EGF. Starved A431 cells were exposed for 40 sec to 15 nm EGF and fixed directly in ice cold EtOH. Activated ErbB1 was detected by immunofluorescence using a phosphotyrosine-specific ErbB1 antibody (pY1068) from cell signaling and a Cy3-labeled secondary antibody. No signal was observed on cells not exposed to EGF. Images were taken on a Zeiss 510 Meta microscope using 532 nm excitation and emission ≥585 nm. Background subtracted. Scale bars, 5 μm.

veer et al. 2000; Brock and Jovin 2003; Friedländer et al. 2005).

Other proteins may activate RTKs. Integrins have been implicated in ErbB signaling even without ligand binding (Yu et al. 2000; Bill et al. 2004), although Alexi et al. (2011) concluded that integrins were not involved but rather promote autocrine signaling. Recently, it was shown that neuregulin 1 binds both ErbB3 and integrin $\alpha6\beta4$, forming a precipitable ternary complex that activates AKT and Erk1/2 (Ieguchi et al. 2010).

The fact that high densities of ErbB1 receptors increase the probability of unliganded dimer formation, as discussed in the first section, led us to investigate superparamagnetic Fe_3O_4 nanoparticles (SPIONS) as possible effectors of ErbB1 activation by a magnetic field (Bharde et al. 2013). Such particles can be conjugated to mAbs or aptamers specific for an ErbB receptor and targeted directly to tumor cells. We have shown that anti-ErbB1-SPIONS (coupled with an antibody against ErbB1 that inhibits ligand binding) can selectively bind to ErbB1-expressing cells. Such complexes are inert but after application of a magnetic field the receptor is activated (Fig. 5) and subsequently internalized. These results suggest that activation of growth factor receptors may be triggered by ligand-independent molecular crowding resulting from overexpression and/or sequestration in membrane microdomains. Because nanoparticles can also be conjugated with drugs they constitute potential therapeutic agents. Somewhat larger nanoparticles also function well in magnetic resonance imaging (MRI).

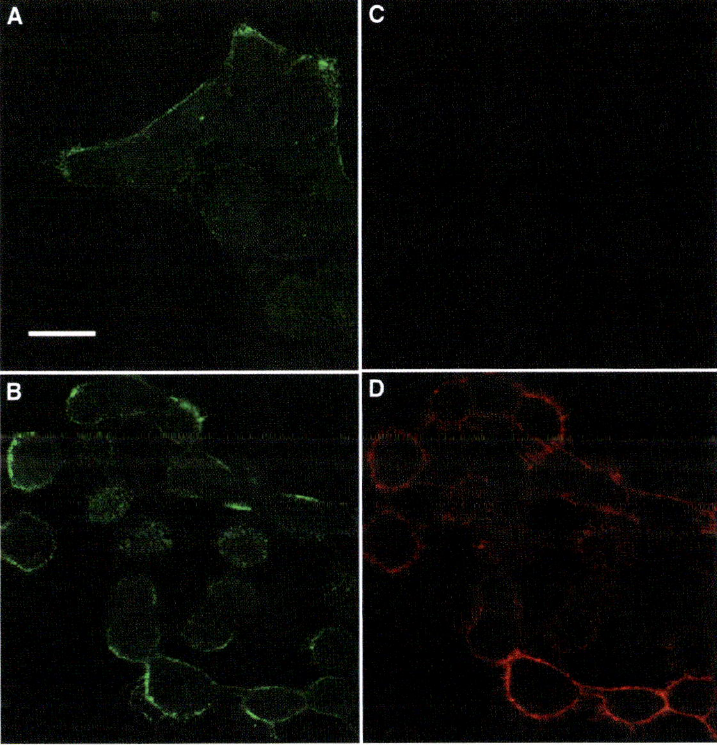

Figure 5. Binding of SPIONS to A431 cells and activation by exposure to a magnet for 180 sec. Panels (*A*) and (*B*) show specific binding of anti-ErbB1-targeted SPIONs on A431 cells after incubation for 15 min at 15°C. Panels (*C*) and (*D*) show the amount of transphosphorylation of ErbB1 on the cells after no magnetic field (*C*), or after application of a magnetic field for 180 sec (*D*) by indirect immunofluorescence staining for anti-ErbB1 pY1068. Data from experiments are similar to those published in Bharde et al. (2013). Scale bar, 20 μm.

CONCLUDING REMARKS

We have reviewed the literature and unpublished data that bear on the disposition and structure of ErbB receptors in living cells. The understanding of the complex interactions and consequent downstream signaling of this key RTK family in the presence of various effectors is increasing, and effective models with testable biomedically relevant hypotheses are emerging. Importantly, there are new techniques available that allow interrogation of individual receptors in living cells, thereby providing new, quantitative estimations of the underlying equilibria and kinetic processes. Furthermore, our understanding of the disposition and activity of RTKs in tissue culture cell systems is assisting the design of more effective cancer therapeutic strategies for the 50% of all human tumors in which the ErbB family of RTKs is functionally up-regulated.

ACKNOWLEDGMENTS

We thank the numerous colleagues that have worked with us on ErbB-mediated signal transduction over many years for their lively discussions, interest, insight, and experimental expertise. Funding for the research in our laboratory is provided by the Max Planck Society, the EU FP6 STREP project FLUOROMAG (EC Contract No. 037465), and the Deutsche Forschungsgemeinschaft (DFG) Research Center for Molecular Physiology of the Brain (CMPB, Excellence Cluster EXC 171–FZT 103 "Microscopy at Nanometer Range").

REFERENCES

*Reference is also in this collection.

Alexi X, Berditchevski F, Odintsova E. 2011. The effect of cell-ECM adhesion on signalling via the ErbB family of growth factor receptors. *Biochem Soc Trans* **39**: 568–573.

Arkhipov A, Shan Y, Das R, Endres NF, Eastwood MP, Wemmer DE, Kuriyan J, Shaw DE. 2013. Architecture and membrane interactions of the EGF receptor. *Cell* **152**: 557–569.

*Barton WA, Dalton AC, Seegar TCM, Himanen JP, Nikolov DB. 2014. Tie2 and Eph receptor tyrosine kinase activation and signaling. *Cold Spring Harb Perspect Biol* doi: 10.1101/cshperspect.a009142.

Berger MB, Mendrola JM, Lemmon MA. 2004. ErbB3/HER3 does not homodimerize upon neuregulin binding at the cell surface. *FEBS Lett* **569**: 332–336.

Bharde AA, Palankar R, Fritsch C, Klaver A, Kanger JS, Jovin TM, Arndt-Jovin DJ. 2013. Magnetic nanoparticles as mediators of ligand-free activation of EGFR signaling. *Plos ONE* **8**: e68879.

Bill HM, Knudsen B, Moores SL, Muthuswamy SK, Rao VR, Brugge JS, Miranti CK. 2004. Epidermal growth factor receptor-dependent regulation of integrin-mediated signaling and cell cycle entry in epithelial cells. *Mol Cell Biol* **24**: 8586–8599.

Brock R, Jovin TM. 2003. Quantitative image analysis of cellular protein translocation induced by magnetic microspheres: Application to the EGF receptor. *Cytometry* **52A**: 1–11.

Brock R, Hamelers IHL, Jovin TM. 1999. Comparison of fixation protocols for adherent cultured cells applied to a GFP fusion protein of the epidermal growth factor receptor. *Cytometry* **35**: 353–362.

Cho I, Cho HS, Mason K, Ramyar KX, Stanley AM, Gabelli SB, Denney DW Jr, Leahy DJ. 2003. Structure of the extracellular region of HER2 alone and in complex with the Herceptin Fab. *Nature* **421**: 756–760.

Chung I, Akita R, Vandlen R, Toomre D, Schlessinger J, Mellman I. 2010. Spatial control of EGF receptor activation by reversible dimerization on living cells. *Nature* **464**: 783–787.

Citri A, Skaria KB, Yarden Y. 2003. The deaf and the dumb: The biology of ErbB-2 and ErbB-3. *Exp Cell Res* **284**: 54–65.

Coskun U, Grzybek M, Drechsel D, Simons K. 2011. Regulation of human EGF receptor by lipids. *Proc Natl Acad Sci* **108**: 9044–9048.

Digman MA, Dalal R, Horwitz AF, Gratton E. 2008. Mapping the number of molecules and brightness in the laser scanning microscope. *Biophys J* **94**: 2320–2332.

Endres NF, Das R, Smith AW, Arkhipov A, Kovacs E, Huang Y, Pelton JG, Shan Y, Shaw DE, Wemmer DE, et al. 2013. Conformational coupling across the plasma membrane in activation of the EGF receptor. *Cell* **152**: 543–556.

Ferguson KM, Berger MB, Mendrola JM, Cho HS, Leahy DJ, Lemmon MA. 2003. EGF activates its receptor by removing interactions that autoinhibit ectodomain dimerization. *Mol Cell* **11**: 507–517.

Friedländer E, Nagy P, Arndt-Jovin DJ, Jovin TM, Szöllösi J, Vereb G. 2005. Signal transduction of erbB receptors in trastuzumab (Herceptin) sensitive and resistant cell lines: Local stimulation using magnetic microspheres as assessed by quantitative digital microscopy. *Cytometry* **67A**: 161–171.

Gadella TWJ Jr, Jovin TM. 1995. Oligomerization of epidermal growth factor receptors on A431 cells studied by time-resolved fluorescence imaging microscopy: A stereochemical model for tyrosine kinase receptor activation. *J Cell Biol* **129**: 1543–1558.

Garcia-Morales P, Minami Y, Luong E, Klausner RD, Samelson LE. 1990. Tyrosine phosphorylation in T cells is regulated by phosphatase activity: Studies with phenylarsine oxide. *Proc Natl Acad Sci* **87**: 9255–9259.

Garrett TPJ, McKern NM, Lou MZ, Elleman TC, Adams TE, Lovrecz GO, Zhu HJ, Walker F, Frenkel MJ, Hoyne PA, et al. 2002. Crystal structure of a truncated epidermal growth factor receptor extracellular domain bound to transforming growth factor α. *Cell* **110:** 763–773.

Garrett TPJ, McKern NM, Lou MZ, Elleman TC, Adams TE, Lovrecz GO, Kofler M, Jorissen RN, Nice EC, Burgess AW, et al. 2003. The crystal structure of a truncated ErbB2 ectodomain reveals an active conformation, poised to interact with other ErbB receptors. *Mol Cell* **11:** 495–505.

George N. 2006. A new method for protein labeling with small molecules based on acyl carrier protein. Chemistry and chemical genetics, 149 pp. PhD thesis, École Polytechnique Fédérale de Lausanne, Lausanne.

Ghatak S, Misra S, Toole BP. 2005. Hyaluronan constitutively regulates ErbB2 phosphorylation and signaling complex formation in carcinoma cells. *J Biol Chem* **280:** 8875–8883.

Gilmour LMR, Macleod KG, McCaig A, Sewell JM, Gullick WJ, Smyth JF, Langdon SP. 2002. Neuregulin expression, function, and signaling in human ovarian cancer cells. *Clin Cancer Res* **8:** 3933–3942.

Grecco HE, Roda-Navarro P, Girod A, Hou J, Frahm T, Truxius DC, Pepperkok R, Squire A, Bastiaens PI. 2010. In situ analysis of tyrosine phosphorylation networks by FLIM on cell arrays. *Nat Methods* **2010:** 9.

Holbro T, Beerli RR, Maurer F, Koziczak M, Barbas CF, Hynes NE. 2003. The ErbB2/ErbB3 heterodimer functions as an oncogenic unit: ErbB2 requires ErbB3 to drive breast tumor cell proliferation. *Proc Natl Acad Sci* **100:** 8933–8938.

* Hunter T. 2014. The genesis of tyrosine phosphorylation. *Cold Spring Harb Perspect Biol* doi: 10.1101/cshperspect.a020644.

Ieguchi K, Fujita M, Ma Z, Davari P, Taniguchi Y, Sekiguchi K, Wang BZ, Takada YK, Takada Y. 2010. Direct binding of the EGF-like domain of neuregulin-1 to integrins (αvβ3 and α6β4) is involved in neuregulin-1/ErbB signaling. *J Biol Chem* **285:** 31388–31398.

Janssen-Heininger YM, Mossman BT, Heintz NH, Forman HJ, Kalyanaraman B, Finkel T, Stamler JS, Rhee SG, van der Vliet A. 2008. Redox-based regulation of signal transduction: Principles, pitfalls, and promises. *Free Radic Biol Med* **45:** 1–17.

Jura N, Endres NF, Engel K, Deindl S, Das R, Lamers MH, Wemmer DE, Zhang X, Kuriyan J. 2009. Mechanism for activation of the EGF receptor catalytic domain by the juxtamembrane segment. *Cell* **137:** 1293–1307.

Kozer N, Henderson C, Jackson JT, Nice EC, Burgess AW, Clayton AH. 2011. Evidence for extended YFP-EGFR dimers in the absence of ligand on the surface of living cells. *Phys Biol* **8:** 066002.

Kucherak OA, Oncul S, Darwich Z, Yushchenko DA, Arntz Y, Didier P, Mely Y, Klymchenko AS. 2010. Switchable nile red-based probe for cholesterol and lipid order at the outer leaflet of biomembranes. *J Am Chem Soc* **132:** 4907–4916.

Kusumi A, Suzuki KG, Kasai RS, Ritchie K, Fujiwara TK. 2011. Hierarchical mesoscale domain organization of the plasma membrane. *Trends Biochem Sci* **36:** 604–615.

Lee S-R, Kwon K-S, Kim S-R, Rhee SG. 1998. Reversible inactivation of protein-tyrosine phosphatase 1B in A431 cells stimulated with epidermal growth factor. *J Biol Chem* **273:** 15366–15372.

Lemmon MA, Bu ZM, Ladbury JE, Zhou M, Pinchasi D, Lax I, Engelman DM, Schlessinger J. 1997. Two EGF molecules contribute additively to stabilization of the EGFR dimer. *EMBO J* **16:** 281–294.

Li S, Schmitz KR, Jeffrey PD, Wiltzius JJ, Kussie P, Ferguson KM. 2005. Structural basis for inhibition of the epidermal growth factor receptor by cetuximab. *Cancer Cell* **7:** 301–311.

Lidke DS, Nagy P, Heintzmann R, Arndt-Jovin DJ, Post JN, Grecco H, Jares-Erijman EA, Jovin TM. 2004. Quantum dot ligands provide new insights into erbB/HER receptor-mediated signal transduction. *Nat Biotechnol* **22:** 198–203.

Lidke DS, Lidke KA, Rieger B, Jovin TM, Arndt-Jovin DJ. 2005. Reaching out for signals: Filopodia sense EGF and respond by directed retrograde transport of activated receptors. *J Cell Biol* **170:** 619–626.

Low-Nam ST, Lidke KA, Cutler PJ, Roovers RC, van Bergen en Henegouwen PMP, Wilson BS, Lidke DS. 2011. ErbB1 dimerization is promoted by domain co-confinement and stabilized by ligand binding. *Nat Struct Mol Biol* **18:** 1244–1249.

Lu C, Mi LZ, Schurpf T, Walz T, Springer TA. 2012. Mechanisms for kinase-mediated dimerization of the epidermal growth factor receptor. *J Biol Chem* **287:** 38244–38253.

Martin-Fernandez M, Clarke DT, Tobin MJ, Jones SV, Jones GR. 2002. Preformed oligomeric epidermal growth factor receptors undergo an ectodomain structure change during signaling. *Biophys J* **82:** 2415–2427.

Michailidis IE, Rusinova R, Georgakopoulos A, Chen Y, Iyengar R, Robakis NK, Logothetis DE, Baki L. 2011. Phosphatidylinositol-4,5-bisphosphate regulates epidermal growth factor receptor activation. *Pflugers Arch* **461:** 387–397.

Moriki T, Maruyama H, Maruyama IN. 2001. Activation of preformed EGF receptor dimers by ligand-induced rotation of the transmembrane domain. *J Mol Biol* **311:** 1011–1026.

Nagy P, Bene L, Balazs M, Hyun WC, Lockett SJ, Chiang NY, Waldman F, Feuerstein BG, Damjanovich S, Szollosi J. 1998. EGF-induced redistribution of erbB2 on breast tumor cells: Flow and image cytometric energy transfer measurements. *Cytometry* **32:** 120–131.

Nagy P, Jenei A, Kirsch AK, Szollosi J, Damjanovich S, Jovin TM. 1999. Activation-dependent clustering of the erbB2 receptor tyrosine kinase detected by scanning near-field optical microscopy. *J Cell Sci* **112:** 1733–1741.

Nagy P, Vereb G, Sebestyen Z, Horvath G, Lockett SJ, Damjanovich S, Park JW, Jovin TM, Szollosi J. 2002. Lipid rafts and the local density of ErbB proteins influence the biological role of homo- and heteroassociations of ErbB2. *J Cell Sci* **115:** 4251–4262.

Nagy P, Friedländer E, Tanner M, Kapanen AI, Carraway KL, Isola J, Jovin TM. 2005. Decreased accessibility and lack of activation of erbB2 in JIMT-1, a Herceptin-resistant, MUC-4-expressing breast cancer cell line. *Cancer Res* **65:** 473–482.

Nagy P, Claus J, Jovin TM, Arndt-Jovin DJ. 2010. Distribution of resting and ligand-bound ErbB1 and ErbB2 receptor tyrosine kinases in living cells using number and brightness analysis. *Proc Natl Acad Sci* **107**: 16524–16529.

Offterdinger M, Bastiaens PI. 2008. Prolonged EGFR signaling by ERBB2-mediated sequestration at the plasma membrane. *Traffic* **9**: 147–155.

Offterdinger M, Georget V, Girod A, Bastiaens PI. 2004. Imaging phosphorylation dynamics of the epidermal growth factor receptor. *J Biol Chem* **279**: 36972–36981.

Ogiso H, Ishitani R, Nureki O, Fukai S, Yamanaka M, Kim JH, Saito K, Sakamoto A, Inoue M, Shirouzu M, et al. 2002. Crystal structure of the complex of human epidermal growth factor and receptor extracellular domains. *Cell* **110**: 775–787.

Ozcan F, Klein P, Lemmon MA, Lax I, Schlessinger J. 2006. On the nature of low- and high-affinity EGF receptors on living cells. **103**: 5735–5740.

Park E, Baron R, Landgraf R. 2008. Higher-order association states of cellular ERBB3 probed with photo-cross-linkable aptamers. *Biochemistry* **47**: 11992–12005.

Red Brewer M, Choi SH, Alvarado D, Moravcevic K, Pozzi A, Lemmon MA, Carpenter G. 2009. The juxtamembrane region of the EGF receptor functions as an activation domain. *Mol Cell* **34**: 641–651.

Reynolds AR, Tischer C, Verveer PJ, Rocks O, Bastiaens PI. 2003. EGFR activation coupled to inhibition of tyrosine phosphatases causes lateral signal propagation. *Nat Cell Biol* **5**: 447–453.

Rhee SG, Bae YS, Lee SR, Kwon J. 2000. Hydrogen peroxide: A key messenger that modulates protein phosphorylation through cysteine oxidation. *Sci STKE* **2000**: e1.

Rubin I, Yarden Y. 2001. The basic biology of HER2. *Ann Oncol* **12**: 3–8.

Saffarian S, Li Y, Elson EL, Pike LJ. 2007. Oligomerization of the EGF receptor investigated by live cell fluorescence intensity distribution analysis. *Biophys J* **93**: 1021–1031.

Sako Y, Minoghchi S, Yanagida T. 2000. Single-molecule imaging of EGFR signalling on the surface of living cells. *Nat Cell Biol* **2**: 168–172.

Salas-Vidal E, Lomeli H. 2004. Imaging filopodia dynamics in the mouse blastocyst. *Dev Biol* **265**: 75–89.

Santos Mde S, Naal RM, Baird B, Holowka D. 2013. Inhibitors of PI(4,5)P2 synthesis reveal dynamic regulation of IgE receptor signaling by phosphoinositides in RBL mast cells. *Mol Pharmacol* **83**: 793–804.

Schmitz KR, Ferguson KM. 2009. Interaction of antibodies with ErbB receptor extracellular regions. *Exp Cell Res* **315**: 659–670.

* Shibuya M. 2013. VEGFR and type-V RTK activation and signaling. *Cold Spring Harb Perspect Biol* doi: 10.1101/cshperspect.a009092.

Smirnova T, Zhou ZN, Flinn RJ, Wyckoff J, Boimel PJ, Pozzuto M, Coniglio SJ, Backer JM, Bresnick AR, Condeelis JS, et al. 2011. Phosphoinositide 3-kinase signaling is critical for ErbB3-driven breast cancer cell motility and metastasis. *Oncogene* **4**: 275.

* Song S, Rosen KM, Corfas G. 2013. Biological function of nuclear receptor tyrosine kinase action. *Cold Spring Harb Perspect Biol* **5**: a009001.

Songtawee N, Gleeson MP, Choowongkomon K. 2013. Computational study of EGFR inhibition: Molecular dynamics studies on the active and inactive protein conformations. *J Mol Model* **19**: 497–509.

Stamos J, Sliwkowski MX, Eigenbrot C. 2002. Structure of the epidermal growth factor receptor kinase domain alone and in complex with a 4-anilinoquinazoline inhibitor. *J Biol Chem* **277**: 46265–46272.

Szabó A, Horváth G, Szöllősi J, Nagy P. 2008. Quantitative characterization of the large-scale association of ErbB1 and ErbB2 by flow cytometric homo-FRET measurements. *Biophys J* **95**: 2086–2096.

Tanaka KA, Suzuki KG, Shirai YM, Shibutani ST, Miyahara MS, Tsuboi H, Yahara M, Yoshimura A, Mayor S, Fujiwara TK, et al. 2010. Membrane molecules mobile even after chemical fixation. *Nat Methods* **7**: 865–866.

Tynan CJ, Clarke DT, Coles BC, Rolfe DJ, Martin-Fernandez ML, Webb SE. 2012. Multicolour single molecule imaging in cells with near infra-red dyes. *PLoS ONE* **7**: e36265.

Tyson DR, Larkin S, Hamai Y, Bradshaw RA. 2003. PC12 cell activation by epidermal growth factor receptor: Role of autophosphorylation sites. *Int J Dev Neurosci* **21**: 63–74.

Verveer PJ, Wouters FS, Reynolds AR, Bastiaens PI. 2000. Quantitative imaging of lateral ErbB1 receptor signal propagation in the plasma membrane. *Science* **290**: 1567–1570.

Vivero-Pol L, George N, Krumm H, Johnsson K, Johnsson N. 2005. Multicolor imaging of cell surface proteins. *J Am Chem Soc* **127**: 12770–12771.

* Volinsky N, Kholodenko BN. 2013. Complexity of receptor tyrosine kinase signaling processing. *Cold Spring Harb Perspect Biol* **5**: a009043.

Wang Z, Zhang L, Yeung TK, Chen X. 1999. Endocytosis deficiency of epidermal growth factor (EGF) receptor-ErbB2 heterodimers in response to EGF stimulation. *Mol Biol Cell* **10**: 1621–1636.

Wang J, Boriskina SV, Wang H, Reinhard BM. 2011. Illuminating epidermal growth factor receptor densities on filopodia through plasmon coupling. *ACS Nano* **5**: 6619–6628.

Yang S, Raymond-Stintz MA, Ying W, Zhang J, Lidke DS, Steinberg SL, Williams L, Oliver JM, Wilson BS. 2007. Mapping ErbB receptors on breast cancer cell membranes during signal transduction. *J Cell Sci* **120**: 2763–2773.

Yarden Y, Schlessinger J. 1987. Self-phosphorylation of epidermal growth factor receptor: Evidence for a model of intermolecular allosteric activation. *Biochemistry* **26**: 1434–1442.

Yin J, Lin A, Golan D, Walsh C. 2006. Site-specific protein labeling by Sfp phosphopantetheinyl transferase. *Nat Protoc* **1**: 280–285.

Yoon SJ, Nakayama K, Hikita T, Handa K, Hakomori SI. 2006. Epidermal growth factor receptor tyrosine kinase is modulated by GM3 interaction with *N*-linked GlcNAc termini of the receptor. *Proc Natl Acad Sci* **103**: 18987–18991.

Yu X, Miyamoto S, Mekada E. 2000. Integrin α2β1-dependent EGF receptor activation at cell-cell contact sites. *J Cell Sci* **113**: 2139–2147.

Zacharias DA, Violin JD, Newton AC, Tsien RY. 2002. Partitioning of lipid-modified monomeric GFPs into membrane microdomains of live cells. *Science* **296**: 913–916.

Zhang Y, Opresko L, Shankaran H, Chrisler WB, Wiley HS, Resat H. 2009. HER/ErbB receptor interactions and signaling patterns in human mammary epithelial cells. *BMC Cell Biol* **10**: 78.

Ziomkiewicz I, Loman A, Klement R, Fritsch C, Klymchenko A, Bunt G, Jovin TM, Arndt-Jovin DJ. 2013. Dynamic conformational transitions of the EGF receptor in living mammalian cells determined by FRET and fluorescence lifetime imaging microscopy. *Cytometry A* **83A**: 794–805.

Structural and Functional Properties of Platelet-Derived Growth Factor and Stem Cell Factor Receptors

Carl-Henrik Heldin and Johan Lennartsson

Ludwig Institute for Cancer Research, Uppsala University, SE-751 24 Uppsala, Sweden

Correspondence: c-h.heldin@licr.uu.se

The receptors for platelet-derived growth factor (PDGF) and stem cell factor (SCF) are members of the type III class of PTK receptors, which are characterized by five Ig-like domains extracellularly and a split kinase domain intracellularly. The receptors are activated by ligand-induced dimerization, leading to autophosphorylation on specific tyrosine residues. Thereby the kinase activities of the receptors are activated and docking sites for downstream SH2 domain signal transduction molecules are created; activation of these pathways promotes cell growth, survival, and migration. These receptors mediate important signals during the embryonal development, and control tissue homeostasis in the adult. Their overactivity is seen in malignancies and other diseases involving excessive cell proliferation, such as atherosclerosis and fibrotic diseases. In cancer, mutations of PDGF and SCF receptors—including gene fusions, point mutations, and amplifications—drive subpopulations of certain malignancies, such as gastrointestinal stromal tumors, chronic myelomonocytic leukemia, hypereosinophilic syndrome, glioblastoma, acute myeloid leukemia, mastocytosis, and melanoma.

The type III tyrosine kinase receptor family consists of platelet-derived growth factor (PDGF) receptor α and β, stem cell factor (SCF) receptor (Kit), colony-stimulating factor-1 (CSF-1) receptor, and Flt-3 (Blume-Jensen and Hunter 2001). Members of this receptor family are characterized by five Ig-like domains in their extracellular part, a single transmembrane domain, and an intracellular part consisting of a rather well-conserved juxtamembrane domain, a tyrosine kinase domain with a characteristic inserted sequence without homology with kinases, and a less well-conserved carboxy-terminal tail. The ligands for these receptors are all dimeric molecules, and on binding they induce receptor dimerization. Although the overall mechanisms for the activation of the type III tyrosine kinase receptors and the signaling pathways they induce are similar, the receptors are expressed on different cell types and thus have different functions in vivo.

Here we will describe the structural and functional properties of the PDGF receptors and Kit.

PDGF RECEPTORS

Ligand-Binding Specificities of PDGF Receptors

The PDGF family consists of five members (i.e., disulfide-bonded dimers of homologous A-, B-, C-, and D-polypeptide chains, and the AB heterodimer) (Heldin and Westermark 1999). The PDGF-α receptor binds all PDGF chains except the D chain, whereas the β receptor binds PDGF-B and -D; thus, the different PDGF isoforms can induce $\alpha\alpha$-, $\alpha\beta$-, or $\beta\beta$-receptor dimers (Fig. 1). The ligand-binding sites are located in Ig-like domains 2 and 3 (Heidaran et al. 1990; Lokker et al. 1997; Miyazawa et al. 1998; Shim et al. 2010); however, ligand-induced receptor dimerization is stabilized by direct receptor–receptor interactions in Ig-like domains 4 and 5 (Omura et al. 1997; Yang et al.

2008). The latter interactions are important because they orient the receptors so that their activation by autophosphorylation in *trans* is facilitated. Binding of vascular endothelial growth factor (VEGF)-A to PDGFR-α and PDGFR-β has been reported (Ball et al. 2007), but the physiological significance of this finding remains to be elucidated.

Ligand stimulation results in homo- as well as heterodimerization of PDGF-α and -β receptors; the different dimeric receptor complexes have overlapping but slightly different signaling capacities (see further below). However, PDGF receptors can also form complexes with other tyrosine kinase receptors, such as the epidermal growth factor (EGF) receptor (Saito et al. 2001) and fibroblast growth factor (FGF) receptor-1 (Faraone et al. 2006), but also with non-kinase receptors, such as integrins (Sundberg

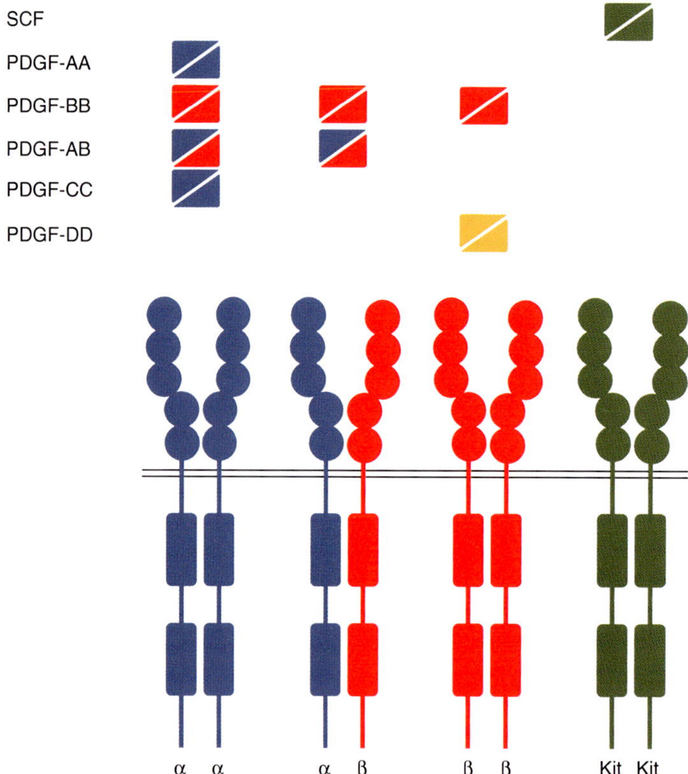

Figure 1. Ligand-binding specificities of PDGF and SCF receptors. The different ligands are depicted above the respective receptor dimers they bind to. Binding of PDGF-CC and PDGF-DD to $\alpha\beta$-heterodimeric PDGF receptors have also been described, but the functional significance of such complexes remains to be determined.

 Cite this article as *Cold Spring Harb Perspect Biol* doi: 10.1101/cshperspect.a009100

and Rubin 1996; Schneller et al. 1997), CD44 (Li et al. 2006), the low-density lipoprotein receptor-related protein (LRP) (Boucher et al. 2002; Loukinova et al. 2002), and the poliovirus receptor Necl-5 (Minami et al. 2010). Such interactions modulate signaling via PDGF receptors.

Activation of PDGF Receptor Kinases

PDGF-induced receptor dimerization leads to autophosphorylation of certain tyrosine residues in the intracellular parts of the receptors. Thus, the α and β receptors have 10 and 11 autophosphorylation sites, respectively (Fig. 2) (Heldin et al. 1998). The autophosphorylation serves two important functions: It leads to changes in the conformation of the intracellular parts of the receptors promoting their activation, and it provides docking sites for SH2-domain-containing signal transduction molecules.

There are at least three mechanisms involved in activation of PDGF receptor kinases.

Like most tyrosine kinase receptors (Hubbard 1997), the PDGF receptors are autophosphorylated in the activation loop of the kinases (residues Tyr849 and Tyr857 in the α and β receptors, respectively). Phosphorylation of this residue of the β receptor has been shown to be necessary for full activation of the receptor kinase (Baxter et al. 1998). Probably phosphorylation causes a change in conformation of the activation loop, which opens up the active site of the kinase and allows access of ATP and protein substrate. Moreover, truncation of the carboxy-terminal tail of the β receptor causes receptor activation (Chiara et al. 2004). This suggests that the carboxyl terminus in the resting state is folded over the kinase domain keeping the kinase inactive; autophosphorylation in the carboxyl terminus is likely to relieve the inhibition. Finally, in the resting state, the juxtamembrane domain of several tyrosine kinase receptors has been shown by X-ray crystallography to be folded over and inhibit the kinase domain; autophosphorylation causes a change in conformation, which

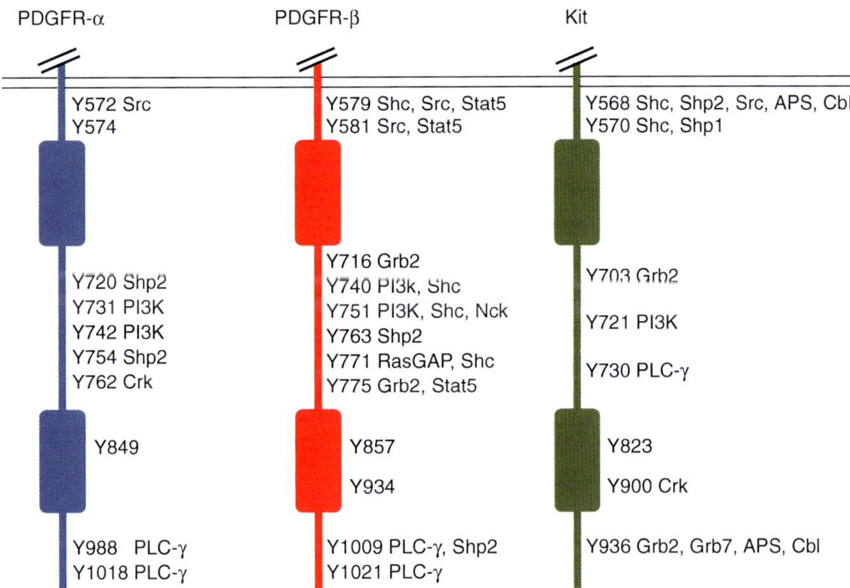

Figure 2. Binding of SH2-containing signaling molecules to phosphorylation sites in PDGF and SCF receptors. The known phosphorylated tyrosine residues and the molecules that bind to them are indicated. Y849, Y857, and Y823 in the α receptor, β receptor, and Kit, respectively, are located in the activation loops of the kinase domains; no molecules are known to bind to these phosphorylation sites. Y934 and Y900 in the β receptor and Kit, respectively, are not autophosphorylation sites, but phosphorylated by Src.

relieves the inhibition (Wybenga-Groot et al. 2001; Hubbard 2004). A similar inhibitory function has been shown in the PDGF-β receptor (Irusta et al. 2002). Moreover, in the context of oncogenic fusion proteins of the PDGF receptors, the juxtamembrane domains have been shown to have an inhibitory function (see further below), supporting the notion that the juxtamembrane domain has an inhibitory function also in full-length PDGF receptors. Together, these mechanisms cooperate to keep the kinase inactive; autophosphorylation of several tyrosine residues is necessary for full activation of the receptor kinase.

Activation of Intracellular Signal Transduction Pathways by PDGF Receptors

The second important function of autophosphorylation of the PDGF receptors is to allow binding of signaling molecules containing SH2 domains, which recognize phosphorylated tyrosine residues (Heldin et al. 1998). Because different SH2 domains have different preferences regarding the three to six amino acid residues downstream from the phosphorylated tyrosine, there is a certain specificity in binding. The PDGF receptors have been reported to bind about 10 different families of SH2-domain-containing molecules, which initiates the activation of different signaling pathways (Fig. 2). Because the autophosphorylation pattern of PDGF-α and -β receptors differ depending on whether the receptors occur in homo- or heterodimeric complexes, each of the three dimeric PDGF receptor complexes have distinct signaling properties (Ekman et al. 1999).

Certain of the SH2-domain signaling molecules that bind to PDGF receptors have intrinsic enzymatic activities (i.e., members of the Src family of tyrosine kinases, the GTPase-activating protein [GAP] for Ras, the tyrosine phosphatase SHP-2, and phospholipase C-γ [PLC-γ]) (reviewed in Heldin and Westermark 1999). The respective enzymatic activities are activated by binding to the receptors or by phosphorylation by the receptor kinases. Alternatively, the enzymes are constitutively active and just

brought to the inner leaflet of the plasma membrane by the activated receptors.

There are also examples of SH2-domain-containing adaptor molecules, including Nck, Shc, and Crk, which bind to activated PDGF receptors. They act by mediating interactions with different downstream signaling molecules. Some of them form stable complexes with enzymes, such as Grb2, which form a complex with the nucleotide exchange molecule SOS1 activating Ras. Moreover, the α- or β-regulatory p85 subunits of the phosphatidylinositol 3′-kinase bind to the receptors and to the α- or β-p110 catalytic subunits. In addition, certain members of the STAT family of transcription factors bind to and are activated by PDGF receptors (reviewed by Heldin and Westermark 1999).

The intracellular parts of the PDGF receptors also interact with certain signaling molecules independent of autophosphorylation (e.g., the PDZ-domain protein NHERF), which binds to the end of the carboxy-terminal tail of the PDGF receptors and enhances receptor signaling (Maudsley et al. 2000; Demoulin et al. 2003; Takahashi et al. 2006; Theisen et al. 2007), and the adaptor molecule Alix, which binds to the region around Tyr1021 of the carboxy-terminal tail and facilitates binding of the ubiquitin ligase Cbl (Lennartsson et al. 2006).

In order for initiation of signaling via PDGF receptors to be efficient, tyrosine phosphatases need to be inactivated (Sundaresan et al. 1995). This is achieved by a PI3-kinase-dependent oxidation of a cysteine residue in the active site of phosphatases (Bae et al. 2000).

Much effort has been put into the elucidation of which signaling pathways mediate the various effects of PDGF on cells (i.e., cell proliferation, survival, chemotaxis, and actin reorganization). Although cell-type differences have been reported, in general, PI3 kinase has been found to be important for the antiapoptotic and motility responses of PDGF. Src via activation of the transcription factor Myc, and Ras via activation of the Erk MAP kinase pathway, are important for the growth-stimulating effects. It should be noted, however, that there is an extensive cross talk between different signaling pathways. Thus, each of the many signaling pathways induced by the activated receptor can, to different extents and in a cell-type-specific manner, contribute to most of the cellular effects of PDGF.

Modulation of Signaling via PDGF Receptors

Both αα- and ββ-receptor complexes induce powerful mitogenic signals. However, whereas ββ-receptor homodimers and αβ heterodimers induce chemotaxis, the αα homodimer inhibits chemotaxis, at least in certain cell types (Eriksson et al. 1992). The αβ-heterodimeric receptor complex appears to induce the most potent mitogenic signal. A mechanism for this difference could be that Tyr771, to which RasGAP binds, is efficiently phosphorylated in the ββ homodimer, but not in the αβ heterodimer (Ekman et al. 1999). Thus, RasGAP, which deactivates Ras, cannot bind to the αβ-receptor heterodimer leading to a more efficient activation of Ras by the αβ-receptor heterodimer.

The simultaneous binding to activated PDGF receptors of the Grb2/SOS1 complex, which activates Ras, and of RasGAP, which inactivates Ras, provides an example of how signaling via PDGF receptors can be modulated. Another example is the binding of the tyrosine phosphatase SHP-2, which dephosphorylates the β receptor and its substrates (Lechleider et al. 1993). However, in addition to negatively modulating PDGF receptor signaling via dephosphorylation, SHP-2 also positively contributes to signaling through dephosphorylation of the carboxy-terminal tyrosine residue in Src, whereby Src is activated, and by functioning as an adaptor, which can bind Grb2/SOS1, thus promoting activation of Ras (Dance et al. 2008).

Internalization and Sorting of PDGF Receptors

Following ligand binding, PDGF receptors are accumulated in coated pit areas at the cell membrane, and then internalized in a clathrin- and dynamin-dependent manner, in a process that partly depends on the kinase activity of the receptors (Sorkin et al. 1991). Signaling continues in endosomes (Wang et al. 2004), until the pH decreases enough to cause dissociation of PDGF

from its receptors. Most of the internalized PDGF receptors are degraded by fusion of endosomes with multivesicular bodies and lysosomes, or by degradation in proteasomes, processes that are promoted by polyubiquitination of the receptors (Heldin et al. 1982; Mori et al. 1992). Ubiquitination of the PDGF-β receptor can be performed by the ubiquitin ligase Cbl (Miyake et al. 1999); degradation of Cbl is promoted by the adaptor molecule Alix, which binds to the PDGF-β receptor, thereby preventing ubiquitination of the receptor and thus stabilizing it (Lennartsson et al. 2006).

Although most of PDGF receptors are degraded after ligand-induced internalization, there are mechanisms that affect sorting and promote recycling of the receptor to the plasma membrane. One such mechanism was shown to involve overactivation of PLC-γ and its downstream effectors PKCα in cells deficient of the tyrosine phosphatase TC-PTP (Karlsson et al. 2006); TC-PTP dephosphorylates preferentially Tyr1021 in the PDGF-β receptor, which is the binding site of PLC-γ. The overactivated PLC-γ was found to promote recycling in a protein kinase C-dependent manner (Hellberg et al. 2009). Another mechanism was found to operate in Ras-transformed fibroblasts; in these cells, PI3-kinase is overactivated leading to internalization of receptors via macropinocytosis, which is accompanied by increased recycling (Schmees et al. 2012). In both these cases, the induction of recycling was associated with an increased PDGF signal strength and duration. Thus, modulation of intracellular sorting mechanisms can affect PDGF signaling.

Function of PDGF and PDGF Receptors during Embryonal Development

Primarily based on gene knockout studies in mice, PDGF and PDGF receptors have been shown to have important roles to promote proliferation, migration, and differentiation of specific cell types during the embryonal development (reviewed by Andrae et al. 2008). A common theme that has emerged from these studies is that PDGF isoforms, secreted by epithelial or endothelial cells, act in a paracrine manner on nearby mesenchymal cells, such as fibroblasts, pericytes, and smooth muscle cells (Hoch and Soriano 2003; Andrae et al. 2008).

Knockout of PDGFR-α and PDGF-A was found to affect early mesenchymal derivatives in both embryo and extraembryo tissues, and a proportion of these mice die before or at embryonic day 10.5 (Hoch and Soriano 2003). Knockout of PDGFR-α also causes defects in neural crest mesenchyme derivatives, including the cardiac outflow tract, the thymus and skeletal components in the facial and other regions, and in the development of the palate and teeth (Soriano 1997; Tallquist et al. 2000; Tallquist and Soriano 2003; Xu et al. 2005).

PDGF-A knockout mice that survive birth develop lung emphysema, because PDGFR-α-positive alveolar myofibroblast precursors do not migrate to the alveolar saccules (Boström et al. 1996; Lindahl et al. 1997b). Knockout of PDGF-A or PDGFR-α in mice also leads to abnormal development of gastrointestinal villi (Karlsson et al. 2000), skin blistering (Soriano 1997), reduced hair development (Karlsson et al. 1999), and defect development of Leydig cells of the testis (Gnessi et al. 2000; Brennan et al. 2003). In each case, the interaction between the PDGFR-α expressed by mesenchymal cells of different kinds, and the ligand expressed by neighboring epithelial cells, is perturbed.

PDGFR-α also mediates signals needed for proliferation and differentiation of oligodendrocyte progenitor cells (Calver et al. 1998) and retinal astrocytes (Fruttiger et al. 1996), and determines the number of oligodendrocyte progenitor cells in the adult brain (Woodruff et al. 2004). PDGF-AA is constitutively secreted from neuronal cell bodies but not from axons (Fruttiger et al. 2000).

PDGFR-β is expressed on pericytes and vascular smooth muscle cells and mediates recruitment of these cells to newly formed vessels in response to PDGF-BB secreted by endothelial cells (Lindahl et al. 1997a; Hellström et al. 1999; Bjarnegård et al. 2004), and in particular by the tip cells that lead the angiogenic sprout (Gerhardt et al. 2003). PDGFR-β knockout embryos progressively develop abnormal glomeruli in the kidney owing to defect development

of mesangial cells (Levéen et al. 1994; Soriano 1994; Lindahl et al. 1998), capillary microaneurysm (Lindahl et al. 1997a), cardiac defects (Van den Akker et al. 2008), and placental defects (Ohlsson et al. 1999), and die at E16–E19 from hemorrhage. A role for PDGFR-β in the early development of hematopoietic/endothelial precursors has also been shown; activation of PDGFR-β on these cells drives differentiation toward endothelial cells (Rolny et al. 2006).

It is thus clear that PDGFR-α and PDGFR-β have different functions during embryonal development. To determine whether the differences are owing to different expression patterns or to different signaling capacities, the intracellular parts of the receptors were swapped (Klinghoffer et al. 2001). Whereas loss of the cytoplasmic part of PDGFR-α could be rescued by the cytoplasmic part of PDGFR-β, loss of the intracellular part of PDGFR-β was only partly rescued by the intracellular part of PDGFR-α. A partial rescue was also obtained when the PDGFR-α kinase domain was replaced with a more distant kinase domain (Hamilton et al. 2003). These findings suggest that both expression patterns and signaling capacities account for the differences in function of the two PDGF receptors (Klinghoffer et al. 2001).

Function of PDGF and PDGF Receptors in the Adult

In the adult, activation of PDGFR-β controls the intestinal fluid pressure of tissues and prevents formation of edema (Rodt et al. 1996). A probable mechanism is that fibroblasts and myofibroblasts, which express PDGFR-β, make contact with extracellular components via their integrins; activation of PDGFR-β induces contraction of the cells, which controls the interstitial fluid pressure (Lidén et al. 2006).

PDGF has also been shown to stimulate wound healing (Robson et al. 1992). PDGF receptors are expressed by several cell types involved in wound healing, such as fibroblasts, smooth muscle cells, neutrophils, and macrophages; on PDGF stimulation these cells are recruited to the wounded area (e.g., by PDGF released from platelets). PDGF also contributes to wound healing by stimulating the production of different matrix molecules (reviewed by Heldin and Westermark 1999).

A specific function for PDGFR-α to promote proliferation of insulin-promoting β cells in juvenile pancreatic islets was recently elucidated (Chen et al. 2011).

Role of PDGF Receptor Activation in Diseases

PDGF receptor activation has been observed in cancer and in other diseases involving excess cell proliferation, such as atherosclerosis and fibrotic conditions.

Overactivity of PDGF Receptors in Tumor Cells

Certain malignancies are characterized by mutations in PDGF receptor genes. Thus, 5% of gastrointestinal stromal tumors (GIST) show point mutations in the PDGFR-α gene (Heinrich et al. 2003); in this tumor type, mutations in the Kit gene are even more common (see below). The mutations affect the control mechanisms involved in keeping the receptor kinase inactive and lead to a constitutively active kinase. Similar activating point mutations in PDGFR-α have also been observed in hypereosinophilic syndrome (Elling et al. 2011).

In chronic myelomonocytic leukemia (CMML) the intracellular part of the PDGFR-β gene has been found to be fused to the TEL gene (Golub et al. 1994) or other genes that have in common that they encode proteins that can dimerize or oligomerize (Magnusson et al. 2001). Similarly, the PDGFR-α gene is fused to the FIP1L1 gene in hypereosinophilic syndrome (Cools et al. 2003; Griffin et al. 2003) and in systemic mastocytosis (Pardanani et al. 2003). The resulting fusion proteins have constitutively active kinases as a result of juxtaposition of the kinases of the receptors, as well as by the loss of regulatory sequences in the juxtamembrane (Stover et al. 2006) and transmembrane (Toffalini et al. 2010) domains.

The PDGFR-α gene has been found to be amplified in subsets of glioblastomas (Fleming et al. 1992; Kumabe et al. 1992; Puputti et al. 2006), anaplastic oligodendrogliomas (Smith et al. 2000), esophageal squamous cell carcinoma (Arai et al. 2003), and pulmonary artery intimal sarcoma (Zhao et al. 2002). The increased number of receptors on such cells makes the cells very sensitive to PDGF stimulation; moreover, at a sufficiently high receptor density, ligand-independent activation may occur. In addition, an activated deletion mutant of PDGFR-α has been found in a human glioblastoma (Clarke and Dirks 2003).

PDGF receptors may be activated also as a consequence of mutation of ligand genes. Thus, in the rare skin tumor dermatofibrosarcoma protuberans (DFSP), the PDGF-B gene is fused to the collagen 1A1 gene (Simon et al. 1997; O'Brien et al. 1998), resulting in the production of large amounts of a fusion protein, which after processing becomes similar to mature PDGF-BB that activates its receptors in an autocrine manner (Shimizu et al. 1999).

Epithelial tumors can undergo epithelial-mesenchymal transition (EMT), whereby they lose their epithelial characteristics and start to express mesenchymal components, such as PDGF receptors (Thiery et al. 2009). Thus, whereas epithelial tumors generally do not respond to PDGF, they may do so after they have undergone EMT (Jechlinger et al. 2003). EMT correlates with increased invasiveness and metastasis; interestingly, inhibition of PDGF signaling was shown to inhibit metastasis in mouse models for breast cancer (Jechlinger et al. 2006), hepatocellular carcinoma (Gotzmann et al. 2006), and prostate cancer (Dolloff et al. 2005; Russell et al. 2009). PDGFR-α appears to be more important than PDGFR-β in the promotion of metastasis of epithelial tumors.

A malignancy-dependent increase in expression of PDGF isoforms and receptors has been observed in glioblastoma tumors. Thus, in this tumor type, PDGF appears to be involved in autocrine and paracrine stimulation, both in the tumor cells via PDGFR-α and in cells of the stroma via PDGFR-β (Hermanson et al. 1992). Moreover, in basal cell carcinoma, activation of the sonic hedgehog signaling pathway induces the expression of PDGFR-α (Xie et al. 2001).

Activation of PDGF Receptors in the Stromal Compartment of Tumors

In addition to direct effects on the tumor cells, PDGF made by tumor cells and other cell types in solid tumors acts in a paracrine manner on various cell types in the vasculature and in the interstitial stroma of tumors (Pietras et al. 2008). Thus, PDGF promotes tumor angiogenesis by stimulating perivascular progenitor cells, pericytes, and vascular smooth muscle cells (Song et al. 2005), and by recruiting protumorigenic inflammatory cells. Moreover, overactivity of PDGF contributes to the increased interstitial fluid pressure that characterizes a majority of solid tumors, which leads to a decreased transcapillary transport and is therefore an obstacle in the treatment of tumors with chemotherapeutical drugs (Heldin et al. 2004).

Activation of PDGF Receptors in Atherosclerosis

PDGF isoforms and receptors occur at increased levels in atherosclerotic lesions. According to the response to injury hypothesis, PDGF isoforms released by platelets at sites of endothelial cell injury stimulate vascular smooth muscle cells to migrate from the vessel media into the intima and to proliferate (Ross 1993). Chronic inflammation involving release of PDGF from different types of immune cells also contributes to the narrowing of the vessel lumen (Hansson 2005). PDGF and other growth factors and cytokines released then maintain the local inflammation in a vicious circle.

Activation of PDGF Receptors in Fibrotic Diseases

Increased PDGF signaling has also been linked to various fibrotic conditions, such as lung fibrosis, myelofibrosis, glomerulonephritis, and liver cirrhosis. During chronic inflammation, activated macrophages secrete PDGF isoforms as well as other inflammatory cytokines, which up-regulate PDGF receptors on mesenchymal cells, thus promoting their proliferation and production of matrix molecules (Bonner 2004). In support of the notion that overactive PDGF signaling drives fibrosis, knockin of constitutively active mutants of PDGFR-α was found to lead to increased connective tissue growth and a progressive fibrotic phenotype in multiple organs (Olson and Soriano 2009). Knockin of a similar PDGFR-β mutant caused an enhanced wound healing response in the skin and the liver (Krampert et al. 2008).

PDGF Receptors as Binders of Viruses

PDGFR-α has been shown to be a receptor for adeno-associated virus type 5 (Di Pasquale et al. 2003) and cytomegalovirus (Soroceanu et al. 2008). Thus, PDGFR-α facilitates infections by these viruses.

DEVELOPMENT AND CLINICAL USE OF PDGF ANTAGONISTS

Because overactivity of PDGF signaling is connected to many serious diseases, efforts have been made to develop antagonists of PDGF signaling. Several types of inhibitors are now available, including DNA aptamers binding PDGF isoforms, inhibitory antibodies against the receptors, and low molecular weight inhibitors of PDGF receptor kinases (e.g., imatinib, sunitinib, and sorafenib) (Demetri 2011). Beneficial effects have been observed by treatment of the rare malignancies that are driven by mutations in genes for PDGF and PDGF receptors (i.e., CMML, DFSP, hypereosinophilic syndrome, and GIST). However, treatment of glioblastoma that also often is characterized by overactivity of PDGF receptors, with PDGF antagonists, has been less successful; probably these tumors have undergone too many other genetic or epigenetic alterations and inhibition of PDGF receptor signaling is therefore not enough. There are indications that anti-PDGF treatment could be of more general application for the targeting of the stroma compartment of solid tumors, to inhibit metastasis (Pietras et al. 2003; Catena et al. 2010), and to lower the interstitial fluid pressure and thereby increase the transcapillary flow and drug delivery (Heldin et al. 2004). The side effects of anti-PDGF therapy have been relatively mild, but include a tendency to develop edema, reflecting the important role of PDGF in regulation of the interstitial fluid pressure (Rodt et al. 1996), and heart failure, reflecting an important role of PDGF in stress-induced cardiac angiogenesis (Chintalgattu et al. 2010). In patients with advanced ovarian cancer, treatment with the inhibitory Fab fragment CDP860 led to the development of significant ascites (Jayson et al. 2005).

STEM CELL FACTOR RECEPTOR (Kit)

Activation of Kit by SCF

The biologically active SCF is a homodimeric protein that is primarily produced by fibroblasts and endothelial cells, and exists both as a soluble and membrane-bound form owing to alternative splicing of exon 6 (Huang et al. 1992). Both forms of SCF are initially produced as transmembrane proteins, but the protein containing exon 6 becomes cleaved into the soluble SCF and the form lacking this exon remains membrane associated. Several proteases have been suggested to be involved in the processing of SCF, including MMP9, members of the ADAM family, and Chymase-1 (Longley et al. 1997; Heissig et al. 2002; Kawaguchi et al. 2007).

The SCF receptor is also called Kit, because it was first identified as an oncoprotein derived from the Hardy-Zuckerman 4 feline sarcoma virus (Besmer et al. 1986). It is expressed in primitive hematopoietic cells and mast cells, but also in melanocytes and basal cells in the skin, germs cells, interstitial cells of Cajal, and in certain regions of the brain. There are four human isoforms of Kit. Two splice forms are characterized by the presence or absence of a four-amino-acid sequence in the extracellular juxtamembrane region (Reith et al. 1991). These two splice forms display different signaling abilities with the shorter form being more strongly tyrosine phosphorylated and more powerful in activation of downstream signaling, whereas the longer Kit splice form signals less intensely but more persistently (Caruana et al. 1999; Voytyuk et al. 2003). Kit isoforms are also characterized by the absence or presence of a serine residue in the kinase insert region (Crosier et al. 1993).

The process by which SCF activates Kit has been studied both at a biochemical and structural level, and occurs in a manner similar to that described above for the PDGF receptor; the dimeric SCF ligand brings two Kit monomers in close proximity of each other, which enables interactions between the extracellular domains of the receptors generating a stable dimer (Fig. 1) (Lemmon et al. 1997; Zhang et al. 2000; Yuzawa et al. 2007).

The crystal structure of the Kit kinase domain revealed that the juxtamembrane region is inserted between the two lobes of the kinase domain thereby suppressing kinase activity, and that this inhibition is released on phosphorylation of tyrosine residues within the juxtamembrane segment (Mol et al. 2003, 2004). A kinetic study of the autophosphorylation process showed that the juxtamembrane region is first to be phosphorylated, consistent with the role of this region in suppressing the kinase activity (DiNitto et al. 2010).

Activation of Intracellular Pathways by Kit

The intracellular region of Kit contains seven autophosphorylation sites that are involved in the activation of the Kit kinase and can act as docking sites for intracellular signaling proteins (Fig. 2). The signaling pathways activated downstream from Kit overlap to a large extent with those described for the PDGF receptor (see above), and include PI3-kinase, Src kinases, MAP kinase pathways, and phospholipase C and D. Furthermore, it is well established that there exists a comprehensive cross talk between different pathways. For example, Src kinases contribute to Erk1/2 activation, and Src in combination with PI3-kinase promotes Jnk activation, in response to SCF (Timokhina et al. 1998; Lennartsson et al. 1999).

Attempts to study the role of individual signaling proteins include mutation of all tyrosine residues in the intracellular region of Kit and then adding them back one at a time. Using this approach, it was found that the Src docking site on Kit is important for the migratory, survival, and, partially, the proliferative response to SCF, whereas restoring coupling to PI3-kinase only had a small effect on the survival and migration of cells (Hong et al. 2004). In addition, by mutating individual tyrosine residues in Kit, it was found that Src and PI3-kinase binding are essential for cell migration (Ueda et al. 2002). These studies identify Src and PI3-kinase signaling as especially important in SCF-induced responses in vitro. Also, in vivo studies have confirmed the role of the Src and PI3-kinase in Kit-mediated functions. Mice expressing Kit lacking the Src binding site display primarily defects in the immune system, whereas lack of the PI3-kinase binding site generated fertility deficiencies (Blume-Jensen et al. 2000; Kissel et al. 2000; Agosti et al. 2004; Kimura et al. 2004).

The absence of clear functions of the other autophosphorylation sites in Kit probably reflects redundancy masking the role of binding of signaling proteins to these sites. Furthermore, the role of individual signaling pathways may vary between different cell types illustrated by the specific defects described above in mice expressing different Kit mutations, in which some tissues are strongly affected, whereas others are not, although they all express the same Kit mutant.

Down-Regulation of Kit

There are several mechanisms that act in parallel to down-regulate Kit signaling ensuring an appropriate signal intensity and duration. These include Kit ubiquitination and internalization, dephosphorylation, and PKC-dependent serine phosphorylation. Because all these processes require Kit activation for their initiation they function as negative-feedback loops limiting the signaling output from the receptor once it has been activated.

As described above for PDGF receptors, the activated Kit is internalized through clathrin-coated pits. Ubiquitination of Kit, which occurs in concert with the internalization process, depends on the ubiquitin ligase Cbl that can bind either directly to Kit or indirectly through adaptor proteins, such as Grb2, p85, or CrkL (Wisniewski et al. 1996; Sattler et al. 1997; Masson et al. 2006). In addition, a SOCS6-containing ubiquitination complex has been found to interact with Kit and promote its down-regulation (Bayle et al. 2004; Zadjali et al. 2011). The internalized Kit is sorted inside the cell toward degradation, which occurs in both lysosomes and proteasomes (Miyazawa et al. 1994; Zeng et al. 2005).

The activity of many kinases is suppressed by the action of phosphatases. In the case of Kit, the phosphatase SHP1 has been shown to associate and negatively regulate the activated Kit (Paulson et al. 1996; Kozlowski et al. 1998).

Protein kinase C (PKC) is activated downstream from Kit and phosphorylates Kit on serine residues in the kinase insert region, which leads to reduced Kit kinase activity (Blume-Jensen et al. 1993, 1995). Moreover, activation of PKC also results in shedding of the extracellular domain of Kit limiting the ability of SCF to stimulate cells (Yee et al. 1994).

Function of Kit

There are a large number of mouse loss-of-function mutations both in the Kit (*W* loci) and SCF (*Sl* loci) genes with varying degrees of severity. From the study of these mutant mice it was found that hematopoiesis, pigmentation, fertility, peristaltic movement, and certain aspects of the nervous system, are regulated by Kit signaling.

Hematopoiesis

Kit has been found to be expressed in primitive hematopoietic cells but the expression declines during differentiation, except for mast cells that retain high Kit expression also as fully differentiated cells (Ogawa et al. 1991; Oliveira and Lukacs 2003). Kit promotes cell survival and proliferation of primitive hematopoietic cells, often in synergy with other cytokines.

Pigmentation

Kit is expressed on melanocytes and has been found to be important for the migration of these cells from the neuronal crest to the dermis during development (Wehrle-Haller 2003). Defects in Kit signaling interferes with melanocyte migration and gives rise to the pigmentation defects found in patients with piebaldism (see below).

Fertility

The reduced fertility of *W* and *Sl* mutant mice was confirmed by knockin experiments with a Kit mutant unable to activate PI3-kinase; this resulted in sterile mice or mice with reduced fertility depending on gender (Blume-Jensen et al. 2000; Kissel et al. 2000). Because Kit signaling through PI3-kinase has been found to be important for survival (Blume-Jensen et al. 1998), a likely reason for the reduced fertility is an inability of mutant Kit to support viability of germ cells. A truncated cytoplasmic form of Kit (tr-Kit) exists in sperm owing to usage of an alternative promoter (Rossi et al. 2003). Tr-Kit has been found to be important for activation of the oocyte after fertilization (Sette et al. 1997).

Peristaltic Movement

Interstitial cells of Cajal (ICC) function as pacemaker cells and can communicate with both smooth muscle cells and nerves. It was found that *W* and *Sl* mice have a constipation phenotype, which is correlated with a reduced number of ICC (Huizinga et al. 1995; Ward et al. 1995). Also, in humans, a loss of ICC has been associated with defective gut movement and constipation (Lyford et al. 2002).

Nervous System

Kit expression has been found in neurons and *W* and *Sl* mutant mice have defects in spatial learning (Motro et al. 1996; Katafuchi et al. 2000). Furthermore, neuroproliferative zones in adult rat have been found to be positive for Kit (Jin et al. 2002).

Role of Kit in Diseases

Acute Myeloid Leukemia

Acute myeloid leukemia (AML) is characterized by an uncontrolled increase in the number of cells from the myeloid lineage that accumulate in the bone marrow and interfere with normal hematopoiesis. Kit expression is found in 60%–80% of AML cells and treatment with SCF results in increased cell proliferation (Ikeda et al. 1991; Odenike et al. 2011). Furthermore, activating mutations in Kit have been found in AML patients, although not very frequently (Odenike et al. 2011).

GIST

GISTs derive from ICC, which are pacemaker cells for peristaltic movement (Wang et al. 2000). These tumors are frequently driven by mutations in Kit, or PDGFR-α, often located in the juxtamembrane region, although mutations in the kinase domain and extracellular region have also been found (Antonescu 2011).

Mastocytosis

Mastocytosis is a disease characterized by abnormal accumulation of mast cells in tissues. Mast cells are among the few differentiated hematopoietic cells that retain Kit expression and patients with mastocytosis frequently harbor activating mutations within the kinase domain (Ustun et al. 2011).

Melanoma

Melanoma is a malignant disease of melanocytes, which can be located in different parts of the body, although most frequently in the skin. Normally, melanocytes express Kit, but the role of this receptor in melanoma is complicated; Kit has been found to be down-regulated in cutaneous melanoma (Easty and Bennett 2000), but activating mutations have been reported in other types of melanoma (Curtin et al. 2006; Ashida et al. 2009).

Small-Cell Lung Cancer

Small-cell lung cancer (SCLC) is an aggressive form of lung cancer that is often associated with smoking (Jackman and Johnson 2005). It was shown that SCLC cells coexpress Kit and SCF suggesting the presence of an autocrine loop (Hibi et al. 1991). However, immunohistochemical staining of tumors for Kit expression yielded varying results both regarding Kit expression and correlation with clinical outcome (Fischer et al. 2007). Furthermore, no activating Kit mutations could be found in SCLC tumors. Although imatinib inhibits growth of SCLC cell lines, it failed to show efficiency in animal models (Wolff et al. 2004). However, combining imatinib with chemotherapy (Decaudin et al. 2005), or using SU5416, which inhibits several kinases including Kit, produced more positive results in experimental models (Litz et al. 2004).

Piebaldism

In cancer, Kit is activated by gain-of-function mutations or by autocrine stimulation. In contrast, piebaldism involves germline loss-of-function mutations in Kit leading to patches of skin and hair that are depigmented owing to lack of melanocytes (Spritz 1994).

Clinically Used Kit Antagonists

Overactivity of Kit has been found in several malignant diseases (see above) and inhibiting Kit kinase activity is an approach to treat these conditions. Imatinib, also discussed in the PDGF receptor context above, inhibits Kit and has been used clinically. Because imatinib does not efficiently inhibit Kit with activating mutations within the kinase domain (Frost et al. 2002; Ma et al. 2002), other drugs have been developed that can inhibit Kit with such mutations, such as dasatinib and nilotinib (Demetri 2011). Thus, depending on the mutational status of Kit, different drugs should be selected. Neither of the drugs mentioned above exclusively inhibit Kit, but inhibit also other kinases and this may contribute to their clinical effectiveness. Other multiselective drugs whose targets include Kit are sunitinib and sorafenib. Masitinib is a potent inhibitor that targets Kit among other kinases and is approved for treating mast cell tumors in a veterinary setting.

FUTURE PERSPECTIVE

Studies over the last 30 years have revealed much about the structural properties, signaling capacities, normal functions, and role in diseases of PDGF and SCF receptors. Moreover, treatment regimens for malignant diseases in which these receptors are overactive are used routinely in the clinic. Future studies will aim at elucidating mechanisms for control of receptor signaling, including internalization, intracellular sorting, and termination of signaling. It is anticipated that such control mechanisms will include posttranslational modifications of the receptors and their signaling mediators, other than the already characterized phosphorylation and ubiquitination. Additional information is also needed regarding cooperation of the receptors with co-receptors at the cell membrane, as well as the importance of interactions with the receptors and intracellular molecules, to enhance or control their signaling capacities. It will also be important to determine exactly where in the cell the receptors induce their different signals: at the plasma membrane, in endosomes, or in other intracellular organelles. In addition, whereas some information is available regarding the involvement of PDGF and SCF receptors in disease, a more detailed analysis of these receptors in malignant as well as nonmalignant diseases is highly warranted, as are more efficient and specific methods to treat diseases in which the receptors are overactive.

ACKNOWLEDGMENT

We thank Ingegärd Schiller for valuable help in the preparation of this manuscript.

REFERENCES

Agosti V, Corbacioglu S, Ehlers I, Waskow C, Sommer G, Berrozpe G, Kissel H, Tucker CM, Manova K, Moore MA, et al. 2004. Critical role for Kit-mediated Src kinase but not PI 3-kinase signaling in pro T and pro B cell development. *J Exp Med* **199:** 867–878.

Andrae J, Gallini R, Betsholtz C. 2008. Role of platelet-derived growth factors in physiology and medicine. *Genes Dev* **22:** 1276–1312.

Antonescu CR. 2011. The GIST paradigm: Lessons for other kinase-driven cancers. *J Pathol* **223:** 251–261.

Arai H, Ueno T, Tangoku A, Yoshino S, Abe T, Kawauchi S, Oga A, Furuya T, Oka M, Sasaki K. 2003. Detection of amplified oncogenes by genome DNA microarrays in human primary esophageal squamous cell carcinoma: Comparison with conventional comparative genomic hybridization analysis. *Cancer Genet Cytogenet* **146:** 16–21.

Ashida A, Takata M, Murata H, Kido K, Saida T. 2009. Pathological activation of KIT in metastatic tumors of acral and mucosal melanomas. *Int J Cancer* **124:** 862–868.

Bae YS, Sung J-Y, Kim O-S, Kim YJ, Hur KC, Kazlauskas A, Rhee SG. 2000. Platelet-derived growth factor-induced H_2O_2 production requires the activation of phosphatidylinositol 3-kinase. *J Biol Chem* **275:** 10527–10531.

Ball SG, Shuttleworth CA, Kielty CM. 2007. Vascular endothelial growth factor can signal through platelet-derived growth factor receptors. *J Cell Biol* **177:** 489–500.

Baxter RM, Secrist JP, Vaillancourt RR, Kazlauskas A. 1998. Full activation of the platelet-derived growth factor β-receptor kinase involves multiple events. *J Biol Chem* **273:** 17050–17055.

Bayle J, Letard S, Frank R, Dubreuil P, De Sepulveda P. 2004. Suppressor of cytokine signaling 6 associates with KIT and regulates KIT receptor signaling. *J Biol Chem* **279:** 12249–12259.

Besmer P, Murphy JE, George PC, Qiu FH, Bergold PJ, Lederman L, Snyder HW Jr, Brodeur D, Zuckerman EE, Hardy WD. 1986. A new acute transforming feline retrovirus and relationship of its oncogene v-kit with the protein kinase gene family. *Nature* **320:** 415–421.

Bjarnegård M, Enge M, Norlin J, Gustafsdottir S, Fredriksson S, Abramsson A, Takemoto M, Gustafsson E, Fässler R, Betsholtz C. 2004. Endothelium-specific ablation of PDGFB leads to pericyte loss and glomerular, cardiac and placental abnormalities. *Development* **131:** 1847–1857.

Blume-Jensen P, Hunter T. 2001. Oncogenic kinase signalling. *Nature* **411:** 355–365.

Blume-Jensen P, Siegbahn A, Stabel S, Heldin C-H, Rönnstrand L. 1993. Increased Kit/SCF receptor induced mitogenicity but abolished cell motility after inhibition of protein kinase C. *EMBO J* **12:** 4199–4209.

Blume-Jensen P, Wernstedt C, Heldin C-H, Rönnstrand L. 1995. Identification of the major phosphorylation sites for protein kinase C in Kit/stem cell factor receptor in vitro and in intact cells. *J Biol Chem* **270:** 14192–14200.

Blume-Jensen P, Janknecht R, Hunter T. 1998. The Kit receptor promotes cell survival via activation of PI 3-kinase and subsequent Akt-mediated phosphorylation of Bad on Ser136. *Curr Biol* **8:** 779–782.

Blume-Jensen P, Jiang G, Hyman R, Lee KF, O'Gorman S, Hunter T. 2000. Kit/stem cell factor receptor-induced activation of phosphatidylinositol 3′-kinase is essential for male fertility. *Nat Genet* **24:** 157–162.

Bonner JC. 2004. Regulation of PDGF and its receptors in fibrotic diseases. *Cytokine Growth Factor Rev* **15:** 255–273.

Boström H, Willetts K, Pekny M, Levéen P, Lindahl P, Hedstrand H, Pekna M, Hellström M, Gebre-Medhin S, Schalling M, et al. 1996. PDGF-A signaling is a critical event in lung alveolar myofibroblast development and alveogenesis. *Cell* **85:** 863–873.

Boucher P, Liu P, Gotthardt M, Hiesberger T, Anderson RG, Herz J. 2002. Platelet-derived growth factor mediates tyrosine phosphorylation of the cytoplasmic domain of the low density lipoprotein receptor-related protein in caveolae. *J Biol Chem* **277:** 15507–15513.

Brennan J, Tilmann C, Capel B. 2003. PDGFR-α mediates testis cord organization and fetal Leydig cell development in the XY gonad. *Genes Dev* **17:** 800–810.

Calver AR, Hall AC, Yu WP, Walsh FS, Heath JK, Betsholtz C, Richardson WD. 1998. Oligodendrocyte population dynamics and the role of PDGF in vivo. *Neuron* **20:** 869–882.

Caruana G, Cambareri AC, Ashman LK. 1999. Isoforms of c-KIT differ in activation of signalling pathways and transformation of NIH3T3 fibroblasts. *Oncogene* **18:** 5573–5581.

Catena R, Luis-Ravelo D, Anton I, Zandueta C, Salazar-Colocho P, Larzabal L, Calvo A, Lecanda F. 2010. PDGFR signaling blockade in marrow stroma impairs lung cancer bone metastasis. *Cancer Res* **71:** 164–174.

Chen H, Gu X, Liu Y, Wang J, Wirt SE, Bottino R, Schorle H, Sage J, Kim SK. 2011. PDGF signalling controls age-dependent proliferation in pancreatic β-cells. *Nature* **478:** 349–355.

Chiara F, Bishayee S, Heldin C-H, Demoulin J-B. 2004. Autoinhibition of the platelet-derived growth factor β receptor tyrosine kinase by its C-terminal tail. *J Biol Chem* **279:** 19732–19738.

Chintalgattu V, Ai D, Langley RR, Zhang J, Bankson JA, Shih TL, Reddy AK, Coombes KR, Daher IN, Pati S, et al. 2010. Cardiomyocyte PDGFR-β signaling is an essential component of the mouse cardiac response to load-induced stress. *J Clin Invest* **120:** 472–484.

Clarke ID, Dirks PB. 2003. A human brain tumor-derived PDGFR-α deletion mutant is transforming. *Oncogene* **22:** 722–733.

Cools J, DeAngelo DJ, Gotlib J, Stover EH, Legare RD, Cortes J, Kutok J, Clark J, Galinsky I, Griffin JD, et al. 2003. A tyrosine kinase created by fusion of the PDGFRA and FIP1L1 genes as a therapeutic target of imatinib in idiopathic hypereosinophilic syndrome. *N Engl J Med* **348:** 1201–1214.

Crosier PS, Ricciardi ST, Hall LR, Vitas MR, Clark SC, Crosier KE. 1993. Expression of isoforms of the human receptor tyrosine kinase c-kit in leukemic cell lines and acute myeloid leukemia. *Blood* **82:** 1151–1158.

Curtin JA, Busam K, Pinkel D, Bastian BC. 2006. Somatic activation of KIT in distinct subtypes of melanoma. *J Clin Oncol* **24:** 4340–4346.

Dance M, Montagner A, Salles JP, Yart A, Raynal P. 2008. The molecular functions of Shp2 in the Ras/Mitogen-activated protein kinase (ERK1/2) pathway. *Cell Signal* **20:** 453–459.

Decaudin D, de Cremoux P, Sastre X, Judde JG, Nemati F, Tran-Perennou C, Freneaux P, Livartowski A, Pouillart P, Poupon MF. 2005. In vivo efficacy of STI571 in xenografted human small cell lung cancer alone or combined with chemotherapy. *Int J Cancer* **113:** 849–856.

Demetri GD. 2011. Differential properties of current tyrosine kinase inhibitors in gastrointestinal stromal tumors. *Semin Oncol* **38:** S10–S19.

Demoulin J-B, Seo JK, Ekman S, Grapengiesser E, Hellman U, Rönnstrand L, Heldin C-H. 2003. Ligand-induced recruitment of Na^+/H^+ exchanger regulatory factor to the PDGF (platelet-derived growth factor) receptor regulates actin cytoskeleton reorganization by PDGF. *Biochem J* **376:** 505–510.

DiNitto JP, Deshmukh GD, Zhang Y, Jacques SL, Coli R, Worrall JW, Diehl W, English JM, Wu JC. 2010. Function of activation loop tyrosine phosphorylation in the mechanism of c-Kit auto-activation and its implication in sunitinib resistance. *J Biochem* **147:** 601–609.

Di Pasquale G, Davidson BL, Stein CS, Martins I, Scudiero D, Monks A, Chiorini JA. 2003. Identification of PDGFR as a receptor for AAV-5 transduction. *Nat Med* **9:** 1306–1312.

Dolloff NG, Shulby SS, Nelson AV, Stearns ME, Johannes GJ, Thomas JD, Meucci O, Fatatis A. 2005. Bone-metastatic potential of human prostate cancer cells correlates with Akt/PKB activation by α platelet-derived growth factor receptor. *Oncogene* **24:** 6848–6854.

Easty DJ, Bennett DC. 2000. Protein tyrosine kinases in malignant melanoma. *Melanoma Res* **10:** 401–411.

Ekman S, Rupp Thuresson E, Heldin C-H, Rönnstrand L. 1999. Increased mitogenicity of an αβ heterodimeric PDGF receptor complex correlates with lack of RasGAP binding. *Oncogene* **18:** 2481–2488.

Elling C, Erben P, Walz C, Frickenhaus M, Schemionek M, Stehling M, Serve H, Cross NC, Hochhaus A, Hofmann WK, et al. 2011. Novel imatinib-sensitive PDGFRA-activating point mutations in hypereosinophilic syndrome induce growth factor independence and leukemia-like disease. *Blood* **117:** 2935–2943.

Eriksson A, Siegbahn A, Westermark B, Heldin C-H, Claesson-Welsh L. 1992. PDGF α- and β-receptors activate unique and common signal transduction pathways. *EMBO J* **11:** 543–550.

Faraone D, Aguzzi MS, Ragone G, Russo K, Capogrossi MC, Facchiano A. 2006. Heterodimerization of FGF-receptor 1 and PDGF-receptor-α: A novel mechanism underlying the inhibitory effect of PDGF-BB on FGF-2 in human cells. *Blood* **107:** 1896–1902.

Fischer B, Marinov M, Arcaro A. 2007. Targeting receptor tyrosine kinase signalling in small cell lung cancer (SCLC): What have we learned so far? *Cancer Treat Rev* **33:** 391–406.

Fleming TP, Saxena A, Clark WC, Robertson JT, Oldfield EH, Aaronson SA, Ali IU. 1992. Amplification and/or overexpression of platelet-derived growth factor receptors and epidermal growth factor receptor in human glial tumors. *Cancer Res* **52:** 4550–4553.

Frost MJ, Ferrao PT, Hughes TP, Ashman LK. 2002. Juxtamembrane mutant V560GKit is more sensitive to Imatinib (STI571) compared with wild-type c-kit whereas the kinase domain mutant D816VKit is resistant. *Mol Cancer Ther* **1:** 1115–1124.

Fruttiger M, Calver AR, Krüger WH, Mudhar HS, Michalovich D, Takakura N, Nishikawa SI, Richardson WD. 1996. PDGF mediates a neuron-astrocyte interaction in the developing retina. *Neuron* **17:** 1117–1131.

Fruttiger M, Calver AR, Richardson WD. 2000. Platelet-derived growth factor is constitutively secreted from neuronal cell bodies but not from axons. *Curr Biol* **10:** 1283–1286.

Gerhardt H, Golding M, Fruttiger M, Ruhrberg C, Lundkvist A, Abramsson A, Jeltsch M, Mitchell C, Alitalo K, Shima D, et al. 2003. VEGF guides angiogenic sprouting utilizing endothelial tip cell filopodia. *J Cell Biol* **161:** 1163–1177.

Gnessi L, Basciani S, Mariani S, Arizzi M, Spera G, Wang C, Bondjers C, Karlsson L, Betsholtz C. 2000. Leydig cell loss and spermatogenic arrest in platelet-derived growth factor (PDGF)-A-deficient mice. *J Cell Biol* **149:** 1019–1026.

Golub TR, Barker GF, Lovett M, Gilliland DG. 1994. Fusion of PDGF receptor β to a novel *ets*-like gene, *tel*, in chronic myelomonocytic leukemia with t(5;12) chromosomal translocation. *Cell* **77:** 307–316.

Gotzmann J, Fischer AN, Zojer M, Mikula M, Proell V, Huber H, Jechlinger M, Waerner T, Weith A, Beug H, et al. 2006. A crucial function of PDGF in TGF-β-mediated cancer progression of hepatocytes. *Oncogene* **25:** 3170–3185.

Griffin JH, Leung J, Bruner RJ, Caligiuri MA, Briesewitz R. 2003. Discovery of a fusion kinase in EOL-1 cells and idiopathic hypereosinophilic syndrome. *Proc Natl Acad Sci* **100:** 7830–7835.

Hamilton TG, Klinghoffer RA, Corrin PD, Soriano P. 2003. Evolutionary divergence of platelet-derived growth factor α receptor signaling mechanisms. *Mol Cell Biol* **23:** 4013–4025.

Hansson GK. 2005. Inflammation, atherosclerosis, and coronary artery disease. *N Engl J Med* **352:** 1685–1695.

Heidaran MA, Pierce JH, Jensen RA, Matsui T, Aaronson SA. 1990. Chimeric α- and β-platelet-derived growth factor (PDGF) receptors define three immunoglobulin-like domains of the α-PDGF receptor that determine PDGF-AA binding specificity. *J Biol Chem* **265:** 18741–18744.

Heinrich MC, Corless CL, Duensing A, McGreevey L, Chen CJ, Joseph N, Singer S, Griffith DJ, Haley A, Town A, et al. 2003. PDGFRA activating mutations in gastrointestinal stromal tumors. *Science* **299:** 708–710.

Heissig B, Hattori K, Dias S, Friedrich M, Ferris B, Hackett NR, Crystal RG, Besmer P, Lyden D, Moore MA, et al. 2002. Recruitment of stem and progenitor cells from the bone marrow niche requires MMP-9 mediated release of kit-ligand. *Cell* **109:** 625–637.

Heldin C-H, Westermark B. 1999. Mechanism of action and in vivo role of platelet-derived growth factor. *Physiol Rev* **79:** 1283–1316.

Heldin C-H, Wasteson Å, Westermark B. 1982. Interaction of platelet-derived growth factor with its fibroblast receptor. Demonstration of ligand degradation and receptor modulation. *J Biol Chem* **257:** 4216–4221.

Heldin C-H, Östman A, Rönnstrand L. 1998. Signal transduction via platelet-derived growth factor receptors. *Biochim Biophys Acta* **1378:** F79–F113.

Heldin C-H, Rubin K, Pietras K, Östman A. 2004. High interstitial fluid pressure—An obstacle in cancer therapy. *Nat Rev Cancer* **4:** 806–813.

Hellberg C, Schmees C, Karlsson S, Åhgren A, Heldin C-H. 2009. Activation of protein kinase C α is necessary for sorting the PDGF β-receptor to Rab4a-dependent recycling. *Mol Biol Cell* **20:** 2856–2863.

Cite this article as *Cold Spring Harb Perspect Biol* doi: 10.1101/cshperspect.a009100

Hellström M, Kalén M, Lindahl P, Abramsson A, Betsholtz C. 1999. Role of PDGF-B and PDGFR-β in recruitment of vascular smooth muscle cells and pericytes during embryonic blood vessel formation in the mouse. *Development* **126:** 3047–3055.

Hermanson M, Funa K, Hartman M, Claesson-Welsh L, Heldin C-H, Westermark B, Nistér M. 1992. Platelet-derived growth factor and its receptors in human glioma tissue: Expression of messenger RNA and protein suggests the presence of autocrine and paracrine loops. *Cancer Res* **52:** 3213–3219.

Hibi K, Takahashi T, Sekido Y, Ueda R, Hida T, Ariyoshi Y, Takagi H. 1991. Coexpression of the stem cell factor and the c-kit genes in small-cell lung cancer. *Oncogene* **6:** 2291–2296.

Hoch RV, Soriano P. 2003. Roles of PDGF in animal development. *Development* **130:** 4769–4784.

Hong L, Munugalavadla V, Kapur R. 2004. c-Kit-mediated overlapping and unique functional and biochemical outcomes via diverse signaling pathways. *Mol Cell Biol* **24:** 1401–1410.

Huang EJ, Nocka KH, Buck J, Besmer P. 1992. Differential expression and processing of two cell associated forms of the kit-ligand: KL-1 and KL-2. *Mol Biol Cell* **3:** 349–362.

Hubbard SR. 1997. Crystal structure of the activated insulin receptor tyrosine kinase in complex with peptide substrate and ATP analog. *EMBO J* **16:** 5572–5581.

Hubbard SR. 2004. Juxtamembrane autoinhibition in receptor tyrosine kinases. *Nat Rev Mol Cell Biol* **5:** 464–471.

Huizinga JD, Thuneberg L, Kluppel M, Malysz J, Mikkelsen HB, Bernstein A. 1995. W/kit gene required for interstitial cells of Cajal and for intestinal pacemaker activity. *Nature* **373:** 347–349.

Ikeda H, Kanakura Y, Tamaki T, Kuriu A, Kitayama H, Ishikawa J, Kanayama Y, Yonezawa T, Tarui S, Griffin JD. 1991. Expression and functional role of the proto-oncogene c-kit in acute myeloblastic leukemia cells. *Blood* **78:** 2962–2968.

Irusta PM, Luo Y, Bakht O, Lai CC, Smith SO, DiMaio D. 2002. Definition of an inhibitory juxtamembrane WW-like domain in the platelet-derived growth factor β receptor. *J Biol Chem* **277:** 38627–38634.

Jackman DM, Johnson BE. 2005. Small-cell lung cancer. *Lancet* **366:** 1385–1396.

Jayson GC, Parker GJ, Mullamitha S, Valle JW, Saunders M, Broughton L, Lawrance J, Carrington B, Roberts C, Issa B, et al. 2005. Blockade of platelet-derived growth factor receptor-β by CDP860, a humanized, PEGylated di-Fab', leads to fluid accumulation and is associated with increased tumor vascularized volume. *J Clin Oncol* **23:** 973–981.

Jechlinger M, Grunert S, Tamir IH, Janda E, Ludemann S, Waerner T, Seither P, Weith A, Beug H, Kraut N. 2003. Expression profiling of epithelial plasticity in tumor progression. *Oncogene* **22:** 7155–7169.

Jechlinger M, Sommer A, Moriggl R, Seither P, Kraut N, Capodiecci P, Donovan M, Cordon-Cardo C, Beug H, Grunert S. 2006. Autocrine PDGFR signaling promotes mammary cancer metastasis. *J Clin Invest* **116:** 1561–1570.

Jin K, Mao XO, Sun Y, Xie L, Greenberg DA. 2002. Stem cell factor stimulates neurogenesis in vitro and in vivo. *J Clin Invest* **110:** 311–319.

Karlsson L, Bondjers C, Betsholtz C. 1999. Roles for PDGF-A and sonic hedgehog in development of mesenchymal components of the hair follicle. *Development* **126:** 2611–2621.

Karlsson L, Lindahl P, Heath JK, Betsholtz C. 2000. Abnormal gastrointestinal development in PDGF-A and PDGFR-α deficient mice implicates a novel mesenchymal structure with putative instructive properties in villus morphogenesis. *Development* **127:** 3457–3466.

Karlsson S, Kowanetz K, Sandin Å, Persson C, Östman A, Heldin C-H, Hellberg C. 2006. Loss of T-cell protein tyrosine phosphatase induces recycling of the platelet-derived growth factor (PDGF) β-receptor but not the PDGF α-receptor. *Mol Biol Cell* **17:** 4846–4855.

Katafuchi T, Li AJ, Hirota S, Kitamura Y, Hori T. 2000. Impairment of spatial learning and hippocampal synaptic potentiation in c-kit mutant rats. *Learn Mem* **7:** 383–392.

Kawaguchi N, Horiuchi K, Becherer JD, Toyama Y, Besmer P, Blobel CP. 2007. Different ADAMs have distinct influences on Kit ligand processing: Phorbol-ester-stimulated ectodomain shedding of Kitl1 by ADAM17 is reduced by ADAM19. *J Cell Sci* **120:** 943–952.

Kimura Y, Jones N, Kluppel M, Hirashima M, Tachibana K, Cohn JB, Wrana JL, Pawson T, Bernstein A. 2004. Targeted mutations of the juxtamembrane tyrosines in the Kit receptor tyrosine kinase selectively affect multiple cell lineages. *Proc Natl Acad Sci* **101:** 6015–6020.

Kissel H, Timokhina I, Hardy MP, Rothschild G, Tajima Y, Soares V, Angeles M, Whitlow SR, Manova K, Besmer P. 2000. Point mutation in kit receptor tyrosine kinase reveals essential roles for kit signaling in spermatogenesis and oogenesis without affecting other kit responses. *EMBO J* **19:** 1312–1326.

Klinghoffer RA, Mueting-Nelsen PF, Faerman A, Shani M, Soriano P. 2001. The two PDGF receptors maintain conserved signaling in vivo despite divergent embryological functions. *Mol Cell* **7:** 343–354.

Kozlowski M, Larose L, Lee F, Le DM, Rottapel R, Siminovitch KA. 1998. SHP-1 binds and negatively modulates the c-Kit receptor by interaction with tyrosine 569 in the c-Kit juxtamembrane domain. *Mol Cell Biol* **18:** 2089–2099.

Krampert M, Heldin C-H, Heuchel R. 2008. A gain-of-function mutation in the PDGFR-β alters the kinetics of injury response in liver and skin. *Lab Invest* **88:** 1204–1214.

Kumabe T, Sohma Y, Kayama T, Yoshimoto T, Yamamoto T. 1992. Amplification of α-platelet-derived growth factor receptor gene lacking an exon coding for a portion of the extracellular region in a primary brain tumor of glial origin. *Oncogene* **7:** 627–633.

Lechleider RJ, Sugimoto S, Bennett AM, Kashishian AS, Cooper JA, Shoelson SE, Walsh CT, Neel BG. 1993. Activation of the SH2-containing phosphotyrosine phosphatase SH-PTP2 by its binding site, phosphotyrosine 1009, on the human platelet-derived growth factor receptor β. *J Biol Chem* **268:** 21478–21481.

Lemmon MA, Pinchasi D, Zhou M, Lax I, Schlessinger J. 1997. Kit receptor dimerization is driven by bivalent binding of stem cell factor. *J Biol Chem* **272:** 6311–6317.

Lennartsson J, Blume-Jensen P, Hermanson M, Pontén E, Carlberg M, Rönnstrand L. 1999. Phosphorylation of Shc by Src family kinases is necessary for stem cell factor receptor/c-*kit* mediated activation of the Ras/MAP kinase pathway and c-*fos* induction. *Oncogene* **18**: 5546–5553.

Lennartsson J, Wardega P, Engström U, Hellman U, Heldin C-H. 2006. Alix facilitates the interaction between c-Cbl and platelet-derived growth factor β-receptor and thereby modulates receptor downregulation. *J Biol Chem* **281**: 39152–39158.

Levéen P, Pekny M, Gebre-Medhin S, Swolin B, Larsson E, Betsholtz C. 1994. Mice deficient for PDGF B show renal, cardiovascular, and hematological abnormalities. *Genes Dev* **8**: 1875–1887.

Li L, Heldin C-H, Heldin P. 2006. Inhibition of platelet-derived growth factor–BB-induced receptor activation and fibroblast migration by hyaluronan activation of CD44. *J Biol Chem* **281**: 26512–26519.

Lidén A, Berg A, Nedrebø T, Reed RK, Rubin K. 2006. Platelet-derived growth factor BB-mediated normalization of dermal interstitial fluid pressure after mast cell degranulation depends on β3 but not β1 integrins. *Circulation Res* **98**: 635–641.

Lindahl P, Johansson BR, Levéen P, Betsholtz C. 1997a. Pericyte loss and microaneurysm formation in PDGF-B-deficient mice. *Science* **277**: 242–245.

Lindahl P, Karlsson L, Hellström M, Gebre-Medhin S, Willetts K, Heath JK, Betsholtz C. 1997b. Alveogenesis failure in PDGF-A-deficient mice is coupled to lack of distal spreading of alveolar smooth muscle cell progenitors during lung development. *Development* **124**: 3943–3953.

Lindahl P, Hellström M, Kalén M, Karlsson L, Pekny M, Pekna M, Soriano P, Betsholtz C. 1998. Paracrine PDGF-B/PDGF-Rβ signaling controls mesangial cell development in kidney glomeruli. *Development* **125**: 3313–3322.

Litz J, Sakuntala Warshamana-Greene G, Sulanke G, Lipson KE, Krystal GW. 2004. The multi-targeted kinase inhibitor SU5416 inhibits small cell lung cancer growth and angiogenesis, in part by blocking Kit-mediated VEGF expression. *Lung Cancer* **46**: 283–291.

Lokker NA, O'Hare JP, Barsoumian A, Tomlinson JE, Ramakrishnan V, Fretto LJ, Giese NA. 1997. Functional importance of platelet-derived growth factor (PDGF) receptor extracellular immunoglobulin-like domains. Identification of PDGF binding site and neutralizing monoclonal antibodies. *J Biol Chem* **272**: 33037–33044.

Longley BJ, Tyrrell L, Ma Y, Williams DA, Halaban R, Langley K, Lu HS, Schechter NM. 1997. Chymase cleavage of stem cell factor yields a bioactive, soluble product. *Proc Natl Acad Sci* **94**: 9017–9021.

Loukinova E, Ranganathan S, Kuznetsov S, Gorlatova N, Migliorini MM, Loukinov D, Ulery PG, Mikhailenko I, Lawrence DA, Strickland DK. 2002. Platelet-derived growth factor (PDGF)-induced tyrosine phosphorylation of the low density lipoprotein receptor-related protein (LRP). Evidence for integrated co-receptor function betwenn LRP and the PDGF. *J Biol Chem* **277**: 15499–15506.

Lyford GL, He CL, Soffer E, Hull TL, Strong SA, Senagore AJ, Burgart LJ, Young-Fadok T, Szurszewski JH, Farrugia G. 2002. Pan-colonic decrease in interstitial cells of Cajal in patients with slow transit constipation. *Gut* **51**: 496–501.

Ma Y, Zeng S, Metcalfe DD, Akin C, Dimitrijevic S, Butterfield JH, McMahon G, Longley BJ. 2002. The c-KIT mutation causing human mastocytosis is resistant to STI571 and other KIT kinase inhibitors; kinases with enzymatic site mutations show different inhibitor sensitivity profiles than wild-type kinases and those with regulatory-type mutations. *Blood* **99**: 1741–1744.

Magnusson MK, Meade KE, Brown KE, Arthur DC, Krueger LA, Barrett AJ, Dunbar CE. 2001. Rabaptin-5 is a novel fusion partner to platelet-derived growth factor β receptor in chronic myelomonocytic leukemia. *Blood* **98**: 2518–2525.

Masson K, Heiss E, Band H, Rönnstrand L. 2006. Direct binding of Cbl to Tyr568 and Tyr936 of the stem cell factor receptor/c-Kit is required for ligand-induced ubiquitination, internalization and degradation. *Biochem J* **399**: 5–67.

Maudsley S, Zamah AM, Rahman N, Blitzer JT, Luttrell LM, Lefkowitz RJ, Hall RA. 2000. Platelet-derived growth factor receptor association with Na^+/H^+ exchanger regulatory factor potentiates receptor activity. *Mol Cell Biol* **20**: 8352–8363.

Minami A, Mizutani K, Waseda M, Kajita M, Miyata M, Ikeda W, Takai Y. 2010. Necl-5/PVR enhances PDGF-induced attraction of growing microtubules to the plasma membrane of the leading edge of moving NIH3T3 cells. *Genes Cells* **15**: 1123–1135.

Miyake S, Mullane-Robinson KP, Lill NL, Douillard P, Band H. 1999. Cbl-mediated negative regulation of platelet-derived growth factor receptor-dependent cell proliferation—A critical role for Cbl tyrosine kinase-binding domain. *J Biol Chem* **274**: 16619–16628.

Miyazawa K, Toyama K, Gotoh A, Hendrie PC, Mantel C, Broxmayer HE. 1994. Ligand-dependent polyubiquitination of c-kit product: A possible mechanism of receptor down regulation in M107e cells. *Blood* **83**: 137–145.

Miyazawa K, Bäckström G, Leppänen O, Persson C, Wernstedt C, Hellman U, Heldin C-H, Östman A. 1998. Role of immunoglobulin-like domains 2-4 of the platelet-derived growth factor α-receptor in ligand-receptor complex assembly. *J Biol Chem* **273**: 25495–25502.

Mol CD, Lim KB, Sridhar V, Zou H, Chien EY, Sang BC, Nowakowski J, Kassel DB, Cronin CN, McRee DE. 2003. Structure of a c-kit product complex reveals the basis for kinase transactivation. *J Biol Chem* **278**: 31461–31464.

Mol CD, Dougan DR, Schneider TR, Skene RJ, Kraus ML, Scheibe DN, Snell GP, Zou H, Sang BC, Wilson KP. 2004. Structural basis for the autoinhibition and STI-571 inhibition of c-Kit tyrosine kinase. *J Biol Chem* **279**: 31655–31663.

Mori S, Heldin C-H, Claesson-Welsh L. 1992. Ligand-induced polyubiquitination of the platelet-derived growth factor β-receptor. *J Biol Chem* **267**: 6429–6434.

Motro B, Wojtowicz JM, Bernstein A, van der Kooy D. 1996. Steel mutant mice are deficient in hippocampal learning but not long-term potentiation. *Proc Natl Acad Sci* **93**: 1808–1813.

O'Brien KP, Seroussi E, Dal Cin P, Sciot R, Mandahl N, Fletcher JA, Turc-Carel C, Dumanski JP. 1998. Various regions within the α-helical domain of the COL1A1 gene are fused to the second exon of the PDGFB gene in dermatofibrosarcomas and giant-cell fibroblastomas. *Gene Chrom Cancer* **23:** 187–193.

Odenike O, Thirman MJ, Artz AS, Godley LA, Larson RA, Stock W. 2011. Gene mutations, epigenetic dysregulation, and personalized therapy in myeloid neoplasia: Are we there yet? *Semin Oncol* **38:** 196–214.

Ogawa M, Matsuzaki Y, Nishikawa S, Hayashi S, Kunisada T, Sudo T, Kina T, Nakauchi H. 1991. Expression and function of c-kit in hemopoietic progenitor cells. *J Exp Med* **174:** 63–71.

Ohlsson R, Falck P, Hellström M, Lindahl P, Boström H, Franklin G, Åhrlund-Richter L, Pollard J, Soriano P, Besholtz C. 1999. PDGFB regulates the development of the labyrinthine layer of the mouse fetal placenta. *Dev Biol* **212:** 124–136.

Oliveira SH, Lukacs NW. 2003. Stem cell factor: A hemopoietic cytokine with important targets in asthma. *Curr Drug Targets Inflamm Allergy* **2:** 313–318.

Olson LE, Soriano P. 2009. Increased PDGFRα activation disrupts connective tissue development and drives systemic fibrosis. *Dev Cell* **16:** 303–313.

Omura T, Heldin C-H, Östman A. 1997. Immunoglobulin-like domain 4-mediated receptor-receptor interactions contribute to platelet-derived growth factor-induced receptor dimerization. *J Biol Chem* **272:** 12676–12682.

Pardanani A, Ketterling RP, Brockman SR, Flynn HC, Paternoster SF, Shearer BM, Reeder TL, Li CY, Cross NC, Cools J, et al. 2003. CHIC2 deletion, a surrogate for FIP1L1-PDGFRA fusion, occurs in systemic mastocytosis associated with eosinophilia and predicts response to imatinib mesylate therapy. *Blood* **102:** 3093–3096.

Paulson RF, Vesely S, Siminovitch KA, Bernstein A. 1996. Signalling by the *W/Kit* receptor tyrosine kinase is negatively regulated in vivo by the protein tyrosine phosphatase Shp1. *Nat Genet* **13:** 309–315.

Pietras K, Sjöblom T, Rubin K, Heldin C-H, Östman A. 2003. PDGF receptors as cancer drug targets. *Cancer Cell* **3:** 439–443.

Pietras K, Pahler J, Bergers G, Hanahan D. 2008. Functions of paracrine PDGF signaling in the proangiogenic tumor stroma revealed by pharmacological targeting. *PLoS Med* **5:** e19.

Puputti M, Tynninen O, Sihto H, Blom T, Maenpaa H, Isola J, Paetau A, Joensuu H, Nupponen NN. 2006. Amplification of KIT, PDGFRA, VEGFR2, and EGFR in gliomas. *Mol Cancer Res* **4:** 927–934.

Reith AD, Ellis C, Lyman SD, Anderson DM, Williams DW, Bernstein A, Pawson T. 1991. Signal transduction by normal isoforms and *W* mutant variants of the Kit receptor tyrosine kinase. *EMBO J* **10:** 2451–2459.

Robson MC, Phillips LG, Thomason A, Robson LE, Pierce GF. 1992. Platelet-derived growth factor BB for the treatment of chronic pressure ulcers. *Lancet* **339:** 23–25.

Rodt SÅ, Åhlén K, Berg A, Rubin K, Reed RK. 1996. A novel physiological function for platelet-derived growth factor-BB in rat dermis. *J Physiol* **495:** 193–200.

Rolny C, Nilsson I, Magnusson P, Armulik A, Jakobsson L, Wentzel P, Lindblom P, Norlin J, Betsholtz C, Heuchel R, et al. 2006. Platelet-derived growth factor receptor-β promotes early endothelial cell differentiation. *Blood* **108:** 1877–1886.

Ross R. 1993. The pathogenesis of atherosclerosis: A perspective for the 1990s. *Nature* **362:** 801–809.

Rossi P, Dolci S, Sette C, Geremia R. 2003. Molecular mechanisms utilized by alternative c-kit gene products in the control of spermatogonial proliferation and sperm-mediated egg activation. *Andrologia* **35:** 71–78.

Russell MR, Jamieson WL, Dolloff NG, Fatatis A. 2009. The α-receptor for platelet-derived growth factor as a target for antibody-mediated inhibition of skeletal metastases from prostate cancer cells. *Oncogene* **28:** 412–421.

Saito Y, Haendeler J, Hojo Y, Yamamoto K, Berk BC. 2001. Receptor heterodimerization: Essential mechanism for platelet-derived growth factor-induced epidermal growth factor receptor transactivation. *Mol Cell Biol* **21:** 6387–6394.

Sattler M, Salgia R, Shrikhande G, Verma S, Pisick E, Prasad KV, Griffin JD. 1997. Steel factor induces tyrosine phosphorylation of CRKL and binding of CRKL to a complex containing c-kit, phosphatidylinositol 3-kinase, and p120(CBL). *J Biol Chem* **272:** 10248–10253.

Schmees C, Villaseñor R, Zheng W, Ma H, Zerial M, Heldin C-H, Hellberg C. 2012. Macropinocytosis of the PDGF β-receptor promotes fibroblast transformation by H-RasG12V. *Mol Biol Cell* **23:** 2571–2582.

Schneller M, Vuori K, Ruoslahti E. 1997. αvβ3 integrin associates with activated insulin and PDGFβ receptors and potentiates the biological activity of PDGF. *EMBO J* **16:** 5600–5607.

Sette C, Bevilacqua A, Bianchini A, Mangia F, Geremia R, Rossi P. 1997. Parthenogenetic activation of mouse eggs by microinjection of a truncated c-kit tyrosine kinase present in spermatozoa. *Development* **124:** 2267–2274.

Shim AH, Liu H, Focia PJ, Chen X, Lin PC, He X. 2010. Structures of a platelet-derived growth factor/propeptide complex and a platelet-derived growth factor/receptor complex. *Proc Natl Acad Sci* **107:** 11307–11312.

Shimizu A, O'Brien KP, Sjöblom T, Pietras K, Buchdunger E, Collins VP, Heldin C-H, Dumanski JP, Östman A. 1999. The dermatofibrosarcoma protuberans-associated collagen type Iα1/platelet derived growth factor (PDGF) B-chain fusion gene generates a transforming protein that is processed to functional PDGF-BB. *Cancer Res* **59:** 3719–3723.

Simon M-P, Pedeutour F, Sirvent N, Grosgeorge J, Minoletti F, Coindre J-M, Terrier-Lacombe M-J, Mandahl N, Craver RD, Blin N, et al. 1997. Deregulation of the platelet-derived growth factor B-chain gene via fusion with collagen gene *COL1A1* in dermatofibrosarcoma protuberans and giant-cell fibroblastoma. *Nat Genet* **15:** 95–98.

Smith JS, Wang XY, Qian J, Hosek SM, Scheithauer BW, Jenkins RB, James CD. 2000. Amplification of the platelet-derived growth factor receptor-A (PDGFRA) gene occurs in oligodendrogliomas with grade IV anaplastic features. *J Neuropathol Exp Neurol* **59:** 495–503.

Song S, Ewald AJ, Stallcup W, Werb Z, Bergers G. 2005. PDGFRβ⁺ perivascular progenitor cells in tumours regulate pericyte differentiation and vascular survival. *Nat Cell Biol* **7:** 870–879.

Soriano P. 1994. Abnormal kidney development and hematological disorders in PDGF β-receptor mutant mice. *Genes Dev* **8**: 1888–1896.

Soriano P. 1997. The PDGFα receptor is required for neural crest cell development and for normal patterning of the somites. *Development* **124**: 2691–2700.

Sorkin A, Westermark B, Heldin C-H, Claesson-Welsh L. 1991. Effect of receptor kinase inactivation on the rate of internalization and degradation of PDGF and the PDGF β-receptor. *J Cell Biol* **112**: 469–478.

Soroceanu L, Akhavan A, Cobbs CS. 2008. Platelet-derived growth factor-α receptor activation is required for human cytomegalovirus infection. *Nature* **455**: 391–395.

Spritz RA. 1994. Molecular basis of human piebaldism. *J Invest Dermatol* **103**: 137S–140S.

Stover EH, Chen J, Folens C, Lee BH, Mentens N, Marynen P, Williams IR, Gilliland DG, Cools J. 2006. Activation of FIP1L1-PDGFRα requires disruption of the juxtamembrane domain of PDGFRα and is FIP1L1-independent. *Proc Natl Acad Sci* **103**: 8078–8083.

Sundaresan M, Yu ZX, Ferrans VJ, Irani K, Finkel T. 1995. Requirement for generation of H_2O_2 for platelet-derived growth factor signal transduction. *Science* **270**: 296–299.

Sundberg C, Rubin K. 1996. Stimulation of β_1 integrins on fibroblasts induces PDGF independent tyrosine phosphorylation of PDGF β-receptors. *J Cell Biol* **132**: 741–752.

Takahashi Y, Morales FC, Kreimann EL, Georgescu MM. 2006. PTEN tumor suppressor associates with NHERF proteins to attenuate PDGF receptor signaling. *EMBO J* **25**: 910–920.

Tallquist MD, Soriano P. 2003. Cell autonomous requirement for PDGFRα in populations of cranial and cardiac neural crest cells. *Development* **130**: 507–518.

Tallquist MD, Weismann KE, Hellstrom M, Soriano P. 2000. Early myotome specification regulates PDGFA expression and axial skeleton development. *Development* **127**: 5059–5070.

Theisen CS, Wahl JK III, Johnson KR, Wheelock MJ. 2007. NHERF links the N-cadherin/catenin complex to the platelet-derived growth factor receptor to modulate the actin cytoskeleton and regulate cell motility. *Mol Biol Cell* **18**: 1220–1232.

Thiery JP, Acloque H, Huang RY, Nieto MA. 2009. Epithelial-mesenchymal transitions in development and disease. *Cell* **139**: 871–890.

Timokhina I, Kissel H, Stella G, Besmer P. 1998. Kit signaling through PI3-kinase and Src kinase pathways: An essential role for Rac1 and JNK activation in mast cell proliferation. *EMBO J* **17**: 6250–6262.

Toffalini F, Hellberg C, Demoulin J-B. 2010. Critical role of the platelet-derived growth factor receptor (PDGFR) β transmembrane domain in the TEL-PDGFRβ cytosolic oncoprotein. *J Biol Chem* **285**: 12268–12278.

Ueda S, Mizuki M, Ikeda H, Tsujimura T, Matsumura I, Nakano K, Daino H, Honda Zi Z, Sonoyama J, Shibayama H, et al. 2002. Critical roles of c-Kit tyrosine residues 567 and 719 in stem cell factor-induced chemotaxis: Contribution of src family kinase and PI3-kinase on calcium mobilization and cell migration. *Blood* **99**: 3342–3349.

Ustun C, DeRemer DL, Akin C. 2011. Tyrosine kinase inhibitors in the treatment of systemic mastocytosis. *Leuk Res* **35**: 1143–1152.

Van den Akker NM, Winkel LC, Nisancioglu MH, Maas S, Wisse LJ, Armulik A, Poelmann RE, Lie-Venema H, Betsholtz C, Gittenberger-de Groot AC. 2008. PDGF-B signaling is important for murine cardiac development: Its role in developing atrioventricular valves, coronaries, and cardiac innervation. *Dev Dyn* **237**: 494–503.

Wang L, Vargas H, French SW. 2000. Cellular origin of gastrointestinal stromal tumors: A study of 27 cases. *Arch Pathol Lab Med* **124**: 1471–1475.

Wang Y, Pennock SD, Chen X, Kazlauskas A, Wang Z. 2004. Platelet-derived growth factor receptor-mediated signal transduction from endosomes. *J Biol Chem* **279**: 8038–8046.

Ward SM, Burns AJ, Torihashi S, Harney SC, Sanders KM. 1995. Impaired development of interstitial cells and intestinal electrical rhythmicity in steel mutants. *Am J Physiol* **269**: C1577–C1585.

Wehrle-Haller B. 2003. The role of Kit-ligand in melanocyte development and epidermal homeostasis. *Pigment Cell Res* **16**: 287–296.

Wisniewski D, Strife A, Clarkson B. 1996. c-kit ligand stimulates tyrosine phosphorylation of the c-Cbl protein in human hematopoietic cells. *Leukemia* **10**: 1436–1442.

Wolff NC, Randle DE, Egorin MJ, Minna JD, Ilaria RL Jr. 2004. Imatinib mesylate efficiently achieves therapeutic intratumor concentrations in vivo but has limited activity in a xenograft model of small cell lung cancer. *Clin Cancer Res* **10**: 3528–3534.

Woodruff RH, Fruttiger M, Richardson WD, Franklin RJ. 2004. Platelet-derived growth factor regulates oligodendrocyte progenitor numbers in adult CNS and their response following CNS demyelination. *Mol Cell Neurosci* **25**: 252–262.

Voytyuk O, Lennartsson J, Mogi A, Caruana G, Courtneidge SA, Ashman LK, Rönnstrand L. 2003. Src family kinases are involved in the differential signaling from two splice forms of c-Kit. *J Biol Chem* **278**: 9159–9166.

Wybenga-Groot LE, Baskin B, Ong SH, Tong J, Pawson T, Sicheri F. 2001. Structural basis for autoinhibition of the EphB2 receptor tyrosine kinase by the unphosphorylated juxtamembrane region. *Cell* **106**: 745–757.

Xie J, Aszterbaum M, Zhang X, Bonifas JM, Zachary C, Epstein E, McCormick F. 2001. A role of PDGFRα in basal cell carcinoma proliferation. *Proc Natl Acad Sci* **98**: 9255–9259.

Xu X, Bringas P Jr, Soriano P, Chai Y. 2005. PDGFR-α signaling is critical for tooth cusp and palate morphogenesis. *Dev Dyn* **232**: 75–84.

Yang Y, Yuzawa S, Schlessinger J. 2008. Contacts between membrane proximal regions of the PDGF receptor ectodomain are required for receptor activation but not for receptor dimerization. *Proc Natl Acad Sci* **105**: 7681–7686.

Yee NS, Hsiau CW, Serve H, Vosseller K, Besmer P. 1994. Mechanism of down-regulation of c-kit receptor. Roles of receptor tyrosine kinase, phosphatidylinositol 3′-kinase, and protein kinase C. *J Biol Chem* **269**: 31991–31998.

Yuzawa S, Opatowsky Y, Zhang Z, Mandiyan V, Lax I, Schlessinger J. 2007. Structural basis for activation of the receptor tyrosine kinase KIT by stem cell factor. *Cell* **130:** 323–334.

Zadjali F, Pike AC, Vesterlund M, Sun J, Wu C, Li SS, Rönnstrand L, Knapp S, Bullock AN, Flores-Morales A. 2011. Structural basis for c-KIT inhibition by the suppressor of cytokine signaling 6 (SOCS6) ubiquitin ligase. *J Biol Chem* **286:** 480–490.

Zeng S, Xu Z, Lipkowitz S, Longley JB. 2005. Regulation of stem cell factor receptor signaling by Cbl family proteins (Cbl-b/c-Cbl). *Blood* **105:** 226–232.

Zhang Z, Zhang R, Joachimiak A, Schlessinger J, Kong XP. 2000. Crystal structure of human stem cell factor: Implication for stem cell factor receptor dimerization and activation. *Proc Natl Acad Sci* **97:** 7732–7737.

Zhao J, Roth J, Bode-Lesniewska B, Pfaltz M, Heitz PU, Komminoth P. 2002. Combined comparative genomic hybridization and genomic microarray for detection of gene amplifications in pulmonary artery intimal sarcomas and adrenocortical tumors. *Gene Chromosome Canc* **34:** 48–57.

VEGFR and Type-V RTK Activation and Signaling

Masabumi Shibuya[1,2]

[1]Jobu University, Isesaki, Gunma 372-8588, Japan

[2]Department of Molecular Oncology, Tokyo Medical and Dental University, Bunkyo-ku,
 Tokyo 113-8519, Japan

Correspondence: shibuya@ims.u-tokyo.ac.jp

Vascular endothelial growth factor receptors (VEGFRs) in vertebrates play essential roles in
the regulation of angiogenesis and lymphangiogenesis. VEGFRs belong to the receptor-type
tyrosine kinase (RTK) supergene family. They consist of a ligand-binding region with seven
immunoglobulin (7 Ig) -like domains, a *trans*-membrane (TM) domain, and a tyrosine kinase
(TK) domain with a long kinase insert (KI) (also known as a type-V RTK). Structurally, VEGFRs
are distantly related to the members of the M-colony stimulating factor receptor/platelet-
derived growth factor receptor (CSFR)/(PDGFR) family, which have five immunoglobulin
(5 Ig)-like domains. However, signal transduction in VEGFRs significantly differs from that
in M-CSFR/PDGFRs. VEGFR2, the major signal transducer for angiogenesis, preferentially
uses the phospholipase Cγ-protein kinase C (PLC-γ-PKC)-MAPK pathway, whereas M-CSFR/
PDGFRs use the PI3 kinase-Ras-MAPK pathway for cell proliferation. In phylogenetic development, the VEGFR-like receptor in nonvertebrates appears to be the ancestor of the 7 Ig-
and 5 Ig-RTK families because most nonvertebrates have only a single *7 Ig-RTK* gene. In
mammals, VEGFRs are deeply involved in pathological angiogenesis, including cancer and
inflammation. Thus, an efficient inhibitor targeting VEGFRs could be useful in suppressing
various diseases.

Angiogenesis, blood vessel formation, is essential for supplying oxygen and nutrients
to the tissues and for removing carbon dioxide
and waste materials (Hanahan and Folkman
1996; Risau 1997). Furthermore, lymph vessels are crucial for the absorption of tissue
fluids and the recovery of tissue-infiltrated
lymphocytes. In the 1980s, two proteins, vascular permeability factor (VPF) and vascular
endothelial growth factor receptors (VEGF),
were independently isolated. However, in 1989,
these proteins were found to be identical and
encoded by a single gene (Ferrara and Davis-
Smith 1997; Dvorak 2002). This protein, named
VEGF (or VEGF-A), has a homodimeric structure with three intramolecular and two intermolecular disulfide (S–S) bonds. However, its
receptor and signaling patterns were yet to be
elucidated.

In 1990, we isolated from human placenta
a novel receptor-type tyrosine kinase (RTK)
cDNA encoding a protein with seven Ig-like domains in its extracellular region. We designated
it Flt-1 (Fms-like tyrosine kinase-1; the first
type-V RTK) because of its similarity with the
Fms (M-CSFR) and PDGFR family (Fig. 1)
(Shibuya et al. 1990; Shibuya 1995). On the basis of this similarity, the ligand of Flt-1 was

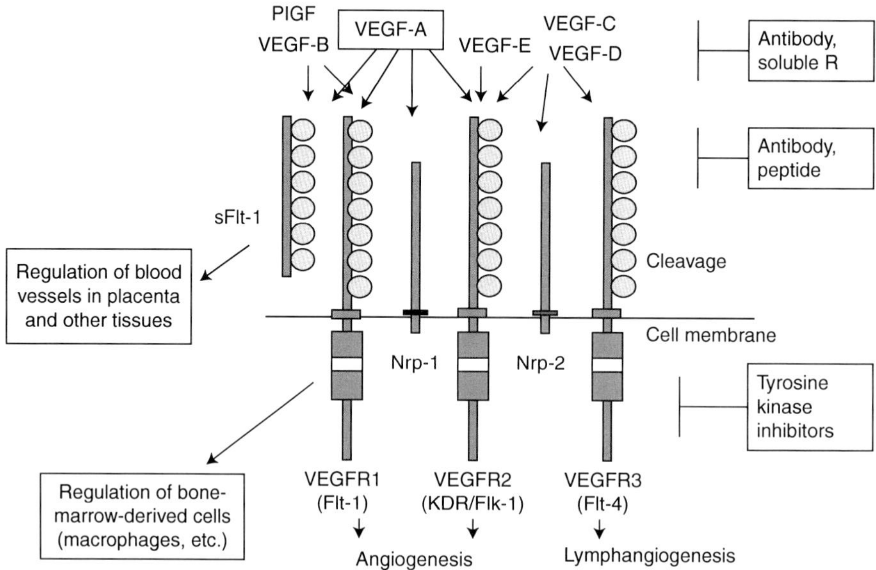

Figure 1. The VEGF-VEGFR system and its inhibitors. VEGF (VEGF-A) and its receptors, VEGFR1 and VEGFR2, play a major role in vasculogenesis and angiogenesis. VEGFR3 regulates not only lymphangiogenesis but also angiogenesis under pathological conditions. Neuropilins (Nrp)-1 and Nrp-2 function as coreceptors for VEGFR1 and VEGFR2, respectively, and efficiently stimulate VEGFR signaling. A heterodimer formed between two receptors, such as VEGFR1/VEGFR2 and VEGFR2/VEGFR3, was reported to regulate angiogenesis as well as lymphangiogenesis. Various inhibitors were developed to suppress this system, and some of them are now widely used clinically.

suggested to be a protein with a homodimeric structure, such as M-CSF and PDGF. In 1992, De Vries et al. (1992) reported tight binding and activation of Flt-1 with VEGF, indicating that Flt-1 is the first receptor for VEGF (VEGFR1).

Several years later, two RTKs structurally related to Flt-1 were isolated: one was KDR/Flk-1 (kinase-insert [KI] domain receptor in humans/fetal liver kinase-1 in mice; VEGFR2), while the other was Flt-4 (VEGFR3) (Matthews et al. 1991; Terman et al. 1991; Alitalo and Carmeliet 2002). VEGF binds VEGFR1 and VEGFR2, but not VEGFR3. Other VEGF family members, VEGF-C and VEGF-D, bind VEGFR3 and activate it for lymphangiogenesis (Joukov et al. 1997; Achen et al. 1998).

When we isolated full-length *flt-1* cDNA, we also found a short 3-kb-long *flt-1* mRNA, encoding a peptide without a TK domain that was highly expressed in normal placenta (Shibuya et al. 1990). This short *flt-1* mRNA was later reported to encode a soluble Flt-1 (sFlt-1) pro-

tein without *trans*-membrane (TM) and TK domains (Kendall and Thomas 1993; Hornig et al. 2000; Helske et al. 2001). Soluble Flt-1 is strongly expressed in the placental trophoblasts and is abnormally expressed in preeclampsia (Koga et al. 2003; Maynard et al. 2003). The main function of sFlt-1 is considered to be trapping of endogenous VEGF. Thus, extreme trapping and suppression of VEGF in various tissues of the body may result in the hypertension and proteinuria observed in preeclampsia patients (Levine et al. 2004).

VEGF heterozygotic knockout mice (*VEGF-A*$^{+/-}$ mice) are embryonic lethal because of immature development of angiogenesis, indicating that the concentration of VEGF is tightly regulated in tissues during embryogenesis (Carmeliet et al. 1996; Ferrara et al. 1996). The VEGF-VEGFR system is the major regulator of angiogenesis. Therefore, this system is an attractive target for antiangiogenic therapy and proangiogenic therapy.

 Cite this article as *Cold Spring Harb Perspect Biol* doi: 10.1101/cshperspect.a009092

STRUCTURAL CHARACTERISTICS OF VEGFRs AND THEIR RELATIONSHIP WITH OTHER RTKs

Mammals possess three *VEGFR* genes, namely, *VEGFR1*, *VEGFR2*, and *VEGFR3*, which produce four proteins, namely, VEGFR1, VEGFR2, VEGFR3, and sFlt-1 (sVEGFR1) (Shibuya and Claesson-Welsh 2006). The three receptors have 7 Ig-like domains that, except for the fourth Ig-like domain, possibly form S–S bonds. Unlike the other domains, the fourth domain does not have cysteines, suggesting that its tertiary structure is different from the others (Shibuya et al. 1990). In addition, the extracellular region of VEGFR3 is cleaved by a protease at the fifth Ig domain, and the amino-terminal and carboxy-terminal peptides are fixed with an S–S bond in the functional receptor (Alitalo and Carmeliet 2002).

The overall structure of the first to fifth Ig domains in VEGFRs is very similar to the extracellular region of the M-CSFR/PDGFR family, the type-III RTKs. The region spanning from the second to third Ig regions of M-CSFR/PDGFRs is the ligand-binding site. Similarly, the same regions in VEGFRs contain the binding regions for VEGF and VEGF-family members (Keyt et al. 1996; Tanaka et al. 1997; Shinkai et al. 1998). For the ligand-dependent activation of VEGFRs, Yang et al. (2010) recently reported that a direct interaction between the extracellular domains of two receptors is required.

VEGFR1: Dual Roles in Angiogenesis and the Antiangiogenic Effect of the Ligand-Binding Domain

VEGFR1 binds VEGF at a very high affinity of $K_d = 1-10$ pmol, and binds two other members of the VEGF family, namely, placental growth factor (PlGF) and VEGF-B (Sawano et al. 1996; Ferrara 2004; Shibuya and Claesson-Welsh 2006). Mice deficient in *flt-1* (*flt-1 $^{-/-}$* mice) die at embryonic day (E) 8.0–8.5 because of overgrowth and disorganization of blood vessels (Fong et al. 1995). This phenotype suggests that VEGFR1 plays a negative role in angiogenesis during early embryogenesis and that it is

needed to maintain an appropriate balance between the positive and negative regulators.

Because the affinity of VEGFR1 for VEGF is very high, the negative role of VEGFR1 suggests two possibilities: the ligand-binding region of VEGFR1 may trap VEGF, or TK activation in VEGFR1 may transduce a negative signal. To distinguish these possibilities, we generated TK domain-deficient mice (*flt-1 TK $^{-/-}$* mice) by a knockout procedure. To our surprise, *flt-1 TK $^{-/-}$* mice were healthy, and angiogenesis was normal in the embryos (Fig. 2) (Hiratsuka et al. 1998). These results clearly indicate that the negative role of VEGFR1 in angiogenesis is localized to the ligand-binding domain of the receptor, and it, most likely, involves trapping VEGF and decreasing its concentration around vascular endothelial cells (EC). In *flt-1 TK $^{-/-}$* mice, VEGF-dependent migration of macrophages was deficient, supporting the idea that VEGFR1 also generates a positive signal, for example, one that stimulates cell migration (Barleon et al. 1996; Clauss et al. 1996; Hiratsuka et al. 1998; LeCouter et al. 2003).

Additional studies addressed whether the ligand-binding region of VEGFR1 needed to be localized on the membrane of vascular endothelial cells (ECs) in order to play a negative role in angiogenesis. To clarify this, knockout mice were generated in which the TM domain as well as the TK domain of VEGFR1 was missing (*flt-1 TM-TK $^{-/-}$* mice). In the mutant mice, approximately half the littermates were embryonic lethal, but the other half were healthy, similar to the *flt-1 TK $^{/}$* mice (Hiratsuka et al. 2005). Therefore, these results suggest that the ligand-binding region of VEGFR1 should be fixed on the cell membrane to function appropriately as a negative regulator of angiogenesis. If the TM domain is missing, the mortality of the embryo may depend on the location of the solubilized VEGFR1. When the peptide is mostly localized near the membrane of vascular ECs, it works properly as a negative regulator just as in *flt-1 TK $^{-/-}$* mice; however, when it is localized far from the membrane, its regulatory activity does not work properly, resulting in disorganization of blood vessels and embryonic lethality.

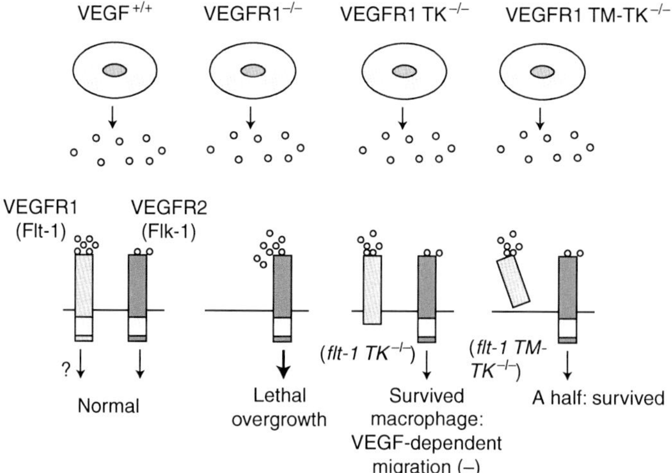

Figure 2. VEGFR1 has dual roles, positive and negative, in angiogenesis. VEGFR1 has a strong binding affinity for VEGF, but its kinase activity is much weaker than that of VEGFR2. VEGFR1 knockout mice (i.e., *flt-1*$^{-/-}$ mice) die during early embryogenesis because of overgrowth of angiogenesis, indicating a negative role for VEGFR1 in angiogenesis. VEGFR1-signal-deficient mice (i.e., *flt-1 TK*$^{-/-}$ mice), organize blood vessel structure in a normal manner. Thus, the negative function of VEGFR1 localizes to the ligand-binding domain. *flt-1 TM-TK*$^{-/-}$ mice indicate that about half of littermates are healthy but the other half are embryonic lethal due to poor development of blood vessels.

VEGFR2: The Major Positive Signal Transducer for Angiogenesis

VEGF$^{+/-}$ heterozygous mice die at the embryonic stage because of insufficient organization of blood vessel formation. Instances of haploid insufficiency like *VEGF* gene are very rare in mammals, indicating that the degree of positive signal from the VEGF-VEGFR system is tightly regulated in the body.

VEGFR2 is the second high-affinity receptor for VEGF: its binding affinity is about one-tenth that of VEGFR1. However, the TK activity of VEGFR2 is about 10-fold stronger than the TK activity of VEGFR1, suggesting that VEGFR2 is the major signal transducer for angiogenesis. Mice deficient in VEGFR2 (*flk-1*$^{-/-}$ mice) were embryonic lethal at E8.5, because of lack of vasculogenesis/angiogenesis, indicating that VEGFR2 is the main positive signal transducer for blood vessel formation in embryogenesis (Shalaby et al. 1995).

VEGFR1 and VEGFR2 are highly homologous at the amino acid level. Particularly, tyrosine residues in the carboxy-terminal region are well conserved: Y1175 in VEGFR2 and Y1169 in VEGFR1 are part of similar amino acid motifs (Shibuya 1995), and appear to be crucial for downstream signaling, as discussed below.

VEGFR3: Regulator of Both Lymph Vessel and Blood Vessel Formation

VEGFR3 is expressed in lymph ECs as well as in venous ECs during early embryogenesis. VEGFR3 binds VEGF-C and VEGF-D (Alitalo and Carmeliet 2002; Lohela et al. 2009). These ligands are synthesized in precursor forms that have limited affinity for the receptor. However, after processing in both amino- and carboxy-terminal portions, the mature ligands bind VEGFR3 with a higher affinity (Joukov et al. 1997; McColl et al. 2007). In addition, mature VEGF-C and VEGF-D bind VEGFR2 at a lower affinity and can activate the receptor. Although VEGF-C and VEGF-D can activate both VEGFR3 and VEGFR2, a mutant of VEGF-C, VEGF-C156S, which binds and activates only VEGFR3, stimulates lymphangiogenesis equally well as wild-type VEGF-C does in the keratin-promoter-driven transgenic mouse system (Veikkola et al. 2001). This suggests that the activation of VEGFR3 is sufficient for signaling toward lymphangiogenesis.

VEGFR3 knockout mice (*flt-4*$^{-/-}$ mice) die because of disorganization of blood vessels at E9.5 when lymphangiogenesis has not yet occurred (Dumont et al. 1998). These results clearly show that VEGFR3 plays an important role not only in lymphangiogenesis but also in blood vessel formation before lymphangiogenesis during early embryogenesis.

Structurally, VEGFR3 is slightly less homologous to VEGFR1 and VEGFR2 than the latter two receptors are to each other. Consequently, VEGFR3 differs from VEGFR1 and VEGFR2 in at least two ways. First, the extracellular domain in VEGFR3 is cleaved and fixed with a S—S bond, whereas the extracellular domains in VEGFR1 and VEGFR2 are not cleaved. Second, in the carboxy-terminal sequences, a tyrosine residue critical for signaling, Y1175 in VEGFR2 and Y1169 in VEGFR1, is not conserved in VEGFR3, suggesting a difference in signaling (Shibuya and Claesson-Welsh 2006).

Makinen et al. (2001) reported that VEGFR3 uses PKC and Ras pathways for lymphangiogenesis. Furthermore, most familial lymphedema patients were found to carry various mutations in the *VEGFR3* gene, particularly in the tyrosine kinase domain. These mutations cause a decrease in kinase activity and signaling, indicating that proper lymphangiogenesis in humans is dependent on the TK activity of VEGFR3 (Irrthum et al. 2000; Karkkainen et al. 2001).

sFlt-1: A Powerful Endogenous Antiangiogenic Protein

sFlt-1 (sVEGFR-1) consists of six Ig-like domains and a 31-amino-acid-long tail (Kendall and Thomas 1993; Shibuya 2011). The *sFlt-1* mRNA is generated by premature termination within intron 13. Surprisingly, the tail sequence encoded in the 5' portion of intron 13 is structurally highly conserved among vertebrates. We found that chicken sFlt-1 and mouse sFlt-1 also have the 31-amino-acid-long tail. The tails of humans, mice, and chickens are highly homologous to each other, approximately 74% iden-

tical at the amino acid level (Kondo et al. 1998; Yamaguchi et al. 2002). This significant homology strongly suggests that the tail sequence has biological significance, and that full-length exon 13 in the *VEGFR1* gene encodes a long form of the peptide that includes the 31-amino-acid-long portion.

Amphibians, but not fish, have mRNA for sFlt-1. Therefore, the system of three VEGFRs and one sFlt-1 was established at an early stage of development in vertebrates (i.e., the amphibian stage). Because sFlt-1 can trap the major angiogenic factor VEGF, it suggests that sFlt-1 is an important suppressor of angiogenesis in various animal tissues.

Furthermore, sFlt-1 might have been a key player in the establishment of the placenta-carrying animals (i.e., mammals). The placenta is essential for communication between the fetal and maternal circulatory systems. Oxygen and nutrients are supplied from the maternal circulation to the fetal circulation, whereas wasted materials are transferred from the fetal circulation to the maternal circulation. Under physiological conditions, sFlt-1 is highly expressed in trophoblasts, which are located between the maternal and fetal blood vessels in the placenta (Helske et al. 2001). Angiogenesis on both fetal and maternal sides in this tissue must be well organized, without overgrowth of blood vessels, to avoid fusion or abnormal hyperpermeability. To obtain such a tight regulation of angiogenesis, sFlt-1, a strong endogenous inhibitor of VEGF, is considered a very efficient molecule. Therefore, it seems reasonable that sFlt-1 plays an important role in the suppression of abnormal angiogenesis in placenta to maintain its function (Clark et al. 1998; Shibuya 2011).

Shorter forms of sFlt-1 that contain only the first three, four, or five Ig domains still maintain a high affinity for VEGF (Tanaka et al. 1997). In contrast, short peptides of VEGFR2, with only the first four or five Ig domains, have a lower affinity for VEGF, 10- to 50-fold lower than full-length VEGFR2 (Shinkai et al. 1998). Recently, Albuquerque et al. (2009) reported that the *VEGFR2* gene also produces sVEGFR2. When sVEGFR2 has six Ig domains, it could bind VEGF or VEGF-C. However, when it is cleaved to less than six Ig domains, it might be difficult to regulate VEGF/VEGF-C levels because of its weaker ligand-trapping activity.

An Intimate Relationship of sFlt-1 with Preeclampsia

In 2003, Koga et al. (2003) and Maynard et al. (2003) reported that sFlt-1 is abnormally high in the serum of preeclampsia patients. Furthermore, Levine et al. (2004) showed a tight relationship between the degree of preeclampsia and the level of sFlt-1 in serum, strongly suggesting that sFlt-1 is an important cause of preeclampsia symptoms such as hypertension and proteinuria. Using a pregnant rat model, Maynard et al. (2003) showed that artificial introduction of sFlt-1 using vector system-induced preeclampsia-like symptoms, specifically hypertension and proteinuria. The molecular mechanism for the overexpression of sFlt-1 in the trophoblasts of preeclampsia patients is not yet fully understood. However, a similar phenomenon was observed with antiangiogenic therapy in the treatment of cancer patients with VEGF-trapping antibody (bevacizumab)-induced adverse effects, including hypertension and proteinuria, which are symptoms observed in preeclampsia (Hurwitz et al. 2004). Therefore, sFlt-1 is an attractive target for future efforts to suppress the symptoms of preeclampsia and improve overall conditions for both mother and baby (Fig. 3) (Thadhani et al. 2011).

Soluble Flt-1 is expressed not only in trophoblasts but also in other cell types, including corneal epithelial cells. Ambati et al. (2006) showed that sFlt-1 expressed in the cornea is important for maintaining the avascularity of the lens and transparent vision.

SIGNAL TRANSDUCTION OF VEGFR: A UNIQUE DEPENDENCY ON THE PLC-γ-PKC PATHWAY

A New Function of sFlt-1 in Kidney

Recently, sFlt-1 has been shown to have another interesting role in vivo. Quaggin's group demonstrated that the podocytes in the glomeruli in

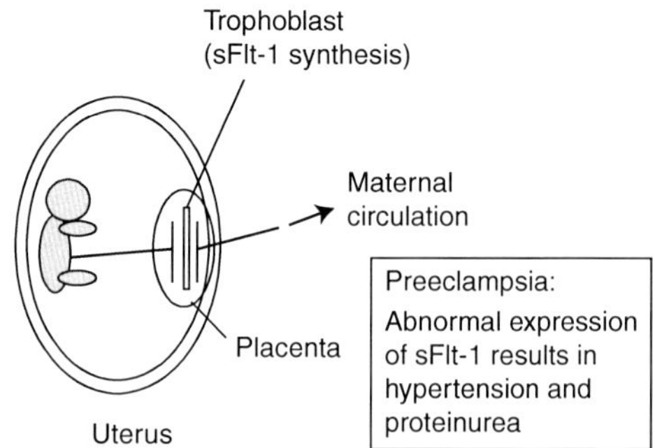

Figure 3. sFlt-1: An endogenous VEGF-inhibitor intimately related to preeclampsia. The soluble form of VEGFR1, sFlt-1, is expressed in normal placenta, particularly in the trophoblasts. Expression of the *sFlt-1* gene is abnormally increased in preeclampsia patients, and the excess sFlt-1 is distributed via the bloodstream to various maternal tissues. sFlt-1 traps endogenous VEGF, resulting in hypertension and proteinuria.

kidney secrete sFlt-1 to regulate glomerular barrier function, and knockdown of sFlt-1 in the podocytes results in chronic proteinurea, similar to the cases of patients with nephrotic syndrome (Jin et al. 2012).

In the 5 Ig-RTKs, such as M-CSFR and PDGFRs, the 60-amino-acid-kinase-insert (KI) sequence located in the middle of the TK domain has several conserved tyrosine residues with YMXM or YXXM motifs. Once phosphorylated at the tyrosine residue, these motifs become strong binding and activation sites for PI3 kinase (PI3K), resulting in the activation of the PI3K-Ras-MAPK pathway for cell proliferation and transformation (Heldin and Westermark 1999). However, interestingly, none of the VEGFRs contains the YXXM motif in their KI regions. Even after activation with VEGF, VEGFR1 and VEGFR2 barely bind the PI3K subunit p85. These differences strongly suggest that the major signaling pathway of VEGF-VEGFR is different from that of M-CSFR/PDGFRs.

Our study and previous studies have found that PKCs, particularly PKC-β, play an important role in VEGFR2 signaling (Xia et al. 1996; Takahashi and Shibuya 1997; Takahashi et al. 1999). After stimulation with VEGF, several tyrosines, including Y1175 and Y1214, were

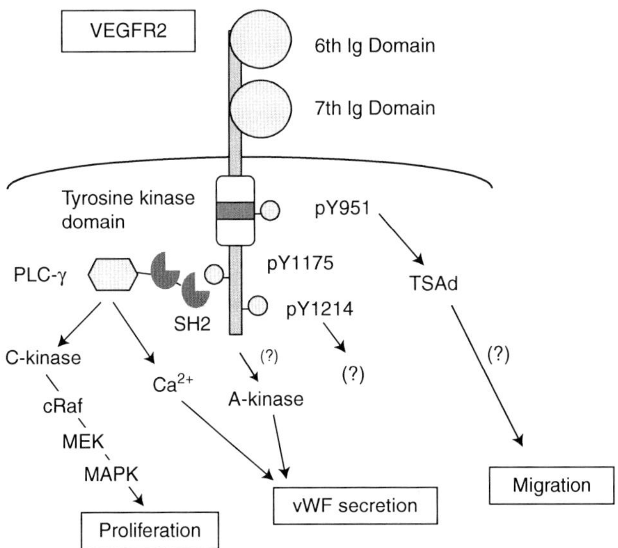

Figure 4. A unique signaling pathway from VEGFR2 toward angiogenesis. The Y1175 site is one of the major autophosphorylation sites on VEGFR2 after stimulation with VEGF. The pY1175-containing motif is the binding and activation site for PLC-γ. It stimulates PKC-Raf-MEK-MAPK and calcium mobilization pathways for EC activation and proliferation.

strongly phosphorylated. We found that phosphorylated Y1175 (pY1175), but not pY1214, is the critical binding site for the carboxy-terminal SH2 domain of PLC-γ and the initiation site of the PLC-γ-PKC-Raf-MEK-MAPK pathway for the proliferation of ECs (Fig. 4) (Takahashi et al. 2001). In addition, activation of PLC-γ stimulates Ca^{2+} mobilization within the cell. Replacement of Y1175 with phenylalanine in VEGFR-2 or intracellular injection of pY1175-blocking antibody (anti-pY1175 peptide Ab) significantly inhibits the activation of the PLC-γ-PKC-MAPK pathway, suppressing the cell proliferation signal in vitro.

To examine whether pY1175 in VEGFR2 is crucial for the angiogenic signal in vivo, we generated a knock-in mouse with a point mutation at amino acid 1175 (1173 in mice) that changes tyrosine to phenylalanine (F). Homozygotic Y1173F/F mice, but not Y1212F/F mice, die because of lack of vasculogenesis similar to *flk-1*$^{-/-}$ mice, even though Y1173F/F mice still express VEGFR2 with other phosphotyrosine sites (Sakurai et al. 2005). These results support the idea that signaling from pY1173 of VEGFR2 to the PLC-γ-PKC-MAPK pathway is essential

for vasculogenesis in embryogenesis. Interestingly, Sase et al. (2009), using a murine embryonic stem (ES) cell system, reported that differentiation of ECs from ES cells strongly depends on the VEGFR2-pY1173 to PLC-γ pathway. Furthermore, a spontaneous mutant of the *PLC-γ1* gene in zebrafish is lethal because of deficiency of arteriogenesis (Lawson et al. 2003). These findings strongly suggest that the VEGFR-PLC-γ-PKC-MAPK pathway is essential for angiogenic signals, not only in mammals but also in fish, the earliest vertebrates.

Recently, Wang et al. (2008) reported that the VEGFR2-PLC-γ-PKC pathway activates the protein kinase D (PKD)-histone-deacetylase 7 (HDAC7) pathway for gene expression toward EC proliferation and migration. In addition, Xiong et al. (2009) found that VEGFR2-pY1175 is essential for the release of von Willebrand factor (vWF) from the ECs to regulate the coagulation system.

VEGF stimulates not only proliferation but also sprouting, survival, migration, and tube formation of ECs. Phosphorylated Y951 was reported to be important for cell migration via recruitment of T-cell-specific adaptor protein

(TSAd) (Matsumoto et al. 2005). Survival signals appear to be generated from the PKC-MAPK pathway and the PI3K pathway. However, the phosphotyrosine(s) on VEGFR2 that is the activation site for PI3K has not yet been identified.

VEGFR1 is expressed not only on vascular ECs but also on macrophages, trophoblasts, tumor cells, and others (Wu et al. 2006; Tsuchida et al. 2008). VEGFR1 is involved in the survival, angiogenic signaling (albeit not strong), and migration of ECs and macrophages (Sawano et al. 2001; LeCouter et al. 2003). VEGFR1 is also involved in bone marrow reconstitution (Niida et al. 2005), and it stimulates tumor growth, metastasis, and inflammation (Hiratsuka et al. 2002; Kaplan et al. 2005; Murakami et al. 2006; Kerber et al. 2008; Muramatsu et al. 2010). However, the signaling of VEGFR1 to promote cell migration has not yet been characterized. Using a stable VEGFR1 overexpressing human EC line, we recently examined this signaling. An intracellular scaffold protein, receptor of activated protein kinase C1 (RACK1), was involved in the migration signal, and stimulated the PI3K-Akt-Rac1 pathway (Wang et al. 2011).

CROSS TALK BETWEEN VEGF-VEGFR SIGNAL AND OTHER SIGNALING PATHWAYS

The VEGF-VEGFR system is reported to communicate with a variety of other signaling systems. In angiogenesis, upon stimulation with VEGF, capillary ECs can shift to two major types, namely, tip cells and stalk cells (Ruhrberg et al. 2002). A tip cell has many filopodia at its front for migration and has low cell proliferation activity. On the other hand, a stalk cell forms a tubular structure and has high proliferation activity. In this system, tip cells express Dll4, a ligand of the Delta family, and bind and activate Notch receptors on stalk cells. Notch signaling in stalk cells suppresses the expression of VEGFR2 to an appropriate level, thereby stimulating cell proliferation (Jakobsson et al. 2009). In addition, sFlt-1 expression in stalk cells regulates tip cell formation in angiogenesis (Chappell et al. 2009).

Recently, Stefater et al. (2011) reported that the Wnt signal cooperates with sFlt-1 to regulate retinal blood vessel formation in perinatal development. The Ang/Tie system is another important regulator of angiogenesis. Ang2, in the presence of VEGF, promotes angiogenesis. However, without VEGF, Ang2 suppresses angiogenesis because of induction of vascular instability. The relationship between the Ang/Tie system and the VEGF/VEGFR system appears to be context dependent, and is yet to be fully understood.

PHYLOGENETIC DEVELOPMENT OF VEGFRS

Genes encoding type-V RTKs (7 Ig-RTKs) have already been identified in nonvertebrates. Examples include the *Drosophila DVR/PVR* gene and the sea squirt *VEGFR*-like receptor gene. Except for *Caenorhabditis elegans*, which possesses three genes encoding type-V RTKs, other nonvertebrates have only one type-V RTK gene. Because nonvertebrates do not have any type-III RTK (5 Ig-RTK) genes, the type-V RTK gene is considered the ancestor of genes in the VEGFR (7 Ig-RTKs) and PDGFR (5 Ig-RTKs) families, including M-CSFR, CSFR, and Flt3 in mammals and other vertebrates. Based on the structural homology, both *cis* gene duplication and *trans* gene duplication may have occurred twice for each during the early development of vertebrates (Fig. 5) (Shibuya 2002). *Drosophila DVR* is mainly used for cell migration and not for cell proliferation (Duchek et al. 2001), and its signaling uses JNK and other pathways (Ishimaru et al. 2004). The exon−intron structure for the KI in VEGFRs differs from that in the PDGFR family (Kondo et al. 1998), suggesting that the KI sequence was replaced during the generation of the 5 Ig-RTKs (PDGFRs) from the 7 Ig-RTKs. During this step, the KI of the 5 Ig-RTKs may have acquired the YXXM motifs that produce strong cell proliferation signals.

ANTIANGIOGENIC THERAPY TARGETING THE VEGF-VEGFR SYSTEM

Solid tumors secrete a variety of angiogenic factors, such as VEGF, fibroblast growth factor

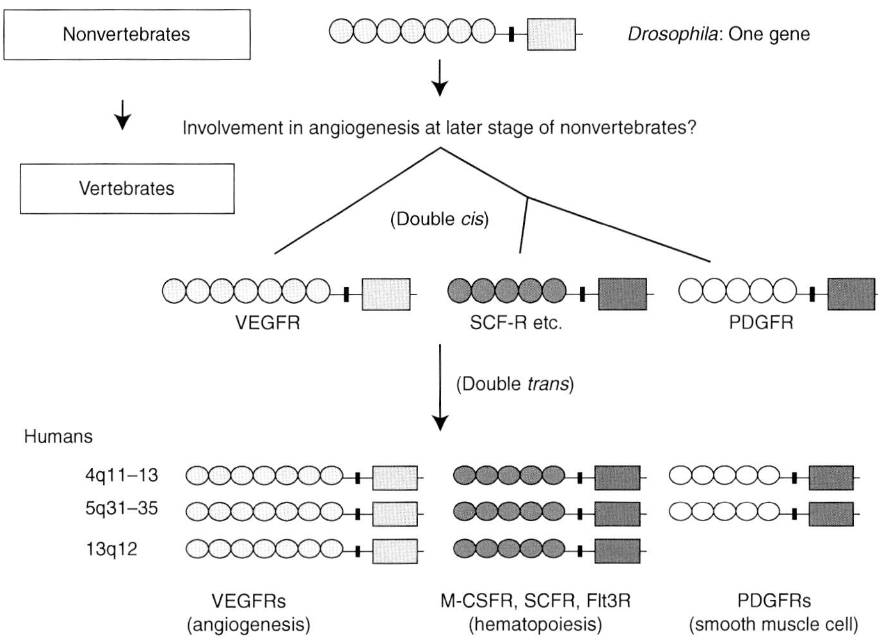

Figure 5. Phylogenetic development of the 7 Ig-RTKs (VEGFRs) and the 5 Ig-RTKs: A hypothesis. Most nonvertebrates carry a single *7 Ig-RTK* gene. Double *cis* gene duplication followed by double *trans* gene duplication appears to have generated the 7 Ig-RTK and 5 Ig-RTK supergene families in vertebrates.

(FGF), hepatocyte growth factor (HGF), and PDGF. However, among these factors, VEGF is thought to play a central role in tumor angiogenesis because blocking VEGF using either anti-VEGF neutralizing antibody or sFlt-1 protein efficiently suppresses solid tumor growth in mouse models (Kim et al. 1993). On the basis of these findings, Genentech Co. developed a human VEGF-neutralizing antibody (bevacizumab, Avastin) for cancer treatment. In 2003, phase III studies of colorectal cancer clearly showed that a combination of chemotherapy with bevacizumab extended overall survival (OS) from 15.6 months (control group; without bevacizumab) to 20.3 months (group with bevacizumab) (Hurwitz et al. 2004). Now, in 2011, this anti-VEGF antibody is approved for use in treating colorectal cancer, lung cancer (non-SCLC) (Cohen et al. 2007), and glioblastoma in many countries. For breast cancer patients, the Food and Drug Administration (FDA) in the United States approved the use of bevacizumab in 2005. However, after reevaluating a recent clinical study, in 2010, the FDA withdrew its

approval, because bevacizumab was less efficacious in improving OS in patients. On the other hand, in 2011, bevacizumab was approved for the treatment of breast cancer patients in several countries because it improved progression-free survival (PFS). Additional clinical studies are necessary to clarify the efficacy of anti-VEGF molecules in various tumors at different settings.

Several chemicals, such as sorafenib and sunitinib, which act as inhibitors of multiple tyrosine kinases, including VEGFRs, were developed; these have been widely used to treat renal cell cancer and hepatocellular cancer because of the increase in the OS.

Several mechanisms have been proposed to explain the efficacy of VEGF-VEGFR inhibitors. These include suppression of preexisting tumor vasculature, suppression of new angiogenesis, and vascular normalization in tumor tissues. Vascular normalization may decrease the abnormal vascular permeability of tumor vessels, resulting in normalized intratumoral pressure and better distribution of anticancer chemicals (Jain 2005).

Although more than five types of solid tumors are sensitive to anti-VEGF-VEGFR medicines, refractoriness or acquired resistance to these drugs might occur after long-term treatment. Expression of other angiogenic factors or receptors such as c-Met after anti-VEGF-VEGFR therapy was reported (You et al. 2011). Furthermore, tumors might acquire a phenotype resistant to conditions of lower availability of oxygen and nutrients through metabolic changes and increases in the phospho-Akt levels (Osawa et al. 2009). The various responses of tumor tissues to antiangiogenic therapy should be clarified to aid the development of more efficient drugs that can improve overall and progression-free survival.

PROANGIOGENIC THERAPY USING THE VEGF-VEGFR SYSTEM

Ischemic diseases, such as cardiac infarction and brain stroke, are major mortal diseases in humans. Because the VEGF-VEGFR system plays a central role in angiogenesis in mammals, VEGF-A is an important candidate for therapeutic angiogenesis. However, a transgenic mouse model using keratin-promoter-driven VEGF-A induced angiogenesis with inflammatory responses. Kunstfeld et al. (2004) and others reported that the skin tissue in *VEGF-A* transgenic mice showed severe inflammation with hyperpermeability and lymphangiogenesis, and could be a model for dermal psoriasis. VEGF binds not only VEGFR2 but also VEGFR1. The activated VEGFR1 on macrophages stimulates cell migration as well as the production of inflammatory cytokines and lymphangiogenic factors like VEGF-C. We and others found that an Orf-virus-genome encoded a VEGF-like molecule (i.e., VEGF-E), which binds and activates only VEGFR2, not VEGFR1 (Lyttle et al. 1994; Ogawa et al. 1998). This characteristic of VEGF-E could promote angiogenesis without stimulating inflammation. We showed that K14-*VEGF-E* transgenic mice have increased angiogenesis under the skin without an inflammatory response (Kiba et al. 2003a). In these mice, vascular permeability was in the normal range, and inflammatory cytokines such as IL-6

and tumor necrosis factor (TNF) were not significantly induced. Thus, the VEGFR2-specific ligand VEGF-E appears be a good candidate for proangiogenic therapy. A possible adverse effect of this molecule might be antigenicity because the *VEGF-E* gene is not present in the human genome. To avoid the antigenicity of VEGF-E, we generated a chimeric molecule consisting of VEGF-E and a human protein, PlGF, that maintains VEGFR2-stimulating activity with less antigenicity (Kiba et al. 2003b; Zheng et al. 2006). Humanized VEGF-E, the Ang1-related protein COMP-Ang1 (Cho et al. 2005), FGF, and HGF (Azuma et al. 2006) are hopeful candidates for the treatment of ischemic diseases.

CONCLUSION

VEGFRs have a 7 Ig-RTK structure and activate a unique signaling pathway for angiogenesis and lymphangiogenesis. Soluble Flt-1 (sFlt-1/sVEGFR-1), a gene product of VEGFR-1, is a strong endogenous VEGF inhibitor, tightly linked with the onset of preeclampsia symptoms. The VEGF-VEGFR system is deeply involved in various diseases, including cancer and ischemic diseases. Thus, further analysis of this system may provide important clinical strategies.

ACKNOWLEDGMENTS

This work is supported by Grant-in-Aid Special Project Research on Cancer-Bioscience 17014020 from the Ministry of Education, Culture, Sports, Science and Technology of Japan.

REFERENCES

Achen MG, Jeltsch M, Kukk E, Mäkinen T, Vitali A, Wilks AF, Alitalo K, Stacker SA. 1998. Vascular endothelial growth factor D (VEGF-D) is a ligand for the tyrosine kinases VEGF receptor 2 (Flk1) and VEGF receptor 3 (Flt4). *Proc Natl Acad Sci* **95:** 548–553.

Albuquerque RJ, Hayashi T, Cho WG, Kleinman ME, Dridi S, Takeda A, Baffi JZ, Yamada K, Kaneko H, Green MG, et al. 2009. Alternatively spliced vascular endothelial growth factor receptor-2 is an essential endogenous inhibitor of lymphatic vessel growth. *Nat Med* **15:** 1023–1030.

Alitalo K, Carmeliet P. 2002. Molecular mechanisms of lymphangiogenesis in health and disease. *Cancer Cell* **1:** 219–227.

Ambati BK, Nozaki M, Singh N, Takeda A, Jani PD, Suthar T, Albuquerque RJ, Richter E, Sakurai E, Newcomb MT, et al. 2006. Corneal avascularity is due to soluble VEGF receptor-1. *Nature* **443**: 993–997.

Azuma J, Taniyama Y, Takeya Y, Iekushi K, Aoki M, Dosaka N, Matsumoto K, Nakamura T, Ogihara T, Morishita R. 2006. Angiogenic and antifibrotic actions of hepatocyte growth factor improve cardiac dysfunction in porcine ischemic cardiomyopathy. *Gene Ther* **13**: 1206–1213.

Barleon B, Sozzani S, Zhou D, Weich HA, Martovani A, Marme D. 1996. Migration of human monocytes in response to vascular endothelilal growth factor (VEGF) is mediated via the VEGF receptor flt-1. *Blood* **87**: 3336–3343.

Carmeliet P, Ferreira V, Breier G, Pollefeyt S, Kleckens L, Gertsenstein M, Fahrig M, Vandenhoeck A, Harpal K, Eberhardt C, et al. 1996. Abnormal blood vessel development and lethality in embryos lacking a single VEGF allele. *Nature* **380**: 435–439.

Chappell JC, Taylor SM, Ferrara N, Bautch VL. 2009. Local guidance of emerging vessel sprouts requires soluble Flt-1. *Dev Cell* **17**: 377–386.

Cho CH, Kim KE, Byun J, Jang HS, Kim DK, Baluk P, Baffert F, Lee GM, Mochizuki N, Kim J, et al. 2005. Long-term and sustained COMP-Ang1 induces long-lasting vascular enlargement and enhanced blood flow. *Circ Res* **97**: 86–94.

Clark DE, Smith SK, He Y, Day KA, Licence DR, Corps AN, Lammoglia R, Charnock-Jones DS. 1998. A vascular endothelial growth factor antagonist is produced by the human placenta and released into the maternal circulation. *Biol Reproduct* **59**: 1540–1548.

Clauss M, Weich H, Breier G, Knies U, Röckl W, Waltenberger J, Risau W. 1996. The vascular endothelial growth factor receptor Flt-1 madiates biological activities. *J Biol Chem* **271**: 17629–17634.

Cohen MH, Gootenberg J, Keegan P, Pazdur R. 2007. FDA drug approval summary: Bevacizumab (Avastin) plus Carboplatin and Paclitaxel as first-line treatment of advanced/metastatic recurrent nonsquamous non-small cell lung cancer. *Oncologist* **12**: 713–718.

De Vries C, Escobedo JA, Ueno H, Houck K, Ferrara N, Williams LT. 1992. The fms-like tyrosine kinase, a receptor for vascular endothelial growth factor. *Science* **255**: 989–991.

Duchek P, Somogyi K, Jekely G, Beccari S, Rorth P. 2001. Guidance of cell migration by the *Drosophila* PDGF/VEGF receptor. *Cell* **107**: 17–26.

Dumont DJ, Jussila L, Taipale J, Lymboussaki A, Mustonen T, Pajusola K, Breitman M, Alitalo K. 1998. Cardiovascular failure in mouse embryos deficient in VEGF receptor-3. *Science* **282**: 946–949.

Dvorak HF. 2002. Vascular permeability factor/vascular endothelial growth factor: A critical cytokine in tumor angiogenesis and a potential target for diagnosis and therapy. *J Clin Oncol* **20**: 4368–4380.

Ferrara N. 2004. Vascular endothelial growth factor: Basic science and clinical progress. *Endocr Rev* **25**: 581–611.

Ferrara N, Davis-Smyth T. 1997. The biology of vascular endothelial growth factor. *Endocrine Rev* **18**: 4–25.

Ferrara N, Carver-Moore K, Chen H, Dowd M, Lu L, O'Shea KS, Powell-Braxton L, Hillan KJ, Moore MW. 1996. Het-erozygous embryonic lethality induced by targeted inactivation of the VEGF gene. *Nature* **380**: 439–442.

Fong GH, Rossant J, Gertsentein M, Breitman ML. 1995. Role of the Flt-1 receptor tyrosine kinase in regulating the assembly of vascular endothelium. *Nature* **376**: 66–70.

Hanahan D, Folkman J. 1996. Patterns and emerging mechanisms of the angiogenic switch during tumorigenesis. *Cell* **86**: 353–364.

Heldin CH, Westermark B. 1999. Mechanism of action and in vivo role of platelet-derived growth factor. *Physiol Rev* **79**: 1283–1316.

Helske S, Vuorela P, Carpen O, Hornig C, Weich H, Halmesmaki E. 2001. Expression of vascular endothelial growth factor receptors 1, 2 and 3 in placentas from normal and complicated pregnancies. *Mol Hum Reprod* **7**: 205–210.

Hiratsuka S, Minowa O, Kuno J, Noda T, Shibuya M. 1998. Flt-1 lacking the tyrosine kinase domain is sufficient for normal development and angiogenesis in mice. *Proc Natl Acad Sci* **95**: 9349–9354.

Hiratsuka S, Nakamura K, Iwai S, Murakami M, Itoh T, Kijima H, Shipley JM, Senior RM, Shibuya M, MS Mincho, et al. 2002. MMP9 induction by vascular endothelial growth factor receptor-1 is involved in lung specific metastasis. *Cancer Cell* **2**: 289–300.

Hiratsuka S, Nakao K, Nakamura K, Katsuki M, Maru Y, Shibuya M. 2005. Membrane-fixation of VRGFR1 ligand-binding domain is important for vasculogenesis/angiogenesis in mice. *Mol Cell Biol* **25**: 346–354.

Hornig C, Barleon B, Ahmad S, Vuorela P, Ahmed A, Weich HA. 2000. Release and complex formation of soluble VEGFR-1 from endothelial cells and biological fluids. *Lab Invest* **80**: 443–454.

Hurwitz H, Fehrenbacher L, Novotny W, Cartwright T, Hainsworth J, Heim W, Berlin J, Baron A, Griffing S, Holmgren E, et al. 2004. Bevacizumab plus irinotecan, fluorouracil, and leucovorin for metastatic colorectal cancer. *N Engl J Med* **350**: 2335–2342.

Irrthum A, Karkkainen MJ, Devriendt K, Alitalo K, Vikkula M. 2000. Congenital hereditary lymphedema caused by a mutation that inactivates VEGFR3 tyrosine kinase. *Am J Hum Genet* **67**: 295–301.

Ishimaru S, Ueda R, Hinohara Y, Ohtani M, Hanafusa H. 2004. PVR plays a critical role via JNK activation in thorax closure during *Drosophila* metamorphosis. *EMBO J* **23**: 3984–3994.

Jain RK. 2005. Normalization of tumor vasculature: An emerging concept in antiangiogenic therapy. *Science* **307**: 58–62.

Jakobsson L, Bentley K, Gerhardt H. 2009. VEGFRs and Notch: A dynamic collaboration in vascular patterning. *Biochem Soc Trans* **37**: 1233–1236.

Jin J, Sison K, Li C, Tian R, Wnuk M, Sung HK, Jeansson M, Zhang C, Tucholska M, Jones N, et al. 2012. Soluble FLT1 binds lipid microdomains in podocytes to control cell morphology and glomerular barrier function. *Cell* **151**: 384–399.

Joukov V, Sorsa T, Kumar V, Jeltsch M, Claesson-Welsh L, Cao Y, Saksela O, Kalkkinen N, Alitalo K. 1997. Proteolytic processing regulates receptor specificity and activity of VEGF-C. *EMBO J* **16**: 3898–3911.

Kaplan RN, Riba RD, Zacharoulis S, Bramley AH, Vincent L, Costa C, MacDonald DD, Jin DK, Shido K, Kerns SA, et al. 2005. VEGFR1-positive haematopoietic bone marrow progenitors initiate the pre-metastatic niche. *Nature* **438:** 820–827.

Karkkainen MJ, Jussila L, Ferrell RE, Finegold DN, Altalo K. 2001. Molecular regulation of lymphangiogenesis and targets for tissue oedema. *Trends Mol Med* **7:** 18–22.

Kendall RL, Thomas KA. 1993. Inhibition of vascular endothelial cell growth factor activity by an endogenously encoded soluble receptor. *Proc Natl Acad Sci* **90:** 10705–10709.

Kerber M, Reiss Y, Wickersheim A, Jugold M, Kiessling F, Heil M, Tchaikovski V, Waltenberger J, Shibuya M, Plate KH, et al. 2008. Flt-1 signaling in macrophages promotes glioma growth in vivo. *Cancer Res* **68:** 7342–7351.

Keyt BA, Nguyen HV, Berleau LT, Duarte CM, Park J, Chen H, Ferrara N. 1996. Identification of vascular endothelial growth factor determinanats for binding KDR and FLT-1 receptors. *J Biol Chem* **271:** 5638–5646

Kiba A, Sagara H, Hara T, Shibuya M. 2003a. VEGFR-2-specific ligand VEGF-E induces non-edematous hypervascularization in mice. *Biochem Biophys Res Commun* **301:** 371–377.

Kiba A, Yabana N, Shibuya M. 2003b. A set of loop-1 and -3 structures in the novel VEGF family member, VEGF-E_{NZ7}, is essential for the activation of VEGFR-2 signaling. *J Biol Chem* **278:** 13453–13461.

Kim KJ, Li B, Winer J, Armanini M, Gillett N, Phillips HS, Ferrara N. 1993. Inhibition of vascular endothelial growth factor-induced angiogenesis suppresses tumour growth in vivo. *Nature* **362:** 841–844.

Koga K, Osuga Y, Yoshino O, Hirota Y, Ruimeng X, Hirata T, Takeda S, Yano T, Tsutsumi O, Taketani Y. 2003. Elevated serum soluble vascular endothelial growth factor receptor 1 (sVEGFR-1) levels in women with preeclampsia. *J Clin Endocrinol Metab* **88:** 2348–2351.

Kondo K, Hiratsuka S, Subbalakshmi E, Matsushime H, Shibuya M. 1998. Genomic organization of the *flt-1* gene encoding for vascular endothelial growth factor (VEGF) receptor-1 suggests an intimate evolutionary relationship between the 7-Ig and the 5-Ig tyrosine kinase receptors. *Gene* **208:** 297–305.

Kunstfeld R, Hirakawa S, Hong YK, Schacht V, Lange-Asschenfeldt B, Velasco P, Lin C, Fiebiger E, Wei X, Wu Y, et al. 2004. Induction of cutaneous delayed-type hypersensitivity reactions in VEGF-A transgenic mice results in chronic skin inflammation associated with persistent lymphatic hyperplasia. *Blood* **104:** 1048–1057.

Lawson ND, Mugford JW, Diamond BA, Weinstein BM. 2003. Phospholipase C γ-1 is required downstream of vascular endothelial growth factor during arterial development. *Genes Dev* **17:** 1346–1351.

LeCouter J, Moritz DR, Li B, Phillips GL, Liang XH, Gerber HP, Hillan KJ, Ferrara N. 2003. Angiogenesis-independent endothelial protection of liver: Role of VEGFR-1. *Science* **299:** 890–893.

Levine RJ, Maynard SE, Qian C, Lim KH, England LJ, Yu KF, Schisterman EF, Thadhani R, Sachs BP, Epstein FH, et al. 2004. Circulating angiogenic factors and the risk of preeclampsia. *N Engl J Med* **350:** 672–683.

Lohela M, Bry M, Tammela T, Alitalo K. 2009. VEGFs and receptors involved in angiogenesis versus lymphangiogenesis. *Curr Opin Cell Biol* **21:** 154–165.

Lyttle DJ, Fraser KM, Fleming SB, Mercer AA, Robinson AJ. 1994. Homologs of vascular endothelial growth factor are encoded by the poxvirus orf virus. *J Virol* **68:** 84–92.

Mäkinen T, Veikkola T, Mustjoki S, Karpanen T, Catimel B, Nice EC, Wise L, Mercer A, Kowalski H, Kerjaschki D, et al. 2001. Isolated lymphatic endothelial cells transduce growth, survival and migratory signals via the VEGF-C/D receptor VEGFR-3. *EMBO J* **20:** 4762–4773.

Matsumoto T, Bohman S, Dixelius J, Berge T, Dimberg A, Magnusson P, Wang L, Wikner C, Qi JH, Wernstedt C, et al. 2005. VEGF receptor-2 Y951 signaling and a role for the adapter molecule TSAd in tumor angiogenesis. *EMBO J* **24:** 2342–2353.

Matthews W, Jordan CT, Gavin M, Jenkins NA, Copeland NG, Lemischka IR. 1991. A receptor tyrosine kinase cDNA isolated from a population of enriched primitive hematopoietic cells and exhibiting close genetic linkage to c-kit. *Proc Natl Acad Sci* **88:** 9026–9030.

Maynard SE, Min JY, Merchan J, Lim KH, Li J, Mondal S, Libermann TA, Morgan JP, Sellke FW, Stillman IE, et al. 2003. Excess placental soluble fms-like tyrosine kinase 1 (sFlt1) may contribute to endothelial dysfunction, hypertension, and proteinuria in preeclampsia. *J Clin Invest* **111:** 649–658.

McColl BK, Paavonen K, Karnezis T, Harris NC, Davydova N, Rothacker J, Nice EC, Harder KW, Roufail S, Hibbs ML, et al. 2007. Proprotein convertases promote processing of VEGF-D, a critical step for binding the angiogenic receptor VEGFR-2. *FASEB J* **21:** 1088–1098.

Murakami M, Iwai S, Hiratsuka S, Yamauchi M, Nakamura K, Iwakura Y, Shibuya M. 2006. Signaling of vascular endothelial growth factor receptor-1 tyrosine kinase promotes rheumatoid arthritis through activation of monocyte/macrophages. *Blood* **108:** 1849–1856.

Muramatsu M, Yamamoto S, Osawa T, Shibuya M. 2010. VEGFR-1 signaling promotes mobilization of macrophage-lineage cells from bone marrow and stimulates solid tumor growth. *Cancer Res* **70:** 8211–8221.

Niida S, Kondo T, Hiratsuka S, Hayashi S-I, Amizuka N, Noda T, Ikeda K, Shibuya M. 2005. Vascular endothelial growth factor receptor-1 signaling is essential for osteoclast development and bone-marrow formation in CSF-1-deficient mice. *Proc Natl Acad Sci* **102:** 14016–14021.

Ogawa S, Oku A, Sawano A, Yamaguchi S, Yazaki Y, Shibuya M. 1998. A novel type of vascular endothelial growth factor: VEGF-E (NZ-7 VEGF) preferentially utilizes KDR/Flk-1 receptor and carries a potent mitotic activity without heparin-binding domain. *J Biol Chem* **273:** 31273–31282.

Osawa T, Muramatsu M, Watanabe M, Shibuya M. 2009. Hypoxia and low nutrition double stress induces aggressiveness in a murine model of melanoma. *Cancer Sci* **100:** 844–851.

Risau W. 1997. Mechanism of angiogenesis. *Nature* **386:** 671–674.

Ruhrberg C, Gerhardt H, Golding M, Watson R, Ioannidou S, Fujisawa H, Betsholtz C, Shima DT. 2002. Spatially restricted patterning cues provided by heparin-binding VEGF-A control blood vessel branching morphogenesis. *Genes Dev* **16:** 2684–2698.

 Cite this article as *Cold Spring Harb Perspect Biol* doi: 10.1101/cshperspect.a009092

Sakurai Y, Ohgimoto K, Kataoka Y, Yoshida N, Shibuya M. 2005. Essential role of Flk-1 (VEGF receptor 2) tyrosine residue 1173 in vasculogenesis in mice. *Proc Natl Acad Sci* **102:** 1076–1081.

Sase H, Watabe T, Kawasaki K, Miyazono K, Miyazawa K. 2009. VEGFR2-PLC-γ1 axis is essential for endothelial specification of VEGFR2+ vascular progenitor cells. *J Cell Sci* **122:** 3303–3311.

Sawano A, Takahashi T, Yamaguchi S, Aonuma T, Shibuya M. 1996. Flt-1 but not KDR/Flk-1 tyrosine kinase is a receptor for placenta growth factor (PlGF), which is related to vascular endothelial growth factor (VEGF). *Cell Growth Diff* **7:** 213–221.

Sawano A, Iwai S, Sakurai Y, Ito M, Shitara K, Nakahata T, Shibuya M. 2001. Vascular endothelial growth factor receptor-1 (Flt-1) is a novel cell surface marker for the lineage of monocyte-macrophages in humans. *Blood* **97:** 785–791

Shalaby F, Rossant J, Yamaguchi TP, Gertsenstein M, Wu X-F, Breitman ML, Schuh AC. 1995. Failure of blood-island formation and vasculogenesis in Flk-1-deficient mice. *Nature* **376:** 62–66.

Shibuya M. 1995. Role of VEGF-Flt receptor system in normal and tumor angiogenesis. *Adv Cancer Res* **67:** 281–316.

Shibuya M. 2002. Vascular endothelial growth factor receptor family genes: When did the three genes phylogenetically segregate? *Biol Chem* **383:** 1573–1579.

Shibuya M. 2011. Involvement of Flt-1 (VEGFR-1) in cancer and preeclampsia. *Proc Jpn Acad Ser B Phys Biol Sci* **87:** 167–178.

Shibuya M, Claesson-Welsh L. 2006. Signal transduction by VEGF receptors in regulation of angiogenesis and lymphangiogenesis. *Exp Cell Res* **312:** 549–560.

Shibuya M, Yamaguchi S, Yamane A, Ikeda T, Tojo A, Matsushime H, Sato M. 1990. Nucleotide sequence and expression of a novel human receptor-type tyrosine kinase gene (flt) closely related to the fms family. *Oncogene* **5:** 519–524.

Shinkai A, Ito M, Anazawa H, Yamaguchi S, Shitara K, Shibuya M. 1998. Mapping of the sites involved in ligand-association and dissociation at the extracellular domain of vascular endothelial growth factor receptor KDR. *J Biol Chem* **273:** 31283–31288

Stefater JA III, Lewkowich I, Rao S, Mariggi G, Carpenter AC, Burr AR, Fan J, Ajima R, Molkentin JD, Williams BO, et al. 2011. Regulation of angiogenesis by a non-canonical Wnt-Flt1 pathway in myeloid cells. *Nature* **474:** 511–515.

Takahashi T, Shibuya M. 1997. The 230 kDa mature form of KDR/Flk-1 (VEGF receptor-2) activates the PLC-γ pathway and partially induces mitotic signals in NIH3T3 fibroblasts. *Oncogene* **14:** 2079–2089.

Takahashi T, Ueno H, Shibuya M. 1999. VEGF activates protein kinase C-dependent, but Ras-independent Raf-MEK-MAP kinase pathway for DNA synthesis in primary endothelial cells. *Oncogene* **18:** 2221–2230.

Takahashi T, Yamaguchi S, Chida K, Shibuya M. 2001. A single autophosphorylation site on KDR/Flk-1 is essential for VEGF-A-dependent activation of PLC-γ and DNA synthesis in vascular endothelial cells. *EMBO J* **20:** 2768–2778.

Tanaka K, Yamaguchi S, Sawano A, Shibuya M. 1997. Characterization of the extracellular domain in the vascular endothelial growth factor receptor-1 (Flt-1 tyrosine kinase). *Jpn J Cancer Res* **88:** 867–876.

Terman BI, Carrion ME, Kovacs E, Rasmussen BA, Eddy R, Shows TB. 1991. Identification of a new endothelial cell growth factor receptor tyrosine kinase. *Oncogene* **6:** 1677–1683.

Thadhani R, Kisner T, Hagmann H, Bossung V, Noack S, Schaarschmidt W, Jank A, Kribs A, Cornely OA, Kreyssig C, et al. 2011. Pilot study of extracorporeal removal of soluble fms-like tyrosine kinase 1 in preeclampsia. *Circulation* **124:** 940–950.

Tsuchida R, Das B, Yeger H, Koren G, Shibuya M, Thorner PS, Baruchel S, Malkin D. 2008. Cisplatin treatment increases survival and expansion of a highly tumorigenic side-population fraction by upregulating VEGF/Flt1 autocrine signaling. *Oncogene* **27:** 3923–3934.

Veikkola T, Jussila L, Makinen T, Karpanen T, Jeltsch M, Petrova TV, Kubo H, Thurston G, McDonald DM, Achen MG, et al. 2001. Signalling via vascular endothelial growth factor receptor-3 is sufficient for lymphangiogenesis in transgenic mice. *EMBO J* **20:** 1223–1231.

Wang S, Li X, Parra M, Verdin E, Basel-Duby R, Olson EN. 2008. Control of endothelial cell proliferation and migration by VEGF signaling to histone deacetylase 7. *Proc Natl Acad Sci* **105:** 7738–7743.

Wang F, Yamauchi M, Muramatsu M, Osawa T, Tsuchida R, Shibuya M. 2011. RACK1 regulates VEGF/Flt1-mediated cell migration via activation of a PI3K/Akt pathway. *J Biol Chem* **286:** 9097–9106.

Wu Y, Hooper AT, Zhong Z, Witte L, Bohlen P, Rafii S, Hicklin DJ. 2006. The vascular endothelial growth factor receptor (VEGFR-1) supports growth and survival of human breast carcinoma. *Int J Cancer* **119:** 1519–1529.

Xia P, Aiello LP, Ishii H, Jiang ZY, Park DJ, Robinson GS, Takagi H, Newsome WP, Jirousek MR, King GL. 1996. Characterization of vascular endothelial growth factor's effect on the activation of protein kinase C, its isoforms, and endothelial cell growth. *J Clin Invest* **98:** 2018–2026.

Xiong Y, Huo Y, Chen C, Zeng H, Lu X, Wei C, Ruan C, Zhang X, Hu Z, Shibuya M, et al. 2009. VEGF receptor-2 Y1175 signaling controls VEGF-induced vWF release from endothelial cells via PLC-γ1- and PKA-dependent pathways. *J Biol Chem* **284:** 23217–23224.

Yamaguchi S, Iwata K, Shibuya M. 2002. Soluble Flt-1 (soluble VEGFR-1), a potent natural anti-angiogenic molecule in mammals, is phylogenetically conserved in avians. *Biochem Biophys Res Commun* **291:** 554–559.

Yang Y, Xie P, Opatowsky Y, Schlessinger J. 2010. Direct contacts between extracellular membrane-proximal domains are required for VEGF receptor activation and cell signaling. *Proc Natl Acad Sci* **107:** 1906–1911.

You WK, Sennino B, Williamson CW, Falcón B, Hashizume H, Yao LC, Aftab DT, McDonald DM. 2011. VEGF and c-Met blockade amplify angiogenesis inhibition in pancreatic islet cancer. *Cancer Res* **71:** 4758–4768.

Zheng Y, Murakami M, Takahashi H, Yamauchi M, Kiba A, Yamaguchi S, Yabana N, Alitalo K, Shibuya M. 2006. Chimeric VEGF-E$_{NZ7}$/PlGF promotes angiogenesis via VEGFR-2 without significant enhancement of vascular permeability and inflammation. *Arterioscler Thromb Vasc Biol* **26:** 2019–2026.

Molecular Mechanisms of Fibroblast Growth Factor Signaling in Physiology and Pathology

Artur A. Belov and Moosa Mohammadi

Department of Biochemistry and Molecular Pharmacology, New York University School of Medicine, New York, New York 10016

Correspondence: Moosa.Mohammadi@nyumc.org

Fibroblast growth factors (FGFs) signal in a paracrine or endocrine fashion to mediate a myriad of biological activities, ranging from issuing developmental cues, maintaining tissue homeostasis, and regulating metabolic processes. FGFs carry out their diverse functions by binding and dimerizing FGF receptors (FGFRs) in a heparan sulfate (HS) cofactor- or Klotho coreceptor-assisted manner. The accumulated wealth of structural and biophysical data in the past decade has transformed our understanding of the mechanism of FGF signaling in human health and development, and has provided novel concepts in receptor tyrosine kinase (RTK) signaling. Among these contributions are the elucidation of HS-assisted receptor dimerization, delineation of the molecular determinants of ligand–receptor specificity, tyrosine kinase regulation, receptor *cis*-autoinhibition, and tyrosine *trans*-autophosphorylation. These structural studies have also revealed how disease-associated mutations highjack the physiological mechanisms of FGFR regulation to contribute to human diseases. In this paper, we will discuss the structurally and biophysically derived mechanisms of FGF signaling, and how the insights gained may guide the development of therapies for treatment of a diverse array of human diseases.

Fibroblast growth factor (FGF) signaling fulfills essential roles in metazoan development and metabolism. A wealth of literature has documented the requirement for FGF signaling in multiple processes during embryogenesis, including implantation (Feldman et al. 1995), gastrulation (Sun et al. 1999), somitogenesis (Dubrulle and Pourquie 2004; Wahl et al. 2007; Lee et al. 2009; Naiche et al. 2011; Niwa et al. 2011), body plan formation (Martin 1998; Rodriguez Esteban et al. 1999; Tanaka et al. 2005; Mariani et al. 2008), morphogenesis (Metzger et al. 2008; Makarenkova et al. 2009), and organogenesis (Goldfarb 1996; Kato and Sekine 1999; Sekine et al. 1999; Sun et al. 1999; Colvin et al. 2001; Serls et al. 2005; Vega-Hernandez et al. 2011). Recent clinical and biochemical data have uncovered unexpected roles for FGF signaling in metabolic processes, including phosphate/vitamin D homeostasis (Consortium 2000; Razzaque and Lanske 2007; Nakatani et al. 2009; Gattineni et al. 2011; Kir et al. 2011), cholesterol/bile acid homeostasis (Yu et al. 2000a; Holt et al. 2003), and glucose/lipid metabolism (Fu et al. 2004; Moyers et al. 2007). Highlighting its diverse biology, deranged FGF signaling contributes to many human diseases, such as congenital craniosynostosis and dwarfism syndromes (Naski et al. 1996; Wilkie et al. 2002, 2005), Kallmann syndrome (Dode et al. 2003; Pitteloud et al. 2006a), hearing loss (Tekin et al. 2007, 2008), and renal phosphate wasting disorders (Shimada et al. 2001; White et al. 2001), as well as many acquired forms of cancers (Rand et al. 2005; Pollock et al. 2007; Gartside et al. 2009; di Martino et al. 2012). Endocrine FGFs have also been implicated in the progression of acquired metabolic disorders, including chronic kidney disease (Fliser et al. 2007), obesity (Inagaki et al. 2007; Moyers et al. 2007; Reinehr et al. 2012), and insulin resistance (Fu et al. 2004; Chen et al. 2008b; Chateau et al. 2010; Huang et al. 2011), giving rise to many opportunities for drug discovery in the field of FGF biology (Beenken and Mohammadi 2012).

Based on sequence homology and phylogeny, the 18 mammalian FGFs are grouped into six subfamilies (Ornitz and Itoh 2001; Popovici et al. 2005; Itoh and Ornitz 2011). Five of these subfamilies act in a paracrine fashion, namely, the *FGF1 subfamily* (FGF1 and FGF2), the *FGF4 subfamily* (FGF4, FGF5, and FGF6), the *FGF7 subfamily* (FGF3, FGF7, FGF10, and FGF22), the *FGF8 subfamily* (FGF8, FGF17, and FGF18), and the *FGF9 subfamily* (FGF9, FGF16, and FGF20). In contrast, the *FGF19 subfamily* (FGF19, FGF21, and FGF23) signals in an endocrine manner (Beenken and Mohammadi 2012). FGFs exert their pleiotropic effects by binding and activating the FGF receptor (FGFR) subfamily of receptor tyrosine kinases that are coded by four genes (FGFR1, FGFR2, FGFR3, and FGFR4) in mammals (Johnson and Williams 1993; Mohammadi et al. 2005b). The extracellular domain of FGFRs consists of three immunoglobulin (Ig)-like domains (D1, D2, and D3), and the intracellular domain harbors the conserved tyrosine kinase domain flanked by the flexible amino-terminal juxtamembrane linker and carboxy-terminal tail (Lee et al. 1989; Dionne et al. 1991; Givol and Yayon 1992). A unique feature of FGFRs is the presence of a contiguous segment of glutamic and aspartic acids in the D1–D2 linker, termed the acid box (AB). The two-membrane proximal D2 and D3 and the intervening D2–

D3 linker are necessary and sufficient for ligand binding/specificity (Dionne et al. 1990; Johnson et al. 1990), whereas D1 and the D1–D2 linker are implicated in receptor autoinhibition (Wang et al. 1995; Roghani and Moscatelli 2007; Kalinina et al. 2012). Alternative splicing and translational initiation further diversify both ligands and receptors. The amino-terminal regions of FGF8 and FGF17 can be differentially spliced to yield FGF8a, FGF8b, FGF8e, FGF8f (Gemel et al. 1996; Blunt et al. 1997), and FGF17a and FGF17b isoforms (Xu et al. 1999), whereas cytosine-thymine-guanine (CTG)-mediated translational initiation gives rise to multiple high molecular weight isoforms of FGF2 and FGF3 (Florkiewicz and Sommer 1989; Prats et al. 1989; Acland et al. 1990). The tissue-specific alternative splicing in D3 of FGFR1, FGFR2, and FGFR3 yields "b" and "c" receptor isoforms which, along with their temporal and spatial expression patterns, is the major regulator of FGF–FGFR specificity/promiscuity (Orr-Urtreger et al. 1993; Ornitz et al. 1996; Zhang et al. 2006). A large body of structural data on FGF–FGFR complexes has begun to reveal the intricate mechanisms by which different FGFs and FGFRs combine selectively to generate quantitatively and qualitatively different intracellular signals, culminating in distinct biological responses. In addition, these structural data have unveiled how pathogenic mutations hijack the normal physiological mechanisms of FGFR regulation to lead to pathogenesis. We will discuss the current state of the structural biology of the FGF–FGFR system, lessons learned from studying the mechanism of action of pathogenic mutations, and how the structural data are beginning to shape and advance the translational research.

STRUCTURE–FUNCTION RELATIONSHIP OF FGFs

FGFs range in size from ~150–300 amino acids (Basilico and Moscatelli 1992; Mohammadi et al. 2005b). Crystal structures with at least one representative from each subfamily, namely FGF1 and FGF2 (Eriksson et al. 1991; Zhang et al. 1991; Zhu et al. 1991), FGF4 (Bellosta et al. 2001), FGF7 (Ye et al. 2001), FGF8b (Olsen et al. 2006), FGF9 (Plotnikov et al. 2001), FGF10 (Yeh et al. 2003), FGF19 (Harmer et al. 2004; Goetz et al. 2007), FGF20 (Kalinina et al. 2009), and FGF23 (Goetz et al. 2007) have been solved. In addition, crystal structures of FGF1, FGF2, FGF8b, and FGF10 in complex with their cognate receptor(s) have been solved, providing valuable insights into the structure–function relationships of the FGF family (Fig. 3). These structures show that the FGF core homology domain (composed of ∼125 amino acids) adopts a conserved β-trefoil fold consisting of 12 antiparallel β strands (β1–β12) in paracrine FGFs (Mohammadi et al. 2005b). Endocrine FGFs lack the β11 strand, and hence have an atypical trefoil fold (Goetz et al. 2007; Beenken and Mohammadi 2011). The FGF trefoil core is flanked by amino- and carboxy-terminal regions that are highly variable in length and sequence among FGFs. These variable amino and carboxyl termini contribute key components in the regulation of distinct biological functions of different FGFs.

Paracrine FGFs exhibit moderate to high affinity for heparan sulfate (HS), a mandatory cofactor in paracrine FGF signaling (Rapraeger et al. 1991; Yayon et al. 1991; Ornitz and Leder 1992). HS is a heterogeneously sulfated glycosaminoglycan that is covalently linked to select serine residues in proteoglycans, such as membrane-anchored syndecans and glypicans and extracellular matrix (ECM) perlecans (Hacker et al. 2005; Iozzo et al. 2009). HS is a polymer of repeating disaccharide units consisting of glucuronic acid and amino acteylglucosamine linked through α-1,4-glycosidic bonds. HS is heterogeneously sulfated on the 2-O position of glucuronic and amino acids and the 6-O position of amino acteylglucosamine (Esko and Lindahl 2001). Attributed to their high affinity for HS, paracrine FGFs can diffuse only a short distance away from their source of secretion, thus acting locally (Asada et al. 2009; Xu et al. 2012). Unique among paracrine FGFs, the FGF9 subfamily undergoes reversible dimerization, whereby the dimer has higher affinity for HS. This provides an additional mechanism for acutely regulating the diffusion radius of the FGF9 subfamily of ligands, further reducing

their signaling dimensionality (Plotnikov et al. 2001; Harada et al. 2009; Kalinina et al. 2009). The HS binding site of FGFs is composed of the β1–β2 loop and the extended β10–β12 region, which provide solvent-exposed basic amino acids and backbone atoms for HS binding (Fig. 1B). Because of the primary sequence variation of the HS binding site, each ligand has discrete affinity for HS, resulting in the formation of FGF-specific morphogenetic gradients that contribute to the distinct biology of FGF (Makarenkova et al. 2009). Despite primary sequence variations, however, the HS binding sites of paracrine FGFs adopt a common topology (Faham et al. 1998; Goetz et al. 2007). This is mainly because of the presence of the paracrine-conserved **GXXXXGXXS/T** motif (Goetz et al. 2007), referred to as the glycine box (Luo et al. 1998). The backbone atoms of these two glycines engage in conserved hydrogen bonds with underlying β strands to facilitate the formation of the β11 strand (Goetz et al. 2007) (Fig. 1B).

The topology of the HS binding region of the endocrine FGFs differs drastically from that of paracrine FGFs because of the lack of the paracrine-conserved glycine box and the truncated β10–β12 region (Goetz et al. 2007) (Fig. 1B). Importantly, the altered topologies of HS binding sites of the endocrine FGFs disallow the interaction of HS with the backbone atoms of the HS binding region. This manifests itself in major reductions in HS affinity of the endocrine FGFs (Yu et al. 2005; Goetz et al. 2007), which allows these ligands to permeate freely through the HS-rich ECM and enter blood circulation (Fig. 1A,C).

STRUCTURAL FEATURES OF FGFRs

Currently, there is no crystal structure of either the intact ectodomain or the D2–D3 ligand-binding region of FGFR in the absence of the ligand. Nuclear magnetic resonance (NMR) solution studies of the D2–D3 region show that D3 is an intrinsically flexible domain (Kalinina et al. 2012). However, in the X-ray structures of ligand-bound D2–D3, D3 adopts a stable Ig-like fold (Plotnikov et al. 1999). To date, there are crystal structures of eight different

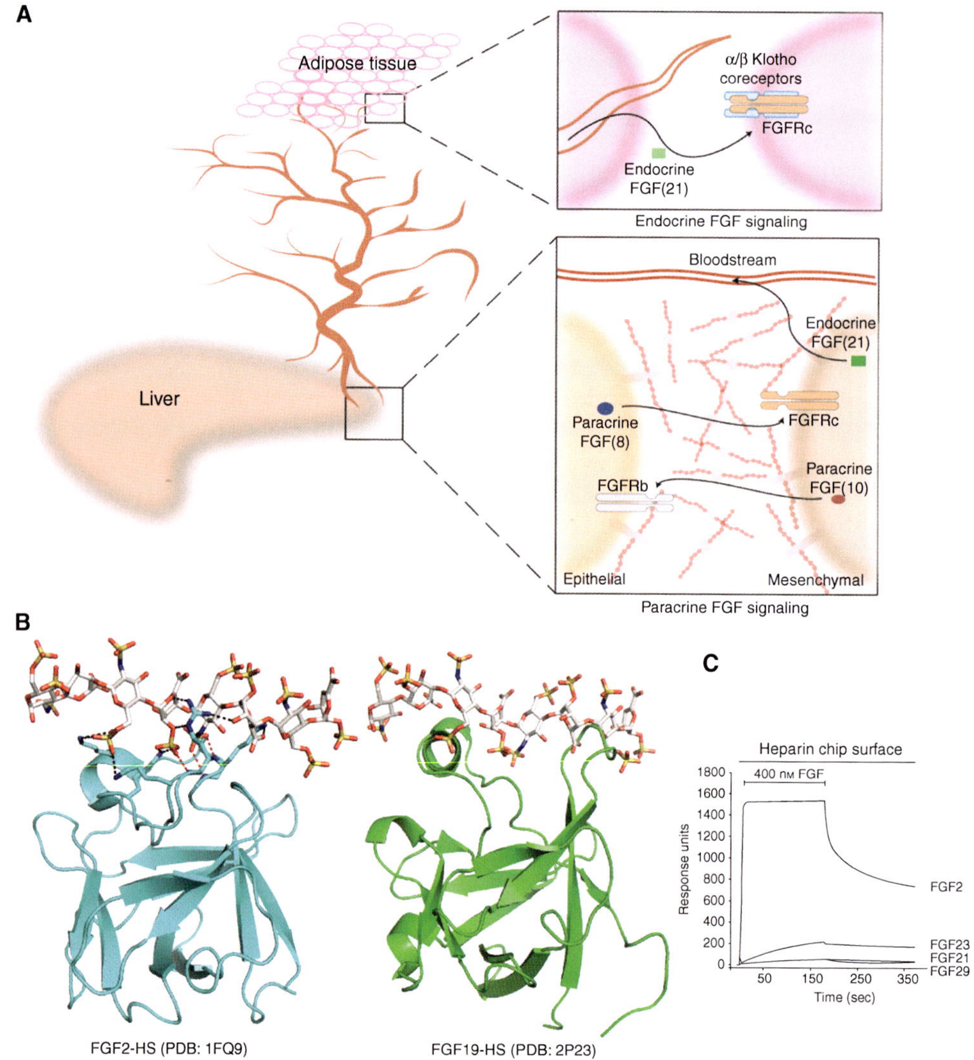

Figure 1. FGF signaling in the liver and adipose tissue. (*A*) The paracrine FGF signaling loop in the liver. FGFs are expressed in both the epithelial or mesenchymal tissue, and signal in a paracrine fashion through their cognate receptors, which are expressed in the opposite tissues. Shown are two examples of paracrine ligands, FGF8 and FGF10, which signal exclusively in an epithelial-to-mesenchyme and mesenchyme-to-epithelial manner, respectively. (*B*) Comparison of the crystal structures of FGF2 (PDB: 1FQ9) and FGF19 (PDB: 2P23) provides the structural basis for the low affinity of endocrine ligands for HS. (*C*) Comparison of the binding interactions of FGF2, FGF19, FGF21, and FGF23 with HS using surface-plasmon resonance spectroscopy. The low affinity of FGF19 family members (such as FGF21) allows them to permeate freely through the HS-dense intercellular space and enter into the blood. This enables them to act as hormones in target tissues in which α/β Klotho coreceptors are expressed (*top* panel in part *A*).

FGF–FGFR complexes that feature unique ligand–receptor combinations, including FGF1 with FGFR1c (Plotnikov et al. 2000; Beenken et al. 2012), FGFR2c (Stauber et al. 2000), FGFR3c (Olsen et al. 2004), FGFR2b (Beenken et al. 2012), FGF2 with FGFR1c (Plotnikov et al. 1999; Schlessinger et al. 2000), FGFR2c (Plotnikov et al. 2000), FGF8b with FGFR2c (Olsen et al. 2006), and FGF10 with FGFR2b (Yeh et al. 2003). In all these structures, the receptor adopts an extended conformation displaying significant differences in the relative orientation of D2 and D3 to one another, suggesting that the overall receptor orientation is dictated by ligand binding (Fig. 3). This is harmonious with the fact that there are no intramolecular contacts between D2 and D3 that would constrain the relative disposition of the two domains in the absence of the ligand.

As anticipated based on sequence homology (Bateman and Chothia 1995), D1 and D2 adopt Ig folds that belong to the I set of the Ig superfamily (Plotnikov et al. 1999; Hung et al. 2005; Kiselyov et al. 2006) (Fig. 2A). In contrast, D3 has an unusual Ig-fold in that the region between

the βC′ and βE strands (referred to as the βC′– βE loop) adopts a different conformation in different complexes, suggesting that it is highly mobile. In all but the FGF8b–FGFR2c structure, D3 lacks the analogous βD strand of the Ig-like domains D1 and D2. Notably, alternative splicing of D3 occurs at the junction between the βC′ strand and the βC′–βE loop, diversifying the primary sequence of the loop (Johnson et al. 1991; Yeh et al. 2003) (Fig. 2A). As we will discuss later, this, together with the inherent flexibility of the loop, plays a major role in determining ligand-binding specificity/promiscuity. The HS binding site of the receptor resides in D2 and is comprised of several surface-exposed residues

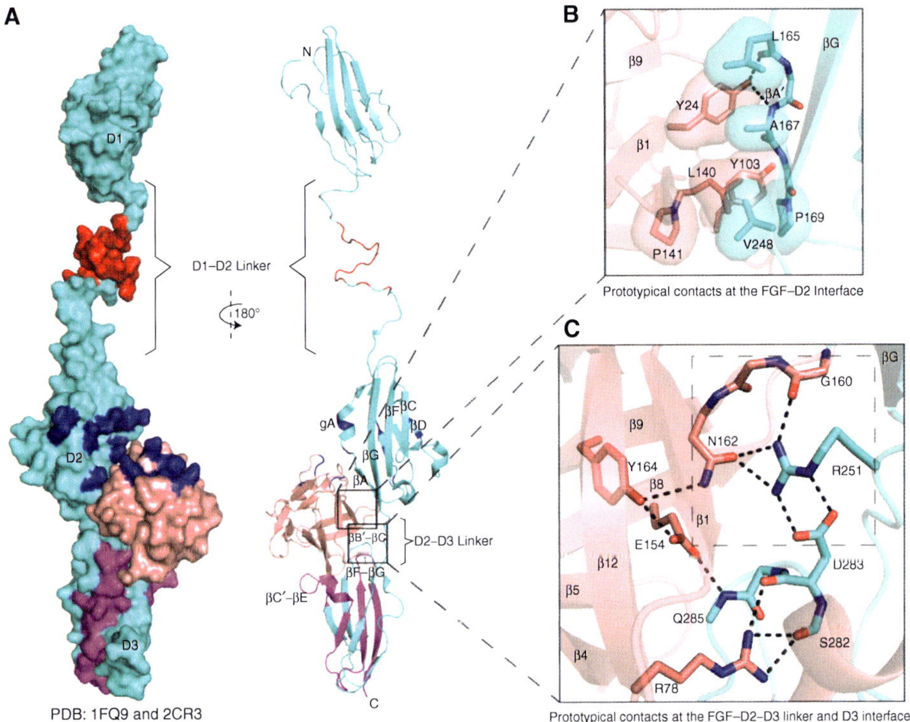

Figure 2. Structural features of a prototypical FGF receptor and FGF-conserved FGF–FGFR contacts. (A) The X-ray structure of FGF2–FGFR1c (PDB: 1FQ9) and an NMR structure of D1 (PDB: 2CR3) were linked arbitrarily with a modeled D1–D2 linker to construct a model of a full-length FGF receptor. The acid box is red and the HS binding region in the FGF2–FGFR1c complex is blue. FGFR1c and FGF2 are cyan and salmon, respectively. The alternatively spliced portion of D3 is magenta. (B) Prototypical contacts (PDB: 1FQ9) between the ligand and receptor D2 are illustrated. Dashed lines denote hydrogen bonds. Hydrophobic contacts are indicated using transparent surfaces. Oxygen and nitrogen atoms are red and blue, respectively, hereafter. (C) The conserved hydrogen bonds at the interface between the D2–D3 linker and D3 of the FGFR and FGF ligand as observed in the FGF10–FGFR2b structure (PDB: 1NUN). The dashed box within panel C highlights the hydrogen bonds between the D2–D3 linker of FGFR and FGF.

emanating from the βB, βE, βD strands, the gA helix, and the loop between the βA and βA′ strands of D2 (Schlessinger et al. 2000). Unlike the FGF–HS interaction, only the side chains of D2 participate in HS coordination, which explains the significantly lower affinity of FGFRs for HS compared to that of FGFs (Powell et al. 2002; Ibrahimi et al. 2004c; Asada et al. 2009; Trueb 2011).

The bottom edge of D2, the D2–D3 linker, and the top portion of D3 compose the ligand-binding pocket (Plotnikov et al. 1999) (Fig 2A). The FGF straddles D3 via the bottom end of the trefoil (the top end being the HS binding site). The βB′–βC and βF–βG loops in D3 are engulfed in a depression formed between the β1, β2, β4, β5, β8, and β9 strands and the intervening loops, whereas the D2–D3 linker cooperates with D3 to further fix the ligand in its observed position (Fig. 2C). The βA′ and βF strands at the bottom edge of D2 sharply engage the β1 through β2 strands, the β3–β4 loop, and β9 and β12 of the ligand (Plotnikov et al. 2000) (Fig. 2A).

THE FGF–FGFR INTERFACE

Conserved Ligand–Receptor Contacts

Each subdomain makes several highly conserved contacts with the ligand, unambiguously demonstrating that the D2–D3 fragment is the minimal binding region of FGFs (Stauber et al. 2000; Olsen et al. 2004; Mohammadi et al. 2005b). At the FGF–D2 interface, two conserved tyrosine residues, one from the β1 strand and another from the β8–β9 loop, along with highly conserved proline in the β12 strand of the ligand engage in hydrophobic and hydrogen-bonding interactions with conserved residues on the βA′ and βG strands in D2 (Fig. 2B). The interface between FGF and the D2–D3 linker is by far the most conserved residue. Here, an invariant arginine from the D2–D3 linker region makes three hydrogen bonds with a side chain of a residue in β9 and a backbone of the β8–β9 loop in the ligand (Fig. 2C). The β9 residue is an asparagine in all FGFs, with the exception of the FGF8 subfamily, which has a threonine instead. Notably

the D2–D3 linker arginine is also engaged in intramolecular hydrogen bonding with a conserved aspartic acid in the βB′–βC loop of D3. This primes the arginine for FGF recognition, thus minimizing the entropic penalty associated with ligand binding. The focal role of these hydrogen bonds in providing general FGF–FGFR affinity is evidenced by the fact that fibroblast homologous factors (FHFs) contain a valine in the equivalent position of β9 asparagine, which hinders them from binding/activating FGFRs (Goldfarb 1996; Olsen et al. 2003; Goetz et al. 2009; Wang et al. 2012). Mutation of the D2–D3 linker arginine to glutamine leads to a loss of function in the Kallmann syndrome (Dode et al. 2003; Pitteloud et al. 2006a,b), further highlighting the importance of these conserved hydrogen bonds in FGF–FGFR binding. The FGF–D3 interface harbors two highly conserved contacts. Here the backbone atoms of the βB′–βC loop are engaged in three strong hydrogen bonds with an arginine from β1 and glutamic acid from β8 of the ligand (Fig. 2C). Together, the aforementioned contacts provide general FGF–FGFR binding affinity, whereas specificity is primarily decided by divergent contacts at the FGF–D3.

ALTERNATIVE SPLICING IN D3 IS A MAIN MECHANISM IN REGULATION OF FGF–FGFR SPECIFICITY

In FGFR1–3, two alternative exons ("b" and "c") code for the second half of D3 (Johnson et al. 1991; Miki et al. 1992; Yayon et al. 1992) that are spliced in tissue-specific fashion (Orr-Urtreger et al. 1993; Wuechner et al. 1996; Beer et al. 2000). Generally, the b-splice variants are expressed in the epithelial tissue, whereas the c-splice isoforms are expressed in the mesenchymal tissue (Orr-Urtreger et al. 1993; McEwen and Ornitz 1997). Paracrine FGFs also show tissue-specific expression patterns with ligands for FGFRb isoforms being expressed in mesenchyme, and ligands for FGFRc isoforms expressed in epithelium (Finch et al. 1989). This results in an epithelial–mesenchymal FGF signaling loop that is crucial for tissue homeostasis and organogenesis (McIntosh et al. 2000; Itoh and Ornitz 2011), as evidenced by the fact that derangements of this signaling loop contribute to human skeletal disorders and cancer (Ibrahimi et al. 2005; Beenken and Mohammadi 2009, 2011). The tissue-specific alternative splicing in D3 is the chief mechanism in the regulation of FGF–FGFR binding specificity (Yayon et al. 1992; Ornitz et al. 1996; Zhang et al. 2006). Crystallographic studies of eight different FGF–FGFR complexes have revealed that this alternative splicing controls FGF–FGFR binding specificity/promiscuity by altering the composition of FGF binding sites in D3, including the βC′–βE and βF–βG loops. Moreover, the βC′–βE loop is inherently flexible, and is capable of adapting uniquely to each ligand. The regions of FGFs that engage the alternatively spliced half of D3, in particular, the amino-terminal region of ligands, is divergent in the primary sequence. These structural data flag the FGF–D3 interface as the key mediator of FGF–FGFR binding specificity/promiscuity. The current structural data on FGF–FGFR complexes have disclosed two distinct modes by which contacts at the FGF–D3 interface mediate FGF–FGFR binding specificity.

GENERAL MODE OF FGF–FGFR SPECIFICITY

The hallmark of the general mode, observed in the FGF1-, FGF2-, and FGF10-receptor complexes, is a cleft in D3 that forms between βB′–βC and the alternatively spliced βC′–βE on ligand binding. This cleft is induced by hydrophobic contacts between residues from the β7 and β8 loops and the β5 strand in the FGF core, and a hydrophobic residue at the apex of the alternatively spliced βC′–βE of the receptor (Fig. 3A–C). Residues from the β4, β5 strands and the intervening loop as well as the amino-terminal tail of the ligand engage the βC′–βE loop, further stabilizing the cleft. The βF–βG loop, which is also alternatively spliced, engages residues from the β4–β5 loop and the β8 strand.

FGF7 Subfamily

FGF7 subfamily members are secreted by the mesenchyme and act exclusively on the FGFR2b

resident in the epithelial tissue to constitute the mesenchymal-to-epithelial arm of the signaling loop (Mason et al. 1994). The FGF10–FGFR2b structure (Yeh et al. 2003) shows that F146, Y131, and A122 in FGF10 form a hydrophobic surface that tethers Ile-317 from the alternatively spliced loop promoting formation of the D3 cleft (Fig 3B). According to the structure, the FGF7 subfamily's preference for FGFR2b can be traced to the highly specific hydrogen bonds between Asp-76, a unique amino-terminal residue in the FGF7 subfamily, and Ser-315 from the alternatively spliced βC′–βE loop in the D3 cleft. A π-cation interaction between Y345 in the βF–βG (a residue unique to b-splice isoforms of receptors) and R155 in the β8 strand of FGF10 further reinforces the specificity (Fig. 3A,B). Interestingly, the substitution of the conserved tyrosine in the β1 strand for phenylalanine in the FGF7 subfamily acts in concert with the above-mentioned specific FGF–D3 contacts to further narrow the specificity of the FGF7 subfamily for FGFR2b. This Y→F substitution disables this subfamily form hydrogen bonding with D2 (Fig. 2B), thus minimizing the contribution of D2 in providing ligand-binding affinity. By primarily relying on the alternatively spliced loops of D3 of the receptor to attain specificity and affinity, the FGF7 subfamily is solely able to bind and activate FGFR2b. In the absence of these hydrogen bonds that impose constrains on the orientation of D3, D2 is observed to rotate about the D2–D3 linker, resulting in a distinct orientation of the HS binding site in D2, relative to the ligand. This structural change has been postulated to play a role in determining the HS selectivity of the FGF7 subfamily FGFR2b complexes (Mohammadi et al. 2005a).

FGF1 Subfamily

Unlike the FGF7 subfamily, whose members share a common receptor specificity profile, the members of the FGF1 subfamily, namely FGF1 and FGF2, have a distinct receptor binding specificity/promiscuity profile (Ibrahimi et al. 2004a,b). Both FGF1 and FGF2 are promiscuous and can bind more than one FGFR isoform.

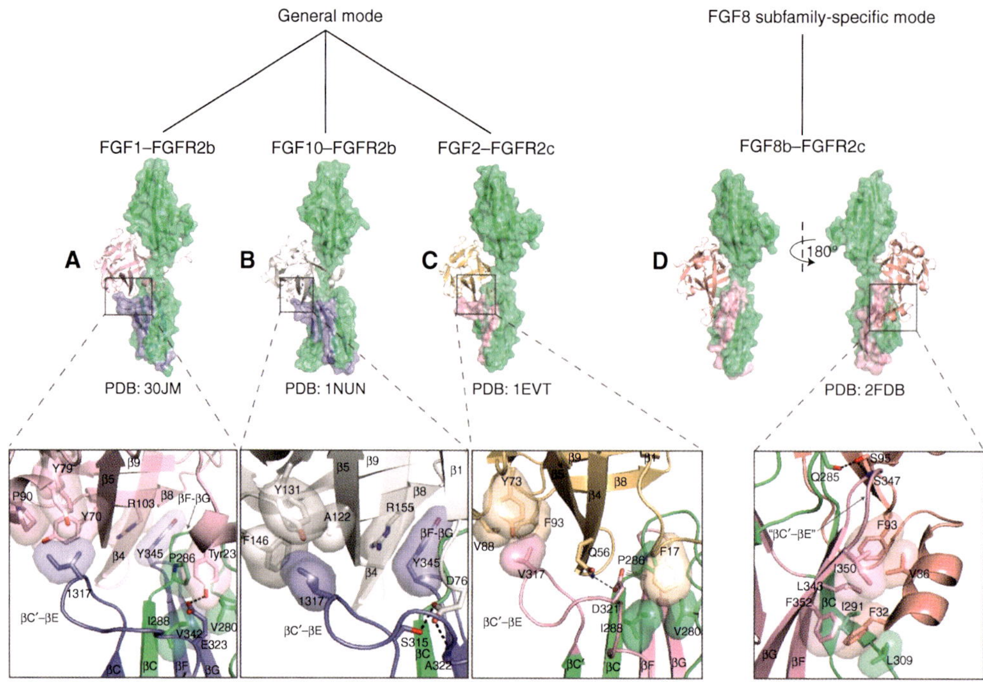

Figure 3. General and FGF8-specific modes of FGF–FGFR binding. FGF1 and FGF10 in complex with FGFR2b are illustrated with the alternatively spliced regions of D3 shown in slate. FGF2 and FGF8b in complex with FGFR2c are also illustrated, with the alternatively spliced regions of D3 in pink. FGF1, FGF10, FGF2, and FGF8b are light pink, gray, wheat, and salmon, respectively. The constant regions of FGFRs are lime green. Each subpanel illustrates the specific contacts the ligands make with D3. In the subpanels *A* and *B*, the β4 strands from FGF1 and FGF10 are made transparent to allow for the visualization of the π-cation interactions at the βF–βG loop outside of the D3 cleft.

FGF2 binds equally well to the "c" isoforms of FGFR1 and FGFR2 but has negligible binding to the "b" isoforms (Ornitz et al. 1996). In contrast, FGF1 overrides the specificity barrier set by alternative splicing and interacts indiscriminately with all seven FGFRs (Zhang et al. 2006).

Crystal structures of FGF2 with both of its cognate receptors have been solved (Plotnikov et al. 1999, 2000), revealing the molecular basis for the FGFR binding specificity/promiscuity of FGF2. Reminiscent of the FGF10–FGFR2b structure, a valine/isoleucine at the apex of the βC′–βE loop makes hydrophobic contacts with Y73 (from the β6 strand) and V88 and F93 (from the β7–β8 loop), resulting in the formation of the D3 cleft (Fig. 3C). The specificity/promiscuity of FGF2 can be traced mostly to specific hydrogen bonds between Q56 from the β4 strand of FGF2 and D321 in the D3 cleft.

F17 from the amino terminus is immersed in a hydrophobic pocket created by I288, P286, and V280 in the D3 cleft, while also engaging in hydrogen-aromatic interactions with D321 (Fig. 3C). The βC′–βE loop of "b-"splice isoforms would not be able to endorse these specific contacts.

The crystal structure of FGF1 in complex with four of its cognate FGFRs, namely, FGFR1c (Plotnikov et al. 2000; Beenken et al. 2012), FGFR2c (Stauber et al. 2000), FGFR3c (Olsen et al. 2004), and FGFR2b (Beenken et al. 2012), have been solved. Analysis of these four FGF1–FGFR structures show that the promiscuity of FGF1 can be traced to the unusual ability of FGF1 to adapt to the alternatively spliced βC′–βE loop on the receptor. The versatility in the interactions of FGF1 with the βC′–βE loop manifests itself in the observed divergent conformation of the βC′–βE loop in the four FGF1–FGFR structures. The FGF1–FGFR1c, FGF1–FGFR2c, and FGF1–FGFR2b complexes all feature the characteristic D3 cleft, whereas the FGF1–FGFR3c structure lacks it. Three amino-terminal residues of FGF1, namely, F16, N22, and Y23 FGF1 make variable interactions with D3, depending on which receptor they interact with (Beenken et al. 2012). Replacing the corresponding three amino-terminal residues in FGF2 with that of FGF1 bestows on FGF2 the ability to bind to the FGFR2b isoform, thus verifying the structural data. As a corollary to FGF10 specificity, subtle changes at the FGF1–D2 contacts augment the promiscuity of FGF1. Notably, FGF1 has an L135 instead of an M142 in FGF2 at the FGF–D2 interface, which enables FGF1 to engage in stronger hydrophobic contacts with D2, thereby gaining more affinity through D2 contacts. This enhances the promiscuity of FGF1 as it reduces the dependency of FGF1 on D3 for receptor binding.

Structural and biochemical studies of pathogenic mutations affecting the extracellular domain of FGFRs strongly support the regulatory mechanisms of FGF–FGFR binding specificity/promiscuity deduced from FGF–FGFR crystal structures (Wilkie 2005). For example, the S252W and P253R mutations in the D2–D3 linker region of FGFR2, responsible for the Apert syndrome (Wilkie et al. 1995), introduce additional conserved contacts with FGFs that result in a generalized increase in affinity of the receptor for all FGFs, thus minimizing the reliance of FGF on specific contacts with D3 for receptor binding (Anderson et al. 1998; Ibrahimi et al. 2001; Yu and Ornitz 2001; Glaser et al. 2003; Yoon et al. 2009). This enables the mesenchymally expressed diseased FGFR2c to illegitimately bind and become activated in an autocrine fashion by mesenchymmal FGF10, thereby short-circuiting the epithelial-to-mesenchyme signaling polarity (Yu et al. 2000b). Likewise, structural studies of the D321A mutation, which maps onto the βC′–βE loop of FGFR2c, show that this mutation removes the electrostatic repulsion and steric conflict that prohibits binding of FGFR2c to FGF10, thereby enabling the illegitimate activation of the "diseased" receptor

by FGF10 in the mesenchyme (Ibrahimi et al. 2004a).

FGF8 SUBFAMILY-SPECIFIC MODE OF FGF–FGFR SPECIFICITY

FGF8 subfamily members are expressed in the epithelial tissue and activate the c-splice isoforms of FGFRs that reside in the underlying mesenchyme to mediate the epithelial-to-mesenchymal arm of the signaling loop (Blunt et al. 1997). Members of the FGF8 subfamily share an overlapping receptor binding specificity/promiscuity profile and bind redundantly to FGFR1c–3c and FGFR4 (Ornitz et al. 1996). The FGF8 subfamily uses a different mode to attain FGFR binding specificity that can be traced to unique prestructured amino termini of this subfamily. In contrast to FGF1, FGF2, and FGF10, which have flexible amino termini, the amino terminus of the FGF8 subfamily adopts a rigid conformation that enables it to wrap around D3 and engage the opposite face of D3 (Olsen et al. 2006) (Fig. 3D). Structural analysis shows that in FGF1, FGF2, and FGF10, the first occurrence of a glycine or proline aborts the β1 strand amino terminally, while also causing the amino terminus to turn away from D3. In the FGF8 subfamily, however, the absence of an amino-terminal proline/glycine allows the β1 strand to continue stranding with β4, which places the amino terminus in an opposite orientation compared to FGF1, FGF2, and FGF10 (Olsen et al. 2006). FGF8b residues F32–S40 form a g helix, which is linked to the β1 strand via an extended loop. The conformation of this loop is stabilized by intramolecular contacts with the FGF8 core. F32, V36 from the gN helix, along with F92 from the β4–β5 loop, engage an extended hydrophobic groove formed by I291 (βC), L309 (βC′), L343 (βF), I350 (βG), and F352 (βG) at the bottom sheet of D3 (Fig. 3D). The βF and βG strands reside in the alternatively spliced half of D3, and the replacement of L343 and I350 by polar residues in b isoforms of FGFR1–3 and FGFR4 explain the subfamily's specificity for FGFRc isoforms and FGFR4 (Olsen et al. 2006). Additional specificity is mediated by the FGF8-specific serine insertion in the β4–β5 loop, which forms a unique network of hydrogen bonds with the backbone atoms of the alternatively spliced βB′–βC and βF–βG loops.

The βC′–βE loop conformation is totally rearranged to make room for the unique FGF8 amino terminus. In fact, a section of this loop forms the canonical Ig-folded βD strand that is connected to βE through a short loop. All the loops connecting the strands of the top sheet localize on one side, and, as a result, the FGF8b–FGFR2c structure lacks the D3 cleft. The unique mode of FGF8b–FGFR2c binding induces a unique D3 rotation that pivots about the D2–D3 linker. Modeling studies show that the membrane insertion points of receptor monomers in the FGF8b–FGFR2c dimer would be closer by ∼15 Å vis-à-vis the FGF2–FGFR2c dimer. These topological differences have been postulated to contribute to the distinct signaling capacity by different FGFs (Olsen et al. 2006). The L341S loss-of-function mutation in FGFR1, which is responsible for the Kallmann syndrome, maps to the D3 groove that the FGF8 subfamily engages (Dode et al. 2003; Pitteloud et al. 2006a; Falardeau et al. 2008). The substitution of leucine for the polar serine in this hydrophobic groove severely impairs the FGF8b binding, thus implicating the FGF8 subfamily in the etiology of the Kallmann syndrome. Indeed, subsequent genetic screening of a cohort of patients led to the identification of loss-of-function mutations in FGF8 and FGF17 (Falardeau et al. 2008; Trarbach et al. 2010; McCabe et al. 2011).

In summary, contacts between FGF and alternatively spliced regions in D3 dictate FGF–FGFR specificity and promiscuity, whereas contacts between FGF and D2 and the D2–D3 linker serve primarily to provide basal ligand-binding affinity. Importantly, differences in the contacts between FGF and D2 and/or D2–D3 contacts can enhance specificity/promiscuity of FGFs by modifying the basal FGF–FGFR affinity. The fidelity of FGF–FGFR binding specificity/promiscuity combined with ligand-dependent differences in receptor orientation would allow for the precise regulation of FGF-induced signaling.

HS-ASSISTED PARACRINE FGF–FGFR DIMERIZATION

HS is a mandatory cofactor in paracrine FGF signaling (Imamura and Mitsui 1987; Rapraeger et al. 1991; Yayon et al. 1991; Olwin and Rapraeger 1992; Ornitz et al. 1992), as documented by the fact that mice and flies with defects in components of FGF signaling or HS biosynthetic enzymes share overlapping phenotypes (Lin et al. 1999; Inatani et al. 2003). Structural data have shown that HS promotes the formation of a symmetric 2:2:2 dimer between FGF, FGFR, and HS, which is required for signal transmission across the plasma membrane (Schlessinger et al. 2000). In the dimeric complex, FGFs engage D2, D3, and the D2–D3 linker of their primary receptor (as discussed in depth above). In addition, residues from the β8–β9 and β11–β12 loops of FGFs interact with the βC′–βD and βE–βF loops in D2 of the neighboring (secondary) receptor (Fig. 4A). Notably, the primary sequence of the β11–β12 loop shows considerable variation among FGFs, indicating that additional FGF–FGFR signaling specificity may be achieved on receptor dimerization. The dimer interface is further fortified by the direct interactions between FGFRs mediated via the bottom end of their D2 domains (Fig. 2A, B). A 2:2:1 FGF–FGFR–HS asymmetric model has also been proposed (Pellegrini et al. 2000); however, analysis of the mechanism of action of pathogenic mutations has lent unbiased support for the symmetric mode of dimerization (Mohammadi et al. 2005a). For example, the A172F gain-of-function mutation that is implicated in the Pfeifer syndrome maps to the D2–D2 receptor interface that has been shown to cause gain of function by facilitating ligand-dependent FGFR dimerization (Ibrahimi et al. 2005).

The HS binding sites of the ligands and receptors are adjacent to one other, forming a continuous basic canyon on the membrane distal end of the dimer. The HS binding residues of FGFs and FGFRs act in concert to recruit two HS molecules in a symmetric fashion (Fig. 4A). Each HS oligosaccharide makes a total of 30 hydrogen bonds with a single FGF and both

 Cite this article as *Cold Spring Harb Perspect Biol* doi: 10.1101/cshperspect.a015958

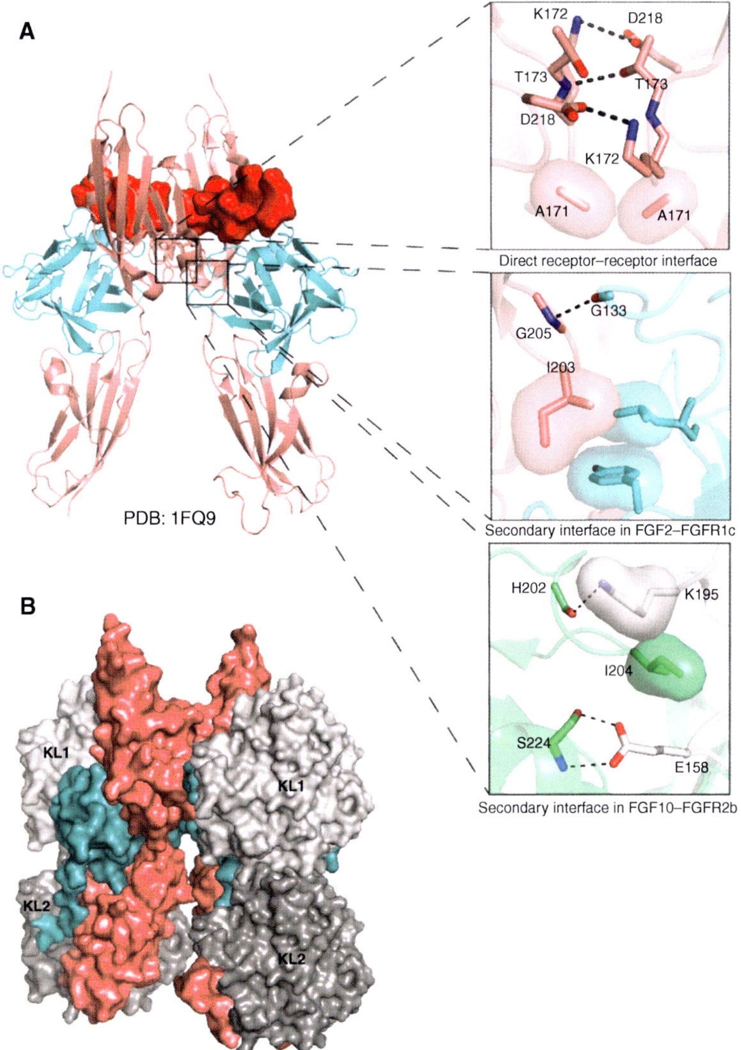

Figure 4. HS/Klotho-assisted FGFR dimerization. (*A*) FGF2–FGFR1c–HS ternary complex (PDB:1FQ9). FGF2 (cyan) and FGFR1c (salmon) are shown in ribbons, while HS is depicted in the surface representation (red). The *top* panel illustrates the direct receptor–receptor contacts. In the middle and bottom panels, secondary receptor–ligand contacts are shown for the 2:2 FGF2–FGFR1c and 2:2 FGF10–FGFR2b dimers, respectively. (*B*) A working model for the endocrine FGF–FGFR–Klotho signaling dimer constructed by the superimposition of the FGF23 structure (PDB: 2P23) onto FGF2 in the FGF2–FGFR1c–HS ternary complex (PDB:1FQ9). The two Klotho (KL) domains, shown in two shades of gray, were modeled using the crystal structure of myrosinase (PDB: 1E6S). The carboxy tail of FGF23 is also modeled to show that it engages a composite site created at the FGFR–Klotho interface.

receptor D2 domains (Schlessinger et al. 2000). Primary sequence differences at the HS binding sites of FGFs and FGFRs are proposed to lead to the formation of distinctly charged canyons destined to bind tissue-specific HS molecules (Mohammadi et al. 2005a; Zhang et al. 2009). Notably, the primary and secondary FGF–FGFR binding sites on D2, the direct FGFR–FGFR binding site, and the HS binding site are adjacent to each other, indicating that HS-mediated dimerization is a cooperative process (Fig. 4A, middle and bottom panels). By engaging ligand and receptors in the dimer, HS promotes the kinetics and thermodynamics of FGF–FGFR binding and dimerization, allowing for the transmission of a sustained and robust intracellular signal as opposed to the transient downstream signaling that is observed in HS-deficient cells (Yayon et al. 1991; Nugent and Edelman 1992; Ornitz et al. 1992; Mathieu et al. 1995; Delehedde et al. 2000).

KLOTHO CORECEPTOR-DEPENDENT ENDOCRINE FGF SIGNALING

In addition to exhibiting a negligible HS binding affinity (Fig. 1C), the endocrine FGFs also have poor affinity for their cognate FGFRs (Goetz et al. 2012b). Modeling studies show key residues at their predicted receptor binding site of endocrine FGFs are substituted for residues that are suboptimal for receptor binding (Goetz et al. 2007) (Fig. 1B). For example, substitution of the conserved arginine in the β1 strand and glutamic acid in the β8 strand with glycine and histidine, respectively, in FGF23 should cause a major reduction in receptor binding affinity of this ligand. As discussed earlier, each of these two residues make conserved hydrogen bonds with D3 to provide general receptor binding affinity (Fig. 2B,C). The poor HS binding affinity, along with negligible FGFR binding affinity renders HS ineffective in promoting endocrine FGF–FGFR binding and dimerization (Goetz et al. 2007; Beenken and Mohammadi 2012). Instead, these ligands must rely on α/β Klotho coreceptors to signal (Urakawa et al. 2006; Kurosu et al. 2007; Ogawa et al. 2007; Kharitonenkov et al. 2008; Suzuki et al. 2008; Kuro-o 2012). Klotho coreceptors are single-pass transmembrane proteins whose ectodomain consists of tandem KL domains, which are homologous to β-glucosidases (Kuro-o et al. 1997; Ito et al. 2000). The Klotho coreceptors have been shown to associate constitutively with the c-splice isoforms of FGFR1–3 and FGFR4 to promote binding and dimerization of endocrine FGF–FGFR complexes (Kurosu et al. 2006, 2007; Goetz et al. 2012a). The Klotho dependency confines the target tissue specificity of endocrine FGFs to those that express α Klotho and β Klotho (Wu et al. 2007; Kurosu and Kuro-o 2008). Signaling specificity is further reinforced by the inherent specificity of FGF19 subfamily members for FGFRs (Goetz et al. 2012a). For example, FGF21 primarily activates the FGFR1c–β Klotho complex (Yie et al. 2012), whereas FGF19 is able to activate both FGFR1c–β Klotho as well as FGFR4–β Klotho. FGF23, on the other hand, binds promiscuously to FGFR1c–α Klotho, FGFR3c–α Klotho, and FGFR4–α Klotho (Yu et al. 2005; Goetz et al. 2012a).

Although the structural basis for ternary complex formation remains to be elucidated, biochemical studies have already provided significant insights into the molecular interactions between the components in the ternary complex (Fig. 4B). These studies have shown that the α Klotho and β Klotho coreceptors employ two different mechanisms to promote ternary complex formation. α Klotho combines with FGFR1c to create a de novo site for the FGF23 carboxy tail, whereas β Klotho uses two distinct sites to bind independently to the FGFR and to either the FGF19 or FGF21 carboxy tail (Wu et al. 2008; Goetz et al. 2012a). Consistent with the key role of the carboxy tail of FGF23 in signaling, the biological activity of FGF23 is downregulated by a naturally occurring proteolytic cleavage at an RXXR motif following the β-trefoil core (Shimada et al. 2001). Interestingly, the proteolytically cleaved carboxy tail can competitively inhibit binding of native FGF23 to the FGFR1c–α Klotho complex (Goetz et al. 2010, 2012a), indicating that this cleavage acts at two levels to inhibit FGF23 signaling: by inactivating FGF23 as well as by generating an endogenous inhibitor of FGF23 signaling. Pathogenic mutations of the RXXR motif abrogate proteolytic cleavage of the ligand (Shimada et al. 2002) and elevate the serum concentration of full-length bioactive FGF23, which accelerates phosphate excretion in the kidney and results in autosomal dominant hypophosphatemic rickets (ADHR) (White et al. 2001). Recent biochemical data show that the mutation of residues that comprises the D3 hydrophobic groove in FGFRc isoforms and FGFR4, which mediates binding of the FGF8 subfamily, also abolishes Klotho binding (Goetz et al. 2012a). Consistent with the overlap between FGF8 and Klotho binding sites on FGFR, the association of the Klotho coreceptor with FGFRs retards the ability of these receptors to respond to FGF8, indicating that endocrine and paracrine FGF signaling impact each other.

The insights gained into the mechanism of endocrine FGF signaling are already being exploited to develop agonists and antagonists of the endocrine FGF system for treating metabolic diseases, including diabetes, obesity, and disorders associated with perturbed phosphate and bile acid homeostasis (Goetz et al. 2010). For example, a novel FGF21 agonist has been engineered by knocking out the HS binding affinity of FGF2 and swapping its carboxy-terminal tail with that of FGF21 or FGF19 (Goetz et al. 2012b). This engineered FGF21 agonist is superior to native FGF21 in its insulin-sensitizing potential and is currently being evaluated in a mouse model for obesity. Using the same approach, FGF2 was converted into an FGF23 agonist, which may be used to treat patients with familial tumoral calcinosis, an inherited disorder, associated with loss-of-function mutations in FGF23 (Goetz et al. 2012b). Conversely, the carboxy-terminal tail of FGF23 is being developed for treatment of renal phosphate wasting disorders, and possibly for combating the cardiovascular morbidity factor in chronic kidney disease that has been shown to be caused by elevated FGF23 serum levels (Goetz et al. 2010).

ALTERNATIVE SPLICING OF D1 AND D1–D2 LINKER CONTROLS RECEPTOR AUTOINHIBITION

In FGFR1–3, a second major alternative splicing event (involving exons encoding D1 and the AB-containing D1–D2 linker regions) generates receptor isoforms lacking D1, the D1–D2 linker, or both (Johnson et al. 1991; Givol and Yayon 1992; Hou et al. 1992; Xu et al. 1992; Shimizu et al. 2001). The loss of D1 and the D1–D2 linker enhances the affinity of FGFR for both HS and FGF, indicating that they play an autoinhibitory role in FGFR regulation (Wang

et al. 1995; Roghani and Moscatelli 2007). The molecular basis by which these regions exert receptor autoinhibition has been interrogated using solution NMR and surface plasmon resonance (SPR) spectroscopies (Kalinina et al. 2012). The data show that the negatively charged AB subregion of the linker electrostatically engages the positively charged HS binding site on D2 in *cis*, thereby directly suppressing the HS affinity of the receptor (illustrated in Fig. 2A). Because of the close proximity of the HS binding site to the primary and secondary ligand-binding sites, as well as the direct receptor–receptor binding sites on D2, the *cis* electrostatic AB:HS interactions also sterically autoinhibits FGF–FGFR binding and dimerization. Consistent with the AB subregion playing a key role in FGFR autoinhibition, the amino acid sequence of the AB subregion is strongly conserved among FGFR orthologs (Kalinina et al. 2012).

MECHANISM OF FGFR KINASE REGULATION

HS or Klotho-dependent FGF–FGFR dimerization juxtaposes the cytoplasmic kinase domains, providing them with sufficient opportunity to *trans*-phosphorylate each other on specific tyrosine residues. A-loop tyrosine phosphorylation leads to a local rearrangement of the A loop into its active state, which in turn stabilizes the active conformation of the kinase globally and culminates in the upregulation of the intrinsic kinase activity (Hubbard 1999). Secondary phosphorylation events on tyrosines in the JM (juxtamembrane), kinase insert, and carboxy-tail regions are then ensued, providing docking sites for SH2-containing downstream signaling substrates, such as PLCγ and CrkL (Eswarakumar et al. 2005; Seo et al. 2009). Crystallographic studies of FGFR kinases have afforded major insights into the molecular mechanisms of FGFR kinase regulation and deregulation in pathological states. The crystal structures of the FGFR1 kinase (FGFR1K) (Mohammadi et al. 1996; Bae et al. 2010) and FGFR2K (Chen et al. 2007) have been solved in both their unphosphorylated (low activity) and A-loop tyrosine phosphorylated (activated) states. Reminiscent of all currently solved A-loop phosphorylated RTK kinase domains, including IRK (Hubbard 1997), IGF1R (Favelyukis et al. 2001; Pautsch et al. 2001), and MUSK (Bergamin et al. 2010), the phosphate moiety of A-loop tyrosine (pY654 in FGFRK1 and pY657 in FGFRK2) engages in two hydrogen bonds with an RTK-invariant arginine in the first half of the A-loop (R646 in FGFR1 and R649 in FGFR2) to stabilize the active conformation of the A-loop (Chen et al. 2007; Bae et al. 2010) (Fig. 5C,D). This in turn triggers a more intimate interaction between the amino lobe and the carboxy lobe of kinase, optimally aligning residues from the A-loop, the catalytic loop, and the αC-helix to catalyze phosphotransfer reactions (Fig. 5D).

In sharp contrast to other unphosphorylated RTKs, in which the ATP and substrate binding pockets are occluded in *cis* by the kinase A-loop or JM region, these sites remain accessible in FGFRKs, although the conformation of the carboxy-terminal end of the A-loop is not optimal for substrate binding (Mohammadi et al. 1996). A crystallographic analysis of FGF-R2Ks harboring pathogenic gain-of-function mutations, responsible for craniosynostosis syndromes and cancers, has proved to be instrumental in unraveling the molecular mechanism

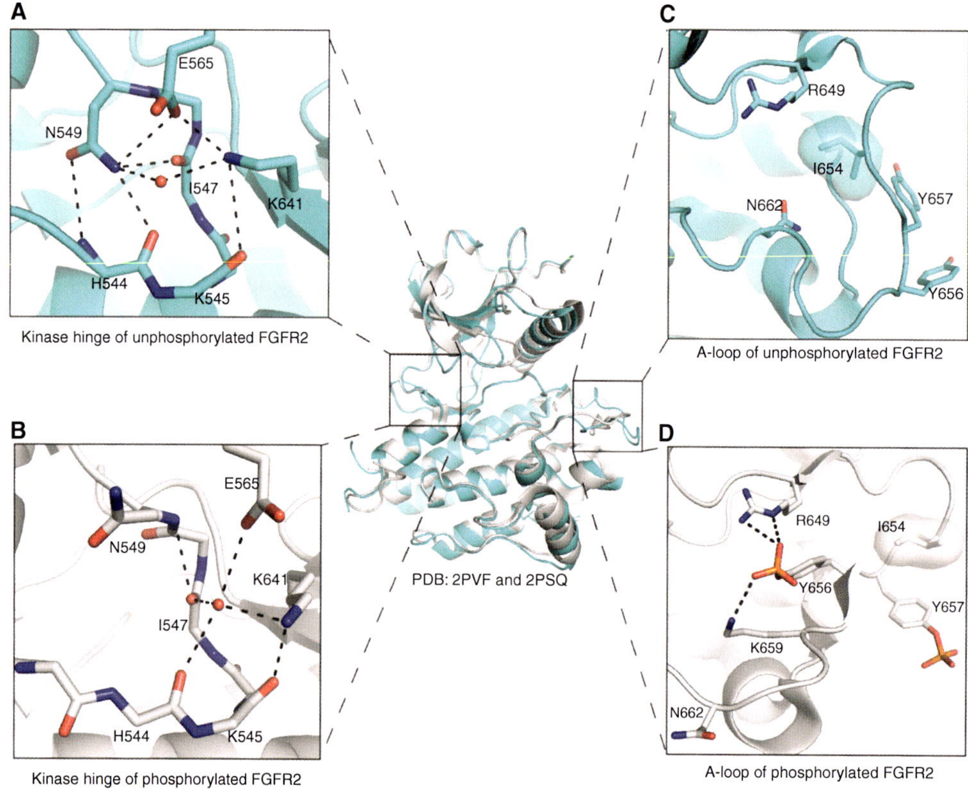

Figure 5. Comparison of FGFR2 kinases in the unphosphorylated and A-loop phosphorylated states. (*A*) The inhibitory network of hydrogen bonds, termed the "molecular brake," at the kinase hinge/interlobe region of the unphosphorylated low-activity FGFR2 kinase (PDB: 2PSQ, cyan). (*B*) Disengagement of the molecular brake in the A-loop phosphorylated activated FGFR2 kinase (PDB: 2PVF, gray). (*C*) The unphosphorylated A-loop conformation and select amino acids are shown. (*D*) The phosphorylated A-loop is held in an active conformation by hydrogen bonds between the phosphate moiety of phosphorylated A-loop tyrosine and basic residues (Arg-?? and Lys-??) within the A-loop. Note that on phosphorylation of A-loop Y656 and Y657, the A-loop undergoes a major conformational change as evidenced by the pistonlike conformational switch in the orientation of I654 and N662.

of FGFRK autoinhibition (Chen et al. 2007). A comparison of these unphosphorylated "diseased" FGFR2Ks with the wild-type unphosphorylated and A-loop phosphorylated FGFRKs shows that kinase autoinhibition is controlled at the level of protein dynamics. Specifically, a network of hydrogen bonds mediated by a triad of residues from the kinase hinge (E565), the αC–β4 loop (N549), and the β8 strand (K641), are engaged in an autoinhibitory network of hydrogen bonds (termed "molecular brake") at the kinase-hinge region that restricts the transition of the kinase into the active state, thus stabilizing the low activity state (Chen et al. 2007) (Fig. 5A,B). A-loop tyrosine phosphorylation or pathogenic gain-of-function mutations disengage this molecular break either directly or indirectly through allosteric communication between the A-loop and kinase hinge, thereby stabilizing the active state (Chen et al. 2007). The constituents of this molecular brake are conserved in other RTKs, including PDGFR, CSFLR, and KIT, and indicate that kinase regulation by a molecular brake may also apply to other RTKs.

Loss-of-function mutations in the tyrosine kinase domains of FGFRs are also implicated in human diseases, such as Kallmann syndrome, lacrimo-auriculo-dento-digital (LADD) syndrome, and cleft lip and palate. The crystal structure of FGFR2K, containing the A628T mutation responsible for LADD syndrome shows that the introduction of the polar and bulkier threonine sterically hinders formation of hydrogen bonds between Arg630 and Asp626 in the catalytic loop, thus hampering substrate tyrosine binding and phosphate-transfer reaction (Lew et al. 2007).

STRUCTURAL BASIS FOR FGFR KINASE *TRANS*-AUTOPHOSPHORYLATION

As introduced earlier, tyrosine *trans*-autophosphorylation fulfills two key roles in RTK signaling: (1) the upregulation of kinase activity, and (2) the generation of docking sites for downstream signaling proteins (Lemmon and Schlessinger 2010). Hence, elucidation of the structural basis for *trans*-phosphorylation is essential

for grasping the mechanism of RTK signal transduction in physiological and pathological states and for identifying new strategies for targeted drug discovery. Two recent studies have captured snapshots of FGFR1 and FGFR2 kinases caught in the act of *trans*-phosphorylation (Chen et al. 2008a; Bae et al. 2010). One kinase acts as the enzyme whereas the other kinase offers a phosphorylatable tyrosine from the kinase insert (Y583F in FGFR1) (Fig. 6B) or the carboxy-tail region (Y769 in FGFR2) (Chen et al. 2008a) (Fig. 6A). The structural data show that tyrosine *trans*-phosphorylation entails both a substantial degree of sequence specificity and structural complementarities. Both *trans*-phosphorylation complexes are asymmetric, whereby the carboxy lobe of the "substrate" engages both the amino lobe and carboxy lobe of the enzyme, burying a 1650 $\mathring{A}^2$ (kinase-insert tyrosine *trans*-phosphorylation complex) and a 1850 $\mathring{A}^2$ (carboxy-tail tyrosine phosphorylation complex) solvent-exposed surface area. The enzyme–substrate interface can be roughly divided into a proximal (vicinity of the catalytic cleft) and a distal (remote from the catalytic cleft) site (Fig. 6A,B). At the proximal interface, the phosphorylatable tyrosine and its immediate surrounding sequences from the substrate engage the catalytic cleft of the enzyme, whereas in the distal interface, the carboxy-lobe regions of the substrate engage the amino lobe of the enzyme remote from the enzyme catalytic cleft.

The structural mode by which the enzyme and the substrate embrace each other is very different between these two *trans*-phosphorylation complexes. Importantly, the kinase-insert tyrosine and the carboxy-tail tyrosine of the substrate dock into the active site of the enzyme from the opposite direction (Fig. 6A,B). The overall topology of the carboxy-tail tyrosine *trans*-phosphorylation complex is dictated by the latching of the carboxyl-terminal end (negative pole) of the αI helix of the substrate onto the amino-terminal end (positive pole) of the αG helix in the substrate at the proximal interface. This interaction assists Y769 (P-0) and its surrounding P-4 (T765) to P + 3 (L772) residues to optimally engage the active site and the P + 1 pocket of the enzyme to mediate the proximal specificity. Prominent contacts at the prox-

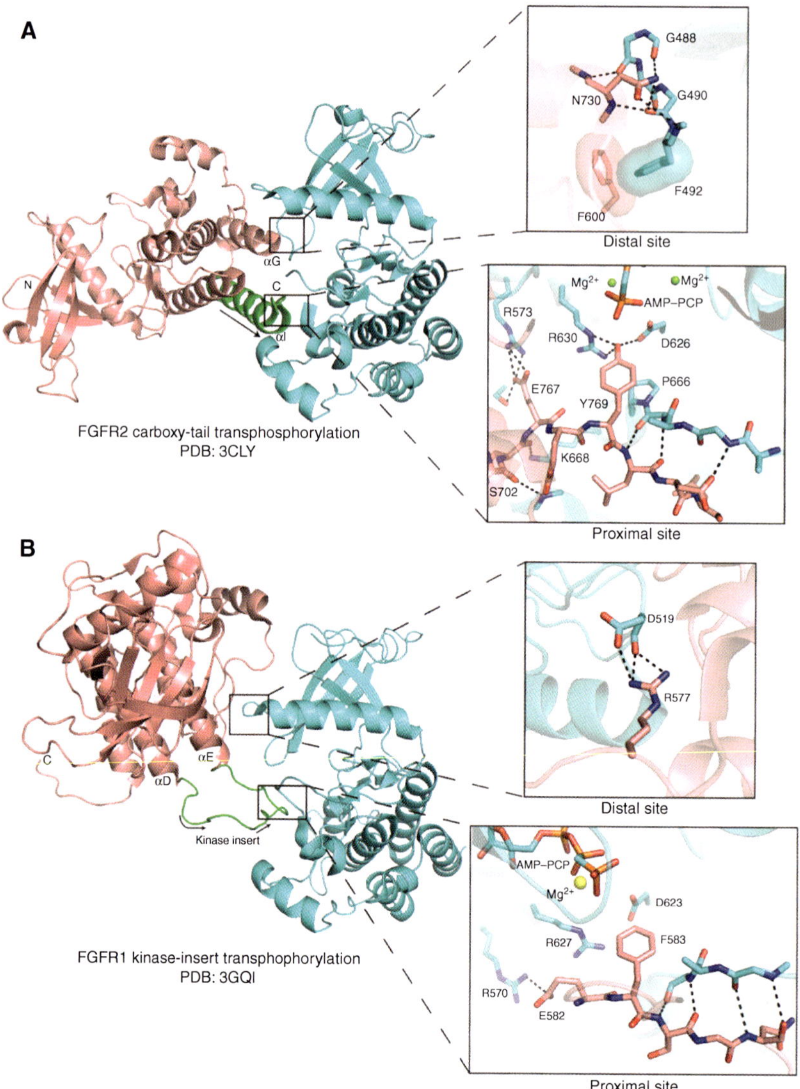

Figure 6. Comparison of the carboxy-tail and kinase-insert *trans*-phosphorylation complexes. (*A*) The crystal structure of FGFR2 kinases caught in the act of *trans*-phosphorylation on carboxy-tail tyrosine Tyr-769 (PDB: 3CLY). The enzyme-acting kinase is cyan, and the substrate kinase is salmon. The αI helix and kinase insert are green, with the amino to carboxy polarity indicated with arrows. (*B*) The kinase-insert *trans*-phosphorylation complex (PDB: 3GQI) is colored as in *A*. Note that the kinase insert Tyr-583 has been mutated to phenylalanine. Compared to the carboxy-tail tyrosine *trans*-phosphorylation complex, there are fewer interactions between the enzyme and the substrate in the kinase-insert tyrosine *trans*-phosphorylation complex.

imal site include the salt bridge between E767 (P-2) of the substrate with R573 of the enzyme, hydrogen bonds of T765 (P-4) and N766 (P-3) with the αG helix of the enzyme, and hydrophobic contacts of L770 (P + 1) and L772 (P + 3) with a P + 1 pocket (Fig. 6A, lower panel). Spe-

cificity is augmented by specific contacts at the distal interface between the carboxy lobe of the substrate and the nucleotide-binding loop in the amino lobe of the enzyme. Prominent contacts at the distal site include perpendicular aromatic interactions between F600 of substrate and F492 of enzyme, and hydrogen bonds between N730 of substrate and backbone carbonyl oxygens of the nucleotide binding loop in the enzyme (Chen et al. 2008a).

In contrast, the enzyme–substrate relationship in the kinase-insert *trans*-phosphorylation complex is determined by contacts between the carboxy-lobe of substrate and the amino-lobe of enzyme at the distal site (Bae et al. 2010). Aside from the hydrogen bonds between R577 (from the amino-terminal end of the kinase insert) and backbone carbonyl oxygen atoms in the β3–αC loop, all the remaining interactions at this distal contact site are of van der Waals nature. Interestingly, as in the carboxy-tail tyrosine *trans*-phosphorylation complex, the nucleotide binding loop of the enzyme partakes in the distal site, indicating that this loop, in addition to coordinating ATP, may also participate in substrate recognition. At the proximal site, the pseudosubstrate Phe583 (Tyr583) occupies a very similar position in the catalytic pocket of the enzyme-acting kinase, even though phenylalanine is not in position to form hydrogen bonds with the catalytic base D623 and R627 (Fig. 6B). Akin to the carboxy-tail tyrosine *trans*-phosphorylation complex, residues P-0 to P + 3 form an antiparallel strand with β10 in the A-loop. Compared to the carboxy-tail tyrosine *trans*-phosphorylation complex, there are significantly fewer interactions at the proximal site in the FGFR1 kinase-insert *trans*-phosphorylation complex. In fact the hydrophobic P + 1 pocket remains unengaged in this complex. Overall, the enzyme interface in the kinase-insert *trans*-phosphorylation complex contains fewer contacts and possesses lower shape complementarity than in the carboxy-tail tyrosine *trans*-phosphorylation complex (0.62 vs. 0.72). These differences are consistent with the kinase autophosphorylation data showing that phosphorylation of carboxy-tail tyrosine precedes that of kinase-insert tyrosine (Furdui et al. 2006; Chen et al. 2008a).

FUTURE PERSPECTIVES

The structural studies have provided the molecular bases for several major signaling events in FGF signaling, including FGF–FGFR specificity, HS-assisted FGF–FGFR dimerization, and FGFR kinase regulation. Many imminent questions in the FGF field remain for structural biologists to address, however. Elucidation of the crystal structure of the endocrine FGF–FGFR–Klotho ternary complex is a top priority, as it will provide a novel mechanism by which FGF–FGFR dimerization is achieved. Crystal structures with a representative from the FGF4 and FGF9 subfamilies in complex with their cognate FGFR should complete our understanding of the molecular mechanisms governing the specificity/promiscuity of this complex system. At the intracellular level, future work should be directed toward obtaining additional structures of *trans*-phosphorylation events, especially of the A-loop tyrosine phosphorylation, the gatekeeping phosphorylation event. Crystal structures of downstream signaling substrates in complex with the intracellular kinase domain will undoubtedly yield valuable information on how FGFR substrates are docked onto the kinase and are phosphorylated. The crystallographic data indicate that intrinsic protein dynamics may also modulate the activity of both the extracellular domain and intracellular domains of FGFRs (Chen et al. 2007; Kalinina et al. 2012). Ideally, the crystallographic snapshots should be complemented by NMR dynamic studies and computational methods to provide a more accurate four-dimensional perspective of FGF signaling. These structural data will be valuable in directing future biochemical interrogations of this complex signaling network, and provide a framework for a comprehensive understanding of the basis of disease and structure-based drug discovery for many human diseases where perturbed FGF signaling is implicated.

ACKNOWLEDGMENTS

We thank Yang Liu for critically reading the manuscript, and Jinghong Ma for assistance in figure preparation. This work is supported by the National Institute of Dental and Craniofacial Research Grant DE13686 (to M.M.). A.A.B. is partially supported by the Macromolecular

A.A. Belov and M. Mohammadi

Structure and Mechanism Training Grant 5T32GM088118-03.

REFERENCES

Acland P, Dixon M, Peters G, Dickson C. 1990. Subcellular fate of the int-2 oncoprotein is determined by choice of initiation codon. *Nature* **343:** 662–665.

Anderson J, Burns HD, Enriquez-Harris P, Wilkie AO, Heath JK. 1998. Apert syndrome mutations in fibroblast growth factor receptor 2 exhibit increased affinity for FGF ligand. *Hum Mol Genet* **7:** 1475–1483.

Asada M, Shinomiya M, Suzuki M, Honda E, Sugimoto R, Ikekita M, Imamura T. 2009. Glycosaminoglycan affinity of the complete fibroblast growth factor family. *Biochim Biophys Acta* **1790:** 40–48.

Bae JH, Boggon TJ, Tome F, Mandiyan V, Lax I, Schlessinger J. 2010. Asymmetric receptor contact is required for tyrosine autophosphorylation of fibroblast growth factor receptor in living cells. *Proc Natl Acad Sci* **107:** 2866–2871.

Basilico C, Moscatelli D. 1992. The FGF family of growth factors and oncogenes. *Adv Cancer Res* **59:** 115–165.

Bateman A, Chothia C. 1995. Outline structures for the extracellular domains of the fibroblast growth factor receptors. *Nat Struct Biol* **2:** 1068–1074.

Beenken A, Mohammadi M. 2009. The FGF family: Biology, pathophysiology and therapy. *Nat Rev Drug Discov* **8:** 235–253.

Beenken A, Mohammadi M. 2011. *Craniosynostoses—Molecular genetics, principles of diagnosis, and treatment.* Karger, Basel, Switzerland.

Beenken A, Mohammadi M. 2012. The structural biology of the FGF19 subfamily. *Adv Exp Med Biol* **728:** 1–24.

Beenken A, Eliseenkova AV, Ibrahimi OA, Olsen SK, Mohammadi M. 2012. Plasticity in interactions of fibroblast growth factor 1 (FGF1) N terminus with FGF receptors underlies promiscuity of FGF1. *J Biol Chem* **287:** 3067–3078.

Beer HD, Vindevoghel L, Gait MJ, Revest JM, Duan DR, Mason I, Dickson C, Werner S. 2000. Fibroblast growth factor (FGF) receptor 1-IIIb is a naturally occurring functional receptor for FGFs that is preferentially expressed in the skin and the brain. *J Biol Chem* **275:** 16091–16097.

Bellosta P, Iwahori A, Plotnikov AN, Eliseenkova AV, Basilico C, Mohammadi M. 2001. Identification of receptor and heparin binding sites in fibroblast growth factor 4 by structure-based mutagenesis. *Mol Cell Biol* **21:** 5946–5957.

Bergamin E, Hallock PT, Burden SJ, Hubbard SR. 2010. The cytoplasmic adaptor protein Dok7 activates the receptor tyrosine kinase MuSK via dimerization. *Mol Cell* **39:** 100–109.

Blunt AG, Lawshe A, Cunningham ML, Seto ML, Ornitz DM, MacArthur CA. 1997. Overlapping expression and redundant activation of mesenchymal fibroblast growth factor (FGF) receptors by alternatively spliced FGF-8 ligands. *J Biol Chem* **272:** 3733–3738.

Chateau MT, Araiz C, Descamps S, Galas S. 2010. Klotho interferes with a novel FGF-signalling pathway and insu-lin/Igf-like signalling to improve longevity and stress resistance in Caenorhabditis elegans. *Aging (Albany NY)* **2:** 567–581.

Chen H, Ma J, Li W, Eliseenkova AV, Xu C, Neubert TA, Miller WT, Mohammadi M. 2007. A molecular brake in the kinase hinge region regulates the activity of receptor tyrosine kinases. *Mol Cell* **27:** 717–730.

Chen H, Xu CF, Ma J, Eliseenkova AV, Li W, Pollock PM, Pitteloud N, Miller WT, Neubert TA, Mohammadi M. 2008a. A crystallographic snapshot of tyrosine *trans*-phosphorylation in action. *Proc Natl Acad Sci* **105:** 19660–19665.

Chen WW, Li L, Yang GY, Li K, Qi XY, Zhu W, Tang Y, Liu H, Boden G. 2008b. Circulating FGF-21 levels in normal subjects and in newly diagnose patients with Type 2 diabetes mellitus. *Exp Clin Endocrinol Diabetes* **116:** 65–68.

Colvin JS, White AC, Pratt SJ, Ornitz DM. 2001. Lung hypoplasia and neonatal death in Fgf9-null mice identify this gene as an essential regulator of lung mesenchyme. *Development* **128:** 2095–2106.

Delehedde M, Seve M, Sergeant N, Wartelle I, Lyon M, Rudland PS, Fernig DG. 2000. Fibroblast growth factor-2 stimulation of p42/44MAPK phosphorylation and IκB degradation is regulated by heparan sulfate/heparin in rat mammary fibroblasts. *J Biol Chem* **275:** 33905–33910.

di Martino E, Tomlinson DC, Knowles MA. 2012. A decade of FGF receptor research in bladder cancer: Past, present, and future challenges. *Adv Urol* **2012:** 429213.

Dionne CA, Crumley G, Bellot F, Kaplow JM, Searfoss G, Ruta M, Burgess WH, Jaye M, Schlessinger J. 1990. Cloning and expression of two distinct high-affinity receptors cross-reacting with acidic and basic fibroblast growth factors. *EMBO J* **9:** 2685–2692.

Dionne CA, Jaye M, Schlessinger J. 1991. Structural diversity and binding of FGF receptors. *Ann NY Acad Sci* **638:** 161–166.

Dode C, Levilliers J, Dupont JM, De Paepe A, Le Du N, Soussi-Yanicostas N, Coimbra RS, Delmaghani S, Compain-Nouaille S, Baverel F, et al. 2003. Loss-of-function mutations in FGFR1 cause autosomal dominant Kallmann syndrome. *Nat Genet* **33:** 463–465.

Dubrulle J, Pourquie O. 2004. fgf8 mRNA decay establishes a gradient that couples axial elongation to patterning in the vertebrate embryo. *Nature* **427:** 419–422.

Eriksson AE, Cousens LS, Weaver LH, Matthews BW. 1991. Three-dimensional structure of human basic fibroblast growth factor. *Proc Natl Acad Sci* **88:** 3441–3445.

Esko JD, Lindahl U. 2001. Molecular diversity of heparan sulfate. *J Clin Invest* **108:** 169–173.

Eswarakumar VP, Lax I, Schlessinger J. 2005. Cellular signaling by fibroblast growth factor receptors. *Cytokine Growth Factor Rev* **16:** 139–149.

Faham S, Linhardt RJ, Rees DC. 1998. Diversity does make a difference: Fibroblast growth factor–heparin interactions. *Curr Opin Struct Biol* **8:** 578–586.

Falardeau J, Chung WC, Beenken A, Raivio T, Plummer L, Sidis Y, Jacobson-Dickman EE, Eliseenkova AV, Ma J, Dwyer A, et al. 2008. Decreased FGF8 signaling causes deficiency of gonadotropin-releasing hormone in humans and mice. *J Clin Invest* **118:** 2822–2831.

Cite this article as *Cold Spring Harb Perspect Biol* doi: 10.1101/cshperspect.a015958

Favelyukis S, Till JH, Hubbard SR, Miller WT. 2001. Structure and autoregulation of the insulin-like growth factor 1 receptor kinase. *Nat Struct Biol* **8:** 1058–1063.

Feldman B, Poueymirou W, Papaioannou VE, DeChiara TM, Goldfarb M. 1995. Requirement of FGF-4 for postimplantation mouse development. *Science* **267:** 246–249.

Finch PW, Rubin JS, Miki T, Ron D, Aaronson SA. 1989. Human KGF is FGF-related with properties of a paracrine effector of epithelial cell growth. *Science* **245:** 752–755.

Fliser D, Kollerits B, Neyer U, Ankerst DP, Lhotta K, Lingenhel A, Ritz E, Kronenberg F, Kuen E, Konig P, et al. 2007. Fibroblast growth factor 23 (FGF23) predicts progression of chronic kidney disease: The mild to moderate kidney disease (MMKD) Study. *J Am Soc Nephrol* **18:** 2600–2608.

Florkiewicz RZ, Sommer A. 1989. Human basic fibroblast growth factor gene encodes four polypeptides: Three initiate translation from non-AUG codons. *Proc Natl Acad Sci* **86:** 3978–3981.

Fu L, John LM, Adams SH, Yu XX, Tomlinson E, Renz M, Williams PM, Soriano R, Corpuz R, Moffat B, et al. 2004. Fibroblast growth factor 19 increases metabolic rate and reverses dietary and leptin-deficient diabetes. *Endocrinology* **145:** 2594–2603.

Furdui CM, Lew ED, Schlessinger J, Anderson KS. 2006. Autophosphorylation of FGFR1 kinase is mediated by a sequential and precisely ordered reaction. *Mol Cell* **21:** 711–717.

Gartside MG, Chen H, Ibrahimi OA, Byron SA, Curtis AV, Wellens CL, Bengston A, Yudt LM, Eliseenkova AV, Ma J, et al. 2009. Loss-of-function fibroblast growth factor receptor-2 mutations in melanoma. *Mol Cancer Res* **7:** 41–54.

Gattineni J, Twombley K, Goetz R, Mohammadi M, Baum M. 2011. Regulation of serum 1,25(OH)$_2$ vitamin D3 levels by fibroblast growth factor 23 is mediated by FGF receptors 3 and 4. *Am J Physiol Renal Physiol* **301:** F371–377.

Gemel J, Gorry M, Ehrlich GD, MacArthur CA. 1996. Structure and sequence of human FGF8. *Genomics* **35:** 253–257.

Givol D, Yayon A. 1992. Complexity of FGF receptors: Genetic basis for structural diversity and functional specificity. *FASEB J* **6:** 3362–3369.

Glaser RL, Broman KW, Schulman RL, Eskenazi B, Wyrobek AJ, Jabs EW. 2003. The paternal-age effect in Apert syndrome is due, in part, to the increased frequency of mutations in sperm. *Am J Hum Genet* **73:** 939–947.

Goetz R, Beenken A, Ibrahimi OA, Kalinina J, Olsen SK, Eliseenkova AV, Xu C, Neubert TA, Zhang F, Linhardt RJ, et al. 2007. Molecular insights into the klotho-dependent, endocrine mode of action of fibroblast growth factor 19 subfamily members. *Mol Cell Biol* **27:** 3417–3428.

Goetz R, Dover K, Laezza F, Shtraizent N, Huang X, Tchetchik D, Eliseenkova AV, Xu CF, Neubert TA, Ornitz DM, et al. 2009. Crystal structure of a fibroblast growth factor homologous factor (FHF) defines a conserved surface on FHFs for binding and modulation of voltage-gated sodium channels. *J Biol Chem* **284:** 17883–17896.

Goetz R, Nakada Y, Hu MC, Kurosu H, Wang L, Nakatani T, Shi M, Eliseenkova AV, Razzaque MS, Moe OW, et al. 2010. Isolated C-terminal tail of FGF23 alleviates hypophosphatemia by inhibiting FGF23–FGFR–Klotho complex formation. *Proc Natl Acad Sci* **107:** 407–412.

Goetz R, Ohnishi M, Ding X, Kurosu H, Wang L, Akiyoshi J, Ma J, Gai W, Sidis Y, Pitteloud N, et al. 2012a. Klotho coreceptors inhibit signaling by paracrine fibroblast growth factor 8 subfamily ligands. *Mol Cell Biol* **32:** 1944–1954.

Goetz R, Ohnishi M, Kir S, Kurosu H, Wang L, Pastor J, Ma J, Gai W, Kuro OM, Razzaque MS, et al. 2012b. Conversion of a paracrine fibroblast growth factor into an endocrine fibroblast growth factor. *J Biol Chem* **287:** 29134–29146.

Goldfarb M. 1996. Functions of fibroblast growth factors in vertebrate development. *Cytokine Growth Factor Rev* **7:** 311–325.

Hacker U, Nybakken K, Perrimon N. 2005. Heparan sulphate proteoglycans: The sweet side of development. *Nat Rev Mol Cell Biol* **6:** 530–541.

Harada M, Murakami H, Okawa A, Okimoto N, Hiraoka S, Nakahara T, Akasaka R, Shiraishi Y, Futatsugi N, Mizutani-Koseki Y, et al. 2009. FGF9 monomer–dimer equilibrium regulates extracellular matrix affinity and tissue diffusion. *Nat Genet* **41:** 289–298.

Harmer NJ, Pellegrini L, Chirgadze D, Fernandez-Recio J, Blundell TL. 2004. The crystal structure of fibroblast growth factor (FGF) 19 reveals novel features of the FGF family and offers a structural basis for its unusual receptor affinity. *Biochem* **43:** 629–640.

Holt JA, Luo G, Billin AN, Bisi J, McNeill YY, Kozarsky KF, Donahee M, Wang DY, Mansfield TA, Kliewer SA, et al. 2003. Definition of a novel growth factor-dependent signal cascade for the suppression of bile acid biosynthesis. *Genes Dev* **17:** 1581–1591.

Hou J, Kan M, Wang F, Xu JM, Nakahara M, McBride G, McKeehan K, McKeehan WL. 1992. Substitution of putative half-cystine residues in heparin-binding fibroblast growth factor receptors. Loss of binding activity in both two and three loop isoforms. *J Biol Chem* **267:** 17804–17808.

Huang Z, Wang H, Lu M, Sun C, Wu X, Tan Y, Ye C, Zhu G, Wang X, Cai L, et al. 2011. A better anti-diabetic recombinant human fibroblast growth factor 21 (rhFGF21) modified with polyethylene glycol. *PLoS ONE* **6:** e20669.

Hubbard SR. 1997. Crystal structure of the activated insulin receptor tyrosine kinase in complex with peptide substrate and ATP analog. *EMBO J* **16:** 5572–5581.

Hubbard SR. 1999. Structural analysis of receptor tyrosine kinases. *Prog Biophys Mol Biol* **71:** 343–358.

Hung KW, Kumar TK, Kathir KM, Xu P, Ni F, Ji HH, Chen MC, Yang CC, Lin FP, Chiu IM, et al. 2005. Solution structure of the ligand binding domain of the fibroblast growth factor receptor: Role of heparin in the activation of the receptor. *Biochem* **44:** 15787–15798.

Ibrahimi OA, Eliseenkova AV, Plotnikov AN, Yu K, Ornitz DM, Mohammadi M. 2001. Structural basis for fibroblast growth factor receptor 2 activation in Apert syndrome. *Proc Natl Acad Sci* **98:** 7182–7187.

Ibrahimi OA, Zhang F, Eliseenkova AV, Itoh N, Linhardt RJ, Mohammadi M. 2004a. Biochemical analysis of pathogenic ligand-dependent FGFR2 mutations suggests distinct pathophysiological mechanisms for craniofacial and limb abnormalities. *Hum Mol Genet* **13:** 2313–2324.

Ibrahimi OA, Zhang F, Eliseenkova AV, Linhardt RJ, Mohammadi M. 2004b. Proline to arginine mutations in FGF receptors 1 and 3 result in Pfeiffer and Muenke craniosynostosis syndromes through enhancement of FGF binding affinity. *Hum Mol Genet* **13:** 69–78.

Ibrahimi OA, Zhang F, Hrstka SC, Mohammadi M, Linhardt RJ. 2004c. Kinetic model for FGF, FGFR, and proteoglycan signal transduction complex assembly. *Biochem* **43:** 4724–4730.

Ibrahimi OA, Chiu ES, McCarthy JG, Mohammadi M. 2005. Understanding the molecular basis of Apert syndrome. *Plas Reconstr Surg* **115:** 264–270.

Imamura T, Mitsui Y. 1987. Heparan sulfate and heparin as a potentiator or a suppressor of growth of normal and transformed vascular endothelial cells. *Exp Cell Res* **172:** 92–100.

Inagaki T, Dutchak P, Zhao G, Ding X, Gautron L, Parameswara V, Li Y, Goetz R, Mohammadi M, Esser V, et al. 2007. Endocrine regulation of the fasting response by PPARα-mediated induction of fibroblast growth factor 21. *Cell Metab* **5:** 415–425.

Inatani M, Irie F, Plump AS, Tessier-Lavigne M, Yamaguchi Y. 2003. Mammalian brain morphogenesis and midline axon guidance require heparan sulfate. *Science* **302:** 1044–1046.

Iozzo RV, Zoeller JJ, Nystrom A. 2009. Basement membrane proteoglycans: Modulators par excellence of cancer growth and angiogenesis. *Mol Cell* **27:** 503–513.

Ito S, Kinoshita S, Shiraishi N, Nakagawa S, Sekine S, Fujimori T, Nabeshima YI. 2000. Molecular cloning and expression analyses of mouse β klotho, which encodes a novel Klotho family protein. *Mech Dev* **98:** 115–119.

Itoh N, Ornitz DM. 2011. Fibroblast growth factors: From molecular evolution to roles in development, metabolism and disease. *J Biochem* **149:** 121–130.

Johnson DE, Williams LT. 1993. Structural and functional diversity in the FGF receptor multigene family. *Adv Cancer Res* **60:** 1–41.

Johnson DE, Lee PL, Lu J, Williams LT. 1990. Diverse forms of a receptor for acidic and basic fibroblast growth factors. *Mol Cell Biol* **10:** 4728–4736.

Johnson DE, Lu J, Chen H, Werner S, Williams LT. 1991. The human fibroblast growth factor receptor genes: A common structural arrangement underlies the mechanisms for generating receptor forms that differ in their third immunoglobulin domain. *Mol Cell Biol* **11:** 4627–4634.

Kalinina J, Byron SA, Makarenkova HP, Olsen SK, Eliseenkova AV, Larochelle WJ, Dhanabal M, Blais S, Ornitz DM, Day LA, et al. 2009. Homodimerization controls the fibroblast growth factor 9 subfamily's receptor binding and heparan sulfate-dependent diffusion in the extracellular matrix. *Mol Cell Biol* **29:** 4663–4678.

Kalinina J, Dutta K, Ilghari D, Beenken A, Goetz R, Eliseenkova AV, Cowburn D, Mohammadi M. 2012. The alternatively spliced acid box region plays a key role in FGF receptor autoinhibition. *Structure* **20:** 77–88.

Kato S, Sekine K. 1999. FGF–FGFR signaling in vertebrate organogenesis. *Cell Mol Biol (Noisy-le-grand)* **45:** 631–638.

Kharitonenkov A, Dunbar JD, Bina HA, Bright S, Moyers JS, Zhang C, Ding L, Micanovic R, Mehrbod SF, Knierman MD, et al. 2008. FGF-21/FGF-21 receptor interaction and activation is determined by β Klotho. *J Cell Physiol* **215:** 1–7.

Kir S, Beddow SA, Samuel VT, Miller P, Previs SF, Suino-Powell K, Xu HE, Shulman GI, Kliewer SA, Mangelsdorf DJ. 2011. FGF19 as a postprandial, insulin-independent activator of hepatic protein and glycogen synthesis. *Science* **331:** 1621–1624.

Kiselyov VV, Bock E, Berezin V, Poulsen FM. 2006. NMR structure of the first Ig module of mouse FGFR1. *Protein Sci* **15:** 1512–1515.

Kuro-o M. 2012. Klotho and β Klotho. *Adv Exp Med Biol* **728:** 25–40.

Kuro-o M, Matsumura Y, Aizawa H, Kawaguchi H, Suga T, Utsugi T, Ohyama Y, Kurabayashi M, Kaname T, Kume E, et al. 1997. Mutation of the mouse klotho gene leads to a syndrome resembling ageing. *Nature* **390:** 45–51.

Kurosu H, Kuro-o M. 2008. The Klotho gene family and the endocrine fibroblast growth factors. *Curr Opin Nephrol Hypertens* **17:** 368–372.

Kurosu H, Ogawa Y, Miyoshi M, Yamamoto M, Nandi A, Rosenblatt KP, Baum MG, Schiavi S, Hu MC, Moe OW, et al. 2006. Regulation of fibroblast growth factor-23 signaling by klotho. *J Biol Chem* **281:** 6120–6123.

Kurosu H, Choi M, Ogawa Y, Dickson AS, Goetz R, Eliseenkova AV, Mohammadi M, Rosenblatt KP, Kliewer SA, Kuro-o M. 2007. Tissue-specific expression of β Klotho and fibroblast growth factor (FGF) receptor isoforms determines metabolic activity of FGF19 and FGF21. *J Biol Chem* **282:** 26687–26695.

Lee PL, Johnson DE, Cousens LS, Fried VA, Williams LT. 1989. Purification and complementary DNA cloning of a receptor for basic fibroblast growth factor. *Science* **245:** 57–60.

Lee HC, Tseng WA, Lo FY, Liu TM, Tsai HJ. 2009. FoxD5 mediates anterior–posterior polarity through upstream modulator Fgf signaling during zebrafish somitogenesis. *Dev Biol* **336:** 232–245.

Lemmon MA, Schlessinger J. 2010. Cell signaling by receptor tyrosine kinases. *Cell* **141:** 1117–1134.

Lew ED, Bae JH, Rohmann E, Wollnik B, Schlessinger J. 2007. Structural basis for reduced FGFR2 activity in LADD syndrome: Implications for FGFR autoinhibition and activation. *Proc Natl Acad Sci* **104:** 19802–19807.

Lin X, Buff EM, Perrimon N, Michelson AM. 1999. Heparan sulfate proteoglycans are essential for FGF receptor signaling during *Drosophila* embryonic development. *Development* **126:** 3715–3723.

Luo Y, Lu W, Mohamedali KA, Jang JH, Jones RB, Gabriel JL, Kan M, McKeehan WL. 1998. The glycine box: A determinant of specificity for fibroblast growth factor. *Biochem* **37:** 16506–16515.

Makarenkova HP, Hoffman MP, Beenken A, Eliseenkova AV, Meech R, Tsau C, Patel VN, Lang RA, Mohammadi M. 2009. Differential interactions of FGFs with heparan sulfate control gradient formation and branching morphogenesis. *Sci Signal* **2:** ra55.

Mariani FV, Ahn CP, Martin GR. 2008. Genetic evidence that FGFs have an instructive role in limb proximal–distal patterning. *Nature* **453:** 401–405.

Cite this article as *Cold Spring Harb Perspect Biol* doi: 10.1101/cshperspect.a015958

Martin GR. 1998. The roles of FGFs in the early development of vertebrate limbs. *Genes Dev* **12**: 1571–1586.

Mason IJ, Fuller-Pace F, Smith R, Dickson C. 1994. FGF-7 (keratinocyte growth factor) expression during mouse development suggests roles in myogenesis, forebrain regionalisation and epithelial–mesenchymal interactions. *Mech Dev* **45**: 15–30.

Mathieu M, Chatelain E, Ornitz D, Bresnick J, Mason I, Kiefer P, Dickson C. 1995. Receptor binding and mitogenic properties of mouse fibroblast growth factor 3. Modulation of response by heparin. *J Biol Chem* **270**: 24197–24203.

McCabe MJ, Gaston-Massuet C, Tziaferi V, Gregory LC, Alatzoglou KS, Signore M, Puelles E, Gerrelli D, Farooqi IS, Raza J, et al. 2011. Novel FGF8 mutations associated with recessive holoprosencephaly, craniofacial defects, and hypothalamo-pituitary dysfunction. *J Clin Endocrinol Metab* **96**: E1709–E1718.

McEwen DG, Ornitz DM. 1997. Determination of fibroblast growth factor expression in mouse, rat and human samples using a single primer pair. *Biotechniques* **22**: 1068–1070.

McIntosh I, Bellus GA, Jab EW. 2000. The pleiotropic effects of fibroblast growth factor receptors in mammalian development. *Cell Struct Funct* **25**: 85–96.

Metzger RJ, Klein OD, Martin GR, Krasnow MA. 2008. The branching programme of mouse lung development. *Nature* **453**: 745–750.

Miki T, Bottaro DP, Fleming TP, Smith CL, Burgess WH, Chan AM, Aaronson SA. 1992. Determination of ligand-binding specificity by alternative splicing: Two distinct growth factor receptors encoded by a single gene. *Proc Natl Acad Sci* **89**: 246–250.

Mohammadi M, Schlessinger J, Hubbard SR. 1996. Structure of the FGF receptor tyrosine kinase domain reveals a novel autoinhibitory mechanism. *Cell* **86**: 577–587.

Mohammadi M, Olsen SK, Goetz R. 2005a. A protein canyon in the FGF–FGF receptor dimer selects from an a la carte menu of heparan sulfate motifs. *Curr Opin Struct Biol* **15**: 506–516.

Mohammadi M, Olsen SK, Ibrahimi OA. 2005b. Structural basis for fibroblast growth factor receptor activation. *Cytokine Growth Factor Rev* **16**: 107–137.

Moyers JS, Shiyanova TL, Mehrbod F, Dunbar JD, Noblitt TW, Otto KA, Reifel-Miller A, Kharitonenkov A. 2007. Molecular determinants of FGF-21 activity-synergy and cross-talk with PPARγ signaling. *J Cell Physiol* **210**: 1–6.

Naiche LA, Holder N, Lewandoski M. 2011. FGF4 and FGF8 comprise the wavefront activity that controls somitogenesis. *Proc Natl Acad Sci* **108**: 4018–4023.

Nakatani T, Sarraj B, Ohnishi M, Densmore MJ, Taguchi T, Goetz R, Mohammadi M, Lanske B, Razzaque MS. 2009. In vivo genetic evidence for klotho-dependent, fibroblast growth factor 23 (Fgf23) -mediated regulation of systemic phosphate homeostasis. *FASEB J* **23**: 433–441.

Naski MC, Wang Q, Xu J, Ornitz DM. 1996. Graded activation of fibroblast growth factor receptor 3 by mutations causing achondroplasia and thanatophoric dysplasia. *Nat Genet* **13**: 233–237.

Niwa Y, Shimojo H, Isomura A, Gonzalez A, Miyachi H, Kageyama R. 2011. Different types of oscillations in Notch and Fgf signaling regulate the spatiotemporal periodicity of somitogenesis. *Genes Dev* **25**: 1115–1120.

Nugent MA, Edelman ER. 1992. Kinetics of basic fibroblast growth factor binding to its receptor and heparan sulfate proteoglycan: A mechanism for cooperactivity. *Biochem* **31**: 8876–8883.

Ogawa Y, Kurosu H, Yamamoto M, Nandi A, Rosenblatt KP, Goetz R, Eliseenkova AV, Mohammadi M, Kuro-o M. 2007. β Klotho is required for metabolic activity of fibroblast growth factor 21. *Proc Natl Acad Sci* **104**: 7432–7437.

Olsen SK, Garbi M, Zampieri N, Eliseenkova AV, Ornitz DM, Goldfarb M, Mohammadi M. 2003. Fibroblast growth factor (FGF) homologous factors share structural but not functional homology with FGFs. *J Biol Chem* **278**: 34226–34236.

Olsen SK, Ibrahimi OA, Raucci A, Zhang F, Eliseenkova AV, Yayon A, Basilico C, Linhardt RJ, Schlessinger J, Mohammadi M. 2004. Insights into the molecular basis for fibroblast growth factor receptor autoinhibition and ligand-binding promiscuity. *Proc Natl Acad Sci* **101**: 935–940.

Olsen SK, Li JY, Bromleigh C, Eliseenkova AV, Ibrahimi OA, Lao Z, Zhang F, Linhardt RJ, Joyner AL, Mohammadi M. 2006. Structural basis by which alternative splicing modulates the organizer activity of FGF8 in the brain. *Genes Dev* **20**: 185–198.

Olwin BB, Rapraeger A. 1992. Repression of myogenic differentiation by aFGF, bFGF, and K-FGF is dependent on cellular heparan sulfate. *J Cell Biol* **118**: 631–639.

Ornitz DM, Itoh N. 2001. Fibroblast growth factors. *Genome Biol* **2**: REVIEWS3005.

Ornitz DM, Leder P. 1992. Ligand specificity and heparin dependence of fibroblast growth factor receptors 1 and 3. *J Biol Chem* **267**: 16305–16311.

Ornitz DM, Yayon A, Flanagan JG, Svahn CM, Levi E, Leder P. 1992. Heparin is required for cell-free binding of basic fibroblast growth factor to a soluble receptor and for mitogenesis in whole cells. *Mol Cell Biol* **12**: 240–247.

Ornitz DM, Xu J, Colvin JS, McEwen DG, MacArthur CA, Coulier F, Gao G, Goldfarb M. 1996. Receptor specificity of the fibroblast growth factor family. *J Biol Chem* **271**: 15292–15297.

Orr Urtreger A, Bedford MT, Burakova T, Arman E, Zimmer Y, Yayon A, Givol D, Lonai P. 1993. Developmental localization of the splicing alternatives of fibroblast growth factor receptor-2 (FGFR2). *Dev Biol* **158**: 475–486.

Pautsch A, Zoephel A, Ahorn H, Spevak W, Hauptmann R, Nar H. 2001. Crystal structure of bisphosphorylated IGF-1 receptor kinase: Insight into domain movements upon kinase activation. *Structure* **9**: 955–965.

Pellegrini L, Burke DF, von Delft F, Mulloy B, Blundell TL. 2000. Crystal structure of fibroblast growth factor receptor ectodomain bound to ligand and heparin. *Nature* **407**: 1029–1034.

Pitteloud N, Acierno JS Jr, Meysing A, Eliseenkova AV, Ma J, Ibrahimi OA, Metzger DL, Hayes FJ, Dwyer AA, Hughes VA, et al. 2006a. Mutations in fibroblast growth factor receptor 1 cause both Kallmann syndrome and normosmic idiopathic hypogonadotropic hypogonadism. *Proc Natl Acad Sci* **103:** 6281–6286.

Pitteloud N, Meysing A, Quinton R, Acierno JS Jr, Dwyer AA, Plummer L, Fliers E, Boepple P, Hayes F, Seminara S, et al. 2006b. Mutations in fibroblast growth factor receptor 1 cause Kallmann syndrome with a wide spectrum of reproductive phenotypes. *Mol Cell Endocrinol* **254–255:** 60–69.

Plotnikov AN, Schlessinger J, Hubbard SR, Mohammadi M. 1999. Structural basis for FGF receptor dimerization and activation. *Cell* **98:** 641–650.

Plotnikov AN, Hubbard SR, Schlessinger J, Mohammadi M. 2000. Crystal structures of two FGF–FGFR complexes reveal the determinants of ligand–receptor specificity. *Cell* **101:** 413–424.

Plotnikov AN, Eliseenkova AV, Ibrahimi OA, Shriver Z, Sasisekharan R, Lemmon MA, Mohammadi M. 2001. Crystal structure of fibroblast growth factor 9 reveals regions implicated in dimerization and autoinhibition. *J Biol Chem* **276:** 4322–4329.

Pollock PM, Gartside MG, Dejeza LC, Powell MA, Mallon MA, Davies H, Mohammadi M, Futreal PA, Stratton MR, Trent JM, et al. 2007. Frequent activating FGFR2 mutations in endometrial carcinomas parallel germline mutations associated with craniosynostosis and skeletal dysplasia syndromes. *Oncogene* **26:** 7158–7162.

Popovici C, Roubin R, Coulier F, Birnbaum D. 2005. An evolutionary history of the FGF superfamily. *BioEssays* **27:** 849–857.

Powell AK, Fernig DG, Turnbull JE. 2002. Fibroblast growth factor receptors 1 and 2 interact differently with heparin/heparan sulfate. Implications for dynamic assembly of a ternary signaling complex. *J Biol Chem* **277:** 28554–28563.

Prats H, Kaghad M, Prats AC, Klagsbrun M, Lelias JM, Liauzun P, Chalon P, Tauber JP, Amalric F, Smith JA, et al. 1989. High molecular mass forms of basic fibroblast growth factor are initiated by alternative CUG codons. *Proc Natl Acad Sci* **86:** 1836–1840.

Rand V, Huang J, Stockwell T, Ferriera S, Buzko O, Levy S, Busam D, Li K, Edwards JB, Eberhart C, et al. 2005. Sequence survey of receptor tyrosine kinases reveals mutations in glioblastomas. *Proc Natl Acad Sci* **102:** 14344–14349.

Rapraeger AC, Krufka A, Olwin BB. 1991. Requirement of heparan sulfate for bFGF-mediated fibroblast growth and myoblast differentiation. *Science* **252:** 1705–1708.

Razzaque MS, Lanske B. 2007. The emerging role of the fibroblast growth factor-23-klotho axis in renal regulation of phosphate homeostasis. *J Endocrinol* **194:** 1–10.

Reinehr T, Woelfle J, Wunsch R, Roth CL. 2012. Fibroblast growth factor 21 (FGF-21) and its relation to obesity, metabolic syndrome, and nonalcoholic fatty liver in children: A longitudinal analysis. *J Clin Endocrinol Metab* **97:** 2143–2150.

Rodriguez Esteban C, Capdevila J, Economides AN, Pascual J, Ortiz A, Izpisua Belmonte JC. 1999. The novel Cer-like protein Caronte mediates the establishment of embryonic left–right asymmetry. *Nature* **401:** 243–251.

Roghani M, Moscatelli D. 2007. Prostate cells express two isoforms of fibroblast growth factor receptor 1 with different affinities for fibroblast growth factor-2. *Prostate* **67:** 115–124.

Schlessinger J, Plotnikov AN, Ibrahimi OA, Eliseenkova AV, Yeh BK, Yayon A, Linhardt RJ, Mohammadi M. 2000. Crystal structure of a ternary FGF–FGFR–heparin complex reveals a dual role for heparin in FGFR binding and dimerization. *Mol Cell* **6:** 743–750.

Sekine K, Ohuchi H, Fujiwara M, Yamasaki M, Yoshizawa T, Sato T, Yagishita N, Matsui D, Koga Y, Itoh N, et al. 1999. Fgf10 is essential for limb and lung formation. *Nat Genet* **21:** 138–141.

Seo JH, Suenaga A, Hatakeyama M, Taiji M, Imamoto A. 2009. Structural and functional basis of a role for CRKL in a fibroblast growth factor 8-induced feed-forward loop. *Mol Cell Biol* **29:** 3076–3087.

Serls AE, Doherty S, Parvatiyar P, Wells JM, Deutsch GH. 2005. Different thresholds of fibroblast growth factors pattern the ventral foregut into liver and lung. *Development* **132:** 35–47.

Shimada T, Mizutani S, Muto T, Yoneya T, Hino R, Takeda S, Takeuchi Y, Fujita T, Fukumoto S, Yamashita T. 2001. Cloning and characterization of FGF23 as a causative factor of tumor-induced osteomalacia. *Proc Natl Acad Sci* **98:** 6500–6505.

Shimada T, Muto T, Urakawa I, Yoneya T, Yamazaki Y, Okawa K, Takeuchi Y, Fujita T, Fukumoto S, Yamashita T. 2002. Mutant FGF-23 responsible for autosomal dominant hypophosphatemic rickets is resistant to proteolytic cleavage and causes hypophosphatemia in vivo. *Endocrinology* **143:** 3179–3182.

Shimizu A, Tada K, Shukunami C, Hiraki Y, Kurokawa T, Magane N, Kurokawa-Seo M. 2001. A novel alternatively spliced fibroblast growth factor receptor 3 isoform lacking the acid box domain is expressed during chondrogenic differentiation of ATDC5 cells. *J Biol Chem* **276:** 11031–11040.

Stauber DJ, DiGabriele AD, Hendrickson WA. 2000. Structural interactions of fibroblast growth factor receptor with its ligands. *Proc Natl Acad Sci* **97:** 49–54.

Sun X, Meyers EN, Lewandoski M, Martin GR. 1999. Targeted disruption of Fgf8 causes failure of cell migration in the gastrulating mouse embryo. *Genes Dev* **13:** 1834–1846.

Suzuki M, Uehara Y, Motomura-Matsuzaka K, Oki J, Koyama Y, Kimura M, Asada M, Komi-Kuramochi A, Oka S, Imamura T. 2008. β Klotho is required for fibroblast growth factor (FGF) 21 signaling through FGF receptor (FGFR) 1c and FGFR3c. *Mol Endocrinol* **22:** 1006–1014.

Tanaka Y, Okada Y, Hirokawa N. 2005. FGF-induced vesicular release of Sonic hedgehog and retinoic acid in leftward nodal flow is critical for left–right determination. *Nature* **435:** 172–177.

Tekin M, Hismi BO, Fitoz S, Ozdag H, Cengiz FB, Sirmaci A, Aslan I, Inceoglu B, Yuksel-Konuk EB, Yilmaz ST, et al. 2007. Homozygous mutations in fibroblast growth factor 3 are associated with a new form of syndromic deafness characterized by inner ear agenesis, microtia, and microdontia. *Am J Hum Genet* **80:** 338–344.

Tekin M, Ozturkmen Akay H, Fitoz S, Birnbaum S, Cengiz FB, Sennaroglu L, Incesulu A, Yuksel Konuk EB, Hasanefendioglu Bayrak A, Senturk S, et al. 2008. Homozygous FGF3 mutations result in congenital deafness with inner ear agenesis, microtia, and microdontia. *Clin Genet* **73:** 554–565.

Trarbach EB, Abreu AP, Silveira LF, Garmes HM, Baptista MT, Teles MG, Costa EM, Mohammadi M, Pitteloud N, Mendonca BB, et al. 2010. Nonsense mutations in FGF8 gene causing different degrees of human gonadotropin-releasing deficiency. *J Clin Endocrinol Metab* **95:** 3491–3496.

Trueb B. 2011. Biology of FGFRL1, the fifth fibroblast growth factor receptor. *Cell Mol Life Sci* **68:** 951–964.

Urakawa I, Yamazaki Y, Shimada T, Iijima K, Hasegawa H, Okawa K, Fujita T, Fukumoto S, Yamashita T. 2006. Klotho converts canonical FGF receptor into a specific receptor for FGF23. *Nature* **444:** 770–774.

Vega-Hernandez M, Kovacs A, De Langhe S, Ornitz DM. 2011. FGF10/FGFR2b signaling is essential for cardiac fibroblast development and growth of the myocardium. *Development* **138:** 3331–3340.

Wahl MB, Deng C, Lewandoski M, Pourquie O. 2007. FGF signaling acts upstream of the NOTCH and WNT signaling pathways to control segmentation clock oscillations in mouse somitogenesis. *Development* **134:** 4033–4041.

Wang F, Kan M, Yan G, Xu J, McKeehan WL. 1995. Alternately spliced NH2-terminal immunoglobulin-like Loop I in the ectodomain of the fibroblast growth factor (FGF) receptor 1 lowers affinity for both heparin and FGF-1. *J Biol Chem* **270:** 10231–10235.

Wang C, Chung BC, Yan H, Lee SY, Pitt GS. 2012. Crystal structure of the ternary complex of a NaV C-terminal domain, a fibroblast growth factor homologous factor, and calmodulin. *Structure* **20:** 1167–1176.

White KE, Evans WE, O'Riordan JLH, Speer MC, Econs MJ, Lorenz-Depiereux B, Grabowski M, Meitinger T, Strom TM. 2000. Autosomal dominant hypophosphataemic rickets is associated with mutations in *FGF23*. *Nat Genet* **26:** 345–348.

White KE, Carn G, Lorenz-Depiereux B, Benet-Pages A, Strom TM, Econs MJ. 2001. Autosomal-dominant hypophosphatemic rickets (ADHR) mutations stabilize FGF-23. *Kidney Int* **60:** 2079–2086.

Wilkie AO. 2005. Bad bones, absent smell, selfish testes: The pleiotropic consequences of human FGF receptor mutations. *Cytokine Growth Factor Rev* **16:** 187–203.

Wilkie AO, Slaney SF, Oldridge M, Poole MD, Ashworth GJ, Hockley AD, Hayward RD, David DJ, Pulleyn LJ, Rutland P, et al. 1995. Apert syndrome results from localized mutations of FGFR2 and is allelic with Crouzon syndrome. *Nat Genet* **9:** 165–172.

Wilkie AO, Patey SJ, Kan SH, van den Ouweland AM, Hamel BC. 2002. FGFs, their receptors, and human limb malformations: clinical and molecular correlations. *Am J Med Genet* **112:** 266–278.

Wu X, Ge H, Gupte J, Weiszmann J, Shimamoto G, Stevens J, Hawkins N, Lemon B, Shen W, Xu J, et al. 2007. Co-receptor requirements for fibroblast growth factor-19 signaling. *J Biol Chem* **282:** 29069–29072.

Wu X, Lemon B, Li X, Gupte J, Weiszmann J, Stevens J, Hawkins N, Shen W, Lindberg R, Chen JL, et al. 2008. C-terminal tail of FGF19 determines its specificity toward Klotho co-receptors. *J Biol Chem* **283:** 33304–33309.

Wuechner C, Nordqvist AC, Winterpacht A, Zabel B, Schalling M. 1996. Developmental expression of splicing variants of fibroblast growth factor receptor 3 (FGFR3) in mouse. *Int J Dev Biol* **40:** 1185–1188.

Xu J, Nakahara M, Crabb JW, Shi E, Matuo Y, Fraser M, Kan M, Hou J, McKeehan WL. 1992. Expression and immunochemical analysis of rat and human fibroblast growth factor receptor (flg) isoforms. *J Biol Chem* **267:** 17792–17803.

Xu J, Lawshe A, MacArthur CA, Ornitz DM. 1999. Genomic structure, mapping, activity and expression of fibroblast growth factor 17. *Mech Dev* **83:** 165–178.

Xu R, Ori A, Rudd TR, Uniewicz KA, Ahmed YA, Guimond SE, Skidmore MA, Siligardi G, Yates EA, Fernig DG. 2012. Diversification of the structural determinants of fibroblast growth factor–heparin interactions: Implications for binding specificity. *J Biol Chem*.

Yayon A, Klagsbrun M, Esko JD, Leder P, Ornitz DM. 1991. Cell surface, heparin-like molecules are required for binding of basic fibroblast growth factor to its high affinity receptor. *Cell* **64:** 841–848.

Yayon A, Zimmer Y, Shen GH, Avivi A, Yarden Y, Givol D. 1992. A confined variable region confers ligand specificity on fibroblast growth factor receptors: Implications for the origin of the immunoglobulin fold. *EMBO J* **11:** 1885–1890.

Ye S, Luo Y, Lu W, Jones RB, Linhardt RJ, Capila I, Toida T, Kan M, Pelletier H, McKeehan WL. 2001. Structural basis for interaction of FGF-1, FGF-2, and FGF-7 with different heparan sulfate motifs. *Biochem* **40:** 14429–14439.

Yeh BK, Igarashi M, Eliseenkova AV, Plotnikov AN, Sher I, Ron D, Aaronson SA, Mohammadi M. 2003. Structural basis by which alternative splicing confers specificity in fibroblast growth factor receptors. *Proc Natl Acad Sci* **100:** 2266–2271.

Yie J, Wang W, Deng L, Tam LT, Stevens J, Chen MM, Li Y, Xu J, Lindberg R, Hecht R, et al. 2012. Understanding the physical interactions in the FGF21/FGFR/β-Klotho complex: Structural requirements and implications in FGF21 signaling. *Chem Biol Drug Des* **79:** 398–410.

Yoon SR, Qin J, Glaser RL, Jabs EW, Wexler NS, Sokol R, Arnheim N, Calabrese P. 2009. The ups and downs of mutation frequencies during aging can account for the Apert syndrome paternal age effect. *PLoS Genet* **5:** e1000558.

Yu K, Ornitz DM. 2001. Uncoupling fibroblast growth factor receptor 2 ligand binding specificity leads to Apert syndrome-like phenotypes. *Proc Natl Acad Sci* **98:** 3641–3643.

Yu C, Wang F, Kan M, Jin C, Jones RB, Weinstein M, Deng CX, McKeehan WL. 2000a. Elevated cholesterol metabolism and bile acid synthesis in mice lacking membrane tyrosine kinase receptor FGFR4. *J Biol Chem* **275:** 15482–15489.

Yu K, Herr AB, Waksman G, Ornitz DM. 2000b. Loss of fibroblast growth factor receptor 2 ligand-binding specificity in Apert syndrome. *Proc Natl Acad Sci* **97:** 14536–14541.

Yu X, Ibrahimi OA, Goetz R, Zhang F, Davis SI, Garringer HJ, Linhardt RJ, Ornitz DM, Mohammadi M, White KE. 2005. Analysis of the biochemical mechanisms for the endocrine actions of fibroblast growth factor-23. *Endocrinology* **146:** 4647–4656.

Zhang JD, Cousens LS, Barr PJ, Sprang SR. 1991. Three-dimensional structure of human basic fibroblast growth factor, a structural homolog of interleukin 1β. *Proc Natl Acad Sci* **88:** 3446–3450.

Zhang X, Ibrahimi OA, Olsen SK, Umemori H, Mohammadi M, Ornitz DM. 2006. Receptor specificity of the fibroblast growth factor family. The complete mammalian FGF family. *J Biol Chem* **281:** 15694–15700.

Zhang F, Zhang Z, Lin X, Beenken A, Eliseenkova AV, Mohammadi M, Linhardt RJ. 2009. Compositional analysis on heparin/heparan sulfate interacting with FGF•FGFR complexes. *Biochemistry* **48:** 8379–8386.

Zhu X, Komiya H, Chirino A, Faham S, Fox GM, Arakawa T, Hsu BT, Rees DC. 1991. Three-dimensional structures of acidic and basic fibroblast growth factors. *Science* **251:** 90–93.

Cite this article as *Cold Spring Harb Perspect Biol* doi: 10.1101/cshperspect.a015958

The Role of MuSK in Synapse Formation and Neuromuscular Disease

Steven J. Burden, Norihiro Yumoto, and Wei Zhang

Molecular Neurobiology Program, Helen L. and Martin S. Kimmel Center for Biology and Medicine at the Skirball Institute of Biomolecular Medicine, NYU Medical School, New York, New York 10016

Correspondence: burden@saturn.med.nyu.edu

Muscle-specific kinase (MuSK) is essential for each step in neuromuscular synapse formation. Before innervation, MuSK initiates postsynaptic differentiation, priming the muscle for synapse formation. Approaching motor axons recognize the primed, or prepatterned, region of muscle, causing motor axons to stop growing and differentiate into specialized nerve terminals. MuSK controls presynaptic differentiation by causing the clustering of Lrp4, which functions as a direct retrograde signal for presynaptic differentiation. Developing synapses are stabilized by neuronal Agrin, which is released by motor nerve terminals and binds to Lrp4, a member of the low-density lipoprotein receptor family, stimulating further association between Lrp4 and MuSK and increasing MuSK kinase activity. In addition, MuSK phosphorylation is stimulated by an inside-out ligand, docking protein-7 (Dok-7), which is recruited to tyrosine-phosphorylated MuSK and increases MuSK kinase activity. Mutations in *MuSK* and in genes that function in the MuSK signaling pathway, including *Dok-7*, cause congenital myasthenia, and autoantibodies to MuSK, Lrp4, and acetylcholine receptors are responsible for myasthenia gravis.

Muscle-specific kinase (MuSK) is a single-pass transmembrane protein that has a critical role in signaling between motor neurons and skeletal muscle. MuSK is expressed in skeletal muscle cells, and once activated, MuSK stimulates pathways that (1) cluster and anchor acetylcholine receptors (AChRs) and additional muscle proteins that are critical for synaptic transmission, (2) enhance transcription of genes encoding synaptic proteins in muscle "synaptic nuclei," and (3) promote the production of retrograde signals that stimulate presynaptic differentiation. In the absence of MuSK, neuromuscular synapses fail to form (DeChiara et al. 1996). Mutations that impair MuSK kinase activity or signaling steps downstream from MuSK cause congenital myasthenia, characterized by structurally and functionally defective synapses, leading to muscle weakness and fatigue (Beeson et al. 2006; Muller et al. 2007; Selcen et al. 2008). In addition, autoantibodies to MuSK, as well as autoantibodies to AChRs and Lrp4, which forms a complex with MuSK (see below), are responsible for autoimmune myasthenia gravis (MG).

Here we describe the structure of MuSK; the mechanisms by which Agrin, a motor neuron–derived signal, stimulates MuSK; the role of Lrp4, the receptor for Agrin and ligand for MuSK, in

"

activating MuSK; and the signaling steps downstream from MuSK.

MuSK IS ESSENTIAL FOR SYNAPSE FORMATION

MuSK, which is expressed by skeletal muscle and not by motor neurons, is essential for the formation and maintenance of neuromuscular synapses (Jennings et al. 1993; Valenzuela et al. 1995; DeChiara et al. 1996). MuSK acts in two phases of synapse formation: (1) prepatterning muscle in the prospective synaptic region before innervation, and (2) responding to neuronal Agrin to form and stabilize synapses.

Before motor innervation, AChRs and additional synaptic proteins are preferentially expressed and clustered in the central, prospective synaptic region of muscle in a manner that is independent of nerve-derived signals but dependent on MuSK (Yang et al. 2000, 2001; Lin et al. 2001; Arber et al. 2002; Burden 2011). This muscle prepatterning has a role in directing where motor axons grow and form synapses (Fig. 1) (Kim and Burden 2008). In the absence

of MuSK, AChRs are expressed uniformly in muscle, and motor axons reach the muscle but fail to terminate and instead grow aimlessly without forming synapses (Fig. 2) (DeChiara et al. 1996). Muscle prepatterning is also dependent on Lrp4, which associates with MuSK and is sufficient to activate MuSK in nonmuscle cells (Weatherbee et al. 2006; Kim et al. 2008; Zhang et al. 2008) (see below).

MuSK itself is prepatterned in muscle (Kim and Burden 2008), but how MuSK expression and activity become confined to the prospective synaptic region is poorly understood. One idea suggests that prepatterning depends on (1) early expression of *Lrp4* and *MuSK* in developing myotubes; (2) activation of MuSK by Lrp4; (3) positive-feedback mechanisms, dependent on MuSK activity, which enhance *MuSK* and *Lrp4* expression in muscle nuclei that are near the site of MuSK activation; (4) clustering of Lrp4 and MuSK at sites where MuSK is activated; and (5) the orderly pattern of muscle growth, achieved by sequential fusion of myoblasts at the ends of growing myotubes (Kim and Burden 2008). The mechanisms that hinder MuSK

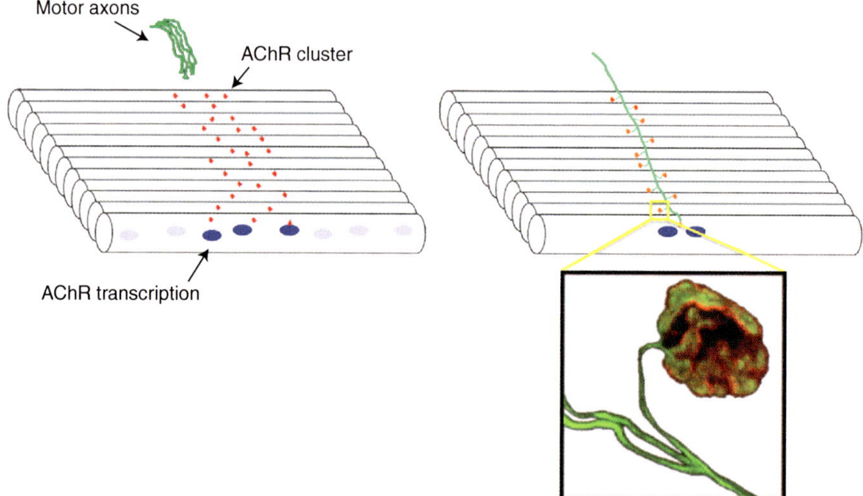

Figure 1. During development, motor axons (green) approach muscle in which *AChR* gene expression (blue) and clustering of AChRs (red) is enhanced in the prospective synaptic region of muscle. Motor axons form synapses in the prepatterned region of muscle. Motor axons release Agrin, which binds Lrp4 and stimulates MuSK, enhancing postsynaptic differentiation; and ACh, which depolarizes muscle, suppressing AChR expression. The combination of the two antagonist signals sharpens the prepattern so that AChR expression is restricted to synaptic sites.

          Cite this article as *Cold Spring Harb Perspect Biol* doi: 10.1101/cshperspect.a009167

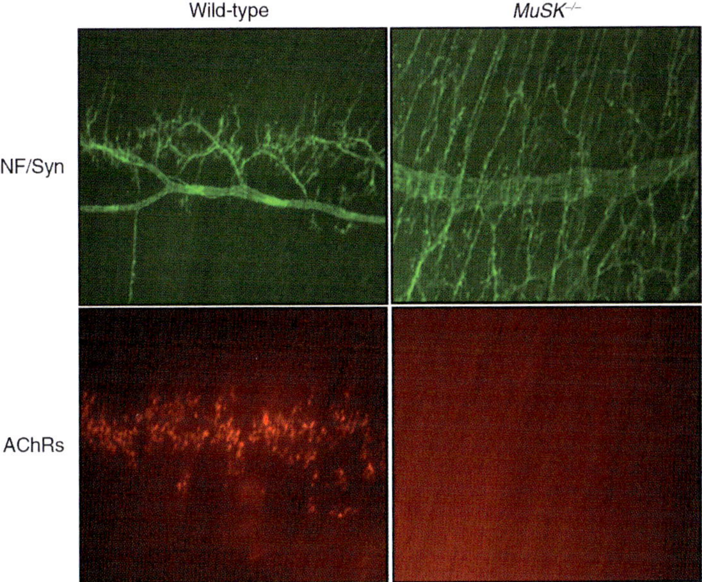

Figure 2. In the absence of MuSK, AChRs (red) fail to cluster and motor axons (green) fail to stop and differentiate. Motor axons and nerve terminals were stained with antibodies to neurofilament (NF) and synaptophysin (Syn), respectively.

activation and AChR clustering in peripheral regions of muscle are not fully understood.

Additional muscle-derived ligands may regulate MuSK activity; notably, Wnt11R, which binds to the Frizzled-like domain in MuSK (see below), can stimulate prepatterning in adaxial muscles of zebrafish (Jing et al. 2009). Further, murine Wnt9a, as well as Wnt11, can bind MuSK and stimulate clustering of AChRs, in a manner that depends on Lrp4 (Zhang et al. 2012a), suggesting that Wnts may serve as muscle-derived ligands that activate MuSK and stimulate muscle prepatterning. Because the hydrophobic pocket in Frizzled, which binds the lipid moiety attached to Wnts (Janda et al. 2012), is not present in the MuSK Frizzled-like domain (Stiegler et al. 2009) (see below), Wnts apparently bind MuSK in a manner that is distinct from the way that Wnts bind to Frizzled. It will be interesting to learn whether the MuSK Frizzled-like domain, which is not required for Agrin to stimulate AChR clustering (Zhou et al. 1999), is required for muscle prepatterning and whether Wnt signaling is required for muscle prepatterning and synapse formation in vivo in mammals.

Motor axons branch and arborize in a stereotyped manner within the prepatterned, central region of muscle, forming synapses in a narrow band adjacent to the main intramuscular nerve. Agrin, released from the tips of motor axons, binds directly to Lrp4, which stimulates further association between Lrp4 and MuSK and substantially increases MuSK phosphorylation (Figs. 3 and 4) (Kim et al. 2008; Zhang et al. 2008, 2011). Once MuSK is phosphorylated, signaling downstream from MuSK leads to clustering of Lrp4 and MuSK (see below), as well as clustering of AChRs and synapse-specific gene expression. Clustered Lrp4 signals in turn to motor axons, stimulating their differentiation (Yumoto et al. 2012). As such, Lrp4 functions bidirectionally, as it not only responds to Agrin, stimulating MuSK and postsynaptic differentiation, but also, once clustered as part of the program for postsynaptic differentiation, signals in a retrograde manner to motor neurons to stimulate presynaptic differentiation. Thus, Lrp4 coordinates synaptic differentiation (Yumoto et al. 2012).

MuSK is not only required to form neuromuscular synapses during embryogenesis but is

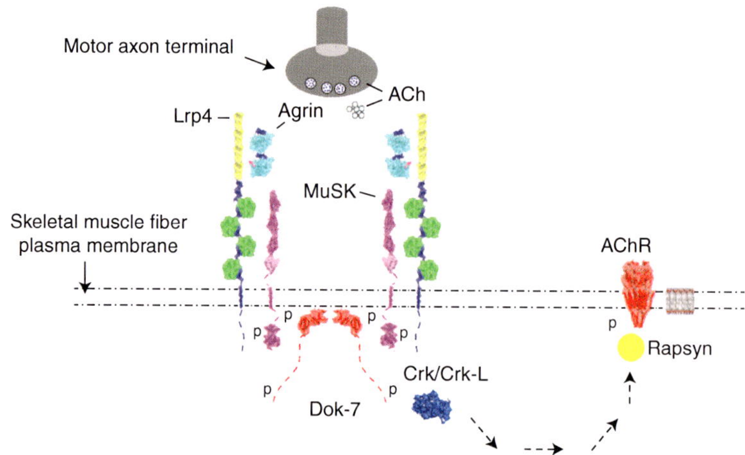

Figure 3. Motor axons release Agrin and ACh. Agrin binds to Lrp4, which stimulates association between Lrp4 and MuSK and MuSK phosphorylation. Once phosphorylated in the juxtamembrane region, MuSK recruits Dok-7, which forms a dimer and stabilizes a dimer of MuSK. Tyrosine phosphorylation of Dok-7 stimulates recruitment of Crk/Crk-L, which activates a poorly understood signaling pathway that leads to the anchoring of Rapsyn and AChRs at sites where MuSK is activated.

also essential to maintain neuromuscular synapses in adults. Reducing *MuSK* expression in adult mice, either by RNA interference or by conditional gene inactivation, leads to disassembly of the postsynaptic membrane and destabilization of synapses (Kong et al. 2004; Hesser et al. 2006). Consistent with the idea that MuSK has a critical role at adult synapses, ∼15% of patients with MG have autoantibodies to MuSK (Vincent and Leite 2005) (see below).

MuSK STRUCTURE

The MuSK Extracellular Region

The extracellular region of MuSK contains three immunoglobulin (Ig)-like domains and a Frizzled-like domain (Fig. 4A). In fish, avians, amphibians, and reptiles, a kringle domain is additionally present in the extracellular region (Jennings et al. 1993; Fu et al. 1999; Ip et al. 2000). The role of this kringle domain is not known, but its presence suggests an evolutionary relationship to Ror1 and Ror2, mammalian kringle-containing kinases, as well as CAM-1 in *Caenorhabditis elegans* and Dnrk and Dror in *Drosophila* (Wilson et al. 1993; Francis et al. 2005; Green et al. 2008).

The first Ig-like (Ig1) domain in MuSK forms a homodimer, and mutation of the hydrophobic dimer interface in Ig1, as well as deletion of the entire Ig1 domain, prevents Agrin from stimulating MuSK activation (Zhou et al. 1999; Stiegler et al. 2006). These findings indicate that MuSK activation requires MuSK dimerization, mediated, at least in part, by the first Ig-like domain. The Ig1 domain is also involved in binding Lrp4, and mutation of a solvent-exposed residue in the Ig1 domain, opposite to the hydrophobic dimer interface, prevents association between MuSK and Lrp4 and blocks Agrin from stimulating MuSK (Stiegler et al. 2006; Zhang et al. 2011). These findings raise the possibility that binding of MuSK to Lrp4 may orient the hydrophobic dimer interface to promote the formation or stability of a MuSK homodimer.

The Frizzled-like domain is not required for Agrin to activate MuSK, but it remains possible that the Frizzled-like domain engages alternative ligands, such as Wnts, that can activate MuSK (Zhou et al. 1999; Jing et al. 2009).

The MuSK Intracellular Region

The intracellular region of MuSK includes a 52-amino-acid juxtamembrane region; a typical

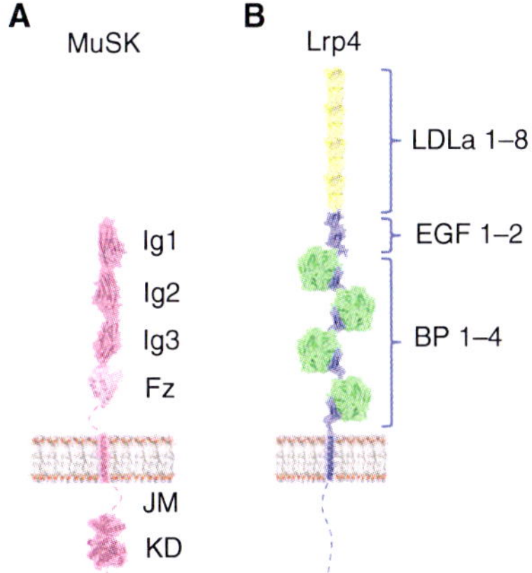

Figure 4. The domain organization of MuSK and Lrp4. (*A*) The extracellular region of MuSK contains three Ig-like domains and a Frizzled-like domain (Fz). The intracellular region of MuSK contains a juxtamembrane region (JM), a kinase domain (KD), and a short cytoplasmic tail. The crystal structures of the first two Ig-like domains (Stiegler et al. 2006), the Frizzled-like domain (Stiegler et al. 2009), and the intracellular region (Till et al. 2002) have been determined, but their arrangement in the context of the full kinase is not known and is depicted here only for illustrative purposes. (*B*) The extracellular region of Lrp4 contains eight LDLa repeats, two EGF-like domains, and four BP domains with embedded EGF-like domains. The crystal structure of the first BP domain in Lrp4 has been determined (Zong et al. 2012), whereas the structures of all other domains in Lrp4 have been inferred from the structures of these domains in other Lrp family members (Rudenko et al. 2002). The arrangement of these domains in the context of the full receptor is not known and is shown only for illustrative purposes.

kinase domain, containing three tyrosine residues within the activation loop; and an eight-amino-acid carboxy-terminal sequence (Jennings et al. 1993; Valenzuela et al. 1995; Till et al. 2002). Four tyrosine residues, the three in the activation loop and one in the juxtamembrane region, are the major sites for tyrosine phosphorylation in MuSK (Watty et al. 2000; Till et al. 2002).

Lrp4 STRUCTURE AND FUNCTION

The extracellular region of Lrp4 contains eight low-density lipoprotein receptor domain class A (LDLa) repeats, two epidermal growth factor

(EGF)-like domains, and four β-propeller (BP) domains, which are joined to an EGF-like domain (Fig. 4B). The last few LDLa repeats together with the first BP domain in Lrp4 are required for maximal binding of neural Agrin (Zhang et al. 2011), whereas the second and third BP domains in Lrp4 are required for Lrp4 to bind MuSK and for Agrin to stimulate MuSK phosphorylation (Zhang et al. 2011). Binding of Agrin to Lrp4 stimulates association between Lrp4 and MuSK, indicating that Agrin causes structural changes in Lrp4 to promote association with MuSK and stimulate MuSK kinase activity (Zhang et al. 2011). Unfortunately, isolation of the extracellular region of Lrp4 has proved difficult, impeding structural studies. A crystal structure of a complex between the first BP domain of Lrp4 and the carboxy-terminal region of Agrin, mini-Agrin, a 21-kDa protein containing the third laminin G-like domain, reveals that residues in the neuronal-specific insert of Agrin interact with the first BP domain in Lrp4. The structure also suggests a 2:2 Agrin/Lrp4 complex, mediated by an Agrin dimer, which could recruit two MuSK molecules into the complex for *trans*-phosphorylation of MuSK and initiation of downstream signaling (Zong et al. 2012).

Mini-Agrin, analyzed in the crystal structure, activates MuSK 100-fold less efficiently than recombinant forms of neural Agrin (e.g., 50 kDa) containing two laminin G-like domains and the last of four EGF-like domains (Gesemann et al. 1995). Although mini-Agrin binds the first BP domain in Lrp4 (Zong et al. 2012), the last few LDLa repeats in Lrp4 are required for maximal binding between Lrp4 and the 50-kDa form of neural Agrin (Zhang et al. 2011). Thus, these LDLa repeats may contact the second laminin G–like domain or the last EGF-like domain in neural Agrin and confer the ~100-fold-greater activity of the larger forms of Agrin.

Although Agrin causes Lrp4 to adopt a configuration that is competent to bind and activate MuSK (Zhang et al. 2011), Lrp4 can bind and activate MuSK independently of Agrin (Kim et al. 2008; Zhang et al. 2011). The ability of Lrp4 to activate MuSK likely underlies the role of Lrp4 in muscle prepatterning. Thus, Lrp4 may be in a dynamic equilibrium, and Agrin may shift this equilibrium to favor a conformation that binds MuSK.

Because Lrp4 self-associates, independently of Agrin, a preformed oligomer of Lrp4 may serve as a template to recruit MuSK and facilitate the formation of a MuSK dimer, promoting MuSK *trans*-phosphorylation (Kim et al. 2008; Zhang et al. 2011). The formation or stability of MuSK dimers may also be enhanced by sequences that promote homodimerization of MuSK, as structural studies reveal that a hydrophobic surface in the first Ig-like domain in MuSK mediates formation of a homodimer, and mutation of this interface prevents Agrin from activating MuSK (Stiegler et al. 2006).

Because Agrin binds Lrp4 rather than MuSK, MuSK is not a traditional receptor tyrosine kinase. Because Lrp4 binds and stimulates MuSK kinase activity, whereas Agrin promotes association between Lrp4 and MuSK, Lrp4 appears to function as a ligand for MuSK, whereas Agrin behaves as an allosteric regulator. In this sense, MuSK resembles Ret and ErbB2, tyrosine kinases that associate with partner proteins, GDNFR and ErbB1, ErbB3, or ErbB4, respectively, which serve as receptors for activating ligands (Zhang et al. 2011).

SIGNALING DOWNSTREAM FROM MuSK

Once the juxtamembrane tyrosine, Y553, becomes phosphorylated, MuSK recruits docking protein-7 (Dok-7), a noncatalytic adaptor protein (Fig. 3) (Okada et al. 2006). Dok-7 contains amino-terminal pleckstrin homology and phosphotyrosine-binding domains, required to bind pY553 MuSK, as well as a carboxy-terminal domain (Okada et al. 2006). In the absence of Dok-7, as in the absence of MuSK, neuromuscular synapses fail to form (Okada et al. 2006). Further, hypomorphic mutations in human *Dok-7*, which impair Dok-7 function, cause congenital myasthenia (Beeson et al. 2006) (see below).

Once recruited to MuSK, Dok-7 stimulates further MuSK phosphorylation and increases MuSK kinase activity (Okada et al. 2006; Inoue et al. 2009; Bergamin et al. 2010), showing that

Dok-7 functions, at least in part, as an inside-out ligand for MuSK. Consistent with this idea, the pleckstrin homology and phosphotyrosine-binding domains of Dok-7 form a dimer, which binds MuSK, promotes the formation of MuSK dimers, and stimulates MuSK phosphorylation in vitro (Bergamin et al. 2010). Moreover, certain disease-causing mutations in the phosphotyrosine binding domain of human Dok-7 impair Dok-7 dimerization, showing that Dok-7 dimerization is critical to activate MuSK in vivo (Bergamin et al. 2010).

A majority of patients with Dok-7 congenital myasthenia carry mutations that truncate the carboxy-terminal domain, indicating that the carboxy-terminal domain has an important function, separate from the role of the amino-terminal domain in stimulating MuSK kinase activity (Muller et al. 2007). Agrin stimulates phosphorylation of two tyrosine residues in the carboxy-terminal domain of Dok-7, which leads to recruitment of two adaptor proteins, Crk and Crk-L (Hallock et al. 2010). Muscle-selective inactivation of Crk and Crk-L causes defects in neuromuscular synapses that are virtually identical to the aberrant synapses found in humans with Dok-7 congenital myasthenia caused by mutations that truncate Dok-7 and prevent phosphorylation of Dok-7 at these two sites (Hallock et al. 2010). Thus, Dok-7 not only stimulates MuSK phosphorylation but also functions downstream from MuSK and transduces MuSK activation to synaptic differentiation. The molecules that are recruited to the Crk/Crk-L/Dok-7/MuSK complex and function in postsynaptic differentiation have not been identified. Ultimately, Dok-7 recruitment and phosphorylation lead to the redistribution and anchoring of critical proteins to the postsynaptic membrane (Fig. 3). This pathway requires Rac, Rho, Actin, and Rapsyn, but additional players are clearly involved (Dai et al. 2000; Weston et al. 2000, 2003; Ramarao et al. 2001; Burden 2011). Rapsyn, an intracellular, peripheral membrane protein, binds directly to AChRs (Neubig et al. 1979; Burden et al. 1983), but how MuSK activation and Dok-7 recruitment regulate the distribution and anchoring of Rapsyn is not currently understood. In addition, MuSK and Dok-7, but not Rapsyn, are required for synapse-specific gene expression, which appears to involve activation of JNK and Ets-domain-containing proteins (Gautam et al. 1995; Fromm and Burden 1998; Schaeffer et al. 1998; Si et al. 1999; Weston et al. 2000; Hippenmeyer et al. 2007). Thus, the pathways downstream from Dok-7 ultimately diverge, as the posttranslational pathway for anchoring postsynaptic proteins requires Rapsyn, whereas the transcriptional pathway for stimulating "synaptic genes" is Rapsyn-independent.

AUTOIMMUNE MG AND MYASTHENIA

Autoantibodies to the AChR, MuSK, or Lrp4 are responsible for MG, an autoimmune disease with a prevalence of 1–2 in 10,000 (Kalb et al. 2002; Murai et al. 2011). Intermittent muscle weakness, which worsens after activity, is the hallmark feature of MG. Autoantibodies to the AChR are responsible for most (~80%) cases of MG. A preponderance of these antibodies are directed to the main immunogenic region in the AChR α subunit, which can block AChR function, increase AChR turnover, and/or stimulate complement-mediated damage (Fig. 5).

Anti-AChR MG is treated by plasmapheresis, systemic immunosuppression, and anticholinesterases, which prolong the action of acetylcholine (ACh) and thereby enhance synaptic transmission.

Approximately 15% of patients with MG have autoantibodies to MuSK, although the percentage varies among different ethnic groups (Evoli et al. 2008; Vincent et al. 2008). Muscle weakness in anti-MuSK MG can be severe, requiring respiratory assistance. Moreover, the disease is recalcitrant to treatments, such as acetylcholinesterase inhibitors, which effectively manage MG caused by autoantibodies to the AChR (Farrugia and Vincent 2010). As such, current treatment for anti-MuSK MG relies on systemic immunosuppression. Unlike autoantibodies to the AChR, the disease-causing autoantibodies to MuSK are largely IgG4, which are functionally monovalent and do not engage complement, indicating that these antibodies interfere with MuSK function rather than causing cell damage (Niks et al. 2008; Klooster et al. 2012). It will be important to learn whether these antibodies obstruct association of MuSK with Lrp4 or interfere with the formation of MuSK dimers.

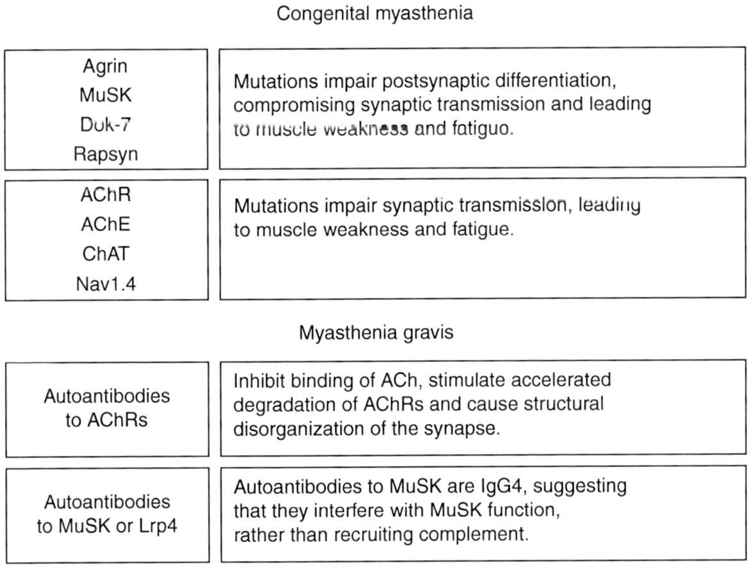

Congenital myasthenia

Agrin MuSK Dok-7 Rapsyn	Mutations impair postsynaptic differentiation, compromising synaptic transmission and leading to muscle weakness and fatigue.
AChR AChE ChAT Nav1.4	Mutations impair synaptic transmission, leading to muscle weakness and fatigue.

Myasthenia gravis

Autoantibodies to AChRs	Inhibit binding of ACh, stimulate accelerated degradation of AChRs and cause structural disorganization of the synapse.
Autoantibodies to MuSK or Lrp4	Autoantibodies to MuSK are IgG4, suggesting that they interfere with MuSK function, rather than recruiting complement.

Figure 5. Neuromuscular diseases: autoimmune myasthenia gravis and congenital myasthenia.

A smaller percentage (0.5%–5%) of MG patients have autoantibodies to Lrp4 (Higuchi et al. 2011; Pevzner et al. 2012; Zhang et al. 2012b). The clinical and pathological features of anti-Lrp4 MG have not been well defined. Autoantibodies to Lrp4 inhibit binding between Lrp4 and Agrin (Higuchi et al. 2011), but may also interfere with synaptic function by preventing oligomerization of Lrp4 or blocking association between Lrp4 and MuSK.

In anti-AChR MG, reduced expression of AChRs in the postsynaptic membrane leads to a decreased responsiveness of the postsynaptic membrane to ACh.

This postsynaptic change is accompanied by an increase in ACh release from motor nerve terminals, indicative of a compensatory mechanism that enhances synaptic transmission. In anti-MuSK MG, the postsynaptic membrane is likewise less responsive to ACh, but ACh release is not enhanced (Klooster et al. 2012). Together, these data suggest that the compensatory mechanism, initiated by a decrease in postsynaptic responsiveness and leading to an increase in transmitter release, requires MuSK.

Congenital myasthenia is caused by mutations in genes that are essential for the formation or function of the synapse (Fig. 5) (Engel et al. 2003). Congenital myasthenia is less common than MG and has an estimated prevalence of 1 in 500,000 in Europe. Mutations in *AChR* subunit genes, *Rapsyn*, and *Dok-7* are major causes of congenital myasthenia. Mutations in *MuSK* are a less common cause for congenital myasthenia. One patient, who is transheterozygous for two mutant *MuSK* alleles, suffers from congenital myasthenia, typified by highly fatigable muscle weakness (Chevessier et al. 2004). One allele is an early-frameshift null mutation, whereas the other allele truncates the kinase domain.

MuSK EXPRESSION AND FUNCTION OUTSIDE THE NEUROMUSCULAR SYNAPSE

Mice and zebrafish deficient in MuSK have defects in neural crest migration (Banerjee et al. 2011). In the absence of MuSK, streaming neural crest cells, which are normally confined to the central region of each somite, are dispersed and present throughout the somite. These data indicate that muscle cells provide guidance clues, which are dependent on MuSK activity, not only to motor axons but also to neural crest cells, thereby confining their movement to the prepatterned zone of muscle.

Skeletal muscle is the major site for *MuSK* expression. *MuSK* expression is regulated during myogenesis, as *MuSK*, like *Lrp4* and *Dok-7*, is expressed in myotubes but not in myoblasts (Valenzuela et al. 1995; Okada et al. 2006; Kim et al. 2008). In adult muscle *MuSK* expression is largely restricted to the synaptic nuclei of muscle (Kim and Burden 2008), which constitute ∼1% of total myofiber nuclei (Merlie and Sanes 1985; Burden 1993).

MuSK is expressed in the central nervous system, where it is concentrated at excitatory synapses, marked by coexpression of PSD-95 and Agrin (Garcia-Osta et al. 2006; Ksiazek et al. 2007). *MuSK* expression is evident in the hippocampus, olfactory bulb, cerebellum, and cortex (see http://www.brainatlas.org). The role of MuSK in the central nervous system is poorly understood, although MuSK has been reported to have a role in memory consolidation (Garcia-Osta et al. 2006).

Further, *MuSK* RNA expression is detected in the neural tube of *Xenopus* tadpoles and in the nervous system of developing and adult chickens (Fu et al. 1999; Ip et al. 2000). Moreover, *MuSK* is expressed transiently in the developing liver of chick and rat and in the eye vesicles, spleen, and lung of *Xenopus* (Fu et al. 1999). The role for MuSK in nonmuscle cells is not currently understood.

CONCLUSIONS

Studies over the past 20 years have revealed that MuSK is a master regulatory kinase that is essential for each step in neuromuscular synapse formation. In the future we are likely to learn (1) how Agrin binding to Lrp4 alters the conformation of Lrp4 to promote association between Lrp4 and MuSK, (2) how the formation of a complex between Lrp4 and MuSK stimulates MuSK kinase activity, (3) which molecules function downstream from Dok-7 and link recruitment of Crk/Crk-L to Rapsyn and to synapse-specific transcription, (4) how auto-antibodies to MuSK cause MG, (5) whether increasing MuSK activity might be therapeutic for diseases in which the neuromuscular synapse is compromised, and (6) the functions of MuSK outside the neuromuscular synapse.

REFERENCES

Arber S, Burden SJ, Harris AJ. 2002. Patterning of skeletal muscle. *Curr Opin Neurobiol* **12:** 100–103.

Banerjee S, Gordon L, Donn TM, Berti C, Moens CB, Burden SJ, Granato M. 2011. A novel role for MuSK and non-canonical Wnt signaling during segmental neural crest cell migration. *Development* **138:** 3287–3296.

Beeson D, Higuchi O, Palace J, Cossins J, Spearman H, Maxwell S, Newsom-Davis J, Burke G, Fawcett P, Motomura M, et al. 2006. Dok-7 mutations underlie a neuromuscular junction synaptopathy. *Science* **313:** 1975–1978.

Bergamin E, Hallock PT, Burden SJ, Hubbard SR. 2010. The cytoplasmic adaptor protein Dok7 activates the receptor tyrosine kinase MuSK via dimerization. *Mol Cell* **39:** 100–109.

Burden SJ. 1993. Synapse-specific gene expression. *Trends Genet* **9:** 12–16.

Burden SJ. 2011. SnapShot: Neuromuscular junction. *Cell* **144:** 826–826.e1.

Burden SJ, DePalma RL, Gottesman GS. 1983. Crosslinking of proteins in acetylcholine receptor-rich membranes: Association between the β-subunit and the 43 kd subsynaptic protein. *Cell* **35:** 687–692.

Chevessier F, Faraut B, Ravel-Chapuis A, Richard P, Gaudon K, Bauche S, Prioleau C, Herbst R, Goillot E, Ioos C, et al. 2004. MUSK, a new target for mutations causing congenital myasthenic syndrome. *Hum Mol Genet* **13:** 3229–3240.

Dai Z, Luo X, Xie H, Peng HB. 2000. The actin driven movement and formation of acetylcholine receptor clusters. *J Cell Biol* **150:** 1321–1334.

DeChiara TM, Bowen DC, Valenzuela DM, Simmons MV, Poueymirou WT, Thomas S, Kinetz E, Compton DL, Rojas E, Park JS, et al. 1996. The receptor tyrosine kinase MuSK is required for neuromuscular junction formation in vivo. *Cell* **85:** 501–512.

Engel AG, Ohno K, Sine SM. 2003. Sleuthing molecular targets for neurological diseases at the neuromuscular junction. *Nat Rev Neurosci* **4:** 339–352.

Evoli A, Bianchi MR, Riso R, Minicuci GM, Batocchi AP, Servidei S, Scuderi F, Bartoccioni E. 2008. Response to therapy in myasthenia gravis with anti-MuSK antibodies. *Ann N Y Acad Sci* **1132:** 76–83.

Farrugia ME, Vincent A. 2010. Autoimmune mediated neuromuscular junction defects. *Curr Opin Neurol* **23:** 489–495.

Francis MM, Evans SP, Jensen M, Madsen DM, Mancuso J, Norman KR, Maricq AV. 2005. The Ror receptor tyrosine kinase CAM-1 is required for ACR-16-mediated synaptic transmission at the *C. elegans* neuromuscular junction. *Neuron* **46:** 581–594.

Fromm L, Burden SJ. 1998. Synapse-specific and neuregulin-induced transcription require an Ets site that binds GABPα/GABPβ. *Genes Dev* **12:** 3074–3083.

Fu AK, Smith FD, Zhou H, Chu AH, Tsim KW, Peng BH, Ip NY. 1999. *Xenopus* muscle-specific kinase: Molecular cloning and prominent expression in neural tissues during early embryonic development. *Eur J Neurosci* **11:** 373–382.

Garcia-Osta A, Tsokas P, Pollonini G, Landau EM, Blitzer R, Alberini CM. 2006. MuSK expressed in the brain mediates cholinergic responses, synaptic plasticity, and memory formation. *J Neurosci* **26:** 7919–7932.

Gautam M, Noakes PG, Mudd J, Nichol M, Chu GC, Sanes JR, Merlie JP. 1995. Failure of postsynaptic specialization to develop at neuromuscular junctions of rapsyn-deficient mice. *Nature* **377:** 232–236.

Gesemann M, Denzer AJ, Ruegg MA. 1995. Acetylcholine receptor-aggregating activity of agrin isoforms and mapping of the active site. *J Cell Biol* **128:** 625–636.

Green JL, Kuntz SG, Sternberg PW. 2008. Ror receptor tyrosine kinases: Orphans no more. *Trends Cell Biol* **18:** 536–544.

Hallock PT, Xu CF, Park TJ, Neubert TA, Curran T, Burden SJ. 2010. Dok-7 regulates neuromuscular synapse formation by recruiting Crk and Crk-L. *Genes Dev* **24:** 2451–2461.

Hesser BA, Henschel O, Witzemann V. 2006. Synapse disassembly and formation of new synapses in postnatal muscle upon conditional inactivation of MuSK. *Mol Cell Neurosci* **31:** 470–480.

Higuchi O, Hamuro J, Motomura M, Yamanashi Y. 2011. Autoantibodies to low-density lipoprotein receptor-related protein 4 in myasthenia gravis. *Ann Neurol* **69:** 418–422.

Hippenmeyer S, Huber RM, Ladle DR, Murphy K, Arber S. 2007. ETS transcription factor Erm controls subsynaptic gene expression in skeletal muscles. *Neuron* **55:** 726–740.

Inoue A, Setoguchi K, Matsubara Y, Okada K, Sato N, Iwakura Y, Higuchi O, Yamanashi Y. 2009. Dok-7 activates the muscle receptor kinase MuSK and shapes synapse formation. *Sci Signal* **2:** ra7.

Ip FC, Glass DG, Gies DR, Cheung J, Lai KO, Fu AK, Yancopoulos GD, Ip NY. 2000. Cloning and characterization of muscle-specific kinase in chicken. *Mol Cell Neurosci* **16:** 661–673.

Janda CY, Waghray D, Levin AM, Thomas C, Garcia KC. 2012. Structural basis of Wnt recognition by Frizzled. *Science* **337:** 59–64.

Jennings CG, Dyer SM, Burden SJ. 1993. Muscle-specific *trk*-related receptor with a kringle domain defines a distinct class of receptor tyrosine kinases. *Proc Natl Acad Sci* **90:** 2895–2899.

Jing L, Lefebvre JL, Gordon LR, Granato M. 2009. Wnt signals organize synaptic prepattern and axon guidance through the zebrafish unplugged/MuSK receptor. *Neuron* **61:** 721–733.

Kalb B, Matell G, Pirskanen R, Lambe M. 2002. Epidemiology of myasthenia gravis: A population-based study in Stockholm, Sweden. *Neuroepidemiology* **21**: 221–225.

Kim N, Burden SJ. 2008. MuSK controls where motor axons grow and form synapses. *Nature Neurosci* **11**: 19–27.

Kim N, Stiegler AL, Cameron TO, Hallock PT, Gomez AM, Huang JH, Hubbard SR, Dustin ML, Burden SJ. 2008. Lrp4 is a receptor for Agrin and forms a complex with MuSK. *Cell* **135**: 334–342.

Klooster R, Plomp JJ, Huijbers MG, Niks EH, Straasheijm KR, Detmers FJ, Hermans PW, Sleijpen K, Verrips A, Losen M, et al. 2012. Muscle-specific kinase myasthenia gravis IgG4 autoantibodies cause severe neuromuscular junction dysfunction in mice. *Brain* **135**: 1081–1101.

Kong XC, Barzaghi P, Ruegg MA. 2004. Inhibition of synapse assembly in mammalian muscle in vivo by RNA interference. *EMBO Rep* **5**: 183–188.

Ksiazek I, Burkhardt C, Lin S, Seddik R, Maj M, Bezakova G, Jucker M, Arber S, Caroni P, Sanes JR, et al. 2007. Synapse loss in cortex of agrin-deficient mice after genetic rescue of perinatal death. *J Neurosci* **27**: 7183–7195.

Lin W, Burgess RW, Dominguez B, Pfaff SL, Sanes JR, Lee KF. 2001. Distinct roles of nerve and muscle in postsynaptic differentiation of the neuromuscular synapse. *Nature* **410**: 1057–1064.

Merlie JP, Sanes JR. 1985. Concentration of acetylcholine receptor mRNA in synaptic regions of adult muscle fibres. *Nature* **317**: 66–68.

Muller JS, Herczegfalvi A, Vilchez JJ, Colomer J, Bachinski LL, Mihaylova V, Santos M, Schara U, Deschauer M, Shevell M, et al. 2007. Phenotypical spectrum of *DOK7* mutations in congenital myasthenic syndromes. *Brain* **130**: 1497–1506.

Murai H, Yamashita N, Watanabe M, Nomura Y, Motomura M, Yoshikawa H, Nakamura Y, Kawaguchi N, Onodera H, Araga S, et al. 2011. Characteristics of myasthenia gravis according to onset-age: Japanese nationwide survey. *J Neurol Sci* **305**: 97–102.

Neubig RR, Krodel EK, Boyd ND, Cohen JB. 1979. Acetylcholine and local anesthetic binding to Torpedo nicotinic postsynaptic membranes after removal of nonreceptor peptides. *Proc Natl Acad Sci* **76**: 690–694.

Niks EH, van Leeuwen Y, Leite MI, Dekker FW, Wintzen AR, Wirtz PW, Vincent A, van Tol MJ, Jol-van der Zijde CM, Verschuuren JJ. 2008. Clinical fluctuations in MuSK myasthenia gravis are related to antigen-specific IgG4 instead of IgG1. *J Neuroimmunol* **195**: 151–156.

Okada K, Inoue A, Okada M, Murata Y, Kakuta S, Jigami T, Kubo S, Shiraishi H, Eguchi K, Motomura M, et al. 2006. The muscle protein Dok-7 is essential for neuromuscular synaptogenesis. *Science* **312**: 1802–1805.

Pevzner A, Schoser B, Peters K, Cosma NC, Karakatsani A, Schalke B, Melms A, Kroger S. 2012. Anti-LRP4 autoantibodies in AChR- and MuSK-antibody-negative myasthenia gravis. *J Neurol* **259**: 427–435.

Ramarao MK, Bianchetta MJ, Lanken J, Cohen JB. 2001. Role of rapsyn tetratricopeptide repeat and coiled-coil domains in self-association and nicotinic acetylcholine receptor clustering. *J Biol Chem* **276**: 7475–7483.

Rudenko G, Henry L, Henderson K, Ichtchenko K, Brown MS, Goldstein JL, Deisenhofer J. 2002. Structure of the LDL receptor extracellular domain at endosomal pH. *Science* **298**: 2353–2358.

Schaeffer L, Duclert N, Huchet-Dymanus M, Changeux JP. 1998. Implication of a multisubunit Ets-related transcription factor in synaptic expression of the nicotinic acetylcholine receptor. *EMBO J* **17**: 3078–3090.

Selcen D, Milone M, Shen XM, Harper CM, Stans AA, Wieben ED, Engel AG. 2008. Dok-7 myasthenia: Phenotypic and molecular genetic studies in 16 patients. *Ann Neurol* **64**: 71–87.

Si J, Wang Q, Mei L. 1999. Essential roles of c-JUN and c-JUN N-terminal kinase (JNK) in neuregulin-increased expression of the acetylcholine receptor ε-subunit. *J Neurosci* **19**: 8498–8508.

Stiegler AL, Burden SJ, Hubbard SR. 2006. Crystal structure of the agrin-responsive immunoglobulin-like domains 1 and 2 of the receptor tyrosine kinase MuSK. *J Mol Biol* **364**: 424–433.

Stiegler AL, Burden SJ, Hubbard SR. 2009. Crystal structure of the Frizzled-like cysteine-rich domain of the receptor tyrosine kinase MuSK. *J Mol Biol* **393**: 1–9.

Till JH, Becerra M, Watty A, Lu Y, Ma Y, Neubert TA, Burden SJ, Hubbard SR. 2002. Crystal structure of the MuSK tyrosine kinase: Insights into receptor autoregulation. *Structure* **10**: 1187–1196.

Valenzuela DM, Stitt TN, DiStefano PS, Rojas E, Mattsson K, Compton DL, Nunez L, Park JS, Stark JL, Gies DR, et al. 1995. Receptor tyrosine kinase specific for the skeletal muscle lineage: Expression in embryonic muscle, at the neuromuscular junction, and after injury. *Neuron* **15**: 573–584.

Vincent A, Leite MI. 2005. Neuromuscular junction autoimmune disease: Muscle specific kinase antibodies and treatments for myasthenia gravis. *Curr Opin Neurol* **18**: 519–525.

Vincent A, Leite MI, Farrugia ME, Jacob S, Viegas S, Shiraishi H, Benveniste O, Morgan BP, Hilton-Jones D, Newsom-Davis J, et al. 2008. Myasthenia gravis seronegative for acetylcholine receptor antibodies. *Ann NY Acad Sci* **1132**: 84–92.

Watty A, Neubauer G, Dreger M, Zimmer M, Wilm M, Burden SJ. 2000. The in vitro and in vivo phosphotyrosine map of activated MuSK. *Proc Natl Acad Sci* **97**: 4585–4590.

Weatherbee SD, Anderson KV, Niswander LA. 2006. LDL-receptor-related protein 4 is crucial for formation of the neuromuscular junction. *Development* **133**: 4993–5000.

Weston C, Yee B, Hod E, Prives J. 2000. Agrin-induced acetylcholine receptor clustering is mediated by the small guanosine triphosphatases Rac and Cdc42. *J Cell Biol* **150**: 205–212.

Weston C, Gordon C, Teressa G, Hod E, Ren XD, Prives J. 2003. Cooperative regulation by Rac and Rho of agrin-induced acetylcholine receptor clustering in muscle cells. *J Biol Chem* **278**: 6450–6455.

Wilson C, Goberdhan DC, Steller H. 1993. *Dror*, a potential neurotrophic receptor gene, encodes a *Drosophila* homolog of the vertebrate Ror family of Trk-related receptor tyrosine kinases. *Proc Natl Acad Sci* **90**: 7109–7113.

Cite this article as *Cold Spring Harb Perspect Biol* doi: 10.1101/cshperspect.a009167

Yang X, Li W, Prescott ED, Burden SJ, Wang JC. 2000. DNA topoisomerase IIβ and neural development. *Science* **287:** 131–134.

Yang X, Arber S, William C, Li L, Tanabe Y, Jessell TM, Birchmeier C, Burden SJ. 2001. Patterning of muscle acetylcholine receptor gene expression in the absence of motor innervation. *Neuron* **30:** 399–410.

Yumoto N, Kim N, Burden SJ. 2012. Lrp4 is a retrograde signal for presynaptic differentiation at neuromuscular synapses. *Nature* **489:** 438–442.

Zhang B, Luo S, Wang Q, Suzuki T, Xiong WC, Mei L. 2008. LRP4 serves as a coreceptor of agrin. *Neuron* **60:** 285–297.

Zhang W, Coldefy AS, Hubbard SR, Burden SJ. 2011. Agrin binds to the N-terminal region of Lrp4 and stimulates association between Lrp4 and the first Ig-like domain in MuSK. *J Biol Chem* **286:** 40624–40630.

Zhang B, Liang C, Bates R, Yin Y, Xiong WC, Mei L. 2012a. Wnt proteins regulate acetylcholine receptor clustering in muscle cells. *Mol Brain* **5:** 7.

Zhang B, Tzartos JS, Belimezi M, Ragheb S, Bealmear B, Lewis RA, Xiong WC, Lisak RP, Tzartos SJ, Mei L. 2012b. Autoantibodies to lipoprotein-related protein 4 in patients with double-seronegative myasthenia gravis. *Arch Neurol* **69:** 445–451.

Zhou H, Glass DJ, Yancopoulos GD, Sanes JR. 1999. Distinct domains of MuSK mediate its abilities to induce and to associate with postsynaptic specializations. *J Cell Biol* **146:** 1133–1146.

Zong Y, Zhang B, Gu S, Lee K, Zhou J, Yao G, Figueiredo D, Perry K, Mei L, Jin R. 2012. Structural basis of agrin-LRP4-MuSK signaling. *Genes Dev* **26:** 247–258.

Structure and Physiology of the RET Receptor Tyrosine Kinase

Carlos F. Ibáñez

Department of Neuroscience, Karolinska Institute, S-17177 Stockholm, Sweden; Life Sciences Institute, Department of Physiology, National University of Singapore 117456, Singapore

Correspondence: carlos.ibanez@ki.se

The identification of the *ret* oncogene by Masahide Takahashi and Geoffrey Cooper in 1985 was both serendipitous and paradigmatic (Takahashi et al. 1985). By transfecting total DNA from a human lymphoma into mouse NIH3T3 cells, they obtained one clone, which in secondary transformants yielded more than 100-fold improvement in transformation efficiency. Subsequent investigations revealed that the *ret* oncogene was not present as such in the primary lymphoma, but was derived by DNA rearrangement during transfection from normal human sequences of the *ret* locus. At the time, activation by DNA rearrangement had not been previously described for a transforming gene with the NIH3T3 transfection assay. The discovery of *ret* opened a field of study that has had a profound impact in cancer research, developmental biology, and neuroscience, and that continues to yield surprises and important insights to this day.

AN UNUSUAL RECEPTOR TYROSINE KINASE WITH CADHERIN REPEATS

Isolation of *ret* cDNA clones revealed a carboxy-terminal domain with high homology with members of the tyrosine kinase gene family preceded by a hydrophobic sequence characteristic of a transmembrane domain, suggesting that the *ret* oncogene encoded a cell-surface receptor (Takahashi and Cooper 1987). The characterization of the human (Takahashi et al. 1988) and mouse (Iwamoto et al. 1993; Pachnis et al. 1993) *ret* proto-oncogenes revealed the full primary structure of the RET protein and the unusual presence in its extracellular region of a sequence with similarity to cadherins, transmembrane proteins that mediate Ca^{2+}-dependent homophilic cell adhesion (Nollet et al.

2000). A molecular modeling study of the extracellular domain of RET later revealed four cadherin repeats—termed cadherin-like domains or CLDs 1–4—each of about 110 residues, and one Ca^{2+}-binding site between CLD2 and CLD3 (Anders et al. 2001). Ca^{2+} binding is required for the functional integrity of the RET protein and for its ability to interact with ligand. Following the four CLDs, the extracellular domain of RET contains a Cys-rich region of 120 residues connected to the transmembrane domain. The intracellular region of RET begins with a juxtamembrane portion of 50 residues, a tyrosine kinase domain split by a 14-residue linker, and a 100-residue-long carboxy-terminal tail, which comes in two flavors as a result of alternative splicing. After position 1063, the "short" RET isoform contains nine unique carboxy-terminal residues (RET9), whereas the "long" contains 51 (RET51).

ONE GENE, MANY DISEASES

Mutations in the *RET* gene have been found in a number of human diseases, including several different cancers of neuroendocrine origin and a gut syndrome characterized by intestinal obstruction known as Hirschsprung's disease. Four different human cancers carry mutations in the *RET* gene, including papillary thyroid carcinoma (PTC) (Grieco et al. 1990), medullary thyroid carcinoma (familial and sporadic) (Donis-Keller et al. 1993; Hofstra et al. 1994), and the multiple endocrine neoplasias type 2A (MEN2A) (Donis-Keller et al. 1993; Mulligan et al. 1993) and 2B (MEN2B) (Hofstra et al. 1994). Dozens of different substitutions and rearrangements in the *RET* gene underlie these syndromes, a complexity that has profound implications for our understating of genotype/phenotype relationships and the molecular mechanisms of signal transduction (for an in-depth review of the cancer biology of *RET*, see the article by Santoro and Carlomagno 2013). Although *RET* mutations that lead to tumor formation have in most cases been described as gain of function, mutations that result in Hirschsprung's disease—of which more than 50 are known so far—invariably result in loss of RET function by targeting its kinase activity (Iwashita et al. 1996; Pelet et al. 1998), docking sites for intracellular signaling effectors (Geneste et al. 1999), or residues in the RET extracellular domain that affect RET processing in the endoplasmic reticulum and prevent RET expression at the cell surface (Iwashita et al. 1996; Cosma et al. 1998; Kjaer and Ibanez 2003b).

A WEALTH OF RET LIGANDS AND CORECEPTORS

The RET ligand has been identified as glial cell line-derived neurotrophic factor (GDNF) (Durbec et al. 1996; Trupp et al. 1996), a dimeric growth factor-like protein distantly related to members of the transforming growth factor-β (TGF-β) superfamily. Three additional proteins highly related in sequence to GDNF, known respectively as Neurturin (Kotzbauer et al. 1996), Persephin (Milbrandt et al. 1998), and Artemin (Baloh et al. 1998b), were subsequently also identified as ligands of RET. However, this ligand–receptor relationship proved to be a little more unusual than initially expected. Neither ligand is able to bind RET on its own, but require a ligand-binding subunit acting as coreceptor, known as the GDNF family receptor α (GFRα) component. Four different GFRαs have been described (GFRα1–4), each with selectivity—although not absolute specificity—for each of the four distinct members of the GDNF ligand family (Jing et al. 1996; Baloh et al. 1997, 1998a; Buj-Bello et al. 1997; Klein et al. 1997; Sanicola et al. 1997; Naveilhan et al. 1998; Trupp et al. 1998; Worby et al. 1998; Masure et al. 2000). GDNF can be chemically cross-linked to RET (Trupp et al. 1996), indicating that it does make direct contact with the receptor, although its binding affinity is too low to stabilize a complex. On the other hand, GDNF has high affinity for GFRα1 independently of RET. A model initially proposed had GDNF forming first a complex with GFRα1 and subsequently recruiting RET to the complex (Massagué 1996). An alternative model, in which GFRα1 and RET are preassociated to some extent before ligand binding, was later proposed based on binding and site-directed mutagenesis studies (Eketjäll et al. 1999; Cik et al. 2000). It should be noted that two additional receptors for GDNF have also been described, namely, the neural cell adhesion molecule NCAM (Paratcha et al. 2003) and syndecan-3 (Bespalov et al. 2011), which are able to transmit GDNF signals independently of RET.

EVOLUTIONARY RELATIONSHIPS

Only one *ret* gene is known to exist in higher organisms. RET, GDNF family ligands, and GFRα proteins have been found in all vertebrate species investigated so far. A *ret* orthologue has also been found in the genome of the cephalochordate *Amphioxus*, along with sequences corresponding to one GDNF-like and one GFRα-like encoded protein product. RET is also found in *Drosophila melanogaster*. Intriguingly, its expression pattern in the fly is to some extent reminiscent of the one found in vertebrates (Hahn and Bishop 2001). However, no GDNF or GFRα proteins appear to be encoded in the fly genome. *Drosophila* RET is unable to interact with GDNF or GFRα1 of mammalian origin, nor is it capable of mediating cell adhesion (Abrescia et al. 2005). A chimeric approach was used to show that *Drosophila* RET contains an active tyrosine kinase that is competent to induce neuronal differentiation on activation in PC12 cells (Abrescia et al. 2005). The physiological function of RET in *Drosophila* remains unknown.

STRUCTURE–FUNCTION STUDIES OF RET EXTRACELLULAR AND KINASE DOMAINS

Loss-of-function mutations in RET cause abnormal development of the enteric nervous system, leading to Hirschsprung's disease. Hirschsprung mutations in the extracellular domain of RET (RET^{ECD}) affect processing in the endoplasmic reticulum (ER) and prevent RET expression at the cell surface. Most Hirschsprung mutations examined prevent the maturation of RET^{ECD} in the ER, indicating defects in protein folding (Kjaer and Ibanez 2003b). Maturation of RET^{ECD} mutants can be rescued by allowing protein expression to proceed at 30°C, a condition known to facilitate protein folding, regaining their ability to bind to the GDNF/GFRα1. Analysis of autonomous folding subunits in the RET^{ECD} has indicated an intrinsic propensity to misfolding in the amino-terminal CLDs 1–3 (Kjaer and Ibanez 2003b), which also concentrate the majority of Hirschsprung mutations affecting the RET^{ECD}. A recent crystal structure of the two amino-terminal CLD1–2 domains of the RET^{ECD} has revealed these two CLDs folded onto each other in a compact clam-shell arrangement distinct from that of classical cadherins (Kjaer et al. 2010). CLD1 structural elements and disulfide composition are unique to mammals, indicating an unexpected structural diversity within higher and lower vertebrate RET CLD regions. The same study identified two unpaired cysteines that predispose human RET to maturation impediments in the ER. The intrin-

sic susceptibility to misfolding of mammalian RET^{ECD} may be the result of a trade-off that helps to avoid an increased incidence of tumors, at the expense of a greater vulnerability to Hirschsprung's disease.

Sequence and functional divergences between the ectodomains of mammalian and amphibian RET molecules have been exploited to map binding determinants in the human RET^{ECD} responsible for its interaction with the GDNF-GFRα1 complex through homolog-scanning mutagenesis. It was found that Xenopus RET^{ECD} was unable to bind to GDNF-GFRα1 complexes of mammalian origin. However, a chimeric molecule containing CLD1, 2, and 3 from human RET^{ECD}, but neither domain alone, had similar binding activity than full-length human RET^{ECD} (Kjaer and Ibanez 2003a). This suggested the existence of an extended ligand-binding surface within the three amino-terminal cadherin-like domains of human RET^{ECD}. Subsequently, a study using chemical cross-linking of a reconstituted GDNF/ GFRα1/RET complex followed by matrix-assisted laser desorption/ionization (MALDI) mass spectrometry analysis, indicated that CLD4 and the carboxy-terminal cysteine-rich domain (CRD) of the RET^{ECD} are in direct contact with GFRα1 in complex with GDNF (Amoresano et al. 2005). This study failed to identify any direct contacts between RET and GDNF, although, as mentioned earlier, those were known to exist from previous cross-linking experiments (Trupp et al. 1996) These discrepant sets of results could be reconciled if the role of the amino-terminal CLD1–3 in ligand binding was indirect, rather than in establishing physical contact with the GDNF/GFRα1 complex. In this scenario, CLD1–3 would be necessary for RET to adopt a conformation that is competent for binding and complex assembly but not directly involved in contacting the RET ligands (Amoresano et al. 2005). At the time of this writing, efforts to solve the three-dimensional structure of the tripartite GDNF/GFRα1/RET complex are still ongoing, but preliminary results would seem to offer support for this latter model.

The crystal structures of the nonphosphorylated (inactive) and phosphorylated (active) RET kinase have been determined and shown to adopt the same active kinase conformation competent to bind ATP and substrate (Knowles et al. 2006). Both structures show a preorganized activation loop conformation that is independent of phosphorylation status. In agreement with the structural data, enzyme kinetic data showed that autophosphorylation produces only a modest increase in activity (Knowles et al. 2006). Longer forms of the RET intracellular domain containing the juxtamembrane domain and carboxy-terminal tail showed similar kinetic behavior as the isolated kinase, indicating the absence of a *cis*-inhibitory mechanism within the RET intracellular domain (Knowles et al. 2006). Unlike the situation of most other receptor tyrosine kinases, these results suggest the existence of alternative inhibitory mechanisms, possibly in *trans*, for the autoregulation of RET kinase activity.

RET SIGNALING MECHANISMS

Dimerization of receptor tyrosine kinases is known to be required, although most likely not sufficient, for ligand-induced kinase transphosphorylation and activation. Several receptor tyrosine kinases are found as preformed homodimers at the plasma membrane independently of ligand binding. Using a transmembrane (TM) domain self-association assay, Kjaer et al. observed strong self-association of the RET-TM in a biological membrane (Kjaer et al. 2006). These investigators found that mutagenesis of specific residues in the RET-TM domain reduced receptor homodimerization and abolished the transforming activity of MEN2A RET, one of the strongest oncogenic variants of the RET protein, suggesting that self-association of RET TM domains contributes to the mechanism of activation of RET (Kjaer et al. 2006).

Like the majority of receptor tyrosine kinases studied, signaling pathways initiated by the RET receptor include the Ras/MAP kinase, PI3 kinase/AKT, and phospholipase C-γ (PLCγ) pathways. On activation, RET undergoes autophosphorylation of intracellular tyrosine residues, which then serve as docking sites for downstream signaling effectors carrying Src ho-

mology 2 (SH2) or phosphotyrosine-binding (PTB) domains. Previous studies have indicated that at least 14 of the 18 tyrosine residues present in the intracellular region of RET can become phosphorylated (Liu et al. 1996; Kawamoto et al. 2004; Knowles et al. 2006). Among those, Tyr^{900} and Tyr^{905} are present in the kinase activation loop and are known to contribute to full kinase activation (Knowles et al. 2006). Autophosphorylation of the key residue Tyr^{1062} is required for activation of Ras/MAP kinase and PI3 kinase/AKT (Besset et al. 2000; Hayashi et al. 2000; Segouffin-Cariou and Billaud 2000; Coulpier et al. 2002). This residue appears to be critical for RET function, and mice with a point mutation in Tyr^{1062} show a severe loss-of-function phenotype (Jijiwa et al. 2004; Wong et al. 2005; Jain et al. 2006a). Phosphorylation of Tyr^{1096}, present only in the long RET51 isoform, also contributes to these pathways. On ligand stimulation, at least two distinct protein complexes assemble on phosphorylated Tyr^{1062} of RET via Shc, one leading to activation of the Ras/MAP kinase pathway through recruitment of Grb2 and Sos, and another to the PI3K/AKT pathway through recruitment of adaptors Grb2 and Gab2 followed by $p85^{PI3K}$ and the SHP2 tyrosine phosphatase (Besset et al. 2000).

The adaptor protein FRS2 can also bind to phosphorylated Tyr^{1062} in the activated RET receptor (Kurokawa et al. 2001; Melillo et al. 2001). FRS2 competes with Shc for binding to Tyr^{1062}, and it has been shown that Shc but not FRS2 is responsible for cell survival effects of RET in neuroblastoma cells (Lundgren et al. 2006). This differential signaling may be mediated from different compartments in the plasma membrane, as RET has been shown to interact with FRS2 in lipid rafts, but with Shc outside lipid rafts (Paratcha et al. 2001). A series of adaptor molecules from the p62dok family have also been shown to interact with the activated RET receptor. Dok-1, -2, -4, -5, and -6 all interact with phosphorylated Tyr^{1062} via their PTB domains (Grimm et al. 2001; Crowder et al. 2004; Kurotsuchi et al. 2010) and are thought to contribute to neuronal differentiation (Grimm et al. 2001; Crowder et al. 2004).

Cite this article as *Cold Spring Harb Perspect Biol* doi: 10.1101/cshperspect.a009134

The Grb2/Gab2 complex can also assemble directly onto phosphorylated Tyr[1096], offering an alternative route to PI3K activation by GDNF. Regarding the remaining autophosphorylation sites, it has been found that phosphorylation of Tyr[1015] leads to activation of PLCγ (Borrello et al. 1996), and phospho-Tyr[981] binds the Src cytoplasmic tyrosine kinase (Encinas et al. 2004). Recently, a yeast-two-hybrid screen led to the identification of the GTPase-activating protein (GAP) for Rap1, Rap1GAP, as a novel RET-binding protein (Jiao et al. 2011). Like Src, Rap1GAP was also found to require phosphorylation of Tyr[981] for RET binding and suppressed GDNF-induced activation of ERK and neurite outgrowth. A recent study has established a biochemical function and a physiological role for the phosphorylation of Tyr[687] in the juxtamembrane region of the RET intracellular domain (Perrinjaquet et al. 2010). Using a phage display strategy, Perrinjaquet et al. found that the phosphotyrosine phosphatase SHP2 binds to phospho-Tyr[687] on ligand-induced RET activation. SHP2 is recruited to activated RET in a cooperative fashion, such that both interaction with Tyr[687] and association with components of the Tyr[1062] signaling complex are required for stable recruitment of SHP2 to the receptor. SHP2 recruitment was found to contribute to the ability of RET to activate the PI3K/AKT pathway and promote survival and neurite outgrowth in primary neurons (Perrinjaquet et al. 2010).

In addition to tyrosine autophosphorylation, RET has been found to undergo serine phosphorylation at Ser[696] by protein kinase A (PKA) (Fukuda et al. 2002). Mutation of Ser[696] affected the ability of RET to activate the small GTPase Rac1 and stimulate formation of cell lamellipodia (Fukuda et al. 2002). Homozygous knock-in mice carrying this mutation lacked neuronal elements of the enteric nervous system in the distal colon, resulting from a migration defect of enteric neural crest cells (Asai et al. 2006), indicating a physiological role for PKA-dependent modification of RET function. Interestingly, the signaling deficits of the Ser[696] RET mutant could be alleviated—at least in vitro—by simultaneous mutation of the nearby residue Tyr[687] (Fukuda et al. 2002). In line with this, activation of PKA by forskolin was found to impair the recruitment of SHP2 to RET and negatively affected ligand-mediated neurite outgrowth (Perrinjaquet et al. 2010). Moreover, mutation of Ser[696] enhanced SHP2 binding to the receptor and eliminated the effect of forskolin on ligand-induced neurite outgrowth. Together, these findings established Tyr[687] as a critical platform for integration of RET and PKA signals.

Among the interactions not mediated by phosphorylation, the PDZ domain-containing Shank3 protein was found to interact with a PDZ-binding motif present in the RET9 but not in the RET51 isoform (Schuetz et al. 2004). Shank3 was shown to mediate sustained MAP kinase and PI3 kinase signaling, and the formation of branched tubular structures in three-dimensional cultures of epithelial cells.

RET FUNCTION IN KIDNEY DEVELOPMENT

Knockout studies have shown that RET inactivation results in renal agenesis or severe hypodysplasia, owing to failure of the ureteric bud to evaginate from the Wolffian duct and branch normally (Schuchardt et al. 1994, 1996). RET expression defines a population of ureteric bud tip cells that proliferate under GDNF stimulation from the metanephric mesenchyme. In the absence of RET, tip cells change fate and instead contribute to the ureteric bud trunk (Shakya et al. 2005). Studies in knock-in mice have provided evidence for differential and isoform-specific roles of RET phospho-Tyr docking sites in kidney development. One earlier study initially reported that mice expressing only RET9 were normal, whereas those expressing only RET51 showed kidney hypodysplasia (de Graaff et al. 2001). In contrast, a later study reported that mice monoisomorphic for either RET9 or RET51 were viable and showed normal kidneys, indicating redundant roles of RET isoforms in kidney development (Jain et al. 2006a). As discussed elsewhere, a possible reason for this discrepancy may lie in the use of chimeric mouse–human knock-in cDNAs in the first study. It has also been reported that wild-type human RET51 and RET9 are both able to promote branching morphogenesis to a similar extent, but mutation of Tyr^{1062} abrogates this activity only in RET9 and not in RET51, presumably because of redundancy through Tyr^{1096} (Jain et al. 2006a). In contrast, mutation of Tyr^{1015} produced clear defects in ureteric bud outgrowth in the context of either isoform, providing evidence for the importance of PLCγ signaling downstream from RET in renal development. Despite the prominent role of RET signaling in kidney development, no human *RET* mutations have yet been uncovered in children suffering from renal tract malformations.

RET FUNCTIONS IN NERVOUS SYSTEM DEVELOPMENT

Enteric Nervous System

Hirschsprung's disease is a genetic disorder of neural crest development characterized by the absence of enteric parasympathetic neurons in the lower regions of the gut. In agreement with a role in neural crest development, RET is expressed in several neuronal subpopulations derived from this structure, including cells in the enteric, sensory, and sympathetic nervous systems (Pachnis et al. 1993). Mice that are homozygotes for a targeted mutation in the RET gene lack enteric neurons throughout the digestive tract (Schuchardt et al. 1994). A subpopulation of enteric neural crest was found to undergo apoptotic cell death specifically in the foregut of embryos lacking the RET receptor (Taraviras et al. 1999). Together with defects in kidney organogenesis, this leads to the death of *RET* null animals at birth. It has more recently been found that conditional ablation of *RET*, or the GFRα1 coreceptor, in postmigratory enteric neurons causes widespread neuronal death in the colon, leading to colonic aganglionosis that resembles Hirschsprung's disease (Uesaka et al. 2007, 2008).

Motoneurons

RET is expressed in all spinal cord motoneurons from the earliest stages of their development

(Pachnis et al. 1993; Trupp et al. 1997; Garcès et al. 2000). GDNF has potent survival activities in spinal motoneurons (Henderson et al. 1994) and is required for their in vivo survival during late embryogenesis (Oppenheim et al. 2000). At early stages of development, RET is required for the topographic projection of hind limb-innervating axons, functioning as an instructive guidance signal for motor axons (Kramer et al. 2006). In this case, RET was shown to cooperate with the EphA4 receptor tyrosine kinase to enforce the precision of this binary choice in motor axon guidance. More recent studies using conditional alleles have shown that the effect of RET on motoneuron survival during programmed cell death is restricted to the early neonatal development of the subpopulation of γ-motoneurons that innervates muscle spindles (Gould et al. 2008). RET signaling would thus appear to have multiple effects on motoneuron survival and connectivity.

Ventral Midbrain Dopaminergic Neurons

GDNF was identified on the basis of its survival-promoting effects on ventral midbrain dopaminergic neurons, which are important in the pathogenesis of Parkinson's disease (Lin et al. 1993). RET is expressed at high levels in adult ventral midbrain dopaminergic neurons of the substantia nigra, and exogenous application of GDNF was shown to protect RET-expressing neurons in this structure from cell death induced by 6-hydroxydopamine, an animal model of Parkinson's disease (Trupp et al. 1996). The robust effects of GDNF/RET signaling on dopaminergic neuron survival in several lesion paradigms naturally raised expectations of an important physiological function for RET in dopaminergic neurons, where it is expressed from very early stages of development. In agreement with this, knock-in of a constitutive allele of *RET* resulted in increased numbers of dopaminergic neurons and profound elevation of brain dopamine concentration, suggesting that RET signaling can have a direct biological effect in the brain dopaminergic system (Mijatovic et al. 2007). Despite the successes of gain-of-function approaches, the results from loss-of-function studies have been less clear-cut with regard to the importance of RET activity in dopaminergic neurons. Conditional ablation of RET in dopaminergic neurons has failed to reveal a prominent role for RET signaling in dopaminergic neuron development or maintenance, at least during the average life span of the mouse (Jain et al. 2006b; Kramer et al. 2007). One of these two studies, however, examined aging mutant mice and found that RET ablation caused progressive, but moderate, adult-onset loss of dopaminergic neurons in the substantia nigra and reduced dopaminergic nerve terminals in striatum, reaching 38% reduction by 2 years of age (Kramer et al. 2007). In contrast to those mild effects, another study induced adult deletion of a conditional allele of *Gdnf* and found widespread deficits in dopaminergic and noradrenergic neurons (Pascual et al. 2008). As discussed elsewhere (Ibanez 2008), the striking discrepancy between these studies could be explained by the presence of alternative receptors for GDNF in dopaminergic neurons, compensatory effects in developing RET-deficient neurons, or the possibility that RET functions as a "dependence receptor" in dopaminergic neurons. Dependence receptors make cells that express them dependent on their ligands, and evidence from transformed cell lines has been provided suggesting that RET might indeed function in this way (Bordeaux et al. 2000). At the time of this writing, independent efforts are underway to replicate several of the above-mentioned studies and so it is likely that the controversy over the physiological role of RET signaling in dopaminergic neuron survival and maintenance will become clarified before long.

CONCLUDING REMARKS

Nearly three decades after its discovery, unique aspects of RET function and physiology continue to fascinate biologists and biochemists. Unlike other receptor tyrosine kinases, autophosphorylation has only a modest effect on kinase activity, and so the mechanism of activation of the RET kinase remains elusive. The possibility of autoinhibition in *trans* is tantalizing and would mechanistically set RET apart from the bulk of other receptor tyrosine kinases. Its ligand system, with the GFRα coreceptors, is also rather unique, and upcoming crystal structures of the full ternary complex promise to reveal the mechanism of complex formation, and perhaps explain how a distant TGF-β superfamily member ended up binding a receptor tyrosine kinase. These insights may also find applications in drug discovery. Indeed, although its physiological significance in dopaminergic neurons remains unclear, the robust effects of RET gain-of-function on the survival and function of these neurons has encouraged efforts to identify agonists for treatment of Parkinson's disease (Aron and Klein 2011). Such compounds may circumvent the intrinsic problems of protein delivery of current GDNF-based approaches.

ACKNOWLEDGMENTS

The author apologizes to all of the colleagues whose work could not be cited owing to space constraints. Work at the author's laboratory is funded by research grants from the Swedish Research Council, Swedish Cancer Society, Swedish Foundation for Strategic Research, European Research Council, and the National Institutes of Health.

REFERENCES

*Reference is also in this collection.

Abrescia C, Sjöstrand D, Kjaer S, Ibanez CF. 2005. *Drosophila* RET contains an active tyrosine kinase and elicits neurotrophic activities in mammalian cells. *FEBS Lett* **579:** 3789–3796.

Amoresano A, Incoronato M, Monti G, Pucci P, de Franciscis V, Cerchia L. 2005. Direct interactions among Ret, GDNF and GFRα1 molecules reveal new insights into the assembly of a functional three-protein complex. *Cell Signal* **17:** 717–727.

Anders J, Kjar S, Ibanez CF. 2001. Molecular modeling of the extracellular domain of the RET receptor tyrosine kinase reveals multiple cadherin-like domains and a calcium-binding site. *J Biol Chem* **276:** 35808–35817.

Aron L, Klein R. 2011. Repairing the parkinsonian brain with neurotrophic factors. *Trends Neurosci* **34:** 88–100.

Asai N, Fukuda T, Wu Z, Enomoto A, Pachnis V, Takahashi M, Costantini F. 2006. Targeted mutation of serine 697 in the Ret tyrosine kinase causes migration defect of enteric neural crest cells. *Development* **133:** 4507–4516.

Baloh RH, Tansey MG, Golden JP, Creedon DJ, Heuckeroth RO, Keck CL, Zimonjic DB, Popescu NC, Johnson EM, Milbrandt J. 1997. TrnR2, a novel receptor that mediates neurturin and GDNF signaling through Ret. *Neuron* **18:** 793–802.

Baloh RH, Gorodinsky A, Golden JP, Tansey MG, Keck CL, Popescu NC, Johnson EM, Milbrandt J. 1998a. GFRα3 is an orphan member of the GDNF/neurturin/persephin receptor family. *Proc Natl Acad Sci* **95:** 5801–5806.

Baloh RH, Tansey MG, Lampe PA, Fahrner TJ, Enomoto H, Simburger KS, Leitner ML, Araki T, Johnson EM, Milbrandt J. 1998b. Artemin, a novel member of the GDNF ligand family, supports peripheral and central neurons and signals through the GFRα3-RET receptor complex. *Neuron* **21:** 1291–1302.

Bespalov MM, Sidorova YA, Tumova S, Ahonen-Bishopp A, Magalhães AC, Kulesskiy E, Paveliev M, Rivera C, Rauvala H, Saarma M. 2011. Heparan sulfate proteoglycan syndecan-3 is a novel receptor for GDNF, neurturin, and artemin. *J Cell Biol* **192:** 153–169.

Besset V, Scott RP, Ibanez CF. 2000. Signaling complexes and protein–protein interactions involved in the activation of the Ras and phosphatidylinositol 3-kinase pathways by the c-Ret receptor tyrosine kinase. *J Biol Chem* **275:** 39159–39166.

Bordeaux MC, Forcet C, Granger L, Corset V, Bidaud C, Billaud M, Bredesen DE, Edery P, Mehlen P. 2000. The RET proto-oncogene induces apoptosis: A novel mechanism for Hirschsprung disease. *EMBO J* **19:** 4056–4063.

Borrello MG, Alberti L, Arighi E, Bongarzone I, Battistini C, Bardelli A, Pasini B, Piutti C, Rizzetti MG, Mondellini P, et al. 1996. The full oncogenic activity of Ret/ptc2 depends on tyrosine 539, a docking site for phospholipase Cγ. *Mol Cell Biol* **16:** 2151–2163.

Buj-Bello A, Adu J, Piñón LG, Horton A, Thompson JF, Rosenthal A, Chinchetru M, Buchman VL, Davies AM. 1997. Neurturin responsiveness requires a GPI-linked receptor and the Ret receptor tyrosine kinase. *Nature* **387:** 721–724.

Cik M, Masure S, Lesage AS, Van der Linden I, Van Gompel P, Pangalos MN, Gordon RD, Leysen JE. 2000. Binding of GDNF and neurturin to human GDNF family receptor α 1 and 2. Influence of cRET and cooperative interactions. *J Biol Chem* **275:** 27505–27512.

Cosma MP, Cardone M, Carlomagno F, Colantuoni V. 1998. Mutations in the extracellular domain cause RET loss of function by a dominant negative mechanism. *Mol Cell Biol* **18:** 3321–3329.

Coulpier M, Anders J, Ibanez CF. 2002. Coordinated activation of autophosphorylation sites in the RET receptor tyrosine kinase: Importance of tyrosine 1062 for GDNF mediated neuronal differentiation and survival. *J Biol Chem* **277:** 1991–1999.

Crowder R, Enomoto H, Yang M, Johnson E, Milbrandt J. 2004. Dok-6, a novel p62 Dok family member, promotes Ret-mediated neurite outgrowth. *J Biol Chem* **279:** 42072–42081.

de Graaff E, Srinivas S, Kilkenny C, D'Agati V, Mankoo BS, Costantini FD, Pachnis V. 2001. Differential activities of the RET tyrosine kinase receptor isoforms during mammalian embryogenesis. *Genes Dev* **15:** 2433–2444.

Donis-Keller H, Dou S, Chi D, Carlson KM, Toshima K, Lairmore TC, Howe JR, Moley JF, Goodfellow P, Wells SA. 1993. Mutations in the RET proto-oncogene are associated with MEN 2A and FMTC. *Hum Mol Genet* **2:** 851–856.

Durbec P, Marcos-Gutierrez CV, Kilkenny C, Grigoriou M, Wartiowaara K, Suvanto P, Smith D, Ponder B, Costantini FD, Saarma M, et al. 1996. GDNF signalling through the Ret receptor tyrosine kinase. *Nature* **381:** 789–793.

Eketjäll S, Fainzilber M, Murray-Rust J, Ibanez CF. 1999. Distinct structural elements in GDNF mediate binding to GFRα1 and activation of the GFRα1-c-Ret receptor complex. *EMBO J* **18:** 5901–5910.

Encinas M, Crowder RJ, Milbrandt J, Johnson EM. 2004. Tyrosine 981, a novel ret autophosphorylation site, binds c-Src to mediate neuronal survival. *J Biol Chem* **279:** 18262–18269.

Fukuda T, Kiuchi K, Takahashi M. 2002. Novel mechanism of regulation of Rac activity and lamellipodia formation by RET tyrosine kinase. *J Biol Chem* **277:** 19114–19121.

Garcès A, Haase G, Airaksinen MS, Livet J, Filippi P, Delapeyrière O. 2000. GFRα 1 is required for development of distinct subpopulations of motoneuron. *J Neurosci* **20:** 4992–5000.

Geneste O, Bidaud C, De Vita G, Hofstra RM, Tartare-Deckert S, Buys CH, Lenoir GM, Santoro M, Billaud M. 1999. Two distinct mutations of the RET receptor causing Hirschsprung's disease impair the binding of signalling effectors to a multifunctional docking site. *Hum Mol Genet* **8:** 1989–1999.

Gould TW, Yonemura S, Oppenheim RW, Ohmori S, Enomoto H. 2008. The neurotrophic effects of glial cell line-derived neurotrophic factor on spinal motoneurons are restricted to fusimotor subtypes. *J Neurosci* **28:** 2131–2146.

Grieco M, Santoro M, Berlingieri MT, Melillo RM, Donghi R, Bongarzone I, Pierotti MA, Porta Della G, Fusco A, Vecchio G. 1990. PTC is a novel rearranged form of the ret proto-oncogene and is frequently detected in vivo in human thyroid papillary carcinomas. *Cell* **60:** 557–563.

Grimm J, Sachs M, Britsch S, Di Cesare S, Schwarz-Romond T, Alitalo K, Birchmeier W. 2001. Novel p62dok family members, dok-4 and dok-5, are substrates of the c-Ret receptor tyrosine kinase and mediate neuronal differentiation. *J Cell Biol* **154:** 345–354.

Hahn M, Bishop J. 2001. Expression pattern of *Drosophila ret* suggests a common ancestral origin between the metamorphosis precursors in insect endoderm and the vertebrate enteric neurons. *Proc Natl Acad Sci* **98:** 1053–1058.

Hayashi H, Ichihara M, Iwashita T, Murakami H, Shimono Y, Kawai K, Kurokawa K, Murakumo Y, Imai T, Funahashi H, et al. 2000. Characterization of intracellular signals via tyrosine 1062 in RET activated by glial cell line-derived neurotrophic factor. *Oncogene* **19:** 4469–4475.

Henderson C, Phillips H, Pollock R, Davies A, Lemeulle C, Armanini M, Simpson L, Moffet B, Vandlen R, Koliatsos V, et al. 1994. Gdnf: A potent survival factor for motoneurons present in peripheral-nerve and muscle. *Science* **266:** 1062–1064.

Hofstra RM, Landsvater RM, Ceccherini I, Stulp RP, Stelwagen T, Luo Y, Pasini B, Höppener JW, van Amstel HK, Romeo G. 1994. A mutation in the RET proto-oncogene associated with multiple endocrine neoplasia type 2B and sporadic medullary thyroid carcinoma. *Nature* **367:** 375–376.

Ibanez CF. 2008. Catecholaminergic neuron survival: Getting hooked on GDNF. *Nat Neurosci* **11:** 735–736.

Iwamoto T, Taniguchi M, Asai N, Ohkusu K, Nakashima I, Takahashi M. 1993. cDNA cloning of mouse ret proto-oncogene and its sequence similarity to the cadherin superfamily. *Oncogene* **8:** 1087–1091.

Iwashita T, Murakami H, Asai N, Takahashi M. 1996. Mechanism of ret dysfunction by Hirschsprung mutations affecting its extracellular domain. *Hum Mol Genet* **5:** 1577–1580.

Jain S, Encinas M, Johnson EM, Milbrandt J. 2006a. Critical and distinct roles for key RET tyrosine docking sites in renal development. *Genes Dev* **20:** 321–333.

Jain S, Golden JP, Wozniak D, Pehek E, Johnson EM, Milbrandt J. 2006b. RET is dispensable for maintenance of midbrain dopaminergic neurons in adult mice. *J Neurosci* **26:** 11230–11238.

Jiao L, Zhang Y, Hu C, Wang Y-G, Huang A, He C. 2011. Rap1GAP interacts with RET and suppresses GDNF-induced neurite outgrowth. *Cell Res* **21:** 327–337.

Jijiwa M, Fukuda T, Kawai K, Nakamura A, Kurokawa K, Murakumo Y, Ichihara M, Takahashi M. 2004. A targeting mutation of tyrosine 1062 in Ret causes a marked decrease of enteric neurons and renal hypoplasia. *Mol Cell Biol* **24:** 8026–8036.

Jing S, Wen D, Yu Y, Holst PL, Luo Y, Fang M, Tamir R, Antonio L, Hu Z, Cupples R, et al. 1996. GDNF-induced activation of the ret protein tyrosine kinase is mediated by GDNFR-α, a novel receptor for GDNF. *Cell* **85:** 1113–1124.

Kawamoto Y, Takeda K, Okuno Y, Yamakawa Y, Ito Y, Taguchi R, Kato M, Suzuki H, Takahashi M, Nakashima I. 2004. Identification of RET autophosphorylation sites by mass spectrometry. *J Biol Chem* **279:** 14213–14224.

Kjaer S, Ibanez CF. 2003a. Identification of a surface for binding to the GDNF-GFR α 1 complex in the first cadherin-like domain of RET. *J Biol Chem* **278:** 47898–47904.

Kjaer S, Ibanez CF. 2003b. Intrinsic susceptibility to misfolding of a hot-spot for Hirschsprung disease mutations in the ectodomain of RET. *Hum Mol Genet* **12:** 2133–2144.

Kjaer S, Kurokawa K, Perrinjaquet M, Abrescia C, Ibanez CF. 2006. Self-association of the transmembrane domain of RET underlies oncogenic activation by MEN2A mutations. *Oncogene* **25:** 7086–7095.

Kjaer S, Heilshorn S, Totty N, McDonald NQ. 2010. Mammal-restricted elements predispose human RET to folding impairment by HSCR mutations. *Nat Struct Biol* **17:** 726–731.

Klein RD, Sherman D, Ho WH, Stone D, Bennett GL, Moffat B, Vandlen R, Simmons L, Gu Q, Hongo JA, et al. 1997. A GPI-linked protein that interacts with Ret to form a candidate neurturin receptor. *Nature* **387:** 717–721.

Knowles PP, Murray-Rust J, Kjaer S, Scott RP, Heilshorn S, Santoro M, Ibanez CF, McDonald NQ. 2006. Structure and chemical inhibition of the RET tyrosine kinase domain. *J Biol Chem* **281:** 33577–33587.

Kotzbauer PT, Lampe PA, Heuckeroth RO, Golden JP, Creedon DJ, Johnson EM, Milbrandt J. 1996. Neurturin, a relative of glial-cell-line-derived neurotrophic factor. *Nature* **384:** 467–470.

Kramer ER, Knott L, Su F, Dessaud E, Krull CE, Helmbacher F, Klein R. 2006. Cooperation between GDNF/Ret and ephrinA/EphA4 signals for motor-axon pathway selection in the limb. *Neuron* **50:** 35–47.

Kramer, Aron L, Ramakers, Seitz, Zhuang, Beyer, Smidt, Klein. 2007. Absence of Ret signaling in mice causes progressive and late degeneration of the nigrostriatal system. *PLoS Biol* **5:** e39.

Kurokawa K, Iwashita T, Murakami H, Hayashi H, Kawai K, Takahashi M. 2001. Identification of SNT/FRS2 docking site on RET receptor tyrosine kinase and its role for signal transduction. *Oncogene* **20:** 1929–1938.

Kurotsuchi A, Murakumo Y, Jijiwa M, Kurokawa K, Itoh Y, Kodama Y, Kato T, Enomoto A, Asai N, Terasaki H, et al. 2010. Analysis of DOK-6 function in downstream signaling of RET in human neuroblastoma cells. *Cancer Sci* **101:** 1147–1155.

Lin LF, Doherty DH, Lile JD, Bektesh S, Collins F. 1993. GDNF: A glial cell line-derived neurotrophic factor for midbrain dopaminergic neurons. *Science* **260:** 1130–1132.

Liu X, Vega QC, Decker RA, Pandey A, Worby CA, Dixon JE. 1996. Oncogenic RET receptors display different autophosphorylation sites and substrate binding specificities. *J Biol Chem* **271:** 5309–5312.

Lundgren TK, Scott RP, Smith M, Pawson T, Ernfors P. 2006. Engineering the recruitment of phosphotyrosine binding domain-containing adaptor proteins reveals distinct roles for RET receptor-mediated cell survival. *J Biol Chem* **281:** 29886–29896.

Massagué J. 1996. Neurotrophic factors. Crossing receptor boundaries. *Nature* **382:** 29–30.

Masure S, Cik M, Hoefnagel E, Nosrat CA, Van der Linden I, Scott RP, Van Gompel P, Lesage AS, Verhasselt P, Ibanez CF, et al. 2000. Mammalian GFRα 4, a divergent member of the GFRα family of coreceptors for glial cell line-derived neurotrophic factor family ligands, is a receptor for the neurotrophic factor persephin. *J Biol Chem* **275:** 39427–39434.

Melillo RM, Santoro M, Ong SH, Billaud M, Fusco A, Hadari YR, Schlessinger J, Lax I. 2001. Docking protein FRS2 links the protein tyrosine kinase RET and its oncogenic forms with the mitogen-activated protein kinase signaling cascade. *Mol Cell Biol* **21:** 4177–4187.

Mijatovic J, Airavaara M, Planken A, Auvinen P, Raasmaja A, Piepponen TP, Costantini FD, Ahtee L, Saarma M. 2007. Constitutive Ret activity in knock-in multiple endocrine neoplasia type B mice induces profound elevation of brain dopamine concentration via enhanced synthesis and increases the number of TH-positive cells in the substantia nigra. *J Neurosci* **27:** 4799–4809.

Milbrandt J, de Sauvage FJ, Fahrner TJ, Baloh RH, Leitner ML, Tansey MG, Lampe PA, Heuckeroth RO, Kotzbauer PT, Simburger KS, et al. 1998. Persephin, a novel neurotrophic factor related to GDNF and neurturin. *Neuron* **20:** 245–253.

Mulligan LM, Kwok JB, Healey CS, Elsdon MJ, Eng C, Gardner E, Love DR, Mole SE, Moore JK, Papi L. 1993. Germ-line mutations of the RET proto-oncogene in multiple endocrine neoplasia type 2A. *Nature* **363:** 458–460.

Naveilhan P, Baudet C, Mikaels A, Shen L, Westphal H, Ernfors P. 1998. Expression and regulation of GFRα3, a glial cell line-derived neurotrophic factor family receptor. *Proc Natl Acad Sci* **95:** 1295–1300.

Nollet F, Kools P, van Roy F. 2000. Phylogenetic analysis of the cadherin superfamily allows identification of six major subfamilies besides several solitary members. *J Mol Biol* **299:** 551–572.

Oppenheim RW, Houenou LJ, Parsadanian AS, Prevette D, Snider WD, Shen L. 2000. Glial cell line-derived neurotrophic factor and developing mammalian motoneurons: Regulation of programmed cell death among motoneuron subtypes. *J Neurosci* **20:** 5001–5011.

Pachnis V, Mankoo B, Costantini FD. 1993. Expression of the c-ret proto-oncogene during mouse embryogenesis. *Development* **119:** 1005–1017.

Paratcha G, Ledda F, Baars L, Coulpier M, Besset V, Anders J, Scott RP, Ibanez CF. 2001. Released GFRα1 potentiates downstream signaling, neuronal survival, and differentiation via a novel mechanism of recruitment of c-Ret to lipid rafts. *Neuron* **29:** 171–184.

Paratcha G, Ledda F, Ibanez CF. 2003. The neural cell adhesion molecule NCAM is an alternative signaling receptor for GDNF family ligands. *Cell* **113:** 867–879.

Pascual A, Hidalgo-Figueroa M, Piruat JI, Pintado CO, Gómez-Díaz R, López-Barneo J. 2008. Absolute requirement of GDNF for adult catecholaminergic neuron survival. *Nat Neurosci* **11:** 755–761.

Pelet A, Geneste O, Edery P, Pasini A, Chappuis S, Atti T, Munnich A, Lenoir G, Lyonnet S, Billaud M. 1998. Various mechanisms cause RET-mediated signaling defects in Hirschsprung's disease. *J Clin Invest* **101:** 1415–1423.

Perrinjaquet M, Vilar M, Ibanez CF. 2010. Protein-tyrosine phosphatase SHP2 contributes to GDNF neurotrophic activity through direct binding to phospho-Tyr687 in the RET receptor tyrosine kinase. *J Biol Chem* **285:** 31867–31875.

Sanicola M, Hession CA, Worley D, Carmillo P, Ehrenfels C, Walus L, Robinson S, Jaworski G, Wei H, Tizard R, et al. 1997. Glial cell line-derived neurotrophic factor-dependent RET activation can be mediated by two different cell-surface accessory proteins. *Proc Natl Acad Sci* **94:** 6238–6243.

* Santoro M, Carlomagno F. 2013. Central role of RET in thyroid cancer. *Cold Spring Harb Perspect Biol* doi: 10.1101/cshperspect.a009233.

Schuchardt A, D'Agati V, Larsson-Blomberg L, Costantini FD, Pachnis V. 1994. Defects in the kidney and enteric nervous system of mice lacking the tyrosine kinase receptor Ret. *Nature* **367:** 380–383.

Schuchardt A, D'Agati V, Pachnis V, Costantini F. 1996. Renal agenesis and hypodysplasia in ret-k- mutant mice result from defects in ureteric bud development. *Development* **122:** 1919–1929.

Schuetz G, Rosário M, Grimm J, Boeckers TM, Gundelfinger ED, Birchmeier W. 2004. The neuronal scaffold protein Shank3 mediates signaling and biological function of the receptor tyrosine kinase Ret in epithelial cells. *J Cell Biol* **167:** 945–952.

Segouffin-Cariou C, Billaud M. 2000. Transforming ability of MEN2A-RET requires activation of the phosphatidylinositol 3-kinase/AKT signaling pathway. *J Biol Chem* **275:** 3568–3576.

Shakya R, Watanabe T, Costantini FD. 2005. The role of GDNF/Ret signaling in ureteric bud cell fate and branching morphogenesis. *Dev Cell* **8:** 65–74.

Takahashi M, Cooper GM. 1987. ret transforming gene encodes a fusion protein homologous to tyrosine kinases. *Mol Cell Biol* **7:** 1378–1385.

Takahashi M, Ritz J, Cooper GM. 1985. Activation of a novel human transforming gene, ret, by DNA rearrangement. *Cell* **42:** 581–588.

Takahashi M, Buma Y, Iwamoto T, Inaguma Y, Ikeda H, Hiai H. 1988. Cloning and expression of the ret proto-oncogene encoding a tyrosine kinase with two potential transmembrane domains. *Oncogene* **3:** 571–578.

Taraviras S, Marcos-Gutierrez CV, Durbec P, Jani H, Grigoriou M, Sukumaran M, Wang LC, Hynes MA, Raisman G, Pachnis V. 1999. Signalling by the RET receptor tyrosine kinase and its role in the development of the mammalian enteric nervous system. *Development* **126:** 2785–2797.

Trupp M, Arenas E, Fainzilber M, Nilsson AS, Sieber B-A, Grigoriou M, Kilkenny C, Salazar-Grueso E, Pachnis V, Arumae U, et al. 1996. Functional receptor for GDNF encoded by the c-ret proto-oncogene. *Nature* **381:** 785–789.

Trupp M, Belluardo N, Funakoshi H, Ibanez CF. 1997. Complementary and overlapping expression of glial cell line-derived neurotrophic factor (GDNF), c-ret proto-oncogene, and GDNF receptor-α indicates multiple mechanisms of trophic actions in the adult rat CNS. *J Neurosci* **17:** 3554–3567.

Trupp M, Raynoschek C, Belluardo N, Ibanez CF. 1998. Multiple GPI-anchored receptors control GDNF-dependent and independent activation of the c-Ret receptor tyrosine kinase. *Mol Cell Neurosci* **11:** 47–63.

Uesaka T, Jain S, Yonemura S, Uchiyama Y, Milbrandt J, Enomoto H. 2007. Conditional ablation of GFRα1 in postmigratory enteric neurons triggers unconventional neuronal death in the colon and causes a Hirschsprung's disease phenotype. *Development* **134:** 2171–2181.

Uesaka T, Nagashimada M, Yonemura S, Enomoto H. 2008. Diminished Ret expression compromises neuronal survival in the colon and causes intestinal aganglionosis in mice. *J Clin Invest* **118:** 1890–1898.

Wong A, Bogni S, Kotka P, de Graaff E, D'Agati V, Costantini FD, Pachnis V. 2005. Phosphotyrosine 1062 is critical for the in vivo activity of the Ret9 receptor tyrosine kinase isoform. *Mol Cell Biol* **25:** 9661–9673.

Worby CA, Vega QC, Chao HH, Seasholtz AF, Thompson RC, Dixon JE. 1998. Identification and characterization of GFRα-3, a novel co-receptor belonging to the glial cell line-derived neurotrophic receptor family. *J Biol Chem* **273:** 3502–3508.

Tie2 and Eph Receptor Tyrosine Kinase Activation and Signaling

William A. Barton[1], Annamarie C. Dalton[1], Tom C.M. Seegar[1], Juha P. Himanen[2], and Dimitar B. Nikolov[2]

[1]Department of Biochemistry and Molecular Biology, School of Medicine, Virginia Commonwealth University, Richmond, Virginia 23298

[2]Structural Biology Program, Memorial Sloan-Kettering Cancer Center, New York, New York 10065

Correspondence: nikolovd@mskcc.org

The Eph and Tie cell surface receptors mediate a variety of signaling events during development and in the adult organism. As other receptor tyrosine kinases, they are activated on binding of extracellular ligands and their catalytic activity is tightly regulated on multiple levels. The Eph and Tie receptors display some unique characteristics, including the requirement of ligand-induced receptor clustering for efficient signaling. Interestingly, both Ephs and Ties can mediate different, even opposite, biological effects depending on the specific ligand eliciting the response and on the cellular context. Here we discuss the structural features of these receptors, their interactions with various ligands, as well as functional implications for downstream signaling initiation. The Eph/ephrin structures are already well reviewed and we only provide a brief overview on the initial binding events. We go into more detail discussing the Tie-angiopoietin structures and recognition.

ANGIOPOIETINS AND TIE2

Vasculogenesis and angiogenesis are distinct cellular processes essential to the creation of the adult vasculature. In early embryonic development, precursor angioblasts differentiate into endothelial cells, migrate, and form the vasculature framework including major primitive blood vessels and the endocardium of the developing heart. This process, known as vasculogenesis, results in a poorly branched and loosely connected capillary plexus. Angiogenesis further remodels the primitive endothelial network into a highly branched microvasculature and results in the intussusception of vessels into some organs (Adams and Alitalo 2007; Huang et al. 2010).

In contrast to vasculogenesis, angiogenesis is continually required in the adult for wound repair and remodeling of reproductive tissues during female menstruation. Importantly, pathological angiogenesis aids solid tumor growth by providing an enriched nutrient and oxygen supply, as well as a mechanism for tumor cell dissemination (metastasis). Thus, understanding the role of receptors and ligands that control angiogenesis is essential for shaping a fundamental understanding of tumor development (Adams and Alitalo 2007; Huang et al. 2010).

Two major endothelial receptor tyrosine kinase signaling pathways are essential for angiogenesis: these include the vascular endothelial growth factor (VEGF) receptor and the Tie2 receptor. Whereas VEGF appears to function as a general regulator of vasculogenesis and angiogenesis, the Ang-Tie system plays a role downstream of VEGF signaling during angiogenesis. Since the initial discovery of the Tie receptors in 1992, a stream of studies have slowly illuminated the role of this signaling pathway in angiogenesis, particularly with regard to its role in the communication between support cells and endothelium (Adams and Alitalo 2007; Huang et al. 2010). However, despite significant molecular developments, high-resolution structural information has only recently become available. Below, we discuss the structural characteristics, and their functional implications, of the unique Tie-angiopoietin signaling system.

Angiopoietin Ligands

The angiopoietins (Ang1-4) modulate the activity of Tie2. These four secreted protein ligands maintain a high level of sequence homology while eliciting distinct responses from their target receptor (Fig. 1) (Davis et al. 1996, 2003; Maisonpierre et al. 1997; Ramsauer and D'Amore 2002). Although the agonist Ang3 and antagonist Ang4 are poorly characterized (Valenzuela et al. 1999), extensive data establishes Ang1 to be a strict agonist of Tie2 activation, leading to prosurvival signaling and quiescence of the endothelium (Davis et al. 1996; Papapetropoulos et al. 2000). In contrast, Ang2 has been shown to competitively inhibit Ang1 activation, suggesting a single ligand-binding site on Tie2 and an antagonistic role for Ang2 (Maisonpierre et al. 1997; Fiedler et al. 2003). The precise role of Ang2 is actually context-dependent, as dimeric Ang2 is capable of activating Tie2 in fibroblasts stably expressing the endothelial-specific receptor (Davis et al. 2003).

Early studies by Davis et al. (1996, 2003) established that Tie2 recognition is predominantly mediated by the angiopoietin conserved carboxy-terminal fibrinogen-like domain (see below); however, it was further shown that the fibrinogen domain alone is not sufficient for activation of the receptor. Instead, activation requires the presence of the central coiled-coil region that enables dimerization of the ligands

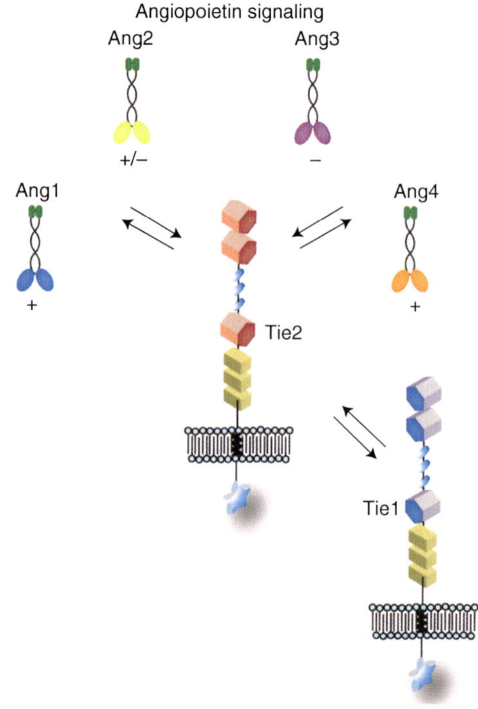

Figure 1. Schematic representation of the Tie receptors and angiopoietin ligands. The Tie receptors are highly homologous endothelial-specific receptor tyrosine kinases. Each receptor consists of three Ig domains (shown in red, green, and blue), three EGF domains (yellow, magenta, orange), and three fibronectin type III repeats (gray) in the ectodomain, followed by a single-pass transmembrane domain, and a split tyrosine kinase domain in the cytoplasm. Tie2 interacts with all four of the structurally similar angiopoietin ligands (Ang1–4), although each ligand is functionally distinct. The angiopoietins contain an amino-terminal super-clustering domain (green), a coiled-coil domain, and a fibrinogen-like receptor-binding domain. Ang1 (blue) and Ang3 (purple) are agonists of Tie2 activation, Ang4 (orange) is an antagonist, and Ang2 (yellow) is a context-dependent antagonist as indicated. Despite the high level of sequence conservation between the two receptors, Tie1 is an orphan receptor, yet is able to heterodimerize with Tie2 on the cell surface.

while further higher order homo-, or potentially hetero-, oligomerization can be induced by the amino-terminal "super-clustering" domain (Fig. 1). Indeed, electron micrographs show both Ang1 and Ang2 as dimers, tetramers, and higher-order multimers, although Ang2 has been shown to exist primarily as a dimer in solution. Biochemical assays show a requirement for tetrameric Ang1 to elicit endogenous Tie2 activation; however, an engineered dimer may also elicit some receptor activation in fibroblasts exogenously expressing Tie2 (Ward and Dumont 2002; Davis et al. 2003; Fiedler et al. 2003).

Crystal structures of the fibrinogen-like receptor-binding domains (RBDs) of both human Ang1 and Ang2 have been determined at 2.7 and 2.4 Å, respectively. Predictably, at 64% sequence homology, very little structural deviation (root-mean-square deviation [rmsd] of 0.77 Å) was observed between the compact Ang-RBD structures ($\sim$50 $\times$ 40 $\times$ 35 Å) (Fig. 2A) (Barton et al. 2005, 2006; Yu et al. 2013). Of the three subdomains, termed A, B, and P, as per human fibrinogen nomenclature, the P domain is the least evolutionarily conserved and is solely responsible for the receptor interactions. It contains little secondary structure but is stabilized by a conserved Ca^{2+}-binding site coordinated by two aspartic acid side chains and main chain oxygen atoms (Barton et al. 2005).

An analysis of the hydrophobicity and electrostatic potential of the Ang-RBDs surfaces yields additional details about the functional differences between the angiopoietin ligands. For example, the A and B domains are largely similar and even the P domain, which contains the majority of the significant structural differences between the ligands, contains mostly conserved surface side chains. The only significant difference is within a three amino acid surface loop containing residues 461-463 in Ang2 and residues 463-465 in Ang1 (Fig. 2, left panel). The loop, which varies from T-A-G in Ang1 to P-Q-R in Ang2, shows differences in both hydrophobic and electrostatic properties. Interestingly, functional differences between Ang1 and Ang2 result from alterations in this sequence as a chimeric Ang2 behaves as a receptor agonist (see below) (Yu et al. 2013).

Tie Receptors

The structurally related Tie1 and Tie2 are type 1 transmembrane receptor tyrosine kinases regulating vessel branching and maintaining endothelial homeostasis (Ramsauer and D'Amore 2002). The Tie receptors consist of three immunoglobulin-like (Ig) domains, three epidermal growth factor (EGF) domains, and three fibronectin type III repeats (FNIII) in the extracellular protein segment (Figs. 1 and 2B) (Barton et al. 2006; Seegar et al. 2010). A single-pass transmembrane domain separates this ectodomain from a structurally conserved split tyrosine kinase domain with homology with fibroblast growth factor receptor (FGFR1) (Shewchuk et al. 2000). Despite extensive homology, Tie1 remains an orphan receptor while Tie2 binds all angiopoietin ligands (Davis et al. 1996; Maisonpierre et al. 1997; Valenzuela et al. 1999).

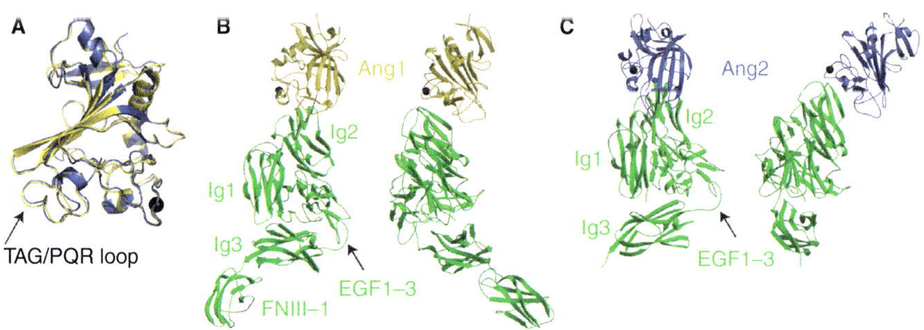

Figure 2. Crystal structures of the angiopoietin receptor binding domains and the Tie2 ectodomain unbound and in complex with Ang1 and Ang2. (*A*) The Ang1 and Ang2 receptor-binding domains superimposed in blue and yellow, respectively. The calcium ion located within the P domain is displayed in space-filling representation in black. The surface loop, which mediates Ang1/Ang2 functional differences, is labeled. (*B, C*) The Ang1-Tie2 and Ang2-Tie2 crystal structures illustrated in two orientations. The Tie2 receptor is shown in green and its domains are labeled. The Ang1-Tie2 model contains an additional fibronectin type-III repeat not present in the Ang2-Tie2 structure.

The Tie2 ectodomain structure was studied by X-ray crystallography and, more recently, by transmission electron microscopy (TEM). A high-resolution crystal structure at 2.5 Å (Fig. 2) reveals a compact architecture with a substantial amount of surface area (3800 Å^2) buried in intradomain interactions in contrast to other family members, such as c-kit and FGFR, which display an extended Ig domain arrangement (Barton et al. 2006). The compact structure ($\sim$90 × 65 × 50 Å) is achieved as flexible linkers allow Ig1 to fold down and interact with the third Ig domain, creating an arrowhead arrangement. This leaves Ig2 as the tip of the arrowhead in position to interact with the angiopoietin ligands. Ig3, Ig1, and EGF3 form the base and two sides, respectively. EGF domains 1 and 2 are mostly buried, forming numerous contacts with the outer 4 domains to stabilize the overall structure. Four N-linked glycosylation sites are found in the crystal structure split between Ig1 and Ig3; however, their role in receptor function, if any, remains unknown (Barton et al. 2006). In agreement, MacDonald et al. (2006) analyzed the structure of the Tie2 ectodomain including the FNIII repeats by rotary shadow-cast TEM. They describe Tie2 as a "lollipop" structure with the compact Ig and EGF domains as the head and the FNIII repeats creating a stalk.

Tie2-Angiopoietin Recognition

In addition to the unique tertiary architecture of the Tie receptors, they are also distinct in their binding of the fibrinogen-like domain of the angiopoietin ligands. Typical Ig superfamily receptors have been shown to interact with members of the cysteine knot, four-helix bundle, and β-trefoil families; thus, the structures of Ang1 and Ang2 bound to the Tie2 receptor ectodomain revealed a novel ligand/receptor interaction (Wiesmann et al. 2000).

The first structure of Ang2-RBD bound to the Tie2 ectodomain (excluding the three FNIII repeats) confirmed that Ig2 of Tie2 and the P domain of the ligand were exclusively responsible for ligand/receptor recognition and binding. The 3.5 Å structure has overall dimensions of 130 × 65 × 50 Å with one Ang2-RBD and one Tie2 ectodomain in the complex, in congruence with the previously reported 1:1 binding stoichiometry (Fig. 2) (Barton et al. 2006). The small, uninterrupted binding interface (1300 Å) is primarily dominated by van der Waals interactions between nonpolar side chains with additional stability resulting from a hydrogen-bonding network and several salt bridges. The binding surface is adjacent to the conserved calcium-binding site of the Ang ligand P domain; however, the calcium ion does not appear to play a direct role in receptor recognition other than to stabilize the P domain. Independent biochemical studies show that disruption of Ca^{2+} binding does, in fact, result in disordered and receptor-binding incompetent ligand (Barton et al. 2005).

The structure of the ligand-receptor interface highlights yet another unique feature of the Ang-Tie system: it uses a lock-and-key mode of recognition similar to that of the antibody-antigen interactions. Both Tie2 and antibodies mediate recognition of their ligands through a molecular surface with complementary electrostatic and chemical properties. Each molecule of the heterodimer undergoes very little structural alteration on binding, primarily restricted to minor side chain rearrangements. Additional parallels between the Ang/Tie2 binding interface and antibody-antigen interactions include a relatively small amount of buried surface area (typically between 700–1150 Å for antigens), an abundance of aromatic residues, and van der Waals contacts, and/or hydrogen bonds mediating recognition (Barton et al. 2006; Sundberg 2009).

Biochemical studies have suggested that Ang1 and Ang2 interact with Tie2 using a common interface and that was confirmed by the crystal structure of the Ang1/Tie2 complex (Barton et al. 2006; Yu et al. 2013). The Ang1/Tie2 complex superimposes on the Ang2/Tie2 structure with few major alterations (rms deviation at 0.574 Å for equivalent Cα positions). Not surprisingly, the additional FNIII repeat, not present in earlier models, does not influence ligand binding and is located on the opposite face of the receptor. However, a small shift in the overall ligand position is observed ($\sim$1.5–6 Å), despite the involvement and preservation of most equivalent contact residues between Ang1 and Ang2 (Barton et al. 2006; Yu et al. 2013).

The unbound and Ang-bound Tie2 structures show few architectural changes. The only major deviation is within a surface loop of Ig2, which shifts slightly in the complex structure to facilitate ligand interactions; yet, the overall packing of the five domains remains essentially identical (Barton et al. 2006). Similarly, the angiopoietin structure undergoes minimal alteration on Tie2 binding. Two small variations include the Ca^{2+} ion-binding loop, which undergoes a small shift of approximately 1.0 Å to create a van der Waals contact, and Ser480, which shifts $\sim$1.8 Å to accommodate a hydrogen bond at the ligand-binding interface (Barton et al. 2005, 2006).

Tie2 Tyrosine Kinase Domain Structure and Signaling

The structures of both wild-type and a nonphosphorylatable mutant of the Tie2 tyrosine kinase catalytic domain (TKD) (residues 808–1124) were determined at 2.1–2.5 Å resolution (Shewchuk et al. 2000). The overall molecular architecture is comparable to previously determined protein kinases and contains a catalytic cleft between the smaller amino and larger carboxyl lobes (Fig. 3). The amino-terminal lobe (residues 808–904) contains two charged residues (K855, E872) and a glycine rich nucleotide-binding loop (residues 831–836) responsible for coordination of the α, β, and γ phosphates of ATP. The carboxy-terminal lobe (residues 905–1124) consists of seven α-helices, four short β-strands, and an extended carboxy-terminal tail. The kinase active site consists of the catalytic loop (residues 962–968), including the essential aspartic residue (D964), and the activation loop (residues 982–1008) containing a single tyrosine residue (Y992).

Despite general similarity with many receptor tyrosine kinase domains, Tie2 TKD most closely resembles FGFR1 with 45% primary sequence identity, and an rmsd for the Cα atoms of the carboxy-terminal and amino-terminal lobes of 0.76 Å and 0.58 Å, respectively. Both kinase structures were determined in an "open" conformation, with relative rotations of 15° between the N and C lobes compared with the "closed" conformation observed, for example, in IRK (Shewchuk et al. 2000).

Activation of many TKDs is thought to occur in *trans* via ligand-induced dimerization of the kinase domains. As opposed to the prototypic receptor tyrosine kinase, Tie2 signal initiation requires receptor tetramerization and/or clustering facilitated by the multimeric an-

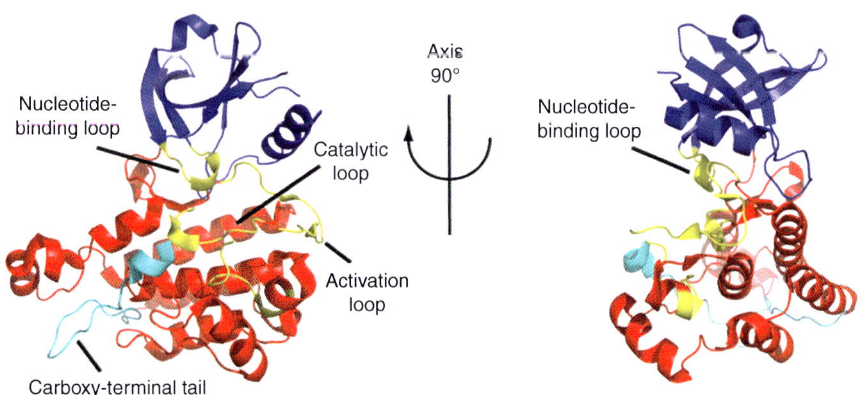

Figure 3. The structure of the Tie2 TKD. Two views rotated 90° about the *y*-axis are shown (PBD 1FVR). The two conserved lobes, amino terminal and carboxy terminal, are colored blue and red, respectively. The three catalytic loops are labeled and colored yellow. The extended carboxy-terminal tail, containing the substrate tyrosines 1101, 1107, and 1112, is colored cyan.

giopoietin ligands (Davis et al. 2003). Autophosphorylation and activation is inhibited primarily through an unproductive conformation of both the nucleotide-binding loop and charged residues responsible for ATP coordination. Interestingly, the activation loop in Tie2 adopts an overall "active conformation" independent of phosphorylation, somewhat analogous to ErbB receptors (Stamos et al. 2002). However, the carboxy-terminal tail adopts an extended conformation into the active site, presumably preventing substrate binding by acting as a substrate mimetic (Shewchuk et al. 2000). In support of this conclusion, a mutant kinase lacking 15 carboxy-terminal residues displays a drastic increase in kinase autophosphorylation, as compared with wild-type Tie2 (Niu et al. 2002).

Within the carboxy-terminal tail are three tyrosine residues, known to undergo reversible phosphorylation during signaling, that serve as important docking sites for various PTB and SH2 domain-containing proteins. Two of these residues, Y1101 and Y1112, are involved in extensive hydrogen bonding and van der Waals interactions with the core of the protein, a conformation that seemingly prevents their post-translational modification. Therefore, a conformational change presumably occurs following activation that exposes Y1101, Y1107, and Y1112 for phosphorylation and subsequent binding events. Phospho-Y1101 has been shown to recruit Grb2 and the p85 subunit of PI3K, promoting cell motility and survival through the MAPK and Akt pathways, respectively (Huang et al. 1995; Kontos et al. 2002). Similarly, Phospho-Y1112 has been reported to recruit the protein tyrosine phosphatase, SH-PTP2, which in turn may negatively regulate Tie2 signaling (Huang et al. 1995). Additional studies aimed at inhibiting the activity of PI3K have shown Y1107 mediating cell mobility by recruitment of Dok-R in a PI3K independent mechanism (Master et al. 2001; Jones et al. 2003). Thus, like most other tyrosine kinase receptors, the Tie2 kinase domain is controlled through a combination of conformational changes involving the activation loop, nucleotide-binding loop, and carboxy-terminal domain.

The Role of Tie1 in Tie2 Signaling

The coreceptor Tie1 is highly homologous to Tie2 yet does not associate with any of the angiopoietin ligands. Examination of the Tie2/Ang2 interface in comparison with a Tie1 homology model illustrates that many of the residues essential for ligand recognition in Tie2 are replaced in Tie1 with residues that would result in highly unfavorable contacts with an incoming ligand (Barton et al. 2006; Seegar et al. 2010). Instead of being directly activated by ligand binding, it appears that Tie1 forms ligand-independent heterodimers with Tie2 at the cell surface. Indeed, Tie1 contains a large basic surface, primarily composed of arginine and lysine residues, that forms a functional electrostatic interaction with a positively charged face of Tie2. Interestingly, the heterodimerization with Tie1 mediates negative regulation of Tie2, inhibiting its phosphorylation and preventing downstream signaling (Fig. 4). This inhibition depends on the relative concentrations of Tie1 and Tie2 in an individual cell, a level of control that may vary between vascular and lymphatic endothelial cells. The different angiopoietin ligands are capable of eliciting various functional responses by either stabilizing or disrupting the Tie1/Tie2 complexes (Hansen et al. 2010; Seegar et al. 2010). A short (three residue) loop in the P domain (see Fig. 2, left panel) adjacent to the Ang/Tie2 binding interface seems to mediate functional differences between the ligands (Yu et al. 2013). This loop in the antagonist Ang2 permits Tie1/Tie2 heterodimerization, whereas the equivalent loop in the agonist Ang1 disrupts the electrostatic interaction, promoting Tie2 clustering and activation of downstream signaling cascades. The context-dependent activation of Tie2 by Ang2 likely occurs when Tie1 is not present in high concentrations within the cell, highlighting the complexity of Tie receptor regulation.

Eph RECEPTORS AND EPHRINS

Eph receptors, the largest family of receptor tyrosine kinases (RTKs) and their ephrin ligands (see also Lisabeth et al. 2013) have central roles in axon pathfinding and in a diverse array of

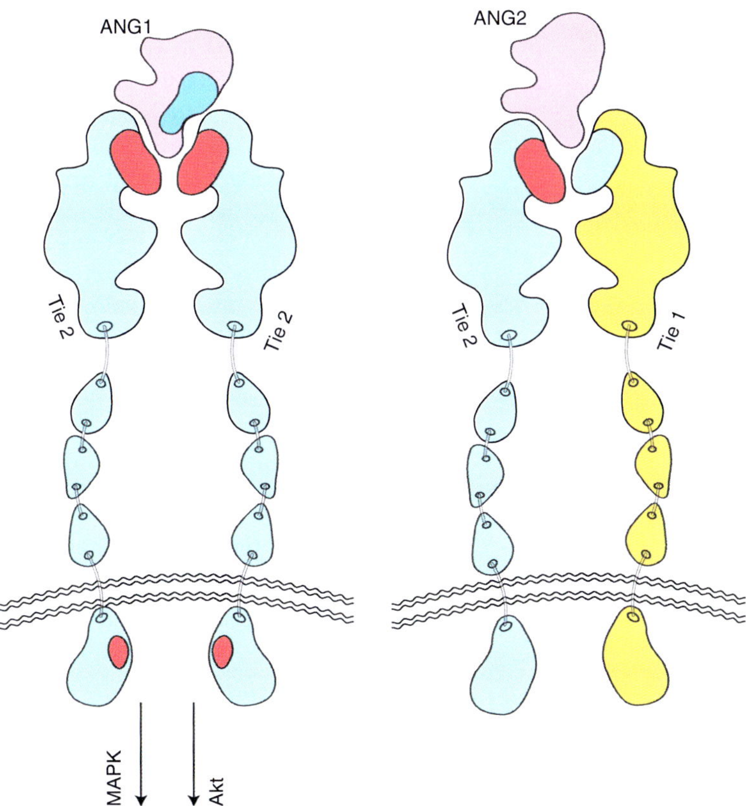

Figure 4. Model of Tie-angiopoietin signaling. The angiopoietin growth factors initiate complex signaling pathways through the Tie receptor tyrosine kinases on the endothelial cell surface. Ang1 activation of the primary receptor, Tie2, leads to prosurvival signaling through pathways such as Akt and MAPK, and results in endothelial cell quiescence and recruitment of surrounding support cells. Disruption of Tie2 signaling gives increase to a leaky vessel wall as support cells detach and endothelial cells begin to migrate. The interaction of Tie2 with its coreceptor Tie1 leads to such a vessel branching phenotype. The context-dependent antagonist Ang2 inactivates the Tie2 receptor by facilitating the inhibitory Tie1/Tie2 interactions; however, if Tie1 is not present in the cell, Ang2 is capable of clustering and activating Tie2 in a manner similar to the agonist Ang1. Tie2 is in cyan, Tie1 is in yellow, and Ang1 and Ang2 are in magenta. Ang1 and Ang2 represent ligand dimers. Blue and red regions indicate electrostatically positive and negative surface regions, including phosphorylation of the Tie2 kinase domain.

other cell–cell interactions, including those of vascular endothelial cells and specialized epithelia (Flanagan and Vanderhaeghen 1998; Klein 2001; Himanen et al. 2007). The 16 Eph receptors and nine ephrins are divided into two subclasses based on sequence homology and binding affinities. The domain organization of Eph receptors and ephrins is shown in Figure 5. As both receptors and ligands are membrane-bound, their interactions at sites of cell–cell contact initiate unique bidirectional signaling cascades (Cowan and Henkemeyer 2001). The

signaling downstream of the Ephs is referred to as "forward" and downstream of the ephrins as "reverse." Whereas the receptor-induced activation of the B-class ephrins is well documented (Cowan and Henkemeyer 2002; Song et al. 2002; Klein 2009; Lee and Daar 2009), the A-class ligands lack a cytoplasmic domain and their downstream signaling mechanism is less clear. It has been suggested that TrkB is a coreceptor for ephrin-A5 and is necessary for A-class reverse signaling (Marler et al. 2008) but the details are still unknown.

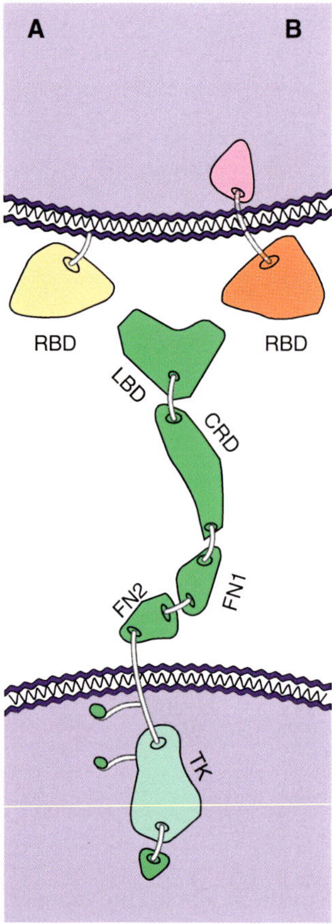

Figure 5. Schematic representation of A and B types of ephrins and Eph receptors. Shown are the receptor-binding domain (RBD) of the ephrins, the Eph's ligand-binding domain (LBD), the kinase domain (TK), the Cys-rich domain (CRD), and the fibronectin III domains (FN3). The A ephrin family are GPI-linked to the membrane, whereas the B ephrin family has a transmembrane domain and a short cytoplasmic tail.

Although Ephs and ephrins were originally identified as axon guidance molecules, they have been implicated in a vast array of cell communication events. Those include bone morphogenesis and homeostasis, immunological and inflammatory host responses, stem cell plasticity, learning and memory, and Alzheimer's disease (Pasquale 2008). However, currently, the most intensely studied function of the Eph/ephrin system is that during development and progression of cancer in multiple cell types. Many A and

B class receptors were shown to be overexpressed in various tumor types (Robinson et al. 1996; Hafner et al. 2004; Noblitt et al. 2005; Sjoblom et al. 2006) and to regulate critical steps of blood vessel formation (vasculogenesis) and remodeling (angiogenesis) and hence tumor growth. Eph receptors are also implicated in tumor invasion and metastasis (Wang 2011). Intriguingly, they have a dual role in tumorous cells, sometimes promoting sometimes suppressing cancer growth (Chen et al. 2008). Given their importance in multiple aspects of cancer progression, it is not surprising that there is widespread interest in developing Eph-targeted anticancer therapeutics (Pasquale 2010).

Eph/Ephrin Recognition and Binding

The structural details of how the Eph receptors and ephrins bind each other have been well studied and documented and have been recently reviewed (Himanen 2012). Consequently, we will give here only a brief overview on the fundamentals of ligand-receptor recognition. Overall, it has been known for more than two decades that the amino-terminal globular domains of Eph receptors and ephrins are necessary and sufficient for binding (Labrador et al. 1997). Later structural studies revealed how the ephrin minimal interaction domain forms an initial high-affinity 1:1 complex with the Eph receptor minimal interaction domain by inserting its long G-H loop into a hydrophobic cavity on the surface of the receptor (Himanen et al. 2001; Himanen and Nikolov 2002). The same mode for initiation of receptor-ligand interactions on cell–cell contact is the same for all investigated cases, regardless of whether the interaction is between A-class or B-class molecules, or whether it is cross-class (Himanen 2012). Individual differences do exist between different complexes, mainly in the intimacy of the contacts surrounding the central G-H-loop/cavity interface and in the conformational changes of the interacting loops, and these differences are responsible for fine-tuning the interactions.

Based on seemingly class-specific differences in the flexibility of the receptor loops forming the sides of the binding cavity, B-class Eph/ephrin binding has been referred to as "induced fit," whereas the A-class binding has been referred to as "lock-and-key" (Himanen et al. 2009). Studies on EphA4, a receptor that is able to bind both classes of ligands, have been particularly informative in terms of explaining the mechanism of initial receptor-ligand recognition and subclass specificity (Bowden et al. 2009; Qin et al. 2010). EphA4 displays significant structural plasticity, especially in the ligand-binding loops and one study reported 16 different conformations, obtained from two EphA4 crystal forms (Qin et al. 2012), highlighting the key role of Eph protein dynamics on recognition and signaling. Although it is now known that other regions, outside the minimal binding domains of the molecules, also participate in the formation of Eph/ephrin complexes (Day et al. 2005; Himanen et al. 2010; Seiradake et al. 2010), there is no evidence that they increase the association rate measured for the minimal binding domains. It is evident, however, that they are essential for the formation of the stable receptor clusters necessary for the full biological activity of these molecules.

Eph/Ephrin Interaction Interfaces as Drug Targets

All known Eph/ephrin structures highlight the importance of the hydrophobic cavity/loop for Eph/ephrin binding. This has prompted an intense investigation toward identifying Eph agonists/antagonists targeting this region that could be developed into therapeutic agents. Several structures were determined to show how the hydrophobic cavity of the receptor, indeed, provides a binding pocket for peptides and small organic molecules (Chrencik et al. 2006, 2007; Qin et al. 2008; Noberini et al. 2012b). For example, an EphB4-specific peptide binds to this channel and blocks EphB4 signaling (Chrencik et al. 2006), whereas other peptides bind selectively to EphA2 with submicromolar Kd and compete with ephrin binding (Koolpe et al. 2002). Remarkably, one of these peptides has ephrin-like activity (i.e., it stimulates EphA2 tyrosine phosphorylation and signaling). It was further revealed that only five peptide residues

might be essential for receptor binding and selectivity (Mitra et al. 2010). Antagonistic peptides that target the ligand-binding pocket of EphA4 were also identified (Lamberto et al. 2012). Recently, lower-resolution NMR structures of two small organic molecules binding in the same hydrophobic cavity of EphA2 and EphA4 were published (Noberini et al. 2008; Qin et al. 2008). The compounds act as competitive inhibitors for ephrin-A5, selectively binding to EphA2 and EphA4. They inhibit ephrin-induced phosphorylation without affecting cell viability or phosphorylation of other receptor tyrosine kinases and, importantly, also inhibit EphA2-dependent retraction of the cell periphery in prostate cancer cells. Moreover, natural compounds such as lithocholic acid (Giorgio et al. 2011) and tea polyphenols (Noberini et al. 2012a), have also been shown to inhibit ephrin binding to EphA4 and several other Eph receptors at low micromolar concentrations. Some of the polyphenols were shown to inhibit tyrosine phosphorylation, which was affected by mutations within the ligand-binding cavity of EphA4. This year, the first small-molecule agonist (doxazosin) for any RTK was identified (Petty et al. 2012). It not only inhibits EphA2-dependent Akt and ERK activation but, remarkably, also reduces metastasis of human prostate cancer cells in a mouse xenograft model.

These studies suggest that small-molecule inhibitors, selected based on their ability to disrupt ephrin-Eph interactions, can display a wide range of biological effects and have clear pharmaceutical potential. What remains to be seen is whether it is possible to also target the other Eph receptor interfaces, those outside of the ephrin-binding domain, which are lower affinity and presumably even easier to disrupt. With the publication of the three-dimensional structures of the complete Eph ectodomains (Himanen et al. 2010 ; Seiradake et al. 2010) (see also below), we expect to see rapid progress in this direction. Finally, it is also possible to target the ephrin ligand, as has been performed by using an ephrin-specific antibody (Abengozar et al. 2012), the systemic administration of which caused a reduction of tumor growth in xenografted mice.

Eph Receptor Clustering and Activation

The structural studies on the minimal binding domains have fairly rapidly given us a comprehensive understanding of the initial Eph/ephrin recognition steps. However, until recently, there were no studies describing the structural rearrangements taking place within the full receptor molecules on ligand binding. This knowledge is essential for understanding the events that trigger the formation of receptor clusters necessary for downstream signaling. During the past couple of years, however, a string of papers have provided considerably better view on these events. They include structural studies of the entire Eph ectodomains, biophysical studies on their transmembrane domains, as well as structural, biophysical, and cell biological studies on their intracytoplasmic regions.

Two papers on the structures of the extracellular domain (ECD) of the Eph receptor (Himanen et al. 2010; Seiradake et al. 2010) have now shed light on ligand-induced Eph clustering. The papers describe structures for the complete or partial EphA2 ECD, either alone or in complex with ephrin-A1 or -A5 ligands. The structures reveal that the Eph-ECD folds into a rigid, rod-like structure that does not significantly change on ligand binding. Thus, ligand-induced conformational changes in the receptor ectodomain do not seem to be the underlying molecular mechanism of Eph signal transduction. The structures further show how Eph receptors use two different interacting surface areas to generate signaling clusters (Fig. 6). The first interface is within the ligand-binding domain of Eph and causes the formation of receptor dimers. The second interacting surface is a novel protein-interaction module within the Cys-rich domain (CRD) that cooperates with ligand-mediated clustering to cause the formation of continuous signaling Eph/Eph assemblies. Thus, once the receptor concentration is high enough, a receptor dimer can associate with two other receptor dimers through the second interacting interfaces. The ensuing receptor hexamer can then again bind other receptor dimers and so on, using a so-called "seeding" mechanism, resulting in the formation of large receptor assemblies (clusters)

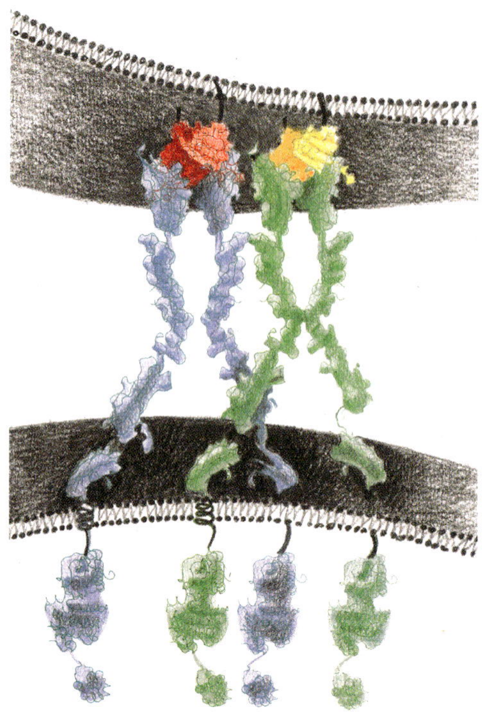

Figure 6. Schematic presentation of the Eph receptors bound to ephrin ligands at the cell–cell contact regions. Eph receptors are in blue and green, the ephrins are in red and yellow. The receptors use a "seeding" mechanism for creating signaling-competent assemblies where the ligand-binding domains first form receptor dimers (2:2 Eph/ephrin complexes), after which they bind other dimers via their cystein-rich domains, thus creating large clusters containing hundreds of molecules. Seemingly, the main role of the ligand is to increase the local receptor concentration so that full downstream signaling can be attained.

that have been visualized on the cell surface. At high receptor concentrations, this process can also happen independently of ligand binding (Wimmer-Kleikamp et al. 2004), potentially leading to transforming phenotypes. Indeed, nearly half of human breast cancers overexpress the EphA2 receptor (Lackmann and Boyd 2008). The presence of highly ordered receptor assemblies on the cell surface is a unique feature of the Eph receptors within the receptor kinase superfamily. The function of ephrin ligands seems to be to increase local receptor concentration so that these Eph/Eph and Eph/ephrin assemblies can be formed. Moreover, EphA and EphB receptors can cocluster so that the assembly of one receptor type promotes the recruitment and activation of the other receptor (Janes et al. 2011). Studies of receptor clustering also include treating cells with antibodies that recognize the Eph ectodomain, consequently inducing Eph receptor activation and initiation of downstream signaling (Vearing et al. 2005).

Transmembrane Domains in Eph Activation

In addition to the ECD, the transmembrane domains (TMDs) of receptor tyrosine kinases play an active role in signaling, contributing to the stability of receptor dimers or maintaining a signaling-competent receptor conformation. Earlier interpretations of the role of TMD were based on biophysical studies using various RTKs (Li and Hristova 2006; Artemenko et al. 2008; Bocharov et al. 2008, 2010; Li and Hristova 2010; Volynsky et al. 2010). Some understanding of the participation of TMD in the biological activity of RTKs came from studies that showed how isolated TMDs of EGFR and other members of the ErbB family dimerize in bacterial membrane (Mendrola et al. 2002). A recent NMR study shows that the membrane-spanning helices of ErbB4 form a parallel dimer in lipid bicelles and undergo a structural adjustment to form a network of intermonomeric polar contacts and provide entropic enhancement for the weak helix–helix interactions (Bocharov et al. 2012). Interestingly, when thousands of peptides, based on the TMD of ErbB2 (Neu) were screened for their dimerization properties, some of the sequences were found to activate the ErbB2 kinase significantly more than the wild-type sequence (He et al. 2011), further highlighting the potential importance of this region. In addition, the isolated FGFR TMD was shown to dimerize in the absence of the extracellular domain or ligands (Li et al. 2005). However, this dimerization propensity is about tenfold weaker than that of glycophorinA (Artemenko et al. 2008), a well-characterized TMD dimer. Recent structures of the EphA1 and EphA2 TMDs also suggest that they mediate lateral movement and functional dimerization of the

receptors (Bocharov et al. 2008, 2010; Volynsky et al. 2010). Interestingly the NMR structure of the dimeric transmembrane domain of EphA2 embedded into lipid bicelle shows a left-handed parallel packing of the transmembrane helices (Bocharov et al. 2010), whereas the same TMD motif of EphA1 favors right-handed helical packing (Bocharov et al. 2008).

How the Eph preceptor TMD mediates, if at all, the phosphorylation of the kinase domain has remained elusive. Studies on other RTKs have suggested two different models. According to the first one, not only the conformation of the ECD is different ("closed") in the absence of ligand, but also the TMDs do not interact productively through their dimerization motifs (Moriki et al. 2001; Schlessinger 2003). Ligand binding causes a rotation of the TMDs, positioning the catalytic domains in a signaling-competent orientation ("open"). For example, in the case of the insulin receptor that is constitutively dimeric, ligand binding would bring the TMDs closer together, allowing a more intimate contact of the catalytic subunits (Ottensmeyer et al. 2000). The other mechanism proposes that the receptor is in the "open" conformation even before ligand binding, but the TMD-mediated receptor dimers are not stable. The dimers would then be stabilized on ligand binding without undergoing major conformational changes (Cho et al. 2003; Li and Hristova 2006). The biological and biophysical studies discussed above suggest that the Eph receptors use the latter mechanism. Furthermore, their ligand-independent autophosphorylation in several cancers and the colocalization of mixed subclasses of Eph receptors in the same lipid rafts (Janes et al. 2011) point to the existence of signaling-competent assemblies even in the absence of ligand stimulation. On the other hand, full downstream signaling that follows full phosphorylation of the receptors requires ligand-induced formation of large-sized clusters (Janes et al. 2012).

Eph Kinase Domain Activation

Although the exact role of the ECD and TMD multimerization and clustering in signaling initiation is still not fully understood, the structural changes that the activation causes in the kinase domain (TKD) are well-studied and reviewed (Wybenga-Groot et al. 2001; Nowakowski et al. 2002; Hubbard 2004; Hubbard and Miller 2007; Lemmon and Schlessinger 2010). The general mechanisms and principals of TKD activation are also used by the Eph receptors. Catalytic activation involves auto-phosphorylation of two key tyrosines in the juxtamembrane segment (JMS) and one tyrosine in the activation loop at the active center of the kinase. In the unphosphorylated, autoinhibitory state, an α-helix in the EphB2 JMS suppresses the catalytic activity by an intimate association with the amino-terminal lobe of the kinase domain. Phosphorylation of the JMS tyrosines induces disorder in the JMS and dissociation from the kinase. This dissociation occurs without major structural changes but with partial ordering of the activation segment. An important Eph characteristic is that kinase phosphorylation also causes increased flexibility between the amino- and the carboxy-terminal TKD lobes (Wiesner et al. 2006), which is correlated with the kinase activity. Later studies on EphA3 have shown that the JMS also affects the activation segment through a separate pathway that includes a tyrosine residue in the active center and a serine in the amino terminus of the activation segment (Davis et al. 2008). These detailed structural studies have prompted efforts to discover structure-based second-generation kinase inhibitors (Choi et al. 2009).

Interestingly, the Eph receptors display some unique TKD phosphorylation and activation features. A recent study used semisynthetic (Singla et al. 2008) Eph receptors to document a sequential and ordered autophosphorylation process in which the carboxy-terminal JMS tyrosine is always phosphorylated first, followed by the amino-terminal JMS tyrosine, and, finally, by the activation loop tyrosine (Singla et al. 2011). This is in contrast with observations for most other receptor kinases in which the kinase activation loop is phosphorylated first (e.g., Furdui et al. 2006). The observed phosphorylation time-dependences coupled with site-directed mutagenesis further revealed that the Eph kinase activity directly correlates with the phosphory-

lation status of the juxtamembrane region, and in particular with the carboxy-terminal JMS tyrosine, and not that of the activation loop tyrosine. This is in contrast to the accepted view for most other characterized protein kinases.

Just as the phosphorylation of the tyrosine residues is important for activation of kinase receptors, dephosphorylation by phosphatases is crucial for their negative control. Protein tyrosine phosphatase (PTP) receptor type O (Ptpro) has been shown to specifically dephosphorylate both A- and B-type Eph receptors (Shintani et al. 2006). Ptpro dephosphorylates the same carboxy-terminal JMS tyrosine residue that is required for receptor activation and, hence, seems to regulate the threshold of the response of Eph receptors to ephrins. Another study suggests that PTP activity switches the response to ephrin from repulsion to adhesion and, thus, may play a role in the pathology of tumors (Wimmer-Kleikamp et al. 2008). Furthermore, it has been shown that there is a direct interaction between a PTP and EphA3 before ligand-stimulation (Nievergall et al. 2010). These studies are paving the way for understanding the precise roles of PTPs in regulating Eph signaling.

Termination of Eph Signaling

The balance between Eph kinase activation and signaling termination is relatively poorly understood. Because the main mechanism for Eph signaling termination is receptor internalization and degradation following ephrin binding and because the Eph/ephrin complexes are extremely stable once formed, mechanisms should exist that allow the endocytosis of Eph/ephrin complexes. Moreover because the signaling is most often repulsive, these mechanisms should allow for quick separation of the interacting cells without damaging the cell membranes at the contact regions. There are currently two models to explain how this occurs—ectodomain cleavage and "transendocytosis." According to the first, on Eph/ephrin complex formation, the Ephs, the ephrins, or both are cleaved from the membrane surface by proteases. Indeed it has been well documented that the GPI-anchored eph-

Cite this article as *Cold Spring Harb Perspect Biol* doi: 10.1101/cshperspect.a009142

rin-A ligands are cleaved off the cell membrane by ADAM10 (A disintegrin and metalloprotease). Originally, ADAM10 was shown to associate with ephrin-A2 *in cis*, with both proteins on the surface of the same cell, while EphA3 on the opposing cell was also required for efficient ephrin cleavage (Hattori et al. 2000). A later study, however, showed that ADAM10 is constitutively associated with EphA3 and ligand-binding repositions ADAM10 to activate the cleavage of ephrin-A5 in *trans* from the membrane of the opposing cell (Janes et al. 2005). The consequence in both cases is ephrin shedding on receptor binding allowing the opposing cells to detach and the Eph/ephrin complexes to be internalized in the Eph-expressing cell. It has also been suggested that activation of Eph receptors cause extension of their intracellular domains away from the cell membrane facilitating a direct physical association with ADAM10 and consequent ligand shedding (Janes et al. 2009).

The second model, transendocytosis, suggests that Eph/ephrin complexes are removed in both interacting cells via endocytic vesicles that are formed rapidly on cell–cell contact (Marston et al. 2003; Zimmer et al. 2003; Lauterbach and Klein 2006). Just as in ADAM-mediated cleavage, the intracellular domains of the proteins regulate these events, with carboxy-terminal Eph truncations being able to change forward to reverse or bidirectional endocytosis. Interestingly, the "transendocytosis" is currently the only internalization/degradation mechanism documented for B-type complexes.

Conclusion and Perspectives

Studies of the Tie-angiopoietin and Eph/ephrin signaling systems during the past two decades have provided significant insight into the activation and function of these vital signaling pathways in the healthy organism and in disease. Structural studies have visualized the initial ligand-receptor interaction events and have provided a foundation for the development of compounds targeting these interactions for the potential treatment of conditions as diverse as cancer, cardiovascular disease, neurological disorders, and spinal cord injury. What is not yet fully understood is how the initial ligand-receptor interactions are translated into kinase domain activation and initiation of downstream signaling. Toward that, novel approaches are being used for the production of full-length receptors for biochemical and structural studies in solution and in liposomes. In addition, high-resolution visualization techniques are being developed to monitor the ligand-receptor interactions and clustering at the surface of live cells, as well as the interactions of the receptors with co-receptors, regulatory molecules such as phosphatases and proteinases, and downstream effectors such as scaffold proteins. Because both Eph receptors and Tie2 require the formation of higher-order clusters as signaling centers, the interplay between kinase and phosphatase activity at these sites is particularly important for the regulation of signaling. Another area of ongoing interest is the regulation of receptor internalization and degradation and/or recycling, including potential receptor signaling from within the endocytic vesicles.

ACKNOWLEDGMENTS

We thank all members of our laboratories for their work on Eph receptors and Ties. This research was supported by grants from the National Institutes of Health 1RO1CA127501 to W.A.B. and 1RO1HL077249 and RO1NS038486 to D.B.N., as well as pilot project funding from the Massey Cancer Center and School of Medicine (VCU) to W.A.B.

REFERENCES

*Reference is also in this collection.

Abengozar MA, de Frutos S, Ferreiro S, Soriano J, Perez-Martinez M, Olmeda D, Marenchino M, Canamero M, Ortega S, Megias D, et al. 2012. Blocking ephrinB2 with highly specific antibodies inhibits angiogenesis, lymphangiogenesis, and tumor growth. *Blood* **119:** 4565–4576.

Adams RH, Alitalo K. 2007. Molecular regulation of angiogenesis and lymphangiogenesis. *Nat Rev* **8:** 464–478.

Artemenko EO, Egorova NS, Arseniev AS, Feofanov AV. 2008. Transmembrane domain of EphA1 receptor forms dimers in membrane-like environment. *Biochim Biophys Acta* **1778:** 2361–2367.

Barton WA, Tzvetkova D, Nikolov DB. 2005. Structure of the angiopoietin-2 receptor binding domain and identification of surfaces involved in Tie2 recognition. *Structure* **13:** 825–832.

Barton WA, Tzvetkova-Robev D, Miranda EP, Kolev MV, Rajashankar KR, Himanen JP, Nikolov DB. 2006. Crystal structures of the Tie2 receptor ectodomain and the angiopoietin-2-Tie2 complex. *Nat Struct Mol Biol* **13:** 524–532.

Bocharov EV, Mayzel ML, Volynsky PE, Goncharuk MV, Ermolyuk YS, Schulga AA, Artemenko EO, Efremov RG, Arseniev AS. 2008. Spatial structure and pH-dependent conformational diversity of dimeric transmembrane domain of the receptor tyrosine kinase EphA1. *J Biol Chem* **283:** 29385–29395.

Bocharov EV, Mayzel ML, Volynsky PE, Mineev KS, Tkach EN, Ermolyuk YS, Schulga AA, Efremov RG, Arseniev AS. 2010. Left-handed dimer of EphA2 transmembrane domain: Helix packing diversity among receptor tyrosine kinases. *Biophys J* **98:** 881–889.

Bocharov EV, Mineev KS, Goncharuk MV, Arseniev AS. 2012. Structural and thermodynamic insight into the process of "weak" dimerization of the ErbB4 transmembrane domain by solution NMR. *Biochim Biophys Acta* **1818:** 2158–2170.

Bowden TA, Aricescu AR, Nettleship JE, Siebold C, Rahman-Huq N, Owens RJ, Stuart DI, Jones EY. 2009. Structural plasticity of eph receptor A4 facilitates cross-class ephrin signaling. *Structure* **17:** 1386–1397.

Chen J, Zhuang G, Frieden L, Debinski W. 2008. Eph receptors and Ephrins in cancer: Common themes and controversies. *Cancer Res* **68:** 10031–10033.

Cho HS, Mason K, Ramyar KX, Stanley AM, Gabelli SB, Denney DW Jr, Leahy DJ. 2003. Structure of the extracellular region of HER2 alone and in complex with the Herceptin Fab. *Nature* **421:** 756–760.

Choi Y, Syeda F, Walker JR, Finerty PJ Jr, Cuerrier D, Wojciechowski A, Liu Q, Dhe-Paganon S, Gray NS. 2009. Discovery and structural analysis of Eph receptor tyrosine kinase inhibitors. *Bioorg Med Chem Lett* **19:** 4467–4470.

Chrencik JE, Brooun A, Recht MI, Kraus ML, Koolpe M, Kolatkar AR, Bruce RH, Martiny-Baron G, Widmer H, Pasquale EB, et al. 2006. Structure and thermodynamic characterization of the EphB4/Ephrin-B2 antagonist peptide complex reveals the determinants for receptor specificity. *Structure* **14:** 321–330.

Chrencik JE, Brooun A, Recht MI, Nicola G, Davis LK, Abagyan R, Widmer H, Pasquale EB, Kuhn P. 2007. Three-dimensional structure of the EphB2 receptor in complex with an antagonistic peptide reveals a novel mode of inhibition. *J Biol Chem* **282:** 36505–36513.

Cowan CA, Henkemeyer M. 2001. The SH2/SH3 adaptor Grb4 transduces B-ephrin reverse signals. *Nature* **413:** 174–179.

Cowan CA, Henkemeyer M. 2002. Ephrins in reverse, park and drive. *Trends Cell Biol* **12:** 339–346.

Davis S, Aldrich TH, Jones PF, Acheson A, Compton DL, Jain V, Ryan TE, Bruno J, Radziejewski C, Maisonpierre PC, et al. 1996. Isolation of angiopoietin-1, a ligand for the TIE2 receptor, by secretion-trap expression cloning. *Cell* **87:** 1161–1169.

Davis S, Papadopoulos N, Aldrich TH, Maisonpierre PC, Huang T, Kovac L, Xu A, Leidich R, Radziejewska E, Rafique A, et al. 2003. Angiopoietins have distinct modular domains essential for receptor binding, dimerization and superclustering. *Nat Struct Biol* **10:** 38–44.

Davis TL, Walker JR, Loppnau P, Butler-Cole C, Allali-Hassani A, Dhe-Paganon S. 2008. Autoregulation by the juxtamembrane region of the human ephrin receptor tyrosine kinase A3 (EphA3). *Structure* **16:** 873–884.

Day B, To C, Himanen JP, Smith FM, Nikolov DB, Boyd AW, Lackmann M. 2005. Three distinct molecular surfaces in ephrin-A5 are essential for a functional interaction with EphA3. *J Biol Chem* **280:** 26526–26532.

Fiedler U, Krissl T, Koidl S, Weiss C, Koblizek T, Deutsch U, Martiny-Baron G, Marme D, Augustin HG. 2003. Angiopoietin-1 and angiopoietin-2 share the same binding domains in the Tie-2 receptor involving the first Ig-like loop and the epidermal growth factor-like repeats. *J Biol Chem* **278:** 1721–1727.

Flanagan JG, Vanderhaeghen P. 1998. The ephrins and Eph receptors in neural development. *Annu Rev Neurosci* **21:** 309–345.

Furdui CM, Lew ED, Schlessinger J, Anderson KS. 2006. Autophosphorylation of FGFR1 kinase is mediated by a sequential and precisely ordered reaction. *Mol Cell* **21:** 711–717.

Giorgio C, Hassan Mohamed I, Flammini L, Barocelli E, Incerti M, Lodola A, Tognolini M. 2011. Lithocholic acid is an Eph-ephrin ligand interfering with Eph-kinase activation. *PloS ONE* **6:** e18128.

Hafner C, Schmitz G, Meyer S, Bataille F, Hau P, Langmann T, Dietmaier W, Landthaler M, Vogt T. 2004. Differential gene expression of Eph receptors and ephrins in benign human tissues and cancers. *Clin Chem* **50:** 490–499.

Hansen TM, Singh H, Tahir TA, Brindle NP. 2010. Effects of angiopoietins-1 and -2 on the receptor tyrosine kinase Tie2 are differentially regulated at the endothelial cell surface. *Cell Signal* **22:** 527–532.

Hattori M, Osterfield M, Flanagan JG. 2000. Regulated cleavage of a contact-mediated axon repellent. *Science* **289:** 1360–1365.

He L, Hoffmann AR, Serrano C, Hristova K, Wimley WC. 2011. High-throughput selection of transmembrane sequences that enhance receptor tyrosine kinase activation. *J Mol Biol* **412:** 43–54.

Himanen JP. 2012. Ectodomain structures of Eph receptors. *Semin Cell Dev Biol* **23:** 35–42.

Himanen JP, Nikolov DB. 2002. Purification, crystallization and preliminary characterization of an Eph-B2/ephrin-B2 complex. *Acta Crystallogr D Biol Crystallogr* **58:** 533–535.

Himanen JP, Rajashankar KR, Lackmann M, Cowan CA, Henkemeyer M, Nikolov DB. 2001. Crystal structure of an Eph receptor-ephrin complex. *Nature* **414:** 933–938.

Himanen JP, Saha N, Nikolov DB. 2007. Cell-cell signaling via Eph receptors and ephrins. *Curr Opin Cell Biol* **19:** 534–542.

Himanen JP, Goldgur Y, Miao H, Myshkin E, Guo H, Buck M, Nguyen M, Rajashankar KR, Wang B, Nikolov DB. 2009. Ligand recognition by A-class Eph receptors: Crystal structures of the EphA2 ligand-binding domain and the EphA2/ephrin-A1 complex. *EMBO Rep* **10:** 722–728.

Himanen JP, Yermekbayeva L, Janes PW, Walker JR, Xu K, Atapattu L, Rajashankar KR, Mensinga A, Lackmann M,

Nikolov DB, et al. 2010. Architecture of Eph receptor clusters. *Proc Natl Acad Sci* **107:** 10860–10865.

Huang L, Turck CW, Rao P, Peters KG. 1995. GRB2 and SH-PTP2: Potentially important endothelial signaling molecules downstream of the TEK/TIE2 receptor tyrosine kinase. *Oncogene* **11:** 2097–2103.

Huang H, Bhat A, Woodnutt G, Lappe R. 2010. Targeting the ANGPT-TIE2 pathway in malignancy. *Nat Rev Cancer* **10:** 575–585.

Hubbard SR. 2004. Juxtamembrane autoinhibition in receptor tyrosine kinases. *Nat Rev Mol Cell Biol* **5:** 464–471.

Hubbard SR, Miller WT. 2007. Receptor tyrosine kinases: mechanisms of activation and signaling. *Curr Opin Cell Biol* **19:** 117–123.

Janes PW, Saha N, Barton WA, Kolev MV, Wimmer-Kleikamp SH, Nievergall E, Blobel CP, Himanen JP, Lackmann M, Nikolov DB. 2005. Adam meets Eph: An ADAM substrate recognition module acts as a molecular switch for ephrin cleavage in trans. *Cell* **123:** 291–304.

Janes PW, Wimmer-Kleikamp SH, Frangakis AS, Treble K, Griesshaber B, Sabet O, Grabenbauer M, Ting AY, Saftig P, Bastiaens PI, et al. 2009. Cytoplasmic relaxation of active Eph controls ephrin shedding by ADAM10. *PLoS Biol* **7:** e1000215.

Janes PW, Griesshaber B, Atapattu L, Nievergall E, Hii LL, Mensinga A, Chheang C, Day BW, Boyd AW, Bastiaens PI, et al. 2011. Eph receptor function is modulated by heterooligomerization of A and B type Eph receptors. *J Cell Biol* **195:** 1033–1045.

Janes PW, Nievergall E, Lackmann M. 2012. Concepts and consequences of Eph receptor clustering. *Semin Cell Dev Biol* **23:** 43–50.

Jones N, Chen SH, Sturk C, Master Z, Tran J, Kerbel RS, Dumont DJ. 2003. A unique autophosphorylation site on Tie2/Tek mediates Dok-R phosphotyrosine binding domain binding and function. *Mol Cell Biol* **23:** 2658–2668.

Klein R. 2001. Excitatory Eph receptors and adhesive ephrin ligands. *Curr Opin Cell Biol* **13:** 196–203.

Klein R. 2009. Bidirectional modulation of synaptic functions by Eph/ephrin signaling. *Nat Neurosci* **12:** 15–20.

Kontos CD, Cha EH, York JD, Peters KG. 2002. The endothelial receptor tyrosine kinase Tie1 activates phosphatidylinositol 3-kinase and Akt to inhibit apoptosis. *Mol Cell Biol* **22:** 1704–1713.

Koolpe M, Dail M, Pasquale EB. 2002. An ephrin mimetic peptide that selectively targets the EphA2 receptor. *J Biol Chem* **277:** 46974–46979.

Labrador JP, Brambilla R, Klein R. 1997. The N-terminal globular domain of Eph receptors is sufficient for ligand binding and receptor signaling. *EMBO J* **16:** 3889–3897.

Lackmann M, Boyd AW. 2008. Eph, a protein family coming of age: more confusion, insight, or complexity? *Sci Signal* **1:** pre2.

Lamberto I, Qin H, Noberini R, Premkumar L, Bourgin C, Riedl SJ, Song J, Pasquale EB. 2012. Distinctive binding of three antagonistic peptides to the ephrin-binding pocket of the EphA4 receptor. *Biochem J* **445:** 47–56.

Lauterbach J, Klein R. 2006. Release of full-length EphB2 receptors from hippocampal neurons to cocultured glial cells. *J Neurosci* **26:** 11575–11581.

Lee HS, Daar IO. 2009. EphrinB reverse signaling in cell-cell adhesion: Is it just par for the course? *Cell Adh Migr* **3:** 250–255.

Lemmon MA, Schlessinger J. 2010. Cell signaling by receptor tyrosine kinases. *Cell* **141:** 1117–1134.

Li E, Hristova K. 2006. Role of receptor tyrosine kinase transmembrane domains in cell signaling and human pathologies. *Biochemistry* **45:** 6241–6251.

Li E, Hristova K. 2010. Receptor tyrosine kinase transmembrane domains: Function, dimer structure and dimerization energetics. *Cell Adh Migr* **4:** 249–254.

Li E, You M, Hristova K. 2005. Sodium dodecyl sulfate-polyacrylamide gel electrophoresis and forster resonance energy transfer suggest weak interactions between fibroblast growth factor receptor 3 (FGFR3) transmembrane domains in the absence of extracellular domains and ligands. *Biochemistry* **44:** 352–360.

* Lisabeth EM, Falivelli G, Pasquale EB. 2013. Eph receptor signaling and ephrins. *Cold Spring Harb Perspect Biol* **5:** a009159.

Macdonald PR, Progias P, Ciani B, Patel S, Mayer U, Steinmetz MO, Kammerer RA. 2006. Structure of the extracellular domain of Tie receptor tyrosine kinases and localization of the angiopoietin-binding epitope. *J Biol Chem* **281:** 28408–28414.

Maisonpierre PC, Suri C, Jones PF, Bartunkova S, Wiegand SJ, Radziejewski C, Compton D, McClain J, Aldrich TH, Papadopoulos N, et al. 1997. Angiopoietin-2, a natural antagonist for Tie2 that disrupts in vivo angiogenesis. *Science* **277:** 55–60.

Marler KJ, Becker-Barroso E, Martinez A, Llovera M, Wentzel C, Poopalasundaram S, Hindges R, Soriano E, Comella J, Drescher U. 2008. A TrkB/EphrinA interaction controls retinal axon branching and synaptogenesis. *J Neurosci* **28:** 12700–12712.

Marston DJ, Dickinson S, Nobes CD. 2003. Rac-dependent trans-endocytosis of ephrinBs regulates Eph-ephrin contact repulsion. *Nat Cell Biol* **5:** 879–888.

Master Z, Jones N, Tran J, Jones J, Kerbel RS, Dumont DJ. 2001. Dok-R plays a pivotal role in angiopoietin-1-dependent cell migration through recruitment and activation of Pak. *EMBO J* **20:** 5919–5928.

Mendrola JM, Berger MB, King MC, Lemmon MA. 2002. The single transmembrane domains of ErbB receptors self-associate in cell membranes. *J Biol Chem* **277:** 4704–4712.

Mitra S, Duggineni S, Koolpe M, Zhu X, Huang Z, Pasquale EB. 2010. Structure-activity relationship analysis of peptides targeting the EphA2 receptor. *Biochemistry* **49:** 6687–6695.

Moriki T, Maruyama H, Maruyama IN. 2001. Activation of preformed EGF receptor dimers by ligand-induced rotation of the transmembrane domain. *J Mol Biol* **311:** 1011–1026.

Nievergall E, Janes PW, Stegmayer C, Vail ME, Haj FG, Teng SW, Neel BG, Bastiaens PI, Lackmann M. 2010. PTP1B regulates Eph receptor function and trafficking. *J Cell Biol* **191:** 1189–1203.

Niu XL, Peters KG, Kontos CD. 2002. Deletion of the carboxyl terminus of Tie2 enhances kinase activity, signaling, and function. Evidence for an autoinhibitory mechanism. *J Biol Chem* **277:** 31768–31773.

Noberini R, Koolpe M, Peddibhotla S, Dahl R, Su Y, Cosford ND, Roth GP, Pasquale EB. 2008. Small molecules can selectively inhibit ephrin binding to the EphA4 and EphA2 receptors. *J Biol Chem* **283:** 29461–29472.

Noberini R, Koolpe M, Lamberto I, Pasquale EB. 2012a. Inhibition of Eph receptor-ephrin ligand interaction by tea polyphenols. *Pharmacol Rese* **66:** 363–373.

Noberini R, Lamberto I, Pasquale EB. 2012b. Targeting Eph receptors with peptides and small molecules: progress and challenges. *Semin Cell Dev Biol* **23:** 51–57.

Noblitt LW, Bangari DS, Shukla S, Mohammed S, Mittal SK. 2005. Immunocompetent mouse model of breast cancer for preclinical testing of EphA2-targeted therapy. *Cancer Gene Ther* **12:** 46–53.

Nowakowski J, Cronin CN, McRee DE, Knuth MW, Nelson CG, Pavletich NP, Rogers J, Sang BC, Scheibe DN, Swanson RV, et al. 2002. Structures of the cancer-related Aurora-A, FAK, and EphA2 protein kinases from nanovolume crystallography. *Structure* **10:** 1659–1667.

Ottensmeyer FP, Beniac DR, Luo RZ, Yip CC. 2000. Mechanism of transmembrane signaling: Insulin binding and the insulin receptor. *Biochemistry* **39:** 12103–12112.

Papapetropoulos A, Fulton D, Mahboubi K, Kalb RG, O'Connor DS, Li F, Altieri DC, Sessa WC. 2000. Angiopoietin-1 inhibits endothelial cell apoptosis via the Akt/survivin pathway. *J Biol Chem* **275:** 9102–9105.

Pasquale EB. 2008. Eph-ephrin bidirectional signaling in physiology and disease. *Cell* **133:** 38–52.

Pasquale EB. 2010. Eph receptors and ephrins in cancer: Bidirectional signalling and beyond. *Nat Rev Cancer* **10:** 165–180.

Petty A, Myshkin E, Qin H, Guo H, Miao H, Tochtrop GP, Hsieh JT, Page P, Liu L, Lindner DJ, et al. 2012. A small molecule agonist of EphA2 receptor tyrosine kinase inhibits tumor cell migration in vitro prostate cancer metastasis in vivo. *PloS ONE* **7:** e42120.

Qin H, Shi J, Noberini R, Pasquale EB, Song J. 2008. Crystal structure and NMR binding reveal that two small molecule antagonists target the high affinity ephrin-binding channel of the EphA4 receptor. *J Biol Chem* **283:** 29473–29484.

Qin H, Noberini R, Huan X, Shi J, Pasquale EB, Song J. 2010. Structural characterization of the EphA4-Ephrin-B2 complex reveals new features enabling Eph-ephrin binding promiscuity. *J Biol Chem* **285:** 644–654.

Qin H, Lim L, Song J. 2012. Protein dynamics at Eph receptor-ligand interfaces as revealed by crystallography, NMR and MD simulations. *BMC Biophys* **5:** 2.

Ramsauer M, D'Amore PA. 2002. Getting Tie(2)d up in angiogenesis. *J Clin Invest* **110:** 1615–1617.

Robinson D, He F, Pretlow T, Kung HJ. 1996. A tyrosine kinase profile of prostate carcinoma. *Proc Natl Acad Sci* **93:** 5958–5962.

Schlessinger J. 2003. Signal transduction. Autoinhibition control. *Science* **300:** 750–752.

Seegar TC, Eller B, Tzvetkova-Robev D, Kolev MV, Henderson SC, Nikolov DB, Barton WA. 2010. Tie1-Tie2 interactions mediate functional differences between angiopoietin ligands. *Mol Cell* **37:** 643–655.

Seiradake E, Harlos K, Sutton G, Aricescu AR, Jones EY. 2010. An extracellular steric seeding mechanism for Eph-ephrin signaling platform assembly. *Nat Struct Mol Biol* **17:** 398–402.

Shewchuk LM, Hassell AM, Ellis B, Holmes WD, Davis R, Horne EL, Kadwell SH, McKee DD, Moore JT. 2000. Structure of the Tie2 RTK domain: Self-inhibition by the nucleotide binding loop, activation loop, and C-terminal tail. *Structure* **8:** 1105–1113.

Shintani T, Ihara M, Sakuta H, Takahashi H, Watakabe I, Noda M. 2006. Eph receptors are negatively controlled by protein tyrosine phosphatase receptor type O. *Nat Neurosci* **9:** 761–769.

Singla N, Himanen JP, Muir TW, Nikolov DB. 2008. Toward the semisynthesis of multidomain transmembrane receptors: Modification of Eph tyrosine kinases. *Protein Sci* **17:** 1740–1747.

Singla N, Erdjument-Bromage H, Himanen JP, Muir TW, Nikolov DB. 2011. A semisynthetic Eph receptor tyrosine kinase provides insight into ligand-induced kinase activation. *Chem Biol* **18:** 361–371.

Sjoblom T, Jones S, Wood LD, Parsons DW, Lin J, Barber TD, Mandelker D, Leary RJ, Ptak J, Silliman N, et al. 2006. The consensus coding sequences of human breast and colorectal cancers. *Science* **314:** 268–274.

Song J, Vranken W, Xu P, Gingras R, Noyce RS, Yu Z, Shen SH, Ni F. 2002. Solution structure and backbone dynamics of the functional cytoplasmic subdomain of human ephrin B2, a cell-surface ligand with bidirectional signaling properties. *Biochemistry* **41:** 10942–10949.

Stamos J, Sliwkowski MX, Eigenbrot C. 2002. Structure of the epidermal growth factor receptor kinase domain alone and in complex with a 4-anilinoquinazoline inhibitor. *J Biol Chem* **77:** 46265–46272.

Sundberg EJ. 2009. Structural basis of antibody-antigen interactions. *Methods Mol Biol* **524:** 23–36.

Valenzuela DM, Griffiths JA, Rojas J, Aldrich TH, Jones PF, Zhou H, McClain J, Copeland NG, Gilbert DJ, Jenkins NA, et al. 1999. Angiopoietins 3 and 4: Diverging gene counterparts in mice and humans. *Proc Natl Acad Sci* **96:** 1904–1909.

Vearing C, Lee FT, Wimmer-Kleikamp S, Spirkoska V, To C, Stylianou C, Spanevello M, Brechbiel M, Boyd AW, Scott AM, et al. 2005. Concurrent binding of anti-EphA3 antibody and ephrin-A5 amplifies EphA3 signaling and downstream responses: Potential as EphA3-specific tumor-targeting reagents. *Cancer Res* **65:** 6745–6754.

Volynsky PE, Mineeva EA, Goncharuk MV, Ermolyuk YS, Arseniev AS, Efremov RG. 2010. Computer simulations and modeling-assisted ToxR screening in deciphering 3D structures of transmembrane alpha-helical dimers: Ephrin receptor A1. *Phys Biol* **7:** 16014.

Wang B. 2011. Cancer cells exploit the Eph-ephrin system to promote invasion and metastasis: Tales of unwitting partners. *Sci Signal* **4:** e28.

Ward NL, Dumont DJ. 2002. The angiopoietins and Tie2/Tek: Adding to the complexity of cardiovascular development. *Semin Cell Dev Biol* **13:** 19–27.

Wiesmann C, Muller YA, de Vos AM. 2000. Ligand-binding sites in Ig-like domains of receptor tyrosine kinases. *J Mol Med (Berl)* **78:** 247–260.

Wiesner S, Wybenga-Groot LE, Warner N, Lin H, Pawson T, Forman-Kay JD, Sicheri F. 2006. A change in conformational dynamics underlies the activation of Eph receptor tyrosine kinases. *EMBO J* **25:** 4686–4696.

Wimmer-Kleikamp SH, Janes PW, Squire A, Bastiaens PI, Lackmann M. 2004. Recruitment of Eph receptors into signaling clusters does not require ephrin contact. *J Cell Biol* **164:** 661–666.

Wimmer-Kleikamp SH, Nievergall E, Gegenbauer K, Adikari S, Mansour M, Yeadon T, Boyd AW, Patani NR, Lackmann M. 2008. Elevated protein tyrosine phosphatase activity provokes Eph/ephrin-facilitated adhesion of pre-B leukemia cells. *Blood* **112:** 721–732.

Wybenga-Groot LE, Baskin B, Ong SH, Tong J, Pawson T, Sicheri F. 2001. Structural basis for autoinhibition of the Ephb2 receptor tyrosine kinase by the unphosphorylated juxtamembrane region. *Cell* **106:** 745–757.

Yu X, Seegar TCM, Dalton AC, Tzvetkova-Robev D, Goldgur Y, Nikolov DB, Barton WA. 2013. Structural basis for angiopoietin-1 mediated signaling initiation. *Proc Natl Acad Sci* **110:** 7205–7210.

Zimmer M, Palmer A, Kohler J, Klein R. 2003. EphB-ephrinB bi-directional endocytosis terminates adhesion allowing contact mediated repulsion. *Nat Cell Biol* **5:** 869–878.

Eph Receptor Signaling and Ephrins

Erika M. Lisabeth[1], Giulia Falivelli[1,2], and Elena B. Pasquale[1,3]

[1]Cancer Center, Sanford-Burnham Medical Research Institute, La Jolla, California 92037

[2]Department of Pharmacology, University of Bologna, Bologna 40100, Italy

[3]Department of Pathology, University of California San Diego, La Jolla, California 92093

Correspondence: elenap@sanfordburnham.org

The Eph receptors are the largest of the RTK families. Like other RTKs, they transduce signals from the cell exterior to the interior through ligand-induced activation of their kinase domain. However, the Eph receptors also have distinctive features. Instead of binding soluble ligands, they generally mediate contact-dependent cell–cell communication by interacting with surface-associated ligands—the ephrins—on neighboring cells. Eph receptor–ephrin complexes emanate bidirectional signals that affect both receptor- and ephrin-expressing cells. Intriguingly, ephrins can also attenuate signaling by Eph receptors coexpressed in the same cell. Additionally, Eph receptors can modulate cell behavior independently of ephrin binding and kinase activity. The Eph/ephrin system regulates many developmental processes and adult tissue homeostasis. Its abnormal function has been implicated in various diseases, including cancer. Thus, Eph receptors represent promising therapeutic targets. However, more research is needed to better understand the many aspects of their complex biology that remain mysterious.

The Eph receptors have the prototypical RTK topology, with a multidomain extracellular region that includes the ephrin ligand-binding domain, a single transmembrane segment, and a cytoplasmic region that contains the kinase domain (Fig. 1). There are nine EphA receptors in the human genome, which promiscuously bind five ephrin-A ligands and five EphB receptors, which promiscuously bind three ephrin-B ligands (Pasquale 2004, 2005). Additionally, EphA4 and EphB2 can also bind ephrins of a different class. Two members of the family, EphA10 and EphB6, have modifications in conserved regions of their kinase domains that prevent kinase activity. Furthermore, a variety of alternatively spliced forms identified for many Eph receptors differ from the prototypical structure and have distinctive functions (Zisch and Pasquale 1997; Pasquale 2010).

Both ephrin classes include a conserved Eph receptor-binding domain, which is connected to the plasma membrane by a linker segment whose length can be affected by alternative splicing (Fig. 1). The ephrin-As are attached to the cell surface by a glycosylphosphatidylinositol (GPI) anchor, although they can also be released to activate EphA receptors at a distance (Bartley et al. 1994; Wykosky et al. 2008), whereas the

Cite this article as *Cold Spring Harb Perspect Biol* doi: 10.1101/cshperspect.a009159

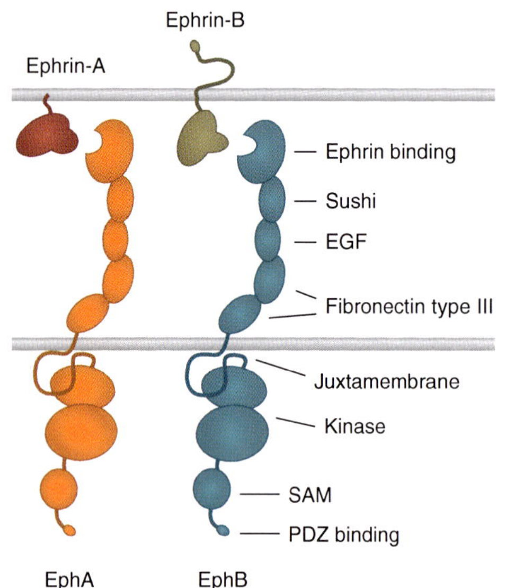

Figure 1. Domain structure of Eph receptors and ephrins.

ephrin-Bs contain a transmembrane segment and a short cytoplasmic region. Ephrin-A3 and ephrin-B3 also bind heparan sulfate proteoglycans through an interaction that involves their extracellular linker region and that, at least in the case of ephrin-A3, potentiates EphA receptor activation and signaling (Irie et al. 2008; Holen et al. 2011).

The Eph receptor family has greatly expanded during evolution, and includes almost one fourth of the 58 human RTKs (Schlessinger and Lemmon 2013). A large number of Eph receptors and ephrins may be required to achieve and maintain the sophisticated tissue organization of higher organisms. Indeed, many are highly expressed in the most complex organ, the brain, particularly during the establishment of its complex architecture and intricate wiring of neuronal connections (Yamaguchi and Pasquale 2004). Besides the brain, Eph receptors and ephrins are also present in most—if not all—other tissues, often in a combinatorial manner and with dynamically changing expression patterns (Pasquale 2005). In some regions, Eph receptors and ephrins are both coexpressed in the same cells, in others they have mutually exclusive expression patterns or they

can be expressed in complementary gradients. These situations likely reflect different signaling modalities with different biological outcomes.

Eph receptors and ephrins engage in a multitude of activities. They typically mediate contact-dependent communication between cells of the same or different types to control cell morphology, adhesion, movement, proliferation, survival, and differentiation (Pasquale 2005). Through these activities, during development, the Eph/ephrin system plays a role in the spatial organization of different cell populations, axon guidance, formation of synaptic connections between neurons, and blood vessel remodeling. In the adult, the Eph/ephrin system regulates remodeling of synapses, epithelial differentiation and integrity, bone remodeling, immune function, insulin secretion, and stem cell self-renewal (Pasquale 2008; Genander and Frisen 2010). In addition, Eph receptors and ephrins are often up-regulated in injured tissues, where they inhibit some regenerative processes but promote angiogenesis, as well as in cancer cells, where they seem to be able to both promote and suppress tumorigenicity (Du et al. 2007; Pasquale 2008, 2010).

Here we provide an overview of Eph receptor and ephrin signaling mechanisms and biological effects, with an emphasis on recent findings. More detailed information on specific aspects of Eph receptor/ephrin biology and downstream signaling networks can be found in other recent reviews (Pasquale 2005, 2008, 2010; Arvanitis and Davy 2008; Lackmann and Boyd 2008; Klein 2009; Genander and Frisen 2010).

EPH RECEPTOR FORWARD SIGNALING

"Forward" signaling corresponds to the prototypical RTK mode of signaling, which is triggered by ligand binding and involves activation of the kinase domain. However, the activation mechanisms of Eph receptors have unique features as compared to other RTK families (Barton et al. 2013). Binding between Eph receptors and ephrins on juxtaposed cell surfaces leads to oligomerization through not only Eph receptor–ephrin interfaces but also receptor–receptor *cis* interfaces located in multiple domains

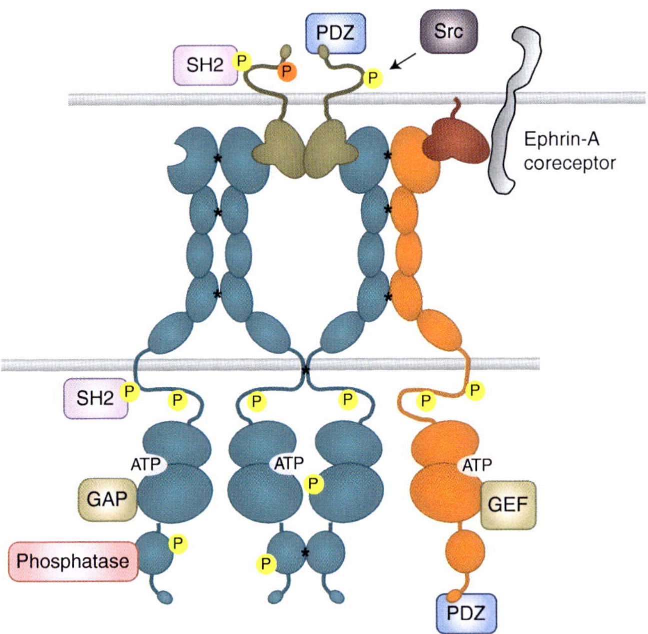

Figure 2. Eph receptor clustering and bidirectional signaling. SH2 and PDZ indicate proteins containing these domains. All types of signaling proteins shown can associate with both EphA and EphB receptors. Asterisks indicate receptor–receptor interactions favoring clustering; yellow circles indicate tyrosine phosphorylation and the orange circle indicates serine phosphorylation.

(Fig. 2) (Himanen et al. 2010; Seiradake et al. 2010). In fact, Eph receptor clusters induced by ephrin binding can enlarge to incorporate Eph receptors that are not bound to ephrins (Wimmer-Kleikamp et al. 2004). The cellular context can also affect Eph receptor clustering ability, which depends on association with the actin cytoskeleton (Salaita and Groves 2010). Given the promiscuity of Eph receptor–ephrin interactions, and also receptor–receptor *cis* interactions, the clusters can include Eph receptors of both A and B classes (Janes et al. 2011).

The proximity of clustered Eph receptor molecules leads to *trans*-phosphorylation. Phosphorylation of two conserved tyrosines in the juxtamembrane domain relieves inhibitory intramolecular interactions with the kinase domain, enabling efficient kinase activity (Binns et al. 2000; Zisch et al. 2000; Wybenga-Groot et al. 2001). Phosphorylation of the conserved tyrosine in the activation loop appears to be less critical for Eph receptor activation than it is for many other RTKs, although it may be important

for maximal activity (Binns et al. 2000; Singla et al. 2011). There are also differences in the kinase domains within the Eph receptor family. For example, the "gatekeeper" residue in the hinge region between the kinase domain lobes, which controls access to a hydrophobic pocket adjacent to the ATP-binding site, is a threonine in most Eph receptors but a valine in EphA6 and an isoleucine in EphA7. Hence, EphA6 and EphA7 likely differ from the other Eph receptors in their sensitivity to kinase inhibitors and possibly substrate specificity (Skaggs et al. 2006; Zhang et al. 2009a).

The Eph receptors modulate many of the same networks of adaptor and effector proteins that also function downstream of other RTK families (Wagner et al. 2013). Various tyrosine autophosphorylation sites in activated Eph receptors—including the two regulatory juxtamembrane sites—enable recruitment of downstream signaling proteins that contain SH2 domains, including nonreceptor tyrosine kinases of the Src and Abl families and adaptors such as Nck and Crk, which are crucial for signal transduction (Fig. 2) (Jorgensen et al. 2009; Pasquale 2010). Binding of PDZ domain-containing proteins to the carboxy-terminal tails of Eph receptors also contributes to signaling. Particularly important effectors are Rho and Ras family GTPases and Akt/mTORC1. Interestingly, whereas most other RTK families use these central regulators of cellular physiology to stimulate cell proliferation, survival, and forward movement, the Eph receptors can use them to inhibit cell growth and achieve cell repulsion. In cancer cells, this can result in tumor suppression.

Signaling by the Eph receptors, however, is not always consistent and can lead to divergent outcomes. The kinase-inactive Eph receptors and alternatively spliced forms lacking the kinase domain can modulate signaling outcome by reducing signal strength in the clusters as well as by contributing distinctive signals. For example, the kinase inactive EphB6 can be phosphorylated by other Eph receptors and subvert the effects of EphB4 in breast cancer cells (Truitt and Freywald 2011). Moreover, a truncated membrane-anchored form of the EphA7 extracellular domain can convert repulsion to cell–cell adhesion in the developing neural tube by decreasing signaling by full-length EphA7 (Holmberg et al. 2000), and a secreted truncated form of EphA7 acts as a tumor suppressor in follicular lymphoma by binding EphA2 and blocking its oncogenic signals (Oricchio et al. 2011). There is also evidence that small and large Eph receptor clusters differ in their ability to recruit certain signaling molecules (Salaita and Groves 2010). Other aspects of the cellular context, and implementation of positive and negative feedback loops, further contribute to the diversity of Eph receptor activities.

Rho Family GTPases

The Eph receptors are well known for their effects on the actin cytoskeleton, which impact cell shape, adhesion, and movement through regulation of the Rho GTPase family, including RhoA, Rac1, and Cdc42 (Pasquale 2008, 2010). GTPases cycle between a GDP bound (inactive) state and a GTP bound (active) state that binds downstream effectors. The Eph receptors can influence these conversions by regulating both guanine nucleotide exchange factors (GEFs, which facilitate GDP to GTP exchange) and GTPase-activating proteins (GAPs, which promote GTP hydrolysis to GDP). Regulation of GEFs and GAPs by Eph receptors can involve constitutive or ephrin-induced association, tyrosine phosphorylation, or even ubiquitination and degradation.

RhoA is mostly involved in the formation of stress fibers and focal adhesions as well as contraction of the actomyosin cytoskeleton, whereas Rac1 and Cdc42 drive the formation of protrusive structures such as lamellipodia and filopodia, respectively (Heasman and Ridley 2008). An increased balance of RhoA versus Rac1/Cdc42 activities has been implicated in the characteristic repulsive effects of Eph receptor forward signaling, including process retraction and inhibition of cell migration/invasiveness (Fig. 3A–C). The collapse or local retraction of neuronal growth cones and dendritic spines (the small protrusions on dendrites bearing excitatory synapses) are well-known repulsive effects of EphA receptors that depend on Rho family GTPases (Fig. 3B,C) (Wahl et al. 2000; Murai et al. 2003; Fu et al. 2007). Growth cone collapse involves RhoA activation, for example, by the GEF Ephexin1 (Shamah et al. 2001; Sahin et al. 2005), and Rac1 inactivation, for example, by the GAP α2-Chimaerin (Beg et al. 2007; Iwasato et al. 2007; Shi et al. 2007; Wegmeyer et al. 2007). However Rac1 activation, which can occur downstream of Vav family GEFs, is also required for growth cone collapse and to process retraction by enabling endocytic removal of adhesive Eph receptor–ephrin complexes from sites of cell–cell contact (Cowan et al. 2005; Yoo et al. 2011). Activation and inactivation of Rho family GTPases may occur with different spatial and/or temporal resolution to achieve growth cone collapse and regulate dendritic spines. In other cell types, Eph repulsive signaling involving Rho family GTPases can lead to mesodermal–ectodermal tissue separation during gastrulation (Park et al. 2011; Rohani et al. 2011), Schwann cell-astrocyte segregation in the in-

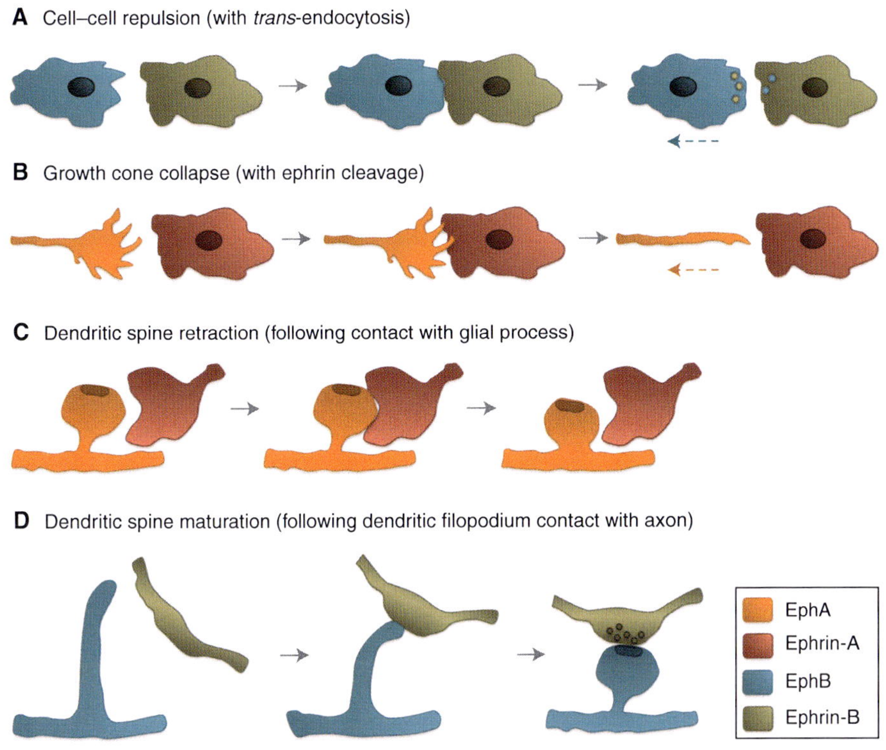

Figure 3. Eph receptor–ephrin repulsive effects and dendritic spine maturation. (*A*) An EphB-expressing cell encounters an ephrin-B-expressing cell and retracts after the internalization of EphB–ephrin-B complexes enables cell separation. (*B*) An EphA-expressing growth cone at the leading edge of an axon encounters an ephrin-A-expressing cell, collapses, and begins to retract after the cleavage of ephrin-A molecules enables cell separation. (*C*) An EphA-expressing spine on a dendrite (bearing an excitatory postsynaptic terminal represented as a darker oval) comes in contact with an ephrin-A-expressing glial process and retracts becoming shorter. (*D*) An EphB-expressing filopodial protrusion on a dendrite acquires an enlarged "head" and shortens following contact with an ephrin-B-expressing axon. The presynaptic terminal also matures following contact.

jured nervous system (Afshari et al. 2010), and contact inhibition of locomotion (Astin et al. 2010).

Regulation of Rho family GTPases by Eph receptors can also control cellular processes beyond repulsion. For example, the maturation of neuronal filopodial protrusions into dendritic spines (Fig. 3D) entails ephrin-B/EphB-dependent activation of the Rac-GEFs Kalirin and Tiam1 and the Cdc42-GEF Intersectin to promote the formation of branched actin filaments that enlarge the distal portion of the filopodial protrusions (Irie and Yamaguchi 2002; Penzes et al. 2003; Tolias et al. 2007), whereas RhoA activation through focal adhesion kinase and a RhoGEF shortens the protrusions (Moeller et al. 2006). Interestingly, EphB receptor forward signaling can also promote synapse formation through ubiquitination and degradation of the Rho-GEF Ephexin5, which decreases RhoA activity without obvious effects on spine morphology (Margolis et al. 2010). Furthermore, EphA2 forward signaling can promote endothelial angiogenic responses by activating Rac1 through Vav family GEFs (Hunter et al. 2006) and enhance epithelial characteristics by inhibiting RhoA through p190RhoGAP (Wakayama et al. 2011). In addition, EphB receptor activation by ephrin-B-expressing stromal cells promotes HGF-dependent invasiveness of metastatic PC3 prostate cancer cells through sustained Cdc42 activation (Astin et al. 2010).

Further work is needed to understand in detail the mechanisms leading to activation versus inhibition of Rho GTPases by Eph receptors and examine the role of the many less well-characterized Rho family members in the biological activities of Eph receptors.

Ras Family GTPases

Perhaps the most prototypical RTK signaling pathway involves activation of the H-Ras GTPase by the GEF Sos, which is recruited by the adaptors Shc and/or Grb2 bound to activated RTKs (McKay and Morrison 2007). H-Ras-GTP triggers a phosphorylation cascade that culminates in activation of the Erk1 and Erk2 serine/threonine kinases. Through phosphorylation of cytoplasmic effectors and nuclear transcription factors, the Ras-Erk pathway regulates many physiological processes—including cell proliferation, survival, differentiation, adhesion, and migration—and its deregulation can cause cancer and other diseases.

Remarkably, Eph receptor forward signaling frequently inhibits the Ras-Erk pathway and can override its activation by other RTKs (Pasquale 2008, 2010). For example, polarized Eph receptor activation in progenitor cells of the ascidian embryo attenuates Erk activation by the FGF RTK, leading to asymmetric division and fate specification (Picco et al. 2007; Shi and Levine 2008). Furthermore, ephrin-A/EphA signaling induced by contact between myoblasts suppresses Erk activation by the IGF-1 RTK, facilitating myogenic differentiation (Minami et al. 2011). In neurons, EphA-dependent Erk inhibition suppresses the effects of the TrkB RTK on growth cone motility and gene expression (Meier et al. 2011) and promotes growth cone collapse (Nie et al. 2010). In cancer cells, ephrin-A/EphA signals that suppress Erk activation by RTKs can inhibit tumorigenicity (Miao et al. 2001; Macrae et al. 2005).

A common mechanism of Eph receptor-dependent Erk inhibition is through p120Ras-GAP, which inactivates H-Ras (Elowe et al. 2001; Minami et al. 2011). Through p120RasGAP, the Eph receptors can also inhibit another Ras family GTPase, R-Ras, causing the reduced integrin

activity that is important for retraction of cell processes and decreased malignancy (Dail et al. 2006). Eph receptors can also negatively regulate Rap1, another member of the Ras family involved in integrin activation, by inhibiting the GEF C3G or activating the GAP SPAR (Riedl et al. 2005; Richter et al. 2007; Huang et al. 2008; Pasquale 2008).

In some cases, however, Eph receptors behave similarly to other RTKs and activate the Ras-Erk pathway. For example, in cultured mouse mesenchymal cells, ephrin-B1-EphB signaling activates Erk to promote proliferation and regulate immediate early gene transcription (Bush and Soriano 2010). In P19 embryonal carcinoma cells and microvascular endothelial cells, ephrin-stimulated EphB1 recruits the adaptors Shc and Grb2 to activate H-Ras and increase cell migration (Vindis et al. 2003). Interestingly, the activation of EphB4 by ephrin-B2 in MCF7 breast cancer cells promotes Erk1/2 activation through an unusual pathway that seems to require the PP2A serine/threonine phosphatase (Xiao et al. 2012). In stably transfected HEK293 cells, EphB2 forward signaling activates Erk to promote cell repulsion (Poliakov et al. 2008). The interplay between Eph receptors and Ras GTPases also involves feedback loops in which Ras-Erk signaling reciprocally influences Eph receptors, for example, by reinforcing ephrin-B1/EphB2 signaling or up-regulating EphA2 gene transcription (Menges and McCance 2008; Poliakov et al. 2008).

Akt

Akt is a serine/threonine kinase that regulates cell size, proliferation, and survival through various downstream effectors such as mTOR complex 1 (mTORC1). RTKs typically activate Akt through PI3 kinase, a lipid kinase that initiates a pathway leading to Akt activation through phosphorylation on T308 and S473 (Manning and Cantley 2007). In contrast, Eph receptor forward signaling can suppress Akt activation. For instance, in a variety of cancer cells, ephrin-dependent EphA2 activation leads to rapid dephosphorylation of Akt T308 and S473, which likely depends on regulation of a phosphatase, leading to mTORC1 inactivation and decreased cell growth and migration (Menges and McCance 2008; Miao et al. 2009; Yang et al. 2011). Remarkably, this can occur even in cancer cells where the PI3 kinase-Akt pathway is activated by oncogenic mutations. EphB3 kinase activation can also inhibit Akt, which leads to suppression of non-small-cell lung cancer migration and metastasis, by promoting the assembly of a complex involving the EphB3-binding partner RACK1 (receptor for activated C-kinase 1), the serine/threonine phosphatase PP2A and Akt itself (Li et al. 2012). However, Eph receptors can also activate Akt, for example, in pancreatic cancer cells stimulated with ephrin-A1 (Chang et al. 2008) or in malignant T lymphocytes where ephrin-B treatment suppresses apoptosis (Maddigan et al. 2011).

Akt signaling can reciprocally influence Eph receptors through feedback loops. For example, phospho-RTK arrays suggest an up-regulation of several tyrosine-phosphorylated Eph receptors in cancer cells treated with Akt inhibitors (Chandarlapaty et al. 2011). Furthermore, Akt can phosphorylate EphA2 on S897 drastically altering receptor function (see below), whereas ephrin-A1 stimulation causes loss of S897 phosphorylation (Miao et al. 2009).

EPHRIN REVERSE SIGNALING

Besides forward signaling, the Eph receptors can also stimulate "reverse" signaling in the ephrin-expressing cells (Fig. 2) (Pasquale 2005, 2010). A central feature enabling signaling by the ephrins, which lack an enzymatic domain, is the activation of Src family kinases. Eph receptor binding causes ephrin-B phosphorylation by Src kinases, creating binding sites for the SH2 domains of signaling proteins such as the adaptor Grb4 (Cowan and Henkemeyer 2001; Palmer et al. 2002). Ephrin-B signaling through Grb4 controls axon pruning, synapse formation and dendritic spine morphogenesis in the developing mouse hippocampus (Segura et al. 2007; Xu and Henkemeyer 2009). Phosphorylation of a serine near the ephrin-B carboxyl terminus, which is also induced by EphB receptor binding, leads to stabilization of AMPA neurotransmit-

ter receptors at synapses (Essmann et al. 2008). This might regulate synaptic plasticity in concert with ephrin-B tyrosine phosphorylation (Bouzioukh et al. 2007).

Recruitment of signaling proteins containing PDZ domains to the ephrin-B carboxyl terminus is also crucial for reverse signaling. For example, the adaptor PDZ-RGS3 connects ephrin-B to G-protein-coupled receptors that control neuronal cell migration and neural progenitor self-renewal (Lu et al. 2001; Qiu et al. 2010). Ephrin-B interaction with PDZ domain proteins also promotes angiogenesis and lymphangiogensis by enabling VEGF receptor endocytosis, and can regulate axon guidance and synaptic plasticity (Makinen et al. 2005; Bouzioukh et al. 2007; Bush and Soriano 2009; Sawamiphak et al. 2010; Wang et al. 2010). Furthermore, ephrin-B signaling controls neuronal migration in the developing mouse brain through cross talk with the secreted glycoprotein Reelin (Senturk et al. 2011), modulates epithelial cell–cell junctions through the Par polarity complex (Lee et al. 2008), disrupts gap junctional communication (Mellitzer et al. 1999; Davy et al. 2006), and enhances glioma cell invasiveness by activating Rac1 (Nakada et al. 2006).

The ephrin-As lack a cytoplasmic domain and it is not well understood how they activate intracellular signaling. Studies in neurons have implicated the p75 neurotrophin receptor and the TrkB and Ret RTKs as transmembrane-binding partners that enable ephrin-A-dependent reverse signals involved in axon guidance and branching (Fig. 2) (Lim et al. 2008; Marler et al. 2008, 2010; Bonanomi et al. 2012). Through these and likely other binding partners, the ephrin-As have diverse signaling activities. Ephrin-A2 reverse signaling can inhibit neural progenitor cell proliferation, perhaps opposing the positive effects of ephrin-B1 (Holmberg et al. 2005). In the adult hippocampus, glial ephrin-A3 functions together with neuronal EphA4 to modulate uptake of the neurotransmitter glutamate by glial cells and, thus, synaptic plasticity (Carmona et al. 2009; Filosa et al. 2009). Ephrin-A4 can inhibit apoptotic cell death in Jurkat immune cells by activating Src family kinases and Akt (Holen et al. 2008).

Ephrin-A5 reverse signaling in pancreatic β cells can stimulate Rac1 activity, which is necessary for insulin secretion after glucose stimulation (Konstantinova et al. 2007). Ephrin-A5 can also increase cell-substrate adhesion in fibroblasts and astrocytes by activating the Src family kinase Fyn and integrins, and seems able to also promote invasiveness (Davy et al. 1999, 2000; Campbell et al. 2006). Furthermore, this ephrin promotes Fyn activation in glioma and HEK293 cells, leading to Cbl-dependent EGF RTK ubiquitination and degradation (Li et al. 2009). Interestingly, Fyn can in turn function in a negative feedback loop to down-regulate cell surface ephrin-A levels by modulating the metabolism of sphingomyelin (Baba et al. 2009).

BEYOND BIDIRECTIONAL SIGNALING

Internalization and Proteolytic Cleavage

Following ligand-dependent activation, RTKs are typically internalized by endocytosis and can continue to signal from intracellular compartments until they are inactivated by dephosphorylation and degradation or traffic back to the cell surface (Goh and Sorkin 2013). For the Eph receptors, this process has unique features as a result of the plasma membrane association of the ephrins (Marston et al. 2003; Zimmer et al. 2003; Pitulescu and Adams 2010). Eph receptor–ephrin complexes can be internalized into either the Eph receptor- or the ephrin-expressing cells through the formation of vesicles containing plasma membrane fragments derived from both cells (Fig. 3A). This Rac1-dependent process, which has been defined "*trans*-endocytosis," is critical for removal of adhesive complexes from cell–cell contact sites to allow cell separation and repulsive effects. Another protein that contributes to Eph receptor internalization and degradation is the ubiquitin ligase Cbl, which can interact with several Eph receptors promoting their ubiquitination (Walker-Daniels et al. 2002; Fasen et al. 2008).

Besides *trans*-endocytosis, Eph receptor–ephrin complexes can convert adhesive interactions into cell repulsion by activating metalloproteases, such as ADAM family members. For example, the transmembrane ADAM10 protease can associate with ephrin-A2 on the same cell surface and cleave it following EphA receptor binding in *trans* to enable repulsive axon guidance (Hattori et al. 2000). ADAM10 can also associate with EphA3, whose active conformation promotes protease activity toward the ephrin in *trans* (Fig. 4A) (Janes et al. 2005, 2009). EphB receptors also interact with ADAM10, as well as the cell-adhesion molecule E-cadherin, and their binding to ephrin-Bs in *trans* provokes shedding of E-cadherin by ADAM10 preferentially in the ephrin-B-expressing cells (Fig. 4A) (Solanas et al. 2011). Cleavage by metalloproteases also plays a role in other Eph receptor/ephrin activities. For example, ephrin-B cleavage by ADAM13 can terminate EphB/ephrin-B signals that inhibit canonical Wnt signaling in the *Xenopus* embryo, thus enabling cranial neural crest induction (Wei et al. 2010). Furthermore, ADAM19 functions independently of its protease activity to stabilize developing neuromuscular junctions by preventing internalization of the complexes between ephrin-A5 on the muscle and EphA4 on the innervating motor neuron (Yumoto et al. 2008).

Ephrin binding and other stimuli can also induce cleavage of the Eph receptor extracellular domain, followed by further intramembrane proteolytic processing via γ-secretase to generate cytoplasmic fragments capable of signaling (Litterst et al. 2007; Inoue et al. 2009; Xu et al. 2009). For example, calcium influx can induce combined metalloprotease/γ-secretase processing of both EphA4 and EphB2 (Litterst et al. 2007; Inoue et al. 2009). The released EphA4 cytoplasmic fragment increases dendritic spine numbers through kinase-independent Rac1 activation (Inoue et al. 2009). Instead, the EphB2 cytoplasmic fragment can phosphorylate NMDA neurotransmitter receptors, which promotes their cell surface localization and may lead to a positive feedback loop by increasing NMDA receptor-mediated calcium currents (Litterst et al. 2007; Xu et al. 2009). Interestingly, stress in mice can also cause cleavage of EphB2 by the extracellular serine protease neuropsin in the amygdala (Attwood et al. 2011). This cleavage results in EphB2 dissociation from the

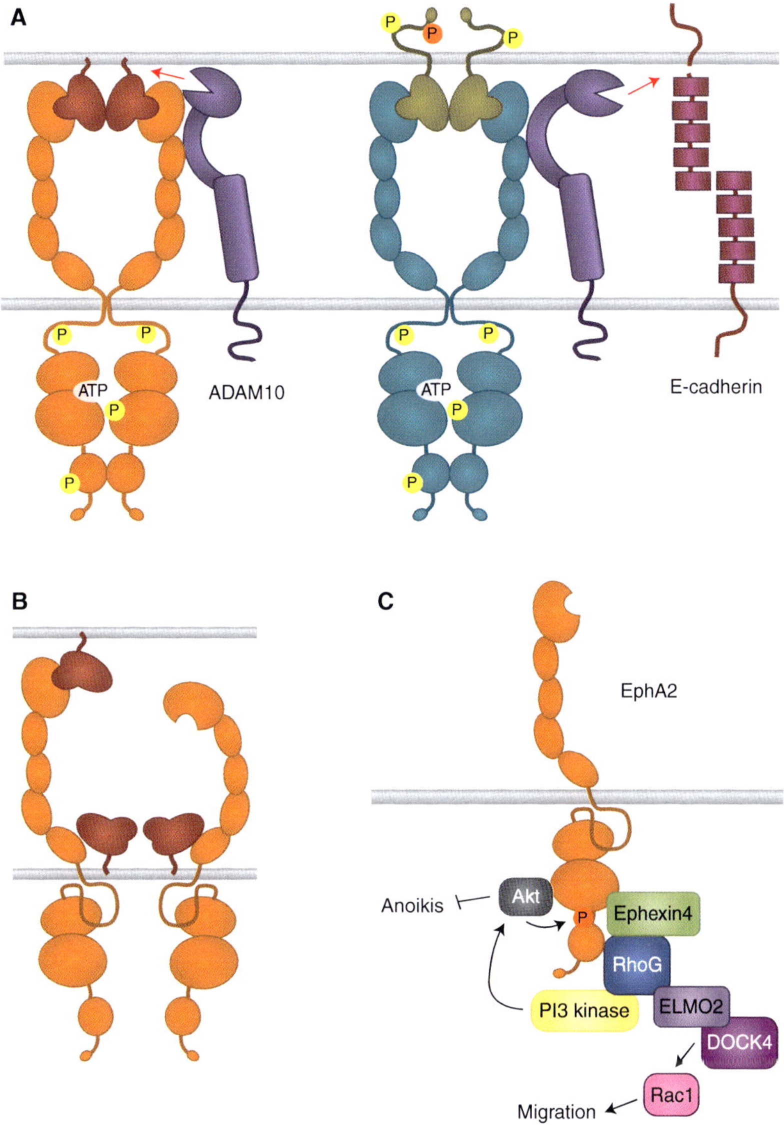

Figure 4. Eph receptor/ephrin signaling activities beyond bidirectional signaling. (*A*) Roles of the ADAM10 metalloprotease in Eph receptor/ephrin signaling. (*B*) Inhibition of Eph receptor forward signaling by *cis* interaction with ephrins. (*C*) Ephrin-independent Eph receptor signaling.

NMDA receptor as well as enhances NMDA receptor ion currents and the behavioral signatures of anxiety. Whether this may also be a consequence of NMDA receptor phosphorylation by a proteolytically released EphB2 cytoplasmic fragment remains to be determined.

Ephrin-B ligands can also undergo metalloprotease/γ-secretase processing following binding to EphB receptors. The released ephrin-B2 cytoplasmic fragment can promote Src activation and Src-dependent phosphorylation of uncleaved ephrin-B2, which is important for reverse signaling (Georgakopoulos et al. 2006). This involves regulating the interplay between Src- and the Csk-binding protein Cbp/PAG, an adaptor that controls Src activity (Georgakopoulos et al. 2011). Moreover, ephrin-B1 cytoplasmic fragments present in the developing mouse brain can associate with the ZHX2 transcriptional repressor and enhance its activity in the nucleus to prevent neural progenitor differentiation (Wu et al. 2009).

Ephrin-Mediated *Cis* Attenuation of Eph Receptor Forward Signaling

Eph receptors and ephrins can be coexpressed in normal and cancer cells (Carvalho et al. 2006; Pasquale 2010; Kao and Kania 2011). In contrast to the autocrine signaling occurring when other RTKs and their soluble ligands are coexpressed (Zwick et al. 2002), a lateral *cis* interaction between Eph receptors and ephrins on the same cell surface can attenuate forward signaling (Fig. 4B) (Bohme et al. 1996; Yin et al. 2004). For example, EphA *cis* attenuation plays a role in topographic mapping of retinal axons (Hornberger et al. 1999; Carvalho et al. 2006) and ephrin-B3 inhibits signaling by EphB2 coexpressed in hippocampal synapses, decreasing tyrosine phosphorylation of NMDA neurotransmitter receptors (Antion et al. 2010). Ephrins also cause *cis* attenuation of EphA and EphB signaling in spinal cord motor neuron populations where they are highly expressed, which is important for proper axon guidance in the limb (Kao and Kania 2011). In contrast, in motor neuron populations where they are present at lower levels, ephrin-As segregate in different membrane microdomains than the coexpressed EphA receptors (Marquardt et al. 2005; Kao and Kania 2011). This segregation allows parallel activation of forward and reverse signaling in the same neurons. Biochemical and structural studies have implicated the second Eph receptor fibronectin type III domain in the *cis* interaction (Fig. 4B) (Carvalho et al. 2006; Seiradake et al. 2010). Consistent with this, an ephrin-A5 mutant that cannot engage the EphA ephrin-binding pocket was shown to still induce *cis* attenuation (Bohme et al. 1996; Carvalho et al. 2006; Kao and Kania 2011). However, how *cis* binding inhibits forward signaling remains unclear. Whereas a mechanism involving decreased Eph receptor cell surface localization seems unlikely (Yin et al. 2004; Carvalho et al. 2006), the association with coexpressed ephrins might induce Eph receptor translocation to an environment rich in phosphotyrosine phosphatases or sterically inhibit Eph receptor clustering, which is necessary for activation. Through *cis* attenuation of Eph receptor forward signaling, coexpressed ephrins can fine-tune the responsiveness of cells to ephrins in *trans* beyond what is achieved by mere regulation of Eph receptor levels.

Ephrin-Independent Activities of Eph Receptors

In addition to their ephrin-dependent activities, the Eph receptors can signal independently of ephrin ligands, for example through cross talk with other receptor systems and cytoplasmic signaling molecules. Ephrin-independent signaling can have opposite effects compared to ephrin-dependent signaling, as exemplified by EphA2. This receptor is widely up-regulated in many cancers, which often correlates with low ephrin-A expression or failure of coexpressed ephrin-As to activate forward signaling (Zelinski et al. 2001; Macrae et al. 2005; Wykosky et al. 2005; Pasquale 2010; Tandon et al. 2011). This is consistent with the ability of EphA2 forward signaling to inhibit the Ras-Erk, Akt-mTORC1 and other oncogenic pathways. However, EphA2 overexpression can induce oncogenic transformation, suggesting that this receptor also has tumor-promoting activities that may not depend on ephrin binding (Zelinski et al. 2001; Tandon et al. 2011; Udayakumar et al. 2011). Recent studies have begun to unravel the mechanism of tumor promotion by EphA2. In cancer cells where Akt is highly activated by oncogenic mutations or growth factor stimulation, EphA2 is phosphorylated at S897 by Akt, which leads to an increase in cell migration/invasion that is independent of both ephrin binding and EphA2 kinase activity (Fig. 4C) (Miao et al. 2009). Other stimuli increasing Akt activation also cause S897 phosphorylation. For example, binding of extracellular Hsp90 to the LRP1 receptor induces Akt-dependent EphA2 S897 phosphorylation and association of EphA2 with LRP1, leading to glioblastoma cell invasiveness (Gopal et al. 2011). Moreover, ephrin-B3 expression in lung cancer cells can enhance the levels of EphA2 in its S897 phosphorylated form, concomitant with increasing resistance to γ-radiation (Stahl et al. 2011). Because ephrin-B3 has poor affinity for the ephrin-binding pocket of EphA2 (Gale et al. 1996), it will be interesting to investigate the connection between ephrin-B3 and EphA2 phosphorylation.

Remarkably, EphA2 seems to be at least in part responsible for the proliferative, migratory and tumorigenic activities of the EGF RTK family, as shown in several cultured cancer cell lines and in a mouse ErbB2 mammary tumor model (Larsen et al. 2007; Brantley-Sieders et al. 2008; Hiramoto-Yamaki et al. 2010; Argenzio et al. 2011). EGF stimulation can promote the association of EphA2 with the Rho-GEF Ephexin4, and it will be interesting to investigate the involvement of S897 phosphorylation in this ephrin-independent association (Hiramoto-Yamaki et al. 2010). The EphA2−Ephexin4 interaction promotes RhoG activation and recruitment of the RhoG-GTP-binding protein ELMO2 and the Rac-GEF DOCK4 to EphA2, leading to Rac1 activation and cancer cell invasiveness (Fig. 4C). The EphA2-Ephexin4-RhoG pathway also suppresses cell death due to detachment from the extracellular matrix (anoikis) in epithelial and cancer cells (Harada et al. 2011). This involves activation of PI3 kinase and Akt, which might also create a positive feedback loop further enhancing EphA2 S897 phosphorylation. Overexpression of EphB3 was also recently shown to promote lung cancer cell tumorigenicity through a kinase-independent mechanism (Ji et al. 2011). It will be important to investigate the full extent of ephrin- and kinase-independent activities of Eph receptors and how they differ from forward signaling.

Dephosphorylation

Phosphotyrosine phosphatases can modulate the Eph receptor/ephrin system by terminating forward signaling and favoring tyrosine phosphorylation-independent activities. For instance, the cytoplasmic phosphotyrosine phosphatase LMW-PTP can dephosphorylate EphA2, thus counteracting the tumor suppressive effects of EphA2 forward signaling and promoting cell transformation (Kikawa et al. 2002; Chiarugi et al. 2004; Parri et al. 2005). Similarly, the cytoplasmic phosphotyrosine phosphatase PTP1B can attenuate ephrin-induced EphA3 phosphorylation, endocytosis, and repulsive effects (Nievergall et al. 2010), while elevated endogenous phosphatase activity in pre-B leukemia cells can switch the EphA3-mediated response to ephrins from repulsion to adhesion (Wimmer-Kleikamp et al. 2008). In addition, PTP1B anchored to the endoplasmic reticulum has been reported to dephosphorylate EphA2 at sites of cell−cell contact where the endoplasmic reticulum comes in close proximity to the plasma membrane (Haj et al. 2012). Phosphatase activity induced by glucose in pancreatic β cells attenuates EphA phosphorylation and forward signaling, which are inhibitory for insulin secretion (Konstantinova et al. 2007). The protein tyrosine phosphatase receptor type O can dephosphorylate both EphA and EphB receptors, and it targets in particular the second of the two conserved phosphotyrosine residues in the juxtamembrane domain, which is the most critical for activation (Shintani et al. 2006). The LAR protein tyrosine phosphatase receptor can dephosphorylate EphB2, and LAR down-regulation by the FGF RTK results in increased ephrin-independent EphB2 tyrosine phosphorylation (Poliakov et al. 2008). LMW-PTP is also involved in EphB receptor signaling, being recruited to EphB clusters to promote cell attachment (Stein et al. 1998). However, it is not known whether this involves EphB receptor dephosphorylation by the phosphatase. Furthermore, the lipid phosphatase Ship2 can interact with EphA2 and decrease its ephrin-dependent tyrosine phosphorylation, internalization, and degradation through a mechanism likely not involving direct receptor dephosphorylation (Zhuang et al. 2007; Lee et al. 2012). Other phosphotyrosine phosphatases, such as the PDZ domain-containing PTP-BL, dephosphorylate ephrin-Bs to terminate reverse signaling (Palmer et al. 2002). Future studies will likely implicate additional phosphatases, including serine/threonine phosphatases (Yang et al. 2011), in the regulation of Eph receptor/ephrin signaling.

GENE MUTATIONS

Given the importance of the Eph receptor/ephrin system in developmental processes and adult tissue homeostasis, it is not surprising that its aberrant functioning has been implicated in a variety of diseases (Pasquale 2008, 2010). In particular, somatic and germline mutations in Eph receptors and ephrin genes are beginning to be linked to cancer and other pathologies. Large-scale sequencing of tumor specimens identified somatic mutations in all the Eph receptors, with frequencies of up to 2%–6% for some Eph receptors in lung cancer and melanoma (Ding et al. 2008; Prickett et al. 2009; Peifer et al. 2012) (www.sanger.ac.uk/genetics/CGP/cosmic). The mutations are scattered throughout the Eph receptor domains (Fig. 5), and their functional consequences are mostly unknown. However, many of the nearly 40 missense mutations identified in EphA3 (Fig. 5), the receptor found to be the most highly mutated in cancer, have been recently shown to cause various degrees of loss-of-function through multiple mechanisms

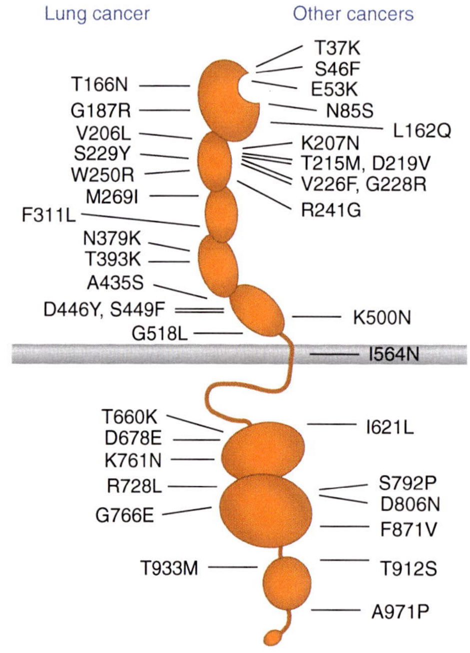

Figure 5. EphA3 receptor somatic mutations in cancer.

(Lisabeth et al. 2012; Zhuang et al. 2012). Most mutations in the ephrin-binding domain and the neighboring sushi domain impair ephrin binding either by directly affecting the high affinity ephrin-binding pocket or by causing overall conformational alterations. Mutations in the EphA3 kinase domain inhibit receptor tyrosine phosphorylation and kinase activity. A further consequence of many EphA3 mutations is a reduction in cell surface localization, which suggests that the mutations cause misfolding and/or alter receptor trafficking.

These findings suggest that the mutations disrupt a tumor suppressive function of EphA3 that depends on ephrin binding and kinase activity and, thus, forward signaling. The EphA3 cancer mutations indeed have different characteristics compared to mutations in other RTK families, which are typically clustered in "hot spots" and promote constitutive activation and tumorigenesis (Lee et al. 2006; Sharma et al. 2007; Greulich and Pollock 2011). Furthermore, wild-type EphA3, but not several mutants identified in tumor specimens, can suppress lung cancer cell growth in mouse xenograft models of lung cancer (Zhuang et al. 2012). Two mutations in EphA6 and EphA7 correspond to inactivating mutations in EphA3, suggesting that these Eph receptors may also suppress tumorigenesis (Lisabeth et al. 2012). EphB2 inactivating mutations identified in prostate cancer also suggested a tumor suppressor role for this receptor, consistent with the growth inhibition induced by EphB2 overexpressed in DU145 prostate cancer cells (Huusko et al. 2004). However, other Eph receptors like EphA2 or EphA4 do not seem to be frequently mutated in cancer, perhaps suggesting differences in the oncogenic activities of these receptors. A mutation in the first FNIII domain of EphA2 identified in lung cancer has indeed been proposed to promote invasiveness and survival (Faoro et al. 2010).

Although classical tumor suppressors are typically inactivated by homozygous mutations, most of the EphA3 inactivating mutations are heterozygous (Lisabeth et al. 2012). Hence, the EphA3 mutants may act as dominant negatives, disrupting the function of the wild-type receptor (Zhuang et al. 2012) and possibly other Eph

Cite this article as *Cold Spring Harb Perspect Biol* doi: 10.1101/cshperspect.a009159

receptors that may be part of the same signaling clusters. Furthermore, concurrent mutations in multiple Eph receptors have been found in a relatively high proportion of tumor samples, suggesting that they may be advantageous for tumor cells by more severely disrupting signaling in Eph receptor clusters than a single mutation (Lisabeth et al. 2012).

Germline mutations in Eph receptors and ephrins also play a role in human disease. For instance, EphA2 mutations enhancing basal receptor activation or possibly increasing EphA2 association with the LMW-PTP phosphatase have been associated with cataract development (Shiels et al. 2008; Jun et al. 2009; Zhang et al. 2009b). Inactivating mutations in the X-linked ephrin-B1 gene cause craniofrontonasal syndrome as a result of inhibition of gap junctional communication and improper tissue boundary formation in the developing skull (Bush and Soriano 2010; Makarov et al. 2010; Zafeiriou et al. 2011). On the other hand, loss-of-function EphA4 mutations in amyotrophic lateral sclerosis patients are associated with long survival (Van Hoecke et al. 2012) and single nucleotide polymorphisms in various Eph receptors and ephrins have been implicated as modifiers in the pathogenesis of amyotrophic lateral sclerosis as well as Parkinson's disease (Lesnick et al. 2008). Furthermore, a common EphA1 polymorphism was recently associated with late-onset Alzheimer's disease (Naj et al. 2011) and an EphA6 polymorphism with responsiveness to an antischizophrenic drug (Ikeda et al. 2010). Studies on the functional effects of Eph receptor and ephrin mutations and polymorphisms will undoubtedly provide a wealth of new information on the physiological and pathological roles of this intriguingly complex signaling system.

CONCLUDING REMARKS

Understanding signaling by the Eph RTK family has been challenging because of the many idiosyncrasies that distinguish it from the other RTK families. The peculiar characteristics of the Eph RTKs include the membrane-bound nature of the ephrins, the bidirectional mode of Eph re-

ceptor-ephrin signaling, the ability of the ephrins to stimulate but also attenuate Eph receptor signaling, and the ability of the Eph receptors to signal without ephrin involvement and even independently of kinase activity. Given the emerging view that different coexpressed Eph receptors signal cooperatively (Janes et al. 2011), to correctly interpret the results of signaling studies it will be important to profile the entire repertoire of Eph receptors present in a biological system (Noberini et al. 2012b) as well as survey their posttranslational modifications, including tyrosine and serine/threonine phosphorylation and ubiquitination. Systems biology approaches are also essential for a comprehensive understanding of the complexities of Eph receptor signaling networks and feedback loops, and the ability of these receptors to produce widely different biological outcomes (Jorgensen et al. 2009; Bush and Soriano 2012). In vivo analysis of Eph receptor/ephrin signaling as well as perturbations by designed or naturally occurring mutations and gene deletions will be critical to elucidate Eph receptor/ephrin physiological functions in the complex in vivo environment. Many fascinating activities of the Eph/ephrin system are only beginning to be appreciated, including key roles in stem cell biology (Genander and Frisen 2010) and in diseases such as Alzheimer's (Cisse et al. 2011; Hollingworth et al. 2011), or their emerging ability to regulate microRNAs (Arvanitis et al. 2010; Bhushan and Kandpal 2011; Khodayari et al. 2011) and gene transcription (Lai et al. 2004; Bong et al. 2007; Bush and Soriano 2010, 2012; Parrinello et al. 2010). Resolution of the paradoxes that plague our understanding of Eph receptor/ephrin function will enable effective exploitation of the many therapeutic opportunities that the Eph/ephrin system offers (Pasquale 2010; Noberini et al. 2012a).

ACKNOWLEDGMENTS

The authors thank R. Noberini, Y. Wallez, and A. Barquilla for helpful comments on the manuscript. Work in the authors' laboratory is supported by grants from the National Institutes of Health and the Tobacco-Related Disease Research Grants Program Office of the University of California.

REFERENCES

*Reference is also in this collection.

Afshari FT, Kwok JC, Fawcett JW. 2010. Astrocyte-produced ephrins inhibit Schwann cell migration via VAV2 signaling. *J Neurosci* **30:** 4246–4255.

Antion MD, Christie LA, Bond AM, Dalva MB, Contractor A. 2010. Ephrin-B3 regulates glutamate receptor signaling at hippocampal synapses. *Mol Cell Neurosci* **45:** 378–388.

Argenzio E, Bange T, Oldrini B, Bianchi F, Peesari R, Mari S, Di Fiore PP, Mann M, Polo S. 2011. Proteomic snapshot of the EGF-induced ubiquitin network. *Mol Syst Biol* **7:** 462.

Arvanitis D, Davy A. 2008. Eph/ephrin signaling: Networks. *Genes Dev* **22:** 416–429.

Arvanitis DN, Jungas T, Behar A, Davy A. 2010. Ephrin-B1 reverse signaling controls a posttranscriptional feedback mechanism via miR-124. *Mol Cell Biol* **30:** 2508–2517.

Astin JW, Batson J, Kadir S, Charlet J, Persad RA, Gillatt D, Oxley JD, Nobes CD. 2010. Competition amongst Eph receptors regulates contact inhibition of locomotion and invasiveness in prostate cancer cells. *Nat Cell Biol* **12:** 1194–1204.

Attwood BK, Bourgognon JM, Patel S, Mucha M, Schiavon E, Skrzypiec AE, Young KW, Shiosaka S, Korostynski M, Piechota M, et al. 2011. Neuropsin cleaves EphB2 in the amygdala to control anxiety. *Nature* **473:** 372–375.

Baba A, Akagi K, Takayanagi M, Flanagan JG, Kobayashi T, Hattori M. 2009. Fyn tyrosine kinase regulates the surface expression of glycosylphosphatidylinositol-linked ephrin via the modulation of sphingomyelin metabolism. *J Biol Chem* **284:** 9206–9214.

Bartley TD, Hunt RW, Welcher AA, Boyle WJ, Parker VP, Lindberg RA, Lu HS, Colombero AM, Elliott RL, Guthrie BA. 1994. B61 is a ligand for the ECK receptor protein-tyrosine kinase. *Nature* **368:** 558–560.

* Barton WA, Dalton AC, Seegar TCM, Himanen JP, Nikolov DB. 2013. Tie2 and Eph receptor tyrosine kinase activation and signaling. *Cold Spring Harb Perspect Biol* doi: 10.1101/cshperspect.a009142.

Beg AA, Sommer JE, Martin JH, Scheiffele P. 2007. α2-Chimaerin is an essential EphA4 effector in the assembly of neuronal locomotor circuits. *Neuron* **55:** 768–778.

Bhushan L, Kandpal RP. 2011. EphB6 receptor modulates micro RNA profile of breast carcinoma cells. *PLoS ONE* **6:** e22484.

Binns KL, Taylor PP, Sicheri F, Pawson T, Holland SJ. 2000. Phosphorylation of tyrosine residues in the kinase domain and juxtamembrane region regulates the biological and catalytic activities of Eph receptors. *Mol Cell Biol* **20:** 4791–4805.

Bohme B, VandenBos T, Cerretti DP, Park LS, Holtrich U, Rubsamen-Waigmann H, Strebhardt K. 1996. Cell–cell adhesion mediated by binding of membrane-anchored ligand LERK-2 to the EPH-related receptor human embryonal kinase 2 promotes tyrosine kinase activity. *J Biol Chem* **271:** 24747–24752.

Bonanomi D, Chivatakarn O, Bai G, Abdesselem H, Lettieri K, Marquardt T, Pierchala BA, Pfaff SL. 2012. Ret is a multifunctional coreceptor that integrates diffusible- and contact-axon guidance signals. *Cell* **148:** 568–582.

Bong YS, Lee HS, Carim-Todd L, Mood K, Nishanian TG, Tessarollo L, Daar IO. 2007. ephrin-B1 signals from the cell surface to the nucleus by recruitment of STAT3. *Proc Natl Acad Sci* **104:** 17305–17310.

Bouzioukh F, Wilkinson GA, Adelmann G, Frotscher M, Stein V, Klein R. 2007. Tyrosine phosphorylation sites in ephrin-B2 are required for hippocampal long-term potentiation but not long-term depression. *J Neurosci* **27:** 11279–11288.

Brantley-Sieders DM, Zhuang G, Hicks D, Fang WB, Hwang Y, Cates JM, Coffman K, Jackson D, Bruckheimer E, Muraoka-Cook RS, et al. 2008. The receptor tyrosine kinase EphA2 promotes mammary adenocarcinoma tumorigenesis and metastatic progression in mice by amplifying ErbB2 signaling. *J Clin Invest* **118:** 64–78.

Bush JO, Soriano P. 2009. Ephrin-B1 regulates axon guidance by reverse signaling through a PDZ-dependent mechanism. *Genes Dev* **23:** 1586–1599.

Bush JO, Soriano P. 2010. Ephrin-B1 forward signaling regulates craniofacial morphogenesis by controlling cell proliferation across Eph-ephrin boundaries. *Genes Dev* **24:** 2068–2080.

Bush JO, Soriano P. 2012. Eph/ephrin signaling: Genetic, phosphoproteomic, and transcriptomic approaches. *Semin Cell Dev Biol* **23:** 26–34.

Campbell TN, Attwell S, Arcellana-Panlilio M, Robbins SM. 2006. Ephrin-A5 expression promotes invasion and transformation of murine fibroblasts. *Biochem Biophys Res Commun* **350:** 623–628.

Carmona MA, Murai KK, Wang L, Roberts AJ, Pasquale EB. 2009. Glial ephrin-A3 regulates hippocampal dendritic spine morphology and glutamate transport. *Proc Natl Acad Sci* **106:** 12524–12529.

Carvalho RF, Beutler M, Marler KJ, Knoll B, Becker-Barroso E, Heintzmann R, Ng T, Drescher U. 2006. Silencing of EphA3 through a *cis* interaction with ephrin-A5. *Nat Neurosci* **9:** 322–330.

Chandarlapaty S, Sawai A, Scaltriti M, Rodrik-Outmezguine V, Grbovic-Huezo O, Serra V, Majumder PK, Baselga J, Rosen N. 2011. AKT inhibition relieves feedback suppression of receptor tyrosine kinase expression and activity. *Cancer Cell* **19:** 58–71.

Chang Q, Jorgensen C, Pawson T, Hedley DW. 2008. Effects of dasatinib on EphA2 receptor tyrosine kinase activity and downstream signalling in pancreatic cancer. *Br J Cancer* **99:** 1074–1082.

Chiarugi P, Taddei ML, Schiavone N, Papucci L, Giannoni E, Fiaschi T, Capaccioli S, Raugei G, Ramponi G. 2004. LMW-PTP is a positive regulator of tumor onset and growth. *Oncogene* **23:** 3905–3914.

Cisse M, Halabisky B, Harris J, Devidze N, Dubal DB, Sun B, Orr A, Lotz G, Kim DH, Hamto P, et al. 2011. Reversing EphB2 depletion rescues cognitive functions in Alzheimer model. *Nature* **469:** 47–52.

Cowan CA, Henkemeyer M. 2001. The SH2/SH3 adaptor Grb4 transduces B-ephrin reverse signals. *Nature* **413:** 174–179.

Cowan CW, Shao YR, Sahin M, Shamah SM, Lin MZ, Greer PL, Gao S, Griffith EC, Brugge JS, Greenberg ME. 2005. Vav family GEFs link activated Ephs to endocytosis and axon guidance. *Neuron* **46:** 205–217.

Dail M, Richter M, Godement P, Pasquale EB. 2006. Eph receptors inactivate R-Ras through different mechanisms to achieve cell repulsion. *J Cell Sci* **119:** 1244–1254.

Davy A, Gale NW, Murray EW, Klinghoffer RA, Soriano P, Feuerstein C, Robbins SM. 1999. Compartmentalized signaling by GPI-anchored ephrin-A5 requires the Fyn tyrosine kinase to regulate cellular adhesion. *Genes Dev* **13:** 3125–3135.

Davy A, Feuerstein C, Robbins SM. 2000. Signaling within a caveolae-like membrane microdomain in human neuroblastoma cells in response to fibroblast growth factor. *J Neurochem* **74:** 676–683.

Davy A, Bush JO, Soriano P. 2006. Inhibition of gap junction communication at ectopic Eph/ephrin boundaries underlies craniofrontonasal syndrome. *PLoS Biol* **4:** e315.

Ding L, Getz G, Wheeler DA, Mardis ER, McLellan MD, Cibulskis K, Sougnez C, Greulich H, Muzny DM, Morgan MB, et al. 2008. Somatic mutations affect key pathways in lung adenocarcinoma. *Nature* **455:** 1069–1075.

Du J, Fu C, Sretavan DW. 2007. Eph/ephrin signaling as a potential therapeutic target after central nervous system injury. *Curr Pharm Des* **13:** 2507–2518.

Elowe S, Holland SJ, Kulkarni S, Pawson T. 2001. Downregulation of the Ras-mitogen-activated protein kinase pathway by the EphB2 receptor tyrosine kinase is required for ephrin-induced neurite retraction. *Mol Cell Biol* **21:** 7429–7441.

Essmann CL, Martinez E, Geiger JC, Zimmer M, Traut MH, Stein V, Klein R, Acker-Palmer A. 2008. Serine phosphorylation of ephrin-B2 regulates trafficking of synaptic AMPA receptors. *Nat Neurosci* **11:** 1035–1043.

Faoro L, Singleton PA, Cervantes GM, Lennon FE, Choong NW, Kanteti R, Ferguson BD, Husain AN, Tretiakova MS, Ramnath N, et al. 2010. EphA2 mutation in lung squamous cell carcinoma promotes increased cell survival, cell invasion, focal adhesions, and mammalian target of rapamycin activation. *J Biol Chem* **285:** 18575–18585.

Fasen K, Cerretti DP, Huynh-Do U. 2008. Ligand binding induces Cbl-dependent EphB1 receptor degradation through the lysosomal pathway. *Traffic* **9:** 251–266.

Filosa A, Paixao S, Honsek SD, Carmona MA, Becker L, Feddersen B, Gaitanos L, Rudhard Y, Schoepfer R, Klopstock T, et al. 2009. Neuron-glia communication via EphA4/ephrin-A3 modulates LTP through glial glutamate transport. *Nat Neurosci* **12:** 1285–1292.

Fu WY, Chen Y, Sahin M, Zhao XS, Shi L, Bikoff JB, Lai KO, Yung WH, Fu AK, Greenberg ME, et al. 2007. Cdk5 regulates EphA4-mediated dendritic spine retraction through an ephexin1-dependent mechanism. *Nat Neurosci* **10:** 67–76.

Gale NW, Flenniken A, Compton DC, Jenkins N, Copeland NG, Gilbert DJ, Davis S, Wilkinson DG, Yancopoulos GD. 1996. Elk-L3, a novel transmembrane ligand for the Eph family of receptor tyrosine kinases, expressed in embryonic floor plate, roof plate and hindbrain segments. *Oncogene* **13:** 1343–1352.

Genander M, Frisen J. 2010. Ephrins and Eph receptors in stem cells and cancer. *Curr Opin Cell Biol* **22:** 611–616.

Georgakopoulos A, Litterst C, Ghersi E, Baki L, Xu C, Serban G, Robakis NK. 2006. Metalloproteinase/Presenilin1 processing of ephrin-B regulates EphB-induced Src phosphorylation and signaling. *EMBO J* **25:** 1242–1252.

Georgakopoulos A, Xu J, Xu C, Mauger G, Barthet G, Robakis NK. 2011. Presenilin1/γ-secretase promotes the EphB2-induced phosphorylation of ephrin-B2 by regulating phosphoprotein associated with glycosphingolipid-enriched microdomains/Csk binding protein. *FASEB J* **25:** 3594–3604.

* Goh LK, Sorkin A. 2013. Endocytosis of receptor tyrosine kinases. *Cold Spring Harb Perspect Biol* **5:** a017459.

Gopal U, Bohonowych JE, Lema-Tome C, Liu A, Garrett-Mayer E, Wang B, Isaacs JS. 2011. A novel extracellular Hsp90 mediated co-receptor function for LRP1 regulates EphA2-dependent glioblastoma cell invasion. *PLoS ONE* **6:** e17649.

Greulich H, Pollock PM. 2011. Targeting mutant fibroblast growth factor receptors in cancer. *Trends Mol Med* **17:** 283–292.

Haj FG, Sabet O, Kinkhabwala A, Wimmer-Kleikamp S, Roukos V, Han HM, Grabenbauer M, Bierbaum M, Antony C, Neel BG, et al. 2012. Regulation of signaling at regions of cell–cell contact by endoplasmic reticulum-bound protein-tyrosine phosphatase 1B. *PLoS ONE* **7:** e36633.

Harada K, Hiramoto-Yamaki N, Negishi M, Katoh H. 2011. Ephexin4 and EphA2 mediate resistance to anoikis through RhoG and phosphatidylinositol 3-kinase. *Exp Cell Res* **317:** 1701–1713.

Hattori M, Osterfield M, Flanagan JG. 2000. Regulated cleavage of a contact-mediated axon repellent. *Science* **289:** 1360–1365.

Heasman SJ, Ridley AJ. 2008. Mammalian Rho GTPases: New insights into their functions from in vivo studies. *Nat Rev Mol Cell Biol* **9:** 690–701.

Himanen JP, Yermekbayeva L, Janes PW, Walker JR, Xu K, Atapattu L, Rajashankar KR, Mensinga A, Lackmann M, Nikolov DB, et al. 2010. Architecture of Eph receptor clusters. *Proc Natl Acad Sci* **107:** 10860–10865.

Hiramoto-Yamaki N, Takeuchi S, Ueda S, Harada K, Fujimoto S, Negishi M, Katoh H. 2010. Ephexin4 and EphA2 mediate cell migration through a RhoG-dependent mechanism. *J Cell Biol* **190:** 461–477.

Holen HL, Shadidi M, Narvhus K, Kjosnes O, Tierens A, Aasheim HC. 2008. Signaling through ephrin-A ligand leads to activation of Src-family kinases, Akt phosphorylation, and inhibition of antigen receptor-induced apoptosis. *J Leukoc Biol* **84:** 1183–1191.

Holen HL, Zernichow L, Fjelland KE, Evenroed IM, Prydz K, Tveit H, Aasheim HC. 2011. Ephrin-B3 binds to a sulfated cell-surface receptor. *Biochem J* **433:** 215–223.

Hollingworth P, Harold D, Sims R, Gerrish A, Lambert JC, Carrasquillo MM, Abraham R, Hamshere ML, Pahwa JS, Moskvina V, et al. 2011. Common variants at ABCA7, MS4A6A/MS4A4E, EPHA1, CD33 and CD2AP are associated with Alzheimer's disease. *Nat Genet* **43:** 429–435.

Holmberg J, Clarke DL, Frisen J. 2000. Regulation of repulsion versus adhesion by different splice forms of an Eph receptor. *Nature* **408:** 203–206.

Holmberg J, Armulik A, Senti KA, Edoff K, Spalding K, Momma S, Cassidy R, Flanagan JG, Frisen J. 2005. Ephrin-A2 reverse signaling negatively regulates neural progenitor proliferation and neurogenesis. *Genes Dev* **19:** 462–471.

Hornberger MR, Dutting D, Ciossek T, Yamada T, Handwerker C, Lang S, Weth F, Huf J, Wessel R, Logan C, et al. 1999. Modulation of EphA receptor function by coexpressed ephrin-A ligands on retinal ganglion cell axons. *Neuron* **22:** 731–742.

Huang X, Wu D, Jin H, Stupack D, Wang JY. 2008. Induction of cell retraction by the combined actions of Abl-CrkII and Rho-ROCK1 signaling. *J Cell Biol* **183:** 711–723.

Hunter SG, Zhuang G, Brantley-Sieders D, Swat W, Cowan CW, Chen J. 2006. Essential role of vav family guanine nucleotide exchange factors in epha receptor-mediated angiogenesis. *Mol Cell Biol* **26:** 4830–4842.

Huusko P, Ponciano-Jackson D, Wolf M, Kiefer JA, Azorsa DO, Tuzmen S, Weaver D, Robbins C, Moses T, Allinen M, et al. 2004. Nonsense-mediated decay microarray analysis identifies mutations of EPHB2 in human prostate cancer. *Nat Genet* **36:** 979–983.

Ikeda M, Tomita Y, Mouri A, Koga M, Okochi T, Yoshimura R, Yamanouchi Y, Kinoshita Y, Hashimoto R, Williams HJ, et al. 2010. Identification of novel candidate genes for treatment response to risperidone and susceptibility for schizophrenia: Integrated analysis among pharmacogenomics, mouse expression, and genetic case-control association approaches. *Biol Psychiatry* **67:** 263–269.

Inoue E, Deguchi-Tawarada M, Togawa A, Matsui C, Arita K, Katahira-Tayama S, Sato T, Yamauchi E, Oda Y, Takai Y. 2009. Synaptic activity prompts γ-secretase-mediated cleavage of EphA4 and dendritic spine formation. *J Cell Biol* **185:** 551–564.

Irie F, Yamaguchi Y. 2002. EphB receptors regulate dendritic spine development via intersectin, Cdc42 and N-WASP. *Nat Neurosci* **5:** 1117–1118.

Irie F, Okuno M, Matsumoto K, Pasquale EB, Yamaguchi Y. 2008. Heparan sulfate regulates ephrin-A3/EphA receptor signaling. *Proc Natl Acad Sci* **105:** 12307–12312.

Iwasato T, Katoh H, Nishimaru H, Ishikawa Y, Inoue H, Saito YM, Ando R, Iwama M, Takahashi R, Negishi M, et al. 2007. Rac-GAP α-chimerin regulates motor-circuit formation as a key mediator of ephrin-B3/EphA4 forward signaling. *Cell* **130:** 742–753.

Janes PW, Saha N, Barton WA, Kolev MV, Wimmer-Kleikamp SH, Nievergall E, Blobel CP, Himanen JP, Lackmann M, Nikolov DB. 2005. Adam meets Eph: An ADAM substrate recognition module acts as a molecular switch for ephrin cleavage in *trans*. *Cell* **123:** 291–304.

Janes PW, Wimmer-Kleikamp SH, Frangakis AS, Treble K, Griesshaber B, Sabet O, Grabenbauer M, Ting AY, Saftig P, Bastiaens PI, et al. 2009. Cytoplasmic relaxation of active Eph controls ephrin shedding by ADAM10. *PLoS Biol* **7:** e1000215.

Janes PW, Griesshaber B, Atapattu L, Nievergall E, Hii LL, Mensinga A, Chheang C, Day BW, Boyd AW, Bastiaens PI, et al. 2011. Eph receptor function is modulated by heterooligomerization of A and B type Eph receptors. *J Cell Biol* **195:** 1033–1045.

Ji XD, Li G, Feng YX, Zhao JS, Li JJ, Sun ZJ, Shi S, Deng YZ, Xu JF, Zhu YQ, et al. 2011. EphB3 is overexpressed in nonsmall-cell lung cancer and promotes tumor metastasis by enhancing cell survival and migration. *Cancer Res* **71:** 1156–1166.

Jorgensen C, Sherman A, Chen GI, Pasculescu A, Poliakov A, Hsiung M, Larsen B, Wilkinson DG, Linding R, Pawson T. 2009. Cell-specific information processing in segregating populations of Eph receptor ephrin-expressing cells. *Science* **326:** 1502–1509.

Jun G, Guo H, Klein BE, Klein R, Wang JJ, Mitchell P, Miao H, Lee KE, Joshi T, Buck M, et al. 2009. EPHA2 is associated with age-related cortical cataract in mice and humans. *PLoS Genet* **5:** e1000584.

Kao TJ, Kania A. 2011. Ephrin-mediated cis-attenuation of Eph receptor signaling is essential for spinal motor axon guidance. *Neuron* **71:** 76–91.

Khodayari N, Mohammed KA, Goldberg EP, Nasreen N. 2011. Ephrin-A1 inhibits malignant mesothelioma tumor growth via let-7 microRNA-mediated repression of the RAS oncogene. *Cancer Gene Ther* **18:** 806–816.

Kikawa KD, Vidale DR, Van Etten RL, Kinch MS. 2002. Regulation of the EphA2 kinase by the low molecular weight tyrosine phosphatase induces transformation. *J Biol Chem* **277:** 39274–39279.

Klein R. 2009. Bidirectional modulation of synaptic functions by Eph/ephrin signaling. *Nat Neurosci* **12:** 15–20.

Konstantinova I, Nikolova G, Ohara-Imaizumi M, Meda P, Kucera T, Zarbalis K, Wurst W, Nagamatsu S, Lammert E. 2007. EphA-Ephrin-A-mediated β cell communication regulates insulin secretion from pancreatic islets. *Cell* **129:** 359–370.

Lackmann M, Boyd AW. 2008. Eph, a protein family coming of age: More confusion, insight, or complexity? *Sci Signal* **1:** re2.

Lai KO, Chen Y, Po HM, Lok KC, Gong K, Ip NY. 2004. Identification of the Jak/Stat proteins as novel downstream targets of EphA4 signaling in muscle: Implications in the regulation of acetylcholinesterase expression. *J Biol Chem* **279:** 13383–13392.

Larsen AB, Pedersen MW, Stockhausen MT, Grandal MV, van Deurs B, Poulsen HS. 2007. Activation of the EGFR gene target EphA2 inhibits epidermal growth factor-induced cancer cell motility. *Mol Cancer Res* **5:** 283–293.

Lee JC, Vivanco I, Beroukhim R, Huang JH, Feng WL, DeBiasi RM, Yoshimoto K, King JC, Nghiemphu P, Yuza Y, et al. 2006. Epidermal growth factor receptor activation in glioblastoma through novel missense mutations in the extracellular domain. *PLoS Med* **3:** e485.

Lee HS, Nishanian TG, Mood K, Bong YS, Daar IO. 2008. Ephrin-B1 controls cell–cell junctions through the Par polarity complex. *Nat Cell Biol* **10:** 979–986.

Lee HJ, Hota PK, Chugha P, Guo H, Miao H, Zhang L, Kim SJ, Stetzik L, Wang BC, Buck M. 2012. NMR Structure of a heterodimeric SAM:SAM complex: Characterization and manipulation of epha2 binding reveal new cellular functions of SHIP2. *Structure* **20:** 41–55.

Lesnick TG, Sorenson EJ, Ahlskog JE, Henley JR, Shehadeh L, Papapetropoulos S, Maraganore DM. 2008. Beyond Parkinson disease: Amyotrophic lateral sclerosis and the axon guidance pathway. *PLoS ONE* **3:** e1449.

Li JJ, Liu DP, Liu GT, Xie D. 2009. Ephrin-A5 acts as a tumor suppressor in glioma by negative regulation of epidermal growth factor receptor. *Oncogene* **28:** 1759–1768.

Li G, Ji XD, Gao H, Zhao JS, Xu JF, Sun ZJ, Deng YZ, Shi S, Feng YX, Zhu YQ, et al. 2012. EphB3 suppresses non-small-cell lung cancer metastasis via a PP2A/RACK1/Akt signalling complex. *Nat Commun* **3:** 667.

Lim YS, McLaughlin T, Sung TC, Santiago A, Lee KF, O'Leary DD. 2008. p75(NTR) mediates ephrin-A reverse signaling required for axon repulsion and mapping. *Neuron* **59:** 746–758.

Lisabeth EM, Fernandez C, Pasquale EB. 2012. Cancer somatic mutations disrupt functions of the EphA3 receptor tyrosine kinase through multiple mechanisms. *Biochemistry* **51:** 1464–1475.

Litterst C, Georgakopoulos A, Shioi J, Ghersi E, Wisniewski T, Wang R, Ludwig A, Robakis NK. 2007. Ligand binding and calcium influx induce distinct ectodomain/γ-secretase-processing pathways of EphB2 receptor. *J Biol Chem* **282:** 16155–16163.

Lu Q, Sun EE, Klein RS, Flanagan JG. 2001. Ephrin-B reverse signaling is mediated by a novel PDZ-RGS protein and selectively inhibits G protein-coupled chemoattraction. *Cell* **105:** 69–79.

Macrae M, Neve RM, Rodriguez-Viciana P, Haqq C, Yeh J, Chen C, Gray JW, McCormick F. 2005. A conditional feedback loop regulates Ras activity through EphA2. *Cancer Cell* **8:** 111–118.

Maddigan A, Truitt L, Arsenault R, Freywald T, Allonby O, Dean J, Narendran A, Xiang J, Weng A, Napper S, et al. 2011. EphB receptors trigger Akt activation and suppress Fas receptor-induced apoptosis in malignant T lymphocytes. *J Immunol* **187:** 5983–5994.

Makarov R, Steiner B, Gucev Z, Tasic V, Wieacker P, Wieland I. 2010. The impact of CFNS-causing EFNB1 mutations on ephrin-B1 function. *BMC Med Genet* **11:** 98.

Makinen T, Adams RH, Bailey J, Lu Q, Ziemiecki A, Alitalo K, Klein R, Wilkinson GA. 2005. PDZ interaction site in ephrin-B2 is required for the remodeling of lymphatic vasculature. *Genes Dev* **19:** 397–410.

Manning BD, Cantley LC. 2007. AKT/PKB signaling: Navigating downstream. *Cell* **129:** 1261–1274.

Margolis SS, Salogiannis J, Lipton DM, Mandel-Brehm C, Wills ZP, Mardinly AR, Hu L, Greer PL, Bikoff JB, Ho HY, et al. 2010. EphB-mediated degradation of the RhoA GEF Ephexin5 relieves a developmental brake on excitatory synapse formation. *Cell* **143:** 442–455.

Marler KJ, Becker-Barroso E, Martinez A, Llovera M, Wentzel C, Poopalasundaram S, Hindges R, Soriano E, Comella J, Drescher U. 2008. A TrkB/Ephrin-A interaction controls retinal axon branching and synaptogenesis. *J Neurosci* **28:** 12700–12712.

Marler KJ, Poopalasundaram S, Broom ER, Wentzel C, Drescher U. 2010. Pro-neurotrophins secreted from retinal ganglion cell axons are necessary for ephrin-A-p75NTR-mediated axon guidance. *Neural Dev* **5:** 30.

Marquardt T, Shirasaki R, Ghosh S, Andrews SE, Carter N, Hunter T, Pfaff SL. 2005. Coexpressed EphA receptors and ephrin-A ligands mediate opposing actions on growth cone navigation from distinct membrane domains. *Cell* **121:** 127–139.

Marston DJ, Dickinson S, Nobes CD. 2003. Rac-dependent *trans*-endocytosis of ephrin-Bs regulates Eph-ephrin contact repulsion. *Nat Cell Biol* **5:** 879–888.

McKay MM, Morrison DK. 2007. Integrating signals from RTKs to ERK/MAPK. *Oncogene* **26:** 3113–3121.

Meier C, Anastasiadou S, Knoll B. 2011. Ephrin-A5 suppresses neurotrophin evoked neuronal motility, ERK activation and gene expression. *PLoS ONE* **6:** e26089.

Mellitzer G, Xu QL, Wilkinson DG. 1999. Eph receptors and ephrins restrict cell intermingling and communication. *Nature* **400:** 77–81.

Menges CW, McCance DJ. 2008. Constitutive activation of the Raf-MAPK pathway causes negative feedback inhibition of Ras-PI3K-AKT and cellular arrest through the EphA2 receptor. *Oncogene* **27:** 2934–2940.

Miao H, Wei BR, Peehl DM, Li Q, Alexandrou T, Schelling JR, Rhim JS, Sedor JR, Burnett E, Wang BC. 2001. Activation of EphA receptor tyrosine kinase inhibits the Ras/MAPK pathway. *Nat Cell Biol* **3:** 527–530.

Miao H, Li DQ, Mukherjee A, Guo H, Petty A, Cutter J, Basilion JP, Sedor J, Wu J, Danielpour D, et al. 2009. EphA2 mediates ligand-dependent inhibition and ligand-independent promotion of cell migration and invasion via a reciprocal regulatory loop with Akt. *Cancer Cell* **16:** 9–20.

Minami M, Koyama T, Wakayama Y, Fukuhara S, Mochizuki N. 2011. Ephrin-A/EphA signal facilitates insulin-like growth factor-I-induced myogenic differentiation through suppression of the Ras/extracellular signal-regulated kinase 1/2 cascade in myoblast cell lines. *Mol Biol Cell* **22:** 3508–3519.

Moeller ML, Shi Y, Reichardt LF, Ethell IM. 2006. EphB receptors regulate dendritic spine morphogenesis through the recruitment/phosphorylation of focal adhesion kinase and RhoA activation. *J Biol Chem* **281:** 1587–1598.

Murai KK, Nguyen LN, Irie F, Yamaguchi Y, Pasquale EB. 2003. Control of hippocampal dendritic spine morphology through ephrin-A3/EphA4 signaling. *Nat Neurosci* **6:** 153–160.

Naj AC, Jun G, Beecham GW, Wang LS, Vardarajan BN, Buros J, Gallins PJ, Buxbaum JD, Jarvik GP, Crane PK, et al. 2011. Common variants at MS4A4/MS4A6E, CD2AP, CD33 and EPHA1 are associated with late-onset Alzheimer's disease. *Nat Genet* **43:** 436–441.

Nakada M, Drake KL, Nakada S, Niska JA, Berens ME. 2006. Ephrin-B3 ligand promotes glioma invasion through activation of Rac1. *Cancer Res* **66:** 8492–8500.

Nie D, Di Nardo A, Han JM, Baharanyi H, Kramvis I, Huynh T, Dabora S, Codeluppi S, Pandolfi PP, Pasquale EB, et al. 2010. Tsc2-Rheb signaling regulates EphA-mediated axon guidance. *Nat Neurosci* **13:** 163–172.

Nievergall E, Janes PW, Stegmayer C, Vail ME, Haj FG, Teng SW, Neel BG, Bastiaens PI, Lackmann M. 2010. PTP1B regulates Eph receptor function and trafficking. *J Cell Biol* **191:** 1189–1203.

Noberini R, Lamberto I, Pasquale EB. 2012a. Targeting Eph receptors with peptides and small molecules: Progress and challenges. *Semin Cell Dev Biol* **23:** 51–57.

Noberini R, Rubio de la Torre E, Pasquale EB. 2012b. Profiling Eph receptor expression in cells and tissues: A targeted mass spectrometry approach. *Cell Adh Migr* **6:** 102–112.

Oricchio E, Nanjangud G, Wolfe AL, Schatz JH, Mavrakis KJ, Jiang M, Liu X, Bruno J, Heguy A, Olshen AB, et al. 2011. The eph-receptor A7 is a soluble tumor suppressor for follicular lymphoma. *Cell* **147:** 554–564.

Palmer A, Zimmer M, Erdmann KS, Eulenburg V, Porthin A, Heumann R, Deutsch U, Klein R. 2002. Ephrin-B phosphorylation and reverse signaling: Regulation by Src kinases and PTP-BL phosphatase. *Mol Cell* **9:** 725–737.

Park EC, Cho GS, Kim GH, Choi SC, Han JK. 2011. The involvement of Eph-Ephrin signaling in tissue separation and convergence during *Xenopus* gastrulation movements. *Dev Biol* **350:** 441–450.

Parri M, Buricchi F, Taddei ML, Giannoni E, Raugei G, Ramponi G, Chiarugi P. 2005. Ephrin-A1 repulsive response is regulated by an EphA2 tyrosine phosphatase. *J Biol Chem* **280:** 34008–34018.

Parrinello S, Napoli I, Ribeiro S, Digby PW, Fedorova M, Parkinson DB, Doddrell RD, Nakayama M, Adams RH, Lloyd AC. 2010. EphB signaling directs peripheral nerve regeneration through Sox2-dependent Schwann cell sorting. *Cell* **143:** 145–155.

Pasquale EB. 2004. Eph-ephrin promiscuity is now crystal clear. *Nat Neurosci* **7:** 417–418.

Pasquale EB. 2005. Eph receptor signalling casts a wide net on cell behaviour. *Nat Rev Mol Cell Biol* **6:** 462–475.

Pasquale EB. 2008. Eph-ephrin bidirectional signaling in physiology and disease. *Cell* **133:** 38–52.

Pasquale EB. 2010. Eph receptors and ephrins in cancer: Bidirectional signalling and beyond. *Nat Rev Cancer* **10:** 165–180.

Peifer M, Fernandez-Cuesta L, Sos ML, George J, Seidel D, Kasper LH, Plenker D, Leenders F, Sun R, Zander T, et al. 2012. Integrative genome analyses identify key somatic driver mutations of small-cell lung cancer. *Nat Genet* **44:** 1104–1110.

Penzes P, Beeser A, Chernoff J, Schiller MR, Eipper BA, Mains RE, Huganir RL. 2003. Rapid induction of dendritic spine morphogenesis by *trans*-synaptic Ephrin-B-EphB receptor activation of the Rho-GEF Kalirin. *Neuron* **37:** 263–274.

Picco V, Hudson C, Yasuo H. 2007. Ephrin-Eph signalling drives the asymmetric division of notochord/neural precursors in Ciona embryos. *Development* **134:** 1491–1497.

Pitulescu ME, Adams RH. 2010. Eph/ephrin molecules—A hub for signaling and endocytosis. *Genes Dev* **24:** 2480–2492.

Poliakov A, Cotrina ML, Pasini A, Wilkinson DG. 2008. Regulation of EphB2 activation and cell repulsion by feedback control of the MAPK pathway. *J Cell Biol* **183:** 933–947.

Prickett TD, Agrawal NS, Wei X, Yates KE, Lin JC, Wunderlich JR, Cronin JC, Cruz P, Rosenberg SA, Samuels Y. 2009. Analysis of the tyrosine kinome in melanoma reveals recurrent mutations in ERBB4. *Nat Genet* **41:** 1127–1132.

Qiu R, Wang J, Tsark W, Lu Q. 2010. Essential role of PDZ-RGS3 in the maintenance of neural progenitor cells. *Stem Cells* **28:** 1602–1610.

Richter M, Murai KK, Bourgin C, Pak D, Pasquale EB. 2007. The EphA4 receptor regulates neuronal morphology through SPAR-mediated inactivation of Rap GTPases. *J Neurosci* **27:** 14205–14215.

Riedl JA, Brandt DT, Batlle E, Price LS, Clevers H, Bos JL. 2005. Down-regulation of Rap1 activity is involved in ephrin-B1-induced cell contraction. *Biochem J* **389:** 465–469.

Rohani N, Canty L, Luu O, Fagotto F, Winklbauer R. 2011. Ephrin-B/EphB signaling controls embryonic germ layer separation by contact-induced cell detachment. *PLoS Biol* **9:** e1000597.

Sahin M, Greer PL, Lin MZ, Poucher H, Eberhart J, Schmidt S, Wright TM, Shamah SM, O'Connell S, Cowan CW, et al. 2005. Eph-dependent tyrosine phosphorylation of ephexin1 modulates growth cone collapse. *Neuron* **46:** 191–204.

Salaita K, Groves JT. 2010. Roles of the cytoskeleton in regulating EphA2 signals. *Commun Integr Biol* **3:** 454–457.

Sawamiphak S, Seidel S, Essmann CL, Wilkinson GA, Pitulescu ME, Acker T, Acker-Palmer A. 2010. Ephrin-B2 regulates VEGFR2 function in developmental and tumour angiogenesis. *Nature* **465:** 487–491.

* Schlessinger J. 2013. Receptor tyrosine kinases: Legacy of the first two decades. *Cold Spring Harb Perspect Biol* doi: 10.1101/cshperspect.a008912.

Segura I, Essmann CL, Weinges S, Acker-Palmer A. 2007. Grb4 and GIT1 transduce ephrin-B reverse signals modulating spine morphogenesis and synapse formation. *Nat Neurosci* **10:** 301–310.

Seiradake E, Harlos K, Sutton G, Aricescu AR, Jones EY. 2010. An extracellular steric seeding mechanism for Eph-ephrin signaling platform assembly. *Nat Struct Mol Biol* **17:** 398–402.

Senturk A, Pfennig S, Weiss A, Burk K, Acker-Palmer A. 2011. Ephrin Bs are essential components of the Reelin pathway to regulate neuronal migration. *Nature* **472:** 356–360.

Shamah SM, Lin MZ, Goldberg JL, Estrach S, Sahin M, Hu L, Bazalakova M, Neve RL, Corfas G, Debant A, et al. 2001. EphA receptors regulate growth cone dynamics through the novel guanine nucleotide exchange factor ephexin. *Cell* **105:** 233–244.

Sharma SV, Bell DW, Settleman J, Haber DA. 2007. Epidermal growth factor receptor mutations in lung cancer. *Nat Rev Cancer* **7:** 169–181.

Shi W, Levine M. 2008. Ephrin signaling establishes asymmetric cell fates in an endomesoderm lineage of the Ciona embryo. *Development* **135:** 931–940.

Shi L, Fu WY, Hung KW, Porchetta C, Hall C, Fu AK, Ip NY. 2007. α2-chimaerin interacts with EphA4 and regulates EphA4-dependent growth cone collapse. *Proc Natl Acad Sci* **104:** 16347–16352.

Shiels A, Bennett TM, Knopf HL, Maraini G, Li A, Jiao X, Hejtmancik JF. 2008. The EPHA2 gene is associated with cataracts linked to chromosome 1p. *Mol Vis* **14:** 2042–2055.

Shintani T, Ihara M, Sakuta H, Takahashi H, Watakabe I, Noda M. 2006. Eph receptors are negatively controlled by protein tyrosine phosphatase receptor type O. *Nat Neurosci* **9:** 761–769.

Singla N, Erdjument-Bromage H, Himanen JP, Muir TW, Nikolov DB. 2011. A semisynthetic Eph receptor tyrosine kinase provides insight into ligand-induced kinase activation. *Chem Biol* **18:** 361–371.

Skaggs BJ, Gorre ME, Ryvkin A, Burgess MR, Xie Y, Han Y, Komisopoulou E, Brown LM, Loo JA, Landaw EM, et al. 2006. Phosphorylation of the ATP-binding loop directs oncogenicity of drug-resistant BCR-ABL mutants. *Proc Natl Acad Sci* **103:** 19466–19471.

Solanas G, Cortina C, Sevillano M, Batlle E. 2011. Cleavage of E-cadherin by ADAM10 mediates epithelial cell sorting downstream of EphB signalling. *Nat Cell Biol* **13:** 1100–1107.

Stahl S, Branca RM, Efazat G, Ruzzene M, Zhivotovsky B, Lewensohn R, Viktorsson K, Lehtio J. 2011. Phosphoproteomic profiling of NSCLC cells reveals that ephrin B3 regulates pro-survival signaling through Akt1-mediated phosphorylation of the EphA2 receptor. *J Proteome Res* **10:** 2566–2578.

Stein E, Lane AA, Cerretti DP, Schoecklmann HO, Schroff AD, Van Etten RL, Daniel TO. 1998. Eph receptors discriminate specific ligand oligomers to determine alternative signaling complexes, attachment, and assembly responses. *Genes Dev* **12:** 667–678.

Tandon M, Vemula SV, Mittal SK. 2011. Emerging strategies for EphA2 receptor targeting for cancer therapeutics. *Expert Opin Ther Targets* **15:** 31–51.

Tolias KF, Bikoff JB, Kane CG, Tolias CS, Hu L, Greenberg ME. 2007. The Rac1 guanine nucleotide exchange factor Tiam1 mediates EphB receptor-dependent dendritic spine development. *Proc Natl Acad Sci* **104:** 7265–7270.

Truitt L, Freywald A. 2011. Dancing with the dead: Eph receptors and their kinase-null partners. *Biochem Cell Biol* **89:** 115–129.

Udayakumar D, Zhang G, Ji Z, Njauw CN, Mroz P, Tsao H. 2011. EphA2 is a critical oncogene in melanoma. *Oncogene* **30:** 4921–4929.

Van Hoecke A, Schoonaert L, Lemmens R, Timmers M, Staats KA, Laird AS, Peeters E, Philips T, Goris A, Dubois B, et al. 2012. EPHA4 is a disease modifier of amyotrophic lateral sclerosis in animal models and in humans. *Nat Med* **18:** 1418–1422.

Vindis C, Cerretti DP, Daniel TO, Huynh-Do U. 2003. EphB1 recruits c-Src and p52Shc to activate MAPK/ERK and promote chemotaxis. *J Cell Biol* **162:** 661–671.

* Wagner MJ, Stacey MM, Liu BA, Pawson T. 2013. Molecular mechanisms of SH2- and PTB-domain containing proteins in receptor tyrosine kinase signaling. *Cold Spring Harb Perspect Biol* doi: 10.1101/cshperspect.a008987.

Wahl S, Barth H, Ciossek T, Aktories K, Mueller BK. 2000. Ephrin-A5 induces collapse of growth cones by activating Rho and Rho kinase. *J Cell Biol* **149:** 263–270.

Wakayama Y, Miura K, Sabe H, Mochizuki N. 2011. Ephrin-A1-EphA2 signal induces compaction and polarization of Madin-Darby canine kidney cells by inactivating ezrin

through negative regulation of RhoA. *J Biol Chem* **286:** 44243–44253.

Walker-Daniels J, Riese DJ 2nd, Kinch MS. 2002. c-Cbl-dependent EphA2 protein degradation is induced by ligand binding. *Mol Cancer Res* **1:** 79–87.

Wang Y, Nakayama M, Pitulescu ME, Schmidt TS, Bochenek ML, Sakakibara A, Adams S, Davy A, Deutsch U, Luthi U, et al. 2010. Ephrin-B2 controls VEGF-induced angiogenesis and lymphangiogenesis. *Nature* **465:** 483–486.

Wegmeyer H, Egea J, Rabe N, Gezelius H, Filosa A, Enjin A, Varoqueaux F, Deininger K, Schnutgen F, Brose N, et al. 2007. EphA4-dependent axon guidance is mediated by the RacGAP α2-chimaerin. *Neuron* **55:** 756–767.

Wei S, Xu G, Bridges LC, Williams P, White JM, DeSimone DW. 2010. ADAM13 induces cranial neural crest by cleaving class B ephrins and regulating Wnt signaling. *Dev Cell* **19:** 345–352.

Wimmer-Kleikamp SH, Janes PW, Squire A, Bastiaens PI, Lackmann M. 2004. Recruitment of Eph receptors into signaling clusters does not require ephrin contact. *J Cell Biol* **164:** 661–666.

Wimmer-Kleikamp SH, Nievergall E, Gegenbauer K, Adikari S, Mansour M, Yeadon T, Boyd AW, Patani NR, Lackmann M. 2008. Elevated protein tyrosine phosphatase activity provokes Eph/ephrin-facilitated adhesion of pre-B leukemia cells. *Blood* **112:** 721–732.

Wu C, Qiu R, Wang J, Zhang H, Murai K, Lu Q. 2009. ZHX2 Interacts with Ephrin-B and regulates neural progenitor maintenance in the developing cerebral cortex. *J Neurosci* **29:** 7404–7412.

Wybenga-Groot LE, Baskin B, Ong SH, Tong JF, Pawson T, Sicheri F. 2001. Structural basis for autoinhibition of the EphB2 receptor tyrosine kinase by the unphosphorylated juxtamembrane region. *Cell* **106:** 745–757.

Wykosky J, Gibo DM, Stanton C, Debinski W. 2005. EphA2 as a novel molecular marker and target in glioblastoma multiforme. *Mol Cancer Res* **3:** 541–551.

Wykosky J, Palma E, Gibo DM, Ringler S, Turner CP, Debinski W. 2008. Soluble monomeric Ephrin-A1 is released from tumor cells and is a functional ligand for the EphA2 receptor. *Oncogene* **27:** 7260–7273.

Xiao Z, Carrasco R, Kinneer K, Sabol D, Jallal B, Coats S, Tice DA. 2012. EphB4 promotes or suppresses Ras/MEK/ERK pathway in a context-dependent manner: Implications for EphB4 as a cancer target. *Cancer Biol Ther* **13:** 630–637.

Xu NJ, Henkemeyer M. 2009. Ephrin-B3 reverse signaling through Grb4 and cytoskeletal regulators mediates axon pruning. *Nat Neurosci* **12:** 268–276.

Xu J, Litterst C, Georgakopoulos A, Zaganas I, Robakis NK. 2009. Peptide EphB2/CTF2 generated by the γ-secretase processing of EphB2 receptor promotes tyrosine phosphorylation and cell surface localization of *N*-methyl-D-aspartate receptors. *J Biol Chem* **284:** 27220–27228.

Yamaguchi Y, Pasquale EB. 2004. Eph receptors in the adult brain. *Curr Opin Neurobiol* **14:** 288–296.

Yang NY, Fernandez C, Richter M, Xiao Z, Valencia F, Tice DA, Pasquale EB. 2011. Cross talk of the EphA2 receptor with a serine/threonine phosphatase suppresses the Akt-mTORC1 pathway in cancer cells. *Cell Signal* **23:** 201–212.

Yin Y, Yamashita Y, Noda H, Okafuji T, Go MJ, Tanaka H. 2004. EphA receptor tyrosine kinases interact with co-expressed ephrin-A ligands in *cis*. *Neurosci Res* **48:** 285–296.

Yoo S, Kim Y, Noh H, Lee H, Park E, Park S. 2011. Endocytosis of EphA receptors is essential for the proper development of the retinocollicular topographic map. *EMBO J* **30:** 1593–1607.

Yumoto N, Wakatsuki S, Kurisaki T, Hara Y, Osumi N, Frisen J, Sehara-Fujisawa A. 2008. Meltrin β/ADAM19 interacting with EphA4 in developing neural cells participates in formation of the neuromuscular junction. *PLoS ONE* **3:** e3322.

Zafeiriou DI, Pavlidou EL, Vargiami E. 2011. Diverse clinical and genetic aspects of craniofrontonasal syndrome. *Pediatr Neurol* **44:** 83–87.

Zelinski DP, Zantek ND, Stewart JC, Irizarry AR, Kinch MS. 2001. EphA2 overexpression causes tumorigenesis of mammary epithelial cells. *Cancer Res* **61:** 2301–2306.

Zhang J, Yang PL, Gray NS. 2009a. Targeting cancer with small molecule kinase inhibitors. *Nat Rev Cancer* **9:** 28–39.

Zhang T, Hua R, Xiao W, Burdon KP, Bhattacharya SS, Craig JE, Shang D, Zhao X, Mackey DA, Moore AT, et al. 2009b. Mutations of the EPHA2 receptor tyrosine kinase gene cause autosomal dominant congenital cataract. *Hum Mutat* **30:** E603–611.

Zhuang G, Hunter S, Hwang Y, Chen J. 2007. Regulation of EphA2 receptor endocytosis by SHIP2 lipid phosphatase via phosphatidylinositol 3-Kinase-dependent Rac1 activation. *J Biol Chem* **282:** 2683–2694.

Zhuang G, Song W, Amato K, Hwang Y, Lee K, Boothby M, Ye F, Guo Y, Shyr Y, Lin L, et al. 2012. Effects of cancer-associated EPHA3 mutations on lung cancer. *J Natl Cancer Inst* **104:** 1182–1197.

Zimmer M, Palmer A, Kohler J, Klein R. 2003. EphB-ephrin-B bi-directional endocytosis terminates adhesion allowing contact mediated repulsion. *Nat Cell Biol* **5:** 869–878.

Zisch AH, Pasquale EB. 1997. The Eph family: A multitude of receptors that mediate cell recognition signals. *Cell Tissue Res* **290:** 217–226.

Zisch AH, Pazzagli C, Freeman AL, Schneller M, Hadman M, Smith JW, Ruoslahti E, Pasquale EB. 2000. Replacing two conserved tyrosines of the EphB2 receptor with glutamic acid prevents binding of SH2 domains without abrogating kinase activity and biological responses. *Oncogene* **19:** 177–187.

Zwick E, Bange J, Ullrich A. 2002. Receptor tyrosine kinases as targets for anticancer drugs. *Trends Mol Med* **8:** 17–23.

The Role of Ryk and Ror Receptor Tyrosine Kinases in Wnt Signal Transduction

Jennifer Green[1,3], Roel Nusse[1], and Renée van Amerongen[2]

[1]Department of Developmental Biology and Howard Hughes Medical Institute, Stanford University, Stanford, California 94305

[2]Division of Molecular Oncology, Netherlands Cancer Institute, 1066 CX Amsterdam, The Netherlands

Correspondence: r.v.amerongen@nki.nl

Receptor tyrosine kinases of the Ryk and Ror families were initially classified as orphan receptors because their ligands were unknown. They are now known to contain functional extracellular Wnt-binding domains and are implicated in Wnt-signal transduction in multiple species. Although their signaling mechanisms still remain to be resolved in detail, both Ryk and Ror control important developmental processes in different tissues. However, whereas many other Wnt-signaling responses affect cell proliferation and differentiation, Ryk and Ror are mostly associated with controlling processes that rely on the polarized migration of cells. Here we discuss what is currently known about the involvement of this exciting class of receptors in development and disease.

The role of the receptor tyrosine kinases Ror and Ryk in Wnt signaling should be considered within the larger group of Wnt signaling receptors. Historically, the Frizzled (FZD) and LRP5/6 molecules were the first proteins implicated as receptors for Wnt ligands (Clevers and Nusse 2012). FZD proteins consist of a seven-pass transmembrane portion and an extracellular cysteine-rich domain (CRD) (Bhanot et al. 1996). Wnt proteins bind with high affinity to the CRD in a fairly promiscuous way: One Wnt will bind to multiple FZDs and conversely, single FZDs can interact with multiple Wnts (Hsieh et al. 1999b; Carmon and Loose 2010). This lack of a high degree of specificity is also borne out by the structure of the Wnt-CRD complex, as recently established for the *Xenopus* Wnt8 protein in a complex with the Frizzled8 CRD. Of the two domains on Wnt that interact with the CRD, one contains a palmitoleic acid modification, presumably present on multiple Wnt proteins, projecting into a pocket in the Frizzled CRD (Janda et al. 2012).

FZD molecules work together with two transmembrane LRP family members, LRP5 and LRP6 in vertebrates (Pinson et al. 2000; Tamai et al. 2000), both homologs of the *Drosophila* Arrow protein (Wehrli et al. 2000). The current model is that a Wnt protein binds to

---

[3]Present address: ActivX Biosciences, La Jolla, CA 92037.

FZD and LRP at the same time, forming a dimeric/multimeric structure. This may result in a conformational change in the receptor molecules, which leads to phosphorylation of the LRP cytoplasmic domain by associated protein kinases. Phosphorylation of the LRP tail (He et al. 2004) takes place on several clusters of serines and threonines, each containing a PPPSP motif. The protein kinases involved include GSK3 and CK1γ. GSK3 targets the PPPSP motif and phosphorylates a serine residue in that motif (Zeng et al. 2005). Residues adjacent to the PPPSP motif are phosphorylated by CK1γ, a CK1 family member with a membrane anchor in the form of a palmitoylation domain (Davidson et al. 2005). The phosphorylation of LRP leads to binding of the Axin protein to the cytoplasmic tail of LRP6 (Mao et al. 2001), an event that increases cytoplasmic levels of the signal transducer β-catenin, which subsequently translocates to the nucleus and induces gene expression in complex with TCF/LEF transcription factors. On the cytoplasmic side, Frizzled interacts with Dishevelled (Dsh) (Chen et al. 2003; Tauriello et al. 2012), which in turn may promote interactions with Axin through the DIX domain that these two proteins have in common (Schwarz-Romond et al. 2007; Fiedler et al. 2011). Although signaling via the FZD and LRP5/6 receptors has occupied our attention for many years, the identification of receptor tyrosine kinases (RTKs) as additional Wnt receptors has opened the door to new and exciting discoveries of Wnt signaling in development and disease.

Ror1 AND Ror2

Ror1 and Ror2 were first identified in PCR-based screens for molecules with resemblance to tyrosine kinases of the Trk family (Masiakowski and Carroll 1992). Indeed, although Ror1 and Ror2 occupy a separate corner in the RTK dendrogram, they are more closely related to Trk and Musk RTKs than to other RTK proteins. However, this conservation is largely restricted to their intracellular tyrosine kinase domains; their divergent extracellular domains suggested

early on that Ror1 and Ror2 might interact with a distinct set of extracellular ligands.

A distinguishing feature in the extracellular portion of the receptors is the presence of a CRD domain that bears close homology with the Wnt-binding domain found in Frizzled transmembrane receptors as well as in secreted Wnt inhibitors of the SFRP family (Saldanha et al. 1998), signifying that Wnt proteins might be the elusive ligands for this class of receptors (Fig. 1). It was not until much later, however, that this was definitively shown to be the case (Oishi et al. 2003; Mikels and Nusse 2006). In addition to the CRD, Ror proteins are further characterized by extracellular Kringle and immunoglobulin domains, whose functions remain enigmatic to this day, and by intracellular proline-rich and serine-threonine-rich domains (Masiakowski and Carroll 1992).

Ror1 and Ror2 in Development

Ror receptors are evolutionarily conserved across vertebrate and invertebrate species. The genomes of most species are reported to harbor two homologs.

Flies, Worms, and Frogs

Two Ror homologs, named *Dror* and *Dnrk*, have been described in *Drosophila*, but apart from their restricted expression pattern in the developing nervous system, no functional role for these receptors has been reported (Wilson et al. 1993; Makino et al. 1997). *CAM-1*, the sole *Caenorhabditis elegans* Ror homolog, is widely expressed in the nervous system, where it functions to regulate neuronal migration, positioning, and neurite outgrowth (Forrester et al. 1999; Zinovyeva et al. 2008; Hayashi et al. 2009; Kennerdell et al. 2009; Song et al. 2010) as well as synaptic transmission (Francis et al. 2005). CAM-1 also regulates the polarity of epithelial cells and stem cells (Green et al. 2008a; Yamamoto et al. 2011). Functions in cell polarity and cell migration are conserved in *Xenopus*, where *XRor2* regulates polarized cell migrations known as convergent extension (Hikasa et al. 2002; Schambony and Wedlich 2007).

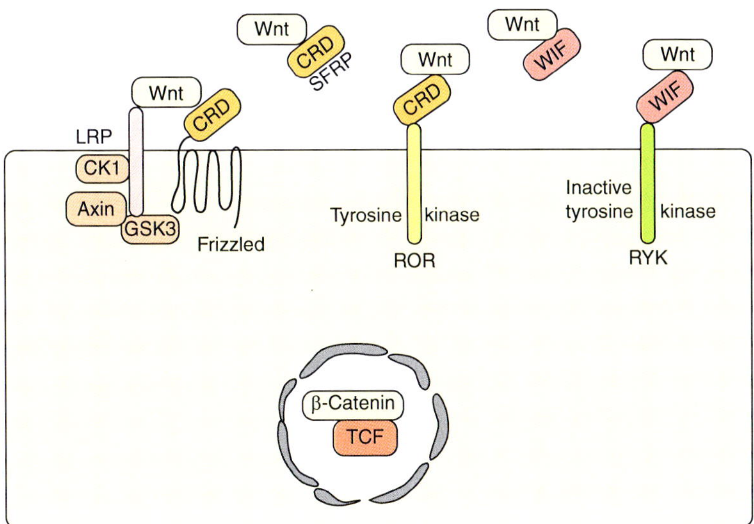

Figure 1. Schematic depicting the use of CRD and WIF domains in Wnt signal transduction. The mammalian genome encodes 19 different Wnts, which can mediate their signaling effects through 10 different FZDs that act in concert with the LRP5 and LRP6 coreceptors. The binding site for Wnt on Fzd is formed by the CRD. This motif is also used as the Wnt-binding site in members of the SFRP family of extracellular Wnt antagonists and in Ror1 and Ror2, both of which are members of the RTK family. A second Wnt-binding module, the so-called WIF domain, is used by extracellular Wnt antagonists of the WIF family, as well as by Ryk, another RTK family member. In spite of possessing functional Wnt-binding domains it remains unclear how Ror and Ryk receptors function in Wnt signal transduction. They have been proposed to function as stand-alone receptors, but also as coreceptors together with Fzd. In addition, they have been implicated in β-catenin-dependent and -independent signaling responses. At present, the molecular mechanisms used by these RTKs to transmit the Wnt-signal are ill understood. Only Ror2 has been shown to have a functional kinase domain, but both Ror1 and Ryk have been proposed to function as pseudokinases. See text for details.

Mammals

Based on the fact that *Ror1* and *Ror2* were first cloned from a human neuroblastoma cell line and the fact that expression of the *Drosophila* homologs appeared to be largely restricted to the developing nervous system, the initial characterization of the expression pattern of the mammalian Ror proteins was also performed in neural tissues (Oishi et al. 1999). Indeed, both Ror1 and Ror2 play a role in maintaining neural progenitor cell fate in the developing mouse brain (Endo et al. 2012). However, *Ror1* and *Ror2* are expressed with partially overlapping expression patterns in a broad range of tissues during mouse embryonic development, including the skeletal system and internal organs (Al-Shawi et al. 2001; Matsuda et al. 2001). The expression of *Ror2* appears to drop

right before birth. In contrast, *Ror1* expression could be detected at the RNA level in a limited number of postnatal tissues, although its expression too drops significantly in the adult (Oishi et al. 1999).

Mice deficient for Ror2 are born with craniofacial abnormalities as well as significantly shortened limbs and tail (DeChiara et al. 2000; Takeuchi et al. 2000), suggesting an important role in skeletal development. The gross appearance of Ror2-deficient and Wnt5a-deficient mice is quite similar, although loss of Wnt5a results in more severe defects than Ror2 (Yamaguchi et al. 1999; Ho et al. 2012). It was in fact this observation that first prompted scientists to investigate a role for Wnt5a in Ror2 signaling (see below) (Oishi et al. 2003).

In humans, mutations in *Ror2* are associated with brachydactyly type B and recessive Robinow syndrome, both of which result in limb malformations (Afzal et al. 2000; Oldridge et al. 2000; Afzal and Jeffery 2003). More recently, mutations in Wnt5a were also discovered in patients with autosomal-dominant Robinow syndrome (Person et al. 2010), suggesting that this signaling axis is disrupted in these patients.

In addition to skeletal malformations, Ror2-null mice have heart and lung defects (Takeuchi et al. 2000), a shortened intestine (Yamada et al. 2010), and aberrant inner ear hair cell orientation (Yamamoto et al. 2008). It should be noted, however, that some of these phenotypes might vary in penetrance and/or might be less pronounced in different mouse strains and/or on a different genetic background, as we have observed for the lung and inner ear defects in *Ror2*-knockout mice (RvA and RN).

The phenotype of mice deficient for *Ror1* is less severe than that of mice lacking *Ror2*. One study reported that these mice died within 24 hours after birth as a result of respiratory failure (Nomi et al. 2001), whereas another study reported more than half of these mice surviving until weaning, with a fair number of them reaching adulthood (Lyashenko et al. 2010). The additional loss of *Ror1* exacerbates the *Ror2*-mutant phenotype, as mice lacking both *Ror* homologs show enhanced skeletal and cardiac abnormalities (Nomi et al. 2001). Of note, it was recently reported that these initial *Ror1*-knockout mice might not represent true null mutants, as a truncated Ror1 protein was still expressed from the targeted allele. Still, newly generated conditional *Ror1*- and *Ror2*-knockout mice largely recapitulate the previously described mutants and *Ror1*/*Ror2* double-knockout mice largely recapitulate the *Wnt5a*-null phenotype (Ho et al. 2012).

Ror1 and Ror2 in Cancer

Both Ror1 and Ror2 have been ascribed oncogenic functions in cancer cells. In particular, siRNA mediated knockdown of Ror1 was shown to induce apoptosis in Hela cells, showing its importance for cell survival (MacKeigan et al. 2005). Ror1 was subsequently found to be overexpressed in B-cell chronic lymphocytic

leukemia (CLL), possibly as a result of enhanced Stat3 signaling (Baskar et al. 2008; Daneshmanesh et al. 2008; Li et al. 2010). It enhances cell viability and might therefore be an attractive candidate for targeted therapy (Fukuda et al. 2008; Hudecek et al. 2010; Yang et al. 2011; Baskar et al. 2012; Daneshmanesh et al. 2012). Its role might extend to other B-cell lymphomas, as a reduction in Ror1 expression affects viability of CLL as well as lymphoblastic leukemia (ALL) cells (Tyner et al. 2009; Choudhury et al. 2010; Dave et al. 2012). Elevated expression of Ror1 was also detected in mantle cell lymphomas (MCLs), marginal zone lymphoma (MZL), and nonhematopoietic malignancies (Barna et al. 2011; Gentile et al. 2011; Zhang et al. 2012).

High levels of Ror2 have been detected in osteosarcoma, melanoma, and renal cell carcinoma cell lines and a reduction in Ror2 expression was shown to decrease cell invasion and motility and extracellular matrix remodeling (Enomoto et al. 2009; Morioka et al. 2009; Wright et al. 2009; O'Connell et al. 2010). Furthermore, Ror2 was recently shown to be a novel prognostic biomarker and potential therapeutic target in patients with leiomyosarcomas or gastrointestinal stromal tumors (Edris et al. 2012). Whereas Ror2 is primarily implicated in cell polarity and migration, it has also been associated with cell survival and proliferation. For example, restoration of *Ror2* expression inhibited tumor cell proliferation in colon cancer, where *Ror2* gene expression is reportedly frequently silenced (Lara et al. 2010). Thus, Ror2 may have tumor-promoting as well as tumor-inhibiting activities depending on the cellular context.

Unfortunately, the biochemical activities of these receptors remain ill defined. As such, there is a large knowledge gap in our understanding of the processes that are controlled by Ror in both healthy and diseased tissues and the molecular mechanisms that underlie them.

Signaling Mechanisms and Effectors

Ror2 has been shown to have an active tyrosine kinase domain and to function as a genuine

RTK under at least some conditions (Masiakowski and Carroll 1992; Mikels et al. 2009). Like other RTKs, forced dimerization induces Ror2 tyrosine phosphorylation, whereas ligand binding can induce either tyrosine or serine/threonine phosphorylation (Liu et al. 2007; Yamamoto et al. 2007; Akbarzadeh et al. 2008; Liu et al. 2008; Mikels et al. 2009; Grumolato et al. 2010). In contrast, Ror1 was found to lack tyrosine kinase activity and most likely functions as a pseudokinase (Masiakowski and Carroll 1992; Gentile et al. 2011).

An alternative splice variant of Ror1, lacking both the extracellular and transmembrane domains, was identified in neural tissues and cancers (Reddy et al. 1996). This is interesting in light of a recent description that, similar to Ryk (see below), the Ror1 receptor might undergo cleavage resulting in a signaling portion that is directed to the nucleus by a signal located in the juxtamembrane domain (Tseng et al. 2010). The biological function of these truncated Ror proteins is unknown, although they have been reported to play a role in cell migration and remodeling of the actin cytoskeleton (Tseng et al. 2011).

It is widely appreciated that Wnt5A is a ligand for Ror2 (reviewed by Green et al. 2008b). At least one potential mechanism underlying Ror2 activation by Wnt5A was proposed recently by the finding that Wnt5A induces the formation of a complex between Ror2 and Frizzled, resulting in Ser/Thr phosphorylation of Ror2 and the recruitment of Dvl, Axin, and GSK3β, the same machinery that mediates the Wnt3A-induced phosphorylation of Lrp5/6 as discussed in the beginning of this article (Yamamoto et al. 2007; Grumolato et al. 2010). According to this model, Wnt3A and Wnt5A compete for binding to Frizzled. The identity of the Wnt ligand determines whether the Frizzled coreceptor will be Lrp5/6 or Ror2, thus dictating whether β-catenin-dependent or -independent signaling will be activated, respectively (Grumolato et al. 2010). In support of this model, Ror proteins have also been reported to bind to Frizzled receptors in *C. elegans*, *Xenopus*, and mammals (Oishi et al. 2003; Li et al. 2008; Nishita et al. 2010; Song et al.

2010). Accessory proteins may promote the use of one type of coreceptor over another. For example, the collagen triple-helix repeat containing protein 1 (Cthrc1) enhances the formation of a complex containing Frizzled and Ror2 (Yamamoto et al. 2008). It remains to be determined, however, whether Ror2 always functions as a coreceptor as proposed by this model.

It is well established that Wnt5a, unlike Wnt3a, has the capacity to inhibit β-catenin-dependent signaling. However, the mechanism underlying this inhibition remains controversial. Proposed explanations include competition with other Wnts for binding to Frizzled (Grumolato et al. 2010; Sato et al. 2010), down-regulation of β-catenin via the E3 ubiquitin ligase Siah2 (Topol et al. 2003), or inhibition downstream of β-catenin (Ishitani et al. 2003; Mikels and Nusse 2006; Verkaar et al. 2010). It has been reported that Wnt5A inhibits Wnt3A-induced β-catenin signaling via Ror2 (Mikels and Nusse 2006; Li et al. 2008); however, others report that Ror2 is not absolutely required for this inhibition (Grumolato et al. 2010; Sato et al. 2010; Ho et al. 2012). Elucidation of the relationship between Ror proteins and canonical Wnt signaling is further complicated by the ability of Ror proteins to modulate Wnt signaling independently of Wnt5A. In *C. elegans*, expression of the Ror extracellular domain (ECD) is sufficient to interfere with Wnt signaling by binding and sequestering Wnt ligands (Forrester et al. 2004; Green et al. 2007). Although a similar sequestration function has not been clearly shown for vertebrate Rors, ectopic expression of a membrane-tethered ECD modulates Wnt signaling in various vertebrate contexts (Hikasa et al. 2002; Nishita et al. 2010). Reports that Ror2 can potentiate canonical Wnt signaling further cloud the issue, although these reports are thus far limited to overexpression studies (Billiard et al. 2005; Li et al. 2008).

In contrast to Ror2, the function of Ror1 in mediating Wnt signals has been explored in very limited detail. Moreover, although Wnt5A has been reported to bind to Ror1 in vitro (Fukuda et al. 2008), Ror1 has also been implicated in various non-Wnt responses. For instance, Ror1 was shown to interact with and be inhibited by

Resistin, a cysteine-rich protein produced by mature adipocytes (Sanchez-Solana et al. 2012). As indicated, Ror1 is generally presumed to be a pseudokinase, its tyrosine phosphorylation either undetectable (Bicocca et al. 2012) or because of transphosphorylation by other kinases (Gentile et al. 2011).

The downstream effector mechanisms of Ror1 remain ill defined. For instance, in B-cell malignancies, cross talk between Ror1 and the pre-B-cell receptor complex has been reported, with Ror1 activating small GTPases to promote survival signaling (Bicocca et al. 2012). In contrast, a dual function for Ror1 was shown in lung adenocarcinoma (Yamaguchi et al. 2012). Here, Ror1 was required to sustain EGFR/ERBB3 signaling in a presumably kinase-independent manner. At the same time, Ror1 was responsible for Src phosphorylation and activation of PI3K/AKT signaling, suggesting that kinase-dependent functions for Ror1 should not be ruled out completely.

Summarizing, at present it remains incompletely understood just how promiscuous the interaction between the Ror receptors and their ligands, both Wnts and others, really is. Similar to the situation observed for Wnts and Fzds, it is unlikely that the Ror CRD will interact with only a single, specific ligand, but just how specificity in ligand/receptor interactions is achieved, particularly in the context of a complex multicellular organism, remains one of the outstanding questions in the Wnt field.

In both *C. elegans* and vertebrates, Ror proteins contribute to the orientation of cells within a tissue, a process known as planar cell polarity (PCP) (Green et al. 2008a; Yamamoto et al. 2008). In the murine limb bud, Wnt5A induces a complex between Ror2 and Vangl2, a core component of the PCP pathway, in a concentration-dependent manner. Thus, Ror proteins transmit directional information provided by a Wnt gradient to the PCP machinery in developing tissues. Likewise, *Xenopus XRor2* regulates convergent extension, a PCP-related process (Hikasa et al. 2002; Schambony and Wedlich 2007).

In addition to modulating canonical Wnt signaling and regulating cell polarity, Ror2 is required for Wnt5A to mediate cell migration, a function that has been reported to involve protein kinase C, JNK, and the actin-binding protein Filamin A (Nishita et al. 2006; Yamamoto et al. 2007; Nomachi et al. 2008; O'Connell et al. 2009). Ror2 signaling also contributes to the invasiveness of various human cancers including those of the skin (O'Connell et al. 2010), prostate (Yamamoto et al. 2010), bone (Enomoto et al. 2009; Yamagata et al. 2012), and kidney (Wright et al. 2009). This increased invasiveness has been associated with the increased expression of extracellular matrix remodeling enzymes called MMPs by Ror2-JNK or Ror2-cSrc signaling. Further complicating matters, Wnt5a was recently also shown to promote cancer cell invasion in a presumably Ror-independent manner by disrupting the association between N-cadherin and β-catenin, thereby promoting β-catenin/TCF signaling (Grossmann et al. 2013).

Many other potential Ror1/2 signaling effectors have been reported in various contexts, but further study is required to determine when and how they are used and in what combinations. In addition to Src and Src family kinases (Akbarzadeh et al. 2008; Enomoto et al. 2009), these include Dlxin-1 (Matsuda et al. 2003), casein kinase Iε (Kani et al. 2004), 14-3-3β (Liu et al. 2007), NF-κB (Fukuda et al. 2008), and BMPR1b (Sammar et al. 2004). It also remains to be determined if and how any of these components are connected to the upstream Wnt signaling machinery. This includes the Dishevelled protein, which plays an important role in both β-catenin-dependent and -independent Wnt-signaling events and which can be phosphorylated in response to Wnt5a/Ror signaling (Ho et al. 2012).

DERAILED/Ryk/lin-18

The Derailed/Ryk/lin18 receptors form a family of conserved transmembrane molecules with an extracellular domain resembling the Wnt inhibitory factor (WIF) protein (Patthy 2000; Angers and Moon 2009; Fradkin et al. 2010). Similar to secreted Frizzled related proteins (SFRPs), WIFs are extracellular Wnt antagonists (Hsieh et al. 1999a). Unlike SFRPs, however, the WIF Wnt-binding domain is different from the classic CRD domain that is observed in both FZD and Ror receptors (Fig. 1). Interestingly, it has been suggested that the WIF domain interacts directly with the lipid modification on the Wnt protein (Malinauskas et al. 2011), similar to the way in which the Fzd CRD domain was recently shown to engage in Wnt binding (Janda et al. 2012). The cytoplasmic portion of Derailed/Ryk/lin-18 has a tyrosine kinase motif, but according to the actual sequence, Derailed and Ryk are not considered to be active tyrosine kinase enzymes. This has been confirmed by functional assays (Yoshikawa et al. 2001; Inoue et al. 2004).

The mammalian Ryk gene was found by homology screens (Stacker et al. 1993), whereas the original identification of *derailed* was based on genetic screens in *Drosophila* for mutants that affect axon path finding in the embryo (Callahan et al. 1995; Bonkowsky et al. 1999). *Derailed* is also required for the specificity of muscle attachment sites (Callahan et al. 1996). A *C. elegans* homolog, *lin-18*, is critical for vulva development, in parallel to the Frizzled family member *lin-17* (Inoue et al. 2004; Wang and Sommer 2011).

Of note, whereas the mammalian genome appears to harbor just a single Ryk gene (in addition to a *Ryk* pseudogene) (Halford et al. 1999), multiple derailed/Ryk/lin-18 family members exist in *Drosophila*. In addition to *Derailed*, these include the *doughnut*, which is itself also involved in muscle attachment formation (Lahaye et al. 2012). Yet another homolog, *Derailed2*, plays a role opposing *Derailed1* during development of the *Drosophila* olfactory system (Sakurai et al. 2009).

A ligand of *Derailed* in these processes was initially not known, but the homology between the cell external domain of *Derailed* with WIF (Patthy 2000) suggested that Wnt proteins could interact with *Derailed*. Indeed, one of the *Drosophila* Wnt family members, *DWnt5* (Fradkin et al. 1995), has genetic interactions with *Derailed* (Yoshikawa et al. 2003; Fradkin et al. 2004). *DWnt5* is expressed in a pattern similar to *Derailed* and the phenotypes of these genes resemble each other as well. In *C. elegans*, the

Wnt proteins LIN-44, MOM-2, and CWN-2 redundantly regulate P7.p patterning, suggesting that Ryk might respond to multiple Wnt ligands. As in *Drosophila*, the mammalian Ryk gene is required for axon guidance and neurite outgrowth (Lu et al. 2004a; Keeble and Cooper 2006). More recently, mammalian Ryk was also shown to interact with the core planar cell polarity (PCP) machinery, and Ryk-deficient mice presented with typical PCP defects such as the misorientation of stereocilia in the inner ear (Macheda et al. 2012).

How do the Derailed/RYK/lin-18 receptors function molecularly? As mentioned above, all Derailed/RYK/lin-18 are catalytically inactive. Hence, Ryk may act as a coreceptor rather than as a primary signal transducing receptor. As a coreceptor, Derailed/RYK/lin-18 molecules may bind Wnts that are then cotargeted to other Wnt receptors. Indeed, whereas the intracellular kinase domains of Derailed/RYK/lin-18 proteins are largely dispensable for function, the extracellular WIF domains are not (Inoue et al. 2004; Taillebourg et al. 2005). Among the candidates for a receptor working with Derailed/RYK/lin18 are, perhaps not surprisingly, the Frizzleds. In neurite outgrowth in mammalian cells, the Ryk molecule stimulates Frizzled activity, both in vivo and in cell culture assays (Lu et al. 2004a). Moreover, in *Xenopus*, Ryk cooperates with Fzd7 and Wnt11 during gastrulation and convergent extension movements (Kim et al. 2008; Lin et al. 2010), suggesting some form of interaction. On the other hand, although the *lin-18* (Ryk) and *lin-17* (a Frizzled) genes in *C. elegans* both operate in vulva development, they are thought to act independently of each other (Inoue et al. 2004). As has been described for Ror2, Ryk might also couple directly to the core planar cell polarity pathway as it has also been shown to form a complex with and promote the stability of Vangl2 (Andre et al. 2012).

When it comes to the mechanism of action of Derailed/RYK/lin-18 receptors at the cytoplasmic side, there is also relatively little known. Surprisingly, Ryk was shown to undergo cleavage and, in response to Wnt-ligand stimulation, the carboxy-terminal fragment translocated to the nucleus (Lyu et al. 2008). Cleavage is mediated by γ-secretase and appears to be essential for the neuronal Ryk-mediated signal transduction events (Lyu et al. 2008). The carboxy-terminal fragment undergoes ubiquitination and degradation by the proteasome, in a way that is inhibited by Cdc37, a subunit of the molecular chaperone Hsp90 complex (Lyu et al. 2009). Inhibition may also require the Src protein kinase (Wouda et al. 2008). In the process of ubiquitination, the RYK molecule also interacts with the E3 ubiquitin ligase Mindbomb 1 (Berndt et al. 2011). The latter work also showed that Ryk can activate β-catenin signaling, suggesting yet another level of interactions among the various Wnt pathways (Berndt et al. 2011).

CONCLUDING REMARKS

In summary, much remains to be learned about the mechanism of action of both the Ror and Ryk tyrosine kinase receptor families. The fact that several of the main classes of Wnt receptors, including the Frizzleds, the RORs, and the RYKs, have ligand-binding domains that exist by themselves as secreted Wnt-binding factors, suggests a complex extracellular landscape of Wnts, secreted inhibitors, and receptors that may compete or assist each other in setting signaling levels and boundaries. But just how Ryk and Ror work as (co)receptors is still far from clear. It should also be stressed that additional non-Frizzled (co)receptors might still be out there. An example is the transmembrane protein PTK7/CCK4, which has an extracellular domain containing immunoglobulin repeats as well as an intracellular tyrosine kinase homology domain and which is required for the establishment of planar cell polarity in mammals (Lu et al. 2004b; Yen et al. 2009). Moreover, in spite of the fact that it does not appear to contain a bona fide Wnt-binding domain, PTK7 has been shown to interact with some, but not other, Wnt ligands and to inhibit Wnt/β-catenin signaling (Peradziryi et al. 2011). Given the multitude of experimental systems that can now be used, it is expected that significant progress will be made in understanding the involvement of these RTKs in development and disease.

REFERENCES

Afzal AR, Jeffery S. 2003. One gene, two phenotypes: ROR2 mutations in autosomal recessive Robinow syndrome and autosomal dominant brachydactyly type B. *Hum Mutat* **22:** 1–11.

Afzal AR, Rajab A, Fenske CD, Oldridge M, Elanko N, Ternes-Pereira E, Tuysuz B, Murday VA, Patton MA, Wilkie AO, et al. 2000. Recessive Robinow syndrome, allelic to dominant brachydactyly type B, is caused by mutation of ROR2. *Nat Genet* **25:** 419–422.

Akbarzadeh S, Wheldon LM, Sweet SM, Talma S, Mardakheh FK, Heath JK. 2008. The deleted in brachydactyly B domain of ROR2 is required for receptor activation by recruitment of Src. *PLoS ONE* **3:** e1873.

Al-Shawi R, Ashton SV, Underwood C, Simons JP. 2001. Expression of the Ror1 and Ror2 receptor tyrosine kinase genes during mouse development. *Dev Genes Evol* **211:** 161–171.

Andre P, Wang Q, Wang N, Gao B, Schilit A, Halford MM, Stacker SA, Zhang X, Yang Y. 2012. The Wnt coreceptor Ryk regulates Wnt/planar cell polarity by modulating the degradation of the core planar cell polarity component Vangl2. *J Biol Chem* **287:** 44518–44525.

Angers S, Moon RT. 2009. Proximal events in Wnt signal transduction. *Nat Rev Mol Cell Biol* **10:** 468–477.

Barna G, Mihalik R, Timar B, Tombol J, Csende Z, Sebestyen A, Bodor C, Csernus B, Reiniger L, Petak I, et al. 2011. ROR1 expression is not a unique marker of CLL. *Hematol Oncol* **29:** 17–21.

Baskar S, Kwong KY, Hofer T, Levy JM, Kennedy MG, Lee E, Staudt LM, Wilson WH, Wiestner A, Rader C. 2008. Unique cell surface expression of receptor tyrosine kinase ROR1 in human B-cell chronic lymphocytic leukemia. *Clin Cancer Res* **14:** 396–404.

Baskar S, Wiestner A, Wilson WH, Pastan I, Rader C. 2012. Targeting malignant B cells with an immunotoxin against ROR1. *MAbs* **4:** 349–361.

Berndt JD, Aoyagi A, Yang P, Anastas JN, Tang L, Moon RT. 2011. Mindbomb 1, an E3 ubiquitin ligase, forms a complex with RYK to activate Wnt/β-catenin signaling. *J Cell Biol* **194:** 737–750.

Bhanot P, Brink M, Samos CH, Hsieh JC, Wang Y, Macke JP, Andrew D, Nathans J, Nusse R. 1996. A new member of the frizzled family from *Drosophila* functions as a Wingless receptor. *Nature* **382:** 225–230.

Bicocca VT, Chang BH, Masouleh BK, Muschen M, Loriaux MM, Druker BJ, Tyner JW. 2012. Crosstalk between ROR1 and the Pre-B cell receptor promotes survival of t(1;19) acute lymphoblastic leukemia. *Cancer Cell* **22:** 656–667.

Billiard J, Way DS, Seestaller-Wehr LM, Moran RA, Mangine A, Bodine PV. 2005. The orphan receptor tyrosine kinase Ror2 modulates canonical Wnt signaling in osteoblastic cells. *Mol Endocrinol* **19:** 90–101.

Bonkowsky JL, Yoshikawa S, O'Keefe DD, Scully AL, Thomas JB. 1999. Axon routing across the midline controlled by the *Drosophila* Derailed receptor. *Nature* **402:** 540–544.

Callahan CA, Muralidhar MG, Lundgren SE, Scully AL, Thomas JB. 1995. Control of neuronal pathway selection by a *Drosophila* receptor protein-tyrosine kinase family member. *Nature* **376:** 171–174.

Callahan CA, Bonkovsky JL, Scully AL, Thomas JB. 1996. Derailed is required for muscle attachment site selection in *Drosophila. Development* **122:** 2761–2767.

Carmon KS, Loose DS. 2010. Development of a bioassay for detection of Wnt-binding affinities for individual frizzled receptors. *Anal Biochem* **401:** 288–294.

Chen W, ten Berge D, Brown J, Ahn S, Hu LA, Miller WE, Caron MG, Barak LS, Nusse R, Lefkowitz RJ. 2003. Dishevelled 2 recruits β-arrestin 2 to mediate Wnt5A-stimulated endocytosis of Frizzled 4. *Science* **301:** 1391–1394.

Choudhury A, Derkow K, Daneshmanesh AH, Mikaelsson E, Kiaii S, Kokhaei P, Osterborg A, Mellstedt H. 2010. Silencing of ROR1 and FMOD with siRNA results in apoptosis of CLL cells. *Br J Haematol* **151:** 327–335.

Clevers H, Nusse R. 2012. Wnt/β-catenin signaling and disease. *Cell* **149:** 1192–1205.

Daneshmanesh AH, Mikaelsson E, Jeddi-Tehrani M, Bayat AA, Ghods R, Ostadkarampour M, Akhondi M, Lagercrantz S, Larsson C, Osterborg A, et al. 2008. Ror1, a cell surface receptor tyrosine kinase is expressed in chronic lymphocytic leukemia and may serve as a putative target for therapy. *Int J Cancer* **123:** 1190–1195.

Daneshmanesh AH, Hojjat-Farsangi M, Khan AS, Jeddi-Tehrani M, Akhondi MM, Bayat AA, Ghods R, Mahmoudi AR, Hadavi R, Osterborg A, et al. 2012. Monoclonal antibodies against ROR1 induce apoptosis of chronic lymphocytic leukemia (CLL) cells. *Leukemia* **26:** 1348–1355.

Dave H, Anver MR, Butcher DO, Brown P, Khan J, Wayne AS, Baskar S, Rader C. 2012. Restricted cell surface expression of receptor tyrosine kinase ROR1 in pediatric B-lineage acute lymphoblastic leukemia suggests targetability with therapeutic monoclonal antibodies. *PLoS ONE* **7:** e52655.

Davidson G, Wu W, Shen J, Bilic J, Fenger U, Stannek P, Glinka A, Niehrs C. 2005. Casein kinase 1 γ couples Wnt receptor activation to cytoplasmic signal transduction. *Nature* **438:** 867–872.

DeChiara TM, Kimble RB, Poueymirou WT, Rojas J, Masiakowski P, Valenzuela DM, Yancopoulos GD. 2000. Ror2, encoding a receptor-like tyrosine kinase, is required for cartilage and growth plate development. *Nat Genet* **24:** 271–274.

Edris B, Espinosa I, Muhlenberg T, Mikels A, Lee CH, Steigen SE, Zhu S, Montgomery KD, Lazar AJ, Lev D, et al. 2012. ROR2 is a novel prognostic biomarker and a potential therapeutic target in leiomyosarcoma and gastrointestinal stromal tumour. *J Pathol* **227:** 223–233.

Endo M, Doi R, Nishita M, Minami Y. 2012. Ror family receptor tyrosine kinases regulate the maintenance of neural progenitor cells in the developing neocortex. *J Cell Sci* **125:** 2017–2029.

Enomoto M, Hayakawa S, Itsukushima S, Ren DY, Matsuo M, Tamada K, Oneyama C, Okada M, Takumi T, Nishita M, et al. 2009. Autonomous regulation of osteosarcoma cell invasiveness by Wnt5a/Ror2 signaling. *Oncogene* **28:** 3197–3208.

Fiedler M, Mendoza-Topaz C, Rutherford TJ, Mieszczanek J, Bienz M. 2011. Dishevelled interacts with the DIX domain polymerization interface of Axin to interfere with its function in down-regulating β-catenin. *Proc Natl Acad Sci* **108:** 1937–1942.

Forrester WC, Dell M, Perens E, Garriga G. 1999. A *C. elegans* Ror receptor tyrosine kinase regulates cell motility and asymmetric cell division. *Nature* **400:** 881–885.

Forrester WC, Kim C, Garriga G. 2004. The *Caenorhabditis elegans* Ror RTK CAM-1 inhibits EGL-20/Wnt signaling in cell migration. *Genetics* **168:** 1951–1962.

Fradkin LG, Noordermeer JN, Nusse R. 1995. The *Drosophila* Wnt protein DWnt-3 is a secreted glycoprotein localized on the axon tracts of the embryonic CNS. *Dev Biol* **168:** 202–213.

Fradkin LG, van Schie M, Wouda RR, de Jong A, Kamphorst JT, Radjkoemar-Bansraj M, Noordermeer JN. 2004. The *Drosophila* Wnt5 protein mediates selective axon fasciculation in the embryonic central nervous system. *Dev Biol* **272:** 362–375.

Fradkin LG, Dura JM, Noordermeer JN. 2010. Ryks: New partners for Wnts in the developing and regenerating nervous system. *Trends Neurosci* **33:** 84–92.

Francis MM, Evans SP, Jensen M, Madsen DM, Mancuso J, Norman KR, Maricq AV. 2005. The Ror receptor tyrosine kinase CAM-1 is required for ACR-16-mediated synaptic transmission at the *C. elegans* neuromuscular junction. *Neuron* **46:** 581–594.

Fukuda T, Chen L, Endo T, Tang L, Lu D, Castro JE, Widhopf GF 2nd, Rassenti LZ, Cantwell MJ, Prussak CE, et al. 2008. Antisera induced by infusions of autologous Ad-CD154-leukemia B cells identify ROR1 as an oncofetal antigen and receptor for Wnt5a. *Proc Natl Acad Sci* **105:** 3047–3052.

Gentile A, Lazzari L, Benvenuti S, Trusolino L, Comoglio PM. 2011. Ror1 is a pseudokinase that is crucial for Met-driven tumorigenesis. *Cancer Res* **71:** 3132–3141.

Green JL, Inoue T, Sternberg PW. 2007. The *C. elegans* ROR receptor tyrosine kinase, CAM-1, non-autonomously inhibits the Wnt pathway. *Development* **134:** 4053–4062.

Green JL, Inoue T, Sternberg PW. 2008a. Opposing Wnt pathways orient cell polarity during organogenesis. *Cell* **134:** 646–656.

Green JL, Kuntz SG, Sternberg PW. 2008b. Ror receptor tyrosine kinases: Orphans no more. *Trends Cell Biol* **18:** 536–544.

Grossmann AH, Yoo JH, Clancy J, Sorensen LK, Sedgwick A, Tong Z, Ostanin K, Rogers A, Grossmann KF, Tripp SR, et al. 2013. The small GTPase ARF6 stimulates β-catenin transcriptional activity during WNT5A-mediated melanoma invasion and metastasis. *Sci Signal* **6:** pra14.

Grumolato L, Liu G, Mong P, Mudbhary R, Biswas R, Arroyave R, Vijayakumar S, Economides AN, Aaronson SA. 2010. Canonical and noncanonical Wnts use a common mechanism to activate completely unrelated coreceptors. *Genes Dev* **24:** 2517–2530.

Halford MM, Oates AC, Hibbs ML, Wilks AF, Stacker SA. 1999. Genomic structure and expression of the mouse growth factor receptor related to tyrosine kinases (Ryk). *J Biol Chem* **274:** 7379–7390.

Hayashi Y, Hirotsu T, Iwata R, Kage-Nakadai E, Kunitomo H, Ishihara T, Iino Y, Kubo T. 2009. A trophic role for Wnt-Ror kinase signaling during developmental pruning in *Caenorhabditis elegans. Nat Neurosci* **12:** 981–987.

He X, Semenov M, Tamai K, Zeng X. 2004. LDL receptor-related proteins 5 and 6 in Wnt/β-catenin signaling: Arrows point the way. *Development* **131**: 1663–1677.

Hikasa H, Shibata M, Hiratani I, Taira M. 2002. The *Xenopus* receptor tyrosine kinase Xror2 modulates morphogenetic movements of the axial mesoderm and neuroectoderm via Wnt signaling. *Development* **129**: 5227–5239.

Ho HY, Susman MW, Bikoff JB, Ryu YK, Jonas AM, Hu L, Kuruvilla R, Greenberg ME. 2012. Wnt5a-Ror-Dishevelled signaling constitutes a core developmental pathway that controls tissue morphogenesis. *Proc Natl Acad Sci* **109**: 4044–4051.

Hsieh JC, Kodjabachian L, Rebbert ML, Rattner A, Smallwood PM, Samos CH, Nusse R, Dawid IB, Nathans J. 1999a. A new secreted protein that binds to Wnt proteins and inhibits their activities. *Nature* **398**: 431–436.

Hsieh JC, Rattner A, Smallwood PM, Nathans J. 1999b. Biochemical characterization of Wnt-frizzled interactions using a soluble, biologically active vertebrate Wnt protein. *Proc Natl Acad Sci* **96**: 3546–3551.

Hudecek M, Schmitt TM, Baskar S, Lupo-Stanghellini MT, Nishida T, Yamamoto TN, Bleakley M, Turtle CJ, Chang WC, Greisman HA, et al. 2010. The B-cell tumor-associated antigen ROR1 can be targeted with T cells modified to express a ROR1-specific chimeric antigen receptor. *Blood* **116**: 4532–4541.

Inoue T, Oz HS, Wiland D, Gharib S, Deshpande R, Hill RJ, Katz WS, Sternberg PW. 2004. *C. elegans* LIN-18 is a Ryk ortholog and functions in parallel to LIN-17/Frizzled in Wnt signaling. *Cell* **118**: 795–806.

Ishitani T, Kishida S, Hyodo-Miura J, Ueno N, Yasuda J, Waterman M, Shibuya H, Moon RT, Ninomiya-Tsuji J, Matsumoto K. 2003. The TAK1-NLK mitogen-activated protein kinase cascade functions in the Wnt-5a/Ca^{2+} pathway to antagonize Wnt/β-catenin signaling. *Mol Cell Biol* **23**: 131–139.

Janda CY, Waghray D, Levin AM, Thomas C, Garcia KC. 2012. Structural basis of Wnt recognition by Frizzled. *Science* **337**: 59–64.

Kani S, Oishi I, Yamamoto H, Yoda A, Suzuki H, Nomachi A, Iozumi K, Nishita M, Kikuchi A, Takumi T, et al. 2004. The receptor tyrosine kinase Ror2 associates with and is activated by casein kinase Iε. *J Biol Chem* **279**: 50102–50109.

Keeble TR, Cooper HM. 2006. Ryk: A novel Wnt receptor regulating axon pathfinding. *Int J Biochem Cell Biol* **38**: 2011–2017.

Kennerdell JR, Fetter RD, Bargmann CI. 2009. Wnt-Ror signaling to SIA and SIB neurons directs anterior axon guidance and nerve ring placement in *C. elegans*. *Development* **136**: 3801–3810.

Kim GH, Her JH, Han JK. 2008. Ryk cooperates with Frizzled 7 to promote Wnt11-mediated endocytosis and is essential for *Xenopus laevis* convergent extension movements. *J Cell Biol* **182**: 1073–1082.

Lahaye LL, Wouda RR, de Jong AW, Fradkin LG, Noordermeer JN. 2012. WNT5 interacts with the Ryk receptors doughnut and derailed to mediate muscle attachment site selection in *Drosophila melanogaster*. *PLoS ONE* **7**: e32297.

Lara E, Calvanese V, Huidobro C, Fernandez AF, Moncada-Pazos A, Obaya AJ, Aguilera O, Gonzalez-Sancho JM, Sanchez L, Astudillo A, et al. 2010. Epigenetic repression of ROR2 has a Wnt-mediated, pro-tumourigenic role in colon cancer. *Mol Cancer* **9**: 170.

Li C, Chen H, Hu L, Xing Y, Sasaki T, Villosis MF, Li J, Nishita M, Minami Y, Minoo P. 2008. Ror2 modulates the canonical Wnt signaling in lung epithelial cells through cooperation with Fzd2. *BMC Mol Biol* **9**: 11.

Li P, Harris D, Liu Z, Liu J, Keating M, Estrov Z. 2010. Stat3 activates the receptor tyrosine kinase like orphan receptor-1 gene in chronic lymphocytic leukemia cells. *PLoS ONE* **5**: e11859.

Lin S, Baye LM, Westfall TA, Slusarski DC. 2010. Wnt5b-Ryk pathway provides directional signals to regulate gastrulation movement. *J Cell Biol* **190**: 263–278.

Liu Y, Ross JF, Bodine PV, Billiard J. 2007. Homodimerization of Ror2 tyrosine kinase receptor induces 14–3-3(β) phosphorylation and promotes osteoblast differentiation and bone formation. *Mol Endocrinol* **21**: 3050–3061.

Liu Y, Rubin B, Bodine PV, Billiard J. 2008. Wnt5a induces homodimerization and activation of Ror2 receptor tyrosine kinase. *J Cell Biochem* **105**: 497–502.

Lu W, Yamamoto V, Ortega B, Baltimore D. 2004a. Mammalian Ryk is a Wnt coreceptor required for stimulation of neurite outgrowth. *Cell* **119**: 97–108.

Lu X, Borchers AG, Jolicoeur C, Rayburn H, Baker JC, Tessier-Lavigne M. 2004b. PTK7/CCK-4 is a novel regulator of planar cell polarity in vertebrates. *Nature* **430**: 93–98.

Lyashenko N, Weissenbock M, Sharir A, Erben RG, Minami Y, Hartmann C. 2010. Mice lacking the orphan receptor ror1 have distinct skeletal abnormalities and are growth retarded. *Dev Dyn* **239**: 2266–2277.

Lyu J, Yamamoto V, Lu W. 2008. Cleavage of the Wnt receptor Ryk regulates neuronal differentiation during cortical neurogenesis. *Dev Cell* **15**: 773–780.

Lyu J, Wesselschmidt RL, Lu W. 2009. Cdc37 regulates Ryk signaling by stabilizing the cleaved Ryk intracellular domain. *J Biol Chem* **284**: 12940–12948.

Macheda ML, Sun WW, Kugathasan K, Hogan BM, Bower NI, Halford MM, Zhang YF, Jacques BE, Lieschke GJ, Dabdoub A, et al. 2012. The Wnt receptor Ryk plays a role in mammalian planar cell polarity signaling. *J Biol Chem* **287**: 29312–29323.

MacKeigan JP, Murphy LO, Blenis J. 2005. Sensitized RNAi screen of human kinases and phosphatases identifies new regulators of apoptosis and chemoresistance. *Nat Cell Biol* **7**: 591–600.

Makino K, Goto Y, Sueyasu M, Futagami K, Kataoka Y, Oishi R. 1997. Micellar electrokinetic capillary chromatography for therapeutic drug monitoring of zonisamide. *J Chromatogr B Biomed Sci Appl* **695**: 417–425.

Malinauskas T, Aricescu AR, Lu W, Siebold C, Jones EY. 2011. Modular mechanism of Wnt signaling inhibition by Wnt inhibitory factor 1. *Nat Struct Mol Biol* **18**: 886–893.

Mao J, Wang J, Liu B, Pan W, Farr GH 3rd, Flynn C, Yuan H, Takada S, Kimelman D, Li L, et al. 2001. Low-density lipoprotein receptor-related protein-5 binds to Axin and regulates the canonical Wnt signaling pathway. *Mol Cell* **7**: 801–809.

Masiakowski P, Carroll RD. 1992. A novel family of cell surface receptors with tyrosine kinase-like domain. *J Biol Chem* **267**: 26181–26190.

Matsuda T, Nomi M, Ikeya M, Kani S, Oishi I, Terashima T, Takada S, Minami Y. 2001. Expression of the receptor tyrosine kinase genes, Ror1 and Ror2, during mouse development. *Mech Dev* **105:** 153–156.

Matsuda T, Suzuki H, Oishi I, Kani S, Kuroda Y, Komori T, Sasaki A, Watanabe K, Minami Y. 2003. The receptor tyrosine kinase Ror2 associates with the melanoma-associated antigen (MAGE) family protein Dlxin-1 and regulates its intracellular distribution. *J Biol Chem* **278:** 29057–29064.

Mikels AJ, Nusse R. 2006. Purified Wnt5a protein activates or inhibits β-catenin-TCF signaling depending on receptor context. *PLoS Biol* **4:** e115.

Mikels A, Minami Y, Nusse R. 2009. Ror2 receptor requires tyrosine kinase activity to mediate Wnt5A signaling. *J Biol Chem* **284:** 30167–30176.

Morioka K, Tanikawa C, Ochi K, Daigo Y, Katagiri T, Kawano H, Kawaguchi H, Myoui A, Yoshikawa H, Naka N, et al. 2009. Orphan receptor tyrosine kinase ROR2 as a potential therapeutic target for osteosarcoma. *Cancer Sci* **100:** 1227–1233.

Nishita M, Yoo SK, Nomachi A, Kani S, Sougawa N, Ohta Y, Takada S, Kikuchi A, Minami Y. 2006. Filopodia formation mediated by receptor tyrosine kinase Ror2 is required for Wnt5a-induced cell migration. *J Cell Biol* **175:** 555–562.

Nishita M, Itsukushima S, Nomachi A, Endo M, Wang Z, Inaba D, Qiao S, Takada S, Kikuchi A, Minami Y. 2010. Ror2/Frizzled complex mediates Wnt5a-induced AP-1 activation by regulating Dishevelled polymerization. *Mol Cell Biol* **30:** 3610–3619.

Nomachi A, Nishita M, Inaba D, Enomoto M, Hamasaki M, Minami Y. 2008. Receptor tyrosine kinase Ror2 mediates Wnt5a-induced polarized cell migration by activating c-Jun N-terminal kinase via actin-binding protein filamin A. *J Biol Chem* **283:** 27973–27981.

Nomi M, Oishi I, Kani S, Suzuki H, Matsuda T, Yoda A, Kitamura M, Itoh K, Takeuchi S, Takeda K, et al. 2001. Loss of mRor1 enhances the heart and skeletal abnormalities in mRor2-deficient mice: Redundant and pleiotropic functions of mRor1 and mRor2 receptor tyrosine kinases. *Mol Cell Biol* **21:** 8329–8335.

O'Connell MP, Fiori JL, Baugher KM, Indig FE, French AD, Camilli TC, Frank BP, Earley R, Hoek KS, Hasskamp JH, et al. 2009. Wnt5A activates the calpain-mediated cleavage of filamin A. *J Invest Dermatol* **129:** 1782–1789.

O'Connell MP, Fiori JL, Xu M, Carter AD, Frank BP, Camilli TC, French AD, Dissanayake SK, Indig FE, Bernier M, et al. 2010. The orphan tyrosine kinase receptor, ROR2, mediates Wnt5A signaling in metastatic melanoma. *Oncogene* **29:** 34–44.

Oishi I, Takeuchi S, Hashimoto R, Nagabukuro A, Ueda T, Liu ZJ, Hatta T, Akira S, Matsuda Y, Yamamura H, et al. 1999. Spatio-temporally regulated expression of receptor tyrosine kinases, mRor1, mRor2, during mouse development: implications in development and function of the nervous system. *Genes Cells* **4:** 41–56.

Oishi I, Suzuki H, Onishi N, Takada R, Kani S, Ohkawara B, Koshida I, Suzuki K, Yamada G, Schwabe GC, et al. 2003. The receptor tyrosine kinase Ror2 is involved in non-canonical Wnt5a/JNK signalling pathway. *Genes Cells* **8:** 645–654.

Oldridge M, Fortuna AM, Maringa M, Propping P, Mansour S, Pollitt C, DeChiara TM, Kimble RB, Valenzuela DM, Yancopoulos GD, et al. 2000. Dominant mutations in ROR2, encoding an orphan receptor tyrosine kinase, cause brachydactyly type B. *Nat Genet* **24:** 275–278.

Patthy L. 2000. The WIF module. *Trends Biochem Sci* **25:** 12–13.

Peradziryi H, Kaplan NA, Podleschny M, Liu X, Wehner P, Borchers A, Tolwinski NS. 2011. PTK7/Otk interacts with Wnts and inhibits canonical Wnt signalling. *EMBO J* **30:** 3729–3740.

Person AD, Beiraghi S, Sieben CM, Hermanson S, Neumann AN, Robu ME, Schleiffarth JR, Billington CJ Jr, van Bokhoven H, Hoogeboom JM, et al. 2010. WNT5A mutations in patients with autosomal dominant Robinow syndrome. *Dev Dyn* **239:** 327–337.

Pinson KI, Brennan J, Monkley S, Avery BJ, Skarnes WC. 2000. An LDL-receptor-related protein mediates Wnt signalling in mice. *Nature* **407:** 535–538.

Reddy UR, Phatak S, Pleasure D. 1996. Human neural tissues express a truncated Ror1 receptor tyrosine kinase, lacking both extracellular and transmembrane domains. *Oncogene* **13:** 1555–1559.

Sakurai M, Aoki T, Yoshikawa S, Santschi LA, Saito H, Endo K, Ishikawa K, Kimura K, Ito K, Thomas JB, et al. 2009. Differentially expressed Drl and Drl-2 play opposing roles in Wnt5 signaling during *Drosophila* olfactory system development. *J Neurosci* **29:** 4972–4980.

Saldanha J, Singh J, Mahadevan D. 1998. Identification of a Frizzled-like cysteine rich domain in the extracellular region of developmental receptor tyrosine kinases. *Protein Sci* **7:** 1632–1635.

Sammar M, Stricker S, Schwabe GC, Sieber C, Hartung A, Hanke M, Oishi I, Pohl J, Minami Y, Sebald W, et al. 2004. Modulation of GDF5/BRI-b signalling through interaction with the tyrosine kinase receptor Ror2. *Genes Cells* **9:** 1227–1238.

Sanchez-Solana B, Laborda J, Baladron V. 2012. Mouse resistin modulates adipogenesis and glucose uptake in 3T3-L1 preadipocytes through the ROR1 receptor. *Mol Endocrinol* **26:** 110–127.

Sato A, Yamamoto H, Sakane H, Koyama H, Kikuchi A. 2010. Wnt5a regulates distinct signalling pathways by binding to Frizzled2. *EMBO J* **29:** 41–54.

Schambony A, Wedlich D. 2007. Wnt-5A/Ror2 regulate expression of XPAPC through an alternative noncanonical signaling pathway. *Dev Cell* **12:** 779–792.

Schwarz-Romond T, Fiedler M, Shibata N, Butler PJ, Kikuchi A, Higuchi Y, Bienz M. 2007. The DIX domain of Dishevelled confers Wnt signaling by dynamic polymerization. *Nat Struct Mol Biol* **14:** 484–492.

Song S, Zhang B, Sun H, Li X, Xiang Y, Liu Z, Huang X, Ding M. 2010. A Wnt-Frz/Ror-Dsh pathway regulates neurite outgrowth in *Caenorhabditis elegans*. *PLoS Genet* **6**.

Stacker SA, Hovens CM, Vitali A, Pritchard MA, Baker E, Sutherland GR, Wilks AF. 1993. Molecular cloning and chromosomal localisation of the human homologue of a receptor related to tyrosine kinases (RYK). *Oncogene* **8:** 1347–1356.

Taillebourg E, Moreau-Fauvarque C, Delaval K, Dura JM. 2005. In vivo evidence for a regulatory role of the kinase activity of the linotte/derailed receptor tyrosine kinase, a *Drosophila* Ryk ortholog. *Dev Genes Evol* **215:** 158–163.

Takeuchi S, Takeda K, Oishi I, Nomi M, Ikeya M, Itoh K, Tamura S, Ueda T, Hatta T, Otani H, et al. 2000. Mouse Ror2 receptor tyrosine kinase is required for the heart development and limb formation. *Genes Cells* **5:** 71–78.

Tamai K, Semenov M, Kato Y, Spokony R, Liu C, Katsuyama Y, Hess F, Saint-Jeannet JP, He X. 2000. LDL-receptor-related proteins in Wnt signal transduction. *Nature* **407:** 530–535.

Tauriello DV, Jordens I, Kirchner K, Slootstra JW, Kruitwagen T, Bouwman BA, Noutsou M, Rudiger SG, Schwamborn K, Schambony A, et al. 2012. Wnt/β-catenin signaling requires interaction of the Dishevelled DEP domain and C terminus with a discontinuous motif in Frizzled. *Proc Natl Acad Sci* **109:** E812–820.

Topol L, Jiang X, Choi H, Garrett-Beal L, Carolan PJ, Yang Y. 2003. Wnt-5a inhibits the canonical Wnt pathway by promoting GSK-3-independent β-catenin degradation. *J Cell Biol* **162:** 899–908.

Tseng HC, Lyu PC, Lin WC. 2010. Nuclear localization of orphan receptor protein kinase (Ror1) is mediated through the juxtamembrane domain. *BMC Cell Biol* **11:** 48.

Tseng HC, Kao HW, Ho MR, Chen YR, Lin TW, Lyu PC, Lin WC. 2011. Cytoskeleton network and cellular migration modulated by nuclear-localized receptor tyrosine kinase ROR1. *Anticancer Res* **31:** 4239–4249.

Tyner JW, Deininger MW, Loriaux MM, Chang BH, Gotlib JR, Willis SG, Erickson H, Kovacsovics T, O'Hare T, Heinrich MC, et al. 2009. RNAi screen for rapid therapeutic target identification in leukemia patients. *Proc Natl Acad Sci* **106:** 8695–8700.

Verkaar F, Blankesteijn WM, Smits JF, Zaman GJ. 2010. β-Galactosidase enzyme fragment complementation for the measurement of Wnt/β-catenin signaling. *FASEB J* **24:** 1205–1217.

Wang X, Sommer RJ. 2011. Antagonism of LIN-17/Frizzled and LIN-18/Ryk in nematode vulva induction reveals evolutionary alterations in core developmental pathways. *PLoS Biol* **9:** e1001110.

Wehrli M, Dougan ST, Caldwell K, O'Keefe L, Schwartz S, Vaizel-Ohayon D, Schejter E, Tomlinson A, DiNardo S. 2000. Arrow encodes an LDL-receptor-related protein essential for Wingless signalling. *Nature* **407:** 527–530.

Wilson C, Goberdhan DC, Steller H. 1993. Dror, a potential neurotrophic receptor gene, encodes a *Drosophila* homolog of the vertebrate Ror family of Trk-related receptor tyrosine kinases. *Proc Natl Acad Sci* **90:** 7109–7113.

Wouda RR, Bansraj MR, de Jong AW, Noordermeer JN, Fradkin LG. 2008. Src family kinases are required for WNT5 signaling through the Derailed/RYK receptor in the *Drosophila* embryonic central nervous system. *Development* **135:** 2277–2287.

Wright TM, Brannon AR, Gordan JD, Mikels AJ, Mitchell C, Chen S, Espinosa I, van de Rijn M, Pruthi R, Wallen E, et al. 2009. Ror2, a developmentally regulated kinase, promotes tumor growth potential in renal cell carcinoma. *Oncogene* **28:** 2513–2523.

Yamada M, Udagawa J, Matsumoto A, Hashimoto R, Hatta T, Nishita M, Minami Y, Otani H. 2010. Ror2 is required for midgut elongation during mouse development. *Dev Dyn* **239:** 941–953.

Yamagata K, Li X, Ikegaki S, Oneyama C, Okada M, Nishita M, Minami Y. 2012. Dissection of Wnt5a-Ror2 signaling leading to matrix metalloproteinase (MMP-13) expression. *J Biol Chem* **287:** 1588–1599.

Yamaguchi TP, Bradley A, McMahon AP, Jones S. 1999. A Wnt5a pathway underlies outgrowth of multiple structures in the vertebrate embryo. *Development* **126:** 1211–1223.

Yamaguchi T, Yanagisawa K, Sugiyama R, Hosono Y, Shimada Y, Arima C, Kato S, Tomida S, Suzuki M, Osada H, et al. 2012. NKX2-1/TITF1/TTF-1-induced ROR1 is required to sustain EGFR survival signaling in lung adenocarcinoma. *Cancer Cell* **21:** 348–361.

Yamamoto H, Yoo SK, Nishita M, Kikuchi A, Minami Y. 2007. Wnt5a modulates glycogen synthase kinase 3 to induce phosphorylation of receptor tyrosine kinase Ror2. *Genes Cells* **12:** 1215–1223.

Yamamoto S, Nishimura O, Misaki K, Nishita M, Minami Y, Yonemura S, Tarui H, Sasaki H. 2008. Cthrc1 selectively activates the planar cell polarity pathway of Wnt signaling by stabilizing the Wnt-receptor complex. *Dev Cell* **15:** 23–36.

Yamamoto H, Oue N, Sato A, Hasegawa Y, Yamamoto H, Matsubara A, Yasui W, Kikuchi A. 2010. Wnt5a signaling is involved in the aggressiveness of prostate cancer and expression of metalloproteinase. *Oncogene* **29:** 2036–2046.

Yamamoto Y, Takeshita H, Sawa H. 2011. Multiple Wnts redundantly control polarity orientation in *Caenorhabditis elegans* epithelial stem cells. *PLoS Genet* **7:** e1002308.

Yang J, Baskar S, Kwong KY, Kennedy MG, Wiestner A, Rader C. 2011. Therapeutic potential and challenges of targeting receptor tyrosine kinase ROR1 with monoclonal antibodies in B-cell malignancies. *PLoS ONE* **6:** e21018.

Yen WW, Williams M, Periasamy A, Conaway M, Burdsal C, Keller R, Lu X, Sutherland A. 2009. PTK7 is essential for polarized cell motility and convergent extension during mouse gastrulation. *Development* **136:** 2039–2048.

Yoshikawa S, Bonkowsky JL, Kokel M, Shyn S, Thomas JB. 2001. The derailed guidance receptor does not require kinase activity in vivo. *J Neurosci* **21:** RC119.

Yoshikawa S, McKinnon RD, Kokel M, Thomas JB. 2003. Wnt-mediated axon guidance via the *Drosophila* Derailed receptor. *Nature* **422:** 583–588.

Zeng X, Tamai K, Doble B, Li S, Huang H, Habas R, Okamura H, Woodgett J, He X. 2005. A dual-kinase mechanism for Wnt co-receptor phosphorylation and activation. *Nature* **438:** 873–877.

Zhang S, Chen L, Cui B, Chuang HY, Yu J, Wang-Rodriguez J, Tang L, Chen G, Basak GW, Kipps TJ. 2012. ROR1 is expressed in human breast cancer and associated with enhanced tumor-cell growth. *PLoS ONE* **7:** e31127.

Zinovyeva AY, Yamamoto Y, Sawa H, Forrester WC. 2008. Complex network of Wnt signaling regulates neuronal migrations during *Caenorhabditis elegans* development. *Genetics* **179:** 1357–1371.

Receptor Tyrosine Kinases in *Drosophila* Development

Richelle Sopko[1] and Norbert Perrimon[1,2]

[1]Department of Genetics, Howard Hughes Medical, Institute Harvard Medical School, Boston, Massachusetts 02115

[2]Howard Hughes Medical, Institute Harvard Medical School, Boston, Massachusetts 02115

Correspondence: perrimon@receptor.med.harvard.edu

Tyrosine phosphorylation plays a significant role in a wide range of cellular processes. The *Drosophila* genome encodes more than 20 receptor tyrosine kinases and extensive studies in the past 20 years have illustrated their diverse roles and complex signaling mechanisms. Although some receptor tyrosine kinases have highly specific functions, others strikingly are used in rather ubiquitous manners. Receptor tyrosine kinases regulate a broad expanse of processes, ranging from cell survival and proliferation to differentiation and patterning. Remarkably, different receptor tyrosine kinases share many of the same effectors and their hierarchical organization is retained in disparate biological contexts. In this comprehensive review, we summarize what is known regarding each receptor tyrosine kinase during *Drosophila* development. Astonishingly, very little is known for approximately half of all *Drosophila* receptor tyrosine kinases.

One of the key strategies that arose during evolution to facilitate the transmission of extracellular information was that of receptor tyrosine kinase (RTK) signaling. This mechanism enables cells to transduce cues from their extracellular environment and thus contributes extensively to developmental processes. Today, we have come to recognize conserved RTK signaling as crucial for most aspects of cell fate determination, differentiation, patterning, proliferation, growth, and survival in metazoans. Activation of RTKs by ligand leads to a canonical deployment of signal transduction involving adaptor proteins, serine/threonine kinases, and transcription factors essential for animal development.

RTKs function reiteratively in different contexts during development to direct, restrain, or alter the commitment of a cell. Genetically tractable model organisms such as *Drosophila melanogaster* have proven instrumental in deciphering the roles of RTKs during development as well as their signaling pathways. Furthermore, extension of this knowledge to mammalian orthologs has substantially broadened our understanding of the function of RTKs in development and cellular transformation. Approximately 20 RTKs are encoded by the *Drosophila* genome, nearly all of which have a mammalian counterpart (Table 1). In this article, we review what we know to date about their functions, illustrating the diversity of cellular

Table 1. Drosophila RTKs/ligands/signaling components/transcription factors

Flybase ID	Symbol	Name	Mammalian homolog	Ligand	Characterized signaling pathway components
FBgn0040505	Alk	Alk	ALK	Jelly belly	mtg, Ras1, rl
FBgn0053531	Ddr	Discoidin domain receptor	DDR1 and DDR2	Collagen?	
FBgn0024245	dnt	Doughnut on 2	RYK	Wnt5?	
FBgn0015380	drl	Derailed	RYK	Wnt5	Src64B
FBgn0033791	Drl-2	Derailed 2	RYK	Wnt5	
FBgn0003731	Egfr	Epidermal growth factor receptor	EGFR	Spitz, Gurken, Vein, Keren	rho, Star, Ras1, Sos, csw, phl, Shc, dos, Gap1, Dsor, drk, ksr, cnk, rl, pnt, aop, ttk, sprouty, kekkon, argos
FBgn0025936	Eph/Dek	Eph receptor tyrosine kinase	EPHA and EPHB	Ephrin, Vap33, Exn	kuz, Exn, cac, Cdc42
FBgn0010389	htl/DFR1/Dtk1	Heartless	FGFR	Pyramus, Thisbe	Ras1, stumps, csw, rl, aop, pnt
FBgn0005592	btl/DFR2/Dtk2	Breathless	FGFR	Branchless	Ras1, stumps, csw, drk, Shc, Sos, ksr, cnk, sprouty, rl, grh, gro, pnt, aop
FBgn0013984	InR	Insulin-like receptor	INSR/IGF1R	Ilp1-7	chico, Sos, Drk, Shc, Ras, Pten, Pi3K92E, Pi3K21B, Pdk1, Tsc1, gigas, Rheb, Tor, Akt1, S6k, foxo
FBgn0038279	CG3837		INSR/IGF1R	Ilps?	
FBgn0032752	CG10702		INSR/IGF1R	Ilps?	
FBgn0032006	Pvr	PDGF- and VEGF-receptor related	VEGFR and PDGFR	PVF1,2,3	Ras1, rl, aop, Rac, mbc, ELMO, Crk, Cdc42
FBgn0011829	Ret	Ret oncogene	RET		
FBgn0010407	Ror	One of two Ror kinases	Ror1 and Ror2	Orphan receptor	
FBgn0020391	Nrk	Neurotropic receptor kinase	MuSK	Orphan receptor	
FBgn0004839	otk/Dtrk	Offtrack	Trk	Wnt4	plexA, dsh
FBgn0003366	sev	Sevenless		Boss	Ras1, Sos, csw, phl, drk, dos, ksr, Gap1, Dsor, rl, aop, Pnt, Lz
FBgn0003733	tor	Torso		Trunk	Torso-like, fs(1)N, fs(1)ph, Ras1, Sos, csw, Shc, dos, Gap1, ksr, phl, Dsor, drk, rl, cic, gro
FBgn0022800	Cad96Ca/Stitcher	Cad96Ca			rl, grh
FBgn0014073	Tie	Tie-like receptor tyrosine kinase			

Cite this article as *Cold Spring Harb Perspect Biol* doi: 10.1101/cshperspect.a009050

processes controlled by RTK signaling as well as the extent of pleiotropy associated with specific RTKs.

Torso: AN RTK DETERMINANT OF ANTERIOR/POSTERIOR PATTERNING AND METAMORPHOSIS

The first RTK to be deployed during *Drosophila* embryogenesis is Torso. Torso is maternally contributed and localized uniformly to the membrane of the syncytial blastocyst. Localized activation of Torso involves the processing of its presumptive ligand, Trunk, at the egg poles, a process requiring at least three genes: *torso-like*, *fs(1)Nasrat (fs(1)N)*, and *fs(1)polehole (fs(1)ph)* (Casanova and Struhl 1989; Sprenger et al. 1989; Stevens et al. 1990; Perrimon et al. 1995; Casali and Casanova 2001). Progeny derived from females lacking *torso, trunk,* or any of the aforementioned "terminal class genes" fail to develop stereotypical head and tail structures (Perrimon et al. 1986; Schupbach and Wieschaus 1986; Nüsslein-Volhard et al. 1987). Gain-of-function alleles of *torso*, on the other hand, drive the opposite phenotype: embryos with an extended posterior domain and minimal thoracic and abdominal regions (Klingler et al. 1988; Casanova and Struhl 1989; Schupbach and Wieschaus 1989; Strecker et al. 1989; Szabad et al. 1989). Screens to uncover suppressors of a *torso* gain-of-function allele identified *Ras1* and *son of sevenless (Sos)*. Further epistasis experiments positioned *corkscrew (csw; SHP2)*, *SHC-adaptor protein(Shc), GTPase-activating protein1(Gap1)*, *kinase suppressor of ras (ksr)*, *leonardo (leo; 14-3-3ζ)*, *polehole (phl; RAF)*, *Downstream of raf1 (Dsor; MEK)*, *downstream of receptor kinases (drk; GRB2)*, and *rolled (rl; ERK)* within the hierarchy responsible for transducing the downstream signal from Torso (Ambrosio et al. 1989a,b; Casanova and Struhl 1989; Stevens et al. 1990; Perkins et al. 1992; Doyle and Bishop 1993; Lu et al. 1993, 1994; Tsuda et al. 1993; Brunner et al. 1994; Hou et al. 1995; Therrien et al. 1995; Li et al. 1997; Luschnig et al. 2000) (Fig. 1).

Torso activation peaks between 1–2 hr of embryonic development (Sprenger and Nüss-

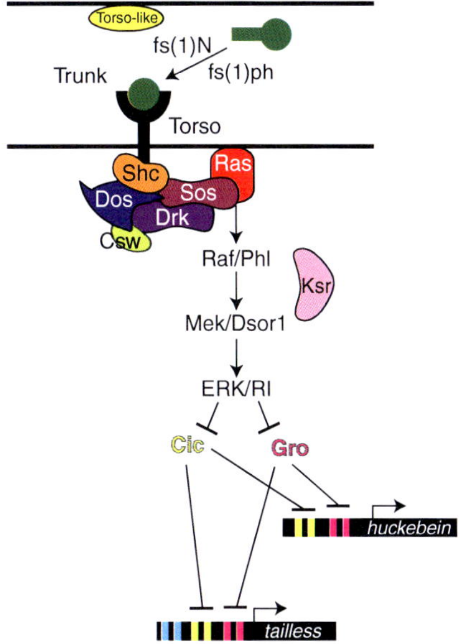

Figure 1. Torso activation in embryogenesis. Processing of the Torso ligand Trunk occurs locally at the anterior and posterior embryonic poles and requires Torso-like, fs(1)N, and fs(1)ph. Engagement of Torso by processed Trunk triggers Torso autophosphorylation and subsequent recruitment of downstream adaptors and effectors. A phosphorylation cascade initiated by Tor activation and involving Raf/Phl, Mek/Dsor1, and ERK/Rl leads to the inhibition of transcriptional repression by Cic and Gro. This permits gap gene (*tailless* and *huckebein*) and subsequent pair-rule gene expression and enables patterning of the developing embryo.

lein-Volhard 1992; Sprenger et al. 1993) results in the expression of *tailless* (*tll*) and *huckebein* (*hkb*), genes encoding for transcriptional repressors, at the embryonic poles (Moran and Jimenez 2006). These "terminal gap genes" demarcate zones of differentiation and embryos deficient for these gene products display phenotypes resembling those deficient for other members of the maternal terminal class (Pignoni et al. 1990; Weigel et al. 1990; Brönner and Jäckle 1991). The terminal class gene, *rl/ERK*, is upstream of *tll*, based on the fact that gain-of-function mutations in *rl/ERK* are unable to rescue *tll* null mutant embryos (Brunner et al. 1994). Phosphorylation of the transcriptional repressor Capicua (Cic) and corepressor Groucho (Gro) by activated ERK relieves transcriptional repression of *tll* and *hkb* at the embryonic posterior pole (Astigarraga et al. 2007; Cinnamon et al. 2008; Helman et al. 2011). At the anterior pole, Torso activation down-regulates the homeodomain transcription factor Bicoid by phosphorylation-independent mechanisms, in addition to inactivating Cic and Gro (Pignoni et al. 1992; Ronchi et al. 1993; Bellaïche et al. 1996; Janody et al. 2001). At both termini, Torso signaling inhibits Gro and permits Tll-dependent suppression of gap gene expression, whereas active Gro in central regions is unaffected and can repress *tll* expression, permitting central gap gene expression and in this way establishing expression "stripes" (Steingrímsson et al. 1991; Moran and Jimenez 2006). Gap genes encode for transcription factors that will activate expression of pair-rule genes. This sequential activation of gene expression enables patterning of the developing embryo (Nasiadka et al. 2002).

Torso also functions as a receptor for the neuropeptide prothoracicotropic hormone (PTTH) in the *Drosophila* brain during metamorphosis (Rewitz et al. 2009). Torso engagement by PTTH in the prothoracic gland (PG), an endocrine organ in insects, triggers Ras/Raf/ERK signaling to drive the production and/or release of the hormone ecdysone. Reduction of *torso* by RNAi specifically in the PG results in developmental delays similar to that resulting from ablation of PTTH-expressing neurons (McBrayer et al. 2007). PTTH shares significant structural homology with the Torso ligand Trunk and can substitute for Trunk in terminal signaling during embryogenesis (Rewitz et al. 2009).

Sevenless: AN RTK SPECIFYING CELL FATE IN THE *Drosophila* EYE AND TESTES

The *Drosophila* compound eye is comprised of 750–800 repetitive units termed ommatidia. Each ommatidium consists of eight photoreceptor neurons (R1–R8) and four lens-secreting cone cells, surrounded by a net of pigment cells that optically insulate each ommatidium from its neighbors (Wolff and Ready 1993). The spectral specificity of photoreceptor subtypes is provided by G-coupled Rhodopsin receptors. Photoreceptor differentiation occurs during the larval stage wherein a progressive "wave" of cell differentiation proceeds from the posterior to anterior region of the eye imaginal disc, the precursor of the adult eye. This wave (the morphogenetic furrow; MF) is visualized as a narrow indentation of epithelial cells contracting in the apical-basal dimension in a concerted fashion. Posterior to the MF, differentiated cells arrange into clusters and adopt the mature ommatidium pattern (Voas and Rebay 2003). A number of signaling pathways initiate MF progression and cell differentiation including epidermal growth factor receptor (EGFR), Notch, Wingless (Wg), Hedgehog (Hh), JAK-STAT, Decapentaplegic (Dpp), and Sevenless (Sev) (Charlton-Perkins et al. 2011).

R7 is the last photoreceptor to be recruited to the ommatidial cluster and specified. Differentiation of R7 relies on signals from neighboring cells within each ommatidial cluster; engagement of the RTK Sev on the surface of R7 by its membrane-associated ligand Bride-of-Sevenless (Boss), expressed exclusively in R8, triggers Ras/Raf/ERK signaling in R7 (Hart et al. 1990; Krämer et al. 1991; Simon et al. 1991). A lack of Sev activity in the R7 precursor (Tomlinson and Ready 1986, 1987; Tomlinson et al. 1987; Basler and Hafen 1988), or a lack of Boss in R8 (Reinke and Zipursky 1988), redirects R7 cell fate to that of a cone cell. Conversely, a cone cell precursor can be directed to become a R7 photoreceptor if the precursor expresses constitutively active Sev (Basler et al. 1991; Dickson et al. 1992; Sprenger and Nüsslein-Volhard 1992). Normally, the activation of Sev in photoreceptors other than R7, is restricted by the activity of *Socs36E*, expressed in all photoreceptors except R7 (Almudi et al. 2009, 2010), and reinforced in R7 by the adaptor protein Drk, specifically expressed in R7 (Olivier et al. 1993; Simon et al. 1993) (Fig. 2). Ras/Raf/Mek downstream from Sev (and EGFR—see below) results in the phosphorylation by Rl/ERK of two transcription factors critical for photoreceptor specification: Anterior open (Aop) and Pointed-P2 (Pnt-P2). Phosphorylation inhibits the repressor activity of Aop (O'Neill et al. 1994) by targeting it for nuclear export (Tootle et al. 2003) and degradation

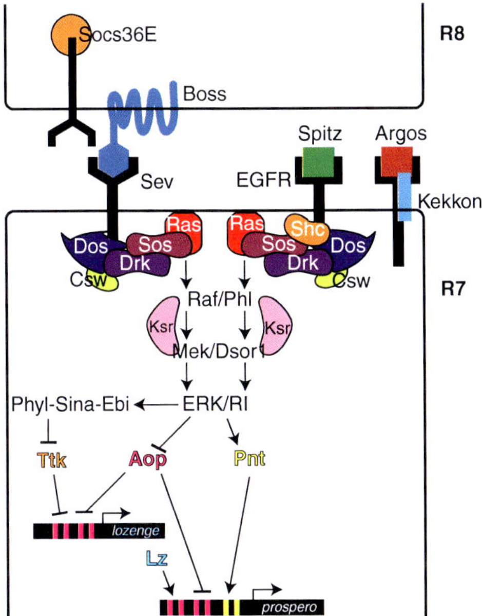

Figure 2. Sevenless and EGFR signaling in photoreceptor specification. Engagement of Sev on the R7 photoreceptor cell by its ligand Boss on R8 activates the Ras/Raf/Mek/ERK signaling cascade in R7. A similar cassette of signaling proteins execute EGFR-dependent functions, following EGFR activation by its ligand Spitz. Sev activation in R8 is prevented by Socs36E and reinforced in R7 by the adaptor protein Drk, whereas Argos and Kekkon limit EGFR activation. Phosphorylated and active ERK/Rl targets the transcriptional repressors Aop and Ttk (via Phyl-Sina-Ebi) for degradation, whereas ERK-dependent phosphorylation stimulates Pnt transcriptional activity. This relieves transcriptional repression at *lozenge* and *prospero* enhancers. Lz functions together with Pnt to activate *pros* expression, thus providing R7 identity by repressing the expression of cone cell and R8-specific rhodopsins.

(Rebay and Rubin 1995). Pnt-P2, on the other hand, requires phosphorylation for its activity (Brunner et al. 1994; O'Neill et al. 1994). These factors play antagonistic roles in regulating the *lozenge* (*lz*) enhancer directly (Xu et al. 2000; Behan et al. 2002; Jackson Behan et al. 2005). High levels of RTK signaling in the prospective R7 cell relieve Tramtrack (Ttk)-dependent repression of *lz* (Daga et al. 1996; Xu et al. 2000; Siddall et al. 2009). Rl/ERK targets Ttk to the E3 ubiquitin ligase complex (comprised of Seven

in absentia, Ebi, and Phyllopod) for degradation (Lai et al. 1997, 2002; Tang et al. 1997; Boulton et al. 2000; Li et al. 2002). Lz regulates *prospero* (*pros*) expression specifically in R7 by binding the *pros* enhancer directly. Lz works with Pnt-P2 at the *pros* enhancer, when Aop is displaced because of high ERK activity (Xu et al. 2000; Jackson Behan et al. 2005; Hayashi et al. 2008; Siddall et al. 2009). Pros functions to repress the expression of cone cell and R8-specific Rhodopsins thereby providing identity to R7 (Cook et al. 2003).

Genetic screening for modifiers of *sev* phenotypes identified many players downstream from Sev, including Ras1, Sos, Raf, Drk/Grb2, Dos, Csw, Gap1, and Rl/ERK (Rogge et al. 1991, 1992; Simon et al. 1991; Dickson et al. 1992; Gaul et al. 1992; Olivier et al. 1993; Biggs et al. 1994; Brunner et al. 1994; Raabe et al. 1996). These experiments were among the first to identify the genetic requirements for RTK signaling and in doing so delineate the hierarchy of canonical RTK signaling.

As is the case of Torso, Sev has also been shown to play a role in an additional cell type, in this case the male testes. Sev is required in a subset of somatic cells of the male embryonic gonad to spatially restrict stem cell niche differentiation. Sev activation in posterior somatic gonadal cells by Boss, presented by adjacent primordial germ cells, represses their differentiation into hub/niche cells and thereby restricts germline stem cell numbers (Kitadate et al. 2007)

EGFR: AN RTK WITH MULTITUDE ROLES

The *Drosophila* EGFR is involved in numerous developmental decisions throughout the *Drosophila* life cycle. For instance, the EGFR pathway has roles in dorsal/ventral patterning of the embryonic ectoderm and the establishment of neuroectoderm, wing development, antennal formation, photoreceptor differentiation, lamina neuron differentiation, and the specification of muscle precursors and invagination of tracheal branches to name a few (Perrimon and Perkins 1997; Shilo 2003). EGFR predominantly mediates short-range signaling that is restricted either to the cells producing EGF or to cells positioned 1–2 cells away. EGFR activates the canonical Ras/Raf/MEK/ERK pathway and visualization of pathway activity has documented the highly dynamic activity of EGFR signaling (Gabay et al. 1997) (Fig. 2). Multiple EGFR ligands and feedback loops are responsible for the complex temporal and spatial regulation of EGFR signaling (Perrimon and McMahon 1999; Shilo 2005). Regarding activation, there are four EGFR ligands in *Drosophila*: Spitz, Keren, Gurken, and Vein, with Vein being the only secreted protein that does not require processing for its activity. Detailed studies of ligand processing, in particular of Spitz, have shown a requirement for two proteins, Star and Rhomboid (Rho). Star is a type II transmembrane protein that associates with Spitz, facilitating its translocation from the ER to a cellular compartment where it is cleaved by the seven-pass transmembrane protein Rhomboid (Rho). Importantly, although Star expression is ubiquitous, Rho is extremely dynamic and thus responsible for controlling EGFR activation in a wide range of tissues. Finally, the transcriptional induction of negative regulators of the pathway restricts the spatial and temporal activation of EGFR signaling. These include the cytoplasmic proteins Sprouty and Cbl, the extracellular secreted molecule Argos, and the extracellular transmembrane protein Kekkon. A comprehensive description of the entirety of EGFR function during *Drosophila* development would exhaust the page limitations of this chapter and so we refer readers to a comprehensive review (Shilo 2005).

FGFR SUPERFAMILY: HEARTLESS AND BREATHLESS

Heartless: An RTK Influencing Mesodermal Cell Migration and Cell Specification

The gene product encoded by *heartless* (*htl*) shares ~60% identity in its kinase domain with vertebrate FGFRs and 80% identity with the other FGFR homolog in *Drosophila*, Breathless (Btl). *htl* is expressed ~2.5 hr postfertilization in presumptive mesoderm at the onset of

gastrulation. *htl* expression persists throughout embryogenesis in somatic muscle precursors including cardiac and pericardial cells, pharyngeal muscle cells, visceral muscle precursors, and additionally in glia of the central nervous system (CNS) (Shishido et al. 1993; Hidalgo and Booth 2000; Egger et al. 2002; Freeman et al. 2003). During larval stages, *htl* is expressed in muscle cell precursors of wing and leg imaginal discs and in neural precursors and glia of the brain and eye imaginal discs (Emori and Saigo 1993; Sato and Kornberg 2002; Butler 2003; Butler et al. 2003; Franzdóttir et al. 2009). During pupal and adult stages, *htl* is expressed in abdominal and thoracic myoblasts (Dutta et al. 2005).

After ventral furrow invagination, the mesoderm primordium undergoes an epithelial-to-mesenchymal transition and spreads dorsally over ectodermal cells to form a monolayer (Schumacher et al. 2004; Wilson 2005; Wilson et al. 2005; Clark et al. 2011). This repositioning is required for the reception by mesodermal cells of patterning cues (Dpp and Wg) from the adjacent ectoderm that specifies the mesoderm lineage into visceral mesoderm, heart tissue, somatic muscle, and the fat body (FB). *htl* null mutant embryos show defects in the bilateral spreading of mesoderm during gastrulation (Murray and Saint 2007; McMahon et al. 2008). *htl* mutants fail to develop visceral mesoderm and heart tissue, whereas somatic muscles are disorganized and reduced because of the absence of differentiated mesodermal subtypes. *htl* mutants show additional defects: failure of CNS glia to migrate and ensheath longitudinal ventral nerve cord (VNC) connectives, and defective salivary gland migration (Beiman et al. 1996; Gisselbrecht et al. 1996; Shishido et al. 1997; Michelson et al. 1998b; Schulz and Gajewski 1999; Mandal et al. 2004).

Expression of activated Ras1 partially rescues mesodermal defects associated with *htl* perturbation (Beiman et al. 1996; Gisselbrecht et al. 1996; Michelson et al. 1998b; Schulz and Gajewski 1999). Ras1 functions downstream or parallel to the adaptor protein Stumps (Carmena et al. 1998; Michelson et al. 1998a; Vincent et al. 1998; Imam et al. 1999) to transduce signals downstream from Htl (Johnson Hamlet and Perkins 2001; Petit et al. 2004; Csiszar et al. 2010). Ectopic expression of activated Aop, a transcriptional repressor downstream from ERK, generates phenotypes similar to that due to *htl* disruption whereas expression of activated Pnt, a transcriptional activator downstream from ERK, increases the number of somatic muscle progenitors (Halfon et al. 2000). These observations are consistent with Htl-dependent ERK activation (Gabay et al. 1997; Wilson et al. 2004).

Thisbe (*ths*) and *pyramus* (*pyr*), ligands for Htl, are expressed in the neurogenic ectoderm coincident with the migration of mesoderm during gastrulation. They are later differentially expressed in other epithelial tissues that flank mesodermal derivatives: the stomadeum, the hindgut, the CNS, and at muscle attachment sites. *ths* and *pyr* mutants are defective in mesodermal cell intercalation and monolayer formation after dorsal spreading. ERK activation at the leading edge of the migrating mesoderm is absent in *ths pyr* double mutants, and expanded as a result of ectopic *ths* expression similar to that attributable to constitutively active Htl expression. Constitutively active Htl partially restores mesodermal differentiation to *ths pyr* mutants (Gryzik and Müller 2004; Stathopoulos et al. 2004; Klingseisen et al. 2009; McMahon et al. 2010; Clark et al. 2011).

The anterior migration of caudal visceral mesoderm, giving rise to the longitudinal muscles that ensheath the gut, is guided by Htl activation. *ths* and *pyr*, expressed in adjacent trunk visceral mesoderm, together promote cell survival and restrict lateral movement of caudal visceral mesoderm cells during their migration along trunk visceral mesoderm (Kadam et al. 2012). In the cardiogenic mesoderm, Pyr plays the major role in activating ERK to maintain cardiogenic lineages (Klingseisen et al. 2009; Grigorian et al. 2011). In the eye imaginal disc, *ths* and *pyr* have different expression patterns that translate to unique contributions to Htl signaling: *pyr* for early glia–glia interactions that promote glial cell proliferation and migration, and *ths* for glial–neuron interactions that inhibit migration and trigger cell differentiation (Franzdóttir et al. 2009).

Htl is additionally required in the gonadal mesoderm for primordial germ cell (PMC) migration. In *htl* mutant embryos, PMCs transverse the posterior midgut but stall at the endoderm/lateral mesoderm border. Those few PMCs that infiltrate the lateral mesoderm fail to navigate toward and associate with somatic gonadal precursors (SGPs)—specialized mesodermal cells that give rise to the somatic portion of the gonad. SGPs of *htl* mutant embryos are reduced in number and deranged in shape (Moore et al. 1998).

htl is required in the *Drosophila* ocellar sensory system (OSS), to direct OSS axon development during puptiation. OSS axons migrate toward their targets in the brain, until metamorphosis when they become detached and reorient. Properties such as the ability to attach, detach, or cross to the brain are lost when dominant-negative Htl is expressed in neurons. Genetic evidence implies Htl functions downstream from Neuroglian—a homophilic cell adhesion molecule required for axon guidance—in this context (García-Alonso et al. 2000).

Htl is necessary for larval cardiac tube remodeling, which occurs without cell migration. Htl is additionally required for the formation of ventral imaginal muscle founders and the differentiation of leg imaginal disc associated myoblasts and abdominal/thoracic adult myoblasts. Modulation of Htl activity alters the number of myoblast founder cells and adult muscle fibers (Dutta et al. 2005; Maqbool et al. 2006; Zeitouni et al. 2007).

Breathless: An RTK Influencing Cell Migration and Patterning, Predominantly in the Tracheal System and CNS

Although they share significant identity, the two FGF receptors in *Drosophila*, Breathless (Btl) and Htl differ in their ligand binding domain structure (Shishido et al. 1993). As such, the Htl ligands Pyr and Ths are unable to activate Btl to influence tracheal branching whereas the Btl ligand, Branchless (Bnl), is unable to influence Htl-dependent mesoderm spreading and differentiation (Kadam et al. 2009).

btl is expressed during embryogenesis in the invaginating tracheal primordia and developing tracheal system, the salivary glands, CNS glia and neurons, cells of the gut and male genitalia primordium (Glazer and Shilo 1991; Klambt et al. 1992; Shishido et al. 1993; Ahmad and Baker 2002). *bnl* is expressed in cells surrounding Btl-expressing cells and prefigures their migratory direction (Sutherland et al. 1996).

btl and *bnl* mutant embryos show defects in tracheal cell migration; however, the specification and proliferation of tracheal precursors is normal. The absence of a tracheal system skeleton in *btl* mutants ensues from an inability of cells to coordinately migrate out from the tracheal placodes in stereotyped directions and then intercalate and elongate to form tubes. Further, the formation and fusion of secondary and terminal branches, each derived from a single cell, is compromised in *btl* mutants because of a failure of tracheal cell fate acquisition. Proper tracheal branching relies on the spatial regulation of Btl activity; localized misexpression of *bnl* can redirect tracheal cell migration and induce branching, through the activation of Btl and a downstream Pnt-dependent gene expression program (Klambt et al. 1992; Reichman-Fried and Shilo 1995; Lee et al. 1996; Samakovlis et al. 1996; Sutherland et al. 1996). In this same manner, oxygen deprivation directs fine terminal branching during larval stages—triggering *bnl* expression and therefore, Btl activation—for oxygen delivery (Jarecki et al. 1999; Ghabrial et al. 2011). Branching relies on an extensive number of factors downstream from Btl (Ghabrial et al. 2011). For instance, Btl autophosphorylation, following Bnl binding (Lee et al. 1996; Sutherland et al. 1996), functions to recruit the adaptor protein Stumps. Phosphorylation of Stumps by Btl induces binding of the phosphatase Csw, another component of the signaling cascade that activates ERK (Michelson et al. 1998a; Vincent et al. 1998; Imam et al. 1999; Wilson et al. 2004) (Fig. 3). ERK function is discharged by the transcriptional activator Grh (Hemphala et al. 2003) and the transcriptional corepressor Gro (Cinnamon et al. 2008). ERK-dependent phosphorylation of these factors as well as of the transcription factor Aop, a

repressor of *btl*, modulates their activity (Ohshiro et al. 2002). The FGFR inhibitor Sprouty limits the range of Bnl signaling and prevents tracheal branch stalk cells from budding ectopically (Hacohen et al. 1998).

Tracheal branch fusion relies on Btl-dependent *Delta* expression in fusion cells of migrating branches. Delta displayed by the fusion cell activates Notch on adjacent cells to limit fusion cell identity to one cell per branch; Notch down-regulates *bnl* expression to restrict fusion cell identity and delimit ERK activation (Ikeya and Hayashi 1999). Notch-mediated lateral inhibition also restricts the number of leading cells in a branch (Ghabrial and Krasnow 2006). Fusion further depends on a single mesodermal cell—the bridge cell—that guides the

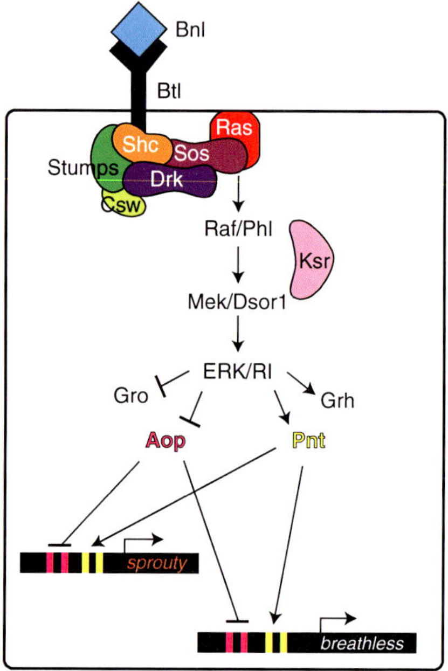

Figure 3. Breathless signaling in the tracheal system. Btl autophosphorylation, on Bnl binding, recruits the adaptor protein Stumps and additional downstream effectors. Ras activation initiates a phosphorylation cascade that culminates in the stimulation of ERK kinase activity. ERK-dependent phosphorylation of the transcriptional activators Grh and Pnt, and the transcriptional repressors Gro and Aop, modulates their activity at promoters. The consequential up-regulation of gene expression induces tracheal cell migration and tracheal branching and fine-tuning of Btl signaling.

extension and fusion of tracheal metameres to generate a continuous dorsal trunk (Wolf and Schuh 2000). Bnl induces filopodial tracheal cell extensions that contact the bridge cell and are essential for dorsal trunk branch fusion (Ribeiro et al. 2002; Wolf et al. 2002).

Btl/Bnl-dependent filopodial extensions are also presented by imaginal tracheoblasts, which proliferate to remodel the adult respiratory system during metamorphosis. The migration of tracheoblasts at the tip of the air sac primordium is dependent on Btl (Sato and Kornberg 2002; Cabernard and Affolter 2005). Btl additionally influences tracheoblast identity in the spiracular branch and the dorsal branch stalk, by inducing tracheoblast differentiation as well as by promoting differentiated cells to reenter the cell cycle, respectively (Guha et al. 2008; Sato et al. 2008; Weaver and Krasnow 2008; Pitsouli and Perrimon 2010).

btl-expressing cells migrate toward distinct populations of *bnl*-expressing cells additionally in the male genital imaginal disc; *bnl* expression by ectodermally-derived genital precursor cells induces *btl*-expressing cell migration into the male disc. The *btl*-expressing mesodermal cells are subsequently converted into epithelia during pupal stages to generate the vascular paragonia and vas deferens. In female genital discs, *bnl* expression is targeted by the female-specific repressor form of the transcription factor Doublesex (Ahmad and Baker 2002).

In the embryonic CNS of *btl* mutants, posterior midline glial cells migrate inappropriately, resulting in irregular commissural patterning (Klambt et al. 1992). Notably, perturbation of *pnt* and *stumps* also generates glial cell migration phenotypes (Klambt 1993; Vincent et al. 1998; Imam et al. 1999). During larval stages of eye patterning Bnl is required for cell adhesion and Hh-dependent apical constriction that enables ommatidial cluster formation. Btl is further required cell autonomously for retinal architectural integrity and noncell-autonomously in directing retinal glia migration (Mukherjee et al. 2012). In the adult brain, Btl is required to mediate axonal retraction rather than guidance specifically in the dorsal cluster neurons (DCN) of the visual system. In this context, Btl is activated in extending DCN axons as they encounter Bnl. Btl signaling activates Rac that in turn inhibits JNK signaling, inducing axonal retraction (Srahna et al. 2006).

INSULIN RECEPTOR RTK SUPERFAMILY: THE INSULIN RECEPTOR AND ANAPLASTIC LYMPHOMA KINASE

The Insulin and Insulin-Like Growth Factor Receptor in *Drosophila*: An RTK Essential for Growth

The insulin-like growth factor receptor in *Drosophila*, *InR*, is ubiquitously expressed throughout embryogenesis, with higher levels accumulating in the brain, midgut primordia, and VNC. A maternally inherited role for *InR* is reflected by abundant *InR* mRNA in nurse cells and mature oocytes (Petruzzelli et al. 1986; Garofalo and Rosen 1988). Accordingly, embryonic lethality is associated with *InR* complete loss-of-function mutations whereas some heteroallelic combinations yield animals that are viable but sterile (Fernandez et al. 1995; Chen et al. 1996; Tatar et al. 2001). Viable embryos lack neuroblasts and are unable to complete germband retraction and dorsal closure. They display abnormal head structures and cuticle. Moreover, mutant embryos display defects in VNC condensation and commissure formation. *InR* expression persists in the nervous system during larval stages, and is detected in all imaginal discs and postsynaptically at neuromuscular junctions (NMJs) (Garofalo and Rosen 1988; Gorczyca et al. 1993; Fernandez et al. 1995). InR is enriched in photoreceptor axons of late larvae (Song et al. 2003). In adults, *InR* mRNA is predominantly localized to the brain cortex, cells of the thoracic and abdominal ganglia, and the gut (Veenstra et al. 2008).

Small animals result from reduced InR activity (Fernandez et al. 1995; Chen et al. 1996; Brogiolo et al. 2001; Tatar et al. 2001). Organismal size is additionally affected by alteration of conserved InR signaling pathway components. Positive regulators downstream from InR include: the insulin receptor substrate (IRS) ortholog Chico (Böhni et al. 1999), the PI3K subunits PI3K92E and PI3K21B (Weinkove et al. 1999), the PI-dependent protein kinase Pdk1 (Rintelen et al. 2001), and the kinases Tor (Oldham et al. 2000), Akt1 (Verdu et al. 1999), and S6K (Montagne et al. 1999) (Fig. 4). Negative regulators downstream from InR include: the phosphatase Pten (Gao et al. 2000; Oldham et al. 2002), the tuberous sclerosis genes *Tsc1* and *Gigas/Tsc2* (Gao and Pan 2001; Potter et al. 2001), and the transcription factor Foxo (Jünger et al. 2003). Cell-autonomous effects of InR on cell size and number rely on kinase activity (Brogiolo et al. 2001) and are antagonized by *Pten* overexpression (Huang et al. 1999), co-expression of *Tsc1* and *gigas* (Potter et al. 2001), or reduced Pdk1 activity (Rintelen et al. 2001). Although *PI3K92E* or *Akt1* overexpression increases cell size it fails to influence cell number or division (Verdu et al. 1999; Weinkove et al. 1999), substantiating the supposition that InR serves two independent functions: to promote proliferation via Ras/ERK and to promote pro-

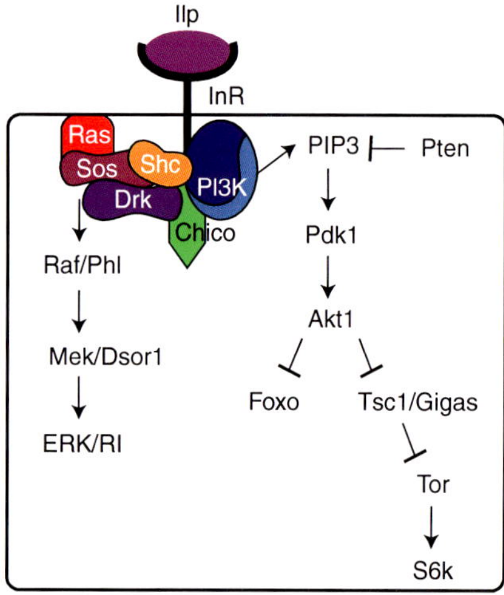

Figure 4. Ubiquitous InR signaling. Interaction of InR with its Ilp ligand induces InR autophosphorylation. Chico interacts directly with the phosphorylated receptor, recruiting PI3K along with a multitude of other factors including adaptor proteins and downstream effectors. InR promotes proliferation via canonical Ras/Raf/Mek/ERK signaling and growth via the activation of Akt1. Akt1 stimulates protein synthesis by activating downstream kinases Tor and S6K, and by inhibiting Foxo nuclear accumulation and transcriptional activity.

tein synthesis through PI3K (Oldham et al. 2002). The list of genes contributing to body size determination and functioning downstream from InR continues to grow and to date includes numerous conserved and *Drosophila*-specific factors.

Seven insulin-like peptides (Ilp1-7), InR ligands, are differentially expressed among *Drosophila* developmental stages and tissues: *Ilp2*, *Ilp4*, and *Ilp7* are expressed in the mesoderm and midgut; *Ilp1*, *Ilp2*, *Ilp3*, and *Ilp5* are expressed in neurosecretory cells of the brain that project to the ring gland, heart and brain lobes, and foregut; *Ilp2* is expressed in imaginal discs and salivary glands; *Ilp7* is expressed in neurons of the ventral ganglion, a subset of which innervates the adult hindgut and another that makes synaptic contact with *Ilp1,2,3,5*-expressing neurosecretory cells; *Ilp5* is expressed in ovarian follicle cells; *Ilp3* is expressed in adult midgut muscle; and *Ilp6* is expressed in the FB, gut, and in CNS surface glia (Gorczyca et al. 1993; Brogiolo et al. 2001; Ikeya et al. 2002; Rulifson et al. 2002; Broughton et al. 2005; Miguel-Aliaga et al. 2008; Veenstra et al. 2008; Okamoto et al. 2009; Slaidina et al. 2009; Chell and Brand 2010; Cognigni et al. 2011). *Ilp* overexpression increases organismal size by stimulating cell growth and division (Brogiolo et al. 2001; Ikeya et al. 2002; Slaidina et al. 2009). FB-specific activation of InR promotes triglyceride storage, by increasing fat cell number and lipid content (DiAngelo and Birnbaum 2009). Reciprocally, ablation of *Drosophila Ilp*-expressing neurons causes developmental delays, a reduction in egg production rates, and a reduction in organismal size owing to decreased cell number and size (Ikeya et al. 2002; Rulifson et al. 2002; Broughton et al. 2005). In addition to developmental disparities, flies devoid of Ilp-producing neurons have elevated levels of glucose and trehalose, like *InR* and *chico* mutants.

Although the InR-signaling pathway autonomously influences growth and proliferation, it nonautonomously influences aging. Long-lived animals result from reduced activity of InR, Chico, Tor, and S6K (Clancy et al. 2001; Tatar et al. 2001; Kapahi et al. 2004). Although nutritional starvation during larval stages influences organismal size and fecundity, it is insufficient to influence aging (Tu and Tatar 2003). Dietary restriction in adults, however, promotes longevity and correlates with reduced *Ilp5* mRNA levels (Min et al. 2008). Ablation of Ilp-producing neurosecretory cells can also extend lifespan (Broughton et al. 2005).

Recently, both the reactivation of proliferation in quiescent embryonic neuroblasts and the elimination of larval neuroblasts postpupariation were shown to correlate with InR signaling. The reception by quiescent embryonic neuroblasts of "nutritional adequacy" signals from the FB elicits *Ilp* expression in glial cells, activating InR and proliferation in quiescent neuroblasts in a paracrine manner (Britton and Edgar 1998). This reactivation of InR relies on Tor and the amino acid transporter Slimfast (Slif) in the FB, and functional InR/PI3K/Akt/Tor in neuroblasts. Consistent with InR promoting proliferation and survival, InR activation in combination with proapoptotic gene ablation delays larval neuroblast elimination. Reduced InR signaling, on the other hand, promotes larval neuroblast elimination postpupariation. In light of observations wherein impaired autophagy enhances caspase-dependent neuroblast survival, Foxo is proposed to activate autophagy in aging neuroblasts (Chell and Brand 2010; Siegrist et al. 2010; Sousa-Nunes et al. 2011). This is consistent with the acknowledged antagonism between autophagy and insulin signaling pathways (Chang et al. 2009) and the proposed mechanism of lifespan extension by Foxo in muscle (Demontis and Perrimon 2010). It is plausible that two other genes in the *Drosophila* genome, *CG3837* and *CG10702*, predicted to encode for insulin-like receptors (Table 1), play a role in this context given that elevated levels of the corresponding proteins have been detected in salivary glands undergoing autophagic programmed cell death (Martin et al. 2007).

In addition to growth and longevity, InR also impinges on reproduction. *InR* mutant eggs chambers, unable to interpret both follicular cell-derived and systemic insulin signals, are developmentally delayed owing to impaired vitellogenesis and cyst growth. InR, deficient germline stem cells (GSCs) show reduced division rates owing to compromised G_2 phase of the GSC division cycle, whereas hindered GSC division in *InR* mutant females reflects additional effects on Notch activity and E-cadherin-mediated adhesion within the niche (Drummond-Barbosa and Spradling 2001; LaFever and Drummond-Barbosa 2005; Hsu et al. 2008; Yu et al. 2009; Hsu and Drummond-Barbosa 2011). An analogous scenario exists in the male germline: insulin signaling in germ and support cells promotes spermatocyte growth and maintains cyst numbers by promoting G_2/M phase progression of GSCs (Ueishi et al. 2009; McLeod et al. 2010).

Ilps are expressed in synaptic boutons at the presynaptic terminals of larval body wall muscles (Gorczyca et al. 1993). Overexpression of *PI3K92E*, *Akt* or *InR*, specifically in larval motor neurons, induces supernumerary synapses projecting to each body wall muscle whereas reduced PI3K or Akt activity reduces synapse number. Similar effects are seen in projection neurons of the brain. Alterations of PI3K activity in motor and projection neurons elicits modified locomotive behaviors (Martin-Pena et al. 2006).

InR mutants display defects in photoreceptor axon path finding reminiscent of animals deficient for *Dystroglycan* and *Dystrophin*—two genes linked to muscular dystrophies. *InR* interacts genetically with *Dystroglycan* and is speculated to function independently of Chico and the PI3K/Akt pathway in this context, and instead associates with the Dg-Dock-Pak pathway to guide neuronal migration (Song et al. 2003; Shcherbata et al. 2007).

Alk: An RTK Involved in the Development of the Visceral Mesoderm, as Well as Motor and Visual Circuitry

Alk is required for the generation of visceral mesoderm—cells that will comprise the inner circular muscles and outer longitudinal muscles that ensheath the intestinal tract. *Alk* deficient embryos are unable to specify visceral mesoderm founder cells and completely lack visceral musculature. *Alk* mutants lack intestinal structures, do not eat and die at the first instar larval stage (Englund et al. 2003; Lee et al. 2003; Lorén et al. 2003).

The derivation of mesoderm in *Drosophila* relies on the coordinated activity of Dpp, Hedgehog (Hh) and Wg. Signaling in these pathways is mediated by Tinman, a conserved homeodomain transcription factor required for somatic, cardiac and visceral mesoderm development (Azpiazu and Frasch 1993; Bodmer 1993; Furlong 2004). A screen for Tinman targets identified *jelly belly (jeb)*. *jeb* is expressed in early ventral and medial somatic mesoderm cells, adjacent to *Alk*-expressing visceral mesoderm cells. Only visceral mesoderm is affected in *jeb* mutants; ergo, the logical gene name— *jelly belly*—referring to a jiggly expansive abdomen and lack of midgut muscles (J Weiss, pers. comm.). The *jeb* mutant phenotype is akin to that of *Alk* and is reflective of a failure of visceral mesodermal cells to differentiate and migrate, because of a lack of either Jeb secretion by ventral somatic mesoderm precursor cells or Jeb engagement by Alk at visceral muscle precursors (Weiss et al. 2001; Stute et al. 2004). Engagement of Alk by Jeb activates ERK in visceral muscle progenitors to establish two separate pools of cells: founder myoblasts and fusion-competent myoblasts. In the absence of *jeb* or *Alk*, all visceral mesoderm progenitors become fusion-competent myoblasts and these cells fuse with the somatic muscle founders because founder myoblasts are nonexistent, resulting in a complete lack of visceral musculature (Stute et al. 2004). Ectopic expression of activated Alk in the visceral mesoderm restores gut morphogenesis to *jeb* mutants by reinstating expression of genes downstream from ERK: *dumbfounded, org-1, sticks and stones, Hand,* and *Dpp* (Englund et al. 2003; Lee et al. 2003; Stute et al. 2004; Varshney and Palmer 2006; Shirinian et al. 2007).

A proposed function for Alk in neuronal development is based on conserved nervous system expression in *C. elegans*, mouse, and chick (Iwahara et al. 1997; Liao et al. 2004; Hurley et al. 2006; Vernersson et al. 2006; Reiner et al. 2008). Like *Alk*, *jeb* is expressed in a subset of neurons distributed throughout the VNC in late *Drosophila* embryogenesis. The accumulation of Jeb in CNS axons is dependent on Alk (Eng-

lund et al. 2003; Lee et al. 2003; Stute et al. 2004; Rohrbough and Broadie 2010). Further, postsynaptic Jeb internalization and Alk accumulation at developing NMJs during embryonic and larval stages is regulated by Mind the gap (Mtg), a neuronally secreted glycoprotein required for synaptic cleft extracellular matrix assembly. Rescue of defective *jeb* mutant larvae locomotion by neuronal-specific *jeb* expression unveiled a neuronal requirement for Jeb (Rohrbough and Broadie 2010).

During pupal stages, Alk functions in the visual system in photoreceptor axon target selection. *jeb* is expressed in photoreceptor axons, whereas *Alk* is expressed in processes of target neurons innervating the optic lobe. Photoreceptor axons secrete Jeb to activate Alk in target neurons within the lamina and medulla. R7 and R8 photoreceptor axons are significantly altered when Alk is removed from target neurons (Bazigou et al. 2007). *jeb* mosaic animals also show photoreceptor projection targeting defects. These photoreceptor navigation errors are reflective of a lack of regular lamina cartridge patterning and defective *duf* expression (Englund et al. 2003; Lee et al. 2003; Stute et al. 2004).

Olfactory learning deficits are associated with Alk activation and surprisingly elevated performance correlates with Alk inactivation. In addition to impaired learning, animals bearing constitutive Alk activation are small and conversely animals deficient for Alk activity are large. These nonautonomous effects on organismal size rely on the activation of Alk in peptidergic and cholinergic neurons and are insulin independent. *Alk* displays genetic interactions with the Ras GTPase activating protein *Nf1*. Consistent with the genetics, Alk activates ERK whereas Nf1 functions as a negative regulator of ERK (Walker et al. 2006). These antagonistic roles for Alk and Nf1 are proposed to regulate the release of GABA neurotransmitter and account for learning and long-term potentiation phenotypes (Ho et al. 2007; Liu et al. 2007).

Alk plays a vital role in protecting the developing CNS from nutrient deprivation. Under these conditions, flies eclose with relatively small bodies. Their heads, however, are of normal size. This is because nutrient deprivation triggers Jeb secretion from glia, activating Alk in the neural progenitors of the brain. Alk stimulates the activation of the InR effectors S6K, Thor, and PI3K, specifically in the brain, salvaging it from nutrient restriction (Cheng et al. 2011). This mechanism is Ras independent and therefore distinct from that described for Alk elsewhere. Of all *Drosophila* RTKs, Alk shares the most sequence similarity with InR. Mammalian ALK binds IRS-1 and SHC via its NPXpY motif. This motif exists in *Drosophila* Alk and offers a mechanism by which Alk activates PI3K when nutrients are limiting (Fujimoto et al. 1996).

THE TRK SUPERFAMILY OF RTKs: Ror, Nrk, AND DDR

Ror: A *Drosophila* Orphan RTK with Roles Exclusively in the Nervous System

Although *Drosophila* Ror is most similar to human Ror1 and Ror2 (~35% identity), it lacks many of the domains found in the human Ror RTKs. *Ror* expression begins ~6.5 hr after fertilization and peaks between 8–12 hr. *Ror* is expressed exclusively in neurons of the CNS and PNS (Wilson et al. 1993). This time in *Drosophila* development coincides with neural differentiation and axonogenesis and as such Ror is suggested to be involved in these processes. Consistent with this, the *C. elegans* Ror ortholog CAM-1 regulates neuronal polarity and the asymmetric division of neurons (Forrester et al. 1999) and vertebrate Rors play roles in neurite outgrowth and synapse formation (Paganoni and Ferreira 2005; Paganoni et al. 2010). A mutant phenotype for *Ror* in *Drosophila*, however, has not been described.

Ror shows a pattern of cysteine residue spacing suggestive of ligand binding like that for other transmembrane receptors such as Frizzled (Fz) (Saldanha et al. 1998). Like Fz, this region binds Wnt ligands in *C. elegans, Xenopus,* and mammalian cell lines (Hikasa et al. 2002; Oishi et al. 2003; Mikels and Nusse 2006; Green et al. 2007). The interaction of Ror proteins with various Wnts has been implicated in diverse contexts including: the migration and asymmetric division of neurons and vulval precursor cells in *C. elegans*, convergent extension in *Xenopus*, mouse embryonic fibroblast migration, and synapse formation in the mouse brain (Green et al. 2008; Grumolato et al. 2010; Paganoni et al. 2010). To date, Ror binding to Wnt has not been reported in *Drosophila*.

Neurotropic Receptor Kinase: Another *Drosophila* Orphan RTK in the Nervous System

Neurotropic receptor kinase (Nrk) is considered a MuSK ortholog, based on extensive homology in its kinase domain (Sossin 2006). Like Ror RTKs, MuSK binds Wnt ligands via the Fz domain (Jing et al. 2009). Although differences exist between Nrk and MuSK extracellular domains, Nrk has an Fz domain and is predicted to bind Wnts. In vertebrates, MuSK induces acetylcholine receptor clustering at NMJs and the stability of clusters relies on the heparan–sulphate proteoglycan Agrin and LRP coreceptor (DeChiara et al. 1996; Bezakova et al. 2001; Kim et al. 2008; Zhang et al. 2008). Nrk, however, lacks the extracellular domains responsible for Agrin binding as well as the intracellular NPXpY motif in MuSK essential for NMJ formation (Herbst and Burden 2000). MuSK has an Agrin-independent role in axon guidance in zebrafish (Zhang et al. 2004) and a noncanonical Wnt-mediated function in neural crest cell migration in mouse (Banerjee et al. 2011). These studies hint at similar roles for Nrk.

Nrk expression begins ~9.5 hr after fertilization, shortly after the determination of neural precursor cells. Expression of *Nrk* is initially detected in the neuroectoderm and becomes restricted to neural progenitor cells situated between epidermal and mesodermal cell layers. Expression persists in the neural cell lineage throughout embryogenesis and peaks again at the pupal stage during restructuring of the nervous system (Oishi et al. 1997). In support of a role for Nrk in the nervous systems, *Nrk* expression was down-regulated in embryos for which neuroectoderm was derived primarily from glial cells, rather than both neurons and glia (Egger et al. 2002).

Discoidin Domain Receptor: A Poorly Characterized RTK in *Drosophila*

Discoidin domain receptors (DDR) are atypical RTKs in that they are activated by collagen rather than secreted factors. Moreover, maximal activation of DDRs occurs several hours after collagen binding, unlike other RTKs that are activated within minutes of receptor engagement (Shrivastava et al. 1997; Vogel et al. 1997). Although *Drosophila* Ddr has not yet been shown to bind collagen, conservation of the discoidin domain suggests that it should possess this ability (Sossin 2006). DDRs have been implicated in cell migration, extracellular matrix remodeling, proliferation and differentiation in a number of tissues including vertebrate lung, skin, GI tract, kidney, heart, liver, mammary gland, endometrium, and brain (Vogel et al. 2006). Like its mammalian counterparts, the sole *Drosophila* Ddr bears an extensive number of tyrosine residues in its cytoplasmic region. In mammals, these residues recruit a number of downstream factors following phosphorylation including Shc, Nck2, Shp-2, PI3K, RasGAP, and Stats (Lemeer et al. 2011). Wnt5a was shown to be required for collagen-induced activation of DDR in a breast cancer cell line (Jonsson and Andersson 2001). It is tempting to speculate that Wnt5A might function as a ligand for DDR, in a manner analogous to that of Ror and Ryk RTKs.

OFFTRACK: A "DEAD" RTK WITH ROLES IN MOTOR AND CNS AXON TARGETING AND EMBRYONIC PATTERNING

Drosophila off-track (also known as *Dtrk*) shares 65% similarity with human Trk. *off-track* (*otk*) is expressed 3–4 hr postfertilization in the anterior midgut primordia, the cephalic furrow, and along the germ band. Expression peaks mid-embryogenesis in neuroectodermal cells and internalized CNS neuroblasts. Otk expression persists in prospective gut and head regions, and eventually accumulates throughout the CNS in segmentally repeated commissures, in axon bundles exiting the CNS, in motor neuron projections innervating muscle fibers, and

in neurons of a subset of sensory organs. Notably, *otk* null animals are embryonic lethal (Pulido et al. 1992; Winberg et al. 2001).

Otk bears structural similarity with cell adhesion molecules of the immunoglobulin (Ig) superfamily expressed in the *Drosophila* nervous system. Like other Ig superfamily members, Otk is glycosylated and permits cell aggregation in vitro (Pulido et al. 1992). This property likely influences neuroblast migration and axon targeting. Accordingly, Otk is implicated in Sema-1a-mediated embryonic motor and CNS axon guidance based on: (1) its physical association with the repulsive axon guidance receptor Plexin A; (2) impaired defasciculation and disrupted axon morphology and targeting in *otk* mutant embryos, similar to that resulting from Sema-1a or PlexA inactivation; and (3) genetic interactions between *otk* and *PlexA* or *Sema-1a* (Pulido et al. 1992; Winberg et al. 2001). Conserved catalytic residues in the kinase domain are altered in Otk and as such Otk belongs to the CCK-4 subfamily of "dead" RTKs (Kroiher et al. 2001). Otk itself is tyrosine phosphorylated, which may serve to recruit signaling molecules to a Sema-1a-PlexA-Otk complex (Pulido et al. 1992; Winberg et al. 2001).

A screen for R1–R6 photoreceptor growth cone targeting identified *otk*; an increased number of R1–R6 photoreceptor axons project through the lamina and inappropriately into the medulla of the developing larval optic lobe in *otk* mosaic heads. The role of Otk in guiding R1–R6 axons appears to be unrelated to that of Otk in Sema-1a signaling (Cafferty et al. 2004). The Otk ligand responsible for R1–R6 targeting is unknown.

Recently, Otk binding to Wnt4 was shown to inhibit canonical β-catenin/TCF signaling in an Fz-dependent but LRP-independent manner. Both *otk* and *Wnt4* mutants show embryonic patterning defects indicative of excessive canonical β-catenin/TCF signaling. In the ventral embryonic epidermis and the adult wing, *otk* and *Wnt4* overexpression reduce canonical β-catenin/TCF signaling. Ectopic *Wnt4* expression phenotypes rely on functional Otk and synergizes with *otk* overexpression. Otk is postulated to direct noncanonical Wnt signaling through interaction with Disheveled, and inhibit canonical Wnt signaling by either occluding LRP or by sequestering canonical Wnts (Peradziryi et al. 2011).

THE RYK SUPERFAMILY OF RTKs: DERAILED AND DOUGHNUT

Derailed: Another "Dead" RTK with Roles in Neuronal Pathway Selection and Muscle Attachment

derailed (drl) expression begins ~6 hr postfertilization, in the embryonic epidermis and in salivary placodes. By 10 hr, Drl is detected in somatic muscle 21–23 precursors and their associated epidermal cells. By 12 hr, *drl* expression is detected in a subset of heterogeneous neurons that project in the anterior commissure (AC) of the VNC (Callahan et al. 1995, 1996; Bonkowsky and Thomas 1999; Harris and Beckendorf 2007). *drl* is expressed later in the larval CNS, as well as the adult brain. Homozygous *drl* mutant animals are viable but uncoordinated because of *drl* axon mistargeting and defasciculation, and the inability of muscles 21–23 to establish functional attachments with the epidermis. *drl* mutant animals show additional learning and memory defects as a consequence of structural defects of the adult brain (Bolwig et al. 1995; Dura et al. 1995; Moreau-Fauvarque et al. 1998; Simon et al. 1998; Grillenzoni et al. 2007; Sakurai et al. 2009).

RYK proteins interact with Wnt5, the cognate ligand for the transmembrane protein Fz. Wnt5 functions as a repulsive signal for Drl during embryogenesis, directing Drl-expressing axons away from the posterior commissure (PC) and toward the AC of each VNC hemisegment (Fradkin et al. 2004). Uncoordinated *wnt5* mutants display axon navigation and axon fasciculation defects similar to *drl* mutant animals. Further, AC loss following *wnt5* overexpression in the midline is dependent on functional Drl. Moreover, both *drl* and *wnt5* mutants show alterations in dendritic branching of CNS serotonergic neurons (Singh et al. 2010). Like Drl, mammalian Wnt5a routes RYK expressing axons through the corpus callosum (Keeble et al. 2006) and corticospinal tract (Liu et al. 2005). Unlike mammal RYK (Lu et al. 2004), however, Drl does not impact on canonical Wnt/β-catenin signaling (Wouda et al. 2008) and the activity of *Drosophila* Wnt5 in PC repulsion appears independent of Fz (Yoshikawa et al. 2003). Interestingly, *wnt5* itself appears to be a target of Drl-activated neurons (Fradkin et al. 2004; Yao et al. 2007).

Drl and Wnt5 positively regulate glutamatergic NMJ development and synaptic transmission. Both *drl* and *wnt5* mutants have reduced numbers of synaptic boutons and reduced NMJ size (Liebl et al. 2008). Growth of the NMJ during larval stages requires coordination between the presynaptic motor neuron and postsynaptic muscle (Zito et al. 1999). *drl* mutant phenotypes at NMJs are rescued by muscle-specific expression of *drl*, whereas Wnt5 functions exclusively in presynaptic motor neurons. Reducing *drl* dosage suppresses NMJ overgrowth resulting from *wnt5* neuronal overexpression, implicating Drl downstream from Wnt5. Further, rescue of *drl* phenotypes with exogenous Drl requires the WIF (Wnt-inhibitory-factor) domain. This evidence suggests that Wnt5 released from the presynaptic boutons binds Drl on postsynaptic muscle to regulate bouton growth and postsynaptic differentiation (Liebl et al. 2008).

Src64B and *Src42A* mutants show defects in mushroom body (MB) anatomy, salivary gland development, and AC formation similar to *Wnt5* and *drl* mutants. *Src64B* interacts genetically with *drl*, as well as *Wnt5*, in MB development, salivary gland migration, and VNC neuron commissure formation (Nicola et al. 2003; Harris and Beckendorf 2007; Wouda et al. 2008). RYK proteins bear alterations in conserved catalytic residues required for phosphotransfer and are therefore inactive (Katso et al. 1999; Yoshikawa et al. 2001). Rather, Src64B functions as the TK responsible for Drl phosphorylation and interaction with Drl stimulates Src64B activity. Src64B is likely responsible for Wnt5/Drl-mediated axon repulsion, given that Drl-dependent axon pathfinding relies on Src64B kinase activity. Drl is proposed to provide substrate specificity for Src proteins (Wouda et al. 2008).

The learning and memory defects of *drl* mutant animals are likely a consequence of Wnt5 regulating olfactory circuitry. A subset of *drl*-expressing cells normally antagonizes Wnt5 produced by olfactory receptor neurons (ORNs) to appropriately pattern and orient glomeruli during antennal lobe development. This function of Drl relies on the WIF domain (Fradkin et al. 2004; Yao et al. 2007). Delineation of the functional requirement of *drl* in the antennal lobe refined cell-type identities to lateral neuroblast and ventral neuroblast-derived neurons (Bolwig et al. 1995; Dura et al. 1995; Moreau-Fauvarque et al. 1998; Simon et al. 1998; Grillenzoni et al. 2007; Sakurai et al. 2009). *drl* mutant phenotypes of the olfactory system resemble those resulting from *wnt5* overexpression in ORNs: aberrant positioning of glomeruli and ectopic targeting of ORN axons to extraneous glomeruli structures. Moreover, enhancement or attenuation of Wnt5 function can either exacerbate or suppress *drl* phenotypes respectively (Fradkin et al. 2004; Yao et al. 2007).

A similar antagonistic relationship between *drl* and *wnt5* exists in MBs. Pan-neuronal *drl* overexpression phenocopies *wnt5* loss-of-function and attenuation of Drl activity can suppress *wnt5* overexpression phenotypes. Neuronal *drl* expression is sufficient to nonautonomously rescue MB defects in *drl* mutants whereas *wnt5* expression in MBs restores MB morphology to *wnt5* mutants. These data suggest that Drl-dependent sequestration of Wnt5 is required to limit MB axonal growth (Bolwig et al. 1995; Dura et al. 1995; Moreau-Fauvarque et al. 1998; Simon et al. 1998; Grillenzoni et al. 2007; Sakurai et al. 2009).

Doughnut on 2: An RTK Involved in Migration

drl and *doughnut on 2* (*dnt*) likely arose by gene duplication because the two genes display more similarity to each other than mammalian RYKs. Although Drl and Dnt share 60% identity, *dnt* cannot completely rescue *drl* mutant phenotypes (Oates et al. 1998). Maximal expression of *dnt* occurs 4–6 hr postfertilization, 2 hr before maximal *drl* expression; however, expression of both persists throughout the *Drosophila* life cycle (Roy et al. 2010). Like Drl, Dnt is considered catalytically inactive, based on substitutions of critical catalytic amino acids in the TK domain.

dnt is expressed initially in the central region and anterior domain of the embryo. Later expression occurs primarily in invaginating cells of the ventral furrow, gut, cephalic, and transverse furrow, and tracheal pits. The name *doughnut* comes from the rings of expression surrounding tracheal primordia (Oates et al. 1998; Savant-Bhonsale et al. 1999). *dnt* plays a minor role, with *drl* and *Drl-2*, in salivary gland cell migration during late embryogenesis (Harris and Beckendorf 2007). Disruption of the *dnt* locus influences multiple body size-related traits including face and head width, thorax length and wing size (Carreira et al. 2008). Although the disparate expression patterns of *dnt* and *drl* are indicative of paralogous function, to date relatively little characterization of *dnt* has substantiated this conjecture.

Derailed 2: An RTK Sharing Overlapping Roles with DRL in Olfactory Circuitry and Salivary Gland Migration

Derailed 2 (Drl-2) shares 35% identity with Drl, yet the two share distinct expression patterns and *drl-2* mutants display relatively mild defects in antennal lobe development compared with *drl*: one of two displaced glomeruli displayed defects similar to *wnt5* and *drl* mutants whereas the other was similar only to that of *wnt5* and opposite to that of *drl* mutants. *drl* and *drl-2* mutant alleles synergize in this context; a *drl drl-2* double mutant displays additional defects resembling *wnt5* mutants, implicating Drl-2 in Wnt5 signaling. Further, ORN-specific overexpression of *wnt5* bears little effect in a *drl-2* mutant or *drl drl-2* double mutant background. Glial-specific expression of *Drl-2* can compensate for loss of *drl* suggesting that these receptors have paralogous functions in Wnt5 signaling dependent on cell context (Bolwig et al. 1995; Dura et al. 1995; Moreau-Fauvarque et al. 1998; Simon et al. 1998; Grillenzoni et al. 2007; Sakurai et al. 2009).

Cite this article as *Cold Spring Harb Perspect Biol* doi: 10.1101/cshperspect.a009050

Wnt5 in the CNS repels *drl*-expressing salivary gland tip cells, thereby dictating salivary gland migration. In this context, *drl-2* mutants show similar defects to *drl* mutants: ventromedial curving of tip cells and a failure of visceral mesoderm attachment. *drl drl-2* double mutant embryos phenocopy *drl* mutants, indicating that Drl-2 plays a minor role in salivary gland morphogenesis (Harris and Beckendorf 2007). Drl-2 has a similar role to Drl in the development of the larval and embryonic musculature; however, in this case *drl-2* expression in specific motor neurons functions in preventing synapse formation with inappropriate ventral muscles (Inaki et al. 2007).

RET: THE HOMOLOG OF THE MAMMALIAN RTK PROTO-ONCOGENE RET

Ret expression begins ~3.5 hr postfertilization in scattered regions throughout the yolk sac and is not detected again until 5–7 hr, in a subset of neuroblasts. At ~10 hr, *Ret* is expressed in midline glia of the VNC, in the somatogastric nervous system anlage, in midgut precursor cells, and transiently in the malpighian tubule anlage. Late expression is observed in the developing PNS and CNS (Sugaya et al. 1994; Hahn and Bishop 2001; Fung et al. 2007). During larval stages, *Ret* is expressed in neuroendocrine cells of the brain and ventral ganglion, as well as in leg, wing, antennal, and eye imaginal discs (Hahn and Bishop 2001; Read et al. 2005). This expression pattern is similar to that of human RET, the closest vertebrate homolog of *Drosophila* Ret (52% identity in the TK domain).

The ligand for vertebrate RET, glial cell line–derived neurotrophic factor (GDNF), does not bind RET, but rather the GPI-linked coreceptor GFR-α. Although *Drosophila* Ret shares homology and structural organization in its extracellular domain with that of vertebrate RET, a homologous GDNF ligand or GPI-linked receptor does not exist in the *Drosophila* genome (Anders et al. 2001). Furthermore, *Drosophila* Ret does not bind GDNF. Rather, four extracellular cadherin-like domains suggest an ancient role for Ret in adhesion, although *Drosophila* Ret is incapable of self-association in vitro (Abrescia et al. 2005).

A Ret transgene with equivalent mutations to that observed in multiple endocrine neoplasia (MEN) expressed in the *Drosophila* eye caused phenotypes analogous to that in vertebrates: excessive proliferation and aberrant neuronal specification. In accordance with a proposed role for Ret in cell adhesion, deficiencies in the adhesion regulators *Moe*, *Pax*, and *Cad-N2*-enhanced Ret^{MEN}-dependent phenotypes (Hahn and Bishop 2001; Read et al. 2005). These phenotypes were further modulated by mutation of Ras, Src, and JNK, consistent with characterized roles for human RET (Arighi et al. 2005). Defective eye development induced by *Ret^{MEN}* expression was altered by DJ-1α/β—proteins linked to Parkinson's disease—likely by modulating Ras/ERK signaling (Aron et al. 2010).

Cad96Ca: A CADHERIN DOMAIN-CONTAINING RTK INVOLVED IN WOUND REPAIR

Like Ret, Cad96Ca (also known as Stitcher) has both cadherin and TK domains (Tepass et al. 2000). *Cad96Ca* is expressed in all ectodermal epithelia during mid- and late embryonic stages but later becomes restricted to the epithelial optic lobe anlagen (Fung et al. 2007). Homozygous *cad96C* null animals die at late pupal stages. Cad96Ca displays TK activity in vitro, and a functional TK domain is required for rescue of *Cad96Ca* mutant animals (Wang et al. 2009).

Cad96Ca is predicted to play a role in nervous system development given that *Cad96Ca* expression was found down-regulated in late-stage embryos for which neuroectoderm was derived from glial cells (Egger et al. 2002). Consistent with this, many classical cadherins are expressed in the developing nervous system and have roles in neurite outgrowth, and axonal patterning and fasciculation (Tepass et al. 2000).

Cad96Ca is expressed in primordia of the spiracle—the external opening of the larval respiratory system—where it reinforces DE-Cad activity in posterior spiracle morphogenesis.

Expression of *Cad96Ca* in this context depends of EGFR and Hh signaling, as well as the transcription factor Cut (Lovegrove et al. 2006; Maurel-Zaffran et al. 2010).

Cad96Ca facilitates embryonic re-epithelialization following wound healing by stimulating actin cable formation and the transcription of cuticle repair genes by the transcription factor Grh (Wang et al. 2009). Moreover, Cad96Ca can induce ERK phosphorylation, which is required for Grh activation (Mace et al. 2005).

Eph: AN RTK INVOLVED IN AXON PATHFINDING

Drosophila Eph displays similarity to both classes of vertebrate Eph receptors (~35% in the extracellular region and 71% in the TK domain), whereas the best characterized ligand for Eph in *Drosophila*, Ephrin, shares ~40% identity in the extracellular ephrin domain with both classes of human ephrin ligands (Bossing and Brand 2002).

Eph is expressed exclusively in the nervous system, initially in the neuroectoderm ~5 hr postfertilization and then after ~10 hr in a subset of neurons of the brain and VNC. Expression persists in the larval CNS and MBs, photoreceptor axonal projections and developing optic ganglia. Eph protein localizes to axons of elongating neurons, with highest concentrations in the growth cones of the earliest differentiating cortical and MB neurons and photoreceptors, and on longitudinal and commissural axons of the VNC (Scully et al. 1999; Dearborn et al. 2002; Boyle et al. 2006). This localization is proximal to that of Ephrin, which is concentrated in neuronal cell bodies along the outer edge of connectives and between commissures (Bossing and Brand 2002).

Disruption of *Eph* function by RNAi results in defective projection of photoreceptor axons as well as the aberrant targeting and loss of medulla and lobular cortical axons. *Eph* function is required specifically at the visual system midline to direct axon targeting in the developing eye and optic ganglia. Eph is predicted to fulfill a comparable role in the VNC; RNAi-mediated disruption of *Eph*, as for *Ephrin*, results in commissure fusion and loss, in addition to connective fragmentation as a consequence of interneuronal axon departure from the CNS longitudinal connectives. *Ephrin* expression at the midline, on the other hand, repels contralateral axon midline crossing and halts axonal growth along connectives, in an Eph-dependent manner. *Eph* null animals are viable and display abnormalities specifically in projection neuron targeting during MB development (Scully et al. 1999; Dearborn et al. 2002; Boyle et al. 2006). Additional phenotypes uncovered for *Eph* in other studies may reflect unintentional RNAi-mediated targeting of homologous targets.

Drosophila Vap33 is a proposed alternative ligand for Eph. Like Ephrin, Vap33 is membrane anchored and *vap33* null mutants display MB defects in late pupae and adult brains identical to those of *Eph*. Moreover, inactivation of *Eph* can suppress muscular defects resulting from neuronal Vap33 expression. Vap33 can bind to the extracellular domain of *C. elegans* VAB-1 Eph receptor. This binding appears conserved among VAP proteins and is proposed to antagonize Ephrin binding (Tsuda et al. 2008).

Finally, *Drosophila* Ephexin (Exn) like its vertebrate Rho-type guanine nucleotide exchange factor counterpart binds Eph, at NMJs. Exn binds the Eph TK domain via its SH3 and Rho-GEF domains. Exn is required at the presynaptic nerve terminal of the NMJ to modulate synaptic vesicle release; disruption of either *Exn* or *Eph* interferes with homeostatic compensatory neurotransmitter release at NMJs. Eph is hypothesized to serve as a presynaptic receptor for a muscle-derived retrograde signal, speculated to be either Ephrin or Vap33 (Frank et al. 2009).

PDGF- AND VEGF-RECEPTOR RELATED: AN RTK WITH ROLES IN TISSUE SCULPTING, CELL MIGRATION, AND SURVIVAL

PDGF- and VEGF-receptor related (Pvr) is expressed ~4 hr postfertilization in the procephalic mesoderm. Expression is later restricted to populations of scattered hemocytes, the hematopoietic cells in *Drosophila*. Three Pvr ligands, Pvf1-3, are expressed along stereotypical routes taken by migrating hemocytes. Although nonessential for differentiation, Pvr is required cell-autonomously for maintaining migrating populations of mature hemocytes. Pvr/Pvf induces ERK activation and hemocyte migration and is dependent on Ras1 (Heino et al. 2001; Cho et al. 2002; Brückner et al. 2004).

Plasmatocytes represent the majority class of hemocytes. They are phagocytic, clearing apoptotic debris generated during the programmed cell death that is necessary for tissue sculpting and metamorphosis (Tepass et al. 1994). *Pvr* mutant embryos show CNS axon scaffolding and glial cell positioning defects, as a consequence of reduced hemocyte numbers and therefore compromised neuron and glial cell elimination (Sears et al. 2003). They also show defective VNC condensation resulting from reduced hemocyte-derived extracellular matrix (Olofsson and Page 2005). Pvr is present in midline glia (MG) and all Pvf ligands localize to midline neurons of the CNS. Pvr/Pvf maintain and direct MG during embryogenesis; expression of activated Pvr at the midline induces enlargement of MG clusters and misallocated supernumerary MG, whereas ectopic *Pvf* expression at the midline or specifically in neurons reroutes MG migration. In the absence of functional Pvr, MG are disorganized or lost because of excessive apoptosis (Learte et al. 2008).

Embryonic plasmatocytes found larval hemocyte populations and self-renewal requires an intact PNS to attract plasmatocytes to a hematopoietic niche (Makhijani et al. 2011). Expression of activated Pvr stimulates larval hemocyte proliferation, whereas dominant negative Pvr has the opposite effect (Zettervall et al. 2004). At the onset of metamorphosis, lymph glands supply large numbers of plasmatocytes to phagocytose unnecessary larval tissue. Pvr is required in the lymph gland to regulate plasmatocyte differentiation and maintain levels of mature hemocytes (Jung et al. 2005).

Pvr is additionally expressed in ovarian border cells and is required for their initial migration in the direction of *Pvf1*, expressed by the developing oocyte (McDonald et al. 2003). Pvf1 engagement by Pvr provides directionality, because Pvr inactivation results in border cell clusters with misallocated and disoriented actin protrusions (Prasad and Montell 2007; Poukkula et al. 2011). The impetus for Pvr-directed border cell migration appears collectively to be the activation of the Rac-Mbc-ELMO complex (Duchek et al. 2001; Bianco et al. 2007; Wang et al. 2010), the accumulation of cortactin and cofilin at the migratory front of the cluster (Somogyi and Rørth 2004; Zhang et al. 2011) and the down-regulation of the transcriptional repressor Aop (Schober et al. 2005).

Pvr has been implicated in the migration of imaginal cells during metamorphosis. Pvr is required for JNK-dependent thorax closure. The Rac effector Crk-Mbc-ELMO complex links Pvr to JNK in this context, similar to border cell migration (Ishimaru et al. 2004). The rotation and dorsal closure of the male genital imaginal disc also relies on Pvr/Pvf1 to activate JNK (Macias et al. 2004). Mbc-ELMO functions downstream from Pvr additionally in epithelial cells stimulated to engulf their oncogenic neighbors. In this environment, JNK is the trigger for both apoptosis in mutant cells and Pvr activation in surrounding wild-type cells (Ohsawa et al. 2011).

Pvr mutants show defects in the anterior projection of renal tubules. Like CNS remodeling, these phenotypes derive from a lack of migrating hemocytes and a consequential lack of collagen secretion/deposition, which normally facilitates Dpp presentation by dorsal epidermal and visceral mesodermal cells and directs renal tubule migration. *Pvf* expression in the renal tubules attracts and activates Pvr-expressing hemocytes (Bunt et al. 2010).

Pvf1 and Pvf3 confine Pvr activity to the apical domain of the wing imaginal disc epithelium. Unrestricted Pvr activity results in a loss of epithelial polarity, ectopic adherens junctions, elevated basolateral actin filament polymerization, and neoplastic overgrowth (Rosin et al. 2004). *Pvr* and *Pvf2* are expressed during heart metamorphosis, in cardiac valve precursors. Cardiac valves, dense accumulations of filamentous actin, are fewer following dominant-negative *Pvr* expression, whereas activated Pvr induces ectopic valve formation (Zeitouni et al. 2007). Congruent with a cardiac requirement for *Pvr*, reduction in *Pvr* function results in aberrant embryonic/larval heart pumping (Wu and Sato 2008). However, this phenotype may reflect defective neural circuitry associated with Pvr loss (Olofsson and Page 2005) rather than defective cardiac valve development (Zeitouni et al. 2007).

Like Ryk RTKs, Pvr is required autonomously to direct salivary gland migration toward the visceral mesoderm. *Pvr* is expressed in the developing salivary gland at the incipient site of placode invagination whereas Pvf1 is abundant at the tip of the migrating gland. *Pvr* mutants, like *Pvf1* and *Pvf2* mutants, have salivary glands that curve ventrally toward the CNS (Harris et al. 2007).

Pvf2 is expressed in the adult midgut in intestinal stem cells (ISCs) and enteroblasts. Expression increases with age and in response to oxidative stress, and correlates with age and stress-dependent increases in ISC populations. Ectopic *Pvr* expression specifically in ISCs and enteroblasts causes lethality. The guts of flies that survive display elevated numbers of proliferating cells that further amplify with age. Both *Pvr* and *Pvf2* expression in ISCs stimulate their division and leads to altered differentiation—increased numbers of enteroendocrine cells at the expense of enterocytes (Choi et al. 2008). These Pvr-dependent effects rely on functional p38b ERK (Park et al. 2009).

Tie-LIKE RECEPTOR TYROSINE KINASE: ANOTHER RTK INVOLVED IN MIGRATION

Drosophila Tie-like RTK (Tie) shares ∼50% identity in its TK domain with the human Tie RTK (Ito et al. 1994). Tie expression peaks at late embryonic and pupal stages (Roy et al. 2010) primarily in the hindgut, salivary gland, and trachea (Chintapalli et al. 2007). *Tie* expression is up-regulated in *Drosophila* egg chambers: specifically in border cells and centripetal cells. This up-regulation is dependent on the basic region/leucine zipper transcription factor Slbo, a C/EBP homolog. Expression of dominant-negative *Tie* exacerbates border cell migration defects resulting from dominant-negative PVR and EGFR. This indicates that Tie has a redundant role with PVR and EGFR in directing border cell cluster migration (Wang et al. 2006).

Tie likely plays a role in development of the *Drosophila* sensory system. *Tie* expression is up-regulated in *Drosophila* imaginal discs on overexpression of the Pax6 homolog Eyeless, a transcription factor directing neuronal differentiation in the retina. *Tie* is a predicted target of Eyeless, based on in silico approaches. *Tie* is expressed in the eye imaginal disc and is predicted to function early in retinal differentiation (Michaut et al. 2003; Ostrin et al. 2006).

CONCLUDING REMARKS

The breadth of scenarios wherein RTK signaling contributes to *Drosophila* development is undeniably impressive. RTKs provide a means of communication between different tissues and cell types that contributes to a robust and highly reproducible developmental program. RTKs vary dramatically in their expression with respect to cell type, ranging from ubiquitous expression, as in the case of InR for instance, to restricted expression in specific subsets of cells, as for Btl. Most interestingly, the activity of a few RTKs is exploited many times over in different developmental contexts, such as for EGFR, although the activity of other RTKs, for instance Torso and Sev, is highly specific to a few specialized functions. Remarkably, many downstream effectors are shared among different RTKs (Table 1) and their hierarchical organization is reiterated in various biological contexts. This implies that specificity in signaling output is likely rendered by a limited number of factors. For example, the insulin receptor substrate Chico is specific for InR signaling (Bohni et al. 1999), whereas the adaptor protein Shc functions downstream from Torso, EGFR, and Btl, but not Sev (Luschnig et al. 2000; Cabernard and Affolter 2005). Additionally, the context and manner in which ligands are presented to the RTK is likely to influence outcome. For instance, the overexpression of *pyr* and *ths* in combination elicits a phenotype opposite to that of individual overexpression of *pyr* or *ths* (Kadam et al. 2012). Furthermore, heparan sulfate proteoglycans are required for delivery and stabilization of ligands with RTKs and therefore maximal activation of signaling (Nybakken and Perrimon 2001). Moreover, the competency and direction by which a cell to responds to RTK activation relies in large part on the transcription factors expressed in that particular cell type. Because of its limited genetic redundancy, *Drosophila* remains an attractive model in terms of RTK pathway component identification, and has successfully served to uncover corresponding conserved vertebrate counterparts. Components rendering fine-tuning functions in many well-characterized RTK pathways continue to be discovered. Astonishingly, relatively little is known regarding downstream signaling evoked by approximately half of all *Drosophila* RTKs. The current availability of genome-wide transgenic RNAi reagents in *Drosophila* will undoubtedly expedite further investigation of these RTKs to alleviate this knowledge gap.

ACKNOWLEDGMENTS

We thank Yanhui Hu for *Drosophila* RTK identification and Chrysoula Pitsouli and Akhila Rajan for helpful comments. R.S. is supported by the Canadian Institutes for Health Research. N.P. is a Howard Hughes Medical Institute investigator.

REFERENCES

Abrescia C, Sjostrand D, Kjaer S, Ibanez CF. 2005. *Drosophila* RET contains an active tyrosine kinase and elicits neurotrophic activities in mammalian cells. *FEBS Lett* **579:** 3789–3796.

Ahmad SM, Baker BS. 2002. Sex-specific deployment of FGF signaling in *Drosophila* recruits mesodermal cells into the male genital imaginal disc. *Cell* **109:** 651–661.

Almudi I, Stocker H, Hafen E, Corominas M, Serras F. 2009. *SOCS36E* specifically interferes with sevenless signaling during *Drosophila* eye development. *Dev Biol* **326:** 212–223.

Almudi I, Corominas M, Serras F. 2010. Competition between *SOCS36E* and Drk modulates sevenless receptor tyrosine kinase activity. *J Cell Sci* **123:** 3857–3862.

Ambrosio L, Mahowald AP, Perrimon N. 1989a. l(1)pole hole is required maternally for pattern formation in the terminal regions of the embryo. *Development* **106:** 145–158.

Ambrosio L, Mahowald AP, Perrimon N. 1989b. Requirement of the *Drosophila* raf homologue for torso function. *Nature* **342:** 288–291.

Anders J, Kjar S, Ibanez CF. 2001. Molecular modeling of the extracellular domain of the RET receptor tyrosine kinase reveals multiple cadherin-like domains and a calcium-binding site. *J Biol Chem* **276:** 35808–35817.

Arighi E, Borrello MG, Sariola H. 2005. RET tyrosine kinase signaling in development and cancer. *Cytokine Growth Factor Rev* **16:** 441–467.

Aron L, Klein P, Pham T-T, Kramer ER, Wurst W, Klein R. 2010. Pro-survival role for Parkinson's associated gene *DJ-1* revealed in trophically impaired dopaminergic neurons. *PLoS Biol* **8:** e1000349.

Astigarraga S, Grossman R, Díaz-Delfín J, Caelles C, Paroush Zae, Jiménez G. 2007. A MAPK docking site is critical for down-regulation of Capicua by Torso and EGFR RTK signaling. *EMBO J* **26:** 668–677.

Azpiazu N, Frasch M. 1993. Tinman and bagpipe: Two homeo box genes that determine cell fates in the dorsal mesoderm of *Drosophila*. *Genes Dev* **7:** 1325–1340.

Banerjee S, Gordon L, Donn TM, Berti C, Moens CB, Burden SJ, Granato M. 2011. A novel role for MuSK and non-canonical Wnt signaling during segmental neural crest cell migration. *Development* **38:** 3287–3296.

Basler K, Hafen E. 1988. Control of photoreceptor cell fate by the sevenless protein requires a functional tyrosine kinase domain. *Cell* **54:** 299–311.

Basler K, Christen B, Hafen E. 1991. Ligand-independent activation of the sevenless receptor tyrosine kinase changes the fate of cells in the developing *Drosophila* eye. *Cell* **64:** 1069–1081.

Bazigou E, Apitz H, Johansson J, Lorén CE, Hirst EMA, Chen P-L, Palmer RH, Salecker I. 2007. Anterograde Jelly belly and Alk receptor tyrosine kinase signaling mediates retinal axon targeting in *Drosophila*. *Cell* **128:** 961–975.

Behan K, Nichols C, Cheung T, Farlow A, Hogan B, Batterham P, Pollock J. 2002. Yan regulates *lozenge* during *Drosophila* eye development. *Dev Genes Evol* **212:** 267–276.

Beiman M, Shilo BZ, Volk T. 1996. Heartless, a *Drosophila* FGF receptor homolog, is essential for cell migration and establishment of several mesodermal lineages. *Genes Dev* **10:** 2993–3002.

Bellaïche Y, Bandyopadhyay R, Desplan C, Dostatni N. 1996. Neither the homeodomain nor the activation domain of Bicoid is specifically required for its down-regulation by the Torso receptor tyrosine kinase cascade. *Development* **122:** 3499–3508.

Bezakova G, Rabben I, Sefland I, Fumagalli G, Lømo T. 2001. Neural agrin controls acetylcholine receptor stability in skeletal muscle fibers. *Proc Natl Acad Sci* **98:** 9924–9929.

Bianco A, Poukkula M, Cliffe A, Mathieu J, Luque CM, Fulga TA, Rørth P. 2007. Two distinct modes of guidance signalling during collective migration of border cells. *Nature* **448:** 362–365.

Biggs WH, Zavitz KH, Dickson B, van der Straten A, Brunner D, Hafen E, Zipursky SL. 1994. The *Drosophila* rolled locus encodes a MAP kinase required in the sevenless signal transduction pathway. *EMBO J* **13:** 1628–1635.

Bodmer R. 1993. The gene tinman is required for specification of the heart and visceral muscles in *Drosophila*. *Development* **118:** 719–729.

Böhni R, Riesgo-Escovar J, Oldham S, Brogiolo W, Stocker H, Andruss BF, Beckingham K, Hafen E. 1999. Autonomous control of cell and organ size by CHICO, a *Drosophila* homolog of vertebrate IRS1–4. *Cell* **97:** 865–875.

Bolwig GM, Del Vecchio M, Hannon G, Tully T. 1995. Molecular cloning of linotte in Drosophila: A novel gene that functions in adults during associative learning. *Neuron* **15:** 829–842.

Bonkowsky JL, Thomas JB. 1999. Cell-type specific modular regulation of derailed in the *Drosophila* nervous system. *Mech Dev* **82:** 181–184.

Bossing T, Brand AH. 2002. Dephrin, a transmembrane ephrin with a unique structure, prevents interneuronal axons from exiting the *Drosophila* embryonic CNS. *Development* **129:** 4205–4218.

Boulton SJ, Brook A, Staehling-Hampton K, Heitzler P, Dyson N. 2000. A role for Ebi in neuronal cell cycle control. *EMBO J* **19:** 5376–5386.

Boyle M, Nighorn A, Thomas JB. 2006. *Drosophila* Eph receptor guides specific axon branches of mushroom body neurons. *Development* **133:** 1845–1854.

Britton JS, Edgar BA. 1998. Environmental control of the cell cycle in *Drosophila*: Nutrition activates mitotic and endoreplicative cells by distinct mechanisms. *Development* **125:** 2149–2158.

Brogiolo W, Stocker H, Ikeya T, Rintelen F, Fernandez R, Hafen E. 2001. An evolutionarily conserved function of the *Drosophila* insulin receptor and insulin-like peptides in growth control. *Curr Biol* **11:** 213–221.

Brönner G, Jäckle H. 1991. Control and function of terminal gap gene activity in the posterior pole region of the *Drosophila* embryo. *Mech Dev* **35:** 205–211.

Broughton SJ, Piper MDW, Ikeya T, Bass TM, Jacobson J, Driege Y, Martinez P, Hafen E, Withers DJ, Leevers SJ, et al. 2005. Longer lifespan, altered metabolism, and stress resistance in *Drosophila* from ablation of cells making insulin-like ligands. *Proc Natl Acad Sci* **102:** 3105–3110.

Brückner K, Kockel L, Duchek P, Luque CM, Rørth P, Perrimon N. 2004. The PDGF/VEGF receptor controls blood cell survival in *Drosophila*. *Dev Cell* **7:** 73–84.

Brunner D, Oellers N, Szabad J, Biggs WH, Zipursky SL, Hafen E. 1994. A gain-of-function mutation in *Drosophila* MAP kinase activates multiple receptor tyrosine kinase signaling pathways. *Cell* **76:** 875–888.

Bunt S, Hooley C, Hu N, Scahill C, Weavers H, Skaer H. 2010. Hemocyte-secreted type IV collagen enhances BMP signaling to guide renal tubule morphogenesis in *Drosophila*. *Dev Cell* **19:** 296–306.

Butler MJ. 2003. Discovery of genes with highly restricted expression patterns in the *Drosophila* wing disc using DNA oligonucleotide microarrays. *Development* **130:** 659–670.

Butler MJ, Jacobsen TL, Cain DM, Jarman MG, Hubank M, Whittle JR, Phillips R, Simcox A. 2003. Discovery of genes with highly restricted expression patterns in the *Drosophila* wing disc using DNA oligonucleotide microarrays. *Development* **130:** 659–670.

Cabernard C, Affolter M. 2005. Distinct roles for two receptor tyrosine kinases in epithelial branching morphogenesis in *Drosophila*. *Dev Cell* **9:** 831–842.

Cafferty P, Yu L, Rao Y. 2004. The receptor tyrosine kinase off-track is required for layer-specific neuronal connectivity in *Drosophila*. *Development* **131:** 5287–5295.

Callahan CA, Muralidhar MG, Lundgren SE, Scully AL, Thomas JB. 1995. Control of neuronal pathway selection by a *Drosophila* receptor protein-tyrosine kinase family member. *Nature* **376:** 171–174.

Callahan CA, Bonkovsky JL, Scully AL, Thomas JB. 1996. Derailed is required for muscle attachment site selection in *Drosophila*. *Development* **122:** 2761–2767.

Carmena A, Gisselbrecht S, Harrison J, Jiménez F, Michelson AM. 1998. Combinatorial signaling codes for the progressive determination of cell fates in the *Drosophila* embryonic mesoderm. *Genes Dev* **12:** 3910–3922.

Carreira VP, Mensch J, Fanara JJ. 2008. Body size in *Drosophila*: Genetic architecture, allometries and sexual dimorphism. *Heredity* **102:** 246–256.

Casali A, Casanova J. 2001. The spatial control of Torso RTK activation: A C-terminal fragment of the Trunk protein acts as a signal for Torso receptor in the *Drosophila* embryo. *Development* **128:** 1709–1715.

Casanova J, Struhl G. 1989. Localized surface activity of torso, a receptor tyrosine kinase, specifies terminal body pattern in *Drosophila*. *Genes Dev* **3:** 2025–2038.

Chang Y-Y, Juhász G, Goraksha-Hicks P, Arsham AM, Mallin DR, Muller LK, Neufeld TP. 2009. Nutrient-dependent regulation of autophagy through the target of rapamycin pathway. *Biochem Soc Trans* **37:** 232.

Charlton-Perkins M, Brown NL, Cook TA. 2011. The lens in focus: A comparison of lens development in *Drosophila* and vertebrates. *Mol Gene Genomics* **286:** 189–213.

Chell JM, Brand AH. 2010. Nutrition-responsive glia control exit of neural stem cells from quiescence. *Cell* **143:** 1161–1173.

Chen C, Jack J, Garofalo RS. 1996. The *Drosophila* insulin receptor is required for normal growth. *Endocrinology* **137:** 846–856.

Cheng LY, Bailey AP, Leevers SJ, Ragan TJ, Driscoll PC, Gould AP. 2011. Anaplastic lymphoma kinase spares organ growth during nutrient restriction in *Drosophila*. *Cell* **146:** 435–447.

Chintapalli VR, Wang J, Dow JAT. 2007. Using FlyAtlas to identify better *Drosophila melanogaster* models of human disease. *Nat Genet* **39:** 715–720.

Cho NK, Keyes L, Johnson E, Heller J, Ryner L, Karim F, Krasnow MA. 2002. Developmental control of blood cell migration by the *Drosophila* VEGF pathway. *Cell* **108:** 865–876.

Choi N-H, Kim J-G, Yang D-J, Kim Y-S, Yoo M-A. 2008. Age-related changes in *Drosophila* midgut are associated with PVF2, a PDGF/VEGF-like growth factor. *Aging Cell* **7:** 318–334.

Cinnamon E, Helman A, Ben-Haroush Schyr R, Orian A, Jimenez G, Paroush Z. 2008. Multiple RTK pathways downregulate Groucho-mediated repression in *Drosophila* embryogenesis. *Development* **135:** 829–837.

Clancy DJ, Gems D, Harshman LG, Oldham S, Stocker H, Hafen E, Leevers SJ, Partridge L. 2001. Extension of lifespan by loss of CHICO, a *Drosophila* insulin receptor substrate protein. *Science* **292:** 104–106.

Cite this article as *Cold Spring Harb Perspect Biol* doi: 10.1101/cshperspect.a009050

Clark IBN, Muha V, Klingseisen A, Leptin M, Muller HAJ. 2011. Fibroblast growth factor signalling controls successive cell behaviors during mesoderm layer formation in *Drosophila*. *Development* **138:** 2705–2715.

Cognigni P, Bailey AP, Miguel-Aliaga I. 2011. Enteric neurons and systemic signals couple nutritional and reproductive status with intestinal homeostasis. *Cell Metab* **13:** 92–104.

Cook T, Pichaud F, Sonneville R, Papatsenko D, Desplan C. 2003. Distinction between color photoreceptor cell fates is controlled by *prospero* in *Drosophila*. *Dev Cell* **4:** 853–864.

Csiszar A, Vogelsang E, Beug H, Leptin M. 2010. A novel conserved phosphotyrosine motif in the *Drosophila* fibroblast growth factor signaling adaptor Dof with a redundant role in signal transmission. *Mol Cell Biol* **30:** 2017–2027.

Daga A, Karlovich CA, Dumstrei K, Banerjee U. 1996. Patterning of cells in the *Drosophila* eye by *lozenge*, which shares homologous domains with AML1. *Genes Dev* **10:** 1194–1205.

Dearborn R, He Q, Kunes S, Dai Y. 2002. Eph receptor tyrosine kinase-mediated formation of a topographic map in the *Drosophila* visual system. *J Neurosci* **22:** 1338–1349.

DeChiara TM, Bowen DC, Valenzuela DM, Simmons MV, Poueymirou WT, Thomas S, Kinetz E, Compton DL, Rojas E, Park JS, et al. 1996. The receptor tyrosine kinase MuSK is required for neuromuscular junction formation in vivo. *Cell* **85:** 501–512.

Demontis F, Perrimon N. 2010. FOXO/4E-BP signaling in *Drosophila* muscles regulates organism-wide proteostasis during aging. *Cell* **143:** 813–825.

DiAngelo JR, Birnbaum MJ. 2009. Regulation of fat cell mass by insulin in *Drosophila* melanogaster. *Mol Cell Biol* **29:** 6341–6352.

Dickson B, Sprenger F, Morrison D, Hafen E. 1992. Raf functions downstream of Ras1 in the sevenless signal transduction pathway. *Nature* **360:** 600–603.

Doyle HJ, Bishop JM. 1993. Torso, a receptor tyrosine kinase required for embryonic pattern formation, shares substrates with the sevenless and EGF-R pathways in *Drosophila*. *Genes Dev* **7:** 633–646.

Drummond-Barbosa D, Spradling AC. 2001. Stem cells and their progeny respond to nutritional changes during *Drosophila* oogenesis. *Dev Biol* **231:** 265–278.

Duchek P, Somogyi K, Jékely G, Beccari S, Rørth P. 2001. Guidance of cell migration by the *Drosophila* PDGF/VEGF receptor. *Cell* **107:** 17–26.

Dura JM, Taillebourg E, Préat T. 1995. The *Drosophila* learning and memory gene linotte encodes a putative receptor tyrosine kinase homologous to the human RYK gene product. *FEBS Lett* **370:** 250–254.

Dutta D, Shaw S, Maqbool T, Pandya H, VijayRaghavan K. 2005. *Drosophila* Heartless acts with Heartbroken/Dof in muscle founder differentiation. *PLoS Biol* **3:** e337.

Egger B, Leemans R, Loop T, Kammermeier L, Fan Y, Radimerski T, Strahm MC, Certa U, Reichert H. 2002. Gliogenesis in *Drosophila*: Genome-wide analysis of downstream genes of glial cells missing in the embryonic nervous system. *Development* **129:** 3295–3309.

Emori Y, Saigo K. 1993. Distinct expression of two *Drosophila* homologs of fibroblast growth factor receptors in imaginal discs. *FEBS Lett* **332:** 111–114.

Englund C, Lorén CE, Grabbe C, Varshney GK, Deleuil F, Hallberg B, Palmer RH. 2003. Jeb signals through the Alk receptor tyrosine kinase to drive visceral muscle fusion. *Nature* **425:** 512–516.

Fernandez R, Tabarini D, Azpiazu N, Frasch M, Schlessinger J. 1995. The *Drosophila* insulin receptor homolog: A gene essential for embryonic development encodes two receptor isoforms with different signaling potential. *EMBO J* **14:** 3373–3384.

Forrester WC, Dell M, Perens E, Garriga G. 1999. A *C elegans* Ror receptor tyrosine kinase regulates cell motility and asymmetric cell division. *Nature* **400:** 881–885.

Fradkin LG, van Schie M, Wouda RR, de Jong A, Kamphorst JT, Radjkoemar-Bansraj M, Noordermeer JN. 2004. The Drosophila Wnt5 protein mediates selective axon fasciculation in the embryonic central nervous system. *Dev Biol* **272:** 362–375.

Frank CA, Pielage J, Davis GW. 2009. A presynaptic homeostatic signaling system composed of the Eph receptor, Ephexin, Cdc42, and CaV2.1 calcium channels. *Neuron* **61:** 556–569.

Franzdóttir SR, Engelen D, Yuva-Aydemir Y, Schmidt I, Aho A, Klämbt C. 2009. Switch in FGF signalling initiates glial differentiation in the Drosophila eye. *Nature* **460:** 758–761.

Freeman MR, Delrow J, Kim J, Johnson E, Doe CQ. 2003. Unwrapping glial biology: Gcm target genes regulating glial development, diversification, and function. *Neuron* **38:** 567–580.

Fujimoto J, Shiota M, Iwahara T, Seki N, Satoh H, Mori S, Yamamoto T. 1996. Characterization of the transforming activity of p80, a hyperphosphorylated protein in a Ki-1 lymphoma cell line with chromosomal translocation t(2;5). *Proc Natl Acad Sci* **93:** 4181–4186.

Fung S, Wang F, Chase M, Godt D, Hartenstein V. 2007. Expression profile of the cadherin family in the developing *Drosophila* brain. *J Comp Neurol* **506:** 469–488.

Furlong EE. 2004. Integrating transcriptional and signalling networks during muscle development. *Curr Opin Genet Devt* **14:** 343–350.

Gabay L, Seger R, Shilo BZ. 1997. In situ activation pattern of *Drosophila* EGF receptor pathway during development. *Science* **277:** 1103–1106.

Gao X, Pan D. 2001. TSC1 and TSC2 tumor suppressors antagonize insulin signaling in cell growth. *Genes Dev* **15:** 1383–1392.

Gao X, Neufeld TP, Pan D. 2000. Drosophila PTEN regulates cell growth and proliferation through PI3K-dependent and -independent pathways. *Dev Biol* **221:** 404–418.

García-Alonso L, Romani S, Jiménez F. 2000. The EGF and FGF receptors mediate neuroglian function to control growth cone decisions during sensory axon guidance in *Drosophila*. *Neuron* **28:** 741–752.

Garofalo RS, Rosen OM. 1988. Tissue localization of *Drosophila melanogaster* insulin receptor transcripts during development. *Mol Cell Biol* **8:** 1638–1647.

Gaul U, Mardon G, Rubin GM. 1992. A putative Ras GTPase activating protein acts as a negative regulator of signaling by the sevenless receptor tyrosine kinase. *Cell* **68:** 1007–1019.

Ghabrial AS, Krasnow MA. 2006. Social interactions among epithelial cells during tracheal branching morphogenesis. *Nature* **441:** 746–749.

Ghabrial AS, Levi BP, Krasnow MA. 2011. A systematic screen for tube morphogenesis and branching genes in the *Drosophila* tracheal system. *PLoS Genet* **7:** e1002087.

Gisselbrecht S, Skeath JB, Doe CQ, Michelson AM. 1996. Heartless encodes a fibroblast growth factor receptor (DFR1/DFGF-R2) involved in the directional migration of early mesodermal cells in the *Drosophila* embryo. *Genes Dev* **10:** 3003–3017.

Glazer L, Shilo BZ. 1991. The *Drosophila* FGF-R homolog is expressed in the embryonic tracheal system and appears to be required for directed tracheal cell extension. *Genes Dev* **5:** 697–705.

Gorczyca M, Augart C, Budnik V. 1993. Insulin-like receptor and insulin-like peptide are localized at neuromuscular junctions in *Drosophila*. *J Neurosci* **13:** 3692–3704.

Green JL, Inoue T, Sternberg PW. 2007. The *C. elegans* ROR receptor tyrosine kinase, CAM-1, non-autonomously inhibits the Wnt pathway. *Development* **134:** 4053–4062.

Green JL, Kuntz SG, Sternberg PW. 2008. Ror receptor tyrosine kinases: Orphans no more. *Trend Cell Biol* **18:** 536–544.

Grigorian M, Mandal L, Hakimi M, Ortiz I, Hartenstein V. 2011. The convergence of Notch and MAPK signaling specifies the blood progenitor fate in the *Drosophila* mesoderm. *Dev Biol* **353:** 105–118.

Grillenzoni N, Flandre A, Lasbleiz C, Dura JM. 2007. Respective roles of the DRL receptor and its ligand WNT5 in *Drosophila* mushroom body development. *Development* **134:** 3089–3097.

Grumolato L, Liu G, Mong P, Mudbhary R, Biswas R, Arroyave R, Vijayakumar S, Economides AN, Aaronson SA. 2010. Canonical and noncanonical Wnts use a common mechanism to activate completely unrelated coreceptors. *Genes Dev* **24:** 2517–2530.

Gryzik T, Müller HAJ. 2004. FGF8-like1 and FGF8-like2 encode putative ligands of the FGF receptor Htl and are required for mesoderm migration in the *Drosophila* gastrula. *Curr Biol* **14:** 659–667.

Guha A, Lin L, Kornberg TB. 2008. Organ renewal and cell divisions by differentiated cells in *Drosophila*. *Proc Natl Acad Sci* **105:** 10832–10836.

Hacohen N, Kramer S, Sutherland D, Hiromi Y, Krasnow MA. 1998. Sprouty encodes a novel antagonist of FGF signaling that patterns apical branching of the *Drosophila* airways. *Cell* **92:** 253–263.

Hahn M, Bishop J. 2001. Expression pattern of *Drosophila* ret suggests a common ancestral origin between the metamorphosis precursors in insect endoderm and the vertebrate enteric neurons. *Proc Natl Acad Sci* **98:** 1053–1058.

Halfon MS, Carmena A, Gisselbrecht S, Sackerson CM, Jiménez F, Baylies MK, Michelson AM. 2000. Ras pathway specificity is determined by the integration of multiple signal-activated and tissue-restricted transcription factors. *Cell* **103:** 63–74.

Harris KE, Beckendorf SK. 2007. Different Wnt signals act through the Frizzled and RYK receptors during *Droso-phila* salivary gland migration. *Development* **134:** 2017–2025.

Harris KE, Schnittke N, Beckendorf SK. 2007. Two ligands signal through the *Drosophila* PDGF/VEGF receptor to ensure proper salivary gland positioning. *Mech Dev* **124:** 441–448.

Hart AC, Krämer H, Van Vactor DL, Paidhungat M, Zipursky SL. 1990. Induction of cell fate in the *Drosophila* retina: The bride of sevenless protein is predicted to contain a large extracellular domain and seven transmembrane segments. *Genes Dev* **4:** 1835–1847.

Hayashi T, Xu C, Carthew RW. 2008. Cell-type-specific transcription of *prospero* is controlled by combinatorial signaling in the *Drosophila* eye. *Development* **135:** 2787–2796.

Heino TI, Kärpänen T, Wahlström G, Pulkkinen M, Eriksson U, Alitalo K, Roos C. 2001. The *Drosophila* VEGF receptor homolog is expressed in hemocytes. *Mech Dev* **109:** 69–77.

Helman A, Cinnamon E, Mezuman S, Hayouka Z, Von Ohlen T, Orian A, Jiménez G, Paroush Zae. 2011. Phosphorylation of groucho mediates RTK feedback inhibition and prolonged pathway target gene expression. *Curr Biol* **21:** 1102–1110.

Hemphala J, Uv A, Cantera R, Bray S, Samakovlis C. 2003. Grainy head controls apical membrane growth and tube elongation in response to branchless/FGF signalling. *Development* **130:** 249–258.

Herbst R, Burden SJ. 2000. The juxtamembrane region of MuSK has a critical role in agrin-mediated signaling. *EMBO J* **19:** 67–77.

Hidalgo A, Booth GE. 2000. Glia dictate pioneer axon trajectories in the *Drosophila* embryonic CNS. *Development* **127:** 393–402.

Hikasa H, Shibata M, Hiratani I, Taira M. 2002. The *Xenopus* receptor tyrosine kinase Xror2 modulates morphogenetic movements of the axial mesoderm and neuroectoderm via Wnt signaling. *Development* **129:** 5227–5239.

Ho IS, Hannan F, Guo HF, Hakker I, Zhong Y. 2007. Distinct functional domains of neurofibromatosis type 1 regulate immediate versus long-term memory formation. *J Neurosci* **27:** 6852–6857.

Hou XS, Chou TB, Melnick MB, Perrimon N. 1995. The torso receptor tyrosine kinase can activate Raf in a Ras-independent pathway. *Cell* **81:** 63–71.

Hsu H-J, Drummond-Barbosa D. 2011. Insulin signals control the competence of the *Drosophila* female germline stem cell niche to respond to Notch ligands. *Dev Biol* **350:** 290–300.

Hsu H-J, LaFever L, Drummond-Barbosa D. 2008. Diet controls normal and tumorous germline stem cells via insulin-dependent and -independent mechanisms in *Drosophila*. *Dev Biol* **313:** 700–712.

Huang H, Potter CJ, Tao W, Li DM, Brogiolo W, Hafen E, Sun H, Xu T. 1999. PTEN affects cell size, cell proliferation and apoptosis during *Drosophila* eye development. *Development* **126:** 5365–5372.

Hurley SP, Clary DO, Copie V, Lefcort F. 2006. Anaplastic lymphoma kinase is dynamically expressed on subsets of motor neurons and in the peripheral nervous system. *J Comp Neurol* **495:** 202–212.

Cite this article as *Cold Spring Harb Perspect Biol* doi: 10.1101/cshperspect.a009050

Ikeya T, Hayashi S. 1999. Interplay of Notch and FGF signaling restricts cell fate and MAPK activation in the *Drosophila* trachea. *Development* 126: 4455–4463.

Ikeya T, Galic M, Belawat P, Nairz K, Hafen E. 2002. Nutrient-dependent expression of insulin-like peptides from neuroendocrine cells in the CNS contributes to growth regulation in *Drosophila*. *Curr Biol* 12: 1293–1300.

Imam F, Sutherland D, Huang W, Krasnow MA. 1999. Stumps, a *Drosophila* gene required for fibroblast growth factor (FGF)-directed migrations of tracheal and mesodermal cells. *Genetics* 152: 307–318.

Inaki M, Yoshikawa S, Thomas JB, Aburatani H, Nose A. 2007. Wnt4 is a local repulsive cue that determines synaptic target specificity. *Curr Biol* 17: 1574–1579.

Ishimaru S, Ueda R, Hinohara Y, Ohtani M, Hanafusa H. 2004. PVR plays a critical role via JNK activation in thorax closure during *Drosophila* metamorphosis. *EMBO J* 23: 3984–3994.

Ito M, Matsui T, Taniguchi T, Chihara K. 1994. Alternative splicing generates two distinct transcripts for the *Drosophila* melanogaster fibroblast growth factor receptor homolog. *Gene* 139: 215–218.

Iwahara T, Fujimoto J, Wen D, Cupples R, Bucay N, Arakawa T, Mori S, Ratzkin B, Yamamoto T. 1997. Molecular characterization of ALK, a receptor tyrosine kinase expressed specifically in the nervous system. *Oncogene* 14: 439–449.

Jackson Behan K, Fair J, Singh S, Bogwitz M, Perry T, Grubor V, Cunningham F, Nichols CD, Cheung TL, Batterham P, et al. 2005. Alternative splicing removes an Ets interaction domain from *lozenge* during *Drosophila* eye development. *Devt Genes Evol* 215: 423–435.

Janody F, Sturny R, Schaeffer V, Azou Y, Dostatni N. 2001. Two distinct domains of Bicoid mediate its transcriptional down-regulation by the Torso pathway. *Development* 128: 2281–2290.

Jarecki J, Johnson E, Krasnow MA. 1999. Oxygen regulation of airway branching in *Drosophila* is mediated by branchless FGF. *Cell* 99: 211–220.

Jing L, Lefebvre JL, Gordon LR, Granato M. 2009. Wnt signals organize synaptic pepattern and axon guidance through the zebrafish unplugged/MuSK receptor. *Neuron* 61: 721–733.

Johnson Hamlet MR, Perkins LA. 2001. Analysis of corkscrew signaling in the *Drosophila* epidermal growth factor receptor pathway during myogenesis. *Genetics* 159: 1073–1087.

Jonsson M, Andersson T. 2001. Repression of Wnt-5a impairs DDR1 phosphorylation and modifies adhesion and migration of mammary cells. *J Cell Sci* 114: 2043–2053.

Jung SH, Evans CJ, Uemura C, Banerjee U. 2005. The *Drosophila* lymph gland as a developmental model of hematopoiesis. *Development* 132: 2521–2533.

Jünger MA, Rintelen F, Stocker H, Wasserman JD, Végh M, Radimerski T, Greenberg ME, Hafen E. 2003. The *Drosophila* forkhead transcription factor FOXO mediates the reduction in cell number associated with reduced insulin signaling. *J Biol* 2: 20.

Kadam S, McMahon A, Tzou P, Stathopoulos A. 2009. FGF ligands in *Drosophila* have distinct activities required to support cell migration and differentiation. *Development* 136: 739–747.

Kadam S, Ghosh S, Stathopoulos A. 2012. Synchronous and symmetric migration of *Drosophila* caudal visceral mesoderm cells requires dual input by two FGF ligands. *Development* 139: 699–708.

Kapahi P, Zid BM, Harper T, Koslover D, Sapin V, Benzer S. 2004. Regulation of lifespan in *Drosophila* by modulation of genes in the TOR signaling pathway. *Curr Biol* 14: 885–890.

Katso RM, Russell RB, Ganesan TS. 1999. Functional analysis of H-Ryk, an atypical member of the receptor tyrosine kinase family. *Mol Cell Biol* 19: 6427–6440.

Keeble TR, Halford MM, Seaman C, Kee N, Macheda M, Anderson RB, Stacker SA, Cooper HM. 2006. The Wnt receptor Ryk is required for Wnt5a-mediated axon guidance on the contralateral side of the corpus callosum. *J Neurosci* 26: 5840–5848.

Kim N, Stiegler AL, Cameron TO, Hallock PT, Gomez AM, Huang JH, Hubbard SR, Dustin ML, Burden SJ. 2008. Lrp4 is a receptor for agrin and forms a complex with MuSK. *Cell* 135: 334–342.

Kitadate Y, Shigenobu S, Arita K, Kobayashi S. 2007. Boss/Sev signaling from germline to soma restricts germline-stem-cell-niche formation in the anterior region of *Drosophila* male gonads. *Dev Cell* 13: 151–159.

Klambt C. 1993. The *Drosophila* gene pointed encodes two ETS-like proteins which are involved in the development of the midline glial cells. *Development* 117: 163–176.

Klambt C, Glazer L, Shilo BZ. 1992. Breathless, a *Drosophila* FGF receptor homolog, is essential for migration of tracheal and specific midline glial cells. *Genes Dev* 6: 1668–1678.

Klingler M, Erdélyi M, Szabad J, Nüsslein-Volhard C. 1988. Function of torso in determining the terminal anlagen of the *Drosophila* embryo. *Nature* 335: 275–277.

Klingseisen A, Clark IBN, Gryzik T, Muller HAJ. 2009. Differential and overlapping functions of two closely related *Drosophila* FGF8-like growth factors in mesoderm development. *Development* 136: 2393–2402.

Krämer H, Cagan RL, Zipursky SL. 1991. Interaction of bride of sevenless membrane-bound ligand and the sevenless tyrosine-kinase receptor. *Nature* 352: 207–212.

LaFever L, Drummond-Barbosa D. 2005. Direct control of germline stem cell division and cyst growth by neural insulin in *Drosophila*. *Science* 309: 1071–1073.

Lai ZC, Fetchko M, Li Y. 1997. Repression of *Drosophila* photoreceptor cell fate through cooperative action of two transcriptional repressors Yan and Tramtrack. *Genetics* 147: 1131–1137.

Learte AR, Forero MG, Hidalgo A. 2008. Gliatrophic and gliatropic roles of PVF/PVR signaling during axon guidance. *Glia* 56: 164–176.

Lee T, Hacohen N, Krasnow M, Montell DJ. 1996. Regulated breathless receptor tyrosine kinase activity required to pattern cell migration and branching in the *Drosophila* tracheal system. *Genes Dev* 10: 2912–2921.

Lee H-H, Norris A, Weiss JB, Frasch M. 2003. Jelly belly protein activates the receptor tyrosine kinase Alk to specify visceral muscle pioneers. *Nature* 425: 507–512.

Lemeer S, Bluwstein A, Wu Z, Leberfinger J, Müller K, Kramer K, Kuster B. 2011. Phosphotyrosine mediated protein interactions of the discoidin domain receptor 1. *J Proteomics* **75:** 3465–3477.

Li W, Skoulakis EM, Davis RL, Perrimon N. 1997. The *Drosophila* 14–3-3 protein Leonardo enhances Torso signaling through D-Raf in a Ras 1-dependent manner. *Development* **124:** 4163–4171.

Li S, Xu C, Carthew RW. 2002. Phyllopod acts as an adaptor protein to link the sina ubiquitin ligase to the substrate protein tramtrack. *Mol Cell Biol* **22:** 6854–6865.

Liao EH, Hung W, Abrams B, Zhen M. 2004. An SCF-like ubiquitin ligase complex that controls presynaptic differentiation. *Nature* **430:** 345–350.

Liebl FLW, Wu Y, Featherstone DE, Noordermeer JN, Fradkin L, Hing H. 2008. Derailed regulates development of the *Drosophila* neuromuscular junction. *Dev Neurobiol* **68:** 152–165.

Liu Y, Shi J, Lu C-C, Wang Z-B, Lyuksyutova AI, Song X, Zou Y. 2005. Ryk-mediated Wnt repulsion regulates posterior-directed growth of corticospinal tract. *Nat Neurosci* **8:** 1151–1159.

Liu X, Krause WC, Davis RL. 2007. GABAA receptor RDL inhibits *Drosophila* olfactory associative learning. *Neuron* **56:** 1090–1102.

Lorén CE, Englund C, Grabbe C, Hallberg B, Hunter T, Palmer RH. 2003. A crucial role for the Anaplastic lymphoma kinase receptor tyrosine kinase in gut development in *Drosophila* melanogaster. *EMBO Rep* **4:** 781–786.

Lovegrove B, Simões S, Rivas ML, Sotillos S, Johnson K, Knust E, Jacinto A, Hombría JC-G. 2006. Coordinated control of cell adhesion, polarity, and cytoskeleton underlies Hox-induced organogenesis in *Drosophila*. *Curr Biol* **16:** 2206–2216.

Lu X, Chou TB, Williams NG, Roberts T, Perrimon N. 1993. Control of cell fate determination by p21ras/Ras1, an essential component of torso signaling in *Drosophila*. *Genes Dev* **7:** 621–632.

Lu X, Melnick MB, Hsu JC, Perrimon N. 1994. Genetic and molecular analyses of mutations involved in *Drosophila* raf signal transduction. *EMBO J* **13:** 2592–2599.

Lu W, Yamamoto V, Ortega B, Baltimore D. 2004. Mammalian Ryk is a wnt coreceptor required for stimulation of neurite outgrowth. *Cell* **119:** 97–108.

Luschnig S, Krauss J, Bohmann K, Desjeux I, Nüsslein-Volhard C. 2000. The *Drosophila* SHC adaptor protein is required for signaling by a subset of receptor tyrosine kinases. *Mol Cell* **5:** 231–241.

Mace KA, Pearson JC, McGinnis W. 2005. An epidermal barrier wound repair pathway in *Drosophila* is mediated by grainy head. *Science* **308:** 381–385.

Macias A, Romero NM, Martin F, Suarez L, Rosa AL, Morata G. 2004. PVF1/PVR signaling and apoptosis promotes the rotation and dorsal closure of the *Drosophila* male terminalia. *Int J Dev Biol* **48:** 1087–1094.

Makhijani K, Alexander B, Tanaka T, Rulifson E, Brückner K. 2011. The peripheral nervous system supports blood cell homing and survival in the *Drosophila* larva. *Development* **138:** 5379–5391.

Mandal L, Dumstrei K, Hartenstein V. 2004. Role of FGFR signaling in the morphogenesis of the *Drosophila* visceral musculature. *Dev Dyn* **231:** 342–348.

Maqbool T, Soler C, Jagla T, Daczewska M, Lodha N, Palliyil S, VijayRaghavan K, Jagla K. 2006. Shaping leg muscles in *Drosophila*: Role of ladybird, a conserved regulator of appendicular myogenesis. *PLoS ONE* **1:** e122.

Martin DN, Balgley B, Dutta S, Chen J, Rudnick P, Cranford J, Kantartzis S, DeVoe DL, Lee C, Baehrecke EH. 2007. Proteomic analysis of steroid-triggered autophagic programmed cell death during *Drosophila* development. *Cell Death Differ* **14:** 916–923.

Martin-Pena A, Acebes A, Rodriguez JR, Sorribes A, de Polavieja GG, Fernandez-Funez P, Ferrus A. 2006. Age-independent synaptogenesis by phosphoinositide 3 kinase. *J Neurosci* **26:** 10199–10208.

Maurel-Zaffran C, Pradel J, Graba Y. 2010. Reiterative use of signalling pathways controls multiple cellular events during *Drosophila* posterior spiracle organogenesis. *Dev Biol* **343:** 18–27.

McBrayer Z, Ono H, Shimell M, Parvy J-P, Beckstead RB, Warren JT, Thummel CS, Dauphin-Villemant C, Gilbert LI, O'Connor MB. 2007. Prothoracicotropic hormone regulates developmental timing and body size in *Drosophila*. *Dev Cell* **13:** 857–871.

McDonald JA, Pinheiro EM, Montell DJ. 2003. PVF1, a PDGF/VEGF homolog, is sufficient to guide border cells and interacts genetically with Taiman. *Development* **130:** 3469–3478.

McLeod CJ, Wang L, Wong C, Jones DL. 2010. Stem cell dynamics in response to nutrient availability. *Curr Biol* **20:** 2100–2105.

McMahon A, Supatto W, Fraser SE, Stathopoulos A. 2008. Dynamic analyses of *Drosophila* gastrulation provide insights into collective cell migration. *Science* **322:** 1546–1550.

McMahon A, Reeves GT, Supatto W, Stathopoulos A. 2010. Mesoderm migration in *Drosophila* is a multi-step process requiring FGF signaling and integrin activity. *Development* **137:** 2167–2175.

Michaut L, Flister S, Neeb M, White KP, Certa U, Gehring WJ. 2003. Analysis of the eye developmental pathway in *Drosophila* using DNA microarrays. *Proc Natl Acad Sci* **100:** 4024–4029.

Michelson AM, Gisselbrecht S, Buff E, Skeath JB. 1998a. Heartbroken is a specific downstream mediator of FGF receptor signalling in *Drosophila*. *Development* **125:** 4379–4389.

Michelson AM, Gisselbrecht S, Zhou Y, Baek KH, Buff EM. 1998b. Dual functions of the heartless fibroblast growth factor receptor in development of the *Drosophila* embryonic mesoderm. *Dev Genet* **22:** 212–229.

Miguel-Aliaga I, Thor S, Gould AP. 2008. Postmitotic specification of *Drosophila* insulinergic neurons from pioneer neurons. *PLoS Biol* **6:** e58.

Mikels AJ, Nusse R. 2006. Purified Wnt5a protein activates or inhibits β-catenin–TCF signaling depending on receptor context. *PLoS Biol* **4:** e115.

Min K-J, Yamamoto R, Buch S, Pankratz M, Tatar M. 2008. Drosophila lifespan control by dietary restriction independent of insulin-like signaling. *Aging Cell* **7:** 199–206.

Montagne J, Stewart MJ, Stocker H, Hafen E, Kozma SC, Thomas G. 1999. *Drosophila* S6 kinase: A regulator of cell size. *Science* **285:** 2126–2129.

Moore LA, Broihier HT, Van Doren M, Lunsford LB, Lehmann R. 1998. Identification of genes controlling germ cell migration and embryonic gonad formation in *Drosophila*. *Development* **125:** 667–678.

Moran E, Jimenez G. 2006. The tailless nuclear receptor acts as a dedicated repressor in the early *Drosophila* embryo. *Mol Cell Biol* **26:** 3446–3454.

Moreau-Fauvarque C, Taillebourg E, Boissoneau E, Mesnard J, Dura JM. 1998. The receptor tyrosine kinase gene linotte is required for neuronal pathway selection in the *Drosophila* mushroom bodies. *Mech Dev* **78:** 47–61.

Mukherjee T, Choi I, Banerjee U, Lipshitz HD. 2012. Genetic analysis of fibroblast growth factor signaling in the *Drosophila* eye. *G3 (Bethesda)* **2:** 23–28.

Murray MJ, Saint R. 2007. Photoactivatable GFP resolves *Drosophila* mesoderm migration behavior. *Development* **134:** 3975–3983.

Nasiadka A, Dietrich BH, Krause HM. 2002. Anterior-posterior patterning in the *Drosophila* embryo. In *Advances in developmental biology and biochemistry* (ed. DePamphilis M), pp. 1–50. Elsevier Science, Amsterdam.

Nicola M, Lasbleiz C, Dura J-M. 2003. Gain-of-function screen identifies a role of the Src64 oncogene in *Drosophila* mushroom body development. *J Neurobiol* **57:** 291–302.

Nüsslein-Volhard C, Frohnhöfer HG, Lehmann R. 1987. Determination of anteroposterior polarity in *Drosophila*. *Science* **238:** 1675–1681.

Oates AC, Bonkovsky JL, Irvine DV, Kelly LE, Thomas JB, Wilks AF. 1998. Embryonic expression and activity of doughnut, a second RYK homolog in *Drosophila*. *Mech Dev* **78:** 165–169.

Ohsawa S, Sugimura K, Takino K, Xu T, Miyawaki A, Igaki T. 2011. Elimination of oncogenic neighbors by JNK-mediated engulfment in *Drosophila*. *Dev Cell* **20:** 315–328.

Ohshiro T, Emori Y, Saigo K. 2002. Ligand-dependent activation of breathless FGF receptor gene in *Drosophila* developing trachea. *Mech Dev* **114:** 3–11.

Oishi I, Sugiyama S, Liu ZJ, Yamamura H, Nishida Y, Minami Y. 1997. A novel *Drosophila* receptor tyrosine kinase expressed specifically in the nervous system. Unique structural features and implication in developmental signaling. *J Biol Chem* **272:** 11916–11923.

Oishi I, Suzuki H, Onishi N, Takada R, Kani S, Ohkawara B, Koshida I, Suzuki K, Yamada G, Schwabe GC, et al. 2003. The receptor tyrosine kinase Ror2 is involved in non-canonical Wnt5a/JNK signalling pathway. *Genes Cells* **8:** 645–654.

Okamoto N, Yamanaka N, Yagi Y, Nishida Y, Kataoka H, Connor MBO, Mizoguchi A. 2009. A fat body-derived IGF-like peptide regulates postfeeding growth in *Drosophila*. *Dev Cell* **17:** 885–891.

Oldham S, Montagne J, Radimerski T, Thomas G, Hafen E. 2000. Genetic and biochemical characterization of dTOR, the *Drosophila* homolog of the target of rapamycin. *Genes Dev* **14:** 2689–2694.

Oldham S, Stocker H, Laffargue M, Wittwer F, Wymann M, Hafen E. 2002. The *Drosophila* insulin/IGF receptor controls growth and size by modulating PtdInsP$_3$ levels. *Development* **129:** 4103–4109.

Olivier JP, Raabe T, Henkemeyer M, Dickson B, Mbamalu G, Margolis B, Schlessinger J, Hafen E, Pawson T. 1993. A *Drosophila* SH2-SH3 adaptor protein implicated in coupling the sevenless tyrosine kinase to an activator of Ras guanine nucleotide exchange, Sos. *Cell* **73:** 179–191.

Olofsson B, Page DT. 2005. Condensation of the central nervous system in embryonic *Drosophila* is inhibited by blocking hemocyte migration or neural activity. *Dev Biol* **279:** 233–243.

O'Neill EM, Rebay I, Tjian R, Rubin GM. 1994. The activities of two Ets-related transcription factors required for *Drosophila* eye development are modulated by the Ras/MAPK pathway. *Cell* **78:** 137–147.

Ostrin EJ, Li Y, Hoffman K, Liu J, Wang K, Zhang L, Mardon G, Chen R. 2006. Genome-wide identification of direct targets of the *Drosophila* retinal determination protein Eyeless. *Genome Res* **16:** 466–476.

Paganoni S, Ferreira A. 2005. Neurite extension in central neurons: A novel role for the receptor tyrosine kinases Ror1 and Ror2. *J Cell Sci* **118:** 433–446.

Paganoni S, Bernstein J, Ferreira A. 2010. Ror1–Ror2 complexes modulate synapse formation in hippocampal neurons. *Neuroscience* **165:** 1261–1274.

Park J-S, Kim Y-S, Yoo M-A. 2009. The role of p38b MAPK in age-related modulation of intestinal stem cell proliferation and differentiation in *Drosophila*. *Aging* **1:** 637–651.

Peradziryi H, Kaplan NA, Podleschny M, Liu X, Wehner P, Borchers A, Tolwinski NS. 2011. PTK7/Otk interacts with Wnts and inhibits canonical Wnt signalling. *EMBO J* **30:** 3729–3740.

Perkins LA, Larsen I, Perrimon N. 1992. corkscrew encodes a putative protein tyrosine phosphatase that functions to transduce the terminal signal from the receptor tyrosine kinase torso. *Cell* **70:** 225–236.

Perrimon N, McMahon AP. 1999. Negative feedback mechanisms and their roles during pattern formation. *Cell* **97:** 13–16.

Perrimon N, Perkins LA. 1997. There must be 50 ways to rule the signal: The case of the *Drosophila* EGF receptor. *Cell* **89:** 13–16.

Perrimon N, Mohler D, Engstrom L, Mahowald AP. 1986. X-linked female-sterile loci in *Drosophila* melanogaster. *Genetics* **113:** 695–712.

Perrimon N, Lu X, Hou XS, Hsu JC, Melnick MB, Chou TB, Perkins LA. 1995. Dissection of the torso signal transduction pathway in *Drosophila*. *Mol Reprod Dev* **42:** 515–522.

Petit V, Nussbaumer U, Dossenbach C, Affolter M. 2004. Downstream-of-FGFR is a fibroblast growth factor-specific scaffolding protein and recruits Corkscrew upon receptor activation. *Mol Cell Biol* **24:** 3769–3781.

Petruzzelli L, Herrera R, Arenas-Garcia R, Fernandez R, Birnbaum MJ, Rosen OM. 1986. Isolation of a *Drosophila* genomic sequence homologous to the kinase domain of the human insulin receptor and detection of the phosphorylated *Drosophila* receptor with an anti-peptide antibody. *Proc Natl Acad Sci* **83:** 4710–4714.

Pignoni F, Baldarelli RM, Steingrímsson E, Diaz RJ, Patapoutian A, Merriam JR, Lengyel JA. 1990. The *Drosophila* gene tailless is expressed at the embryonic termini and is a member of the steroid receptor superfamily. *Cell* **62:** 151–163.

Pignoni F, Steingrímsson E, Lengyel JA. 1992. Bicoid and the terminal system activate tailless expression in the early *Drosophila* embryo. *Development* **115:** 239–251.

Pitsouli C, Perrimon N. 2010. Embryonic multipotent progenitors remodel the *Drosophila* airways during metamorphosis. *Development* **137:** 3615–3624.

Potter CJ, Huang H, Xu T. 2001. *Drosophila* Tsc1 functions with Tsc2 to antagonize insulin signaling in regulating cell growth, cell proliferation, and organ size. *Cell* **105:** 357–368.

Poukkula M, Cliffe A, Changede R, Rørth P. 2011. Cell behaviors regulated by guidance cues in collective migration of border cells. *J Cell Biol* **192:** 513–524.

Prasad M, Montell DJ. 2007. Cellular and molecular mechanisms of border cell migration analyzed using time-lapse live-cell imaging. *Dev Cell* **12:** 997–1005.

Pulido D, Campuzano S, Koda T, Modolell J, Barbacid M. 1992. Dtrk, a *Drosophila* gene related to the trk family of neurotrophin receptors, encodes a novel class of neural cell adhesion molecule. *EMBO J* **11:** 391–404.

Raabe T, Riesgo-Escovar J, Liu X, Bausenwein BS, Deak P, Maröy P, Hafen E. 1996. DOS, a novel pleckstrin homology domain-containing protein required for signal transduction between sevenless and Ras1 in *Drosophila*. *Cell* **85:** 911–920.

Read RD, Goodfellow PJ, Mardis ER, Novak N, Armstrong JR, Cagan RL. 2005. A *Drosophila* model of multiple endocrine neoplasia type 2. *Genetics* **171:** 1057–1081.

Rebay I, Rubin GM. 1995. Yan functions as a general inhibitor of differentiation and is negatively regulated by activation of the Ras1/MAPK pathway. *Cell* **81:** 857–866.

Reichman-Fried M, Shilo BZ. 1995. Breathless, a *Drosophila* FGF receptor homolog, is required for the onset of tracheal cell migration and tracheole formation. *Mech Dev* **52:** 265–273.

Reiner DJ, Ailion M, Thomas JH, Meyer BJ. 2008. *C elegans* anaplastic lymphoma kinase ortholog SCD-2 controls dauer formation by modulating TGF-β signaling. *Curr Biol* **18:** 1101–1109.

Reinke R, Zipursky SL. 1988. Cell–cell interaction in the *Drosophila* retina: The bride of sevenless gene is required in photoreceptor cell R8 for R7 cell development. *Cell* **55:** 321–330.

Rewitz KF, Yamanaka N, Gilbert LI, O'Connor MB. 2009. The insect neuropeptide PTTH activates receptor tyrosine kinase torso to initiate metamorphosis. *Science* **326:** 1403–1405.

Ribeiro C, Ebner A, Affolter M. 2002. In vivo imaging reveals different cellular functions for FGF and Dpp signaling in tracheal branching morphogenesis. *Dev Cell* **2:** 677–683.

Rintelen F, Stocker H, Thomas G, Hafen E. 2001. PDK1 regulates growth through Akt and S6K in *Drosophila*. *Proc Natl Acad Sci* **98:** 15020–15025.

Rogge RD, Karlovich CA, Banerjee U. 1991. Genetic dissection of a neurodevelopmental pathway: Son of sevenless functions downstream of the sevenless and EGF receptor tyrosine kinases. *Cell* **64:** 39–48.

Rogge R, Cagan R, Majumdar A, Dulaney T, Banerjee U. 1992. Neuronal development in the *Drosophila* retina: The sextra gene defines an inhibitory component in the developmental pathway of R7 photoreceptor cells. *Proc Natl Acad Sci* **89:** 5271–5275.

Rohrbough J, Broadie K. 2010. Anterograde Jelly belly ligand to Alk receptor signaling at developing synapses is regulated by Mind the gap. *Development* **137:** 3523–3533.

Ronchi E, Treisman J, Dostatni N, Struhl G, Desplan C. 1993. Down-regulation of the *Drosophila* morphogen bicoid by the torso receptor-mediated signal transduction cascade. *Cell* **74:** 347–355.

Rosin D, Schejter E, Volk T, Shilo BZ. 2004. Apical accumulation of the *Drosophila* PDGF/VEGF receptor ligands provides a mechanism for triggering localized actin polymerization. *Development* **131:** 1939–1948.

Roy S, Ernst J, Kharchenko PV, Kheradpour P, Negre N, Eaton ML, Landolin JM, Bristow CA, Ma L, Lin MF, et al. 2010. Identification of functional elements and regulatory circuits by *Drosophila* modENCODE. *Science* **330:** 1787–1797.

Rulifson EJ, Kim SK, Nusse R. 2002. Ablation of insulin-producing neurons in flies: Growth and diabetic phenotypes. *Science* **296:** 1118–1120.

Sakurai M, Aoki T, Yoshikawa S, Santschi LA, Saito H, Endo K, Ishikawa K, Kimura Ki, Ito K, Thomas JB, et al. 2009. Differentially expressed Drl and Drl-2 play opposing roles in Wnt5 signaling during Drosophila olfactory system development. *J Neurosci* **29:** 4972–4980.

Saldanha J, Singh J, Mahadevan D. 1998. Identification of a Frizzled-like cysteine rich domain in the extracellular region of developmental receptor tyrosine kinases. *Protein Sci* **7:** 1632–1635.

Samakovlis C, Hacohen N, Manning G, Sutherland DC, Guillemin K, Krasnow MA. 1996. Development of the *Drosophila* tracheal system occurs by a series of morphologically distinct but genetically coupled branching events. *Development* **122:** 1395–1407.

Sato M, Kornberg TB. 2002. FGF is an essential mitogen and chemoattractant for the air sacs of the Drosophila tracheal system. *Dev Cell* **3:** 195–207.

Sato M, Kitada Y, Tabata T. 2008. Larval cells become imaginal cells under the control of homothorax prior to metamorphosis in the *Drosophila* tracheal system. *Dev Biol* **318:** 247–257.

Savant-Bhonsale S, Friese M, McCoon P, Montell DJ. 1999. A *Drosophila* derailed homolog, doughnut, expressed in invaginating cells during embryogenesis. *Gene* **231:** 155–161.

Schober M, Rebay I, Perrimon N. 2005. Function of the ETS transcription factor Yan in border cell migration. *Development* **132:** 3493–3504.

Schulz RA, Gajewski K. 1999. Ventral neuroblasts and the heartless FGF receptor are required for muscle founder cell specification in *Drosophila*. *Oncogene* **18:** 6818–6823.

Schumacher S, Gryzik T, Tannebaum S, Muller HA. 2004. The RhoGEF Pebble is required for cell shape changes during cell migration triggered by the *Drosophila* FGF receptor heartless. *Development* **131:** 2631–2640.

Schupbach T, Wieschaus E. 1986. Germline autonomy of maternal-effect mutations altering the embryonic body pattern of *Drosophila*. *Dev Biol* **113:** 443–448.

Schupbach T, Wieschaus E. 1989. Female sterile mutations on the second chromosome of *Drosophila* melanogaster. I. Maternal effect mutations. *Genetics* **121:** 101–117.

Scully AL, McKeown M, Thomas JB. 1999. Isolation and characterization of Dek, a *Drosophila* eph receptor protein tyrosine kinase. *Mol Cell Neurosci* **13:** 337–347.

Sears HC, Kennedy CJ, Garrity PA. 2003. Macrophage-mediated corpse engulfment is required for normal *Drosophila* CNS morphogenesis. *Development* **130:** 3557–3565.

Shcherbata HR, Yatsenko AS, Patterson L, Sood VD, Nudel U, Yaffe D, Baker D, Ruohola-Baker H. 2007. Dissecting muscle and neuronal disorders in a *Drosophila* model of muscular dystrophy. *EMBO J* **26:** 481–493.

Shilo B. 2003. Signaling by the *Drosophila* epidermal growth factor receptor pathway during development. *Exp Cell Res* **284:** 140–149.

Shilo BZ. 2005. Regulating the dynamics of EGF receptor signaling in space and time. *Development* **132:** 4017–4027.

Shirinian M, Varshney G, Lorén CE, Grabbe C, Palmer RH. 2007. *Drosophila* anaplastic lymphoma kinase regulates Dpp signalling in the developing embryonic gut. *Differentiation* **75:** 418–426.

Shishido E, Higashijima S, Emori Y, Saigo K. 1993. Two FGF-receptor homologues of *Drosophila*: One is expressed in mesodermal primordium in early embryos. *Development* **117:** 751–761.

Shishido E, Ono N, Kojima T, Saigo K. 1997. Requirements of DFR1/heartless, a mesoderm-specific *Drosophila* FGF-receptor, for the formation of heart, visceral and somatic muscles, and ensheathing of longitudinal axon tracts in CNS. *Development* **124:** 2119–2128.

Shrivastava A, Radziejewski C, Campbell E, Kovac L, McGlynn M, Ryan TE, Davis S, Goldfarb MP, Glass DJ, Lemke G, et al. 1997. An orphan receptor tyrosine kinase family whose members serve as nonintegrin collagen receptors. *Mol Cell* **1:** 25–34.

Siddall NA, Hime GR, Pollock JA, Batterham P. 2009. Ttk69-dependent repression of *lozenge* prevents the ectopic development of R7 cells in the *Drosophila* larval eye disc. *BMC Dev Biol* **9:** 64.

Siegrist SE, Haque NS, Chen C-H, Hay BA, Hariharan IK. 2010. Inactivation of both foxo and reaper promotes long-term adult neurogenesis in *Drosophila*. *Curr Biol* **20:** 643–648.

Simon MA, Bowtell DD, Dodson GS, Laverty TR, Rubin GM. 1991. Ras1 and a putative guanine nucleotide exchange factor perform crucial steps in signaling by the sevenless protein tyrosine kinase. *Cell* **67:** 701–716.

Simon MA, Dodson GS, Rubin GM. 1993. An SH3-SH2-SH3 protein is required for p21Ras1 activation and binds to sevenless and Sos proteins in vitro. *Cell* **73:** 169–177.

Simon AF, Boquet I, Synguélakis M, Préat T. 1998. The *Drosophila* putative kinase linotte (derailed) prevents central brain axons from converging on a newly described interhemispheric ring. *Mech Dev* **76:** 45–55.

Singh AP, VijayRaghavan K, Rodrigues V. 2010. Dendritic refinement of an identified neuron in the *Drosophila* CNS is regulated by neuronal activity and Wnt signaling. *Development* **137:** 1351–1360.

Slaidina M, Delanoue R, Gronke S, Partridge L, Léopold P. 2009. A *Drosophila* insulin-like peptide promotes growth during nonfeeding states. *Dev Cell* **17:** 874–884.

Somogyi K, Rørth P. 2004. Cortactin modulates cell migration and ring canal morphogenesis during *Drosophila* oogenesis. *Mech Dev* **121:** 57–64.

Song J, Wu L, Chen Z, Kohanski RA, Pick L. 2003. Axons guided by insulin receptor in *Drosophila* visual system. *Science* **300:** 502–505.

Sossin WS. 2006. Tracing the evolution and function of the Trk superfamily of receptor tyrosine kinases. *Brain Behav Evol* **68:** 145–156.

Sousa-Nunes R, Yee LL, Gould AP. 2011. Fat cells reactivate quiescent neuroblasts via TOR and glial insulin relays in *Drosophila*. *Nature* **471:** 508–512.

Sprenger F, Nüsslein-Volhard C. 1992. Torso receptor activity is regulated by a diffusible ligand produced at the extracellular terminal regions of the *Drosophila* egg. *Cell* **71:** 987–1001.

Sprenger F, Stevens LM, Nüsslein-Volhard C. 1989. The *Drosophila* gene torso encodes a putative receptor tyrosine kinase. *Nature* **338:** 478–483.

Sprenger F, Trosclair MM, Morrison DK. 1993. Biochemical analysis of torso and D-raf during *Drosophila* embryogenesis: Implications for terminal signal transduction. *Mol Cell Biol* **13:** 1163–1172.

Srahna M, Leyssen M, Choi CM, Fradkin LG, Noordermeer JN, Hassan BA. 2006. A signaling network for patterning of neuronal connectivity in the *Drosophila* brain. *PLoS Biol* **4:** e348.

Stathopoulos A, Tam B, Ronshaugen M, Frasch M, Levine M. 2004. pyramus and thisbe: FGF genes that pattern the mesoderm of *Drosophila* embryos. *Genes Dev* **18:** 687–699.

Steingrímsson E, Pignoni F, Liaw GJ, Lengyel JA. 1991. Dual role of the *Drosophila* pattern gene tailless in embryonic termini. *Science* **254:** 418–421.

Stevens LM, Frohnhöfer HG, Klingler M, Nüsslein Volhard C. 1990. Localized requirement for torso-like expression in follicle cells for development of terminal anlagen of the *Drosophila* embryo. *Nature* **346:** 660–663.

Strecker TR, Halsell SR, Fisher WW, Lipshitz HD. 1989. Reciprocal effects of hyper- and hypoactivity mutations in the *Drosophila* pattern gene torso. *Science* **243:** 1062–1066.

Stute C, Schimmelpfeng K, Renkawitz-Pohl R, Palmer RH, Holz A. 2004. Myoblast determination in the somatic and visceral mesoderm depends on Notch signalling as well as on milliways (mili(Alk)) as receptor for Jeb signalling. *Development* **131:** 743–754.

Sugaya R, Ishimaru S, Hosoya T, Saigo K, Emori Y. 1994. A *Drosophila* homolog of human proto-oncogene ret transiently expressed in embryonic neuronal precursor cells including neuroblasts and CNS cells. *Mech Dev* **45:** 139–145.

Sutherland D, Samakovlis C, Krasnow MA. 1996. branchless encodes a *Drosophila* FGF homolog that controls tracheal cell migration and the pattern of branching. *Cell* **87:** 1091–1101.

Szabad J, Erdelyi M, Hoffmann G, Szidonya J, Wright TR. 1989. Isolation and characterization of dominant female sterile mutations of *Drosophila* melanogaster. II. Mutations on the second chromosome. *Genetics* **122:** 823–835.

Tang AH, Neufeld TP, Kwan E, Rubin GM. 1997. PHYL acts to down-regulate TTK88, a transcriptional repressor of neuronal cell fates, by a SINA-dependent mechanism. *Cell* **90:** 459–467.

Tatar M, Kopelman A, Epstein D, Tu MP, Yin CM, Garofalo RS. 2001. A mutant *Drosophila* insulin receptor homolog that extends life-span and impairs neuroendocrine function. *Science* **292:** 107–110.

Tepass U, Fessler LI, Aziz A, Hartenstein V. 1994. Embryonic origin of hemocytes and their relationship to cell death in *Drosophila*. *Development* **120:** 1829–1837.

Tepass U, Truong K, Godt D, Ikura M, Peifer M. 2000. Cadherins in embryonic and neural morphogenesis. *Nat Rev Mol Cell Biol* **1:** 91–100.

Therrien M, Chang HC, Solomon NM, Karim FD, Wassarman DA, Rubin GM. 1995. KSR, a novel protein kinase required for RAS signal transduction. *Cell* **83:** 879–888.

Tomlinson A, Ready DF. 1986. Sevenless: A cell-specific homeotic mutation of the *Drosophila* eye. *Science* **231:** 400–402.

Tomlinson A, Ready DF. 1987. Cell fate in the *Drosophila* ommatidium. *Dev Biol* **123:** 264–275.

Tomlinson A, Bowtell DD, Hafen E, Rubin GM. 1987. Localization of the sevenless protein, a putative receptor for positional information, in the eye imaginal disc of *Drosophila*. *Cell* **51:** 143–150.

Tootle TL, Lee PS, Rebay I. 2003. CRM1-mediated nuclear export and regulated activity of the receptor tyrosine kinase antagonist YAN require specific interactions with MAE. *Development* **130:** 845–857.

Tsuda L, Inoue YH, Yoo MA, Mizuno M, Hata M, Lim YM, Adachi-Yamada T, Ryo H, Masamune Y, Nishida Y. 1993. A protein kinase similar to MAP kinase activator acts downstream of the raf kinase in *Drosophila*. *Cell* **72:** 407–414.

Tsuda H, Han SM, Yang Y, Tong C, Lin YQ, Mohan K, Haueter C, Zoghbi A, Harati Y, Kwan J, et al. 2008. The amyotrophic lateral sclerosis 8 Protein VAPB is cleaved, secreted, and acts as a ligand for Eph receptors. *Cell* **133:** 963–977.

Tu M-P, Tatar M. 2003. Juvenile diet restriction and the aging and reproduction of adult *Drosophila* melanogaster. *Aging Cell* **2:** 327–333.

Ueishi S, Shimizu H, Inoue YH. 2009. Male germline stem cell division and spermatocyte growth require insulin signaling in *Drosophila*. *Cell Struct Funct* **34:** 61–69.

Varshney GK, Palmer RH. 2006. The bHLH transcription factor Hand is regulated by Alk in the *Drosophila* embryonic gut. *Biochem Biophys Res Commun* **351:** 839–846.

Veenstra JA, Agricola H-J, Sellami A. 2008. Regulatory peptides in fruit fly midgut. *Cell Tissue Res* **334:** 499–516.

Verdu J, Buratovich MA, Wilder EL, Birnbaum MJ. 1999. Cell-autonomous regulation of cell and organ growth in *Drosophila* by Akt/PKB. *Nat Cell Biol* **1:** 500–506.

Vernersson E, Khoo NKS, Henriksson ML, Roos G, Palmer RH, Hallberg B. 2006. Characterization of the expression of the ALK receptor tyrosine kinase in mice. *Gene Expr Patterns* **6:** 448–461.

Vincent S, Wilson R, Coelho C, Affolter M, Leptin M. 1998. The *Drosophila* protein Dof is specifically required for FGF signaling. *Mol Cell* **2:** 515–525.

Voas MG, Rebay I. 2003. Signal integration during development: Insights from the *Drosophila* eye. *Dev Dyn* **229:** 162–175.

Vogel W, Gish GD, Alves F, Pawson T. 1997. The discoidin domain receptor tyrosine kinases are activated by collagen. *Mol Cell* **1:** 13–23.

Vogel WF, Abdulhussein R, Ford CE. 2006. Sensing extracellular matrix: An update on discoidin domain receptor function. *Cell Signal* **18:** 1108–1116.

Walker JA, Tchoudakova AV, McKenney PT, Brill S, Wu D, Cowley GS, Hariharan IK, Bernards A. 2006. Reduced growth of *Drosophila* neurofibromatosis 1 mutants reflects a non-cell-autonomous requirement for GTPase-activating protein activity in larval neurons. *Genes Dev* **20:** 3311–3323.

Wang X, Bo J, Bridges T, Dugan KD, Pan T-c, Chodosh LA, Montell DJ. 2006. Analysis of cell migration using whole-genome expression profiling of migratory cells in the *Drosophila* ovary. *Dev Cell* **10:** 483–495.

Wang S, Tsarouhas V, Xylourgidis N, Sabri N, Tiklová K, Nautiyal N, Gallio M, Samakovlis C. 2009. The tyrosine kinase Stitcher activates Grainy head and epidermal wound healing in *Drosophila*. *Nature* **11:** 890–895.

Wang X, He L, Wu YI, Hahn KM, Montell DJ. 2010. Light-mediated activation reveals a key role for Rac in collective guidance of cell movement in vivo. *Nature* **12:** 591–597.

Weaver M, Krasnow MA. 2008. Dual origin of tissue-specific progenitor cells in *Drosophila* tracheal remodeling. *Science* **321:** 1496–1499.

Weigel D, Jürgens G, Klingler M, Jäckle H. 1990. Two gap genes mediate maternal terminal pattern information in *Drosophila*. *Science* **248:** 495–498.

Weinkove D, Neufeld TP, Twardzik T, Waterfield MD, Leevers SJ. 1999. Regulation of imaginal disc cell size, cell number and organ size by *Drosophila* class I(A) phosphoinositide 3-kinase and its adaptor. *Curr Biol* **9:** 1019–1029.

Weiss JB, Suyama KL, Lee HH, Scott MP. 2001. Jelly belly: A *Drosophila* LDL receptor repeat-containing signal required for mesoderm migration and differentiation. *Cell* **107:** 387–398.

Wilson R. 2005. FGF signalling and the mechanism of mesoderm spreading in *Drosophila* embryos. *Development* **132:** 491–501.

Wilson C, Goberdhan DC, Steller H. 1993. Dror, a potential neurotrophic receptor gene, encodes a *Drosophila* homolog of the vertebrate Ror family of Trk-related receptor tyrosine kinases. *Proc Natl Acad Sci* **90:** 7109–7113.

Wilson R, Battersby A, Csiszar A, Vogelsang E, Leptin M. 2004. A functional domain of Dof that is required for fibroblast growth factor signaling. *Mol Cell Biol* **24:** 2263–2276.

Wilson R, Vogelsang E, Leptin M. 2005. FGF signalling and the mechanism of mesoderm spreading in *Drosophila* embryos. *Development* **132:** 491–501.

Winberg ML, Tamagnone L, Bai J, Comoglio PM, Montell D, Goodman CS. 2001. The transmembrane protein Off-track associates with Plexins and functions downstream of Semaphorin signaling during axon guidance. *Neuron* **32:** 53–62.

Wolf C, Schuh R. 2000. Single mesodermal cells guide outgrowth of ectodermal tubular structures in *Drosophila*. *Genes Dev* **14:** 2140–2145.

Wolf C, Gerlach N, Schuh R. 2002. *Drosophila* tracheal system formation involves FGF-dependent cell extensions contacting bridge-cells. *EMBO Rep* **3:** 563–568.

Wolff T, Ready DF. 1993. Pattern formation in the *Drosophila* retina. In *The development of Drosophila melanogaster*, pp. 1277–1325. Cold Spring Harbor Laboratory Press, Cold Spring Harbor, New York.

Wouda RR, Bansraj MRKS, de Jong AWM, Noordermeer JN, Fradkin LG. 2008. Src family kinases are required for WNT5 signaling through the Derailed/RYK receptor in the *Drosophila* embryonic central nervous system. *Development* **135:** 2277–2287.

Wu M, Sato TN. 2008. On the mechanics of cardiac function of *Drosophila* embryo. *PLoS ONE* **3:** e4045.

Xu C, Kauffmann RC, Zhang J, Kladny S, Carthew RW. 2000. Overlapping activators and repressors delimit transcriptional response to receptor tyrosine kinase signals in the *Drosophila* eye. *Cell* **103:** 87–97.

Yao Y, Wu Y, Yin C, Ozawa R, Aigaki T, Wouda RR, Noordermeer JN, Fradkin LG, Hing H. 2007. Antagonistic roles of Wnt5 and the Drl receptor in patterning the *Drosophila* antennal lobe. *Nat Neurosci* **10:** 1423–1432.

Yoshikawa S, Bonkowsky JL, Kokel M, Shyn S, Thomas JB. 2001. The derailed guidance receptor does not require kinase activity in vivo. *J Neurosci* **21:** RC119.

Yoshikawa S, McKinnon RD, Kokel M, Thomas JB. 2003. Wnt-mediated axon guidance via the *Drosophila* derailed receptor. *Nature* **422:** 583–588.

Yu JY, Reynolds SH, Hatfield SD, Shcherbata HR, Fischer KA, Ward EJ, Long D, Ding Y, Ruohola-Baker H. 2009. Dicer-1-dependent Dacapo suppression acts downstream of insulin receptor in regulating cell division of *Drosophila* germline stem cells. *Development* **136:** 1497–1507.

Zeitouni B, Sénatore S, Séverac D, Aknin C, Sémériva M, Perrin L. 2007. Signalling pathways involved in adult heart formation revealed by gene expression profiling in *Drosophila*. *PLoS Genet* **3:** e174.

Zettervall C-J, Anderl I, Williams MJ, Palmer R, Kurucz E, Ando I, Hultmark D. 2004. A directed screen for genes involved in *Drosophila* blood cell activation. *Proc Natl Acad Sci* **101:** 14192–14197.

Zhang J, Lefebvre JL, Zhao S, Granato M. 2004. Zebrafish unplugged reveals a role for muscle-specific kinase homologs in axonal pathway choice. *Nat Neurosci* **7:** 1303–1309.

Zhang B, Luo S, Wang Q, Suzuki T, Xiong WC, Mei L. 2008. LRP4 serves as a coreceptor of agrin. *Neuron* **60:** 285–297.

Zhang L, Luo J, Wan P, Wu J, Laski F, Chen J. 2011. Regulation of cofilin phosphorylation and asymmetry in collective cell migration during morphogenesis. *Development* **138:** 455–464.

Zito K, Parnas D, Fetter RD, Isacoff EY, Goodman CS. 1999. Watching a synapse grow: Noninvasive confocal imaging of synaptic growth in *Drosophila*. *Neuron* **22:** 719–729.

Biology of the TAM Receptors

Greg Lemke

Molecular Neurobiology Laboratory, Immunobiology and Microbial Pathogenesis Laboratory, The Salk Institute, La Jolla, California 92037

Correspondence: lemke@salk.edu

The TAM receptors—Tyro3, Axl, and Mer—comprise a unique family of receptor tyrosine kinases, in that as a group they play no essential role in embryonic development. Instead, they function as homeostatic regulators in adult tissues and organ systems that are subject to continuous challenge and renewal throughout life. Their regulatory roles are prominent in the mature immune, reproductive, hematopoietic, vascular, and nervous systems. The TAMs and their ligands—Gas6 and Protein S—are essential for the efficient phagocytosis of apoptotic cells and membranes in these tissues; and in the immune system, they act as pleiotropic inhibitors of the innate inflammatory response to pathogens. Deficiencies in TAM signaling are thought to contribute to chronic inflammatory and autoimmune disease in humans, and aberrantly elevated TAM signaling is strongly associated with cancer progression, metastasis, and resistance to targeted therapies.

The name of the TAM family is derived from the first letter of its three constituents—Tyro3, Axl, and Mer (Prasad et al. 2006). As detailed in Figure 1, members of this receptor tyrosine kinase (RTK) family were independently identified by several different groups and appear in the early literature under multiple alternative names. However, Tyro3, Axl, and Mer (officially c-Mer or MerTK for the protein, *Mertk* for the gene) have now been adopted as the NCBI designations. The TAMs were first grouped into a distinct RTK family (the Tyro3/7/12 cluster) in 1991, through PCR cloning of their kinase domains (Lai and Lemke 1991). The isolation of full-length cDNAs for Axl (O'Bryan et al. 1991), Mer (Graham et al. 1994), and Tyro3 (Lai et al. 1994) confirmed their segregation into a structurally distinctive family of orphan RTKs (Manning et al. 2002b). The two ligands that bind and activate the TAMs—Gas6 and Protein S (Pros1)—were identified shortly thereafter (Ohashi et al. 1995; Stitt et al. 1995; Mark et al. 1996; Nagata et al. 1996).

Subsequent progress on elucidating the biological roles of the TAM receptors was considerably slower and ultimately required the derivation of mouse loss-of-function mutants (Camenisch et al. 1999; Lu et al. 1999). The fact that $Tyro3^{-/-}$, $Axl^{-/-}$, and $Mer^{-/-}$ mice are all viable and fertile permitted the generation of a complete TAM mutant series that included all possible double mutants and even triple mutants that lack all three receptors (Lu et al. 1999). Remarkably, these $Tyro3^{-/-} Axl^{-/-} Mer^{-/-}$ triple knockouts (TAM TKOs) are viable, and for the first 2–3 wk after birth, superficially indistinguishable from their wild-type counterparts (Lu et al. 1999). Because many

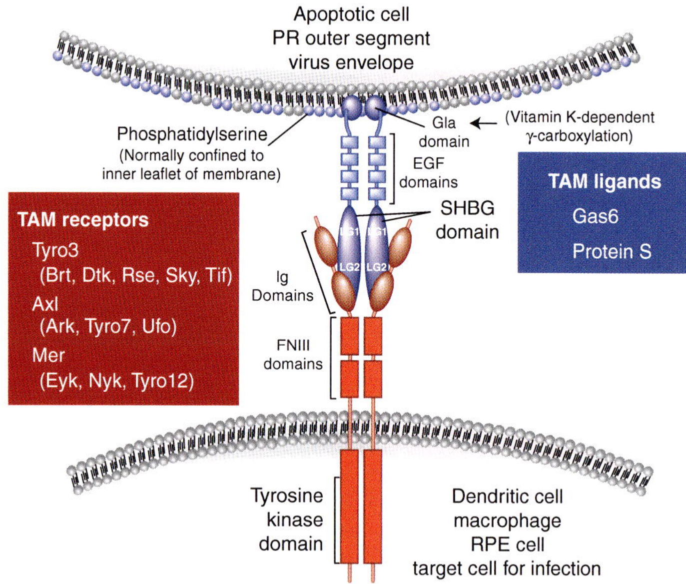

Figure 1. TAM receptors and ligands. The TAM receptors (red) are Tyro3 (Lai and Lemke 1991; Lai et al. 1994)—also designated Brt (Fujimoto and Yamamoto 1994), Dtk (Crosier et al. 1994), Rse (Mark et al. 1994), Sky (Ohashi et al. 1994), and Tif (Dai et al. 1994); Axl (O'Bryan et al. 1991)—also designated Ark (Rescigno et al. 1991), Tyro7 (Lai and Lemke 1991), and Ufo (Janssen et al. 1991); and Mer (Graham et al. 1994)—also designated Eyk (Jia and Hanafusa 1994), Nyk (Ling and Kung 1995), and Tyro12 (Lai and Lemke 1991). The TAMs are widely expressed by cells of the mature immune, nervous, vascular, and reproductive systems. The TAM ligands (blue) are Gas6 and Protein S (Pros1). The carboxy-terminal SHBG domains of the ligands bind to the immunoglobulin (Ig) domains of the receptors, induce dimerization, and activate the TAM tyrosine kinases. When γ-carboxylated in a vitamin-K-dependent reaction, the amino-terminal Gla domains of the dimeric ligands bind to the phospholipid phosphatidylserine expressed on the surface on an apposed apoptotic cell or enveloped virus. See text for details. (From Lemke and Burstyn-Cohen 2010; adapted, with permission, from the authors.)

RTKs play essential roles in embryonic development, even single loss-of-function mutations in RTK genes often result in an embryonic-lethal phenotype (Gassmann et al. 1995; Lee et al. 1995; Soriano 1997; Arman et al. 1998). The postnatal viability of mice in which an entire RTK family is ablated completely—the TAM TKOs can survive for more than a year (Lu et al. 1999)—is therefore highly unusual. Their viability notwithstanding, the TAM mutants go on to develop a plethora of phenotypes, some of them debilitating (Camenisch et al. 1999; Lu et al. 1999; Lu and Lemke 2001; Scott et al. 2001; Duncan et al. 2003; Prasad et al. 2006). Almost without exception, these phenotypes are degenerative in nature and reflect the loss of TAM signaling activities in adult tissues that are subject to regular challenge, renewal, and remodeling. These activities are the subject of this review.

TAM RECEPTOR/LIGAND STRUCTURE AND SIGNALING FEATURES

The extracellular domains of TAM receptors are composed of two structural modules that are used repeatedly in other RTK ectodomains, but that are configured in a defining two-plus-two combination in the TAMs (Fig. 1). The amino-terminal regions of these ectodomains carry tandem immunoglobulin-related domains that mediate ligand binding (Heiring et al. 2004; Sasaki et al. 2006), which are followed by tandem fibronectin type III repeats (O'Bryan et al. 1991;

Graham et al. 1994; Lai et al. 1994; Lemke and Rothlin 2008). All three TAM receptors have a single-pass *trans*-membrane domain, and all carry a catalytically competent protein-tyrosine kinase (Fig. 1). High-resolution crystal structures have been determined for the Tyro3 (Powell et al. 2012) and Mer kinase domains (Huang et al. 2009; Liu et al. 2012).

In many cells, the activation of this tyrosine kinase is coupled to the downstream activation of the phosphoinositide 3 kinase (PI3K)/AKT pathway. Most of this downstream PI3K signaling is nucleated through a TAM-autophosphorylated Grb2-binding site, which is located 18 residues carboxy terminal to the kinase domain and is conserved in all three TAMs (Fig. 2) (Fridell et al. 1996; Ling et al. 1996; Braunger et al. 1997; Goruppi et al. 1997; Georgescu et al. 1999; Lan et al. 2000; Ming Cao et al. 2001; Son et al. 2007; Tibrewal et al. 2008; Weinger et al. 2008). Coupling to phospholipase C, ERK1/2, Ras, and MAP kinase activation have also been described in many different cells (Keating et al. 2010; Lijnen et al. 2011; Ou et al. 2011). These TAM-activated signaling pathways (Fig. 2), which involve what might be called "the usual suspects" downstream from RTK activation, operate in all TAM-expressing cells. Macrophages, dendritic cells, and other sentinel cells of the immune system, however, also express cytokine receptor signaling systems—in particular, the type I interferon (IFN) receptor—that are directly coupled to, interact with, and are codependent on the TAM receptors. In these cells, the TAM-activated PI3K/AKT pathway is often dominated and obscured by a stronger TAM-activated JAK/STAT signaling pathway (Zong et al. 1996; Rothlin et al. 2007; Lemke and Rothlin 2008). Differential TAM activation of PI3K/AKT versus JAK/STAT signaling may be important for the differential activation of distinct TAM-regulated bioactivities (Fig. 2).

TAM receptors are among the last RTKs to have appeared during evolution (Manning et al. 2002a,b). Unlike the FGFR, EGFR, or ROR families, for example, there are no TAM representatives in either *Drosophila* or *Caenorhabditis elegans*. A single TAM-like receptor gene and a single Gas6/Pros1-like ligand gene are first seen in the genomes of prevertebrate urochordates such as *Ciona* (Kulman et al. 2006; Lemke and Rothlin 2008), coincident with the first appearance of type I and type II cytokines (e.g., interferons) and cytokine receptors.

The two TAM ligands—Gas6 and Pros1 (Manfioletti et al. 1993; Stitt et al. 1995; Mark et al. 1996)—are large ($\sim$80-kDa) proteins that are $\sim$42% identical and share the same multidomain arrangement (Fig. 1). They have two unusual structural features that are key to their bioactivities. The first is a carboxy-terminally positioned "sex hormone-binding globulin" (SHBG) domain composed of two laminin G domains (Fig. 1). This SHBG domain binds to the Ig domains of the receptors and induces their dimerization and subsequent kinase activation (Nyberg et al. 1997; Tanabe et al. 1997; Evenas et al. 2000; Sasaki et al. 2002, 2006). The second is a so-called Gla domain positioned at the very amino terminus of both Gas6 and Pros1 (Stitt et al. 1995; Ishimoto et al. 2000; Rajotte et al. 2008). (The SHBG and Gla domains are separated by four EGF-related domains.) This $\sim$60-amino-acid Gla domain is rich in glutamic acid residues whose γ-hydroxyl groups are posttranslationally carboxylated in a vitamin K-dependent modification (Huang et al. 2003; Li et al. 2004; Bandyopadhyay 2008). Gas6 and Pros1 share Gla domains with several proteins of the blood coagulation cascade, such as factors VII, IX, and X (Dahlback 2000; Stafford 2005). Indeed, in addition to acting as a TAM ligand, Pros1 also functions as an anticoagulant in this cascade (Dahlback 2000; Burstyn-Cohen et al. 2009).

γ-Carboxylation of Gla domains allows them to bind to phosphatidylserine (PtdSer). In most cells, the activity of a set of P_4-ATPases—so-called flippases—ensures that this phospholipid is confined to the inner, cytoplasm-facing leaflet of the plasma membrane (van Meer et al. 2008). In activated platelets and apoptotic cells (among other sites), these flippases are disabled such that PtdSer is displayed on the extracellular membrane surface as well. For apoptotic cells (ACs), extracellularly displayed PtdSer is among the most potent "eat-me" signals by which these dead cells are

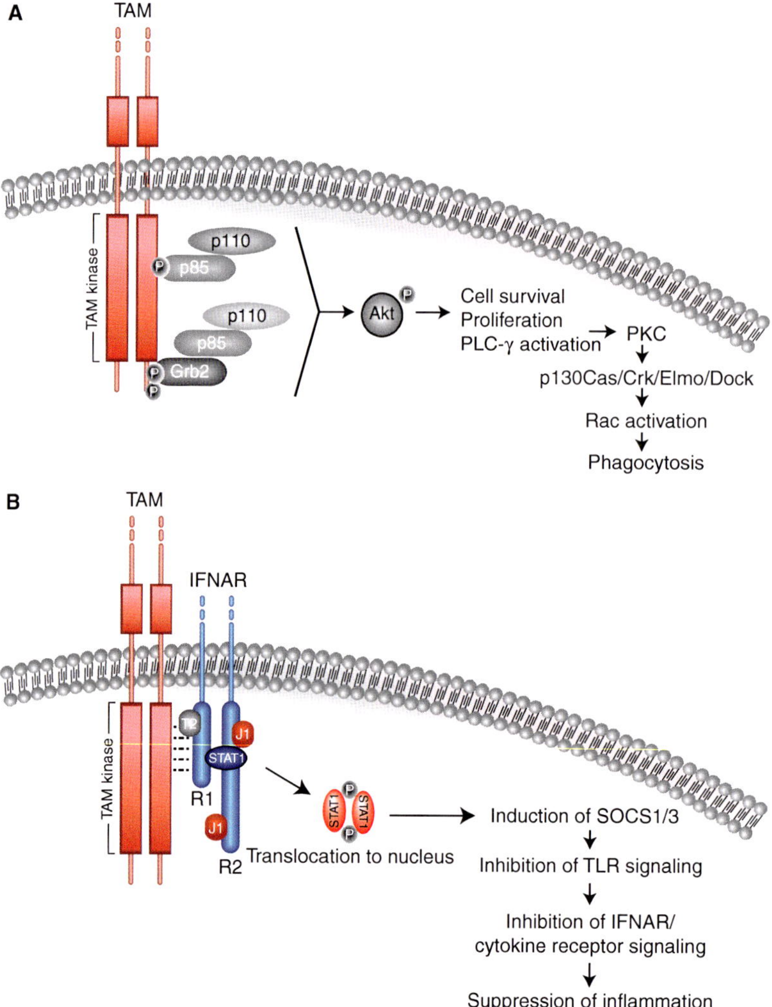

Figure 2. TAM receptor signaling pathways. (*A*) Free TAMs. As receptor dimers, activated TAM proteins drive a conventional RTK signaling pathway that is dominated by the phosphorylation and activation of Akt. The positions of major tyrosine autophosphorylation sites shared between Tyro3, Axl, and Mer are indicated (P). The tyrosine immediately downstream from the kinase domain (Y821 in human Axl) is bound by the SH2 domain of Grb2, which recruits the p85 subunit of PI3 kinase through an SH3 (Grb2)-proline-rich domain (p85) interaction. Alternatively, p85 can bind this phosphotyrosine directly using its own SH2 domain. P85 also binds to the indicated phosphotyrosine within the kinase domain (see, e.g., Weinger et al. 2008). Mobilization of the joint p85/p110 PI3K complex results in the downstream phosphorylation and activation of Akt. Mer activation has also been found to drive the downstream activation of PLC-γ, by a mechanism that is not delineated biochemically (Tibrewal et al. 2008). These pathways are required for TAM regulation of cell survival and the mobilization of the actin cytoskeleton required for the engulfment of apoptotic cells by phagocytes. (*B*) TAM receptors complexed with the type I interferon receptor (IFNAR). In dendritic cells, TAM receptors— when activated by the binding of a TAM ligand—form a coimmunoprecipitable complex specifically with the R1 (or α) chain of the IFNAR (Rothlin et al. 2007). This may be associated with the activation of Jak1 (J1) (Zong et al. 1996). Direct activation of the hybrid TAM-IFNAR receptor by the addition of Gas6 leads to the rapid tyrosine phosphorylation and activation of Stat1. This dimeric transcription factor then translocates to the nucleus, where it drives the expression of the cytoplasmic cytokine inhibitors SOCS1 and 3. This pathway is required for the inhibition of inflammatory responses in dendritic cells (Rothlin et al. 2007; Lemke and Rothlin 2008). See text for details.

recognized by phagocytes (Ravichandran 2010). Gla-domain-containing proteins can therefore bind the surface membranes of ACs. As discussed below, the interaction of the γ-carboxylated amino-terminal Gla domains of Gas6 and Pros1 with a PtdSer-containing membrane is a crucial feature of their activation of TAM receptors, and it is possible that in vivo these ligands *always* signal in the context of membrane association.

As depicted in Figure 1, Gas6 and Pros1 appear to bind to TAM receptors as dimers, and for Pros1, multimerization is required for TAM activation (Uehara and Shacter 2008). Apart from this, receptor–ligand pairing relationships and signaling interactions for the TAM system remain incompletely understood. For example, we do not know the extent to which Pros1 and Gas6 may heterodimerize, and if this occurs, how receptor binding and activation profiles of the heterodimer may differ from those of Gas6 or Pros1 homodimers. Similarly, the extent to which individual TAM receptors may heterodimerize in different cellular settings is also poorly understood. The preponderance of evidence indicates that Gas6 functions as a ligand for all three TAM receptors, with reduced binding affinity for Mer relative to Axl and Tyro3 (Ohashi et al. 1995; Stitt et al. 1995; Mark et al. 1996; Nagata et al. 1996; Chen et al. 1997; Lemke and Rothlin 2008). Pros1, in contrast, appears to bind and activate Tyro3 and Mer, with little or no affinity for Axl (Stitt et al. 1995; Prasad et al. 2006; Lemke and Rothlin 2008; Uehara and Shacter 2008; Zhong et al. 2010). In an active area of research, the extent to which Gas6 and/or Pros1 contribute to the observed activity of specific TAM receptors has just begun to be dissected genetically. As discussed below, the first example of such a differential genetic analysis has recently been reported for Gas6 and Pros1 action in Mer-expressing retinal pigment epithelial cells of the eye (Burstyn-Cohen et al. 2012).

In some settings, the biologically relevant cellular sources of Gas6 and/or Pros1 required for TAM activation also remain to be determined. In several cell types, TAM signaling appears to be autocrine/paracrine, in that a TAM-positive cell has frequently been found to express Pros1 and/or Gas6 (Lu et al. 1999; Prasad et al. 2006; Rothlin et al. 2007). Pros1 is expressed at ~300 nM in the blood, into which it is secreted by hepatocytes and vascular endothelial cells (Burstyn-Cohen et al. 2009). (In contrast, Gas6 is present at ≤0.2 nM in serum, and nearly all of this is complexed with soluble Axl ectodomain [Ekman et al. 2010].) Tyro3- and Mer-expressing cells that transit through the circulation are therefore exposed to saturating levels of Pros1. In the immune system, an important source of Pros1 for TAM-expressing macrophages and dendritic cells (see below) may be activated T cells (Smiley et al. 1997).

TAM MEDIATION OF THE PHAGOCYTOSIS OF APOPTOTIC CELLS

TAM receptor signaling plays an especially important role in the engulfment and phagocytic clearance of apoptotic cells (ACs) and membranes in adult tissues (Lemke and Rothlin 2008; Lemke and Burstyn-Cohen 2010). In this process, a TAM ligand, Gas6 or Pros1, serves as a "bridging molecule" that physically links a TAM receptor, generally Mer or Axl, expressed on the surface of the phagocyte, to PtdSer, which is displayed on the surface of the AC that will be engulfed (Fig. 1) (Wu et al. 2006; Nagata et al. 2010). At the same time, this ligand must also activate the tyrosine kinase activity of the TAM receptor for the process of phagocytosis to go forward (Scott et al. 2001; Mahajan and Earp 2003; Tibrewal et al. 2008; Todt et al. 2008; Lemke and Burstyn-Cohen 2010).

The first phenotype described in the TAM TKOs was male infertility, which is tied to the degenerative death of nearly all germ cells in the testes (Lu et al. 1999). This cell death results from a dramatic pileup of AC corpses in the seminiferous tubules and is degenerative rather than developmental in nature (Lu et al. 1999); this is due to the loss of TAM receptor function in Sertoli cells (Lu et al. 1999; Chen et al. 2009; Sun et al. 2010). These somatic support cells are phagocytes; among their most important roles is the PtdSer-dependent clearance of the enormous number of apoptotic germ cells that are generated during meiosis (Kawasaki et al. 2002).

It has been estimated that more than half of the meiotic population dies during each cycle of mammalian spermatogenesis, and thus the clearance of these AC corpses (on the order of 10^8/d in a human male) by Sertoli cells is critical. This process is TAM dependent; Sertoli cells express all three TAMs and both TAM ligands, and in the absence of TAM signaling, the phagocytosis of apoptotic germ cells in the testes is significantly attenuated (Lu et al. 1999).

A similarly dramatic phenotype is seen in the retina of both TAM TKOs and $Mer^{-/-}$ single mutants. These mutants are born with normal retinas, but by 2 mo after birth, most of their photoreceptors (PRs) have died (Lu et al. 1999; Duncan et al. 2003). This is a nonautonomous phenotype with respect to PRs, in that these cells do not express the TAMs. Rather, both Mer and Tyro3 are expressed by cells of the retinal pigment epithelium (RPE) (Prasad et al. 2006). Like Sertoli cells in the testes, RPE cells are phagocytes (Sparrow et al. 2010). Unlike Sertoli cells, however, they do not engulf ACs. Rather, the apical microvilli of these cells engulf and metabolize only part of a living cell—the distal ends of PR outer segments. These outer segments (OS) are the rhodopsin-containing organelles in which light is detected. PRs synthesize and insert new membrane at the proximal base of their OS every day, and the distal tips of these organelles are phagocytozed by RPE cells—also on a daily basis—to remove toxic oxidative products generated by phototransduction and to maintain a constant OS length (Prasad et al. 2006; Coleman et al. 2009; Strick et al. 2009; Nandrot and Dufour 2010). In $Mer^{-/-}$ mice, RPE cells differentiate normally but fail to perform this phagocytosis, which leads to the apoptotic death of nearly all PRs (Feng et al. 2002; Duncan et al. 2003). Unlike the situation with germ cells in the testes, PR apoptosis does not occur normally but is instead triggered by the failure of mutant RPE cells to phagocytose PR OS. Consistent with the phenotype of the $Mer^{-/-}$ mice, the PR degeneration seen in the *Royal College of Surgeons* rat, a decades-old model of retinitis pigmentosa (Bourne et al. 1938; Edwards and Szamier 1977), has been found to be due to mutation of the rat *Mertk* gene

(D'Cruz et al. 2000; Nandrot and Dufour 2010); and in humans, 12 distinct pathogenic sequence variants in the *Mertk* gene lead to inherited forms of retinitis pigmentosa and retinal dystrophy (Gal et al. 2000; Ostergaard et al. 2011). Gas6 and Pros1 have been found to function as independent and interchangeable Mer ligands in this system. Mouse mutants in which all Gas6 or all Pros1 are singly eliminated from the retina have a wild-type number of PRs, but mice in which both Gas6 and Pros1 are removed display PR degeneration that perfectly phenocopies the degeneration seen in $Mer^{-/-}$ mice (Burstyn-Cohen et al. 2012).

The TAMs play similarly critical roles in AC clearance by phagocytes of the immune system—most prominently macrophages (Scott et al. 2001). In humans, $>10^9$ ACs are generated every day, but at steady state, these dead cells are nearly impossible to detect. This is because they are almost immediately cleared by macrophages and other phagocytes. In many settings, these cells rely on the eat-me signal PtdSer to recognize dead cells as targets for engulfment (Ravichandran 2010). Phagocytic removal of ACs is also prominent during the resolution phase of inflammation, when large numbers of infiltrating granulocytes and lymphocytes undergo apoptosis and must be cleared to terminate an inflammatory response (Elliott and Ravichandran 2010; Nagata et al. 2010). Incomplete phagocytosis of ACs leads to the accumulation of secondary necrotic cells, which constitute a source of self-antigens. Not surprisingly then, defects in these TAM-dependent processes are associated with the development of human autoimmune diseases (Gaipl et al. 2007; Shao and Cohen 2011), and autoimmune phenotypes are prominent features of the TAM mouse mutants (Scott et al. 2001; Seitz et al. 2007; Ait-Oufella et al. 2008; Thorp et al. 2008; Shao et al. 2009; Lemke and Burstyn-Cohen 2010).

TAM REGULATION OF THE INNATE IMMUNE RESPONSE

Mechanistically linked to their role in the phagocytosis of ACs is the role that the TAMs play in the feedback inhibition of the innate immune response to pathogens. This important regulatory activity has been studied in both macrophages and dendritic cells (DCs), although the mechanism of inhibition is known in detail only in the latter (Rothlin et al. 2007). DCs and other sentinel cells use Toll-like receptors (TLRs) and other pattern recognition receptors to detect the presence of invariant molecular patterns, such as lipopolysaccharide and double-stranded RNA, which are associated with bacteria, viruses, and other pathogens (Akira 2006; Beutler et al. 2006). Activation of these receptors leads to the production of proinflammatory cytokines such as tumor necrosis factor (TNF) α, interleukin (IL)-6, and type I interferons (Fig. 3) (Iwasaki and Medzhitov 2004). Although these cytokines are required to combat infection, they are powerful agents that must be controlled after the innate immune response is mobilized, because unrestrained cytokine signaling results in chronic inflammation and can lead to a response against self (Marshak-Rothstein 2006).

In DCs, the *Axl* gene is expressed at a modest steady-state level before pathogen encounter but is strongly induced by TLR activation and subsequently by type I IFNs through a JAK-Stat1-dependent mechanism (Rothlin et al. 2007). The up-regulated Axl protein then binds to and co-opts the type I IFN receptor (IFNAR) by forming a complex with the R1 chain of this receptor (Figs. 2 and 3). In so doing, Axl switches the IFNAR signaling modality from proinflammatory to immunosuppressive, by driving the activation of the genes encoding the suppressor of cytokine signaling (SOCS) 1 and 3 (Rothlin et al. 2007; Yoshimura et al. 2007). An SH2 domain of these cytoplasmic inhibitors binds to phosphotyrosine residues in JAK kinases that are associated with the IFNAR and other cytokine receptors (and to phosphorylated tyrosine within the receptors themselves), and a carboxy-terminal SOCS box then mediates proteosomal degradation of associated proteins. The amino-terminal regions of SOCS1 and SOCS3 also contain a kinase-inhibitory region that acts as a JAK pseudosubstrate (Yoshimura et al. 2007; Croker et al. 2008). In this way, the induced SOCS proteins, whose expression in DCs is very largely dependent on activation of the TAM-IFNAR multimeric complex, terminate the inflammatory response to pathogens (Fig. 3) (Rothlin et al. 2007).

This pathway is an important inhibitor of inflammation in DCs and macrophages. The induction of SOCS 1 and 3 by type I IFNs is markedly blunted in Axl-deficient DCs. At the same time, the induction of these proteins by direct activation of the TAM receptors—through addition of Gas6—is equally dependent on the presence of both the IFNAR and associated Stat1 (Rothlin et al. 2007). The codependence of the TAM and IFNAR receptor systems for immunosuppression provides an explanation for the long-standing conundrum that type I IFNs can be *both* proinflammatory and immunosuppressive in sentinel cells of the immune system. If a type I IFN binds to the free IFNAR, it delivers a proinflammatory stimulus, but if it binds to the TAM-IFNAR receptor complex, it drives an immunosuppressive response (Fig. 3) (Sharif et al. 2006; Rothlin et al. 2007; Lemke and Rothlin 2008). The provision of an immune stimulus—for example, through activation of TLR4 with LPS—to a TAM-deficient cell or mouse inevitably leads to a hyperelevated inflammatory response (Camenisch et al. 1999; Lu and Lemke 2001; Rothlin et al. 2007). This means that deficiencies in TAM signaling are always associated with sustained immune activation and chronic inflammation.

TAM SIGNALING AND AUTOIMMUNE DISEASE

It is therefore not surprising that mouse mutants in TAM receptor genes eventually develop broad-spectrum autoimmune disease (Lu and Lemke 2001; Scott et al. 2001; Radic et al. 2006; Wallet et al. 2008; Rothlin and Lemke 2010; Shao et al. 2010). This disease, which is particularly severe in $Axl^{-/-}Mer^{-/-}$ double mutants and in TAM TKOs (Lu and Lemke 2001), has clinical features of both systemic lupus erythematosus (SLE) and rheumatoid arthritis (RA), and is characterized by swollen joints, IgG deposits in the kidneys and other tissues, and pro-

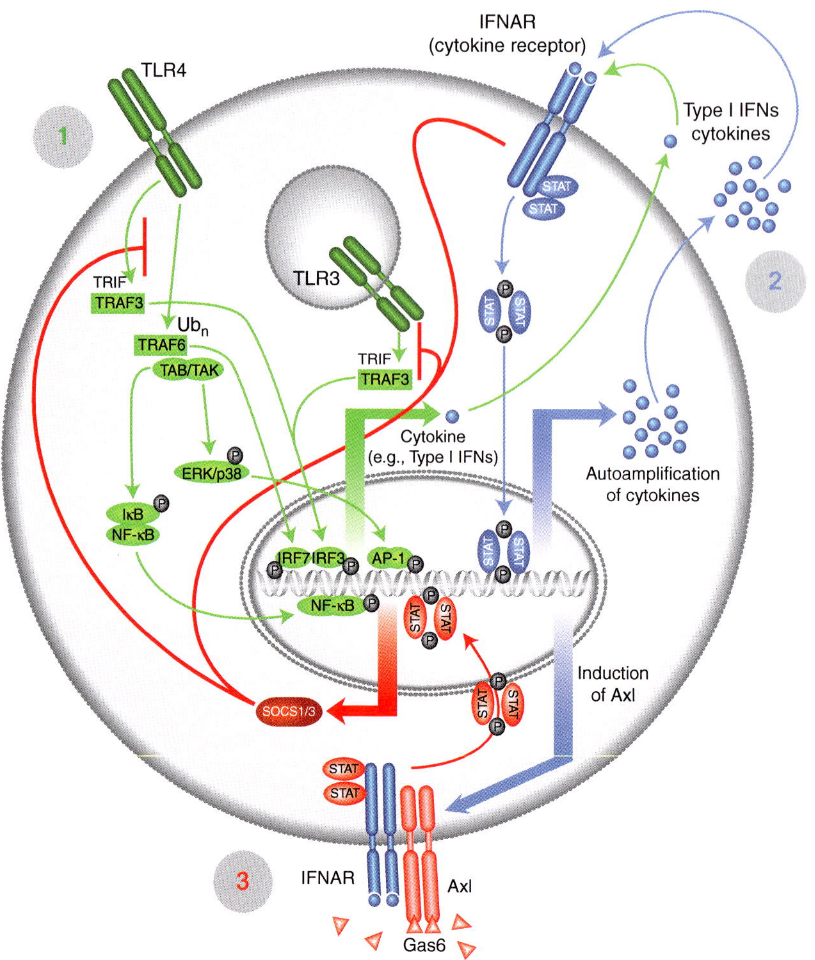

Figure 3. A TAM-regulated cycle of inflammation in dendritic cells. An initial recognition phase (1, green), mediated by Toll-like receptors (e.g., TLR4 on the cell surface and TLR3 in endosomes) and other pattern recognition receptors triggers a kinase cascade that leads to the activation of transcription factors (IRF3/7, AP-1, NF-κB) that drive the production of an initial bolus of type I interferons (IFNs) and other proinflammatory cytokines. In a second response phase (2, blue), the levels of these cytokines are elevated via a feed-forward, JAK-STAT-dependent amplification loop. This same JAK-SAT pathway drives the transcription of the Axl gene. In a final resolution phase (3, red), the induced Axl protein binds to the R1 chain of the type I IFN receptor (IFNAR). The hybrid TAM-IFNAR receptor activates a Stat1 dimer that drives the transcription of the genes encoding SOCS1 and SOCS3. These proteins inhibit both TLR and cytokine receptor signaling, and thereby return the dendritic cell to baseline (Rothlin et al. 2007; Lemke and Rothlin 2008). The features of this self-limiting cycle predict that the provision of an immune stimulus to a dendritic cell with diminished TAM signaling will always result in a hyperelevated inflammatory response. See text for details. (From Rothlin et al. 2007; adapted and reprinted, with permission, from the author.)

nounced splenomegaly and lymphadenopathy. TAM mutant mice also display relatively high titers of antibodies to autoantigens, including double-stranded DNA, phospholipids, and ribonucleoproteins (Lu and Lemke 2001; Scott et al. 2001; Radic et al. 2006). Crossing *Mer* or *Axl* mutants into existing mouse models of autoimmune disease has generally been found to exacerbate disease (Weinger et al. 2011; Ye et al. 2011).

Defects in the clearance of apoptotic cells and unabated type I IFN signaling—both of which are direct consequences of TAM deficiency—are also both thought to drive the development of human autoimmune diseases, including SLE, RA, and inflammatory bowel diseases (IBDs) (Gaipl et al. 2007; Ronnblom and Pascual 2008; Nagata et al. 2010). Correspondingly, several lines of evidence suggest that diminished TAM signaling may contribute to human autoimmunity (Rothlin and Lemke 2010). There is an anecdotal medical literature that ties low circulating levels of Pros1 to IBDs (Song et al. 2000; Zezos et al. 2007; Cakal et al. 2010; Diakou et al. 2011) and a much larger literature that establishes an association between low Pros1 and SLE (Song et al. 2000; Brouwer et al. 2004; Meesters et al. 2007). A recent analysis of a 107-patient SLE cohort found that levels of free protein S—but not Gas6—were significantly lower in SLE patients with a history of serositis, neurologic disorder, hematologic disorder, and immunologic disorder, and that low Pros1 levels were correlated with other disease-associated risk factors such as reductions in the complement proteins C3 and C4 (Suh et al. 2010). Polymorphisms in the *Mertk* gene have been tied to SLE (Cheong et al. 2007), and a clear genetic link has also been made with respect to the development of multiple sclerosis (MS). Here, a large genome-wide association study identified polymorphisms in the *Mertk* gene as risk factors for the development of MS (Ma et al. 2011; Sawcer et al. 2011).

The most widely prescribed drugs used to treat the chronic inflammation associated with many human autoimmune diseases—namely, glucocorticoids (GCs) such as prednisone and prednisolone—have recently been shown to potentiate TAM signaling. One well-described immunosuppressive activity of GCs is their ability to stimulate the phagocytosis of ACs by macrophages (Liu et al. 1999). Agonists for the liver-X-receptor (LXR) family of nuclear hormone receptors display this same activity (A-Gonzalez et al. 2009). Remarkably, the ability of both GCs and LXR agonists to stimulate macrophage phagocytosis of ACs has recently been shown to be due entirely to their ability to up-regulate expression of Mer (A-Gonzalez et al. 2009; McColl et al. 2009). These and related findings suggest that activation of TAM signaling may be therapeutic in the context of autoimmune disease. In this regard, in vivo delivery of adenoviruses expressing either Gas6 or Pros1 has been found to significantly diminish disease symptoms in a mouse model of collagen-induced arthritis (van den Brand et al. 2013).

TAM RECEPTORS AS TARGETS FOR VIRAL INFECTION

A very active area of current research relates to the role of TAM receptors in infection by viruses. In a process termed "apoptotic mimicry" (Mercer and Helenius 2010; Mercer 2011), the eat-me signal PtdSer has been found to be displayed on the extracellular membrane surface of several enveloped viruses, including vaccinia virus, cytomegalovirus, Lassa fever virus, and HIV (Callahan et al. 2003; Mercer and Helenius 2008; Soares et al. 2008). In facilitating infection, the TAMs do not function as direct virus receptors. Rather, Gas6 and Pros1 again serve as "bridging molecules"—this time between a membrane that surrounds a virus capsid and the cell that the virus will infect (Fig. 1).

Tyro3, Axl, and Mer have been found to function as entry factors for the Ebola/Marburg family of hemorrhagic fever filoviruses (Shimojima et al. 2006, 2007). Closely related cell lines that show marked differences in infectivity with pseudotyped viruses containing Ebola or Marburg envelope glycoproteins, or in infectivity with the Zaire or Reston strains of bona fide Ebola, were distinguished principally by TAM expression. Lines with high levels of Axl and/or Tyro3 were more readily infected than TAM-negative cells. Introduction of either Tyro3 or Axl into filovirus-resistant cells rendered these cells susceptible to infection, and anti-Axl antibodies antagonized infection of Axl-expressing cells with MLV viruses pseudotyped with Ebola glycoprotein (Shimojima et al. 2006).

Similar results have been obtained using infection of endothelial cells with lentiviral vectors pseudotyped with Sindbis virus glycoproteins (Morizono et al. 2011). These experiments also identified Gas6 and Pros1 as "bridging factors" that link PtdSer on the external surface of the viral envelope to Axl on the target cell. Depending on culture conditions, introduction of TAM cDNAs potentiated virus titers in infection-resistant cell lines ≥50-fold. This has also recently been observed for infection by the Dengue (DENV) and West Nile viruses (WNV) of the flavivirus family (Meertens et al. 2012). An unbiased cDNA transfection screen in human 293T cells, which are resistant to infection by these viruses, identified Tyro3 and Axl as factors that greatly enhance infection. Consistent with the interaction model depicted in Figure 1, this enhancement was found to be entirely dependent on the presence of a TAM ligand and also on the expression of PtdSer on the extracellular leaflet of the virus envelope (Meertens et al. 2012). In general, the above findings have been interpreted to suggest that TAM receptors serve as docking sites for TAM-ligand-bound virus particles. However, mutational analyses also indicate that tyrosine kinase activity is required for Axl potentiation of infection by Ebola (Shimojima et al. 2007) and DENV (Meertens et al. 2012), and thus active TAM signaling appears to be required for the potentiation of virus infection just as it is required for the phagocytosis of ACs. Given that (1) Axl activation potently suppresses type I IFN signaling in DCs and macrophages (Sharif et al. 2006; Rothlin et al. 2007; Shao et al. 2010), (2) type I IFNs are strong antiviral agents (Diamond 2003), and (3) suppression of type I IFN signaling is a mechanism that viruses exploit repeatedly as a means of immune evasion (Diamond 2003; Bonjardim et al. 2009; Versteeg and Garcia-Sastre 2010), the activation of TAM receptor signaling by viruses may prove to be an exceptionally effective mechanism of viral infection.

TAM RECEPTORS AND CANCER

There is a long association of TAM receptors with cancer—the first cDNAs for Axl and Mer were cloned from myeloid leukemia and lymphoblastoid lines, respectively (O'Bryan et al. 1991; Graham et al. 1994), and a truncated form of Mer (designated v-eyk) was identified initially as an avian retroviral oncogene (Jia et al. 1992; Jia and Hanafusa 1994). Axl was named from the Greek *anexelekto*, meaning "uncontrolled." Over the ensuing two decades, hundreds of papers have appeared that link TAM receptor and ligand expression to various forms of cancer (Linger et al. 2008; Verma et al. 2011). In general, these studies have reported overexpression or up-regulation of Axl, Mer, Tyro3, and/or Gas6. In many settings, however, a definitive demonstration that overexpression is causal for particular features of cancer development or progression has not been made. Elevated expression of TAM signaling components has been reported for leukemias (Graham et al. 1994, 2006; Hong et al. 2008), gliomas (Hutterer et al. 2008; Keating et al. 2010), colorectal carcinomas (Craven et al. 1995), breast cancers (Berclaz et al. 2001; Gjerdrum et al. 2010), gastrointestinal stromal tumors (Mahadevan et al. 2007), hepatocellular carcinoma (He et al. 2010), melanoma (Quong et al. 1994; Koorstra et al. 2009; Zhu et al. 2009), pancreatic adenocarcinoma (Song et al. 2010), and prostate cancer (Wu et al. 2004; Sainaghi et al. 2005), among several others.

Expression of Axl is correlated with an adverse prognosis in acute myeloid leukemia (Rochlitz et al. 1999), glioblastoma multiforme (Hutterer et al. 2008), pancreatic cancer (Koorstra et al. 2009), and esophageal adenocarcinoma (Hector et al. 2010). Axl up-regulation and activation have also been found to be a clinically significant feature of resistance to EGF receptor inhibitor and PI3K inhibitor therapies for non-small-cell lung cancer (Zhang et al. 2012; Byers et al. 2013). In many settings, Axl and/or Gas6 expression is most prominently associated with tumor metastasis, rather than growth of the primary tumor (Gjerdrum et al. 2010; Song et al. 2010). Consistent with this association, a small-molecule inhibitor of the Axl tyrosine kinase has shown efficacy primarily with respect to a reduced metastatic burden, rather than primary tumor growth, in mouse models of breast cancer metastasis (Holland et al. 2010). The link between TAM receptor expression and tumor metastasis is interesting in light of the importance of Axl and Tyro3 in the migration of gonadotropin-releasing hormone (GnRH) neurons from the olfactory placode to hypothalamus of the brain (Allen et al. 2002a; Pierce et al. 2008). This migration of GnRH neurons involves Gas6 activation of the same downstream signaling pathways—PI3 kinase, ERK1/2, and Rac via Ras (Allen et al. 2002a,b; Nielsen-Preiss et al. 2007)—that are engaged downstream from TAMs in tumor cells.

TAM REGULATION OF THE VASCULATURE SYSTEM

TAM signaling plays an important role in the homeostatic regulation of blood vessel integrity and permeability. The TAM ligands Gas6 and Pros1 were first identified in and purified from aortic endothelial cells (Stitt et al. 1995), and studies using conditional Pros1 knockouts have shown that vascular endothelial cells are a major source of the Pros1 that appears in the circulation (Burstyn-Cohen et al. 2009). Axl and Tyro3 are also expressed by the vascular smooth muscle cells that surround these endothelia, and Gas6 and Pros1 have potent trophic effects on these cells, both in vitro and in vivo (Gasic et al. 1992; Fridell et al. 1998; Melaragno et al. 1999; Collett et al. 2007; Son et al. 2007; Cavet et al. 2008). The PI3 kinase/Akt pathway is again implicated as a key effector of TAM signaling in smooth muscle.

Damage to blood vessels results in the upregulation of both Axl and Gas6 (Melaragno et al. 1998), and a complex pattern of differential regulation of Axl, Mer, Gas6, and Pros1 has been reported in human atherosclerotic plaques (Hurtado et al. 2011). Defects in the clearance of apoptotic cells from these plaques are linked to progression of advanced atherosclerotic lesions, and the role of compromised TAM signaling in cardiovascular disease is a subject of active study (Ait-Oufella et al. 2008; Thorp 2010). $Pros1^{+/-}$ mice with a 50% reduction in Pros1 display vessel breaches, with leakage of blood into the parenchyma of tissues (Burstyn-Cohen et al. 2009). Pros1 has also been linked to vascular integrity in the brain. Pros1, signaling through Tyro3, has been implicated in maintenance of the blood–brain barrier and has been found to ameliorate hypoxic/ischemic blood–brain barrier disruption (Zhu et al. 2010).

In addition to these direct activities, TAM signaling has been shown to affect vascular integrity indirectly, through the regulation of platelet function. Loss of one or more TAM receptors has been observed to impair stabilization of platelet aggregates, at least in part by reducing platelet granule secretion. Gas6 activates PI3K/Akt signaling in platelets and stimulates tyrosine phosphorylation of β3 integrin, thereby amplifying thrombus formation (Angelillo-Scherrer et al. 2001, 2005).

PROSPECTS

In addition to the biological settings outlined above, there is a significant body of literature to suggest that TAM signaling may play regulatory roles in the nervous system. Microglia, the tissue macrophages of the brain, also express Axl and Mer (Gautier et al. 2012), and there is evidence that TAM signaling through these receptors controls the phagocytosis of ACs and the inhibition of inflammation in the CNS just as it does in macrophages and DCs in the periphery (Grommes et al. 2008; Weinger et al. 2011). Tyro3 is also prominently expressed by many CNS neurons (Lai and Lemke 1991; Lai et al. 1994; Prieto et al. 2000, 2007). Its role in these neurons has for the most part remained obscure, however, and therefore this area is wide open for future study.

As noted at the outset, TAM receptor and ligand functions are, in the main, devoted to the homeostatic regulation of phenomena that are regular, cyclic, and circadian. These signaling proteins operate in adult, fully differentiated tissues that are subject to constant challenge and regular renewal. These features, together with the fact that the TAMs are RTKs expressed on the cell surface, make the TAM system a particularly favorable target for therapeutic intervention.

REFERENCES

A-Gonzalez N, Bensinger SJ, Hong C, Beceiro S, Bradley MN, Zelcer N, Deniz J, Ramirez C, Diaz M, Gallardo G, et al. 2009. Apoptotic cells promote their own clearance and immune tolerance through activation of the nuclear receptor LXR. *Immunity* **31:** 245–258.

Ait-Oufella H, Pouresmail V, Simon T, Blanc-Brude O, Kinugawa K, Merval R, Offenstadt G, Leseche G, Cohen PL, Tedgui A, et al. 2008. Defective mer receptor tyrosine kinase signaling in bone marrow cells promotes apoptotic cell accumulation and accelerates atherosclerosis. *Arterioscler Thromb Vasc Biol* **28:** 1429–1431.

Akira S. 2006. TLR signaling. *Curr Top Microbiol Immunol* **311:** 1–16.

Allen MP, Linseman DA, Udo H, Xu M, Schaack JB, Varnum B, Kandel ER, Heidenreich KA, Wierman ME. 2002a. Novel mechanism for gonadotropin-releasing hormone neuronal migration involving Gas6/Ark signaling to p38 mitogen-activated protein kinase. *Mol Cell Biol* **22:** 599–613.

Allen MP, Xu M, Linseman DA, Pawlowski JE, Bokoch GM, Heidenreich KA, Wierman ME. 2002b. Adhesion-related kinase repression of gonadotropin-releasing hormone gene expression requires Rac activation of the extracellular signal-regulated kinase pathway. *J Biol Chem* **277:** 38133–38140.

Angelillo-Scherrer A, de Frutos P, Aparicio C, Melis E, Savi P, Lupu F, Arnout J, Dewerchin M, Hoylaerts M, Herbert J, et al. 2001. Deficiency or inhibition of Gas6 causes platelet dysfunction and protects mice against thrombosis. *Nat Med* **7:** 215–221.

Angelillo-Scherrer A, Burnier L, Flores N, Savi P, DeMol M, Schaeffer P, Herbert J-M, Lemke G, Goff SP, Matsushima GK, et al. 2005. Role of Gas6 receptors in platelet signaling during thrombus stabilization and implications for antithrombotic therapy. *J Clin Invest* **115:** 237–246.

Arman E, Haffner-Krausz R, Chen Y, Heath JK, Lonai P. 1998. Targeted disruption of fibroblast growth factor (FGF) receptor 2 suggests a role for FGF signaling in pregastrulation mammalian development. *Proc Natl Acad Sci* **95:** 5082–5087.

Bandyopadhyay PK. 2008. Vitamin K-dependent γ-glutamylcarboxylation: An ancient posttranslational modification. *Vitam Horm* **78:** 157–184.

Berclaz G, Altermatt HJ, Rohrbach V, Kieffer I, Dreher E, Andres AC. 2001. Estrogen dependent expression of the receptor tyrosine kinase axl in normal and malignant human breast. *Ann Oncol* **12:** 819–824.

Beutler B, Jiang Z, Georgel P, Crozat K, Croker B, Rutschmann S, Du X, Hoebe K. 2006. Genetic analysis of host resistance: Toll-like receptor signaling and immunity at large. *Annu Rev Immunol* **24:** 353–389.

Bonjardim CA, Ferreira PC, Kroon EG. 2009. Interferons: Signaling, antiviral and viral evasion. *Immunol Lett* **122:** 1–11.

Bourne MC, Campbell DA, Tansley K. 1938. Hereditary degeneration of the rat retina. *Br J Ophthalmol* **22:** 613–623.

Braunger J, Schleithoff L, Schulz AS, Kessler H, Lammers R, Ullrich A, Bartram CR, Janssen JW. 1997. Intracellular signaling of the Ufo/Axl receptor tyrosine kinase is mediated mainly by a multi-substrate docking-site. *Oncogene* **14:** 2619–2631.

Brouwer JL, Bijl M, Veeger NJ, Kluin-Nelemans HC, van der Meer J. 2004. The contribution of inherited and acquired thrombophilic defects, alone or combined with antiphospholipid antibodies, to venous and arterial thromboembolism in patients with systemic lupus erythematosus. *Blood* **104:** 143–148.

Burstyn-Cohen T, Heeb MJ, Lemke G. 2009. Lack of protein S in mice causes embryonic lethal coagulopathy and vascular dysgenesis. *J Clin Invest* **119:** 2942–2953.

Burstyn-Cohen T, Lew ED, Través PG, Burrola PG, Hash JC, Lemke G. 2012. Genetic dissection of TAM receptor-ligand interaction in retinal pigment epithelial cell phagocytosis. *Neuron* **76:** 1123–1132.

Byers LA, Diao L, Wang J, Saintigny P, Girard L, Peyton M, Shen L, Fan Y, Giri U, Tumula PK, et al. 2013. An epithelial–mesenchymal transition gene signature predicts resistance to EGFR and PI3K inhibitors and identifies Axl as a therapeutic target for overcoming EGFR inhibitor resistance. *Clin Cancer Res* **19:** 279–290.

Cakal B, Gokmen A, Yalinkilic M, Cakal E, Ayaz S, Nadir I, Ozin Y, Dagli U, Ulker A. 2010. Natural anticoagulant protein levels in Turkish patients with inflammatory bowel disease. *Blood Coagul Fibrinolysis* **21:** 118–121.

Callahan MK, Popernack PM, Tsutsui S, Truong L, Schlegel RA, Henderson AJ. 2003. Phosphatidylserine on HIV envelope is a cofactor for infection of monocytic cells. *J Immunol* **170:** 4840–4845.

Camenisch TD, Koller BH, Earp HS, Matsushima GK. 1999. A novel receptor tyrosine kinase, Mer, inhibits TNF-α production and lipopolysaccharide-induced endotoxic shock. *J Immunol* **162:** 3498–3503.

Cavet ME, Smolock EM, Ozturk OH, World C, Pang J, Konishi A, Berk BC. 2008. Gas6–Axl receptor signaling is regulated by glucose in vascular smooth muscle cells. *Arterioscler Thromb Vasc Biol* **28:** 886–891.

Chen J, Carey K, Godowski PJ. 1997. Identification of Gas6 as a ligand for Mer, a neural cell adhesion molecule related receptor tyrosine kinase implicated in cellular transformation. *Oncogene* **14:** 2033–2039.

Chen Y, Wang H, Qi N, Wu H, Xiong W, Ma J, Lu Q, Han D. 2009. Functions of TAM RTKs in regulating spermatogenesis and male fertility in mice. *Reproduction* **138:** 655–666.

Cheong HS, Lee SO, Choi CB, Sung YK, Shin HD, Bae SC. 2007. MERTK polymorphisms associated with risk of haematological disorders among Korean SLE patients. *Rheumatology (Oxford)* **46:** 209–214.

Coleman JA, Kwok MC, Molday RS. 2009. Localization, purification, and functional reconstitution of the P4-ATPase Atp8a2, a phosphatidylserine flippase in photoreceptor disc membranes. *J Biol Chem* **284:** 32670–32679.

Collett GD, Sage AP, Kirton JP, Alexander MY, Gilmore AP, Canfield AE. 2007. Axl/phosphatidylinositol 3-kinase signaling inhibits mineral deposition by vascular smooth muscle cells. *Circ Res* **100:** 502–509.

Craven RJ, Xu LH, Weiner TM, Fridell YW, Dent GA, Srivastava S, Varnum B, Liu ET, Cance WG. 1995. Receptor tyrosine kinases expressed in metastatic colon cancer. *Int J Cancer* **60:** 791–797.

Croker BA, Kiu H, Nicholson SE. 2008. SOCS regulation of the JAK/STAT signalling pathway. *Semin Cell Dev Biol* **19:** 414–422.

Cite this article as *Cold Spring Harb Perspect Biol* doi: 10.1101/cshperspect.a009076

Crosier PS, Lewis PM, Hall LR, Vitas MR, Morris CM, Beier DR, Wood CR, Crosier KE. 1994. Isolation of a receptor tyrosine kinase (DTK) from embryonic stem cells: Structure, genetic mapping and analysis of expression. *Growth Factors* **11:** 125–136.

Dahlback B. 2000. Blood coagulation. *Lancet* **355:** 1627–1632.

Dai W, Pan H, Hassanain H, Gupta SL, Murphy MJ Jr. 1994. Molecular cloning of a novel receptor tyrosine kinase, tif, highly expressed in human ovary and testis. *Oncogene* **9:** 975–979.

D'Cruz PM, Yasumura D, Weir J, Matthes MT, Abderrahim H, LaVail MM, Vollrath D. 2000. Mutation of the receptor tyrosine kinase gene Mertk in the retinal dystrophic RCS rat. *Hum Mol Genet* **9:** 645–651.

Diakou M, Kostadima V, Giannopoulos S, Zikou AK, Argyropoulou MI, Kyritsis AP. 2011. Cerebral venous thrombosis in an adolescent with ulcerative colitis. *Brain Dev* **33:** 49–51.

Diamond MS. 2003. Evasion of innate and adaptive immunity by flaviviruses. *Immunol Cell Biol* **81:** 196–206.

Duncan HJ, LaVail MM, Yasumura D, Matthes MT, Yang H, Trautmann N, Chappelow AV, Feng W, Earp HS, Matsushima GK, et al. 2003. An RCS-like retinal dystrophy phenotype in mer knockout mice. *Invest Ophthalmol Visual Sci* **44:** 826–838.

Edwards RB, Szamier RB. 1977. Defective phagocytosis of isolated rod outer segments by RCS rat retinal pigment epithelium in culture. *Science* **197:** 1001–1003.

Ekman C, Stenhoff J, Dahlback B. 2010. Gas6 is complexed to the soluble tyrosine kinase receptor Axl in human blood. *J Thromb Haemost* **8:** 838–844.

Elliott MR, Ravichandran KS. 2010. Clearance of apoptotic cells: Implications in health and disease. *J Cell Biol* **189:** 1059–1070.

Evenas P, Dahlback B, Garcia de Frutos P. 2000. The first laminin G-type domain in the SHBG-like region of protein S contains residues essential for activation of the receptor tyrosine kinase sky. *Biol Chem* **381:** 199–209.

Feng W, Yasumura D, Matthes MT, LaVail MM, Vollrath D. 2002. Mertk triggers uptake of photoreceptor outer segments during phagocytosis by cultured retinal pigment epithelial cells. *J Biol Chem* **277:** 17016–17022.

Fridell YW, Jin Y, Quilliam LA, Burchert A, McCloskey P, Spizz G, Varnum B, Der C, Liu ET. 1996. Differential activation of the Ras/extracellular-signal-regulated protein kinase pathway is responsible for the biological consequences induced by the Axl receptor tyrosine kinase. *Mol Cell Biol* **16:** 135–145.

Fridell YW, Villa J Jr, Attar EC, Liu ET. 1998. GAS6 induces Axl-mediated chemotaxis of vascular smooth muscle cells. *J Biol Chem* **273:** 7123–7126.

Fujimoto J, Yamamoto T. 1994. *brt*, a mouse gene encoding a novel receptor-type protein-tyrosine kinase, is preferentially expressed in the brain. *Oncogene* **9:** 693–698.

Gaipl US, Munoz LE, Grossmayer G, Lauber K, Franz S, Sarter K, Voll RE, Winkler T, Kuhn A, Kalden J, et al. 2007. Clearance deficiency and systemic lupus erythematosus (SLE). *J Autoimmun* **28:** 114–121.

Gal A, Li Y, Thompson DA, Weir J, Orth U, Jacobson SG, Apfelstedt-Sylla E, Vollrath D. 2000. Mutations in MERTK, the human orthologue of the RCS rat retinal dystrophy gene, cause retinitis pigmentosa. *Nat Genet* **26:** 270–271.

Gasic GP, Arenas CP, Gasic TB, Gasic GJ. 1992. Coagulation factors X, Xa, and protein S as potent mitogens of cultured aortic smooth muscle cells. *Proc Natl Acad Sci* **89:** 2317–2320.

Gassmann M, Casagranda F, Orioli D, Simon H, Lai C, Klein R, Lemke G. 1995. Aberrant neural and cardiac development in mice lacking the ErbB4 neuregulin receptor. *Nature* **378:** 390–394.

Gautier EL, Shay T, Miller J, Greter M, Jakubzick C, Ivanov S, Helft J, Chow A, Elpek KG, Gordonov S, et al. 2012. Gene-expression profiles and transcriptional regulatory pathways that underlie the identity and diversity of mouse tissue macrophages. *Nat Immunol* **13:** 1118–1128.

Georgescu MM, Kirsch KH, Shishido T, Zong C, Hanafusa H. 1999. Biological effects of c-Mer receptor tyrosine kinase in hematopoietic cells depend on the Grb2 binding site in the receptor and activation of NF-κB. *Mol Cell Biol* **19:** 1171–1181.

Gjerdrum C, Tiron C, Hoiby T, Stefansson I, Haugen H, Sandal T, Collett K, Li S, McCormack E, Gjertsen BT, et al. 2010. Axl is an essential epithelial-to-mesenchymal transition-induced regulator of breast cancer metastasis and patient survival. *Proc Natl Acad Sci* **107:** 1124–1129.

Goruppi S, Ruaro E, Varnum B, Schneider C. 1997. Requirement of phosphatidylinositol 3-kinase-dependent pathway and Src for Gas6-Axl mitogenic and survival activities in NIH 3T3 fibroblasts. *Mol Cell Biol* **17:** 4442–4453.

Graham DK, Dawson TL, Mullaney DL, Snodgrass HR, Earp HS. 1994. Cloning and mRNA expression analysis of a novel human protooncogene, *c-mer*. *Cell Growth Differ* **5:** 647–657.

Graham DK, Salzberg DB, Kurtzberg J, Sather S, Matsushima GK, Keating AK, Liang X, Lovell MA, Williams SA, Dawson TL, et al. 2006. Ectopic expression of the protooncogene *Mer* in pediatric T-cell acute lymphoblastic leukemia. *Clin Cancer Res* **12:** 2662–2669.

Grommes C, Lee CY, Wilkinson BL, Jiang Q, Koenigsknecht-Talboo JL, Varnum B, Landreth GE. 2008. Regulation of microglial phagocytosis and inflammatory gene expression by Gas6 acting on the Axl/Mer family of tyrosine kinases. *J Neuroimmune Pharmacol* **3:** 130–140.

He L, Zhang J, Jiang L, Jin C, Zhao Y, Yang G, Jia L. 2010. Differential expression of Axl in hepatocellular carcinoma and correlation with tumor lymphatic metastasis. *Mol Carcinog* **49:** 882–891.

Hector A, Montgomery EA, Karikari C, Canto M, Dunbar KB, Wang JS, Feldmann G, Hong SM, Haffner MC, Meeker AK, et al. 2010. The Axl receptor tyrosine kinase is an adverse prognostic factor and a therapeutic target in esophageal adenocarcinoma. *Cancer Biol Ther* **10:** 1009–1018.

Heiring C, Dahlback B, Muller YA. 2004. Ligand recognition and homophilic interactions in Tyro3: Structural insights into the Axl/Tyro3 receptor tyrosine kinase family. *J Biol Chem* **279:** 6952–6958.

Holland SJ, Pan A, Franci C, Hu Y, Chang B, Li W, Duan M, Torneros A, Yu J, Heckrodt TJ, et al. 2010. R428, a selective small molecule inhibitor of Axl kinase, blocks tumor spread and prolongs survival in models of metastatic breast cancer. *Cancer Res* **70:** 1544–1554.

Hong CC, Lay JD, Huang JS, Cheng AL, Tang JL, Lin MT, Lai GM, Chuang SE. 2008. Receptor tyrosine kinase AXL is induced by chemotherapy drugs and overexpression of AXL confers drug resistance in acute myeloid leukemia. *Cancer Lett* **268:** 314–324.

Huang M, Rigby AC, Morelli X, Grant MA, Huang G, Furie B, Seaton B, Furie BC. 2003. Structural basis of membrane binding by Gla domains of vitamin K-dependent proteins. *Nat Struct Biol* **10:** 751–756.

Huang X, Finerty P Jr, Walker JR, Butler-Cole C, Vedadi M, Schapira M, Parker SA, Turk BE, Thompson DA, Dhe-Paganon S. 2009. Structural insights into the inhibited states of the Mer receptor tyrosine kinase. *J Struct Biol* **165:** 88–96.

Hurtado B, Munoz X, Recarte-Pelz P, Garcia N, Luque A, Krupinski J, Sala N, Garcia de Frutos P. 2011. Expression of the vitamin K-dependent proteins GAS6 and protein S and the TAM receptor tyrosine kinases in human atherosclerotic carotid plaques. *Thromb Haemost* **105:** 873–882.

Hutterer M, Knyazev P, Abate A, Reschke M, Maier H, Stefanova N, Knyazeva T, Barbieri V, Reindl M, Muigg A, et al. 2008. Axl and growth arrest-specific gene 6 are frequently overexpressed in human gliomas and predict poor prognosis in patients with glioblastoma multiforme. *Clin Cancer Res* **14:** 130–138.

Ishimoto Y, Ohashi K, Mizuno K, Nakano T. 2000. Promotion of the uptake of PS liposomes and apoptotic cells by a product of growth arrest-specific gene, *gas6*. *J Biochem* **127:** 411–417.

Iwasaki A, Medzhitov R. 2004. Toll-like receptor control of the adaptive immune responses. *Nat Immunol* **5:** 987–995.

Janssen JW, Schulz AS, Steenvoorden AC, Schmidberger M, Strehl S, Ambros PF, Bartram CR. 1991. A novel putative tyrosine kinase receptor with oncogenic potential. *Oncogene* **6:** 2113–2120.

Jia R, Hanafusa H. 1994. The proto-oncogene of *v-eyk* (*v-ryk*) is a novel receptor-type protein tyrosine kinase with extracellular Ig/GN-III domains. *J Biol Chem* **269:** 1839–1844.

Jia R, Mayer BJ, Hanafusa T, Hanafusa H. 1992. A novel oncogene, *v-ryk*, encoding a truncated receptor tyrosine kinase is transduced into the RPL30 virus without loss of viral sequences. *J Virol* **66:** 5975–5987.

Kawasaki Y, Nakagawa A, Nagaosa K, Shiratsuchi A, Nakanishi Y. 2002. Phosphatidylserine binding of class B scavenger receptor type I, a phagocytosis receptor of testicular sertoli cells. *J Biol Chem* **277:** 27559–27566.

Keating AK, Kim GK, Jones AE, Donson AM, Ware K, Mulcahy JM, Salzberg DB, Foreman NK, Liang X, Thorburn A, et al. 2010. Inhibition of Mer and Axl receptor tyrosine kinases in astrocytoma cells leads to increased apoptosis and improved chemosensitivity. *Mol Cancer Ther* **9:** 1298–1307.

Koorstra JB, Karikari CA, Feldmann G, Bisht S, Rojas PL, Offerhaus GJ, Alvarez H, Maitra A. 2009. The Axl receptor tyrosine kinase confers an adverse prognostic influence in pancreatic cancer and represents a new therapeutic target. *Cancer Biol Ther* **8:** 618–626.

Kulman JD, Harris JE, Nakazawa N, Ogasawara M, Satake M, Davie EW. 2006. Vitamin K-dependent proteins in *Ciona intestinalis*, a basal chordate lacking a blood coagulation cascade. *Proc Natl Acad Sci* **103:** 15794–15799.

Lai C, Lemke G. 1991. An extended family of protein-tyrosine kinase genes differentially expressed in the vertebrate nervous system. *Neuron* **6:** 691–704.

Lai C, Gore M, Lemke G. 1994. Structure, expression, and activity of Tyro 3, a neural adhesion-related receptor tyrosine kinase. *Oncogene* **9:** 2567–2578.

Lan Z, Wu H, Li W, Wu S, Lu L, Xu M, Dai W. 2000. Transforming activity of receptor tyrosine kinase Tyro3 is mediated, at least in part, by the PI3 kinase-signaling pathway. *Blood* **95:** 633–638.

Lee KF, Simon H, Chen H, Bates B, Hung MC, Hauser C. 1995. Requirement for neuregulin receptor erbB2 in neural and cardiac development. *Nature* **378:** 394–398.

Lemke G, Burstyn-Cohen T. 2010. TAM receptors and the clearance of apoptotic cells. *Ann NY Acad Sci* **1209:** 23–29.

Lemke G, Rothlin CV. 2008. Immunobiology of the TAM receptors. *Nat Rev Immunol* **8:** 327–336.

Li T, Chang CY, Jin DY, Lin PJ, Khvorova A, Stafford DW. 2004. Identification of the gene for vitamin K epoxide reductase. *Nature* **427:** 541–544.

Lijnen HR, Christiaens V, Scroyen L. 2011. Growth arrest-specific protein 6 receptor antagonism impairs adipocyte differentiation and adipose tissue development in mice. *J Pharmacol Exp Ther* **337:** 457–464.

Ling L, Kung HJ. 1995. Mitogenic signals and transforming potential of Nyk, a newly identified neural cell adhesion molecule-related receptor tyrosine kinase. *Mol Cell Biol* **15:** 6582–6592.

Ling L, Templeton D, Kung HJ. 1996. Identification of the major autophosphorylation sites of Nyk/Mer, an NCAM-related receptor tyrosine kinase. *J Biol Chem* **271:** 18355–18362.

Linger RM, Keating AK, Earp HS, Graham DK. 2008. TAM receptor tyrosine kinases: Biologic functions, signaling, and potential therapeutic targeting in human cancer. *Adv Cancer Res* **100:** 35–83.

Liu Y, Cousin JM, Hughes J, Van Damme J, Seckl JR, Haslett C, Dransfield I, Savill J, Rossi AG. 1999. Glucocorticoids promote nonphlogistic phagocytosis of apoptotic leukocytes. *J Immunol* **162:** 3639–3646.

Liu J, Yang C, Simpson C, Deryckere D, Van Deusen A, Miley MJ, Kireev D, Norris-Drouin J, Sather S, Hunter D, et al. 2012. Discovery of novel small molecule Mer kinase inhibitors for the treatment of pediatric acute lymphoblastic leukemia. *ACS Med Chem Lett* **3:** 129–134.

Lu Q, Lemke G. 2001. Homeostatic regulation of the immune system by receptor tyrosine kinases of the Tyro 3 family. *Science* **293:** 306–311.

Lu Q, Gore M, Zhang Q, Camenisch T, Boast S, Casagranda F, Lai C, Skinner MK, Klein R, Matsushima GK, et al. 1999. Tyro-3 family receptors are essential regulators of mammalian spermatogenesis. *Nature* **398:** 723–728.

Ma GZ, Stankovich J, Kilpatrick TJ, Binder MD, Field J. 2011. Polymorphisms in the receptor tyrosine kinase *MERTK* gene are associated with multiple sclerosis susceptibility. *PLoS ONE* **6:** e16964.

Mahadevan D, Cooke L, Riley C, Swart R, Simons B, Della Croce K, Wisner L, Iorio M, Shakalya K, Garewal H, et al. 2007. A novel tyrosine kinase switch is a mechanism of imatinib resistance in gastrointestinal stromal tumors. *Oncogene* **26**: 3909–3919.

Mahajan NP, Earp HS. 2003. An SH2 domain-dependent, phosphotyrosine-independent interaction between Vav1 and the Mer receptor tyrosine kinase: A mechanism for localizing guanine nucleotide-exchange factor action. *J Biol Chem* **278**: 42596–42603.

Manfioletti G, Brancolini C, Avanzi G, Schneider C. 1993. The protein encoded by a growth arrest-specific gene (*gas6*) is a new member of the vitamin K-dependent proteins related to protein S, a negative coregulator in the blood coagulation cascade. *Mol Cell Biol* **13**: 4976–4985.

Manning G, Plowman GD, Hunter T, Sudarsanam S. 2002a. Evolution of protein kinase signaling from yeast to man. *Trends Biochem Sci* **27**: 514–520.

Manning G, Whyte DB, Martinez R, Hunter T, Sudarsanam S. 2002b. The protein kinase complement of the human genome. *Science* **298**: 1912–1934.

Mark MR, Scadden DT, Wang Z, Gu Q, Goddard A, Godowski PJ. 1994. *rse*, a novel receptor-type tyrosine kinase with homology to Axl/Ufo, is expressed at high levels in the brain. *J Biol Chem* **269**: 10720–10728.

Mark MR, Chen J, Hammonds RG, Sadick M, Godowsk PJ. 1996. Characterization of Gas6, a member of the super-family of G domain-containing proteins, as a ligand for Rse and Axl. *J Biol Chem* **271**: 9785–9789.

Marshak-Rothstein A. 2006. Toll-like receptors in systemic autoimmune disease. *Nat Rev Immunol* **6**: 823–835.

McColl A, Bournazos S, Franz S, Perretti M, Morgan BP, Haslett C, Dransfield I. 2009. Glucocorticoids induce protein S–dependent phagocytosis of apoptotic neutrophils by human macrophages. *J Immunol* **183**: 2167–2175.

Meertens L, Carnec X, Lecoin MP, Ramdasi R, Guivel-Benhassine F, Lew E, Lemke G, Schwartz O, Amara A. 2012. The TIM and TAM families of phosphatidylserine receptors mediate Dengue virus entry. *Cell Host Microbe* **12**: 544–557.

Meesters EW, Hansen H, Spronk HM, Hamulyak K, Rosing J, Rowshani AT, ten Berge IJ, ten Cate H. 2007. The inflammation and coagulation cross-talk in patients with systemic lupus erythematosus. *Blood Coagul Fibrinolysis* **18**: 21–28.

Melaragno MG, Wuthrich DA, Poppa V, Gill D, Lindner V, Berk BC, Corson MA. 1998. Increased expression of Axl tyrosine kinase after vascular injury and regulation by G protein-coupled receptor agonists in rats. *Circ Res* **83**: 697–704.

Melaragno MG, Fridell YW, Berk BC. 1999. The Gas6/Axl system: A novel regulator of vascular cell function. *Trends Cardiovasc Med* **9**: 250–253.

Mercer J. 2011. Viral apoptotic mimicry party: P.S. Bring your own Gas6. *Cell Host Microbe* **9**: 255–257.

Mercer J, Helenius A. 2008. Vaccinia virus uses macropinocytosis and apoptotic mimicry to enter host cells. *Science* **320**: 531–535.

Mercer J, Helenius A. 2010. Apoptotic mimicry: Phosphatidylserine-mediated macropinocytosis of vaccinia virus. *Ann NY Acad Sci* **1209**: 49–55.

Ming Cao W, Murao K, Imachi H, Sato M, Nakano T, Kodama T, Sasaguri Y, Wong NC, Takahara J, Ishida T. 2001. Phosphatidylinositol 3-OH kinase-Akt/protein kinase B pathway mediates Gas6 induction of scavenger receptor A in immortalized human vascular smooth muscle cell line. *Arterioscler Thromb Vasc Biol* **21**: 1592–1597.

Morizono K, Xie Y, Olafsen T, Lee B, Dasgupta A, Wu AM, Chen IS. 2011. The soluble serum protein Gas6 bridges virion envelope phosphatidylserine to the TAM receptor tyrosine kinase Axl to mediate viral entry. *Cell Host Microbe* **9**: 286–298.

Nagata K, Ohashi K, Nakano T, Arita H, Zong C, Hanafusa H, Mizuno K. 1996. Identification of the product of growth arrest-specific gene 6 as a common ligand for Axl, Sky, and Mer receptor tyrosine kinases. *J Biol Chem* **271**: 30022–30027.

Nagata S, Hanayama R, Kawane K. 2010. Autoimmunity and the clearance of dead cells. *Cell* **140**: 619–630.

Nandrot EF, Dufour EM. 2010. Mertk in daily retinal phagocytosis: A history in the making. *Adv Exp Med Biol* **664**: 133–140.

Nielsen-Preiss SM, Allen MP, Xu M, Linseman DA, Pawlowski JE, Bouchard RJ, Varnum BC, Heidenreich KA, Wierman ME. 2007. Adhesion-related kinase induction of migration requires phosphatidylinositol-3-kinase and Ras stimulation of Rac activity in immortalized gonadotropin-releasing hormone neuronal cells. *Endocrinology* **148**: 2806–2814.

Nyberg P, He X, Hardig Y, Dahlback B, Garcia de Frutos P. 1997. Stimulation of Sky tyrosine phosphorylation by bovine protein S–domains involved in the receptor–ligand interaction. *Eur J Biochem* **246**: 147–154.

O'Bryan JP, Frye RA, Cogswell PC, Neubauer A, Kitch B, Prokop C, Espinosa R III, Le Beau MM, Earp HS, Liu ET. 1991. *axl*, a transforming gene isolated from primary human myeloid leukemia cells, encodes a novel receptor tyrosine kinase. *Mol Cell Biol* **11**: 5016–5031.

Ohashi K, Mizuno K, Kuma K, Miyata T, Nakamura T. 1994. Cloning of the cDNA for a novel receptor tyrosine kinase, Sky, predominantly expressed in brain. *Oncogene* **9**: 699–705.

Ohashi K, Nagata K, Toshima J, Nakano T, Arita H, Tsuda H, Suzuki K, Mizuno K. 1995. Stimulation of sky receptor tyrosine kinase by the product of growth arrest-specific gene 6. *J Biol Chem* **270**: 22681–22684.

Ostergaard E, Duno M, Batbayli M, Vilhelmsen K, Rosenberg T. 2011. A novel MERTK deletion is a common founder mutation in the Faroe Islands and is responsible for a high proportion of retinitis pigmentosa cases. *Mol Vis* **17**: 1485–1492.

Ou WB, Corson JM, Flynn DL, Lu WP, Wise SC, Bueno R, Sugarbaker DJ, Fletcher JA. 2011. AXL regulates mesothelioma proliferation and invasiveness. *Oncogene* **30**: 1643–1652.

Pierce A, Bliesner B, Xu M, Nielsen-Preiss S, Lemke G, Tobet S, Wierman ME. 2008. Axl and Tyro3 modulate female reproduction by influencing gonadotropin-releasing hormone neuron survival and migration. *Mol Endocrinol* **22**: 2481–2495.

Powell NA, Kohrt JT, Filipski KJ, Kaufman M, Sheehan D, Edmunds JE, Delaney A, Wang Y, Bourbonais F, Lee DY, et al. 2012. Novel and selective spiroindoline-based inhibitors of Sky kinase. *Bioorg Med Chem Lett* **22**: 190–193.

Prasad D, Rothlin CV, Burrola P, Burstyn-Cohen T, Lu Q, Garcia de Frutos P, Lemke G. 2006. TAM receptor function in the retinal pigment epithelium. *Mol Cell Neurosci* **33**: 96–108.

Prieto AL, Weber JL, Lai C. 2000. Expression of the receptor protein-tyrosine kinases Tyro-3, Axl, and Mer in the developing rat central nervous system. *J Comp Neurol* **425**: 295–314.

Prieto AL, O'Dell S, Varnum B, Lai C. 2007. Localization and signaling of the receptor protein tyrosine kinase Tyro3 in cortical and hippocampal neurons. *Neuroscience* **150**: 319–334.

Quong RY, Bickford ST, Ing YL, Terman B, Herlyn M, Lassam NJ. 1994. Protein kinases in normal and transformed melanocytes. *Melanoma Res* **4**: 313–319.

Radic MZ, Shah K, Lu Q, Lemke G, Hillard GM. 2006. Heterogeneous nuclear ribonucleoprotein P2 is an autoantibody target in mice deficient for Mer, Axl, and Tyro3 receptor tyrosine kinases. *J Immunology* **176**: 68–74.

Rajotte I, Hasanbasic I, Blostein M. 2008. Gas6-mediated signaling is dependent on the engagement of its γ-carboxyglutamic acid domain with phosphatidylserine. *Biochem Biophys Res Commun* **376**: 70–73.

Ravichandran KS. 2010. Find-me and eat-me signals in apoptotic cell clearance: Progress and conundrums. *J Exp Med* **207**: 1807–1817.

Rescigno J, Mansukhani A, Basilico C. 1991. A putative receptor tyrosine kinase with unique structural topology. *Oncogene* **6**: 1909–1913.

Rochlitz C, Lohri A, Bacchi M, Schmidt M, Nagel S, Fopp M, Fey MF, Herrmann R, Neubauer A. 1999. Axl expression is associated with adverse prognosis and with expression of Bcl-2 and CD34 in de novo acute myeloid leukemia (AML): Results from a multicenter trial of the Swiss Group for Clinical Cancer Research (SAKK). *Leukemia* **13**: 1352–1358.

Ronnblom L, Pascual V. 2008. The innate immune system in SLE: Type I interferons and dendritic cells. *Lupus* **17**: 394–399.

Rothlin CV, Lemke G. 2010. TAM receptor signaling and autoimmune disease. *Curr Opin Immunol* **22**: 740–746.

Rothlin CV, Ghosh S, Zuniga EI, Oldstone MB, Lemke G. 2007. TAM receptors are pleiotropic inhibitors of the innate immune response. *Cell* **131**: 1124–1136.

Sainaghi PP, Castello L, Bergamasco L, Galletti M, Bellosta P, Avanzi GC. 2005. Gas6 induces proliferation in prostate carcinoma cell lines expressing the Axl receptor. *J Cell Physiol* **204**: 36–44.

Sasaki T, Knyazev PG, Cheburkin Y, Gohring W, Tisi D, Ullrich A, Timpl R, Hohenester E. 2002. Crystal structure of a C-terminal fragment of growth arrest-specific protein Gas6. Receptor tyrosine kinase activation by laminin G-like domains. *J Biol Chem* **277**: 44164–44170.

Sasaki T, Knyazev PG, Clout NJ, Cheburkin Y, Gohring W, Ullrich A, Timpl R, Hohenester E. 2006. Structural basis for Gas6–Axl signalling. *EMBO J* **25**: 80–87.

Sawcer S, Hellenthal G, Pirinen M, Spencer CC, Patsopoulos NA, Moutsianas L, Dilthey A, Su Z, Freeman C, Hunt SE, et al. 2011. Genetic risk and a primary role for cell-mediated immune mechanisms in multiple sclerosis. *Nature* **476**: 214–219.

Scott RS, McMahon EJ, Pop SM, Reap EA, Caricchio R, Cohen PL, Earp HS, Matsushima GK. 2001. Phagocytosis and clearance of apoptotic cells is mediated by MER. *Nature* **411**: 207–211.

Seitz HM, Camenisch TD, Lemke G, Earp HS, Matsushima GK. 2007. Macrophages and dendritic cells use different Axl/Mertk/Tyro3 receptors in clearance of apoptotic cells. *J Immunol* **178**: 5635–5642.

Shao WH, Cohen PL. 2011. Disturbances of apoptotic cell clearance in systemic lupus erythematosus. *Arthritis Res Ther* **13**: 202.

Shao WH, Zhen Y, Eisenberg RA, Cohen PL. 2009. The Mer receptor tyrosine kinase is expressed on discrete macrophage subpopulations and mainly uses Gas6 as its ligand for uptake of apoptotic cells. *Clin Immunol* **133**: 138–144.

Shao WH, Kuan AP, Wang C, Abraham V, Waldman MA, Vogelgesang A, Wittenburg G, Choudhury A, Tsao PY, Miwa T, et al. 2010. Disrupted Mer receptor tyrosine kinase expression leads to enhanced MZ B-cell responses. *J Autoimmun* **35**: 368–374.

Sharif MN, Sosic D, Rothlin CV, Kelly E, Lemke G, Olson EN, Ivashkiv LB. 2006. Twist mediates suppression of inflammation by type I IFNs and Ax1. *J Exp Med* **203**: 1891–1901.

Shimojima M, Takada A, Ebihara H, Neumann G, Fujioka K, Irimura T, Jones S, Feldmann H, Kawaoka Y. 2006. Tyro3 family-mediated cell entry of Ebola and Marburg viruses. *J Virol* **80**: 10109–10116.

Shimojima M, Ikeda Y, Kawaoka Y. 2007. The mechanism of Axl-mediated Ebola virus infection. *J Infect Dis* **196**: S259–263.

Smiley ST, Boyer SN, Heeb MJ, Griffin JH, Grusby MJ. 1997. Protein S is inducible by interleukin 4 in T cells and inhibits lymphoid cell procoagulant activity. *Proc Natl Acad Sci* **94**: 11484–11489.

Soares MM, King SW, Thorpe PE. 2008. Targeting inside-out phosphatidylserine as a therapeutic strategy for viral diseases. *Nat Med* **14**: 1357–1362.

Son BK, Kozaki K, Iijima K, Eto M, Nakano T, Akishita M, Ouchi Y. 2007. Gas6/Axl-PI3K/Akt pathway plays a central role in the effect of statins on inorganic phosphate-induced calcification of vascular smooth muscle cells. *Eur J Pharmacol* **556**: 1–8.

Song KS, Park YS, Kim HK. 2000. Prevalence of anti-protein S antibodies in patients with systemic lupus erythematosus. *Arthritis Rheum* **43**: 557–560.

Song X, Wang H, Logsdon CD, Rashid A, Fleming JB, Abbruzzese JL, Gomez HF, Evans DB. 2010. Overexpression of receptor tyrosine kinase Axl promotes tumor cell invasion and survival in pancreatic ductal adenocarcinoma. *Cancer* **117**: 734–743.

Soriano P. 1997. The PDGF α receptor is required for neural crest cell development and for normal patterning of the somites. *Development* **124**: 2691–2700.

Cite this article as *Cold Spring Harb Perspect Biol* doi: 10.1101/cshperspect.a009076

Sparrow JR, Hicks D, Hamel CP. 2010. The retinal pigment epithelium in health and disease. *Curr Mol Med* **10:** 802–823.

Stafford DW. 2005. The vitamin K cycle. *J Thromb Haemost* **3:** 1873–1878.

Stitt TN, Conn G, Gore M, Lai C, Bruno J, Radziejewski C, Mattsson K, Fisher J, Gies DR, Jones PF, et al. 1995. The anticoagulation factor protein S and its relative, Gas6, are ligands for the Tyro 3/Axl family of receptor tyrosine kinases. *Cell* **80:** 661–670.

Strick DJ, Feng W, Vollrath D. 2009. Mertk drives myosin II redistribution during retinal pigment epithelial phagocytosis. *Invest Ophthalmol Vis Sci* **50:** 2427–2435.

Suh CH, Hilliard B, Li S, Merrill JT, Cohen PL. 2010. TAM receptor ligands in lupus: Protein S but not Gas6 levels reflect disease activity in systemic lupus erythematosus. *Arthritis Res Ther* **12:** R146.

Sun B, Qi N, Shang T, Wu H, Deng T, Han D. 2010. Sertoli cell-initiated testicular innate immune response through Toll-like receptor-3 activation is negatively regulated by Tyro3, Axl, and Mer receptors. *Endocrinology* **151:** 2886–2897.

Tanabe K, Nagata K, Ohashi K, Nakano T, Arita H, Mizuno K. 1997. Roles of γ-carboxylation and a sex hormone-binding globulin-like domain in receptor-binding and in biological activities of Gas6. *FEBS Lett* **408:** 306–310.

Thorp EB. 2010. Mechanisms of failed apoptotic cell clearance by phagocyte subsets in cardiovascular disease. *Apoptosis* **15:** 1124–1136.

Thorp E, Cui D, Schrijvers DM, Kuriakose G, Tabas I. 2008. Mertk receptor mutation reduces efferocytosis efficiency and promotes apoptotic cell accumulation and plaque necrosis in atherosclerotic lesions of $Apoe^{-/-}$ mice. *Arterioscler Thromb Vasc Biol* **28:** 1421–1428.

Tibrewal N, Wu Y, D'Mello V, Akakura R, George TC, Varnum B, Birge RB. 2008. Autophosphorylation docking site Tyr-867 in Mer receptor tyrosine kinase allows for dissociation of multiple signaling pathways for phagocytosis of apoptotic cells and down-modulation of lipopolysaccharide-inducible NF-κB transcriptional activation. *J Biol Chem* **283:** 3618–3627.

Todt JC, Hu B, Curtis JL. 2008. The scavenger receptor SR-A I/II (CD204) signals via the receptor tyrosine kinase Mertk during apoptotic cell uptake by murine macrophages. *J Leukoc Biol* **84:** 510–518.

Uchara H, Shacter E. 2008. Auto-oxidation and oligomerization of protein S on the apoptotic cell surface is required for Mer tyrosine kinase-mediated phagocytosis of apoptotic cells. *J Immunol* **180:** 2522–2530.

van den Brand BT, Abdollahi-Roodsaz S, Vermeij EA, Bennink MB, Arntz OJ, Rothlin CV, van den Berg WB, van de Loo FA. 2013. Therapeutic efficacy of Tyro3, Axl, and Mer tyrosine kinase agonists in collagen-induced arthritis. *Arthritis Rheum* **65:** 671–680.

van Meer G, Voelker DR, Feigenson GW. 2008. Membrane lipids: Where they are and how they behave. *Nat Rev Mol Cell Biol* **9:** 112–124.

Verma A, Warner SL, Vankayalapati H, Bearss DJ, Sharma S. 2011. Targeting Axl and Mer kinases in cancer. *Mol Cancer Ther* **10:** 1763–1773.

Versteeg GA, Garcia-Sastre A. 2010. Viral tricks to grid-lock the type I interferon system. *Curr Opin Microbiol* **13:** 508–516.

Wallet MA, Sen P, Flores RR, Wang Y, Yi Z, Huang Y, Mathews CE, Earp HS, Matsushima G, Wang B, et al. 2008. MerTK is required for apoptotic cell-induced T cell tolerance. *J Exp Med* **205:** 219–232.

Weinger JG, Gohari P, Yan Y, Backer JM, Varnum B, Shafit-Zagardo B. 2008. In brain, Axl recruits Grb2 and the p85 regulatory subunit of PI3 kinase; in vitro mutagenesis defines the requisite binding sites for downstream Akt activation. *J Neurochem* **106:** 134–146.

Weinger JG, Brosnan CF, Loudig O, Goldberg MF, Macian F, Arnett HA, Prieto AL, Tsiperson V, Shafit-Zagardo B. 2011. Loss of the receptor tyrosine kinase Axl leads to enhanced inflammation in the CNS and delayed removal of myelin debris during experimental autoimmune encephalomyelitis. *J Neuroinflammation* **8:** 49.

Wu YM, Robinson DR, Kung HJ. 2004. Signal pathways in up-regulation of chemokines by tyrosine kinase MER/NYK in prostate cancer cells. *Cancer Res* **64:** 7311–7320.

Wu Y, Tibrewal N, Birge RB. 2006. Phosphatidylserine recognition by phagocytes: A view to a kill. *Trends Cell Biol* **16:** 189–197.

Ye F, Han L, Lu Q, Dong W, Chen Z, Shao H, Kaplan HJ, Li Q. 2011. Retinal self-antigen induces a predominantly Th1 effector response in Axl and Mertk double-knockout mice. *J Immunol* **187:** 4178–4186.

Yoshimura A, Naka T, Kubo M. 2007. SOCS proteins, cytokine signalling and immune regulation. *Nat Rev Immunol* **7:** 454–465.

Zezos P, Papaioannou G, Nikolaidis N, Vasiliadis T, Giouleme O, Evgenidis N. 2007. Thrombophilic abnormalities of natural anticoagulants in patients with ulcerative colitis. *Hepatogastroenterology* **54:** 1417–1421.

Zhang Z, Lee JC, Lin L, Olivas V, Au V, LaFramboise T, Abdel-Rahman M, Wang X, Levine AD, Rho JK, et al. 2012. Activation of the AXL kinase causes resistance to EGFR-targeted therapy in lung cancer. *Nat Genet* **44:** 852–860.

Zhong Z, Wang Y, Guo H, Sagare A, Fernandez JA, Bell RD, Barrett TM, Griffin JH, Freeman RS, Zlokovic BV. 2010. Protein S protects neurons from excitotoxic injury by activating the TAM receptor Tyro3–phosphatidylinositol 3–kinase–Akt pathway through its sex hormone-binding globulin-like region. *J Neurosci* **30:** 15521–15534.

Zhu S, Wurdak H, Wang Y, Galkin A, Tao H, Li J, Lyssiotis CA, Yan F, Tu BP, Miraglia L, et al. 2009. A genomic screen identifies TYRO3 as a MITF regulator in melanoma. *Proc Natl Acad Sci* **106:** 17025–17030.

Zhu D, Wang Y, Singh I, Bell RD, Deane R, Zhong Z, Sagare A, Winkler EA, Zlokovic BV. 2010. Protein S controls hypoxic/ischemic blood–brain barrier disruption through the TAM receptor Tyro3 and sphingosine 1-phosphate receptor. *Blood* **115:** 4963–4972.

Zong C, Yan R, August A, Darnell JE Jr, Hanafusa H. 1996. Unique signal transduction of Eyk: Constitutive stimulation of the JAK–STAT pathway by an oncogenic receptor-type tyrosine kinase. *EMBO J* **15:** 4515–4525.

Receptor Tyrosine Kinase-Mediated Angiogenesis

Michael Jeltsch[1,2], Veli-Matti Leppänen[1,2], Pipsa Saharinen[1,2], and Kari Alitalo[1,2,3,4]

[1]Wihuri Research Institute, Biomedicum Helsinki, University of Helsinki, FIN-00014 Helsinki, Finland

[2]Translational Cancer Biology Program, Biomedicum Helsinki, University of Helsinki, FIN-00014 Helsinki, Finland

[3]Institute for Molecular Medicine Finland, Biomedicum Helsinki, University of Helsinki, FIN-00014 Helsinki, Finland

[4]Helsinki University Central Hospital, Biomedicum Helsinki, University of Helsinki, FIN-00014 Helsinki, Finland

Correspondence: kari.alitalo@helsinki.fi

The endothelial cell is the essential cell type forming the inner layer of the vasculature. Two families of receptor tyrosine kinases (RTKs) are almost completely endothelial cell specific: the vascular endothelial growth factor (VEGF) receptors (VEGFR1-3) and the Tie receptors (Tie1 and Tie2). Both are key players governing the generation of blood and lymphatic vessels during embryonic development. Because the growth of new blood and lymphatic vessels (or the lack thereof) is a central element in many diseases, the VEGF and the Tie receptors provide attractive therapeutic targets in various diseases. Indeed, several drugs directed to these RTK signaling pathways are already on the market, whereas many are in clinical trials. Here we review the VEGFR and Tie families, their involvement in developmental and pathological angiogenesis, and the different possibilities for targeting them to either block or enhance angiogenesis and lymphangiogenesis.

ANGIOGENESIS IN DEVELOPMENT

Basics of Vasculature

Blood and lymph are the two major bodily fluids that transport and distribute, via the blood and lymphatic vessels, molecules and cells throughout the body. The essential building blocks of all vessels are endothelial cells (ECs). In the smallest vessels (capillaries), the vascular wall consists almost exclusively of ECs, whereas in larger vessels, and in particular the arteries, the vascular wall is multilayered, with the ECs forming the innermost layer (endothelium or the intimal layer). This is followed by a layer of smooth muscle/mural cells embedded in elastic connective tissue (media) and by the outer adventitial layer, which consists mainly of connective tissue (Boulpaep 2009). The growth of new blood vessels in the initially avascular embryo occurs via vasculogenesis, in which precursor cells from the mesoderm (hemangioblasts) aggregate and differentiate. In the yolk sac, cells in the periphery of the blood islands become ECs, whereas cells in the center differentiate into blood cells. A primitive vascular network is then established via sprouting angiogenesis (the growth of new blood vessels from pre-existing vessels) and extensive remodeling (Lugus et al. 2005). The contribution of vasculogenesis in the development of the lymphatic system has so far been shown only in birds (Papoutsi et al. 2001) and frogs (Ny et al. 2005), whereas in mammals, the lymphatic system may arise predominantly by a lymphangiogenic sprouting process that starts from the large veins (Wigle and Oliver 1999).

Vasculogenesis is largely restricted to early embryonic development, and angiogenesis is the major mechanism of vascular growth in later embryogenesis and in the adult. When new vessels start to grow, the ECs execute a complex program. They have to switch from their quiescent, immotile state into a proliferating, migrating state. A major trigger for this switch is hypoxia (insufficient oxygen concentration), which is often created by tissue expansion. An oxygen-sensing transcriptional system activates a genetic regulatory master switch, which engages the angiogenesis machinery.

Two RTK families—the VEGF receptors (VEGFRs) and the Tie receptors (Fig. 1)—are largely restricted to the ECs in vertebrates, although they also are expressed in a few other cell types, notably in some hematopoietic cells (listed in the supplementary information S2 in Olsson et al. 2006). From the three VEGFRs, only two (VEGFR-2 and VEGFR-3) drive angiogenesis, whereas VEGFR-1 mostly acts to restrict the angiogenic response (Ho et al. 2012) and to recruit macrophages for tissue remodeling (Pipp et al. 2003). Under normal conditions, stimulation of VEGFR-2 results in angiogenesis of blood vascular ECs (BECs), whereas stimulation of VEGFR-3 elicits a similar response in lymphatic ECs (LECs). The Tie receptors have context-dependent roles in EC survival and in the stabilization and remodeling of blood and lymphatic vessels.

Two other RTK families play important roles in angiogenesis, namely the platelet-derived growth factor (PDGF) receptors and Eph receptors. The PDGF receptors are important for the stabilization of the vascular wall by mural cells, such as pericytes and smooth muscle cells (Andrae et al. 2008), and the Eph receptors are involved in determining arterial versus venous identity (Adams and Eichmann 2010). ECs normally do not express PDGF receptors, and the Eph receptors and their membrane-bound ligands (ephrins) are not exclusively expressed by endothelial cells. This paper will focus on the relatively EC-specific VEGF and Tie receptors.

Blood versus Lymphatic Vasculature

The cardiovascular system is a high-pressure system from which blood plasma continuously leaks out into tissues. The primary function of the lymphatic vessels is to drain this fluid and to return it into the blood circulation. On its path, the lymph, including its cellular elements, passes through one or several lymph nodes where it is scanned for foreign antigens; hence, the lymphatic system is essential for an efficient immune defense. In the intestine, lymphatic vessels serve yet another specialized function. They absorb long chain dietary triglycerides and other lipophilic compounds after their digestion and transport these in the form of chylomicrons to other parts of the body. The blood and lymphatic vessels are also structurally different; with the exception of fenestrated and discontinuous endothelia, BECs connect to each other via both tight and adherens junctions and form a continuous basement membrane on their abluminal side. LECs, on the other hand, are only loosely connected to each other and have a discontinuous basement membrane, relying on so-called anchoring filaments to connect to the pericellular matrix (Jeltsch et al. 2003).

Several molecular markers differ between BECs and LECs. Although both BECs and LECs carry panendothelial markers, such as platelet–endothelial cell adhesion molecules (PECAM-1), they express different subsets of RTKs. VEGFR-1 and VEGFR-2 are expressed by BECs, whereas VEGFR-2 and VEGFR-3 are expressed by LECs. However, VEGFR-3 is also expressed in discontinuous or fenestrated blood vascular endothelium (Partanen et al. 2000). Also the high endothelial venules (HEVs) that carry out important immune functions (Kaipainen et al.

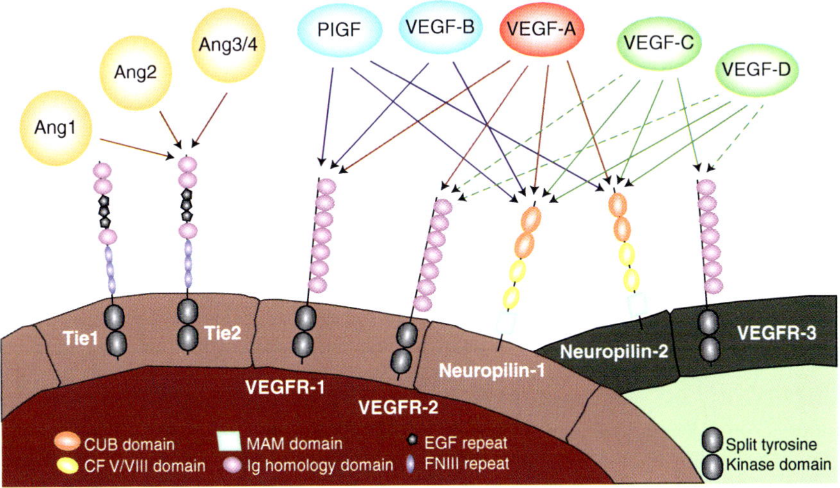

Figure 1. Schematic presentation of Tie and VEGF receptors and their ligands. There are five VEGFs and three angiopoietins in mammals (the mouse ortholog of Ang4 is also referred to as Ang3). Dotted lines indicate that the ligand–receptor interaction is weak or nonexistent for some isoforms of the ligand (Joukov et al. 1997; Baldwin et al. 2001; Leppanen et al. 2011). CUB, Clr/Cls, urchin EGF-like protein, and bone morphogenetic protein I; CF, coagulation factor; MAM, meprin/A5-protein/PTPμ; Ig, immunoglobulin; EGF, epidermal growth factor; FN, fibronectin.

1995; Lacorre et al. 2004) and the abnormal vasculature of tumors can express VEGFR-3 (Tammela et al. 2008).

Tip and Stalk Cells Express Different RTKs

Some of the mechanisms of vessel sprouting in angiogenesis and lymphangiogenesis are surprisingly similar to neuronal axon guidance and the formation of the tracheal system in the fruit fly. The cell leading the EC migration toward a chemotactic stimulus, called the tip cell, sends out numerous filopodia in order to integrate attractive and repellent signals. Repulsive signals are mediated, for example, by the EphB4 RTK and its transmembrane ephrinB2 ligand; these inhibit the direct interaction between arterial and venous endothelial cells (Füller et al. 2003). Such behavior is analogous to functioning of the growth cones of nerve cells. Migration signals are mediated via VEGFR-2 and VEGFR-3 expressed by the tip cells. The Notch ligand Dll4 in the tip cells mediates lateral inhibition of sprouting to the adjacent cells (stalk cells) via Notch signaling. This in turn induces up-regulation of VEGFR-1 and down-regulation of both VEGFR-2 and VEGFR-3 in the stalk cells (Fig. 2)

(Jakobsson et al. 2009). Tip cells also express neuropilin-1, which appears essential for the tip cell functions. Neuropilin-1 acts as a coreceptor for VEGFR-2 and neuropilin-2 acts as a coreceptor for both VEGFR-2 and VEGFR-3 (Favier et al. 2006; Karpanen et al. 2006; Pan et al. 2007; Caunt et al. 2008; Xu et al. 2010). In order to achieve directed, hierarchical sprouting, a VEGF-A gradient is established around the cells producing VEGF-A. Different VEGF-A isoforms vary in their gradient-forming ability due to their different affinities to heparan sulfate proteoglycans on the cell surface and in the pericellular matrix (Ruhrberg et al. 2002). Eventually, tip cells need to detect and connect to neighboring sprouts to form perfused loops. This process is regulated by neuropilin signaling and macrophages that accompany angiogenic processes (Gerhardt et al. 2004; Fantin et al. 2010).

MOLECULAR MECHANISMS OF ANGIOGENESIS

VEGF Receptors and Their Ligands

The ligands of the VEGFRs, the mammalian vascular endothelial growth factors (VEGF-

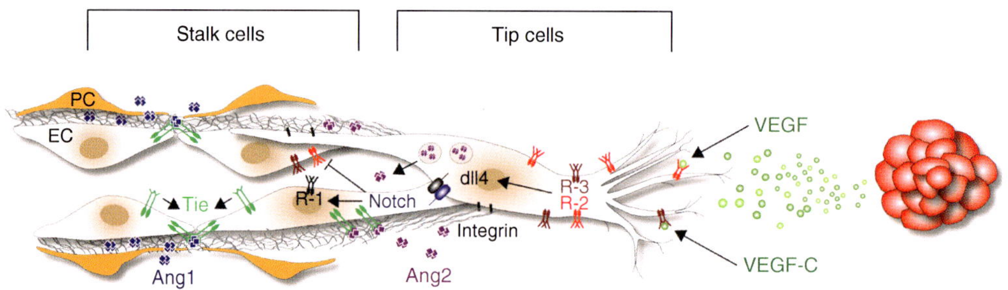

Figure 2. Schematic presentation of the involvement of RTKs in sprouting angiogenesis. VEGF-A and VEGF-C activate VEGFR-2 and VEGFR-3 in the tip cells of angiogenic sprouts, which leads to the migratory cell phenotype. Dll4, which is expressed in the tip cells, interacts with Notch on stalk cells to down-regulate VEGFR-2 and VEGFR-3 and to up-regulate VEGFR-1. Tip cells express Ang2, and may release Ang2 from the Weibel-Palade bodies promoting angiogenesis. Pericyte-produced and matrix-associated Ang1 stabilizes the stalk cells via cell–cell junctional Tie receptor complexes, promoting cell survival, matrix interactions, and endothelial barrier function. In the stalk cells, Ang2 may compete with Ang1, promoting vessel destabilization.

A, VEGF-B, VEGF-C, VEGF-D, and placenta growth factor [PlGF]) (Fig. 1) are crucial regulators of angiogenesis. Their activities are modulated through binding to the heparan sulfate proteoglycan and neuropilin coreceptors. VEGFRs are type-V RTKs comprising a family of transmembrane receptors with an extracellular part of seven immunoglobulin (Ig) homology domains (Fig. 1). VEGFRs utilize distinct Ig homology domains for ligand binding and dimerization. The VEGF family members are antiparallel homodimeric, secreted glycoproteins with multiple isoforms that are generated by alternative splicing and posttranslational processing. Characteristic for the VEGFRs and for most other RTKs is the dimerization of the extracellular domains upon ligand binding. The interactions of the membrane-proximal domains ensure correct positioning of the intracellular domains resulting in their autophosphorylation and downstream signaling (Koch et al. 2011). VEGFR-2 and VEGFR-3 contribute to Erk1,2 activation, whereas Akt activation is mostly induced by VEGFR-3 (Tvorogov et al. 2010; Koch et al. 2011).

Ligand binding to VEGFR-2 induces a robust tyrosine phosphorylation and results in a strong angiogenesis response (Waltenberger et al. 1994), whereas VEGFR-1 has only weak tyrosine kinase activity and seems to modulate angiogenesis as a decoy receptor (Ferrara 2004; Shibuya and Claesson-Welsh 2006; Ho et al.

2012). VEGF-A is a ligand for both VEGFR-2 and VEGFR-1, whereas VEGF-B and PlGF are VEGFR-1-specific ligands. Alternative splicing of both VEGFR-1 and VEGFR-2 gives rise to secreted receptor variants, which are able to bind their respective ligands and may inhibit angiogenesis and lymphangiogenesis (Kendall and Thomas 1993; Albuquerque et al. 2009).

VEGFR-3 and its primary ligand VEGF-C play important roles in the formation of the lymphatic vascular system (Tammela and Alitalo 2010). Upon removal of the propeptides, the VEGFR-3 ligands VEGF-C and VEGF-D acquire a strong binding affinity also for VEGFR-2 and become angiogenic (Joukov et al. 1997; Stacker et al. 1999; Anisimov et al. 2009; Leppanen et al. 2011). VEGF-C activation of VEGFR-3 and ligand-induced VEGFR-3/VEGFR-2 heterodimers are also important in sprouting angiogenesis (Tammela et al. 2008; Nilsson et al. 2010). Furthermore, VEGF-C-induced VEGFR-3 activation contributes to the phenotypic conversion of endothelial cells at fusion points of vessel sprouts (Tammela et al. 2011b).

Tie Receptors and Their Ligands

In addition to the VEGF-VEGFR system, the angiopoietin–Tie system is the second endothelial-specific RTK pathway involved in blood and lymphatic vessel development. The intracellular parts of Tie1 and Tie2 RTKs have a similar split tyrosine kinase domain as the VEGFRs (Partanen et al. 1992), whereas their extracellular domains are composed of two Ig homology domains, followed by three epidermal growth factor (EGF) homology domains, a third Ig homology domain, and three fibronectin type-III domains (Fig. 1) (Barton et al. 2006; Macdonald et al. 2006).

The angiopoietin growth factors (Ang1, Ang2, and Ang4) (Fig. 1) function as ligands for Tie2 (Davis et al. 1996; Maisonpierre et al. 1997; Kim et al. 1999; Valenzuela et al. 1999; Lee et al. 2004). The angiopoietins comprise a structurally unique family of growth factor ligands that multimerize to form various-sized multimers with multiple receptor-binding domains (Kim et al. 2005). The angiopoietins do not directly bind to Tie1, and therefore Tie1 is considered an orphan receptor. However, Ang1 and Ang4 activate Tie1 in heteromeric receptor complexes in endothelial cells, which most likely involve the interaction between Tie1 and Tie2 (Saharinen et al. 2005; Seegar et al. 2010).

Activation of Tie2 occurs via a unique mechanism, which is not used by other soluble growth factor ligands, and thus differs from, for example, VEGF-induced receptor activation (Fukuhara et al. 2008; Saharinen et al. 2008). When contacting ECs are stimulated with angiopoietins, the Tie2 receptors are rapidly translocated to EC–EC junctions to form homomeric Tie2 complexes that associate in *trans* across the EC–EC junctions (Fig. 2) (Fukuhara et al. 2008; Saharinen et al. 2008). Tie1 is associated with Tie2 and traffics also to these complexes (Saharinen et al. 2008). The junctional Ang1–Tie2 complexes activate preferentially the phosphatidyl-inositol-3 kinase (PI3K)/Akt kinase pathway to promote cell survival, EC stability, and barrier function (Fukuhara et al. 2008; Saharinen et al. 2008). The junctional Tie2 complexes are likely activated in the quiescent vasculature. In mobile ECs, matrix-bound Ang1 activates Tie2 in EC-matrix contacts to induce EC-matrix adhesion and cell migration via activation of the extracellular regulated kinases (Erk) (Fukuhara et al. 2008) and the adaptor protein DokR (Saharinen et al. 2008).

Ang2 stimulation of ECs results in Tie2 translocation to EC–EC junctions, but, whereas Ang1 induces Tie2 activation, Ang2 induces only weak Tie2 tyrosine phosphorylation (Maisonpierre et al. 1997; Saharinen et al. 2008). Thus, Ang2 in some conditions acts as an antagonist to inhibit the more robust Ang1-induced Tie2 activation. Ang2 is stored in endothelial cell Weibel-Palade bodies, from where it is rapidly released in response to various stimuli (Fiedler et al. 2004). Ang2 levels are increased in tumor patients and in numerous diseases characterized by vascular leakage and inflammation, for example, in sepsis (Parikh et al. 2006). In the latter situations, Ang2 may compete with Ang1 in Tie2 binding, promoting EC destabilization and vessel regression. In tumors, Ang2 promotes angiogenesis and vascular sprouting, possibly acting as a Tie2 agonist (Holash et al. 1999; Hashizume et al. 2010; Daly et al. 2013).

Angiopoietins may also directly bind to integrins, and this may regulate the functions of tip cells, which express high levels of Ang2 but low Tie2 levels (del Toro et al. 2010; Felcht et al. 2012). The Tie2 receptor also interacts with integrins when endothelial cells adhere on fibronectin, sensitizing cells for Ang1 signaling (Cascone et al. 2005). Ang2 regulates Tie2 localization in an integrin-dependent manner (Pietila et al. 2012), and induces clustering of Tie2 and β3 integrins in cell–cell junctions stimulating integrin turnover (Thomas et al. 2010).

Endothelial Cell-Specific RTKs in Development

VEGF Receptors

VEGFR-2 is expressed in the early mesoderm in cells that undergo angioblast differentiation, and mice lacking VEGFR-2 die at embryonic day E8.5–9.5 as a result of defects in the development of hematopoietic and endothelial cells (Quinn et al. 1993; Shalaby et al. 1995). Mice lacking a single VEGF-A allele display abnormal blood vessel development and embryonic lethality before E9.5 (Carmeliet et al. 1996; Ferrara et al. 1996). VEGF-A and VEGFR-2 expression decrease postnatally, but expression is again up-regulated in tissues undergoing physiological or pathological angiogenesis (Ferrara 2004). In hypoxic cells, VEGF-A expression is up-regulated via the oxygen sensor system that works via the hypoxia-inducible transcription factor (HIF) (Germain et al. 2010).

Deletion of *VEGFR-1* from mouse embryos results in a severely disorganized vasculature, which results from an increased commitment of mesenchymal precursor cells to the hemangioblast lineage (Fong et al. 1995, 1999). Similarly, *VEGFR-1* deletion in adult vessels leads to increased angiogenesis (Ho et al. 2012). However, deletion of the tyrosine kinase domain of VEGFR-1 does not affect vascular development (Hiratsuka et al. 1998), nor does the deletion of its specific ligands VEGF-B or PlGF compromise mouse viability or lead to significant phenotypes in physiological conditions (Bellomo et al. 2000; Carmeliet et al. 2001). This has led to the view that VEGFR-1 and its soluble isoform (sVEGFR-1) act primarily by sequestering excess VEGF-A from its major receptor VEGFR-2 modulating endothelial cell proliferation (Kearney et al. 2002) and sprout formation (Chappell et al. 2009).

During early embryonic development, VEGFR-3 is expressed widely in the ECs but thereafter it becomes gradually restricted to the developing lymphatic vessels (Kaipainen et al. 1995). VEGFR-3-deficient mice die at E9.5 due to defective remodeling and maturation of blood vessels prior to the development of lymphatic vessels (Dumont et al. 1998). VEGF-C-deficient mice die about 3 days later due to edema upon failure of the lymphatic vessel development (Karkkainen et al. 2004). In contrast, the lymphatic vascular development is not affected in VEGF-D-deficient mice (Baldwin et al. 2005). Strikingly, loss of both VEGFR-3 ligands, VEGF-C and VEGF-D, fails to reproduce the early embryonic lethality of VEGFR-3-deficient mice (Haiko et al. 2008). One explanation for this may be that the VEGFR-3-matrix/integrin interactions are sufficient for the early embryonic development (Zhang et al. 2005; Galvagni et al. 2010). In addition, the collagen and calcium-binding EGF domains 1 protein (CCBE1) enhances VEGF-C-induced lymphangiogene-

sis and is also important for lymphatic vascular development (Hogan et al. 2009; Bos et al. 2011; Hagerling et al. 2013).

The Ang/Tie System

The Ang-Tie system is required for the remodeling of the developing blood and lymphatic vasculatures after their initial assembly regulated by the VEGFs (Augustin et al. 2009).

$Tie2^{-/-}$ mouse embryos show severely impaired cardiac development, hemorrhages, and reduced numbers of endothelial cells, resulting in the death of the embryos by E10.5 (Dumont et al. 1994). The gene-targeted embryos deficient of the Tie2 ligand Ang1 have a very similar phenotype, including impaired cardiac development and defective remodeling of the primary vascular plexus, resulting in embryonic lethality (Suri et al. 1996). In addition, the ECs of the $Ang1^{-/-}$ embryos are rounded and poorly associated with basement membranes. Ang1 is dispensable in the adult vasculature during normal homeostasis, but it is required to limit angiogenesis in pathological processes. Conditionally, Ang1-targeted mice develop excessive tissue fibrosis upon vascular stress, suggesting that Ang1 limits perivascular fibrosis, perhaps counteracting TGF-β signaling (Jeansson et al. 2011).

The $Tie1^{-/-}$ mouse embryos have impaired endothelial integrity and hemorrhages, resulting in lethality starting around E13.5, and the deletion of both *Tie1* and *Tie2* results in a similar, but more severe phenotype than that of the $Tie2^{-/-}$ embryos (Puri et al. 1995; Sato et al. 1995). Embryos chimeric for the gene-targeted Tie1 and Tie2 alleles showed that Tie1 and Tie2 are cell-autonomously required for EC survival in the microvasculature during late embryogenesis and in essentially all blood vessels in the adult (Partanen et al. 1996; Puri et al. 1999). Tie1 is also critical for lymphatic development; the jugular lymph sacs of $Tie1^{-/-}$ mouse embryos appear malformed, and the embryos are swollen, before any signs of blood vascular defects (D'Amico et al. 2010; Qu et al. 2010).

The $Ang2^{-/-}$ mice die postnatally or, depending on the background, survive until adulthood. The surviving $Ang2^{-/-}$ mice accumulate

Cite this article as *Cold Spring Harb Perspect Biol* doi: 10.1101/cshperspect.a009183

chylous ascites due to malfunctioning lymphatic vessels, which are abnormally attached to the smooth muscle cells (Gale et al. 2002; Dellinger et al. 2008). The blood vascular defects of $Ang2^{-/-}$ mice are limited to the vitreous vessels. However, ectopic overexpression of Ang2 in developing mouse embryos results in embryonic lethality and a vascular phenotype similar to that of $Ang1^{-/-}$ embryos, suggesting that at least in certain circumstances Ang2 may act as an antagonist of Ang1 (Maisonpierre et al. 1997) despite being an agonist of Tie2 in others (Daly et al. 2013).

ANGIOGENESIS IN DISEASE

The endothelial cells of healthy adult organisms are largely quiescent. Notable exceptions—accompanied by local increases in VEGF-A and VEGF-A receptor expression—include wound healing and tissue repair (Tonnesen et al. 2000), exercise-induced angiogenesis in the heart and skeletal muscle (Prior et al. 2004), the hair cycle (Yano et al. 2001), and, in the female, reproductive cycle and placenta development (Augustin 2005). Pathological angiogenesis differs from physiological angiogenesis, which is tightly controlled and spatially and temporally limited. Pathological angiogenesis plays an important role in many diseases, notably in tumor development.

Angiogenesis and Tumor Development

Tumor Angiogenesis and Tumor Lymphangiogenesis

The concept that tumors are angiogenesis dependent has led to multiple attempts of therapeutic intervention. In several tumor models, the initial stages of tumor growth occur without the involvement of blood vessels until oxygen diffusion becomes limiting and the hypoxic tumor cells start to stimulate vessel growth. This event is called the "angiogenic switch" and it can occur already in the premalignant stages of tumor development (Hanahan et al. 1996; Baeriswyl and Christofori 2009).

VEGF-A/VEGFR-2 is the major growth factor axis involved in tumor angiogenesis (Fer-

rara et al. 2007). The role of VEGF-A in tumor angiogenesis was validated by experiments showing that angiogenesis in tumors and, subsequently, tumor growth could be inhibited by VEGF-A-blocking antibody therapy (Kim et al. 1993). The anti-VEGF-A monoclonal antibody Bevacizumab (Avastin) was the first antiangiogenic cancer drug targeting this axis. However, not all tumors are sensitive to antiangiogenic therapy (intrinsic resistance), and most sensitive tumors will eventually become resistant (evasive resistance) (Abdullah and Perez-Soler 2011; Sennino and McDonald 2012; Singh and Ferrara 2012). A tumor could competitively circumvent VEGF-A inhibition by increasing its VEGF-A production or by switching to an alternative ligand or receptor. For example, VEGFR-3, which is expressed in the tumor vasculature, can be activated by VEGF-C or VEGF-D and after their proteolytic processing, these factors can additionally activate VEGFR-2 (Tammela et al. 2008; Anisimov et al. 2009). Whether PlGF can drive tumor angiogenesis has been controversial, also due to the fact that the PlGF receptor VEGFR-1 appears to be a negative regulator of angiogenesis (Bais et al. 2010; Van de Veire et al. 2010). However, VEGFR-1 is expressed in certain tumor types, and in a fraction of mononuclear phagocytes that contribute other signals for tumor angiogenesis (Bais et al. 2010). Additional mechanisms of resistance to VEGF-A blocking include vessel co-option (Huang et al. 2009), recruitment of myeloid cells from the bone marrow (Ferrara 2010), and other angiogenic factors, such as fibroblast growth factors (FGFs) (Nissen et al. 2007). Targeting multiple angiogenic pathways may lead to improved efficacy of antiangiogenic therapy.

Tumor vasculature is far from normal; it is very heterogeneous and leaky, and shows a lack of hierarchical organization. Hence, blood flow is slow and the interstitial pressure inside the tumor is high, making it difficult for cytostatic drugs to reach their target cells. The current antiangiogenic drugs reduce tumor blood vessels, resulting in the "normalization" of the remaining vasculature, with less vascular leakiness, and improved pericyte coating and blood flow (Jain 2005).

Cite this article as *Cold Spring Harb Perspect Biol* doi: 10.1101/cshperspect.a009183

The angiopoietin-Tie system is also involved in tumor vascularization (see section on Targeting Angiopoietin Signaling). Ang1 has been implicated in the process of tumor vessel normalization during VEGF or Ang2-blocking therapies in mouse models (Winkler et al. 2004; Falcon et al. 2009), whereas Ang2 appears to directly stimulate tumor vascularization. Ang2 is thought to destabilize tumor co-opted blood vessels, resulting in hypoxia and enhanced VEGF-A expression, and the initiation of a neo-angiogenic switch (Holash et al. 1999). Ang2 also stimulates sprouting angiogenesis together with VEGF-A (Hashizume et al. 2010).

Hematogenous and Lymphatic Metastasis

Tumors can metastasize via the blood vessels (hematogenous metastasis) or the lymphatic vessels (lymphatic metastasis). Tumor vascularization correlates with hematogenous metastasis (Weidner et al. 1991; Graeber et al. 1996) and VEGF-A or VEGFR-2 have provided prognostic markers for metastasis in some studies (Bremnes et al. 2006). VEGF-C and VEGF-D stimulate intratumoral and tumor-associated lymphangiogenesis and lymphatic metastasis (Stacker et al. 2002; He et al. 2004; Alitalo 2011). Intratumoral lymphatic vessels are rare and mostly nonfunctional, but they can be detected in several human tumors and at least in some studies their number correlated with VEGF-C and VEGF-D expression and clinical outcome (Achen and Stacker 2008). Most of the tumor-associated lymphatic vessels develop in the tumor periphery; they are recruited from the surrounding lymphatic vasculature (Karpanen et al. 2001; He et al. 2004, 2005), and facilitate tumor dissemination via the lymphatic vessels. Furthermore, VEGF-A and VEGF-C both promote collecting lymph vessel enlargement, increased lymph flow, lymph node lymphangiogenesis, and lymphatic metastasis (He et al. 2002; Roberts et al. 2006; Tammela et al. 2011a).

Recently, Ang2 was shown to increase tumor lymphangiogenesis (Holopainen et al. 2012), as well as metastasis to lymph nodes and distant organs, and its blocking using anti-bodies inhibited the metastasis (Holopainen et al. 2012). The mechanisms likely involve tumor-associated Tie2-expressing macrophages (TEMs) (Mazzieri et al. 2011), and direct effects on endothelial integrity (Holopainen et al. 2012).

Insufficient Angiogenesis in Ischemia

Angiogenesis is one of the hallmarks of malignant tumors and an etiological agent in diseases such as diabetic retinopathy and age-related macular degeneration (AMD). On the other hand, chronic ischemic diseases are characterized by a surprising absence or insufficiency of the angiogenic response to local hypoxia. For example, in coronary artery disease, the formation of new collateral blood vessels to bypass the sclerotic arteries would obviously be beneficial for the patient; however, this process, arteriogenesis, and angiogenesis appear insufficient in most of such cases.

Several trials have attempted to stimulate collateral vessel growth and repair (Zachary and Morgan 2011). The most straightforward approach has been to apply growth factors that directly activate VEGFRs and thus stimulate EC proliferation. Early on, clinical trials appeared to demonstrate the beneficial effect of both VEGF-A and VEGF-C therapy in cardiovascular disease. However, this strategy to improve cardiovascular function has not been translated into clinical practice (Losordo et al. 2002; Reilly et al. 2005; Stewart et al. 2009; Vuorio et al. 2012). The VEGFR-1 ligands PlGF and VEGF-B have both been proposed to play important roles during ischemic revascularization, especially in the heart. In mice, PlGF deletion impaired angiogenesis during ischemia (Carmeliet et al. 2001; Luttun et al. 2002). VEGF-B-stimulated coronary artery growth in transgenic rats (Bry et al. 2010), and deletion of VEGF-B in mice resulted in reduced heart size and vascular dysfunction after transient coronary occlusion as well as impaired recovery from experimentally induced myocardial ischemia (Bellomo et al. 2000; Li et al. 2008b) and VEGF-B-stimulated coronary artery growth in transgenic rats (Bry et al. 2010).

It is possible that other growth factor and cytokine signals are required in addition to VEGF signals to achieve significant therapeutic angiogenesis. Supporting this concept, a designed VEGF-angiopoietin growth factor chimera was superior to VEGF in inducing therapeutic angiogenesis with improved perfusion and less leakage in a mouse model of limb ischemia (Anisimov et al. 2013). On the other hand, the failures could be due to shortcomings in study design or technology (Ylä-Herttuala 2009). In order to activate a multiplicity of proangiogenic stimuli, gene therapy with upstream angiogenic master switches, such as HIF, has been proposed (Kim et al. 2009).

Lack of Lymphangiogenesis in Lymphedema

Lymphedema (insufficient lymphatic drainage) may be caused by deficient lymphangiogenesis, or lymphatic valve deficiency (Alitalo 2011; Norrmén et al. 2011). A relatively common cause of type-I hereditary lymphedema (Milroy disease) is lymphatic vessel hypoplasia due to decreased signaling by a VEGFR-3 allele having a missense point mutation inactivating the kinase domain (Karkkainen et al. 2000). Interestingly, although all ECs carry the mutant allele, it mostly affects the peripheral lymphatic capillaries, which are severely hypoplastic or absent (Bollinger et al. 1983). In one pedigree, a mutation in the coding region of the *VEGF-C* gene has been shown to cosegregate with the lymphedema phenotype (Gordon et al. 2013). Another rare cause of lymphedema is a mutation in the *CCBE1* gene (Alders et al. 2009), which results in compromised VEGF-C-induced lymphangiogenesis (Hogan et al. 2009; Bos et al. 2011). A somewhat similar phenotype is observed in mice when VEGFR-3 signaling is inhibited (Makinen et al. 2001). However, therapeutic dosing of VEGF-C resulted in sufficient activation of VEGFR-3 and rescued the disease phenotype in mice (Karkkainen et al. 2001). Treating lymphedema by activating the primarily lymphangiogenic VEGF-C/VEGFR-3 axis has been also proposed as a rational therapy for secondary lymphedema, which is commonly caused by lymphatic vessel damage by surgery, infectious diseases, or traumas (Tammela et al. 2007; Alitalo 2011).

Role of Ang2 Signals in Inflammation

Tissue inflammation commonly stimulates angiogenesis and lymphangiogenesis. Blood and lymphatic vessels are conduits for immune cell trafficking and, when activated, the inflammatory cells produce a plethora of cytokines, including VEGFs, proteases, and growth factors that directly act on ECs promoting vascular permeability and angiogenesis. Not surprisingly, angiogenesis and often also lymphangiogenesis characterize a variety of chronic inflammatory diseases (e.g., rheumatoid arthritis, psoriasis, and some autoimmune and other disorders with inflammatory features such as obesity and diabetic retinopathy) (Costa et al. 2007; Alitalo 2011). Although angiogenesis is not the underlying cause in these disorders, antiangiogenic intervention could prove beneficial.

Ang2 is locally released from the secretory storage granules at sites of endothelial activation, and it acts as a proinflammatory molecule by sensitizing the endothelium, for example, to the tumor necrosis factor-α-induced expression of endothelial cell adhesion molecules (Fiedler et al. 2006). In a mouse model of chronic airway inflammation, Ang2-blocking agents decreased the remodeling of mucosal capillaries into venules, the amount of leukocyte influx, and disease severity (Tabruyn et al. 2010).

In the pathogenesis of diabetic retinopathy, Ang2 is up-regulated in the endothelium and seems to induce the loss of pericytes from the retinal vasculature. Reducing the Ang2 levels inhibited diabetes-induced pericyte loss and decreased the number of acellular capillary segments (Hammes et al. 2004).

Ang2 may also directly act on endothelial cell–cell junctions to promote vascular leakage. The levels of Ang2 increase in a number of human diseases associated with inflammation and characterized by microvascular leakage, including sepsis and the acute respiratory distress syndrome. Notably, Ang2 induction precedes and contributes to the adverse outcomes in sepsis, suggesting that it is possible to target Ang2 in this condition (Parikh et al. 2006; David et al. 2012). In murine models, Ang1 has beneficial effects on alleviating vascular dysfunction, and some of the effects of Ang1 are likely mediated via its ability to reduce vessel permeability by reducing gaps between the ECs (Baffert et al. 2006; Li et al. 2008a). Furthermore, Ang1 also inhibits excessive angiogenesis, and pathological tissue fibrosis in various murine disease models (Jeansson et al. 2011).

Endothelial Cell-Derived Tumors and Vascular Malformations

A common EC-derived tumor is infantile hemangioma, in which rapidly proliferating endothelial cells give rise to a vascular lesion that can vary from small and innocuous to large and deforming. However, infantile hemangiomas regress spontaneously within several years and treatment is reserved for cases of significant cosmetic and functional interference and complications (North et al. 2006). Although their pathogenesis is still poorly understood, local dysregulation of angiogenesis involving hypoxia-mediated changes in VEGF and VEGFR expression seem to have an important etiological role in hemangiomas (Kleinman et al. 2007; Przewratil et al. 2010). Lymphangiomas, on the other hand, are vascular malformations involving lymphatic vessels, and they seem to be exclusively sporadic unlike capillary and venous malformations, which exist in both sporadic and hereditary forms (Brouillard and Vikkula 2007).

Mutations that increase the ligand-independent autophosphorylation of Tie2 have been identified as an important cause of inherited cutaneomucosal venous malformations, whereas somatic mutations cause more common sporadic lesions (Vikkula et al. 1996; Limaye et al. 2009). Contrary to epithelial cell-derived tumors (carcinomas), endothelial cell-derived tumors are mostly benign and the pathogenesis of their rare malignant counterparts (hemangiosarcoma or lymphangiosarcoma) is unclear. Interestingly, such tumors occur as a complication of postmastectomy edema (Janse et al. 1995). Kaposi's sarcoma is a more frequent tumor that is thought to arise from human herpesvirus-8-induced reprogramming of lymphatic ECs (Liu et al. 2010).

ANTIANGIOGENIC THERAPIES

Inhibitors Targeting the VEGF Signaling Pathways

Up-regulation of certain VEGF family members and their receptors has been implicated in the pathogenesis of the majority of human tumors. Blocking antibodies against VEGF-A have shown the clinical utility of VEGF targeting in antiangiogenic tumor therapy (Hurwitz et al. 2004). Tumor growth inhibition has also been demonstrated with small molecule inhibitors of VEGFRs, anti-VEGFR antibodies, and soluble VEGF-A decoy receptors. However, despite a clear tumor control benefit in clinical trials the response to antiangiogenic therapy is often limited, and relapse of the disease follows (Sennino and McDonald 2012; Singh and Ferrara 2012). In such cases simultaneous targeting of multiple angiogenic pathways may provide improved therapeutic efficacy in the future. Table 1 lists antiangiogenic drugs in clinical use, or in clinical trials targeting VEGFs, VEGFRs, and angiopoietins, and Figure 3 shows a schematic view of the drugs and the drug targets.

Small Molecule VEGFR Tyrosine Kinase Inhibitors

Small molecule receptor tyrosine kinase inhibitors (TKIs) are typically ATP analogs with limited specificity. A number of small molecule inhibitors, including sorafenib, sunitinib, and pazopanib, have been described that inhibit the intrinsic tyrosine kinase activity of VEGFRs (Bhargava and Robinson 2011). However, these multitargeted agents inhibit a wide range of kinases in addition to the VEGFRs, and they are associated with adverse effects, some of which may be unrelated to efficient blocking of VEGF signaling. More selective, second-generation VEGFR TKIs tivozanib, axitinib, and cediranib are currently under evaluation. Extensive preclinical and clinical studies using VEGFR TKIs have shown that these second-

Table 1. Antiangiogenic therapeutics

Drug	Mechanism	Status	Application area	References
Bevacizumab (Avastin)	VEGF-A antibody	Approved	Multiple types of cancer	Ferrara et al. 2005; Van Meter and Kim 2010
Ranibizumab (Lucentis)	Pan-VEGF-A antibody	Approved	Age-related macular degeneration	Rosenfeld et al. 2006; Nguyen et al. 2012
Pegaptanib (Macugen)	VEGF-A-neutralizing aptamer	Approved	Age-related macular degeneration	Ng et al. 2006; Virgili et al. 2012
Sorafenib	Multikinase inhibitor	Approved	Liver and kidney cancer	Escudier et al. 2007
Sunitinib	Multikinase inhibitor	Approved	Multiple cancer types	Motzer et al. 2007
Pazopanib	Multikinase inhibitor	Approved	Renal cell carcinoma, soft tissue sarcoma	Sleijfer et al. 2009
VEGF-A Trap (aflibercept)	VEGF-A-neutralizing receptor fragment	Approved	Age-related macular degeneration, oxaliplatin-refractory metastatic colorectal cancer	Holash et al. 2002; Teng et al. 2010
Axitinib	VEGFR kinase inhibitor	Phase III	Renal cell carcinoma	Bhargava and Robinson 2011
Tivozanib	VEGFR/PDGFR kinase inhibitor	Phase III	Renal cell carcinoma	De Luca and Normanno 2010; Bhargava and Robinson 2011
AMG 386	Ang1/2-inhibiting peptibody	Phase III	Multiple types of cancer	Coxon et al. 2010
Cediranib	Kinase inhibitor	Phase I/II	Multiple types of cancer	Lindsay et al. 2009; Bhargava and Robinson 2011
CovX-body (COV-060)	Ang2 Trap	Phase I/II	Renal cell carcinoma	Huang et al. 2011
MEDI3617	Ang2 antibody	Phase I/II	Multiple types of cancer	Brown et al. 2010
VGX-100	VEGF-C antibody	Phase I	Multiple types of cancer	Goyal et al. 2012
CEP-11981	Tie/VEGF-AR kinase inhibitor	Phase I	Multiple types of cancer	Hudkins et al. 2012

generation VEGFR TKIs are capable of slowing the growth of primary tumors while having reduced off-target toxicities (Bhargava and Robinson 2011).

VEGFR Ligand Inhibitors

Targeting the VEGF axis for antiangiogenic tumor therapy was initially demonstrated by using a monoclonal antibody specific for VEGF-A, which inhibited the tumor growth by decreasing the density of tumor blood vessels (Kim et al. 1993). Subsequently, the humanized anti-VEGF-A antibody bevacizumab was developed as the first antiangiogenic agent approved for clinical use in combination with chemotherapy for the treatment of metastatic colorectal cancer (Hurwitz et al. 2004; Ferrara et al. 2005). Ranibizumab, a recombinant humanized monoclonal antibody Fab fragment against VEGF-A, derived from bevacizumab, is being used for the treatment of age-related macular degeneration (AMD) (Rosenfeld et al. 2006). Antibodies against PlGF have been tested in preclinical models, but the results are controversial concerning the role of PlGF in tumor growth control (Bais et al. 2010; Van de Veire et al. 2010; Snuderl et al. 2013). Antibodies blocking VEGF-C have shown tumor growth inhibition in preclinical models and VEGF-C-blocking an-

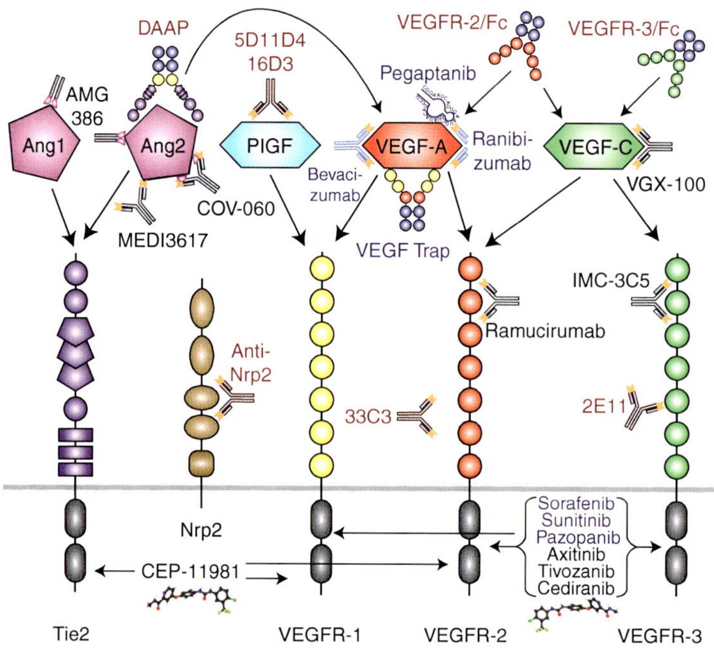

Figure 3. Schematic presentation of inhibitors of the VEGF and Tie pathways. Inhibitors approved by the FDA for clinical use are indicated in blue. Inhibitors that are currently tested in clinical trials are indicated in black and preclinical and nonclinical inhibitors in red.

tibodies are currently being tested in phase-I clinical trials (VGX-100) (Goyal et al. 2012).

Soluble decoy receptors represent a complementary way to block VEGF signaling pathways by preventing the ligands from binding to their cell-surface receptors. One decoy receptor with a high affinity for VEGF-A is a fusion of the VEGFR-1 ligand-binding domains to an Ig constant region, Flt(1-3)-IgG (Gerber et al. 2000). Also, a VEGFR-1/VEGFR-2 fusion protein, VEGF-A-Trap, was described as a potent anti-VEGF-A suppressor of tumor growth and vascularization in vivo (Holash et al. 2002). In recent comparisons of VEGF-A inhibitors, VEGF-A-Trap and Flt(1-3)-IgG had very similar potencies in the bioassays tested and they were both over 10-fold more potent than bevacizumab (Yu et al. 2011; Papadopoulos et al. 2012). In 2011, VEGF-A-Trap received FDA approval for the treatment of wet AMD and recently also for the treatment of oxaliplatin-refractory metastatic colorectal cancer (see www.fda.gov/Drugs/InformationOnDrugs/ApprovedDrugs).

Results of studies using mouse tumor models have suggested that the inhibition of tumor-derived VEGF-C with antibodies or a soluble VEGFR-3 decoy receptor represents another possible strategy for blocking of lymphatic metastasis (Lin et al. 2005). A soluble VEGFR-2 RNA splice variant has also been described as a VEGF-C-specific endogenous inhibitor of lymphatic vessels (Albuquerque et al. 2009).

VEGF Receptor-Blocking Antibodies

VEGFR activation requires ligand binding and dimerization. In principle, VEGFR activation and signaling can be inhibited either by blocking the ligand-binding site or by blocking receptor dimerization. Current VEGFR-2- and VEGFR-3-blocking antibodies under clinical trials are potent inhibitors of ligand binding (Lu et al. 2003; Persaud et al. 2004). In theory, targeting VEGF receptors (which have multiple ligands) should be superior to blocking a single ligand. Receptor-blocking antibodies need to compete with the ligand for binding to the VEGFR. This

Cite this article as *Cold Spring Harb Perspect Biol* doi: 10.1101/cshperspect.a009183

is less of a problem with receptor-blocking antibodies that prevent receptor dimerization (Tvorogov et al. 2010; Kendrew et al. 2011; Hyde et al. 2012), and some of these may also promote incremental inhibition in combination with antibodies targeted against the VEGFR ligand-binding site. Recent results have furthermore indicated that simultaneous VEGFR-2 and VEGFR-3 inhibition results in additive inhibition of tumor angiogenesis (Tammela et al. 2008). Neuropilin coreceptors have been suggested as additional targets for tumor therapy (Caunt et al. 2008).

Targeting Angiopoietin Signaling

Because of its intricate and context-dependent mechanism of action, the Tie-Ang axis has only recently become a target of antiangiogenic drug development. Tumor and stromal cells express Ang1, whereas Ang2 is predominantly produced by activated ECs (Augustin et al. 2009). Ang2 is readily induced in ECs of tumor co-opted blood vessels (Holash et al. 1999), and in the absence of VEGF-A, Ang2 induces endothelial destabilization and vessel regression, leading to an avascular tumor that is highly hypoxic. This results in increased expression of both Ang2 and VEGF-A, which promote the angiogenic switch in the tumor. Interestingly, one recent study showed that low Ang2 expression levels in the tumor stroma of metastatic colorectal cancer patients were associated with better response rates to VEGF-blocking antibodies combined with chemotherapy (Goede et al. 2010).

A number of studies carried out in various preclinical murine models have shown that blocking Ang2 or both Ang1 and Ang2 significantly inhibits tumor growth and angiogenesis (for original articles, see Saharinen et al. 2011). Blocking Ang2 resulted in normalized tumor vessels with increased levels of adhesion molecules in EC–EC junctions, increased pericyte coverage, and reduced EC sprouting and vascular remodeling, resulting in smaller, more uniform vessels (Falcon et al. 2009). In line with these results, tumors grown in $Ang2^{-/-}$ mice displayed a more mature vascular phenotype with increased numbers of pericytes and a small-

er vessel diameter than tumors in wild-type mice (Nasarre et al. 2009). In addition, Ang2 overexpression induced a switch of the vascular phenotype of a breast cancer xenograft by inducing intratumoral hemorrhages and nonfunctional and abnormal blood vessels with increased EC apoptosis and decreased PC coverage (Reiss et al. 2009). At least one study suggests that in combination with Ang2 inhibition, Ang1-blocking agents may prevent tumor vessel normalization (Falcon et al. 2009). Accordingly, COMP-Ang1, a recombinant Ang1 protein, induced vessel normalization and improved vessel perfusion and thereby potentiated the effects of chemotherapy in a Lewis lung carcinoma isograft model (Hwang et al. 2009). The tumor vessel normalization that occurs during VEGF-A blockage may also be mediated in part by Ang1 (Winkler et al. 2004).

In addition to its autocrine functions in the ECs, Ang2 modulates the proangiogenic properties of TEMs, which have been shown to contribute to tumor progression (Mazzieri et al. 2011). Thus, the protumorigenic and proangiogenic functions of Ang2 may involve multiple mechanisms.

Although agents that block the VEGF-A pathway were recently reported to induce invasive growth of some tumors in mice (Ebos et al. 2009; Pàez-Ribes et al. 2009), no such effects have been observed in studies testing Ang2-blocking antibodies. In contrast, Ang2-blocking antibodies inhibited the metastatic colonization of the lungs by tumor cells in mouse models, and subsequent metastatic growth (Mazzieri et al. 2011; Holopainen et al. 2012). The former was at least partially due to enhanced endothelial integrity and improved cell–cell junctions of pulmonary capillaries in Ang2-blocking antibody-treated mice (Holopainen et al. 2012).

The combination of Ang2 inhibition with cytotoxic drugs, or agents targeting the VEGF-A pathway has shown significantly enhanced efficacy when compared to monotherapy (Brown et al. 2010; Hashizume et al. 2010; Huang et al. 2011; Daly et al. 2013). Interestingly, Ang2 overexpression in a mouse model of glioma inhibited the beneficial effects of anti-VEGFR-2 treatment on tumor vessel normalization, brain edema, and animal survival by increasing vascular permeability (Chae et al. 2010). This suggests that high Ang2 levels may compromise the efficacy of anti-VEGF-A therapy and that combinatorial inhibition of Ang2 and VEGF-A signaling should be further investigated. To this end, a chimeric decoy receptor, DAAP (double antiangiogenic protein), simultaneously capable of blocking mouse VEGF-A and angiopoietins, was shown to induce regression of tumor vessels, inhibition of metastasis, reduced ascites formation, and vascular leakage (Koh et al. 2010).

Other specific angiopoietin neutralizing agents that have been used in preclinical models include the soluble Tie2 ectodomain fused to the Fc domain of human immunoglobulin, which binds both Ang1 and Ang2 ligands (Lin et al. 1998), and an Ang2-specific aptamer (Sarraf-Yazdi et al. 2008); however, their clinical applicability is not yet clear.

Currently, several agents targeting the Ang-Tie pathway are in clinical trials (see www.clinicaltrials.gov). A randomized, placebo-controlled phase-II trial showed that an Ang1 and Ang2 dual-specific peptibody (AMG 386) in combination with chemotherapy increased progression-free survival of patients with ovarian cancer (Karlan et al. 2012). The toxicity profile of AMG 386 was mild and distinct from that of VEGF-A inhibitors. Other Ang2-blocking agents, including a fully human anti-Ang2 monoclonal antibody MEDI 3617 (Leow et al. 2012), are currently being tested for safety, dose escalation, and efficacy in phase I/II clinical trials. CEP-11981, a Tie2 and VEGFR tyrosine kinase inhibitor with potential antiangiogenic and antineoplastic activities, is also being tested in clinical phase-I trials, but no information about its potential in tumor growth inhibition is yet available (Hudkins et al. 2012).

CONCLUSIONS

The endothelial cell-specific RTKs are the central effector molecules that orchestrate the development of the vasculature. Because of its omnipresence the vasculature is intimately involved in and affected by many disease processes. VEGF and Tie receptors have gathered most of the attention due to their central roles in tumor growth and progression. Detailed knowledge of these receptors and their functions has allowed the development of additional therapeutic agents against cancer and other diseases where angiogenesis is a contributor. Several drugs have been approved for the treatment of human cancers and certain eye diseases. However, the current data suggest that the existing drugs do not max out the possibilities of antiangiogenic therapy in cancer. Therefore, improved ("second-generation") antiangiogenic drugs are in development. On the flip side, the development of proangiogenic therapies, which could be beneficial in a variety of diseases characterized by insufficient angiogenesis, has not progressed to routine clinical applications. Because pro- and antiangiogenic therapies aim at antagonistic goals, there might be intrinsic limits to their applicability. Would the proangiogenic therapy of coronary heart disease increase the incidence of some types of tumors? Or would long-term antiangiogenic tumor therapy (e.g., to prevent reoccurrence) compromise the vascular health in a patient? To answer these questions we will have to learn more about the molecular basis of vascular biology and this knowledge in turn will allow at least incremental improvements of the existing therapeutic strategies as well as the design of new ones.

REFERENCES

Abdullah SE, Perez-Soler R. 2011. Mechanisms of resistance to vascular endothelial growth factor blockade. *Cancer* **118:** 3455–3467.

Achen MG, Stacker SA. 2008. Molecular control of lymphatic metastasis. *Ann NY Acad Sci* **1131:** 225–234.

Adams RH, Eichmann A. 2010. Axon guidance molecules in vascular patterning. *Cold Spring Harb Perspect Biol* **2:** a001875.

Albuquerque RJC, Hayashi T, Cho WG, Kleinman ME, Dridi S, Takeda A, Baffi JZ, Yamada K, Kaneko H, Green MG, et al. 2009. Alternatively spliced vascular endothelial growth factor receptor-2 is an essential endogenous inhibitor of lymphatic vessel growth. *Nat Med* **15:** 1023–1030.

Alders M, Hogan BM, Gjini E, Salehi F, Al-Gazali L, Hennekam EA, Holmberg EE, Mannens MM, Mulder MF, Offerhaus GJ, et al. 2009. Mutations in CCBE1 cause generalized lymph vessel dysplasia in humans. *Nat Genet* **41:** 1272–1274.

Alitalo K. 2011. The lymphatic vasculature in disease. *Nat Med* **17:** 1371–1380.

Cite this article as *Cold Spring Harb Perspect Biol* doi: 10.1101/cshperspect.a009183

Andrae J, Gallini R, Betsholtz C. 2008. Role of platelet-derived growth factors in physiology and medicine. *Genes Dev* **22:** 1276–1312.

Anisimov A, Alitalo A, Korpisalo P, Soronen J, Kaijalainen S, Leppänen V-M, Jeltsch M, Ylä-Herttuala S, Alitalo K. 2009. Activated forms of VEGF-C and VEGF-D provide improved vascular function in skeletal muscle. *Circ Res* **104:** 1302–1312.

Anisimov A, Tvorogov D, Alitalo A, Leppanen VM, An Y, Han EC, Orsenigo F, Gaal EI, Holopainen T, Koh YJ, et al. 2013. Vascular endothelial growth factor-angiopoietin chimera with improved properties for therapeutic angiogenesis. *Circulation* **127:** 424–434.

Augustin HG. 2005. Angiogenesis in the female reproductive system. *EXS* **94:** 35–52.

Augustin HG, Koh GY, Thurston G, Alitalo K. 2009. Control of vascular morphogenesis and homeostasis through the angiopoietin-Tie system. *Nat Rev Mol Cell Biol* **10:** 165–177.

Baeriswyl V, Christofori G. 2009. The angiogenic switch in carcinogenesis. *Semin Cancer Biol* **19:** 329–337.

Baffert F, Le T, Thurston G, McDonald DM. 2006. Angiopoietin-1 decreases plasma leakage by reducing number and size of endothelial gaps in venules. *Am J Physiol Heart Circ Physiol* **290:** H107–H118.

Bais C, Wu X, Yao J, Yang S, Crawford Y, McCutcheon K, Tan C, Kolumam G, Vernes JM, Eastham-Anderson J, et al. 2010. PlGF blockade does not inhibit angiogenesis during primary tumor growth. *Cell* **141:** 166–177.

Baldwin ME, Catimel B, Nice EC, Roufail S, Hall NE, Stenvers KL, Karkkainen MJ, Alitalo K, Stacker SA, Achen MG. 2001. The specificity of receptor binding by vascular endothelial growth factor-D is different in mouse and man. *J Biol Chem* **276:** 19166–19171.

Baldwin ME, Halford MM, Roufail S, Williams RA, Hibbs ML, Grail D, Kubo H, Stacker SA, Achen MG. 2005. Vascular endothelial growth factor D is dispensable for development of the lymphatic system. *Mol Cell Biol* **25:** 2441–2449.

Barton WA, Tzvetkova-Robev D, Miranda EP, Kolev MV, Rajashankar KR, Himanen JP, Nikolov DB. 2006. Crystal structures of the Tie2 receptor ectodomain and the angiopoietin-2-Tie2 complex. *Nat Struct Mol Biol* **13:** 524–532.

Bellomo D, Headrick JP, Silins GU, Paterson CA, Thomas PS, Gartside M, Mould A, Cahill MM, Tonks ID, Grimmond SM, et al. 2000. Mice lacking the vascular endothelial growth factor-B gene (*Vegfb*) have smaller hearts, dysfunctional coronary vasculature, and impaired recovery from cardiac ischemia. *Circ Res* **86:** E29–E35.

Bhargava P, Robinson MO. 2011. Development of second-generation VEGFR tyrosine kinase inhibitors: Current status. *Curr Oncol Rep* **13:** 103–111.

Bollinger A, Isenring G, Franzeck UK, Brunner U. 1983. Aplasia of superficial lymphatic capillaries in hereditary and connatal lymphedema (Milroy's disease). *Lymphology* **16:** 27–30.

Bos FL, Caunt M, Peterson-Maduro J, Planas-Paz L, Kowalski J, Karpanen T, van Impel A, Tong R, Ernst JA, Korving J, et al. 2011. CCBE1 is essential for mammalian lymphatic vascular development and enhances the lymphangio-

genic effect of vascular endothelial growth factor-C in vivo. *Circ Res* **109:** 486–491.

Boulpaep EL. 2009. Arteries and veins. The microcirculation. *Medical physiology: A cellular and molecular approach* (ed. Boron WF, Boulpaep EL), pp. 467–503. Saunders Elsevier, Philadelphia.

Bremnes RM, Camps C, Sirera R. 2006. Angiogenesis in non-small cell lung cancer: The prognostic impact of neoangiogenesis and the cytokines VEGF and bFGF in tumours and blood. *Lung Cancer* **51:** 143–158.

Brouillard P, Vikkula M. 2007. Genetic causes of vascular malformations. *Hum Mol Genet* **16** (Spec No. 2): R140–R149.

Brown JL, Cao ZA, Pinzon-Ortiz M, Kendrew J, Reimer C, Wen S, Zhou JQ, Tabrizi M, Emery S, McDermott B, et al. 2010. A human monoclonal anti-ANG2 antibody leads to broad antitumor activity in combination with VEGF inhibitors and chemotherapy agents in preclinical models. *Mol Cancer Ther* **9:** 145–156.

Bry M, Kivelä R, Holopainen T, Anisimov A, Tammela T, Soronen J, Silvola J, Saraste A, Jeltsch M, Korpisalo P, et al. 2010. Vascular endothelial growth factor-B acts as a coronary growth factor in transgenic rats without inducing angiogenesis, vascular leak, or inflammation. *Circulation* **122:** 1725–1733.

Carmeliet P, Ferreira V, Breier G, Pollefeyt S, Kieckens L, Gertsenstein M, Fahrig M, Vandenhoeck A, Harpal K, Eberhardt C, et al. 1996. Abnormal blood vessel development and lethality in embryos lacking a single VEGF allele. *Nature* **380:** 435–439.

Carmeliet P, Moons L, Luttun A, Vincenti V, Compernolle V, De Mol M, Wu Y, Bono F, Devy L, Beck H, et al. 2001. Synergism between vascular endothelial growth factor and placental growth factor contributes to angiogenesis and plasma extravasation in pathological conditions. *Nat Med* **7:** 575–583.

Cascone I, Napione L, Maniero F, Serini G, Bussolino F. 2005. Stable interaction between α5β1 integrin and Tie2 tyrosine kinase receptor regulates endothelial cell response to Ang-1. *J Cell Biol* **170:** 993–1004.

Caunt M, Mak J, Liang W-C, Stawicki S, Pan Q, Tong RK, Kowalski J, Ho C, Reslan HB, Ross J, et al. 2008. Blocking neuropilin-2 function inhibits tumor cell metastasis. *Cancer Cell* **13:** 331–342.

Chae S-S, Kamoun WS, Farrar CT, Kirkpatrick ND, Niemeyer E, de Graaf AMA, Sorensen AG, Munn LL, Jain RK, Fukumura D. 2010. Angiopoietin-2 interferes with anti-VEGFR2-induced vessel normalization and survival benefit in mice bearing gliomas. *Clin Cancer Res* **16:** 3618–3627.

Chappell JC, Taylor SM, Ferrara N, Bautch VL. 2009. Local guidance of emerging vessel sprouts requires soluble Flt-1. *Dev Cell* **17:** 377–386.

Costa C, Incio Jo, Soares R. 2007. Angiogenesis and chronic inflammation: Cause or consequence? *Angiogenesis* **10:** 149–166.

Coxon A, Bready J, Min H, Kaufman S, Leal J, Yu D, Lee TA, Sun JR, Estrada J, Bolon B, et al. 2010. Context-dependent role of angiopoietin-1 inhibition in the suppression of angiogenesis and tumor growth: Implications for AMG 386, an angiopoietin-1/2-neutralizing peptibody. *Mol Cancer Ther* **9:** 2641–2651.

Daly C, Eichten A, Castanaro C, Pasnikowski E, Adler A, Lalani AS, Papadopoulos N, Kyle AH, Minchinton AI, Yancopoulos GD, et al. 2013. Angiopoietin-2 functions as a tie2 agonist in tumor models, where it limits the effects of VEGF inhibition. *Cancer Res* **73:** 108–118.

D'Amico G, Korhonen EA, Waltari M, Saharinen P, Laakkonen P, Alitalo K. 2010. Loss of endothelial Tie1 receptor impairs lymphatic vessel development-brief report. *Arterioscler Thromb Vasc Biol* **30:** 207–209.

David S, Mukherjee A, Ghosh CC, Yano M, Khankin EV, Wenger JB, Karumanchi SA, Shapiro NI, Parikh SM. 2012. Angiopoietin-2 may contribute to multiple organ dysfunction and death in sepsis. *Crit Care Med* **40:** 3034–3041.

Davis S, Aldrich TH, Jones PF, Acheson A, Compton DL, Jain V, Ryan TE, Bruno J, Radziejewski C, Maisonpierre PC, et al. 1996. Isolation of angiopoietin-1, a ligand for the TIE2 receptor, by secretion-trap expression cloning. *Cell* **87:** 1161–1169.

Dellinger M, Hunter R, Bernas M, Gale N, Yancopoulos G, Erickson R, Witte M. 2008. Defective remodeling and maturation of the lymphatic vasculature in Angiopoietin-2 deficient mice. *Dev Biol* **319:** 309–320.

del Toro R, Prahst C, Mathivet T, Siegfried G, Kaminker JS, Larrivee B, Breant C, Duarte A, Takakura N, Fukamizu A, et al. 2010. Identification and functional analysis of endothelial tip cell-enriched genes. *Blood* **116:** 4025–4033.

De Luca A, Normanno N. 2010. Tivozanib, a pan-VEGFR tyrosine kinase inhibitor for the potential treatment of solid tumors. *IDrugs* **13:** 636–645.

Dumont DJ, Gradwohl G, Fong GH, Puri MC, Gertsenstein M, Auerbach A, Breitman ML. 1994. Dominant-negative and targeted null mutations in the endothelial receptor tyrosine kinase, tek, reveal a critical role in vasculogenesis of the embryo. *Genes Dev* **8:** 1897–1909.

Dumont DJ, Jussila L, Taipale J, Lymboussaki A, Mustonen T, Pajusola K, Breitman M, Alitalo K. 1998. Cardiovascular failure in mouse embryos deficient in VEGF receptor-3. *Science* **282:** 946–949.

Ebos JML, Lee CR, Cruz-Munoz W, Bjarnason GA, Christensen JG, Kerbel RS. 2009. Accelerated metastasis after short-term treatment with a potent inhibitor of tumor angiogenesis. *Cancer Cell* **15:** 232–239.

Escudier B, Eisen T, Stadler WM, Szczylik C, Oudard S, Siebels M, Negrier S, Chevreau C, Solska E, Desai AA, et al. 2007. Sorafenib in advanced clear-cell renal-cell carcinoma. *N Engl J Med* **356:** 125–134.

Falcon BL, Hashizume H, Koumoutsakos P, Chou J, Bready JV, Coxon A, Oliner JD, McDonald DM. 2009. Contrasting actions of selective inhibitors of angiopoietin-1 and angiopoietin-2 on the normalization of tumor blood vessels. *Am J Pathol* **175:** 2159–2170.

Fantin A, Vieira JM, Gestri G, Denti L, Schwarz Q, Prykhozhij S, Peri F, Wilson SW, Ruhrberg C. 2010. Tissue macrophages act as cellular chaperones for vascular anastomosis downstream of VEGF-mediated endothelial tip cell induction. *Blood* **116:** 829–840.

Favier B, Alam A, Barron P, Bonnin J, Laboudie P, Fons P, Mandron M, Herault J-P, Neufeld G, Savi P, et al. 2006. Neuropilin-2 interacts with VEGFR-2 and VEGFR-3 and promotes human endothelial cell survival and migration. *Blood* **108:** 1243–1250.

Felcht M, Luck R, Schering A, Seidel P, Srivastava K, Hu J, Bartol A, Kienast Y, Vettel C, Loos EK, et al. 2012. Angiopoietin-2 differentially regulates angiogenesis through TIE2 and integrin signaling. *J Clin Invest* **122:** 1991–2005.

Ferrara N. 2004. Vascular endothelial growth factor: Basic science and clinical progress. *Endocr Rev* **25:** 581–611.

Ferrara N. 2010. Role of myeloid cells in vascular endothelial growth factor-independent tumor angiogenesis. *Curr Opin Hematol* **17:** 219–224.

Ferrara N, Carver-Moore K, Chen H, Dowd M, Lu L, O'Shea KS, Powell-Braxton L, Hillan KJ, Moore MW. 1996. Heterozygous embryonic lethality induced by targeted inactivation of the VEGF gene. *Nature* **380:** 439–442.

Ferrara N, Hillan KJ, Novotny W. 2005. Bevacizumab (Avastin), a humanized anti-VEGF monoclonal antibody for cancer therapy. *Biochem Biophys Res Commun* **333:** 328–335.

Ferrara N, Mass RD, Campa C, Kim R. 2007. Targeting VEGF-A to treat cancer and age-related macular degeneration. *Annu Rev Med* **58:** 491–504.

Fiedler U, Scharpfenecker M, Koidl S, Hegen A, Grunow V, Schmidt JM, Kriz W, Thurston G, Augustin HG. 2004. The Tie-2 ligand angiopoietin-2 is stored in and rapidly released upon stimulation from endothelial cell Weibel-Palade bodies. *Blood* **103:** 4150–4156.

Fiedler U, Reiss Y, Scharpfenecker M, Grunow V, Koidl S, Thurston G, Gale NW, Witzenrath M, Rosseau S, Suttorp N, et al. 2006. Angiopoietin-2 sensitizes endothelial cells to TNF-α and has a crucial role in the induction of inflammation. *Nat Med* **12:** 235–239.

Fong GH, Rossant J, Gertsenstein M, Breitman ML. 1995. Role of the Flt-1 receptor tyrosine kinase in regulating the assembly of vascular endothelium. *Nature* **376:** 66–70.

Fong GH, Zhang L, Bryce DM, Peng J. 1999. Increased hemangioblast commitment, not vascular disorganization, is the primary defect in flt-1 knock-out mice. *Development* **126:** 3015–3025.

Fukuhara S, Sako K, Minami T, Noda K, Kim HZ, Kodama T, Shibuya M, Takakura N, Koh GY, Mochizuki N. 2008. Differential function of Tie2 at cell-cell contacts and cell-substratum contacts regulated by angiopoietin-1. *Nat Cell Biol* **10:** 513–526.

Füller T, Korff T, Kilian A, Dandekar G, Augustin HG. 2003. Forward EphB4 signaling in endothelial cells controls cellular repulsion and segregation from ephrinB2 positive cells. *J Cell Sci* **116:** 2461–2470.

Gale NW, Thurston G, Hackett SF, Renard R, Wang Q, McClain J, Martin C, Witte C, Witte MH, Jackson D, et al. 2002. Angiopoietin-2 is required for postnatal angiogenesis and lymphatic patterning, and only the latter role is rescued by Angiopoietin-1. *Dev Cell* **3:** 411–423.

Galvagni F, Pennacchini S, Salameh A, Rocchigiani M, Neri F, Orlandini M, Petraglia F, Gotta S, Sardone GL, Matteucci G, et al. 2010. Endothelial cell adhesion to the extracellular matrix induces c-Src-dependent VEGFR-3 phosphorylation without the activation of the receptor intrinsic kinase activity. *Circ Res* **106:** 1839–1848.

Gerber HP, Kowalski J, Sherman D, Eberhard DA, Ferrara N. 2000. Complete inhibition of rhabdomyosarcoma xenograft growth and neovascularization requires blockade of both tumor and host vascular endothelial growth factor. *Cancer Res* **60:** 6253–6258.

Gerhardt H, Ruhrberg C, Abramsson A, Fujisawa H, Shima D, Betsholtz C. 2004. Neuropilin-1 is required for endothelial tip cell guidance in the developing central nervous system. *Dev Dyn* **231:** 503–509.

Germain S, Monnot C, Muller L, Eichmann A. 2010. Hypoxia-driven angiogenesis: Role of tip cells and extracellular matrix scaffolding. *Curr Opin Hematol* **17:** 245–251.

Goede V, Coutelle O, Neuneier J, Reinacher-Schick A, Schnell R, Koslowsky TC, Weihrauch MR, Cremer B, Kashkar H, Odenthal M, et al. 2010. Identification of serum angiopoietin-2 as a biomarker for clinical outcome of colorectal cancer patients treated with bevacizumab-containing therapy. *Br J Cancer* **103:** 1407–1414.

Gordon K, Schulte D, Brice G, Simpson MA, Roukens MG, van Impel A, Connell F, Kalidas K, Jeffery S, Mortimer PS, et al. 2013. Mutation in vascular endothelial growth factor-C, a ligand for vascular endothelial growth factor receptor-3, is associated with autosomal dominant milroy-like primary lymphedema. *Circ Res* **112:** 956–960.

Goyal S, Chauhan SK, Dana R. 2012. Blockade of prolymphangiogenic vascular endothelial growth factor C in dry eye disease. *Arch Ophthalmol* **130:** 84–89.

Graeber TG, Osmanian C, Jacks T, Housman DE, Koch CJ, Lowe SW, Giaccia AJ. 1996. Hypoxia-mediated selection of cells with diminished apoptotic potential in solid tumours. *Nature* **379:** 88–91.

Hagerling R, Pollmann C, Andreas M, Schmidt C, Nurmi H, Adams RH, Alitalo K, Andresen V, Schulte-Merker S, Kiefer F. 2013. A novel multistep mechanism for initial lymphangiogenesis in mouse embryos based on ultramicroscopy. *EMBO J* **32:** 629–644.

Haiko P, Makinen T, Keskitalo S, Taipale J, Karkkainen MJ, Baldwin ME, Stacker SA, Achen MG, Alitalo K. 2008. Deletion of vascular endothelial growth factor C (VEGF-C) and VEGF-D is not equivalent to VEGF receptor 3 deletion in mouse embryos. *Mol Cell Biol* **28:** 4843–4850.

Hammes H-P, Lin J, Wagner P, Feng Y, Vom Hagen F, Krzizok T, Renner O, Breier G, Brownlee M, Deutsch U. 2004. Angiopoietin-2 causes pericyte dropout in the normal retina: Evidence for involvement in diabetic retinopathy. *Diabetes* **53:** 1104–1110.

Hanahan D, Christofori G, Naik P, Arbeit J. 1996. Transgenic mouse models of tumour angiogenesis: The angiogenic switch, its molecular controls, and prospects for preclinical therapeutic models. *Eur J Cancer* **32A:** 2386–2393.

Hashizume H, Falcon BL, Kuroda T, Baluk P, Coxon A, Yu D, Bready JV, Oliner JD, McDonald DM. 2010. Complementary actions of inhibitors of angiopoietin-2 and VEGF on tumor angiogenesis and growth. *Cancer Res* **70:** 2213–2223.

He Y, Kozaki K-I, Karpanen T, Koshikawa K, Yla-Herttuala S, Takahashi T, Alitalo K. 2002. Suppression of tumor lymphangiogenesis and lymph node metastasis by blocking vascular endothelial growth factor receptor 3 signaling. *J Natl Cancer Inst* **94:** 819–825.

He Y, Rajantie I, Ilmonen M, Makinen T, Karkkainen MJ, Haiko P, Salven P, Alitalo K. 2004. Preexisting lymphatic endothelium but not endothelial progenitor cells are es-

sential for tumor lymphangiogenesis and lymphatic metastasis. *Cancer Res* **64:** 3737–3740.

He Y, Rajantie I, Pajusola K, Jeltsch M, Holopainen T, Yla-Herttuala S, Harding T, Jooss K, Takahashi T, Alitalo K. 2005. Vascular endothelial cell growth factor receptor 3-mediated activation of lymphatic endothelium is crucial for tumor cell entry and spread via lymphatic vessels. *Cancer Res* **65:** 4739–4746.

Hiratsuka S, Minowa O, Kuno J, Noda T, Shibuya M. 1998. Flt-1 lacking the tyrosine kinase domain is sufficient for normal development and angiogenesis in mice. *Proc Natl Acad Sci* **95:** 9349–9354.

Ho VC, Duan LJ, Cronin C, Liang BT, Fong GH. 2012. Elevated vascular endothelial growth factor receptor-2 abundance contributes to increased angiogenesis in vascular endothelial growth factor receptor-1-deficient mice. *Circulation* **126:** 741–752.

Hogan BM, Bos FL, Bussmann J, Witte M, Chi NC, Duckers HJ, Schulte-Merker S. 2009. Ccbe1 is required for embryonic lymphangiogenesis and venous sprouting. *Nat Genet* **41:** 396–398.

Holash J, Maisonpierre PC, Compton D, Boland P, Alexander CR, Zagzag D, Yancopoulos GD, Wiegand SJ. 1999. Vessel cooption, regression, and growth in tumors mediated by angiopoietins and VEGF. *Science* **284:** 1994–1998.

Holash J, Davis S, Papadopoulos N, Croll SD, Ho L, Russell M, Boland P, Leidich R, Hylton D, Burova E, et al. 2002. VEGF-Trap: A VEGF blocker with potent antitumor effects. *Proc Natl Acad Sci* **99:** 11393–11398.

Holopainen T, Saharinen P, D'Amico G, Lampinen A, Eklund L, Sormunen R, Anisimov A, Zarkada G, Lohela M, Helotera H, et al. 2012. Effects of angiopoietin-2-blocking antibody on endothelial cell–cell junctions and lung metastasis. *J Natl Cancer Inst* **104:** 461–475.

Huang J, Bae J-O, Tsai JP, Kadenhe-Chiweshe A, Papa J, Lee A, Zeng S, Kornfeld ZN, Ullner P, Zaghloul N, et al. 2009. Angiopoietin-1/Tie-2 activation contributes to vascular survival and tumor growth during VEGF blockade. *Int J Oncol* **34:** 79–87.

Huang H, Lai J-Y, Do J, Liu D, Li L, Del Rosario J, Doppalapudi VR, Pirie-Shepherd S, Levin N, Bradshaw C, et al. 2011. Specifically targeting angiopoietin-2 inhibits angiogenesis, Tie2-expressing monocyte infiltration, and tumor growth. *Clin Cancer Res* **17:** 1001–1011.

Hudkins RL, Becknell NC, Zulli AL, Underiner TL, Angeles TS, Aimone LD, Albom MS, Chang H, Miknyoczki SJ, Hunter K, et al. 2012. Synthesis and biological profile of the pan-vascular endothelial growth factor receptor/tyrosine kinase with immunoglobulin and epidermal growth factor-like homology domains 2 (VEGF-R/TIE-2) inhibitor 11-(2-methylpropyl)-12,13-dihydro-2-methyl-8-(pyrimidin-2-ylamino)-4H-inda zolo[5,4-a]pyrrolo[3,4-c]carbazol-4-one (CEP-11981): A novel oncology therapeutic agent. *J Med Chem* **55:** 903–913.

Hurwitz H, Fehrenbacher L, Novotny W, Cartwright T, Hainsworth J, Heim W, Berlin J, Baron A, Griffing S, Holmgren E, et al. 2004. Bevacizumab plus irinotecan, fluorouracil, and leucovorin for metastatic colorectal cancer. *N Engl J Med* **350:** 2335–2342.

Hwang J-A, Lee EH, Kim H-W, Park JB, Jeon BH, Cho C-H. 2009. COMP-Ang1 potentiates the antitumor activity of

5-fluorouracil by improving tissue perfusion in murine Lewis lung carcinoma. *Mol Cancer Res* **7:** 1920–1927.

Hyde CA, Giese A, Stuttfeld E, Abram Saliba J, Villemagne D, Schleier T, Binz HK, Ballmer-Hofer K. 2012. Targeting extracellular domains D4 and D7 of vascular endothelial growth factor receptor 2 reveals allosteric receptor regulatory sites. *Mol Cell Biol* **32:** 3802–3813.

Jain RK. 2005. Normalization of tumor vasculature: An emerging concept in antiangiogenic therapy. *Science* **307:** 58–62.

Jakobsson L, Bentley K, Gerhardt H. 2009. VEGFRs and Notch: A dynamic collaboration in vascular patterning. *Biochem Soc Trans* **37:** 1233–1236.

Janse AJ, van Coevorden F, Peterse H, Keus RB, van Dongen JA. 1995. Lymphedema-induced lymphangiosarcoma. *Eur J Surg Oncol* **21:** 155–158.

Jeansson M, Gawlik A, Anderson G, Li C, Kerjaschki D, Henkelman M, Quaggin SE. 2011. Angiopoietin-1 is essential in mouse vasculature during development and in response to injury. *J Clin Invest* **121:** 2278–2289.

Jeltsch M, Tammela T, Alitalo K, Wilting J. 2003. Genesis and pathogenesis of lymphatic vessels. *Cell Tissue Res* **314:** 69–84.

Joukov V, Sorsa T, Kumar V, Jeltsch M, Claesson-Welsh L, Cao Y, Saksela O, Kalkkinen N, Alitalo K. 1997. Proteolytic processing regulates receptor specificity and activity of VEGF-C. *EMBO J* **16:** 3898–3911.

Kaipainen A, Korhonen J, Mustonen T, van Hinsbergh VW, Fang GH, Dumont D, Breitman M, Alitalo K. 1995. Expression of the fms-like tyrosine kinase 4 gene becomes restricted to lymphatic endothelium during development. *Proc Natl Acad Sci* **92:** 3566–3570.

Karkkainen MJ, Ferrell RE, Lawrence EC, Kimak MA, Levinson KL, McTigue MA, Alitalo K, Finegold DN. 2000. Missense mutations interfere with VEGFR-3 signalling in primary lymphoedema. *Nat Genet* **25:** 153–159.

Karkkainen MJ, Saaristo A, Jussila L, Karila KA, Lawrence EC, Pajusola K, Bueler H, Eichmann A, Kauppinen R, Kettunen MI, et al. 2001. A model for gene therapy of human hereditary lymphedema. *Proc Natl Acad Sci* **98:** 12677–12682.

Karkkainen MJ, Haiko P, Sainio K, Partanen J, Taipale J, Petrova TV, Jeltsch M, Jackson DG, Talikka M, Rauvala H, et al. 2004. Vascular endothelial growth factor C is required for sprouting of the first lymphatic vessels from embryonic veins. *Nat Immunol* **5:** 74–80.

Karlan BY, Oza AM, Richardson GE, Provencher DM, Hansen VL, Buck M, Chambers SK, Ghatage P, Pippitt CH Jr, Brown JV III, et al. 2012. Randomized, double-blind, placebo-controlled phase II study of AMG 386 combined with weekly paclitaxel in patients with recurrent ovarian cancer. *J Clin Oncol* **30:** 362–371.

Karpanen T, Egeblad M, Karkkainen MJ, Kubo H, Ylä-Herttuala S, Jäättelä M, Alitalo K. 2001. Vascular endothelial growth factor C promotes tumor lymphangiogenesis and intralymphatic tumor growth. *Cancer Res* **61:** 1786–1790.

Karpanen T, Wirzenius M, Mäkinen T, Veikkola T, Haisma HJ, Achen MG, Stacker SA, Pytowski B, Ylä-Herttuala S, Alitalo K. 2006. Lymphangiogenic growth factor responsiveness is modulated by postnatal lymphatic vessel maturation. *Am J Pathol* **169:** 708–718.

Kearney JB, Ambler CA, Monaco KA, Johnson N, Rapoport RG, Bautch VL. 2002. Vascular endothelial growth factor receptor Flt-1 negatively regulates developmental blood vessel formation by modulating endothelial cell division. *Blood* **99:** 2397–2407.

Kendall RL, Thomas KA. 1993. Inhibition of vascular endothelial cell growth factor activity by an endogenously encoded soluble receptor. *Proc Natl Acad Sci* **90:** 10705–10709.

Kendrew J, Eberlein C, Hedberg B, McDaid K, Smith NR, Weir HM, Wedge SR, Blakey DC, Foltz I, Zhou J, et al. 2011. An antibody targeted to VEGFR-2 Ig domains 4–7 inhibits VEGFR-2 activation and VEGFR-2-dependent angiogenesis without affecting ligand binding. *Mol Cancer Ther* **10:** 770–783.

Kim KJ, Li B, Winer J, Armanini M, Gillett N, Phillips HS, Ferrara N. 1993. Inhibition of vascular endothelial growth factor-induced angiogenesis suppresses tumour growth in vivo. *Nature* **362:** 841–844.

Kim I, Kwak HJ, Ahn JE, So JN, Liu M, Koh KN, Koh GY. 1999. Molecular cloning and characterization of a novel angiopoietin family protein, angiopoietin-3. *FEBS Lett* **443:** 353–356.

Kim KT, Choi HH, Steinmetz MO, Maco B, Kammerer RA, Ahn SY, Kim HZ, Lee GM, Koh GY. 2005. Oligomerization and multimerization are critical for angiopoietin-1 to bind and phosphorylate Tie2. *J Biol Chem* **280:** 20126–20131.

Kim HA, Mahato RI, Lee M. 2009. Hypoxia-specific gene expression for ischemic disease gene therapy. *Adv Drug Deliv Rev* **61:** 614–622.

Kleinman ME, Greives MR, Churgin SS, Blechman KM, Chang EI, Ceradini DJ, Tepper OM, Gurtner GC. 2007. Hypoxia-induced mediators of stem/progenitor cell trafficking are increased in children with hemangioma. *Arterioscler Thromb Vasc Biol* **27:** 2664–2670.

Koch S, Tugues S, Li X, Gualandi L, Claesson-Welsh L. 2011. Signal transduction by vascular endothelial growth factor receptors. *Biochem J* **437:** 169–183.

Koh YJ, Kim H-Z, Hwang S-I, Lee JE, Oh N, Jung K, Kim M, Kim KE, Kim H, Lim N-K, et al. 2010. Double antiangiogenic protein, DAAP, targeting VEGF-A and angiopoietins in tumor angiogenesis, metastasis, and vascular leakage. *Cancer Cell* **18:** 171–184.

Lacorre D-A, Baekkevold ES, Garrido I, Brandtzaeg P, Haraldsen G, Amalric Fo, Girard J-P. 2004. Plasticity of endothelial cells: Rapid dedifferentiation of freshly isolated high endothelial venule endothelial cells outside the lymphoid tissue microenvironment. *Blood* **103:** 4164–4172.

Lee HJ, Cho C-H, Hwang S-J, Choi H-H, Kim K-T, Ahn SY, Kim J-H, Oh J-L, Lee GM, Koh GY. 2004. Biological characterization of angiopoietin-3 and angiopoietin-4. *FASEB J* **18:** 1200–1208.

Leow CC, Coffman K, Inigo I, Breen S, Czapiga M, Soukharev S, Gingles N, Peterson N, Fazenbaker C, Woods R, et al. 2012. MEDI3617, a human anti-angiopoietin 2 monoclonal antibody, inhibits angiogenesis and tumor growth in human tumor xenograft models. *Int J Oncol* **40:** 1321–1330.

Leppanen VM, Jeltsch M, Anisimov A, Tvorogov D, Aho K, Kalkkinen N, Toivanen P, Yla-Herttuala S, Ballmer-Hofer K, Alitalo K. 2011. Structural determinants of vascular endothelial growth factor-D receptor binding and specificity. *Blood* **117:** 1507–1515.

Li X, Stankovic M, Bonder CS, Hahn CN, Parsons M, Pitson SM, Xia P, Proia RL, Vadas MA, Gamble JR. 2008a. Basal and angiopoietin-1-mediated endothelial permeability is regulated by sphingosine kinase-1. *Blood* **111:** 3489–3497.

Li X, Tjwa M, Van Hove I, Enholm B, Neven E, Paavonen K, Jeltsch M, Juan TD, Sievers RE, Chorianopoulos E, et al. 2008b. Reevaluation of the role of VEGF-B suggests a restricted role in the revascularization of the ischemic myocardium. *Arterioscler Thromb Vasc Biol* **28:** 1614–1620.

Limaye N, Wouters V, Uebelhoer M, Tuominen M, Wirkkala R, Mulliken JB, Eklund L, Boon LM, Vikkula M. 2009. Somatic mutations in angiopoietin receptor gene TEK cause solitary and multiple sporadic venous malformations. *Nat Genet* **41:** 118–124.

Lin P, Buxton JA, Acheson A, Radziejewski C, Maisonpierre PC, Yancopoulos GD, Channon KM, Hale LP, Dewhirst MW, George SE, et al. 1998. Antiangiogenic gene therapy targeting the endothelium-specific receptor tyrosine kinase Tie2. *Proc Natl Acad Sci* **95:** 8829–8834.

Lin J, Lalani AS, Harding TC, Gonzalez M, Wu W-W, Luan B, Tu GH, Koprivnikar K, VanRoey MJ, He Y, et al. 2005. Inhibition of lymphogenous metastasis using adeno-associated virus-mediated gene transfer of a soluble VEGFR-3 decoy receptor. *Cancer Res* **65:** 6901–6909.

Lindsay CR, MacPherson IR, Cassidy J. 2009. Current status of cediranib: The rapid development of a novel anti-angiogenic therapy. *Future Oncol* **5:** 421–432.

Liu R, Li X, Tulpule A, Zhou Y, Scehnet JS, Zhang S, Lee JS, Chaudhary PM, Jung J, Gill PS. 2010. KSHV-induced notch components render endothelial and mural cell characteristics and cell survival. *Blood* **115:** 887–895.

Losordo DW, Vale PR, Hendel RC, Milliken CE, Fortuin FD, Cummings N, Schatz RA, Asahara T, Isner JM, Kuntz RE. 2002. Phase 1/2 placebo-controlled, double-blind, dose-escalating trial of myocardial vascular endothelial growth factor 2 gene transfer by catheter delivery in patients with chronic myocardial ischemia. *Circulation* **105:** 2012–2018.

Lu D, Shen J, Vil MD, Zhang H, Jimenez X, Bohlen P, Witte L, Zhu Z. 2003. Tailoring in vitro selection for a picomolar affinity human antibody directed against vascular endothelial growth factor receptor 2 for enhanced neutralizing activity. *J Biol Chem* **278:** 43496–43507.

Lugus JJ, Park C, Choi K. 2005. Developmental relationship between hematopoietic and endothelial cells. *Immunol Res* **32:** 57–74.

Luttun A, Tjwa M, Moons L, Wu Y, Angelillo-Scherrer A, Liao F, Nagy JA, Hooper A, Priller J, De Klerck B, et al. 2002. Revascularization of ischemic tissues by PlGF treatment, and inhibition of tumor angiogenesis, arthritis and atherosclerosis by anti-Flt1. *Nat Med* **8:** 831–840.

Macdonald PR, Progias P, Ciani B, Patel S, Mayer U, Steinmetz MO, Kammerer RA. 2006. Structure of the extracellular domain of Tie receptor tyrosine kinases and localization of the angiopoietin-binding epitope. *J Biol Chem* **281:** 28408–28414.

Maisonpierre PC, Suri C, Jones PF, Bartunkova S, Wiegand SJ, Radziejewski C, Compton D, McClain J, Aldrich TH, Papadopoulos N, et al. 1997. Angiopoietin-2, a natural antagonist for Tie2 that disrupts in vivo angiogenesis. *Science* **277:** 55–60.

Makinen T, Jussila L, Veikkola T, Karpanen T, Kettunen MI, Pulkkanen KJ, Kauppinen R, Jackson DG, Kubo H, Nishikawa S, et al. 2001. Inhibition of lymphangiogenesis with resulting lymphedema in transgenic mice expressing soluble VEGF receptor-3. *Nat Med* **7:** 199–205.

Mazzieri R, Pucci F, Moi D, Zonari E, Ranghetti A, Berti A, Politi LS, Gentner B, Brown JL, Naldini L, et al. 2011. Targeting the ANG2/TIE2 axis inhibits tumor growth and metastasis by impairing angiogenesis and disabling rebounds of proangiogenic myeloid cells. *Cancer Cell* **19:** 512–526.

Motzer RJ, Hutson TE, Tomczak P, Michaelson MD, Bukowski RM, Rixe O, Oudard S, Negrier S, Szczylik C, Kim ST, et al. 2007. Sunitinib versus interferon alfa in metastatic renal-cell carcinoma. *N Engl J Med* **356:** 115–124.

Nasarre P, Thomas M, Kruse K, Helfrich I, Wolter V, Deppermann C, Schadendorf D, Thurston G, Fiedler U, Augustin HG. 2009. Host-derived angiopoietin-2 affects early stages of tumor development and vessel maturation but is dispensable for later stages of tumor growth. *Cancer Res* **69:** 1324–1333.

Ng EW, Shima DT, Calias P, Cunningham ET Jr, Guyer DR, Adamis AP. 2006. Pegaptanib, a targeted anti-VEGF aptamer for ocular vascular disease. *Nat Rev Drug Discov* **5:** 123–132.

Nguyen QD, Brown DM, Marcus DM, Boyer DS, Patel S, Feiner L, Gibson A, Sy J, Rundle AC, Hopkins JJ, et al. 2012. Ranibizumab for diabetic macular edema: Results from 2 phase III randomized trials: RISE and RIDE. *Ophthalmology* **119:** 789–801.

Nilsson I, Bahram F, Li X, Gualandi L, Koch S, Jarvius M, Soderberg O, Anisimov A, Kholova I, Pytowski B, et al. 2010. VEGF receptor 2/-3 heterodimers detected in situ by proximity ligation on angiogenic sprouts. *EMBO J* **29:** 1377–1388.

Nissen LJ, Cao R, Hedlund E-M, Wang Z, Zhao X, Wetterskog D, Funa K, Brakenhielm E, Cao Y. 2007. Angiogenic factors FGF2 and PDGF-BB synergistically promote murine tumor neovascularization and metastasis. *J Clin Invest* **117:** 2766–2777.

Norrmén C, Tammela T, Petrova TV, Alitalo K. 2011. Biological basis of therapeutic lymphangiogenesis. *Circulation* **123:** 1335–1351.

North PE, Waner M, Buckmiller L, James CA, Mihm MCJ. 2006. Vascular tumors of infancy and childhood: Beyond capillary hemangioma. *Cardiovasc Pathol* **15:** 303–317.

Ny A, Koch M, Schneider M, Neven E, Tong RT, Maity S, Fischer C, Plaisance S, Lambrechts D, Heligon C, et al. 2005. A genetic Xenopus laevis tadpole model to study lymphangiogenesis. *Nat Med* **11:** 998–1004.

Olsson A-K, Dimberg A, Kreuger J, Claesson-Welsh L. 2006. VEGF receptor signalling—in control of vascular function. *Nat Rev Mol Cell Biol* **7:** 359–371.

Pàez-Ribes M, Allen E, Hudock J, Takeda T, Okuyama H, Viñals F, Inoue M, Bergers G, Hanahan D, Casanovas O. 2009. Antiangiogenic therapy elicits malignant progression of tumors to increased local invasion and distant metastasis. *Cancer Cell* **15:** 220–231.

Pan Q, Chanthery Y, Liang W-C, Stawicki S, Mak J, Rathore N, Tong RK, Kowalski J, Yee SF, Pacheco G, et al. 2007. Blocking neuropilin-1 function has an additive effect with anti-VEGF to inhibit tumor growth. *Cancer Cell* **11:** 53–67.

Papadopoulos N, Martin J, Ruan Q, Rafique A, Rosconi MP, Shi E, Pyles EA, Yancopoulos GD, Stahl N, Wiegand SJ. 2012. Binding and neutralization of vascular endothelial growth factor (VEGF) and related ligands by VEGF Trap, ranibizumab and bevacizumab. *Angiogenesis* **15:** 171–185.

Papoutsi M, Tomarev SI, Eichmann A, Prols F, Christ B, Wilting J. 2001. Endogenous origin of the lymphatics in the avian chorioallantoic membrane. *Dev Dyn* **222:** 238–251.

Parikh SM, Mammoto T, Schultz A, Yuan HT, Christiani D, Karumanchi SA, Sukhatme VP. 2006. Excess circulating angiopoietin-2 may contribute to pulmonary vascular leak in sepsis in humans. *PLoS Med* **3:** e46.

Partanen J, Armstrong E, Mäkelä TP, Korhonen J, Sandberg M, Renkonen R, Knuutila S, Huebner K, Alitalo K. 1992. A novel endothelial cell surface receptor tyrosine kinase with extracellular epidermal growth factor homology domains. *Mol Cell Biol* **12:** 1698–1707.

Partanen J, Puri MC, Schwartz L, Fischer KD, Bernstein A, Rossant J. 1996. Cell autonomous functions of the receptor tyrosine kinase TIE in a late phase of angiogenic capillary growth and endothelial cell survival during murine development. *Development* **122:** 3013–3021.

Partanen TA, Arola J, Saaristo A, Jussila L, Ora A, Miettinen M, Stacker SA, Achen MG, Alitalo K. 2000. VEGF-C and VEGF-D expression in neuroendocrine cells and their receptor, VEGFR-3, in fenestrated blood vessels in human tissues. *FASEB J* **14:** 2087–2096.

Persaud K, Tille J-C, Liu M, Zhu Z, Jimenez X, Pereira DS, Miao H-Q, Brennan LA, Witte L, Pepper MS, et al. 2004. Involvement of the VEGF receptor 3 in tubular morphogenesis demonstrated with a human anti-human VEGFR-3 monoclonal antibody that antagonizes receptor activation by VEGF-C. *J Cell Sci* **117:** 2745–2756.

Pietila R, Natynki M, Tammela T, Kangas J, Pulkki KH, Limaye N, Vikkula M, Koh GY, Saharinen P, Alitalo K, et al. 2012. Ligand oligomerization state controls Tie2 receptor trafficking and angiopoietin-2-specific responses. *J Cell Sci* **125:** 2212–2223.

Pipp F, Heil M, Issbrucker K, Ziegelhoeffer T, Martin S, van den Heuvel J, Weich H, Fernandez B, Golomb G, Carmeliet P, et al. 2003. VEGFR-1-selective VEGF homologue PlGF is arteriogenic: Evidence for a monocyte-mediated mechanism. *Circ Res* **92:** 378–385.

Prior BM, Yang HT, Terjung RL. 2004. What makes vessels grow with exercise training? *J Appl Physiol* **97:** 1119–1128.

Przewratil Pa, Sitkiewicz A, Andrzejewska E. 2010. Local serum levels of vascular endothelial growth factor in infantile hemangioma: Intriguing mechanism of endothelial growth. *Cytokine* **49:** 141–147.

Puri MC, Rossant J, Alitalo K, Bernstein A, Partanen J. 1995. The receptor tyrosine kinase TIE is required for integrity and survival of vascular endothelial cells. *EMBO J* **14:** 5884–5891.

Puri MC, Partanen J, Rossant J, Bernstein A. 1999. Interaction of the TEK and TIE receptor tyrosine kinases during cardiovascular development. *Development* **126:** 4569–4580.

Qu X, Tompkins K, Batts LE, Puri M, Baldwin S. 2010. Abnormal embryonic lymphatic vessel development in Tie1 hypomorphic mice. *Development* **137:** 1285–1295.

Quinn TP, Peters KG, De Vries C, Ferrara N, Williams LT. 1993. Fetal liver kinase 1 is a receptor for vascular endothelial growth factor and is selectively expressed in vascular endothelium. *Proc Natl Acad Sci* **90:** 7533–7537.

Reilly JP, Grise MA, Fortuin FD, Vale PR, Schaer GL, Lopez J, VAN Camp JR, Henry T, Richenbacher WE, Losordo DW, et al. 2005. Long-term (2-year) clinical events following transthoracic intramyocardial gene transfer of VEGF-2 in no-option patients. *J Interv Cardiol* **18:** 27–31.

Reiss Y, Knedla A, Tal AO, Schmidt MHH, Jugold M, Kiessling F, Burger AM, Wolburg H, Deutsch U, Plate KH. 2009. Switching of vascular phenotypes within a murine breast cancer model induced by angiopoietin-2. *J Pathol* **217:** 571–580.

Roberts N, Kloos B, Cassella M, Podgrabinska S, Persaud K, Wu Y, Pytowski B, Skobe M. 2006. Inhibition of VEGFR-3 activation with the antagonistic antibody more potently suppresses lymph node and distant metastases than inactivation of VEGFR-2. *Cancer Res* **66:** 2650–2657.

Rosenfeld PJ, Rich RM, Lalwani GA. 2006. Ranibizumab: Phase III clinical trial results. *Ophthalmol Clin North Am* **19:** 361–372.

Ruhrberg C, Gerhardt H, Golding M, Watson R, Ioannidou S, Fujisawa H, Betsholtz C, Shima DT. 2002. Spatially restricted patterning cues provided by heparin-binding VEGF-A control blood vessel branching morphogenesis. *Genes Dev* **16:** 2684–2698.

Saharinen P, Kerkelä K, Ekman N, Marron M, Brindle N, Lee GM, Augustin H, Koh GY, Alitalo K. 2005. Multiple angiopoietin recombinant proteins activate the Tie1 receptor tyrosine kinase and promote its interaction with Tie2. *J Cell Biol* **169:** 239–243.

Saharinen P, Eklund L, Miettinen J, Wirkkala R, Anisimov A, Winderlich M, Nottebaum A, Vestweber D, Deutsch U, Koh GY, et al. 2008. Angiopoietins assemble distinct Tie2 signalling complexes in endothelial cell-cell and cell-matrix contacts. *Nat Cell Biol* **10:** 527–537.

Saharinen P, Eklund L, Pulkki K, Bono P, Alitalo K. 2011. VEGF and angiopoietin signaling in tumor angiogenesis and metastasis. *Trends Mol Med* **17:** 347–362.

Sarraf-Yazdi S, Mi J, Moeller BJ, Niu X, White RR, Kontos CD, Sullenger BA, Dewhirst MW, Clary BM. 2008. Inhibition of in vivo tumor angiogenesis and growth via systemic delivery of an angiopoietin 2-specific RNA aptamer. *J Surg Res* **146:** 16–23.

Sato TN, Tozawa Y, Deutsch U, Wolburg-Buchholz K, Fujiwara Y, Gendron-Maguire M, Gridley T, Wolburg H, Risau W, Qin Y. 1995. Distinct roles of the receptor tyrosine kinases Tie-1 and Tie-2 in blood vessel formation. *Nature* **376:** 70–74.

Seegar TC, Eller B, Tzvetkova-Robev D, Kolev MV, Henderson SC, Nikolov DB, Barton WA. 2010. Tie1–Tie2 interactions mediate functional differences between angiopoietin ligands. *Mol Cell* **37:** 643–655.

Sennino B, McDonald DM. 2012. Controlling escape from angiogenesis inhibitors. *Nat Rev Cancer* **12:** 699–709.

Shalaby F, Rossant J, Yamaguchi TP, Gertsenstein M, Wu XF, Breitman ML, Schuh AC. 1995. Failure of blood-island formation and vasculogenesis in Flk-1-deficient mice. *Nature* **376:** 62–66.

Shibuya M, Claesson-Welsh L. 2006. Signal transduction by VEGF receptors in regulation of angiogenesis and lymphangiogenesis. *Exp Cell Res* **312:** 549–560.

Singh M, Ferrara N. 2012. Modeling and predicting clinical efficacy for drugs targeting the tumor milieu. *Nat Biotechnol* **30:** 648–657.

Sleijfer S, Ray-Coquard I, Papai Z, Le Cesne A, Scurr M, Schoffski P, Collin F, Pandite L, Marreaud S, De Brauwer A, et al. 2009. Pazopanib, a multikinase angiogenesis inhibitor, in patients with relapsed or refractory advanced soft tissue sarcoma: A phase II study from the European organization for research and treatment of cancer-soft tissue and bone sarcoma group (EORTC study 62043). *J Clin Oncol* **27:** 3126–3132.

Snuderl M, Batista A, Kirkpatrick ND, Ruiz de Almodovar C, Riedemann L, Walsh EC, Anolik R, Huang Y, Martin JD, Kamoun W, et al. 2013. Targeting placental growth factor/neuropilin 1 pathway inhibits growth and spread of medulloblastoma. *Cell* **152:** 1065–1076.

Stacker SA, Stenvers K, Caesar C, Vitali A, Domagala T, Nice E, Roufail S, Simpson RJ, Moritz R, Karpanen T, et al. 1999. Biosynthesis of vascular endothelial growth factor-D involves proteolytic processing which generates non-covalent homodimers. *J Biol Chem* **274:** 32127–32136.

Stacker SA, Achen MG, Jussila L, Baldwin ME, Alitalo K. 2002. Lymphangiogenesis and cancer metastasis. *Nat Rev Cancer* **2:** 573–583.

Stewart DJ, Kutryk MJB, Fitchett D, Freeman M, Camack N, Su Y, Della Siega A, Bilodeau L, Burton JR, Proulx G, et al. 2009. VEGF gene therapy fails to improve perfusion of ischemic myocardium in patients with advanced coronary disease: Results of the NORTHERN trial. *Mol Ther* **17:** 1109–1115.

Suri C, Jones PF, Patan S, Bartunkova S, Maisonpierre PC, Davis S, Sato TN, Yancopoulos GD. 1996. Requisite role of angiopoietin-1, a ligand for the TIE2 receptor, during embryonic angiogenesis. *Cell* **87:** 1171–1180.

Tabruyn SP, Colton K, Morisada T, Fuxe J, Wiegand SJ, Thurston G, Coyle AJ, Connor J, McDonald DM. 2010. Angiopoietin-2-driven vascular remodeling in airway inflammation. *Am J Pathol* **177:** 3233–3243.

Tammela T, Alitalo K. 2010. Lymphangiogenesis: Molecular mechanisms and future promise. *Cell* **140:** 460–476.

Tammela T, Saaristo A, Holopainen T, Lyytikkä J, Kotronen A, Pitkonen M, Abo-Ramadan U, Ylä-Herttuala S, Petrova TV, Alitalo K. 2007. Therapeutic differentiation and maturation of lymphatic vessels after lymph node dissection and transplantation. *Nat Med* **13:** 1458–1466.

Tammela T, Zarkada G, Wallgard E, Murtomaki A, Suchting S, Wirzenius M, Waltari M, Hellstrom M, Schomber T, Peltonen R, et al. 2008. Blocking VEGFR-3 suppresses angiogenic sprouting and vascular network formation. *Nature* **454:** 656–660.

Tammela T, Saaristo A, Holopainen T, Ylä-Herttuala S, Andersson LC, Virolainen S, Immonen I, Alitalo K. 2011a. Photodynamic ablation of lymphatic vessels and intra-

lymphatic cancer cells prevents metastasis. *Sci Transl Med* **3:** 69ra11.

Tammela T, Zarkada G, Nurmi H, Jakobsson L, Heinolainen K, Tvorogov D, Zheng W, Franco CA, Murtomaki A, Aranda E, et al. 2011b. VEGFR-3 controls tip to stalk conversion at vessel fusion sites by reinforcing Notch signalling. *Nat Cell Biol* **13:** 1202–1213.

Teng LS, Jin KT, He KF, Zhang J, Wang HH, Cao J. 2010. Clinical applications of VEGF-trap (aflibercept) in cancer treatment. *J Chin Med Assoc* **73:** 449–456.

Thomas M, Felcht M, Kruse K, Kretschmer S, Deppermann C, Biesdorf A, Rohr K, Benest AV, Fiedler U, Augustin HG. 2010. Angiopoietin-2 stimulation of endothelial cells induces αvβ3 integrin internalization and degradation. *J Biol Chem* **285:** 23842–23849.

Tonnesen MG, Feng X, Clark RA. 2000. Angiogenesis in wound healing. *J Investig Dermatol Symp Proc* **5:** 40–46.

Tvorogov D, Anisimov A, Zheng W, Leppänen V-M, Tammela T, Laurinavicius S, Holnthoner W, Heloterä H, Holopainen T, Jeltsch M, et al. 2010. Effective suppression of vascular network formation by combination of antibodies blocking VEGFR ligand binding and receptor dimerization. *Cancer Cell* **18:** 630–640.

Valenzuela DM, Griffiths JA, Rojas J, Aldrich TH, Jones PF, Zhou H, McClain J, Copeland NG, Gilbert DJ, Jenkins NA, et al. 1999. Angiopoietins 3 and 4: Diverging gene counterparts in mice and humans. *Proc Natl Acad Sci* **96:** 1904–1909.

Van de Veire S, Stalmans I, Heindryckx F, Oura H, Tijeras-Raballand A, Schmidt T, Loges S, Albrecht I, Jonckx B, Vinckier S, et al. 2010. Further pharmacological and genetic evidence for the efficacy of PlGF inhibition in cancer and eye disease. *Cell* **141:** 178–190.

Van Meter ME, Kim ES. 2010. Bevacizumab: Current updates in treatment. *Curr Opin Oncol* **22:** 586–591.

Vikkula M, Boon LM, Carraway KLr, Calvert JT, Diamonti AJ, Goumnerov B, Pasyk KA, Marchuk DA, Warman ML, Cantley LC, et al. 1996. Vascular dysmorphogenesis caused by an activating mutation in the receptor tyrosine kinase TIE2. *Cell* **87:** 1181–1190.

Virgili G, Parravano M, Menchini F, Brunetti M. 2012. Anti-angiogenic therapy with anti-vascular endothelial growth factor modalities for diabetic macular oedema. *Cochrane Database Syst Rev* **12:** CD007419.

Vuorio T, Jauhiainen S, Yla-Herttuala S. 2012. Pro- and anti-angiogenic therapy and atherosclerosis with special emphasis on vascular endothelial growth factors. *Exp Opin Biol Ther* **12:** 79–92.

Waltenberger J, Claesson-Welsh L, Siegbahn A, Shibuya M, Heldin CH. 1994. Different signal transduction properties of KDR and Flt1, two receptors for vascular endothelial growth factor. *J Biol Chem* **269:** 26988–26995.

Weidner N, Semple JP, Welch WR, Folkman J. 1991. Tumor angiogenesis and metastasis—correlation in invasive breast carcinoma. *N Engl J Med* **324:** 1–8.

Wigle JT, Oliver G. 1999. Prox1 function is required for the development of the murine lymphatic system. *Cell* **98:** 769–778.

Winkler F, Kozin SV, Tong RT, Chae SS, Booth MF, Garkavtsev I, Xu L, Hicklin DJ, Fukumura D, di Tomaso E, et al.

2004. Kinetics of vascular normalization by VEGFR2 blockade governs brain tumor response to radiation: Role of oxygenation, angiopoietin-1, and matrix metalloproteinases. *Cancer Cell* **6:** 553–563.

Xu Y, Yuan L, Mak J, Pardanaud L, Caunt M, Kasman I, Larrivée B, Del Toro R, Suchting S, Medvinsky A, et al. 2010. Neuropilin-2 mediates VEGF-C-induced lymphatic sprouting together with VEGFR3. *J Cell Biol* **188:** 115–130.

Yano K, Brown LF, Detmar M. 2001. Control of hair growth and follicle size by VEGF-mediated angiogenesis. *J Clin Invest* **107:** 409–417.

Ylä-Herttuala S. 2009. Gene therapy with vascular endothelial growth factors. *Biochem Soc Trans* **37:** 1198–1200.

Yu L, Liang XH, Ferrara N. 2011. Comparing protein VEGF inhibitors: In vitro biological studies. *Biochem Biophys Res Commun* **408:** 276–281.

Zachary I, Morgan RD. 2011. Therapeutic angiogenesis for cardiovascular disease: Biological context, challenges, prospects. *Heart* **97:** 181–189.

Zhang X, Groopman JE, Wang JF. 2005. Extracellular matrix regulates endothelial functions through interaction of VEGFR-3 and integrin $\alpha 5\beta 1$. *J Cell Physiol* **202:** 205–214.

Insulin Receptor Signaling in Normal and Insulin-Resistant States

Jérémie Boucher[1,2], André Kleinridders[1,2], and C. Ronald Kahn[1]

[1]Section on Integrative Physiology and Metabolism, Joslin Diabetes Center and Department of Medicine, Brigham and Women's Hospital and Harvard Medical School, Boston, Massachusetts 02115

Correspondence: c.ronald.kahn@joslin.harvard.edu

In the wake of the worldwide increase in type-2 diabetes, a major focus of research is understanding the signaling pathways impacting this disease. Insulin signaling regulates glucose, lipid, and energy homeostasis, predominantly via action on liver, skeletal muscle, and adipose tissue. Precise modulation of this pathway is vital for adaption as the individual moves from the fed to the fasted state. The positive and negative modulators acting on different steps of the signaling pathway, as well as the diversity of protein isoform interaction, ensure a proper and coordinated biological response to insulin in different tissues. Whereas genetic mutations are causes of rare and severe insulin resistance, obesity can lead to insulin resistance through a variety of mechanisms. Understanding these pathways is essential for development of new drugs to treat diabetes, metabolic syndrome, and their complications.

Insulin and IGF-1 control a wide variety of biological processes by acting on two closely related tyrosine kinase receptors. Receptor activation initiates a cascade of phosphorylation events that leads to the activation of enzymes that control many aspects of metabolism and growth. Insulin/IGF-1 signaling contains many different points of regulation or critical nodes, controlled both positively and negatively, to ensure proper signal duration and intensity (see the schematic in Fig. 1). Perturbations in these signaling pathways can lead to insulin resistance. Here we review the insulin-signaling network, its critical nodes, and how these are perturbed in insulin-resistant states.

INSULIN AND IGF-1 RECEPTORS

Insulin and IGF-1 mediate their biological effects via the insulin and IGF-1 receptors (IR and IGF-1R). These highly homologous tyrosine kinase receptors are members of a family that also includes the orphan insulin receptor-related receptor (IRR), which has been suggested to play a role in testis determination (Nef et al. 2003) and act as an extracellular alkali sensor (Deyev et al. 2011). Although insulin and IGF-1 preferentially bind to their own receptors, both ligands can also bind to the alternate receptor with reduced affinity (Belfiore et al. 2009).

The IR, IGF-1R, and IRR are tetrameric proteins that consist of two extracellular α subunits

---

[2]These authors contributed equally to this article.

Cite this article as *Cold Spring Harb Perspect Biol* doi: 10.1101/cshperspect.a009191

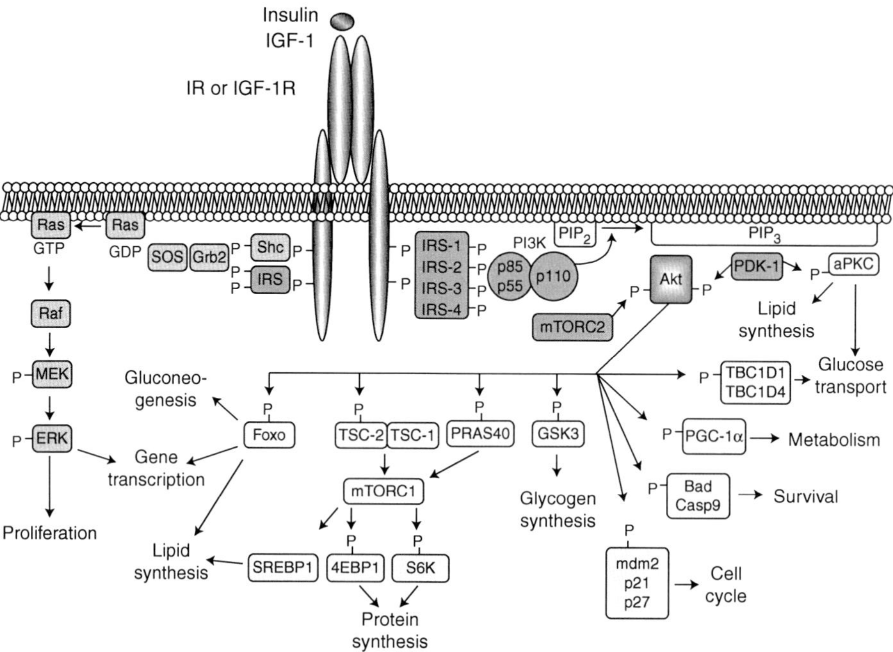

Figure 1. Insulin- and IGF-1-signaling pathways. Activation of insulin and IGF-1 receptors by their ligands initiates a cascade of phosphorylation events. A conformational change and autophosphorylation of the receptors occur at the time of ligand binding, leading to the recruitment and phosphorylation of receptor substrates such as IRS and Shc proteins. Shc activates the Ras-MAPK pathway, whereas IRS proteins mostly activate the PI3K-Akt pathway by recruiting and activating PI3K, leading to the generation of second messenger PIP₃. Membrane-bound PIP₃ recruits and activates PDK-1, which phosphorylates and activates Akt and atypical PKCs. Akt mediates most of insulin's metabolic effects, regulating glucose transport, lipid synthesis, gluconeogenesis, and glycogen synthesis. Akt also plays a role in the control of cell cycle and survival. The Shc-Grb2-Sos-Ras-Raf-MAPK pathway controls cellular proliferation and gene transcription.

and two transmembrane β subunits that are joined by disulfide bonds. Both subunits are generated from a single large precursor by proteolytic cleavage. The IR messenger RNA (mRNA) undergoes alternative splicing of exon 11 to yield two isoforms that differ by exclusion (isoform A) or inclusion (isoform B) of a 12-amino-acid sequence in the carboxy-terminal part of the α subunit (Mosthaf et al. 1990). IR-A is predominantly expressed in fetal tissues and in the brain, has higher affinity for both insulin and IGF-2, has a higher rate of internalization than the type-B isoform, and tends to be up-regulated in cancer (Frasca et al. 1999), whereas IR-B expression is highest in the liver. Heterotetramers composed of an α/β dimer of IR and an α/β dimer of IGF-1R can form hybrid receptor complexes that bind preferentially

IGF-1 and IGF-2 over insulin (Benyoucef et al. 2007). Their formation appears to occur randomly in cells expressing both receptors and depends on the relative expression level of each type of receptor (Bailyes et al. 1997; Pandini et al. 1999). Insulin and IGF-1 differential effects in vivo reflect mostly the hormone concentration and relative expression level of receptors in different tissues rather than the capacity of IR and IGF-1R to convey different signaling (Boucher et al. 2010; Siddle 2012).

INSULIN RECEPTOR SUBSTRATES

At the time of ligand binding to the α subunits, IR and IGF-1R undergo a conformational change-inducing activation of the kinase activity in the β subunits. This results in transphosphorylation among β subunits, further activating the kinase and allowing the recruitment of receptor substrates. The best characterized substrates are members of the insulin receptor substrate (IRS) family of proteins, simply referred to as IRS-1 through IRS-6, which act as scaffolds to organize and mediate signaling complexes (Sun et al. 1991, 1995; Lavan et al. 1997a,b; Cai et al. 2003; White 2006; Shaw 2011). IRS proteins are recruited to the membrane and the activated receptors through both pleckstrin homology (PH) and phosphotyrosine binding (PTB) domains in their amino terminus (Voliovitch et al. 1995). They are subsequently phosphorylated by the activated receptors on multiple tyrosine residues that form binding sites for intracellular molecules that contain Src-homology 2 (SH2) domains (Sun et al. 1993).

Although these substrates have similar tyrosine phosphorylation motifs, they clearly have different functions in vivo. IRS-1 knockout (KO) mice show growth retardation and impaired insulin action, especially in muscle (Araki et al. 1994), but have normal glucose tolerance. IRS-2 KO mice display growth reduction in only selective tissues, such as certain neurons and islet cells, but also have defective insulin signaling in the liver, which when combined with the loss of β cells results in the development of diabetes (Withers et al. 1998). At the cellular level, IRS-1 KO preadipocytes show defects in differentiation, whereas IRS-2 KO preadipocytes differentiate normally, but have impaired insulin-stimulated glucose transport (Miki et al. 2001; Tseng et al. 2005). In skeletal muscle cells, IRS-1, but not IRS-2, is required for myoblast differentiation and glucose metabolism, whereas IRS-2 is important for lipid metabolism and ERK activation (Huang et al. 2005; Bouzakri et al. 2006).

IRS-3 and IRS-4 show a more restricted tissue distribution pattern. In rodents, IRS-3 is most abundant in adipocytes, liver, and lung (Sciacchitano and Taylor 1997), whereas in humans, the *IRS-3* gene is a pseudogene, so no protein is produced at all (Bjornholm et al. 2002). In mice, disruption of the gene for *IRS-3* alone does not result in abnormalities, but leads to a severe defect in adipogenesis when combined with deletion of IRS-1 (Laustsen et al. 2002). IRS-4 mRNA is present in skeletal muscle, liver, heart, brain, and kidney (Fantin et al. 1999), and IRS-4 KO mice show only very minimal growth retardation and glucose intolerance (Fantin et al. 2000). IRS-5 (also called DOK4) and IRS-6 (DOK5) have limited tissue expression (Cai et al. 2003) and are relatively poor IR substrates (Versteyhe et al. 2010).

In addition to the IRS proteins, the insulin and IGF-1 receptors can phosphorylate several other substrates (Siddle 2012). Shc proteins are tyrosine phosphorylated by IR and IGF-1R, and participate in the activation of the Ras/ERK pathway. Grb2-associated binder (GAB) proteins are also substrates for a variety of receptors, including IR and IGF-1R. GAB proteins resemble IRS proteins, but lack a protein tyrosine phosphatase (PTP) domain, and could play a role in insulin/IGF-1 signaling in cells expressing low IRS protein levels. APS (SHB2) and Cbl are IR/IGF-1R substrates that recruit other proteins, such as the Cbl-associated protein (CAP), to the insulin-signaling complex. The latter participates in the control of insulin-stimulated glucose uptake (Baumann et al. 2000). SH2B1 directly binds to insulin receptors and IRS proteins and enhances insulin sensitivity by promoting insulin receptor catalytic activity and by inhibiting tyrosine dephosphorylation of IRS proteins.

PHOSPHATIDYLINOSITOL (3,4,5)-TRIPHOSPHATE AND PHOSPHOINOSITIDE 3-KINASE

The critical pathway linking IRS proteins to the metabolic actions of insulin is the PI3-kinase (PI3K) and Akt pathway. The class Ia PI3-kinases are heterodimers consisting of a regulatory and catalytic subunit, each of which occurs in several isoforms (Vadas et al. 2011). Recruitment and activation of the PI3K depends on the binding of the two SH2 domains in the regulatory subunits to tyrosine-phosphorylated IRS proteins (Myers et al. 1992; Shaw 2011). This results in activation of the catalytic subunit, which rapidly phosphorylates phosphatidylinositol 4,5-bisphosphate (PIP$_2$) to generate the lipid second messenger phosphatidylinositol (3,4,5)-triphosphate (PIP$_3$). The latter recruits Akt to the plasma membrane, where it is activated by phosphorylation and induces downstream signaling.

The different isoforms of the regulatory subunit of PI3K are encoded by three distinct genes. *Pik3r1* encodes 65%–75% of all regulatory subunits, mostly in the form of p85α, but also the splice variants p55α and p50α. *Pik3r2* encodes p85β and accounts for ∼20% of the regulatory subunits. *Pik3r3* encodes p55γ, which is similar in structure to p55α, but expressed at low levels in most tissues.

The three different catalytic subunits—p110α, β, and δ—are derived from three different genes. Binding of a regulatory to a catalytic subunit increases the catalytic subunit stability and maintains it in an inhibited state. This is relieved by binding of the regulatory subunit to specific phosphotyrosine motifs in IRS proteins, resulting in its activation (Yu et al. 1998; Burke et al. 2011; Zhang et al. 2011). Liver-specific ablation of p110α, and to a lesser extent p110β, in mice results in glucose intolerance and insulin resistance (Jia et al. 2008; Sopasakis et al. 2010). Surprisingly, knockouts of the regulatory subunits of PI3K, including a heterozygous deletion of p85α, p85β KO, or p50α/p55α double KO, all display increased insulin sensitivity (Terauchi et al. 1999; Ueki et al. 2002). Different mechanisms by which reducing concentration of regulatory subunits can increase insulin action have been identified. Regulatory subunits typically are in excess concentration to catalytic subunits and thus compete with the enzymatically competent p85/p110 heterodimer for binding to IRS proteins. The p85α monomer has also been linked to regulation of the phosphatase and tensin homolog (PTEN) (Taniguchi et al. 2010). More recently, p85α has been shown to bind to the transcription factor XBP-1 and to modify the unfolded protein response, which contributes to insulin resistance (Park et al. 2010; Winnay et al. 2010).

In addition to PI3K, IRS proteins recruit other proteins potentially contributing to insulin and IGF-1 action. Proteomics analysis of the phosphotyrosine interactome of IRS-1 and IRS-2 indicates that most interacting proteins bind to both substrates, such as adaptor proteins Grb2 or Crk, or tyrosine phosphatase SHP2. However, other interaction partners seem to bind exclusively to IRS-1 (Csk) or IRS-2 (Shc, DOCK-6, and DOCK-7) (Hanke and Mann 2009).

ACTIVATION OF DOWNSTREAM KINASES

Most of the physiological effects of PI3K-generated PIP$_3$ are mediated by a subset of AGC protein kinase family members, which include isoforms of Akt/protein kinase B (PKB), p70 ribosomal S6 kinase (S6K), serum- and glucocorticoid-induced protein kinase (SGK), as well as several isoforms of protein kinase C (PKC), particularly the atypical PKCs. AGC kinase family members share similar structure and mechanisms of activation via phosphorylation of two serine and threonine residues (Pearce et al. 2010). PDK-1 (3-phosphoinositide-dependent protein kinase 1) is the major upstream kinase responsible for the phosphorylation and activation of the AGC kinase members regulated by PI3K (Bayascas 2010). PDK-1 contains a PH domain that binds to membrane-bound PIP$_3$, triggering PDK-1 activation. PDK-1 phosphorylates and activates AGC protein kinases at serine/threonine residues, such as Thr-308 for Akt (Alessi et al. 1997). However, Akt phosphorylation at Ser-473 is required for full activation, and this is accomplished by the mammalian target of rapamycin complex 2 (mTORC2) (Sarbassov et al. 2005; Oh and Jacinto 2011). DNA-dependent protein kinase (DNA PK) has also been described to phosphorylate and activate Akt in response to DNA damage (Bozulic et al. 2008), and is involved in insulin regulation of metabolic genes such as fatty acid synthase (Wong et al. 2009).

The Akt/PKB family of proteins consists of three different isoforms of serine/threonine protein kinases encoded by different genes (Schultze et al. 2011). All isoforms possess a PH domain, allowing interaction with PIP$_3$ and recruitment to the plasma membrane. Akt2 is most abundant in insulin-sensitive tissues and seems to play a predominant role in

Cite this article as *Cold Spring Harb Perspect Biol* doi: 10.1101/cshperspect.a009191

mediating insulin action on metabolism. Thus, Akt2 KO mice are insulin resistant and develop diabetes (Cho et al. 2001), whereas Akt1 and Akt3 KO mice do not.

ACTIONS OF INSULIN DOWNSTREAM FROM AKT

Activation of Akt by PDK-1 and mTORC2 allows the phosphorylation and activation of many downstream targets. Akt phosphorylates tuberous sclerosis complex protein 2 (TSC-2), inducing the degradation of the tumor suppressor complex that consists of TSC-2 and TSC-1, which activates the mTORC1 complex. Akt-induced activation of mTORC1 can also be achieved by phosphorylation of proline-rich Akt substrate 40 KDa (PRAS40), an inhibitor of mTORC1, thereby relieving the inhibition. The mTORC1 complex then phosphorylates and inhibits 4E-binding protein 1 (4E-BP1), activates ribosomal protein S6 kinases S6K1 and S6K2 and SREBP1, and leads to the regulation of a network of genes controlling metabolism, protein synthesis, and cell growth (Duvel et al. 2010).

Transcription factors of the Forkhead box O (Foxo) family control the expression of lipogenic and gluconeogenic genes. Akt phosphorylates Foxos at several sites which provides docking sites for binding proteins of the 14-3-3 family. This interaction leads to the exclusion of Foxo from the nucleus, thus blocking its transcriptional activity (Tzivion et al. 2011). Interestingly, although mice lacking Akt1 and Akt2 show severe hepatic insulin resistance and high levels of hepatic glucose production, these defects are normalized when Foxo1 is concomitantly ablated in the liver. This indicates that an additional pathway exists in the control of hepatic glucose metabolism beyond the Akt/Foxo1 axis, which allows for insulin-mediated regulation of hepatic glucose production (Lu et al. 2012).

There are multiple other substrates of Akt involved in insulin action. The GTPase-activating protein Akt substrate of 160 kDa (AS160), also called TBC1D4, and its homolog TBC1D1, are phosphorylated by Akt and are involved in insulin- and contraction-mediated glucose uptake (Sano et al. 2003; Sakamoto and Holman 2008; Taylor et al. 2008; An et al. 2010). Akt also phosphorylates and inactivates glycogen synthase kinase 3, resulting in glycogen synthase activation and glycogen accumulation in liver (Kim et al. 2004b). Akt-dependent phosphorylation of PGC-1α impairs the ability of PGC-1α to promote gluconeogenesis and fatty acid oxidation (Li et al. 2007). Phosphorylation of phosphodiesterase 3B (PDE3B) by Akt results in its activation and in a decrease in cyclic AMP levels (Kitamura et al. 1999), which plays important roles in the effect of insulin to inhibit lipolysis in adipocytes and insulin secretion in β cells (Degerman et al. 2011).

OTHER ACTIONS OF INSULIN DOWNSTREAM FROM PI3K

Akt plays a central role in mediating many other insulin actions by regulating the expression and activity of a wide range of proteins, including enzymes, transcription factors, cell cycle regulating proteins, or apoptosis and survival proteins (Manning and Cantley 2007). Murine double minute 2 (Mdm2) is phosphorylated by Akt, which inhibits p53-mediated apoptosis and contributes to tumorigenesis (Cheng et al. 2010). Akt phosphorylates cell cycle inhibitors p21Cip1/WAF1 and p27Kip1, resulting in cytoplasmic localization, cell growth, and inhibition of apoptosis (Zhou et al. 2001; Motti et al. 2004). Akt also phosphorylates and inhibits Bax, Bad, and caspase-9, which promotes cell survival (Datta et al. 1997; Cardone et al. 1998; Yamaguchi and Wang 2001; Gardai et al. 2004). Akt can phosphorylate and activate IκB kinase (IKK), leading to NF-κB activation (Bai et al. 2009). Akt phosphorylates and activates endothelial nitric oxide synthase (eNOS), which catalyzes the production of the vasodilator and anti-inflammatory molecule nitric oxide (NO), providing a potential link between insulin resistance and cardiovascular disease (Dimmeler et al. 1999; Fulton et al. 1999; Yu et al. 2011a). Although less well studied in insulin action, the serum- and glucocorticoid-induced protein family of kinases (SGK) are highly homologous to Akt, are also activated by dual phosphorylation by PDK-1 and mTORC2 in a PI3K dependent manner, and have many downstream substrates in common with Akt (Bruhn et al. 2010).

PROTEIN KINASES C

PKC isoforms are both mediators and modifiers of insulin's metabolic action. Of the three major classes of PKC, the atypical PKCs (aPKCs), PKC-ζ and PKC-λ/ι, are activated via phosphorylation by PDK-1. aPKCs play an important role in insulin-stimulated glucose transport and regulation of lipid synthesis, and their expression and/or activation is decreased in muscle from obese and diabetic humans (Farese and Sajan 2010). Both PKC-λ and PKC-ζ have been shown to function interchangeably in mediating insulin-stimulated glucose transport (Sajan et al. 2006). Muscle-specific deletion of PKC-λ in mice leads to impairment in insulin-induced glucose uptake and insulin resistance (Farese et al. 2007). Mice with liver-specific deletion of PKC-λ display decreased insulin-induced expression of SREBP1c and triglyceride content in the liver, resulting in increased insulin sensitivity in these mice (Matsumoto et al. 2003).

THE GRB2-SOS-RAS-MAPK PATHWAY

A second essential branch of the insulin/IGF-1-signaling pathway is the Grb2-SOS-Ras-MAPK pathway, which is activated independently of PI3K/Akt. Activated receptors and IRS proteins both possess docking sites for adaptor molecules that contain SH2 domains such as Grb2 and Shc. The carboxy-terminal SH3 domain of Grb2 binds to proteins such as Gab-1, whereas the amino-terminal SH3 domain binds to proline-rich regions of proteins such as son-of-sevenless (SOS). SOS is a guanine nucleotide exchange factor (GEF) for Ras, catalyzing the switch of membrane-bound Ras from an inactive, GDP-bound form (Ras-GDP) to an active, GTP-bound form (Ras-GTP). Ras-GTP then interacts with and stimulates downstream effectors, such as the Ser/Thr kinase Raf, which stimulates its downstream target MEK1 and 2 that phosphorylate and activate the MAP kinases ERK1 and 2. Stimulated ERK1/2 play

a direct role in cell proliferation or differentiation, regulating gene expression or extra-nuclear events, such as cytoskeletal reorganization, through phosphorylation and activation of targets in the cytosol and nucleus.

NEGATIVE REGULATORS OF INSULIN SIGNALING

Insulin and IGF-1 signaling are tightly controlled because uncontrolled activity of the downstream pathways could lead to severe perturbations in metabolism and tumorigenesis. Intensity and duration of the signal play an important role in determining the specificity of the response to their pleiotropic effects. Therefore, the ability to turn off the insulin signal in a rapid manner at different levels is critical (Fig. 2). On the other hand, some of these inhibitory mechanisms can be altered in pathophysiological conditions and participate in the development of insulin resistance.

Phosphoprotein Phosphatases as Negative Regulators of Insulin Action

Both cytoplasmic protein tyrosine phosphatases, such as PTP1B, and transmembrane phosphatases, such as LAR, have been shown to dephosphorylate the tyrosine residues on activated IR and IGF-1R, as well as IRS proteins, thereby reducing their activity (Goldstein et al. 1998). Although the role of LAR in the control of insulin signaling in vivo remains controversial, PTP1B is an essential component of insulin action. PTP1B KO mice show enhanced insulin sensitivity, increased IR phosphorylation in muscle and liver, and are also resistant to high-fat-diet-induced obesity and associated insulin resistance (Elchebly et al. 1999; Klaman et al. 2000).

The serine/threonine phosphatase protein phosphatase 1 (PP1) has been implicated in the regulation of several rate-limiting enzymes in both glucose and lipid metabolism, including glycogen synthase, hormone-sensitive lipase, or acetyl CoA carboxylase (Brady and Saltiel 2001). Protein phosphatase 2A (PP2A), which accounts for $\sim$80% of serine/threonine phos-

Cite this article as *Cold Spring Harb Perspect Biol* doi: 10.1101/cshperspect.a009191

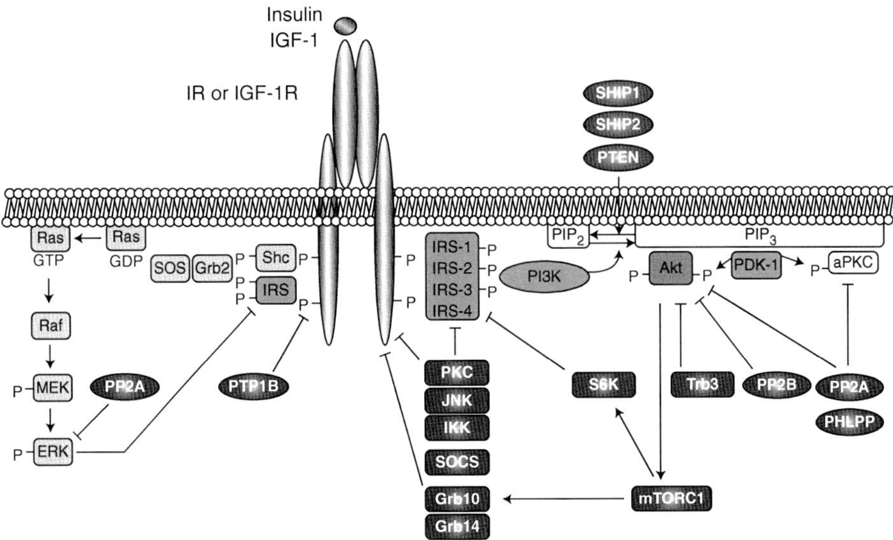

Figure 2. Negative modulators of insulin and IGF-1 signaling. Intensity and duration of insulin and IGF-1 signaling play an important role in determining the specificity and the nature of the response to these hormones. Signaling is attenuated by action of several phosphatases, which dephosphorylate the receptors, IRS proteins, PKCs, and ERK or PIP₃. In addition, stress kinases such as JNK, IKK, and ERK, as well as PKCs or S6K, inhibit insulin/IGF-1 signaling by inducing inhibitory serine/threonine phosphorylation of IR/IGFR and IRS proteins. Trb3 inhibits Akt, and adaptor proteins such as SOCS and Grb bind to the receptors and IRS proteins and inhibit signaling by competition.

phatase activity in cells, also regulates the activities of many protein kinases involved in insulin action, including Akt, PKC, S6K, ERK, cyclin-dependent kinases, and IKK (Millward et al. 1999). Several studies indicate that PP2A is hyperactivated in diabetic states (Kowluru and Matti 2012).

Other serine/threonine phosphatases have been implicated in insulin action. Protein phosphatases 2B (PP2B), also known as calcineurin, has been shown to dephosphorylate Akt (Ni et al. 2007). Two novel members of the PP2C family involved in regulation of insulin action are the PH domain leucine-rich repeat protein phosphatases PHLPP-1 and -2, which dephosphorylate both Akt and PKCs (Brognard and Newton 2008). Overexpression of PHLPP1 in cells impairs Akt and glycogen synthase kinase 3 activity, resulting in decreased glycogen synthesis and glucose transport (Andreozzi et al. 2011). Elevated levels of PHLPP1 have been found in adipose tissue and skeletal muscle of obese and/or diabetic patients and correlate

with decreased Akt2 phosphorylation (Cozzone et al. 2008; Andreozzi et al. 2011).

Lipid Phosphatases as Negative Regulators of Insulin Action

Lipid phosphatases can regulate insulin signaling by modulating PIP₃ levels. PTEN dephosphorylates PIP₃, thus antagonizing PI3K signaling in cells (Cantley and Neel 1999; Carracedo and Pandolfi 2008). Muscle, adipose tissue, or liver-specific deletion of PTEN in mice increases insulin sensitivity (Stiles et al. 2004; Kurlawalla-Martinez et al. 2005; Wijesekara et al. 2005), and mice with whole-body PTEN haploinsufficiency show improved glucose tolerance and increased insulin sensitivity (Wong et al. 2007). Interestingly, the p85α regulatory subunit of PI3K has recently been shown to bind directly to and enhance PTEN activity, creating a unique interface between the generation and degradation of PIP₃ (Taniguchi et al. 2006b; Chagpar et al. 2010).

SH2 domain-containing inositol 5-phosphatases (SHIP) 1 and 2 also dephosphorylate PIP_3. SHIP1 expression is restricted to hematopoietic cells, whereas SHIP2 is ubiquitously expressed and plays a role in insulin signaling (Suwa et al. 2010). SHIP2 deficiency in mice results in hypoglycemia, enhanced insulin-induced Akt activation, and resistance to high-fat-diet-induced obesity, indicating that SHIP2 is a key regulator of glucose and energy homeostasis in vivo (Clement et al. 2001; Sleeman et al. 2005). Conversely, SHIP2-overexpressing mice show reduced insulin-induced Akt activation in the liver, fat, and skeletal muscle (Kagawa et al. 2008).

Other Negative Modulators (Grb, SOCS, Trb3, IP7)

Grb10 and Grb14 are cytoplasmic adaptor proteins that decrease IR and to a lesser extent IGF-1R activity, and prevent access of substrates to the activated receptors (Holt and Siddle 2005). Deletion of the *Grb10* gene in mice leads to increased growth, enhanced insulin signaling, and increased glucose tolerance (Smith et al. 2007; Wang et al. 2007). Grb10 overexpression, on the other hand, results in impaired growth, glucose intolerance, and insulin resistance (Shiura et al. 2005). Grb14 expression is increased in adipose tissue of insulin-resistant animal models and type-2 diabetic patients (Cariou et al. 2004), and Grb14 KO mice display increased glucose tolerance and insulin sensitivity, consistent with an inhibitory role of Grb14 on insulin signaling (Cooney et al. 2004). Grb10 and Grb14 share similar mechanisms as insulin signaling is not further increased in mice with deletion of both proteins (Holt et al. 2009).

Proteins of the suppressor of cytokine signaling (SOCS) family are adaptor proteins that act as negative regulators of cytokine and growth factor signaling. In addition, SOCS proteins, in particular SOCS1 and SOCS3, negatively regulate insulin signaling and thus link cytokine signaling to insulin resistance. Their expression is increased in obesity, and they induce insulin resistance via either inhibition of the tyrosine kinase activity of the IR, competi-

tion for binding of the IRS proteins to the receptor, or targeting the IRS proteins to degradation (Emanuelli et al. 2000, 2001; Rui et al. 2002; Ueki et al. 2004a,b; Sachithanandan et al. 2010).

Tribbles homolog 3 (Trb3) is a member of the family of pseudokinases that is thought to function as adaptor proteins. Trb3 expression is induced in liver in fasting and diabetes, and disrupts insulin signaling by binding to Akt and blocking its activation. Trb3 knockdown in mice improves glucose tolerance (Du et al. 2003; Koo et al. 2004). In cultured cells, insulin-stimulated S6K activation is decreased when Trb3 is overexpressed, and increased when Trb3 levels are reduced (Matsushima et al. 2006). Trb3 action in adipose tissue seems to be independent of Akt. Thus, whereas insulin promotes lipogenesis, Trb3 stimulates lipolysis by triggering the ubiquitination and degradation of acetyl-CoA carboxylase. Transgenic mice overexpressing Trb3 in adipose tissue are protected from diet-induced obesity because of enhanced fatty acid oxidation and display increased insulin sensitivity (Qi et al. 2006).

A novel negative regulator of insulin signaling is the inositol phosphate IP7. It was recently shown that insulin and IGF-1 increase IP7 levels, which in turn inhibits Akt translocation to the plasma membrane and subsequent activation, creating a potential feedback mechanism that attenuates insulin signaling (Chakraborty et al. 2010). Deletion of the enzyme that catalyzes IP7 formation in mice causes increased insulin responsiveness. Further studies will be needed to elucidate the contribution of this pathway in normal or pathological conditions.

Regulation by Inhibitory Serine and Threonine Phosphorylation

Tyrosine phosphorylation is essential for IR/IGF-1R and IRS activation. On the other hand, serine and threonine phosphorylation of the receptors or IRS proteins is primarily involved in turning the insulin signal down (Fig. 3). Increased inhibitory Ser/Thr phosphorylation of IR and especially IRS-1 and -2 occurs in response to cytokines, fatty acids, hyperglycemia,

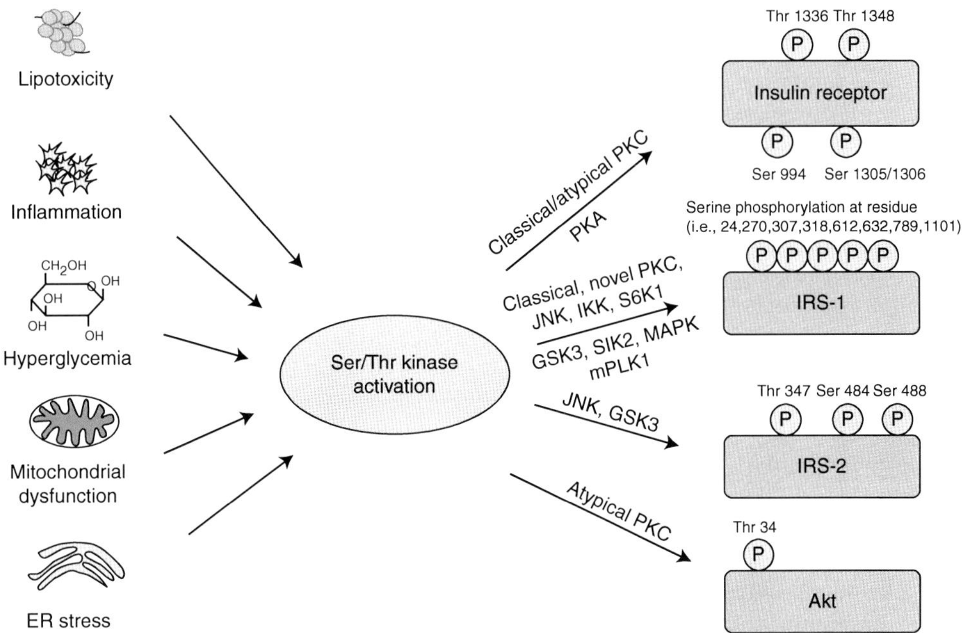

Figure 3. Activation of Ser/Thr kinases causes inhibitory phosphorylation on insulin-signaling molecules. Lipotoxicity, inflammation, hyperglycemia, and subsequently oxidative stress, as well as mitochondrial dysfunction and ER stress, all converge on activation of Ser/Thr kinases, inducing inhibitory Ser/Thr phosphorylation of IR, IRS proteins, and Akt on multiple residues, causing insulin resistance.

mitochondrial dysfunction, and ER stress, and insulin itself via activation of multiple kinases, predominantly by c-Jun amino-terminal kinase (JNK), IKK, conventional and novel PKCs, but also mTORC1/S6K and MAPK (De Fea and Roth 1997a; Aguirre et al. 2000; Gao et al. 2002; Gual et al. 2003; Li et al. 2004; Zhang et al. 2008a; Boura-Halfon and Zick 2009). Increased IR serine phosphorylation associated with decreased tyrosine kinase activity has been observed in insulin-resistant states, both in rodents and in humans (Karasik et al. 1990; Dunaif et al. 1995; Zhou et al. 1999; Shao et al. 2000). An increase in cAMP concentration also induces inhibitory serine phosphorylation of IR in a PKA-dependent manner (Stadtmauer and Rosen 1986; Roth and Beaudoin 1987).

Although inhibitory IRS-1 serine phosphorylation occurs at many different sites (Boura-Halfon and Zick 2009), the best studied of these modifications occurs at Ser-307 (Aguirre et al. 2002). IRS-1 Ser-307 phosphorylation is increased in obese and diabetic mice (Hirosumi

et al. 2002; Um et al. 2004). Although this is widely believed to contribute to insulin resistance by inhibiting insulin receptor kinase activity, recent studies have made this association less clear. Thus, insulin itself can stimulate phosphorylation of IRS-1 on Ser-307 in humans (Yi et al. 2007), and mice with a knock-in of IRS-1 Ser307Ala mutant developed more severe insulin resistance than control mice when fed a high-fat diet, indicating that Ser-307 is required to maintain normal insulin signaling (Copps et al. 2010). Thus, increased IRS-1 Ser-307 phosphorylation observed in insulin-resistance states may be associated with, but not cause, insulin resistance.

Lipids, through their metabolic product diacylglycerols, can activate classical (α, β, γ) and novel PKC members (δ, θ, ε) and impair insulin signaling by inducing multiple serine phosphorylation of IRS proteins and IR specifically at Thr-1336, Thr-1348, and Ser-1305/1306 (Bollag et al. 1986; Karasik et al. 1990; Lewis et al. 1990; Chin et al. 1993; De Fea and Roth

1997b; Turban and Hajduch 2011). Thus, deletion of any member of the novel PKC family prevents the development of insulin resistance in skeletal muscle and liver by decreasing IRS-1 Ser-307 phosphorylation (Kim et al. 2004a; Samuel et al. 2007; Mack et al. 2008; Bezy et al. 2011). Atypical PKC-ζ also inhibits insulin signaling by inducing serine phosphorylation of IRS-1 (Ravichandran et al. 2001) and Thr-34 phosphorylation of Akt, thereby inhibiting its recruitment to the plasma membrane (Powell et al. 2003, 2004).

Another component of the negative feedback loops in insulin signaling is mTORC1. Activation of mTOR and S6K is not only downstream from insulin signaling, but also inhibits it by increasing serine phosphorylation and reducing IRS tyrosine phosphorylation. This is illustrated by the phenotype of S6K null mice, which are lean and display enhanced insulin sensitivity (Um et al. 2004). In addition, IRS-1 is hyperphosphorylated and degraded in TSC-2 KO fibroblasts, which show constitutive S6K activation (Harrington et al. 2004; Shah et al. 2004). mTORC1 also mediates phosphorylation and stabilization of Grb10, leading to feedback inhibition of insulin signaling (Hsu et al. 2011; Yu et al. 2011b).

MECHANISMS OF INSULIN RESISTANCE

A central feature of type-2 diabetes is insulin resistance, a condition in which cells cannot respond properly to insulin. This occurs primarily at the level of so-called insulin-sensitive tissues, such as liver, muscle, and fat, and can be caused by multiple mechanisms (Fig. 3 and Table 1).

Genetic Causes of Insulin Resistance

Insulin Receptor

Mutations in the insulin receptor gene have been identified in several rare forms of severe insulin resistance, including leprechaunism, Rabson-Mendenhall syndrome, or the type-A syndrome of insulin resistance. These patients often require a hundredfold or more insulin than a typical diabetic patient (Kahn et al. 1976; Cochran et al. 2005). Most of these patients have nonsense or missense mutations in the extracellular ligand-binding domain or intracellular tyrosine kinase domain of the receptor, which leads to severely reduced insulin binding, altered kinetics of insulin binding, or reduced tyrosine kinase activity, but some also have presumed promoter defects leading to reduced receptor mRNA expression (Taylor et al. 1991; Haruta et al. 1995). Insulin receptor mutations have not been observed in patients with routine type-2 diabetes (T2D).

Insulin Receptor Substrate Proteins

The G972R polymorphism of IRS-1 is observed with higher frequency in patients with T2D and leads to decreased insulin signaling, mostly decreasing PI3K activity (Almind et al. 1996; Hribal et al. 2008). Although this finding has not been confirmed in all large-scale population analyses (Florez et al. 2004; van Dam et al. 2004), recent studies have continued to show an association between a single-nucleotide polymorphism (SNP) in IRS-1 and T2D (Burguete-Garcia et al. 2010; Martinez-Gomez et al. 2011). A T608R missense mutation in IRS-1 resulting in decreased insulin signaling has been reported in a patient with T2D, but appears to be very rare (Esposito et al. 2003). Numerous polymorphisms have been identified in the human *IRS-2* gene, but a clear association between these polymorphisms and T2D has not been found (Bernal et al. 1998).

Phosphoinositide 3-Kinase

An M326I polymorphism in the p85α regulatory subunit of the PI3K was identified in Pima Indian women and is associated with decreased prevalence for T2D (Baier et al. 1998). However, this M326I mutation only has modest effects on insulin signaling in vitro by decreasing p85α binding to IRS-1 and increasing p85α degradation (Almind et al. 2002). Another polymorphism in p85α (SNP42) is associated with fasting hyperglycemia, but its molecular mechanism so far remains elusive (Barroso et al. 2003).

Table 1. Molecular mechanisms of insulin resistance

Cause	Mechanism	Effect	References
Lipotoxicity Inflammation Hyperglycemia Mitochondrial dysfunction ER stress	Activation of Ser/ Thr kinases	Inhibitory phosphorylation of insulin-signaling molecules	Bollag et al. 1986; Karasik et al. 1990; Lewis et al. 1990; Chin et al. 1993; Powell et al. 2003; Boura-Halfon and Zick, 2009
Genetic mutations	Point mutations in IR and insulin-signaling molecules	Increased protein turnover Reduced expression and ligand affinity Decreased signaling capacities	Kahn et al. 1976; Taylor et al. 1991; Haruta et al. 1995; Almind et al. 1996; George et al. 2004; Prudente et al. 2005; Hribal et al. 2008; Dash et al. 2009; Prudente et al. 2009
	SNP causing increased gene expression	Increased PTEN action leading to reduced PIP_3 levels	Ishihara et al. 2003
Lipotoxicity	Hyperactivation of protein phosphatase PP2A	Reduced phosphorylation of IR and insulin-signaling molecules	Kowluru and Matti 2012
Inflammation	Cytokine-induced SOCS3 activation	Inhibition or IR tyrosine kinase activity Competition for IRS binding to IR Increased IRS degradation	Emanuelli et al. 2000, 2001; Rui et al. 2002; Ueki et al. 2004a,b; Steppan et al. 2005
	Cytokine-induced reduction in gene expression	Decreased expression of insulin-signaling molecules	Rotter et al. 2003; Jager et al. 2007
Hyperglycemia	Glycation of insulin-signaling molecules	Reduced affinity for IR Decreased DNA-binding capacities of transcription factors	Federici et al. 1999; Riboulet-Chavey et al. 2006; Housley et al. 2008
	Hyperactivation of protein phosphatase PP2A	Reduced phosphorylation of IR and insulin-signaling molecules	Kowluru and Matti 2012
Hyperinsulinemia	Hyperactivation of PHLPP1 and Grb14	Decreased AKT Ser 473 phosphorylation Competition for IRS binding to IR	Cariou et al. 2004; Cozzone et al. 2008; Andreozzi et al. 2011

Multiple molecular mechanisms of insulin resistance have been described, in addition to inhibitory Ser/Thr phosphorylation on insulin-signaling molecules (Fig. 3). Genetic mutations, dephosphorylation events, posttranslational modifications, and formation of inhibitory complexes have all been shown to cause insulin resistance.

Phosphatase and Tensin Homolog

In diabetes, mutations of PTEN have not been reported yet. However, three Japanese type-2 diabetic subjects have been identified with polymorphisms in the *PTEN* gene, one of which was associated with T2D. This SNP caused a higher expression rate of PTEN and reduced insulin-induced Akt activation in cells (Ishihara et al. 2003). Very recently, it has been found that individuals with PTEN haploinsufficiency are both obese and insulin sensitive, with a decreased risk of T2D but increased risk of cancer (Pal et al. 2012). The impact of this in the general population is unknown.

AKT and Related Targets

A rare missense mutation (R274H) in Akt2 leading to loss of kinase activity has been identified in a patient with diabetes (George et al. 2004). Two other missense mutations (R208 K and R467W) have also been identified in diabetic patients, but surprisingly, these mutant forms display unaltered insulin-stimulated kinase activities in vitro (Tan et al. 2007). In type-2 diabetic patients, a gain-of-function mutation (Q84R) in Trb3 has been associated with insulin resistance and decreased insulin-stimulated Akt phosphorylation (Prudente et al. 2005, 2009). A mutation in AS160 at position 363, resulting in a premature stop codon, was identified in a patient with severe postprandial hyperinsulinemia, and acts in a dominant-negative manner to reduce glucose transport (Dash et al. 2009).

Lipotoxicity

One feature of metabolic syndrome is ectopic accumulation of lipids, especially fatty acids (FA), which is believed to cause insulin resistance via multiple mechanisms. Tissue-specific increase in lipid content in nonadipose tissues provides direct evidence of lipotoxicity. Increased hydrolysis of circulating triglycerides owing to muscle-specific overexpression of lipoprotein lipase leads to skeletal muscle insulin resistance (Ferreira et al. 2001), whereas increased lipid transport in heart or liver leads to lipotoxic cardiomyopathy and nonalcoholic fatty liver disease, respectively (Chiu et al. 2005; Koonen et al. 2007). Besides the effect of increased lipid flux on insulin sensitivity, multiple lipid intermediates have been shown to promote insulin resistance.

Elevated circulating free fatty acids (FFA) are observed in obesity and induce activation of JNK, IKK, and PKC and IRS-1 Ser-307 phosphorylation (Schenk et al. 2008). The fatty acid palmitate plays a particular role in promoting insulin resistance as it induces endoplasmic reticulum (ER) stress, cytokine production, and activates JNK (Ozcan et al. 2004; Shi et al. 2006). In addition, palmitate activates NF-κB

signaling while inhibition of this pathway reverses lipid-induced insulin resistance (Kim et al. 2001a; Sinha et al. 2004). Interestingly, the detrimental effect of palmitate on skeletal muscle insulin resistance can be reversed by coinfusion with oleate, thereby changing its conversion from phospholipids and diacylglycerol (DAG) to triglycerides (Peng et al. 2011). This indicates that FFA induces insulin resistance through multiple mechanisms, and combinations of FA can influences insulin signaling and highlight the crucial interplay of lipids with respect to dietary interventions.

The lipid metabolite DAG has also been shown to induce insulin resistance. Increased muscle DAG (intramyocellular lipid) leads to muscle insulin resistance by activating PKC-θ and inducing IRS-1 Ser-307 phosphorylation (Yu et al. 2002). Conversely, reducing DAG levels in skeletal muscle and liver protects mice against high-fat-diet-induced insulin resistance (Liu et al. 2007; Ahmadian et al. 2009; Samuel et al. 2010).

Increased plasma concentration of the sphingolipid ceramide is observed in obese and diabetic patients and is associated with severe insulin resistance (Haus et al. 2009). Ceramide has been shown to induce insulin resistance via PKC and JNK activation (Westwick et al. 1995; Schenk et al. 2008) and, thus, inhibition of ceramide synthesis ameliorates insulin resistance (Holland et al. 2007). Ceramides also inhibit Akt activation by increasing the interaction of PP2A with Akt, and phosphorylation of Akt at Thr-34 by PKC ζ, resulting in reduced binding of PIP_3 to Akt (Teruel et al. 2001; Powell et al. 2003; Blouin et al. 2010).

In addition to effects on kinases, alteration of membrane–lipid composition affects insulin signaling. An increase in the saturated-to-unsaturated FA ratio is observed in type-2 diabetic patients and is thought to reduce membrane fluidity and insulin sensitivity (Field et al. 1990; Bakan et al. 2006). Moreover, an increase in the phosphatidylcholine (PC) to phosphatidylethanolamine (PE) ratio in endoplasmic reticulum leads to the activation of ER stress and is associated with insulin resistance (Fu et al. 2011).

Inflammation

Obesity is characterized by the development of a chronic low-grade inflammatory state, which is considered a key component in promoting obesity-associated insulin resistance (Osborn and Olefsky 2012). Adipose tissue expansion occurs in response to caloric overload, and is associated with an increase in immune cell infiltration and a subsequent proinflammatory response (Sun et al. 2011). Two cell types are especially important in this scenario: adipocytes and macrophages, both of them capable of secreting proinflammatory cytokines and inducing insulin resistance. Increased secretion of the chemokine MCP-1 by adipocytes drives macrophage accumulation into adipose tissues and induces insulin resistance (Kamei et al. 2006). Deletion of MCP-1 or its receptor CCR2 improves insulin sensitivity and ameliorates inflammation in mice (Kanda et al. 2006; Weisberg et al. 2006). Increased secretion of cytokines, such as TNF-α, IL1β, or IL-6, by both immune cells and adipocytes is observed with obesity and induces insulin resistance via multiple mechanisms, including activation of Ser/Thr kinases (Ozes et al. 2001; Yuan et al. 2001; Hirosumi et al. 2002; Zhang et al. 2008a; Fan et al. 2010), decreasing IRS-1, GLUT4, and PPARγ expression (Rotter et al. 2003; Jager et al. 2007), or activation of SOCS3 in adipocytes (Steppan et al. 2005).

Another driving factor in obesity-associated inflammation is caused by activation of Toll-like receptor (TLR), especially activation of TLR-2 and -4. TLRs belong to the innate immune system and are generally activated by pathogen-associated molecular patterns such as LPS, and induce inflammation via activation of the NF-κB pathway (Akira and Takeda 2004). TLRs are ubiquitously expressed and TLR-4 is elevated in skeletal muscle (Reyna et al. 2008) and adipose tissue (Shi et al. 2006) with obesity. Interestingly, saturated FA can also activate this pathway (Lee et al. 2001; Shi et al. 2006), indicating a potential role for these receptors in obesity-driven inflammation. Thus, mice with reduced TLR-2- or TLR-4-signaling proteins (Shi et al. 2006; Kleinridders et al. 2009; Himes

and Smith 2010) are protected from obesity and obesity-associated insulin resistance.

Negative Regulation by Hyperglycemia

Glucose itself, at supraphysiological concentrations, is able to alter insulin sensitivity in muscle and fat, as well as decrease insulin secretion from β cells (Leahy et al. 1986; Hager et al. 1991). Hyperglycemia induced by decreased glucose transport in skeletal muscle impairs adipose and hepatic insulin action (Zisman et al. 2000; Kim et al. 2001b) and induces insulin resistance through several pathways, which are all believed to be linked to oxidative stress (Evans et al. 2005). Advanced glycosylation end products (AGE) inhibit insulin signaling by increasing Ser-307 phosphorylation of IRS-1 and forming methylglyoxal-IRS-1 adducts (Riboulet-Chavey et al. 2006).

Hyperglycemia increases the flux through the polyol pathway, which causes JNK activation and increases the hexosamine-biosynthetic pathway. This has been shown to promote insulin resistance in adipose tissue, skeletal muscle, liver, and pancreas in part by O-GlcNAcylation of IRS proteins (Marshall et al. 1991; Patti et al. 1999; McClain 2002; McClain et al. 2002). Furthermore, hyperglycemia also leads to O-GlcNAcylation of IR, which impairs receptor dimerization (Federici et al. 1999), and of Foxo1 leading to increased gluconeogenic gene expression (Housley et al. 2008).

Hyperglycemia also activates the PKC pathway by inducing de novo synthesis of DAG (Xia et al. 1994) and causes insulin resistance by forming a multimolecular complex, including receptor of AGE/IRS-1/Src, thereby activating PKC-α and increasing IRS-1 Ser-307 phosphorylation (Miele et al. 2003; Cassese et al. 2008).

Mitochondrial Dysfunction and ROS Formation

Although low levels of reactive oxygen species (ROS) can enhance insulin action (Krieger-Brauer et al. 1992; Mahadev et al. 2001), high concentration of ROS causes oxidative stress when unresolved. ROS formation occurs as a by-product of the electron transport chain and is a major consequence of mitochondrial dysfunction (Chang and Chuang 2010). Increased ROS levels have been observed in obese and diabetic states and can be caused by an increased metabolite flux into mitochondria, alterations in mitochondrial proteins, and reduced expression of antioxidant enzymes (West 2000; Rosen et al. 2001; Evans et al. 2005; Fridlyand and Philipson 2006). Increased oxidative stress leads to the activation of stress kinases that induce insulin resistance by serine phosphorylation of IRS proteins (Rudich et al. 1998; Evans et al. 2005; Dokken et al. 2008). Besides the aspect of ROS-mediated insulin resistance, altered mitochondrial dynamics in the form of increased mitochondrial fission leads to insulin resistance and can be rescued by inhibiting fission, which decreases the activity of p38 MAP kinase and increases IRS-1 and Akt activation (Jheng et al. 2012). Impairment of mitochondrial FA oxidation in liver can also lead to elevated DAG content, resulting in PKC-ε activation and decreased IRS-2 phosphorylation and PI3-kinase activity (Koh et al. 2005; Zhang et al. 2007).

ER Stress

The ER stress response, also known as unfolded protein response (UPR), is an adaptive process to ensure proper protein folding, maturation, and quality control in the ER. The three crucial pathways of the UPR (PERK, IRE1α, and ATF6) are all activated with obesity and act together to reduce the burden of unfolded proteins (Hotamisligil 2010). Obese mice display enhanced PERK and IRE1α activity in adipose tissue and liver, causing JNK and IKK activation and insulin resistance by phosphorylation of IRS-1 on Ser-307 (Ozcan et al. 2004, 2009; Hu et al. 2006; Zhang et al. 2008b). The transcription factor XBP-1 is activated by splicing during ER stress and increases gene expression of molecular chaperones to restore ER homeostasis. Overexpression of spliced XBP-1 reduces ER stress response, decreases activation of JNK, and increases insulin signaling by decreasing IRS-1 serine phosphorylation (Ozcan et al. 2004).

CONCLUDING REMARKS

Insulin and IGF-1 acting via specific tyrosine kinase receptors propagate signals via two main branches: the PI3K-PDK-1-Akt and the Grb2-SOS-Ras-MAPK pathways that control proliferation, differentiation, and survival at the cellular level, and growth and metabolism in organisms. These signaling pathways contain several points of regulation, signal divergence, and cross talk with other signaling cascades that define critical nodes (Taniguchi et al. 2006a). The complexity of this signaling system is essential to mediate the variety of insulin and IGF-1 biological responses. Many steps are negatively regulated by action of phosphatases or inhibitory proteins. One of the great challenges remaining is deciphering the complexity of insulin-resistance pathogenesis. Causes of insulin resistance are numerous and the mechanisms are multifactorial. In rare cases, the cause is genetic, but in most others, insulin resistance is triggered by cellular perturbations, such as lipotoxicity, inflammation, glucotoxicity, mitochondrial dysfunction, and ER stress, which lead to deregulation of genes and inhibitory protein modifications, resulting in impaired insulin and IGF-1 action. Identifying new molecules that impact insulin signaling and new levels of control, as well as better understanding the causes and mechanisms leading to insulin resistance, will be essential for a more effective treatment of type-2 diabetes and associated diseases.

REFERENCES

Aguirre V, Uchida T, Yenush L, Davis R, White MF. 2000. The c-Jun NH(2)-terminal kinase promotes insulin resistance during association with insulin receptor substrate-1 and phosphorylation of Ser(307). *J Biol Chem* **275**: 9047–9054.

Aguirre V, Werner ED, Giraud J, Lee YH, Shoelson SE, White MF. 2002. Phosphorylation of Ser307 in insulin receptor substrate-1 blocks interactions with the insulin receptor and inhibits insulin action. *J Biol Chem* **277**: 1531–1537.

Ahmadian M, Duncan RE, Varady KA, Frasson D, Hellerstein MK, Birkenfeld AL, Samuel VT, Shulman GI, Wang Y, Kang C, et al. 2009. Adipose overexpression of desnutrin promotes fatty acid use and attenuates diet-induced obesity. *Diabetes* **58**: 855–866.

Akira S, Takeda K. 2004. Toll-like receptor signalling. *Nat Rev Immunol* **4**: 499–511.

Alessi DR, James SR, Downes CP, Holmes AB, Gaffney PR, Reese CB, Cohen P. 1997. Characterization of a 3-phosphoinositide-dependent protein kinase, which phosphorylates and activates protein kinase Bα. *Curr Biol* **7:** 261–269.

Almind K, Inoue G, Pedersen O, Kahn CR. 1996. A common amino acid polymorphism in insulin receptor substrate-1 causes impaired insulin signaling. Evidence from transfection studies. *J Clin Invest* **97:** 2569–2575.

Almind K, Delahaye L, Hansen T, Van Obberghen E, Pedersen O, Kahn CR. 2002. Characterization of the Met326Ile variant of phosphatidylinositol 3-kinase p85α. *Proc Natl Acad Sci* **99:** 2124–2128.

An D, Toyoda T, Taylor EB, Yu H, Fujii N, Hirshman MF, Goodyear LJ. 2010. TBC1D1 regulates insulin- and contraction-induced glucose transport in mouse skeletal muscle. *Diabetes* **59:** 1358–1365.

Andreozzi F, Procopio C, Greco A, Mannino GC, Miele C, Raciti GA, Iadicicco C, Beguinot F, Pontiroli AE, Hribal ML, et al. 2011. Increased levels of the Akt-specific phosphatase PH domain leucine-rich repeat protein phosphatase (PHLPP)-1 in obese participants are associated with insulin resistance. *Diabetologia* **54:** 1879–1887.

Araki E, Lipes MA, Patti ME, Bruning JC, Haag B III, Johnson RS, Kahn CR. 1994. Alternative pathway of insulin signalling in mice with targeted disruption of the IRS-1 gene. *Nature* **372:** 186–190.

Bai D, Ueno L, Vogt PK. 2009. Akt-mediated regulation of NF-κB and the essentialness of NF-κB for the oncogenicity of PI3K and Akt. *Int J Cancer* **125:** 2863–2870.

Baier LJ, Wiedrich C, Hanson RL, Bogardus C. 1998. Variant in the regulatory subunit of phosphatidylinositol 3-kinase (p85α): Preliminary evidence indicates a potential role of this variant in the acute insulin response and type 2 diabetes in Pima women. *Diabetes* **47:** 973–975.

Bailyes EM, Nave BT, Soos MA, Orr SR, Hayward AC, Siddle K. 1997. Insulin receptor/IGF-I receptor hybrids are widely distributed in mammalian tissues: Quantification of individual receptor species by selective immunoprecipitation and immunoblotting. *Biochem J* **327:** 209–215.

Bakan E, Yildirim A, Kurtul N, Polat MF, Dursun H, Cayir K. 2006. Effects of type 2 diabetes mellitus on plasma fatty acid composition and cholesterol content of erythrocyte and leukocyte membranes. *Acta Diabetol* **43:** 109–113.

Barroso I, Luan J, Middelberg RP, Harding AH, Franks PW, Jakes RW, Clayton D, Schafer AJ, O'Rahilly S, Wareham NJ. 2003. Candidate gene association study in type 2 diabetes indicates a role for genes involved in β-cell function as well as insulin action. *PLoS Biol* **1:** E20.

Baumann CA, Ribon V, Kanzaki M, Thurmond DC, Mora S, Shigematsu S, Bickel PE, Pessin JE, Saltiel AR. 2000. CAP defines a second signalling pathway required for insulin-stimulated glucose transport. *Nature* **407:** 202–207.

Bayascas JR. 2010. PDK1: The major transducer of PI 3-kinase actions. *Curr Top Microbiol Immunol* **346:** 9–29.

Belfiore A, Frasca F, Pandini G, Sciacca L, Vigneri R. 2009. Insulin receptor isoforms and insulin receptor/insulin-like growth factor receptor hybrids in physiology and disease. *Endocr Rev* **30:** 586–623.

Benyoucef S, Surinya KH, Hadaschik D, Siddle K. 2007. Characterization of insulin/IGF hybrid receptors: Contributions of the insulin receptor L2 and Fn1 domains and the alternatively spliced exon 11 sequence to ligand binding and receptor activation. *Biochem J* **403:** 603–613.

Bernal D, Almind K, Yenush L, Ayoub M, Zhang Y, Rosshani L, Larsson C, Pedersen O, White MF. 1998. IRS-2 amino acid polymorphisms are not associated with random type 2 diabetes amoung caucasians. *Diabetes* **47:** 976–979.

Bezy O, Tran TT, Pihlajamaki J, Suzuki R, Emanuelli B, Winnay J, Mori MA, Haas J, Biddinger SB, Leitges M, et al. 2011. PKCδ regulates hepatic insulin sensitivity and hepatosteatosis in mice and humans. *J Clin Invest* **121:** 2504–2517.

Bjornholm M, He AR, Attersand A, Lake S, Liu SCH, Lienhard GE, Taylor S, Arner P, Zierath JR. 2002. Absence of functional insulin receptor substrate-3 (IRS-3) gene in humans. *Diabetologia* **45:** 1697–1702.

Blouin CM, Prado C, Takane KK, Lasnier F, Garcia-Ocana A, Ferre P, Dugail I, Hajduch E. 2010. Plasma membrane subdomain compartmentalization contributes to distinct mechanisms of ceramide action on insulin signaling. *Diabetes* **59:** 600–610.

Bollag GE, Roth RA, Beaudoin J, Mochly Rosen D, Koshland DE Jr. 1986. Protein kinase C directly phosphorylates the insulin receptor in vitro and reduces its protein-tyrosine kinase activity. *Proc Natl Acad Sci* **83:** 5822–5824.

Boucher J, Tseng YH, Kahn CR. 2010. Insulin and insulin-like growth factor-1 receptors act as ligand-specific amplitude modulators of a common pathway regulating gene transcription. *J Biol Chem* **285:** 17235–17245.

Boura-Halfon S, Zick Y. 2009. Phosphorylation of IRS proteins, insulin action, and insulin resistance. *Am J Physiol Endocrinol Metab* **296:** E581–E591.

Bouzakri K, Zachrisson A, Al Khalili L, Zhang BB, Koistinen HA, Krook A, Zierath JR. 2006. siRNA-based gene silencing reveals specialized roles of IRS-1/Akt2 and IRS-2/Akt1 in glucose and lipid metabolism in human skeletal muscle. *Cell Metab* **4:** 89–96.

Bozulic L, Surucu B, Hynx D, Hemmings BA. 2008. PKBα/Akt1 acts downstream of DNA-PK in the DNA double-strand break response and promotes survival. *Mol Cell* **30:** 203–213.

Brady MJ, Saltiel AR. 2001. The role of protein phosphatase-1 in insulin action. *Recent Prog Horm Res* **56:** 157–173.

Brognard J, Newton AC. 2008. PHLiPPing the switch on Akt and protein kinase C signaling. *Trends Endocrinol Metab* **19:** 223–230.

Bruhn MA, Pearson RB, Hannan RD, Sheppard KE. 2010. Second AKT: The rise of SGK in cancer signalling. *Growth Factors* **28:** 394–408.

Burguete-Garcia AI, Cruz-Lopez M, Madrid-Marina V, Lopez-Ridaura R, Hernandez-Avila M, Cortina B, Gomez RE, Velasco-Mondragon E. 2010. Association of Gly972Arg polymorphism of IRS1 gene with type 2 diabetes mellitus in lean participants of a national health survey in Mexico: A candidate gene study. *Metabolism* **59:** 38–45.

Burke JE, Vadas O, Berndt A, Finegan T, Perisic O, Williams RL. 2011. Dynamics of the phosphoinositide 3-kinase p110δ interaction with p85α and membranes reveals aspects of regulation distinct from p110α. *Structure* **19:** 1127–1137.

Cai D, Dhe-Paganon S, Melendez PA, Lee J, Shoelson SE. 2003. Two new substrates in insulin signaling, IRS5/DOK4 and IRS6/DOK5. *J Biol Chem* **278**: 25323–25330.

Cantley LC, Neel BG. 1999. New insights into tumor suppression: PTEN suppresses tumor formation by restraining the phosphoinositide 3-kinase/AKT pathway. *Proc Natl Acad Sci* **96**: 4240–4245.

Cardone MH, Roy N, Stennicke HR, Salvesen GS, Franke TF, Stanbridge E, Frisch S, Reed JC. 1998. Regulation of cell death protease caspase-9 by phosphorylation. *Science* **282**: 1318–1321.

Cariou B, Capitaine N, Le M, Vega VN, Bereziat V, Kergoat M, Laville M, Girard J, Vidal H, Burnol AF. 2004. Increased adipose tissue expression of Grb14 in several models of insulin resistance. *FASEB J* **18**: 965–967.

Carracedo A, Pandolfi PP. 2008. The PTEN-PI3K pathway: Of feedbacks and cross-talks. *Oncogene* **27**: 5527–5541.

Cassese A, Esposito I, Fiory F, Barbagallo AP, Paturzo F, Mirra P, Ulianich L, Giacco F, Iadicicco C, Lombardi A, et al. 2008. In skeletal muscle advanced glycation end products (AGEs) inhibit insulin action and induce the formation of multimolecular complexes including the receptor for AGEs. *J Biol Chem* **283**: 36088–36099.

Chagpar RB, Links PH, Pastor MC, Furber LA, Hawrysh AD, Chamberlain MD, Anderson DH. 2010. Direct positive regulation of PTEN by the p85 subunit of phosphatidylinositol 3-kinase. *Proc Natl Acad Sci* **107**: 5471–5476.

Chakraborty A, Koldobskiy MA, Bello NT, Maxwell M, Potter JJ, Juluri KR, Maag D, Kim S, Huang AS, Dailey MJ, et al. 2010. Inositol pyrophosphates inhibit Akt signaling, thereby regulating insulin sensitivity and weight gain. *Cell* **143**: 897–910.

Chang YC, Chuang LM. 2010. The role of oxidative stress in the pathogenesis of type 2 diabetes: From molecular mechanism to clinical implication. *Am J Transl Res* **2**: 316–331.

Cheng X, Xia W, Yang JY, Hsu JL, Lang JY, Chou CK, Du Y, Sun HL, Wyszomierski SL, Mills GB, et al. 2010. Activation of murine double minute 2 by Akt in mammary epithelium delays mammary involution and accelerates mammary tumorigenesis. *Cancer Res* **70**: 7684–7689.

Chin JE, Dickens M, Tavare JM, Roth RA. 1993. Overexpression of protein kinase C isoenzymes α, β I, γ, and ε in cells overexpressing the insulin receptor. Effects on receptor phosphorylation and signaling. *J Biol Chem* **268**: 6338–6347.

Chiu HC, Kovacs A, Blanton RM, Han X, Courtois M, Weinheimer CJ, Yamada KA, Brunet S, Xu H, Nerbonne JM, et al. 2005. Transgenic expression of fatty acid transport protein 1 in the heart causes lipotoxic cardiomyopathy. *Circ Res* **96**: 225–233.

Cho H, Mu J, Kim JK, Thorvaldsen JL, Chu Q, Crenshaw EB III, Kaestner KH, Bartolomei MS, Shulman GI, Birnbaum MJ. 2001. Insulin resistance and a diabetes mellitus-like syndrome in mice lacking the protein kinase Akt2 (PKB β). *Science* **292**: 1728–1731.

Clement S, Krause U, Desmedt F, Tanti JF, Behrends J, Pesesse X, Sasaki T, Penninger J, Doherty M, Malaisse W, et al. 2001. The lipid phosphatase SHIP2 controls insulin sensitivity. *Nature* **409**: 92–97.

Cochran E, Musso C, Gorden P. 2005. The use of U-500 in patients with extreme insulin resistance. *Diabetes Care* **28**: 1240–1244.

Cooney GJ, Lyons RJ, Crew AJ, Jensen TE, Molero JC, Mitchell CJ, Biden TJ, Ormandy CJ, James DE, Daly RJ. 2004. Improved glucose homeostasis and enhanced insulin signalling in Grb14-deficient mice. *EMBO J* **23**: 582–593.

Copps KD, Hancer NJ, Opare-Ado L, Qiu W, Walsh C, White MF. 2010. Irs1 serine 307 promotes insulin sensitivity in mice. *Cell Metab* **11**: 84–92.

Cozzone D, Frojdo S, Disse E, Debard C, Laville M, Pirola L, Vidal H. 2008. Isoform-specific defects of insulin stimulation of Akt/protein kinase B (PKB) in skeletal muscle cells from type 2 diabetic patients. *Diabetologia* **51**: 512–521.

Dash S, Sano H, Rochford JJ, Semple RK, Yeo G, Hyden CS, Soos MA, Clark J, Rodin A, Langenberg C, et al. 2009. A truncation mutation in TBC1D4 in a family with acanthosis nigricans and postprandial hyperinsulinemia. *Proc Natl Acad Sci* **106**: 9350–9355.

Datta SR, Dudek H, Tao X, Masters S, Fu H, Gotoh Y, Greenberg ME. 1997. Akt phosphorylation of BAD couples survival signals to the cell-intrinsic death machinery. *Cell* **91**: 231–241.

De Fea K, Roth RA. 1997a. Modulation of insulin receptor substrate-1 tyrosine phosphorylation and function by mitogen-activated protein kinase. *J Biol Chem* **272**: 31400–31406.

De Fea K, Roth RA. 1997b. Protein kinase C modulation of insulin receptor substrate-1 tyrosine phosphorylation requires serine 612. *Biochemistry* **36**: 12939–12947.

Degerman E, Ahmad F, Chung YW, Guirguis E, Omar B, Stenson L, Manganiello V. 2011. From PDE3B to the regulation of energy homeostasis. *Curr Opin Pharmacol* **11**: 676–682.

Deyev IE, Sohet F, Vassilenko KP, Serova OV, Popova NV, Zozulya SA, Burova EB, Houillier P, Rzhevsky DI, Berchatova AA, et al. 2011. Insulin receptor-related receptor as an extracellular alkali sensor. *Cell Metab* **13**: 679–689.

Dimmeler S, Fleming I, Fisslthaler B, Hermann C, Busse R, Zeiher AM. 1999. Activation of nitric oxide synthase in endothelial cells by Akt-dependent phosphorylation. *Nature* **399**: 601–605.

Dokken BB, Saengsirisuwan V, Kim JS, Teachey MK, Henriksen EJ. 2008. Oxidative stress-induced insulin resistance in rat skeletal muscle: Role of glycogen synthase kinase-3. *Am J Physiol Endocrinol Metab* **294**: E615–E621.

Du K, Herzig S, Kulkarni RN, Montminy M. 2003. TRB3: A tribbles homolog that inhibits Akt/PKB activation by insulin in liver. *Science* **300**: 1574–1577.

Dunaif A, Xia J, Book CB, Schenker E, Tang Z. 1995. Excessive insulin receptor serine phosphorylation in cultured fibroblasts and in skeletal muscle. A potential mechanism for insulin resistance in the polycystic ovary syndrome. *J Clin Invest* **96**: 801–810.

Duvel K, Yecies JL, Menon S, Raman P, Lipovsky AI, Souza AL, Triantafellow E, Ma Q, Gorski R, Cleaver S, et al. 2010. Activation of a metabolic gene regulatory network downstream of mTOR complex 1. *Mol Cell* **39**: 171–183.

Elchebly M, Payette P, Michaliszyn E, Cromlish W, Collins S, Loy AL, Normandin DCA, imms-Hagen J, Chan C, Ramachandran C, et al. 1999. Increased insulin sensitivity and obesity resistance in mice lacking the protein tyrosine phosphatase-1B gene. *Science* **283**: 1544–1548.

Emanuelli B, Peraldi P, Filloux C, Sawka-Verhelle D, Hilton D, Van OE. 2000. SOCS-3 is an insulin-induced negative regulator of insulin signaling. *J Biol Chem* **275**: 15985–15991.

Emanuelli B, Peraldi P, Filloux C, Chavey C, Freidinger L, Hilton DJ, Hotamisligil GS, Van OE. 2001. SOCS-3 inhibits insulin signaling and is up-regulated in response to tumor necrosis factor-α in the adipose tissue of obese mice. *J Biol Chem* **276**: 47944–47949.

Esposito DL, Li Y, Vanni C, Mammarella S, Veschi S, Della LF, Mariani-Costantini R, Battista P, Quon MJ, Cama A. 2003. A novel T608R missense mutation in insulin receptor substrate-1 identified in a subject with type 2 diabetes impairs metabolic insulin signaling. *J Clin Endocrinol Metab* **88**: 1468–1475.

Evans JL, Maddux BA, Goldfine ID. 2005. The molecular basis for oxidative stress-induced insulin resistance. *Antioxid Redox Signal* **7**: 1040–1052.

Fan Y, Yu Y, Shi Y, Sun W, Xie M, Ge N, Mao R, Chang A, Xu G, Schneider MD, et al. 2010. Lysine 63-linked polyubiquitination of TAK1 at lysine 158 is required for tumor necrosis factor α- and interleukin-1β-induced IKK/NF-κB and JNK/AP-1 activation. *J Biol Chem* **285**: 5347–5360.

Fantin VR, Lavan BE, Wang Q, Jenkins NA, Gilbert DJ, Copeland NG, Keller SR, Lienhard GE. 1999. Cloning, tissue expression, and chromosomal location of the mouse insulin receptor substrate 4 gene. *Endocrinology* **140**: 1329–1337.

Fantin VR, Wang Q, Lienhard GE, Keller SR. 2000. Mice lacking insulin receptor substrate 4 exhibit mild defects in growth, reproduction, and glucose homeostasis. *Am J Physiol Endocrinol Metab* **278**: E127–E133.

Farese RV, Sajan MP. 2010. Metabolic functions of atypical protein kinase C: "Good" and "bad" as defined by nutritional status. *Am J Physiol Endocrinol Metab* **298**: E385–E394.

Farese RV, Sajan MP, Yang H, Li P, Mastorides S, Gower WR Jr, Nimal S, Choi CS, Kim S, Shulman GI, et al. 2007. Muscle-specific knockout of PKC-λ impairs glucose transport and induces metabolic and diabetic syndromes. *J Clin Invest* **117**: 2289–2301.

Federici M, Giaccari A, Hribal ML, Giovannone B, Lauro D, Morviducci L, Pastore L, Tamburrano G, Lauro R, Sesti G. 1999. Evidence for glucose/hexosamine in vivo regulation of insulin/IGF-I hybrid receptor assembly. *Diabetes* **48**: 2277–2285.

Ferreira LD, Pulawa LK, Jensen DR, Eckel RH. 2001. Overexpressing human lipoprotein lipase in mouse skeletal muscle is associated with insulin resistance. *Diabetes* **50**: 1064–1068.

Field CJ, Ryan EA, Thomson AB, Clandinin MT. 1990. Diet fat composition alters membrane phospholipid composition, insulin binding, and glucose metabolism in adipocytes from control and diabetic animals. *J Biol Chem* **265**: 11143–11150.

Florez JC, Sjogren M, Burtt N, Orho-Melander M, Schayer S, Sun M, Almgren P, Lindblad U, Tuomi T, Gaudet D, et al. 2004. Association testing in 9,000 people fails to confirm the association of the insulin receptor substrate-1 G972R polymorphism with type 2 diabetes. *Diabetes* **53**: 3313–3318.

Frasca F, Pandini G, Scalia P, Sciacca L, Mineo R, Costantino A, Goldfine ID, Belfiore A, Vigneri R. 1999. Insulin receptor isoform A, a newly recognized, high-affinity insulin-like growth factor II receptor in fetal and cancer cells. *Mol Cell Biol* **19**: 3278–3288.

Fridlyand LE, Philipson LH. 2006. Reactive species and early manifestation of insulin resistance in type 2 diabetes. *Diabetes Obes Metab* **8**: 136–145.

Fu S, Yang L, Li P, Hofmann O, Dicker L, Hide W, Lin X, Watkins SM, Ivanov AR, Hotamisligil GS. 2011. Aberrant lipid metabolism disrupts calcium homeostasis causing liver endoplasmic reticulum stress in obesity. *Nature* **473**: 528–531.

Fulton D, Gratton JP, McCabe TJ, Fontana J, Fujio Y, Walsh K, Franke TF, Papapetropoulos A, Sessa WC. 1999. Regulation of endothelium-derived nitric oxide production by the protein kinase Akt. *Nature* **399**: 597–601.

Gao Z, Hwang D, Bataille F, Lefevre M, York D, Quon M, Ye J. 2002. Serine phosphorylation of insulin receptor substrate 1 (IRS-1) by inhibitor KB kinase (IKK) complex. *J Biol Chem* **277**: 48115–48121.

Gardai SJ, Hildeman DA, Frankel SK, Whitlock BB, Frasch SC, Borregaard N, Marrack P, Bratton DL, Henson PM. 2004. Phosphorylation of Bax Ser184 by Akt regulates its activity and apoptosis in neutrophils. *J Biol Chem* **279**: 21085–21095.

George S, Rochford JJ, Wolfrum C, Gray SL, Schinner S, Wilson JC, Soos MA, Murgatroyd PR, Williams RM, Acerini CL, et al. 2004. A family with severe insulin resistance and diabetes due to a mutation in AKT2. *Science* **304**: 1325–1328.

Goldstein BJ, Ahmad F, Ding W, Li PM, Zhang WR. 1998. Regulation of the insulin signalling pathway by cellular protein-tyrosine phosphatases. *Mol Cell Biochem* **182**: 91–99.

Gual P, Gonzalez T, Gremeaux T, Barres R, Marchand-Brustel Y, Tanti JF. 2003. Hyperosmotic stress inhibits insulin receptor substrate-1 function by distinct mechanisms in 3T3-L1 adipocytes. *J Biol Chem* **278**: 26550–26557.

Hager SR, Jochen AL, Kalkhoff RK. 1991. Insulin resistance in normal rats infused with glucose for 72 h. *Am J Physiol* **260**: E353–E362.

Hanke S, Mann M. 2009. The phosphotyrosine interactome of the insulin receptor family and its substrates IRS-1 and IRS-2. *Mol Cell Proteomics* **8**: 519–534.

Harrington LS, Findlay GM, Gray A, Tolkacheva T, Wigfield S, Rebholz H, Barnett J, Leslie NR, Cheng S, Shepherd PR, et al. 2004. The TSC1-2 tumor suppressor controls insulin-PI3K signaling via regulation of IRS proteins. *J Cell Biol* **166**: 213–223.

Haruta T, Imamura T, Iwanishi M, Egawa K, Goji K, Kobayashi M. 1995. Amplification and analysis of promoter region of insulin receptor gene in a patient with leprechaunism associated with severe insulin resistance. *Metabolism* **44**: 430–437.

Haus JM, Kashyap SR, Kasumov T, Zhang R, Kelly KR, DeFronzo RA, Kirwan JP. 2009. Plasma ceramides are elevated in obese subjects with type 2 diabetes and correlate with the severity of insulin resistance. *Diabetes* **58:** 337–343.

Himes RW, Smith CW. 2010. Tlr2 is critical for diet-induced metabolic syndrome in a murine model. *FASEB J* **24:** 731–739.

Hirosumi J, Tuncman G, Chang L, Gorgun CZ, Uysal KT, Maeda K, Karin M, Hotamisligil GS. 2002. A central role for JNK in obesity and insulin resistance. *Nature* **420:** 333–336.

Holland WL, Brozinick JT, Wang LP, Hawkins ED, Sargent KM, Liu Y, Narra K, Hoehn KL, Knotts TA, Siesky A, et al. 2007. Inhibition of ceramide synthesis ameliorates glucocorticoid-, saturated-fat-, and obesity-induced insulin resistance. *Cell Metab* **5:** 167–179.

Holt LJ, Siddle K. 2005. Grb10 and Grb14: Enigmatic regulators of insulin action—and more? *Biochem J* **388:** 393–406.

Holt LJ, Lyons RJ, Ryan AS, Beale SM, Ward A, Cooney GJ, Daly RJ. 2009. Dual ablation of Grb10 and Grb14 in mice reveals their combined role in regulation of insulin signaling and glucose homeostasis. *Mol Endocrinol* **23:** 1406–1414.

Hotamisligil GS. 2010. Endoplasmic reticulum stress and the inflammatory basis of metabolic disease. *Cell* **140:** 900–917.

Housley MP, Rodgers JT, Udeshi ND, Kelly TJ, Shabanowitz J, Hunt DF, Puigserver P, Hart GW. 2008. O-GlcNAc regulates FoxO activation in response to glucose. *J Biol Chem* **283:** 16283–16292.

Hribal ML, Tornei F, Pujol A, Menghini R, Barcaroli D, Lauro D, Amoruso R, Lauro R, Bosch F, Sesti G, et al. 2008. Transgenic mice overexpressing human G972R IRS-1 show impaired insulin action and insulin secretion. *J Cell Mol Med* **12:** 2096–2106.

Hsu PP, Kang SA, Rameseder J, Zhang Y, Ottina KA, Lim D, Peterson TR, Choi Y, Gray NS, Yaffe MB, et al. 2011. The mTOR-regulated phosphoproteome reveals a mechanism of mTORC1-mediated inhibition of growth factor signaling. *Science* **332:** 1317–1322.

Hu P, Han Z, Couvillon AD, Kaufman RJ, Exton JH. 2006. Autocrine tumor necrosis factor α links endoplasmic reticulum stress to the membrane death receptor pathway through IRE1α-mediated NF-κB activation and down-regulation of TRAF2 expression. *Mol Cell Biol* **26:** 3071–3084.

Huang C, Thirone AC, Huang X, Klip A. 2005. Differential contribution of insulin receptor substrates 1 versus 2 to insulin signaling and glucose uptake in l6 myotubes. *J Biol Chem* **280:** 19426–19435.

Ishihara H, Sasaoka T, Kagawa S, Murakami S, Fukui K, Kawagishi Y, Yamazaki K, Sato A, Iwata M, Urakaze M, et al. 2003. Association of the polymorphisms in the 5′-untranslated region of PTEN gene with type 2 diabetes in a Japanese population. *Growth Regul* **554:** 450–454.

Jager J, Gremeaux T, Cormont M, Le Marchand-Brustel Y, Tanti JF. 2007. Interleukin-1β-induced insulin resistance in adipocytes through down-regulation of insulin receptor substrate-1 expression. *Endocrinology* **148:** 241–251.

Jheng HF, Tsai PJ, Guo SM, Kuo LH, Chang CS, Su IJ, Chang CR, Tsai YS. 2012. Mitochondrial fission contributes to mitochondrial dysfunction and insulin resistance in skeletal muscle. *Mol Cell Biol* **32:** 309–319.

Jia S, Liu Z, Zhang S, Liu P, Zhang L, Lee SH, Zhang J, Signoretti S, Loda M, Roberts TM, et al. 2008. Essential roles of PI3K-p110β in cell growth, metabolism and tumorigenesis. *Nature* **454:** 776–779.

Kagawa S, Soeda Y, Ishihara H, Oya T, Sasahara M, Yaguchi S, Oshita R, Wada T, Tsuneki H, Sasaoka T. 2008. Impact of transgenic overexpression of SH2-containing inositol 5′-phosphatase 2 on glucose metabolism and insulin signaling in mice. *Endocrinology* **149:** 642–650.

Kahn CR, Flier JS, Bar RS, Archer JA, Gorden P, Martin MM, Roth J. 1976. The syndromes of insulin resistance and acanthosis nigricans. Insulin-receptor disorders in man. *N Engl J Med* **294:** 739–745.

Kamei N, Tobe K, Suzuki R, Ohsugi M, Watanabe T, Kubota N, Ohtsuka-Kowatari N, Kumagai K, Sakamoto K, Kobayashi M, et al. 2006. Overexpression of monocyte chemoattractant protein-1 in adipose tissues causes macrophage recruitment and insulin resistance. *J Biol Chem* **281:** 26602–26614.

Kanda H, Tateya S, Tamori Y, Kotani K, Hiasa K, Kitazawa R, Kitazawa S, Miyachi H, Maeda S, Egashira K, et al. 2006. MCP-1 contributes to macrophage infiltration into adipose tissue, insulin resistance, and hepatic steatosis in obesity. *J Clin Invest* **116:** 1494–1505.

Karasik A, Rothenberg PL, Yamada K, White MF, Kahn CR. 1990. Increased protein kinase C activity is linked to reduced insulin receptor autophosphorylation in liver of starved rats. *J Biol Chem* **265:** 10226–10231.

Kim JK, Kim YJ, Fillmore JJ, Chen Y, Moore I, Lee J, Yuan M, Li ZW, Karin M, Perret P, et al. 2001a. Prevention of fat-induced insulin resistance by salicylate. *J Clin Invest* **108:** 437–446.

Kim JK, Zisman A, Fillmore JJ, Peroni OD, Kotani K, Perret P, Zong H, Kahn CR, Kahn BB, Shulman GI. 2001b. Glucose toxicity and the development of diabetes in mice with muscle-specific inactivation of GLUT4. *J Clin Invest* **108:** 153–160.

Kim JK, Fillmore JJ, Sunshine MJ, Albrecht B, Higashimori T, Kim DW, Liu ZX, Soos TJ, Cline GW, O'Brien WR, et al. 2004a. PKC-θ knockout mice are protected from fat-induced insulin resistance. *J Clin Invest* **114:** 823–827.

Kim KH, Song JJ, Yoo EJ, Choe SS, Park SD, Kim JB. 2004b. Regulatory role of glycogen synthase kinase 3 for criptional activity of ADD1/SREBP1c. *J Biol Chem* **279:** 51999–52006.

Kitamura T, Kitamura Y, Kuroda S, Hino Y, Ando M, Kotani K, Konishi H, Matsuzaki H, Kikkawa U, Ogawa W, et al. 1999. Insulin-induced phosphorylation and activation of cyclic nucleotide phosphodiesterase 3B by the serine-threonine kinase Akt. *Mol Cell Biol* **19:** 6286–6296.

Klaman LD, Boss O, Peroni OD, Kim JK, Martino JL, Zabolotny JM, Moghal N, Lubkin M, Kim YB, Sharpe AH, et al. 2000. Increased energy expenditure, decreased adiposity, and tissue-specific insulin sensitivity in protein-tyrosine phosphatase 1B-deficient mice. *Mol Cell Biol* **20:** 5479–5489.

Kleinridders A, Schenten D, Konner AC, Belgardt BF, Mauer J, Okamura T, Wunderlich FT, Medzhitov R, Bruning JC.

2009. MyD88 signaling in the CNS is required for development of fatty acid-induced leptin resistance and diet-induced obesity. *Cell Metab* **10:** 249–259.

Koh EH, Lee WJ, Kim MS, Park JY, Lee IK, Lee KU. 2005. Intracellular fatty acid metabolism in skeletal muscle and insulin resistance. *Curr Diabetes Rev* **1:** 331–336.

Koo SH, Satoh H, Herzig S, Lee CH, Hedrick S, Kulkarni R, Evans RM, Olefsky J, Montminy M. 2004. PGC-1 promotes insulin resistance in liver through PPAR-α-dependent induction of TRB-3. *Nat Med* **10:** 530–534.

Koonen DP, Jacobs RL, Febbraio M, Young ME, Soltys CL, Ong H, Vance DE, Dyck JR. 2007. Increased hepatic CD36 expression contributes to dyslipidemia associated with diet-induced obesity. *Diabetes* **56:** 2863–2871.

Kowluru A, Matti A. 2012. Hyperactivation of protein phosphatase 2A in models of glucolipotoxicity and diabetes: Potential mechanisms and functional consequences. *Biochem Pharmacol* **84:** 591–597.

Krieger-Brauer HI Kather H. 1992. Human fat cells possess a plasma membrane-bound H_2O_2-generating system that is activated by insulin via a mechanism bypassing the receptor kinase. *J Clin Invest* **89:** 1006–1013.

Kurlawalla-Martinez C, Stiles B, Wang Y, Devaskar SU, Kahn BB, Wu H. 2005. Insulin hypersensitivity and resistance to streptozotocin-induced diabetes in mice lacking PTEN in adipose tissue. *Mol Cell Biol* **25:** 2498–2510.

Laustsen PG, Michael MD, Crute BE, Cohen SE, Ueki K, Kulkarni RN, Keller SR, Lienhard GE, Kahn CR. 2002. Lipoatrophic diabetes in $Irs1^{-/-}/Irs3^{-/-}$ double knockout mice. *Genes Dev* **16:** 3213–3222.

Lavan BE, Fantin VR, Chang ET, Lane WS, Keller SR, Lienhard GE. 1997a. A novel 160-kDa phosphotyrosine protein in insulin-treated embryonic kidney cells is a new member of the insulin receptor substrate family. *J Biol Chem* **272:** 21403–21407.

Lavan BE, Lane WS, Lienhard GE. 1997b. The 60-kDa phosphotyrosine protein in insulin-treated adipocytes is a new member of the insulin receptor substrate family. *J Biol Chem* **272:** 11439–11443.

Leahy JL, Cooper HE, Deal DA, Weir DG. 1986. Chronic hyperglycemia is associated with impaired glucose influence on insulin secretion. A study in normal rats using chronic in vivo glucose infusions. *J Clin Invest* **77:** 908–915.

Lee JY, Sohn KH, Rhee SH, Hwang D. 2001. Saturated fatty acids, but not unsaturated fatty acids, induce the expression of cyclooxygenase-2 mediated through Toll-like receptor 4. *J Biol Chem* **276:** 16683–16689.

Lewis RE, Cao L, Perregaux D, Czech MP. 1990. Threonine 1336 of the human insulin receptor is a major target for phosphorylation by protein kinase C. *Biochemistry* **29:** 1807–1813.

Li Y, Soos TJ, Li X, Wu J, Degennaro M, Sun X, Littman DR, Birnbaum MJ, Polakiewicz RD. 2004. Protein kinase C θ inhibits insulin signaling by phosphorylating IRS1 at Ser(1101). *J Biol Chem* **279:** 45304–45307.

Li X, Monks B, Ge Q, Birnbaum MJ. 2007. Akt/PKB regulates hepatic metabolism by directly inhibiting PGC-1α cription coactivator. *Nature* **447:** 1012–1016.

Liu L, Zhang Y, Chen N, Shi X, Tsang B, Yu BH. 2007. Upregulation of myocellular DGAT1 augments triglycer-

ide synthesis in skeletal muscle and protects against fat-induced insulin resistance. *J Clin Invest* **117:** 1679–1689.

Lu M, Wan M, Leavens KF, Chu Q, Monks BR, Fernandez S, Ahima RS, Ueki K, Kahn CR, Birnbaum MJ. 2012. Insulin regulates liver metabolism in vivo in the absence of hepatic Akt and Foxo1. *Nat Med* **18:** 388–395.

Mack E, Ziv E, Reuveni H, Kalman R, Niv MY, Jorns A, Lenzen S, Shafrir E. 2008. Prevention of insulin resistance and β-cell loss by abrogating PKCε-induced serine phosphorylation of muscle IRS-1 in Psammomys obesus. *Diabetes Metab Res Rev* **24:** 577–584.

Mahadev K, Zilbering A, Zhu L, Goldstein BJ. 2001. Insulin-stimulated hydrogen peroxide reversibly inhibits protein-tyrosine phosphatase 1b in vivo and enhances the early insulin action cascade. *J Biol Chem* **276:** 21938–21942.

Manning BD, Cantley LC. 2007. AKT/PKB signaling: Navigating downstream. *Cell* **129:** 1261–1274.

Marshall S, Bacote V, Traxinger RR. 1991. Discovery of a metabolic pathway mediating glucose-induced desensitization of the glucose transport system. Role of hexosamine biosynthesis in the induction of insulin resistance. *J Biol Chem* **266:** 4706–4712.

Martinez-Gomez LE, Cruz M, Martinez-Nava GA, Madrid-Marina V, Parra E, Garcia-Mena J, Espinoza-Rojo M, Estrada-Velasco BI, Piza-Roman LF, Aguilera P, et al. 2011. A replication study of the IRS1, CAPN10, TCF7L2, and PPARG gene polymorphisms associated with type 2 diabetes in two different populations of Mexico. *Ann Hum Genet* **75:** 612–620.

Matsumoto M, Ogawa W, Akimoto K, Inoue H, Miyake K, Furukawa K, Hayashi Y, Iguchi H, Matsuki Y, Hiramatsu R, et al. 2003. PKCλ in liver mediates insulin-induced SREBP-1c expression and determines both hepatic lipid content and overall insulin sensitivity. *J Clin Invest* **112:** 935–944.

Matsushima R, Harada N, Webster NJ, Tsutsumi YM, Nakaya Y. 2006. Effect of TRB3 on insulin and nutrient-stimulated hepatic p70 S6 kinase activity. *J Biol Chem* **281:** 29719–29729.

McClain DA. 2002. Hexosamines as mediators of nutrient sensing and regulation in diabetes. *J Diabetes Complications* **16:** 72–80.

McClain DA, Lubas WA, Cooksey RC, Hazel M, Parker GJ, Love DC, Hanover JA. 2002. Altered glycan-dependent signaling induces insulin resistance and hyperleptinemia. *Proc Natl Acad Sci* **99:** 10695–10699.

Miele C, Riboulet A, Maitan MA, Oriente F, Romano C, Formisano P, Giudicelli J, Beguinot F, Van Obberghen E. 2003. Human glycated albumin affects glucose metabolism in L6 skeletal muscle cells by impairing insulin-induced insulin receptor substrate (IRS) signaling through a protein kinase C α-mediated mechanism. *J Biol Chem* **278:** 47376–47387.

Miki H, Yamauchi T, Suzuki R, Komeda K, Tsuchida A, Kubota N, Terauchi Y, Kamon J, Kaburagi Y, Matsui J, et al. 2001. Essential role of insulin receptor substrate 1 (IRS-1) and IRS-2 in adipocyte differentiation. *Mol Cell Biol* **21:** 2521–2532.

Millward TA, Zolnierowicz S, Hemmings BA. 1999. Regulation of protein kinase cascades by protein phosphatase 2A. *Trends Biochem Sci* **24:** 186–191.

Mosthaf L, Grako K, Dull TJ, Coussens L, Ullrich A, McClain DA. 1990. Functionally distinct insulin receptors generated by tissue-specific alternative splicing. *EMBO J* **9**: 2409–2413.

Motti ML, De MC, Califano D, Fusco A, Viglietto G. 2004. Akt-dependent T198 phosphorylation of cyclin-dependent kinase inhibitor p27kip1 in breast cancer. *Cell Cycle* **3**: 1074–1080.

Myers MG Jr, Backer JM, Sun XJ, Shoelson S, Hu P, Schlessinger J, Yoakim M, Schaffhausen B, White MF. 1992. IRS-1 activates phosphatidylinositol 3′-kinase by associating with src homology 2 domains of p85. *Proc Natl Acad Sci* **89**: 10350–10354.

Nef S, Verma-Kurvari S, Merenmies J, Vassalli JD, Efstratiadis A, Accili D, Parada LF. 2003. Testis determination requires insulin receptor family function in mice. *Nature* **426**: 291–295.

Ni YG, Wang N, Cao DJ, Sachan N, Morris DJ, Gerard RD, Kuro O, Rothermel BA, Hill JA. 2007. FoxO cription factors activate Akt and attenuate insulin signaling in heart by inhibiting protein phosphatases. *Proc Natl Acad Sci* **104**: 20517–20522.

Oh WJ, Jacinto E. 2011. mTOR complex 2 signaling and functions. *Cell Cycle* **10**: 2305–2316.

Osborn O, Olefsky JM. 2012. The cellular and signaling networks linking the immune system and metabolism in disease. *Nat Med* **18**: 363–374.

Ozcan U, Cao Q, Yilmaz E, Lee AH, Iwakoshi NN, Ozdelen E, Tuncman G, Gorgun C, Glimcher LH, Hotamisligil GS. 2004. Endoplasmic reticulum stress links obesity, insulin action, and type 2 diabetes. *Science* **306**: 457–461.

Ozcan L, Ergin AS, Lu A, Chung J, Sarkar S, Nie D, Myers MG Jr, Ozcan U. 2009. Endoplasmic reticulum stress plays a central role in development of leptin resistance. *Cell Metab* **9**: 35–51.

Ozes ON, Akca H, Mayo LD, Gustin JA, Maehama T, Dixon JE, Donner DB. 2001. A phosphatidylinositol 3-kinase/Akt/mTOR pathway mediates and PTEN antagonizes tumor necrosis factor inhibition of insulin signaling through insulin receptor substrate-1. *Proc Natl Acad Sci* **98**: 4640–4645.

Pal A, Barber TM, Van de Bunt M, Rudge SA, Zhang Q, Lachlan KL, Cooper NS, Linden H, Levy JC, Wakelam MJ, et al. 2012. PTEN mutations as a cause of constitutive insulin sensitivity and obesity. *N Engl J Med* **367**: 1002–1011.

Pandini G, Vigneri R, Costantino A, Frasca F, Ippolito A, Fujita-Yamaguchi Y, Siddle K, Goldfine ID, Belfiore A. 1999. Insulin and insulin-like growth factor-I (IGF-I) receptor overexpression in breast cancers leads to insulin/IGF-I hybrid receptor overexpression: Evidence for a second mechanism of IGF-I signaling. *Clin Cancer Res* **5**: 1935–1944.

Park SW, Zhou Y, Lee J, Lu A, Sun C, Chung J, Ueki K, Ozcan J. 2010. The regulatory subunits of PI3K, p85α and p85β, interact with XBP-1 and increase its nuclear translocation. *Nat Med* **16**: 429–437.

Patti ME, Virkamaki A, Landaker EJ, Kahn CR, Yki-Jarvinen H. 1999. Activation of the hexosamine pathway by glucosamine in vivo induces insulin resistance of early postreceptor insulin signaling events in skeletal muscle. *Diabetes* **48**: 1562–1571.

Pearce LR, Komander D, Alessi DR. 2010. The nuts and bolts of AGC protein kinases. *Nat Rev Mol Cell Biol* **11**: 9–22.

Peng G, Li L, Liu Y, Pu J, Zhang S, Yu J, Zhao J, Liu P. 2011. Oleate blocks palmitate-induced abnormal lipid distribution, endoplasmic reticulum expansion and stress, and insulin resistance in skeletal muscle. *Endocrinology* **152**: 2206–2218.

Powell DJ, Hajduch E, Kular G, Hundal HS. 2003. Ceramide disables 3-phosphoinositide binding to the pleckstrin homology domain of protein kinase B (PKB)/Akt by a PKCζ-dependent mechanism. *Mol Cell Biol* **23**: 7794–7808.

Powell DJ, Turban S, Gray A, Hajduch E, Hundal HS. 2004. Intracellular ceramide synthesis and protein kinase Cζ activation play an essential role in palmitate-induced insulin resistance in rat L6 skeletal muscle cells. *Biochem J* **382**: 619–629.

Prudente S, Hribal ML, Flex E, Turchi F, Morini E, De Cosmo S, Bacci S, Tassi V, Cardellini M, Lauro R, et al. 2005. The functional Q84R polymorphism of mammalian Tribbles homolog TRB3 is associated with insulin resistance and related cardiovascular risk in Caucasians from Italy. *Diabetes* **54**: 2807–2811.

Prudente S, Scarpelli D, Chandalia M, Zhang YY, Morini E, Del Guerra S, Perticone F, Li R, Powers C, Andreozzi F, et al. 2009. The TRIB3 Q84R polymorphism and risk of early-onset type 2 diabetes. *J Clin Endocrinol Metab* **94**: 190–196.

Qi L, Heredia JE, Altarejos JY, Screaton R, Goebel N, Niessen S, Macleod IX, Liew cW, Kulkarni RN, Bain J, et al. 2006. TRB3 links the E3 ubiquitin ligase COP1 to lipid metabolism. *Science* **312**: 1763–1766.

Ravichandran LV, Esposito DL, Chen J, Quon MJ. 2001. Protein kinase C-ζ phosphorylates insulin receptor substrate-1 and impairs its ability to activate phosphatidylinositol 3-kinase in response to insulin. *J Biol Chem* **276**: 3543–3549.

Reyna SM, Ghosh S, Tantiwong P, Meka CS, Eagan P, Jenkinson CP, Cersosimo E, DeFronzo RA, Coletta DK, Sriwijitkamol A, et al. 2008. Elevated toll-like receptor 4 expression and signaling in muscle from insulin-resistant subjects. *Diabetes* **57**: 2595–2602.

Riboulet-Chavey A, Pierron A, Durand I, Murdaca J, Giudicelli J, Van OE. 2006. Methylglyoxal impairs the insulin signaling pathways independently of the formation of intracellular reactive oxygen species. *Diabetes* **55**: 1289–1299.

Rosen P, Nawroth PP, King G, Moller W, Tritschler HJ, Packer L. 2001. The role of oxidative stress in the onset and progression of diabetes and its complications: A summary of a Congress Series sponsored by UNESCO-MCBN, the American Diabetes Association and the German Diabetes Society. *Diabetes Metab Res Rev* **17**: 189–212.

Roth RA, Beaudoin J. 1987. Phosphorylation of purified insulin receptor by cAMP kinase. *Diabetes* **36**: 123–126.

Rotter V, Nagaev I, Smith U. 2003. Interleukin-6 (IL-6) induces insulin resistance in 3T3-L1 adipocytes and is, like IL-8 and tumor necrosis factor-α, overexpressed in human fat cells from insulin-resistant subjects. *J Biol Chem* **278**: 45777–45784.

Rudich A, Tirosh A, Potashnik R, Hemi R, Kanety H, Bashan N. 1998. Prolonged oxidative stress impairs insulin-induced GLUT4 translocation in 3T3-L1 adipocytes. *Diabetes* **47:** 1562–1569.

Rui L, Yuan M, Frantz D, Shoelson S, White MF. 2002. SOCS-1 and SOCS-3 block insulin signaling by ubiquitin-mediated degradation of IRS1 and IRS2. *J Biol Chem* **277:** 42394–42398.

Sachithanandan N, Fam BC, Fynch S, Dzamko N, Watt MJ, Wormald S, Honeyman J, Galic S, Proietto J, Andrikopoulos S, et al. 2010. Liver-specific suppressor of cytokine signaling-3 deletion in mice enhances hepatic insulin sensitivity and lipogenesis resulting in fatty liver and obesity. *Hepatology* **52:** 1632–1642.

Sajan MP, Rivas J, Li P, Standaert ML, Farese RV. 2006. Repletion of atypical protein kinase C following RNA interference-mediated depletion restores insulin-stimulated glucose transport. *J Biol Chem* **281:** 17466–17473.

Sakamoto K, Holman GD. 2008. Emerging role for AS160/TBC1D4 and TBC1D1 in the regulation of GLUT4 traffic. *Am J Physiol Endocrinol Metab* **295:** E29–E37.

Samuel VT, Liu ZX, Wang A, Beddow SA, Geisler JG, Kahn M, Zhang XM, Monia BP, Bhanot S, Shulman GI. 2007. Inhibition of protein kinase Cε prevents hepatic insulin resistance in nonalcoholic fatty liver disease. *J Clin Invest* **117:** 739–745.

Samuel VT, Petersen KF, Shulman GI. 2010. Lipid-induced insulin resistance: Unravelling the mechanism. *Lancet* **375:** 2267–2277.

Sano H, Kane S, Sano E, Miinea CP, Asara JM, Lane WS, Garner CW, Lienhard GE. 2003. Insulin-stimulated phosphorylation of a Rab GTPase-activating protein regulates GLUT4 translocation. *J Biol Chem* **278:** 14599–14602.

Sarbassov DD, Guertin DA, Ali SM, Sabatini DM. 2005. Phosphorylation and regulation of Akt/PKB by the rictor-mTOR complex. *Science* **307:** 1098–1101.

Schenk S, Saberi M, Olefsky JM. 2008. Insulin sensitivity: Modulation by nutrients and inflammation. *J Clin Invest* **118:** 2992–3002.

Schultze SM, Jensen J, Hemmings BA, Tschopp O, Niessen M. 2011. Promiscuous affairs of PKB/AKT isoforms in metabolism. *Arch Physiol Biochem* **117:** 70–77.

Sciacchitano S, Taylor SI. 1997. Cloning, tissue expression, and chromosomal localization of the mouse IRS-3 gene. *Endocrinology* **138:** 4931–4940.

Shah OJ, Wang Z, Hunter T. 2004. Inappropriate activation of the TSC/Rheb/mTOR/S6K cassette induces IRS1/2 depletion, insulin resistance, and cell survival deficiencies. *Curr Biol* **14:** 1650–1656.

Shao J, Catalano PM, Yamashita H, Ruyter I, Smith S, Youngren J, Friedman JE. 2000. Decreased insulin receptor tyrosine kinase activity and plasma cell membrane glycoprotein-1 overexpression in skeletal muscle from obese women with gestational diabetes mellitus (GDM): Evidence for increased serine/threonine phosphorylation in pregnancy and GDM. *Diabetes* **49:** 603–610.

Shaw LM. 2011. The insulin receptor substrate (IRS) proteins: At the intersection of metabolism and cancer. *Cell Cycle* **10:** 1750–1756.

Shi H, Kokoeva MV, Inouye K, Tzameli I, Yin H, Flier JS. 2006. TLR4 links innate immunity and fatty acid-induced insulin resistance. *J Clin Invest* **116:** 3015–3025.

Shiura H, Miyoshi N, Konishi A, Wakisaka-Saito N, Suzuki K, Muguruma R, Kohda T, Wakana S, Yokoyama M, Ishino F, et al. 2005. Meg1/Grb10 overexpression causes postnatal growth retardation and insulin resistance via negative modulation of the IGF1R and IR cascades. *Biochem Biophys Res Commun* **329:** 909–916.

Siddle K. 2012. Molecular basis of signaling specificity of insulin and IGF receptors: Neglected corners and recent advances. *Front Endocrinol (Lausanne)* **3:** 34.

Sinha S, Perdomo G, Brown NF, O'Doherty RM. 2004. Fatty acid-induced insulin resistance in L6 myotubes is prevented by inhibition of activation and nuclear localization of nuclear factor κ B. *J Biol Chem* **279:** 41294–41301.

Sleeman MW, Wortley KE, Lai KM, Gowen LC, Kintner J, Kline WO, Garcia K, Stitt TN, Yancopoulos GD, Wiegand SJ, et al. 2005. Absence of the lipid phosphatase SHIP2 confers resistance to dietary obesity. *Nat Med* **11:** 199–205.

Smith FM, Holt LJ, Garfield AS, Charalambous M, Koumanov F, Perry M, Bazzani R, Sheardown SA, Hegarty BD, Lyons RJ, et al. 2007. Mice with a disruption of the imprinted Grb10 gene exhibit altered body composition, glucose homeostasis, and insulin signaling during postnatal life. *Mol Cell Biol* **27:** 5871–5886.

Sopasakis VR, Liu P, Suzuki R, Kondo T, Winnay J, Tran TT, Asano T, Smyth G, Sajan MP, Farese RV, et al. 2010. Specific roles of the p110α isoform of phosphatidylinositol 3-kinase in hepatic insulin signaling and metabolic regulation. *Cell Metab* **11:** 220–230.

Stadtmauer LA, Rosen OM. 1986. Increasing the cAMP content of IM-9 cells alters the phosphorylation state and protein kinase activity of the insulin receptor. *J Biol Chem* **261:** 3402–3407.

Steppan CM, Wang J, Whiteman EL, Birnbaum MJ, Lazar MA. 2005. Activation of SOCS-3 by resistin. *Mol Cell Biol* **25:** 1569–1575.

Stiles B, Wang Y, Stahl A, Bassilian S, Lee WP, Kim YJ, Sherwin R, Devaskar S, Lesche R, Magnuson MA, et al. 2004. Liver-specific deletion of negative regulator Pten results in fatty liver and insulin hypersensitivity [corrected]. *Proc Natl Acad Sci* **101:** 2082–2087.

Sun XJ, Rothenberg PL, Kahn CR, Backer JM, Araki E, Wilden PA, Cahill DA, Goldstein BJ, White MF. 1991. Structure of the insulin receptor substrate IRS-1 defines a unique signal duction protein. *Nature* **352:** 73–77.

Sun XJ, Crimmins DL, Myers MR Jr, Miralpeix M, White MF. 1993. Pleiotropic insulin signals are engaged by multisite phosphorylation of IRS-1. *Mol Cell Biol* **13:** 7418–7428.

Sun XJ, Wang LM, Zhang Y, Yenush L, Myers MG Jr, Glasheen E, Lane WS, Pierce JH, White MF. 1995. Role of IRS-2 in insulin and cytokine signalling. *Nature* **377:** 173–177.

Sun K, Kusminski CM, Scherer PE. 2011. Adipose tissue remodeling and obesity. *J Clin Invest* **121:** 2094–2101.

Suwa A, Kurama T, Shimokawa T. 2010. SHIP2 and its involvement in various diseases. *Expert Opin Ther Targets* **14:** 727–737.

Tan K, Kimber WA, Luan J, Soos MA, Semple RK, Wareham NJ, O'Rahilly S, Barroso I. 2007. Analysis of genetic variation in Akt2/PKB-β in severe insulin resistance, lipodystrophy, type 2 diabetes, and related metabolic phenotypes. *Diabetes* **56:** 714–719.

Taniguchi CM, Emanuelli B, Kahn CR. 2006a. Critical nodes in signalling pathways: Insights into insulin action. *Nat Rev Mol Cell Biol* **7:** 85–96.

Taniguchi CM, Tran TT, Kondo T, Luo J, Ueki K, Cantley LC, Kahn CR. 2006b. Phosphoinositide 3-kinase regulatory subunit p85α suppresses insulin action via positive regulation of PTEN. *Proc Natl Acad Sci* **103:** 12093–12097.

Taniguchi CM, Winnay J, Kondo T, Bronson RT, Guimaraes AR, Aleman JO, Luo J, Stephanopoulos G, Weissleder R, Cantley LC, et al. 2010. The phosphoinositide 3-kinase regulatory subunit p85α can exert tumor suppressor properties through negative regulation of growth factor signaling. *Cancer Res* **70:** 5305–5315.

Taylor SI, Accili D, Cama A, Kadowaki H, Kadowaki T, Imano E, Sierra ML. 1991. Mutations in the insulin receptor gene in patients with genetic syndromes of insulin resistance. *Adv Exp Med Biol* **293:** 197–213.

Taylor EB, An D, Kramer HF, Yu H, Fujii NL, Roeckl KS, Bowles N, Hirshman MF, Xie J, Feener EP, et al. 2008. Discovery of TBC1D1 as an insulin-, AICAR-, and contraction-stimulated signaling nexus in mouse skeletal muscle. *J Biol Chem* **283:** 9787–9796.

Terauchi Y, Tsuji Y, Satoh S, Minoura H, Murakami K, Okuno A, Inukai K, Asano T, Kaburagi Y, Ueki K, et al. 1999. Increased insulin sensitivity and hypoglycaemia in mice lacking the p85 α subunit of phosphoinositide 3-kinase. *Nat Genet* **21:** 230–235.

Teruel T, Hernandez R, Lorenzo M. 2001. Ceramide mediates insulin resistance by tumor necrosis factor-α in brown adipocytes by maintaining Akt in an inactive dephosphorylated state. *Diabetes* **50:** 2563–2571.

Tseng YH, Butte AJ, Kokkotou E, Yechoor VK, Taniguchi CM, Kriauciunas KM, Cypess AM, Niinobe M, Yoshikawa K, Patti ME, et al. 2005. Prediction of preadipocyte differentiation by gene expression reveals role of insulin receptor substrates and necdin. *Nat Cell Biol* **7:** 601–611.

Turban S, Hajduch E. 2011. Protein kinase C isoforms: Mediators of reactive lipid metabolites in the development of insulin resistance. *Growth Regul* **585:** 269–274.

Tzivion G, Dobson M, Ramakrishnan G. 2011. FoxO cription factors; Regulation by AKT and 14–3-3 proteins. *Biochim Biophys Acta* **1813:** 1938–1945.

Ueki K, Yballe CM, Brachmann SM, Vicent D, Watt JM, Kahn CR, Cantley LC. 2002. Increased insulin sensitivity in mice lacking p85β subunit of phosphoinositide 3-kinase. *Proc Natl Acad Sci* **99:** 419–424.

Ueki K, Kondo T, Kahn CR. 2004a. Suppressor of cytokine signaling 1 (SOCS-1) and SOCS-3 cause insulin resistance through inhibition of tyrosine phosphorylation of insulin receptor substrate proteins by discrete mechanisms. *Mol Cell Biol* **24:** 5434–5446.

Ueki K, Kondo T, Tseng YH, Kahn CR. 2004b. Central role of suppressors of cytokine signaling proteins in hepatic steatosis, insulin resistance, and the metabolic syndrome in the mouse. *Proc Natl Acad Sci* **101:** 10422–10427.

Um SH, Frigerio F, Watanabe M, Picard F, Joaquin M, Sticker M, Fumagalli S, Allegrini PR, Kozma SC, Auwerx J, et al. 2004. Absence of S6K1 protects against age- and diet-induced obesity while enhancing insulin sensitivity. *Nature* **431:** 200–205.

Vadas O, Burke JE, Zhang X, Berndt A, Williams RL. 2011. Structural basis for activation and inhibition of class I phosphoinositide 3-kinases. *Sci Signal* **4:** re2.

van Dam RM, Hoebee B, Seidell JC, Schaap MM, Blaak EE, Feskens EJ. 2004. The insulin receptor substrate-1 Gly972Arg polymorphism is not associated with type 2 diabetes mellitus in two population-based studies. *Diabet Med* **21:** 752–758.

Versteyhe S, Blanquart C, Hampe C, Mahmood S, Christeff N, De MP, Gray SG, Issad T. 2010. Insulin receptor substrates-5 and -6 are poor substrates for the insulin receptor. *Mol Med Report* **3:** 189–193.

Voliovitch H, Schindler DG, Hadari YR, Taylor SI, Accili D, Zick Y. 1995. Tyrosine phosphorylation of insulin receptor substrate-1 in vivo depends upon the presence of its pleckstrin homology region. *J Biol Chem* **270:** 18083–18087.

Wang L, Balas B, Christ-Roberts CY, Kim RY, Ramos FJ, Kikani CK, Li C, Deng C, Reyna S, Musi N, et al. 2007. Peripheral disruption of the Grb10 gene enhances insulin signaling and sensitivity in vivo. *Mol Cell Biol* **27:** 6497–6505.

Weisberg SP, Hunter D, Huber R, Lemieux J, Slaymaker S, Vaddi K, Charo I, Leibel RL, Ferrante AW Jr. 2006. CCR2 modulates inflammatory and metabolic effects of high-fat feeding. *J Clin Invest* **116:** 115–124.

West IC. 2000. Radicals and oxidative stress in diabetes. *Diabet Med* **17:** 171–180.

Westwick JK, Bielawska AE, Dbaibo G, Hannun YA, Brenner DA. 1995. Ceramide activates the stress-activated protein kinases. *J Biol Chem* **270:** 22689–22692.

White MF. 2006. Regulating insulin signaling and β-cell function through IRS proteins. *Can J Physiol Pharmacol* **84:** 725–737.

Wijesekara N, Konrad D, Eweida M, Jefferies C, Liadis N, Giacca A, Crackower M, Suzuki A, Mak TW, Kahn CR, et al. 2005. Muscle-specific Pten deletion protects against insulin resistance and diabetes. *Mol Cell Biol* **25:** 1135–1145.

Winnay JN, Boucher J, Mori MA, Ueki K, Kahn CR. 2010. A regulatory subunit of phosphoinositide 3-kinase increases the nuclear accumulation of X-box-binding protein-1 to modulate the unfolded protein response. *Nat Med* **16:** 438–445.

Withers DJ, Gutierrez JS, Towery H, Burks DJ, Ren JM, Previs S, Zhang Y, Bernal D, Pons S, Shulman GI, et al. 1998. Disruption of IRS-2 causes type 2 diabetes in mice. *Nature* **391:** 900–904.

Wong JT, Kim PT, Peacock JW, Yau TY, Mui AL, Chung SW, Sossi V, Doudet D, Green D, Ruth TJ, et al. 2007. Pten (phosphatase and tensin homologue gene) haploinsufficiency promotes insulin hypersensitivity. *Diabetologia* **50:** 395–403.

Wong RH, Chang I, Hudak CS, Hyun S, Kwan HY, Sul HS. 2009. A role of DNA-PK for the metabolic gene regulation in response to insulin. *Cell* **136:** 1056–1072.

Xia P, Inoguchi T, Kern TS, Engerman RL, Oates PJ, King GL. 1994. Characterization of the mechanism for the chronic activation of diacylglycerol-protein kinase C pathway in diabetes and hypergalactosemia. *Diabetes* **43:** 1122–1129.

Yamaguchi H, Wang HG. 2001. The protein kinase PKB/Akt regulates cell survival and apoptosis by inhibiting Bax conformational change. *Oncogene* **20:** 7779–7786.

Yi Z, Langlais P, De Filippis EA, Luo M, Flynn CR, Schroeder S, Weintraub ST, Mapes R, Mandarino LJ. 2007. Global assessment of regulation of phosphorylation of insulin receptor substrate-1 by insulin in vivo in human muscle. *Diabetes* **56:** 1508–1516.

Yu J, Zhang Y, McIlroy J, Rordorf-Nikolic T, Orr GA, Backer JM. 1998. Regulation of the p85/p110 phosphatidylinositol 3′-kinase: Stabilization and inhibition of the p110α catalytic subunit by the p85 regulatory subunit. *Mol Cell Biol* **18:** 1379–1387.

Yu C, Chen Y, Zong H, Wang Y, Bergeron R, Kim JK, Cline GW, Cushman SW, Cooney GJ, Atcheson B, et al. 2002. Mechanism by which fatty acids inhibit insulin activation of IRS-1 associated phosphatidylinositol 3-kinase activity in muscle. *J Biol Chem* **277:** 50230–50236.

Yu Q, Gao F, Ma XL. 2011a. Insulin says NO to cardiovascular disease. *Cardiovasc Res* **89:** 516–524.

Yu Y, Yoon SO, Poulogiannis G, Yang Q, Ma XM, Villen J, Kubica N, Hoffman GR, Cantley LC, Gygi SP, et al. 2011b. Phosphoproteomic analysis identifies Grb10 as an mTORC1 substrate that negatively regulates insulin signaling. *Science* **332:** 1322–1326.

Yuan M, Konstantopoulos N, Lee J, Hansen L, Li ZW, Karin M, Shoelson SE. 2001. Reversal of obesity- and diet-induced insulin resistance with salicylates or targeted disruption of Ikkβ. *Science* **293:** 1673–1677.

Zhang D, Liu ZX, Choi CS, Tian L, Kibbey R, Dong J, Cline GW, Wood PA, Shulman GI. 2007. Mitochondrial dysfunction due to long-chain Acyl-CoA dehydrogenase deficiency causes hepatic steatosis and hepatic insulin resistance. *Proc Natl Acad Sci* **104:** 17075–17080.

Zhang J, Gao Z, Yin J, Quon MJ, Ye J. 2008a. S6K directly phosphorylates IRS-1 on Ser-270 to promote insulin resistance in response to TNF-α signaling through IKK2. *J Biol Chem* **283:** 35375–35382.

Zhang X, Zhang G, Zhang H, Karin M, Bai H, Cai D. 2008b. Hypothalamic IKK-β/NF-κB and ER stress link overnutrition to energy imbalance and obesity. *Cell* **135:** 61–73.

Zhang X, Vadas O, Perisic O, Anderson KE, Clark J, Hawkins PT, Stephens LR, Williams RL. 2011. Structure of lipid kinase p110β/p85β elucidates an unusual SH2-domain-mediated inhibitory mechanism. *Mol Cell* **41:** 567–578.

Zhou Q, Dolan PL, Dohm GL. 1999. Dephosphorylation increases insulin-stimulated receptor kinase activity in skeletal muscle of obese Zucker rats. *Mol Cell Biochem* **194:** 209–216.

Zhou BP, Liao Y, Xia W, Zou Y, Spohn B, Hung MC. 2001. HER-2/neu induces p53 ubiquitination via Akt-mediated MDM2 phosphorylation. *Nat Cell Biol* **3:** 973–982.

Zisman A, Peroni OD, Abel ED, Michael MD, Mauvais-Jarvis F, Lowell BB, Wojtaszewski JF, Hirshman MF, Virkamaki A, Goodyear LJ, et al. 2000. Targeted disruption of the glucose transporter 4 selectively in muscle causes insulin resistance and glucose intolerance. *Nat Med* **6:** 924–928.

MET: A Critical Player in Tumorigenesis and Therapeutic Target

Carrie R. Graveel[1], David Tolbert[2], and George F. Vande Woude[1]

[1]Molecular Oncology, Van Andel Research Institute, Grand Rapids, Michigan 49503

[2]Structural Sciences, Van Andel Research Institute, Grand Rapids, Michigan 49503

Correspondence: carrie.graveel@vai.org

Since its discovery more than 25 years ago, numerous studies have established that the MET receptor is unique among tyrosine kinases. Signaling through MET is necessary for normal development and for the progression of a wide range of human cancers. MET activation has been shown to drive numerous signaling pathways; however, it is not clear how MET signaling mediates diverse cellular responses such as motility, invasion, growth, and angiogenesis. Great strides have been made in understanding the pleotropic aspects of MET signaling using three-dimensional molecular structures, cell culture systems, human tumors, and animal models. These combined approaches have driven the development of MET-targeted therapeutics that have shown promising results in the clinic. Here we examine the unique features of MET and hepatocyte growth factor/scatter factor (HGF/SF) structure and signaling, mutational activation, genetic mouse models of MET and HGF/SF, and MET-targeted therapeutics.

Since the discovery of the receptor tyrosine kinase MET and its ligand, hepatocyte growth factor/scatter factor (HGF/SF), numerous studies have established the significant role of this receptor/ligand pair in tumor growth and metastasis. The last 25 years of work has made it clear that MET is unique among receptor tyrosine kinases (RTKs), yet the MET receptor activates many signaling pathways that are common to other RTKs. Why MET activation produces growth in one cell type and invasion in another is still unclear, yet the variety of responses to MET signaling is what makes this receptor so frequently associated with malignant growth.

Understanding the relationships between MET activation and its downstream signaling effectors is critical to the development of successful therapeutics for a wide range of malignancies.

The *MET* oncogene was first identified in the early 1980s in a human osteosarcoma tumor cell line that was exposed to N-methyl-N'-nitro-N-nitrosoguanidine, which produced a chromosomal translocation and a novel fusion protein, between a region called the translocated promoter region (TPR) on chromosome 1 and MET kinase domain on chromosome 7. Here the activation of the MET tyrosine kinase domain occurs through the dimerization domain from the TPR (Cooper et al. 1984; Park et al. 1986). Isolation of the full-length proto-oncogene revealed that MET was a unique receptor tyrosine kinase (Park et al. 1986).

The ligand was discovered first as a mitogenic factor of liver cells called hepatocyte growth factor (HGF), and it was shortly after determined to be the same as the motogenic factor called scatter factor (SF). The ligand, commonly referred to as HGF/SF, is the only ligand for the MET receptor (Stoker et al. 1987; Nakamura et al. 1989; Bottaro et al. 1991; Weidner et al. 1991; Gherardi et al. 2006). Under normal physiological conditions, HGF/SF is predominantly produced by mesenchymal cells and acts in a paracrine fashion on MET-expressing epithelial cells (Jeffers et al. 1996). The proliferative and motogenic effects observed in these early studies were some of the first indications of the varied roles that MET signaling has in tumor growth and metastasis.

Embryonic development and tissue regeneration are normal physiological processes that parallel the mechanisms of growth and invasion that occur during tumor progression. Several studies have shown that MET-HGF/SF signaling is essential for embryonic development and regeneration. Depending on the cellular context, MET signaling induces cell proliferation, motility, scattering, angiogenesis, or invasion. These pleotropic attributes are what make MET signaling essential in both normal development and tumor progression. During development, paracrine MET signaling drives the epithelial-to-mesenchymal transition (EMT) of myogenic progenitor cells and is crucial for placenta and liver development (Bladt et al. 1995; Schmidt et al. 1995; Uehara et al. 1995). MET signaling is also critical for liver regeneration and wound repair in skin (Chmielowiec et al. 2007). The signaling networks that drive the developmental processes of EMT, wound healing, and invasion are exploited in tumor cells to promote invasive growth.

The expression and/or activation of MET and HGF/SF have been implicated in the development of numerous human cancers (www.vai.org/met), including carcinomas (breast, colon, gastric, renal, pancreatic, bladder, liver, lung, prostate, ovarian, etc.), sarcomas (osteosarcoma, rhabdomyosarcoma), hematopoietic malignancies (multiple myeloma, lymphoma, chronic myeloid leukemia), melanomas, and central nervous system tumors (glioblastomas and astrocytomas) (Birchmeier et al. 2003; Corso et al. 2005; Gherardi et al. 2012). Uncontrolled MET signaling can occur through overexpression of HGF/SF or MET, mutational activation of MET, autocrine signaling, or gene amplification. Numerous in vitro and in vivo studies have shown that MET signaling plays a key role in tumorigenic growth, metastasis, and therapeutic resistance. It is crucial that we develop an in-depth understanding of how MET signaling regulates both normal and tumorigenic cell processes to develop successful therapeutic strategies. In this review, we will discuss the unique features of MET and HGF/SF structure and signaling, mutational activation, genetic mouse models of MET and HGF/SF, and MET-targeted therapeutics.

STRUCTURAL CHARACTERISTICS OF MET AND HGF/SF

MET is a disulfide-linked heterodimer that is created by cleavage of the precursor into a shorter extracellular α chain and a longer β chain (Fig. 1A). The extracellular portion consists of a seven-bladed β-propeller Sema domain atop four immunoglobulin-like (Ig-like) repeated domains (Love et al. 2003). The Sema domain, which is also present in plexins, semaphorins, and integrins, contains the binding sites for HGF/SF and is linked to the Ig-like repeats by a short cysteine-rich PSI domain (Stamos et al. 2004; Holmes et al. 2007). The remainder of the receptor is composed of a juxtamembrane domain, a tyrosine kinase domain, and a carboxy-terminal tail that is involved in downstream signaling. Structurally, the MET receptor is similar to RON (recepteur d'origine nantais) in humans, Stk (RON murine homolog), f-Stk (RON feline homolog), and c-sea, a cell-surface receptor in chickens (Ronsin et al. 1993; Camp et al. 2005).

HGF/SF is also a polyprotein that structurally belongs to the serine protease family and is most closely related to macrophage stimulating

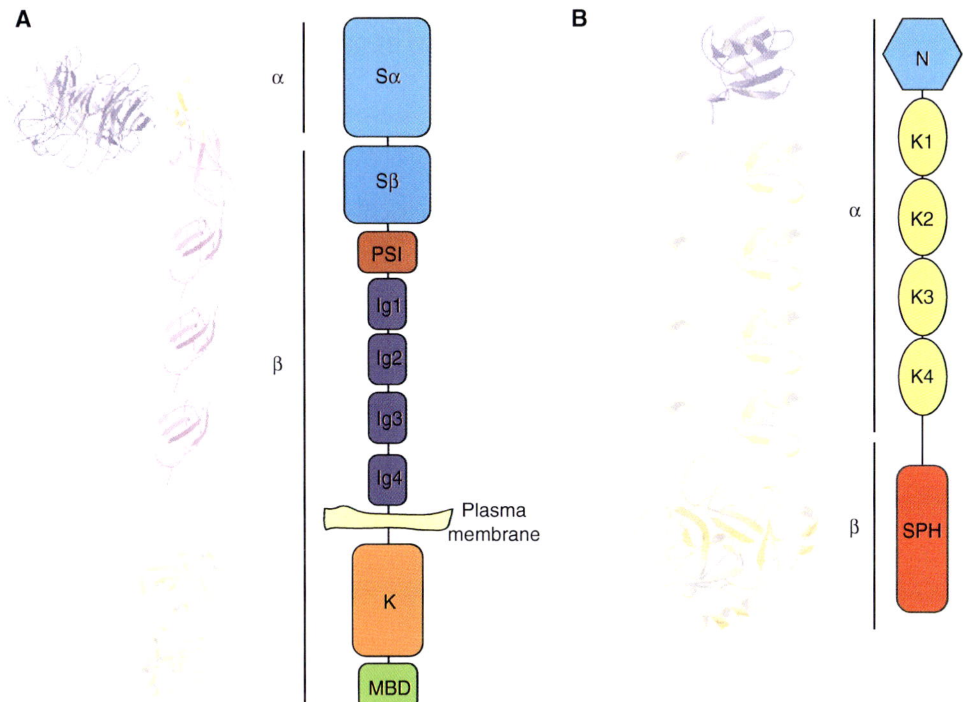

Figure 1. Structural characteristics of MET and HGF/SF. (*A*) Structure of the MET receptor (α and β refer to subunits present after proteolytic cleavage). MET is expressed at the plasma membrane: The extracellular portion consists of the sema domain, a PSI domain, and four immunoglobulin-like (Ig-like) repeated domains; the intracellular region contains the tyrosine kinase domain and the multifunctional binding domain. The three-dimensional models (*left*) were generated using the following coordinates from the Protein Data Bank: 1SHY (sema domain), 2UZY (PSI and Ig1, and 2), and 1R1W (kinase domain). The Ig3 and Ig4 domains are modeled as copies of the Ig2 domain. (*B*) Functional domains of HGF/SF. HGF/SF contains an amino-terminal domain (N), four tandem repeats of kringle domains (K1–K4), and a serine protease homology domain (SPH). The three-dimensional models (*left*) were generated using the following coordinates from the Protein Data Bank: 1NK1 (N and K1) and 1SHY (HGF/SF β chain). K2–4 are represented as copies of K1.

factor (MSP) and plasminogen. HGF/SF is composed of an α chain that contains the amino-terminal domain, four tandem repeats of kringle domains, and a serine protease-like β chain that lacks catalytic activity (Fig. 1B) (Donate et al. 1994). The amino-terminal domain also contains a high-affinity binding site for heparin and dermatan sulfate (Deakin and Lyon 1999a,b; Lietha et al. 2001; Lyon et al. 2004). Similar to plasminogen and MSP, HGF/SF is made as an inactive, single-chain precursor that is proteolytically converted into an active heterodimer. HGF/SF can be activated by a number of proteases, including hepatocyte growth factor activator, plasma kallikrein, and

coagulation factor XIa (Shimomura et al. 1993; Peek et al. 2002), and it can bind with high affinity to MET in an active or inactive state (Gherardi et al. 2006). Activation of HGF/SF occurs by cleavage at Arg494, which produces a sulfhydryl-linked heterodimer of the NK4α domain with the β subunit serine protease domain.

The newly created amino terminus of the serine protease β domain is an integral part of the catalytic site and prevents protease activity from other serine proteases. After cleavage, the available amino terminus inserts into the "activation pocket" of the protease domain, enabling the HGF/SF β chain to bind to the Sema domain of MET (Kirchhofer et al. 2004; Carafoli et al. 2005). Binding of HGF/SF to MET results in receptor oligomerization at the plasma membrane and subsequent autophosphorylation of the activation loop (tyrosines 1234 and 1235) within the intracellular kinase domain (Gonzatti-Haces et al. 1988; Blume-Jensen and Hunter 2001; Bardella et al. 2004). The autophosphorylation destabilizes the loop, allowing substrate access and phosphorylation of tyrosines 1349 and 1356 within the carboxy-terminal domain (Schiering et al. 2003; Wang et al. 2005). These carboxy-terminal tyrosines serve as the multifunctional docking sites for various signaling adaptors, as discussed below.

MET activation by HGF/SF is a complex process; for example, HGF/SF can bind to MET in either an active or an inactive state, and it can be activated by a number of proteases. In fact, the uncleavable mutant form of HGF/SF (arginine 494 mutated to glutamic acid) can even function as a competitive inhibitor to wild-type HGF/SF (Gherardi et al. 2006). In addition, the α chain of HGF/SF is susceptible to proteolysis by coagulation factor Xa (and possibly by other proteases), generating N-domain and kringle fragments that possess their own intrinsic activity in some circumstances (Rubin et al. 2001; Pediaditakis et al. 2002; Shen et al. 2008). HGF/SF also has two shorter splice variants, the amino-terminal domain plus either one or two kringle domains (NK1 and NK2) (Stahl et al. 1997), which can act as agonists or antagonists depending on the conditions and assay used (Montesano et al. 1998). Heparin/dermatan sulfate can also have a dominant effect on NK1, NK2, and the affinity of the α chain for MET (Deakin and Lyon 1999b; Lyon et al. 2004). Two possible HGF/SF dimer interfaces, an α-chain- and a β-chain-based dimer, may induce MET oligomerization and activation (Chirgadze et al. 1999; Stamos et al. 2004). The discovery of these alternative ligand–receptor interactions has led to the development of a number of HGF/SF-based agonist/antagonist derivatives that may have therapeutic value in wound healing or cancer treatment (Lietha et al. 2001; Tolbert et al. 2007; Youles et al. 2008).

MET ACTIVATION, SIGNALING, AND REGULATION

MET activation induces complex signaling events that depend on the cellular context and produce a variety of cellular responses. For example, treatment of epithelial cell lines with HGF/SF can induce proliferation, anchorage-independent growth, cell scattering, survival, or tubulogenesis. Tubulogenesis is a complex process that involves cellular proliferation, motility, differentiation, and polarization (Birchmeier et al. 1997). Induction of these biological responses through MET requires the cooperation of several intracellular adaptors and signaling effectors. The strength and duration of these signals are tightly controlled through interactions with diverse signal modifiers, expression levels, and subcellular localization and degradation.

For ligand-mediated receptor activation, HGF/SF binds to the extracellular domain of MET, resulting in receptor dimerization and trans-phosphorylation of tyrosines Y1234 and Y1235 within the kinase domain. Phosphorylation of these residues initiates a conformational change within the receptor, resulting in the phosphorylation of Y1349 and Y1356. These carboxy-terminal residues are part of a unique bidentate docking site that is necessary for MET signaling. Upon MET activation, the multifunctional docking site is capable of interacting with several adaptor proteins and direct kinase substrates, including Gab1, Grb2, phosphatidylinositol 3-kinase (PI3K), Shc, Src, Shp2, Ship1, and Stat3 (Fig. 2). Scaffolding adaptors contain several phosphorylation sites that facilitate the recruitment of proteins having Src-homology-2 (SH2) or phosphotyrosine-binding (PTB) domains. The binding of specific substrates to the MET carboxy-terminal docking site creates a scaffold of signaling effectors that govern cellular responses to MET activation.

Gab1 is the major substrate for MET in epithelial cells, and several studies have shown its necessity for downstream signaling through MET (Maroun et al. 1999; Sachs et al. 2000). Embryos that are nullizygous for Gab1 have the same lethal defects as animals nullizygous

                    Cite this article as *Cold Spring Harb Perspect Biol* doi: 10.1101/cshperspect.a009209

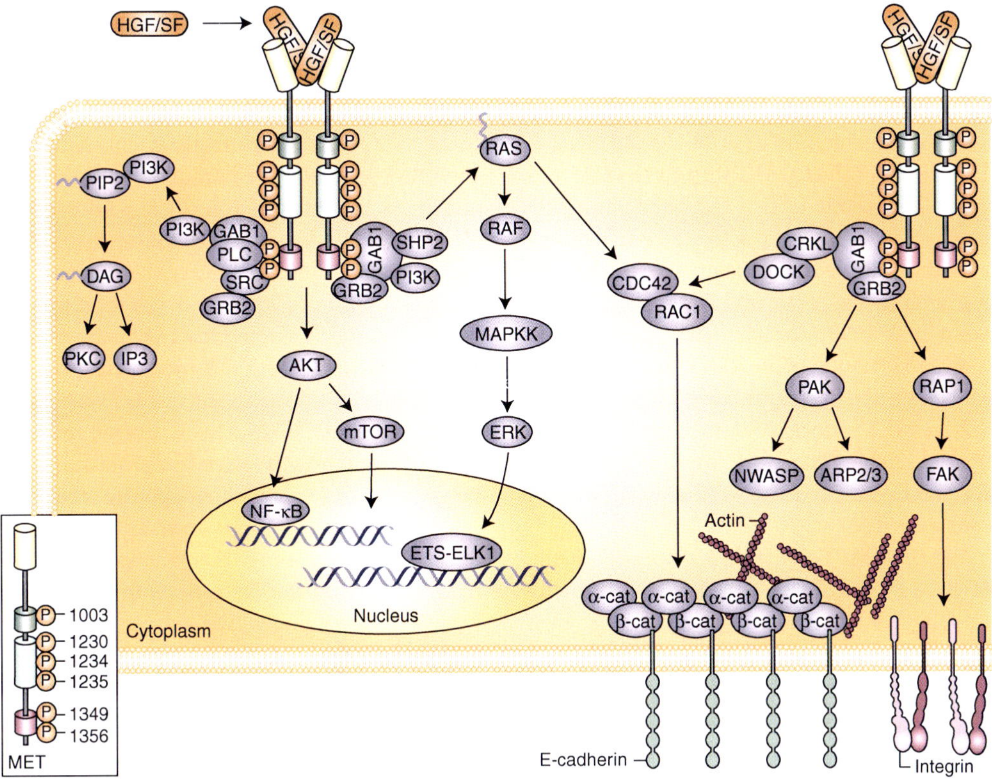

Figure 2. MET-HGF/SF signaling. The MET receptor is activated at the plasma membrane by binding of HGF/SF to the extracellular region of MET. Upon dimerization and activation, tyrosine phosphorylation occurs at Tyr1003 in the juxtamembrane CBL-binding site (shown in green), Tyr1230, Tyr1234, and Tyr1235 in the active site of the kinase (shown in light green), and Tyr1349 and Tyr1356 in the bidentate docking site (shown in pink). MET can then mediate several intracellular signaling pathways through a diverse array of adaptors and downstream effectors. The Gab1 and Grb2 adaptor proteins are critical mediators of MET activation and signaling through RAS-MAPK, PI3K-AKT, RAC1, and PAK pathways drive distinct cellular responses including proliferation, cell survival, and migration. (From Gherardi et al. 2006; reprinted, with permission, from Nature Publishing Group, © 2012.)

for MET and HGF/SF (Itoh et al. 2000; Sachs et al. 2000). The interaction between Gab1 and MET is distinctive in that it can occur either directly or indirectly through Grb2. Direct interaction occurs through a 13-amino-acid MET-binding domain (MBD) on Gab1 that binds to Y1349 in the MET multifunctional docking site (Schaeper et al. 2000). The MBD is a unique motif that is not conserved in other Gab family members, yet it promotes sustained interaction and phosphorylation on MET activation (Maroun et al. 2000). This interaction leads to the activation of several signaling cascades through Gab1, including Shp2, PI3K, PLCγ, and Crk.

Activation of these pathways is able to induce diverse cellular responses necessary for normal and tumorigenic cell growth. For instance, Gab1-mediated activation of PI3K and AKT is known to promote cell survival and cell migration (Rosario and Birchmeier 2003), whereas Crk couples with Rap1 and Rac to mediate cell motility and branching morphogenesis (Lamorte et al. 2002a,b; Rodrigues et al. 2005). The binding of the phosphatase Shp2 to Gab1 is known to up-regulate the Ras/ERK/MAPK pathway, leading to branching morphogenesis and proliferation (Maroun et al. 2000; Schaeper et al. 2000).

In addition to Gab1, the Grb2 and Shc adaptor proteins are critical mediators of MET activation. Grb2 and Shc associate with MET and other RTKs through their respective SH2 and PTB domains (Furge et al. 2000). Grb2 is able to directly bind to MET through its SH2 and SH3 domains, but is also recruited indirectly through Shc. Grb2 links activated MET receptors with multiple downstream signaling pathways, such as Ras/ERK and PI3K/AKT (Lowenstein et al. 1992; Rozakis-Adcock et al. 1992; Gu et al. 2000; Ong et al. 2001). Recent studies have shown that recruitment of Shc, but not Grb2, to MET induces VEGF expression (Saucier et al. 2004). Therefore, binding of Shc to MET may be a crucial angiogenic switch during tumor growth.

Although MET signaling is primarily mediated by the Ras-MAPK and PI3K-AKT pathways (Bertotti et al. 2009), other downstream effectors such as NF-κB, β-catenin, and STAT3 have been linked to MET. NF-κB contributes to HGF/SF-mediated proliferation and tubulogenesis through ERK1/2 and p38 MAPK (Muller et al. 2002). Another study showed that MET activation of NF-κB is able to protect cells from apoptosis through AKT (Fan et al. 2005). STAT3 (signal transducer and activator of transcription 3) is also associated with MET. We have found that MET-mediated STAT3 activation in SK-LMS-1 (human leiomyosarcoma cells) and Madin-Darby canine kidney (MDCK) epithelial cells is required for anchorage-independent growth and tumorigenesis, whereas it has no effect on branching morphogenesis (Zhang et al. 2002). Others have reported that STAT3 mediates anchorage-independent growth and is required for branching morphogenesis, but we found no such activity (Boccaccio et al. 1998; Zhang et al. 2002).

MET is also able to initiate biological responses indirectly by interacting with proteins at the plasma membrane. Crosstalk occurs between MET and the developmental Wnt-β-catenin pathway in which MET directly interacts with E-cadherin and induces nuclear localization of β-catenin (Monga et al. 2002; Apte et al. 2006; Reshetnikova et al. 2007). Recent work has shown that in colon cancer, HGF-producing myofibroblasts activate β-catenin-dependent transcription and stimulate cancer stem cell (CSC) populations (Vermeulen et al. 2010). Further, MET interaction with the hyaluron receptor CD44 creates a complex with the ERM proteins (ezrin, radixin, and moesin) and the actin cytoskeleton (Orian-Rousseau et al. 2002), and in some cell lines, interaction with CD44 is required for Ras/MAPK signaling. MET also interacts with the death receptor Fas, which maintains homeostasis in many tissues. MET is able to prevent Fas-mediated apoptosis in hepatocytes by sequestering Fas (Wang et al. 2002; Zou et al. 2007). This is a novel relationship between growth factor receptors and cytokine receptors in controlling apoptosis.

Tumor angiogenesis is an essential response to hypoxic conditions that allows tumors to progress beyond the limitations of the normal vasculature. Angiogenesis is largely mediated by the vascular endothelial growth factor receptor (VEGFR) family and the hypoxia-inducible factors (HIF). Several studies have shown that MET signaling can promote angiogenesis through induction of VEGFA expression and suppression of thrombospondin 1 (TSP1), a negative regulator of angiogenesis (Bussolino et al. 1992; Grant et al. 1993; Zhang et al. 2003; Abounader and Laterra 2005). By inducing angiogenic promoters and inhibiting angiogenic suppressors, MET ensures angiogenesis during tumor progression. In recent years, it has been shown that hypoxic conditions induce *MET* transcription and amplify MET signaling and MET-mediated invasion in several types of carcinomas (Pennacchietti et al. 2003). These observations have significant clinical implications: for instance, will antiangiogenic therapies induce hypoxia-mediated MET activation? On the other hand, the combined inhibition of angiogenesis and MET is being evaluated in both preclinical and clinical studies.

Regulation of RTK activation is necessary for preventing prolonged activation of downstream signaling cascades. Termination of RTK activation can occur through receptor dephosphorylation, sequestration, and degradation, or through signaling antagonists (e.g., Sprouty, Mimp, and Mig-6). Defective receptor trafficking and degradation can result in increased signaling and, ultimately, malignant transformation (Mosesson et al. 2008). Termination of MET signaling is predominantly controlled through receptor internalization and degradation. Several studies have shown that the ubiquitin ligase Cbl is central to down-regulation of the MET receptor (Peschard and Park 2007). Cbl is recruited to MET through Grb2, but it is also able to bind directly to MET through phosphorylated Y1003 (Fig. 3). Binding of Cbl to Y1003 results in receptor ubiquitination, internalization, and degradation. MET ubiquitina-

tion is crucial to maintaining physiological MET activation levels, because it has been shown that mutations within the Cbl-binding domain are oncogenic (Abella et al. 2005; Peschard and Park 2007). Further studies are needed in both cellular and animal models for us to have a complete understanding of the intricacies of MET regulation and downstream signaling.

MUTATIONAL ACTIVATION OF *MET*

The first activating mutations found within *MET* were identified through a genome-wide

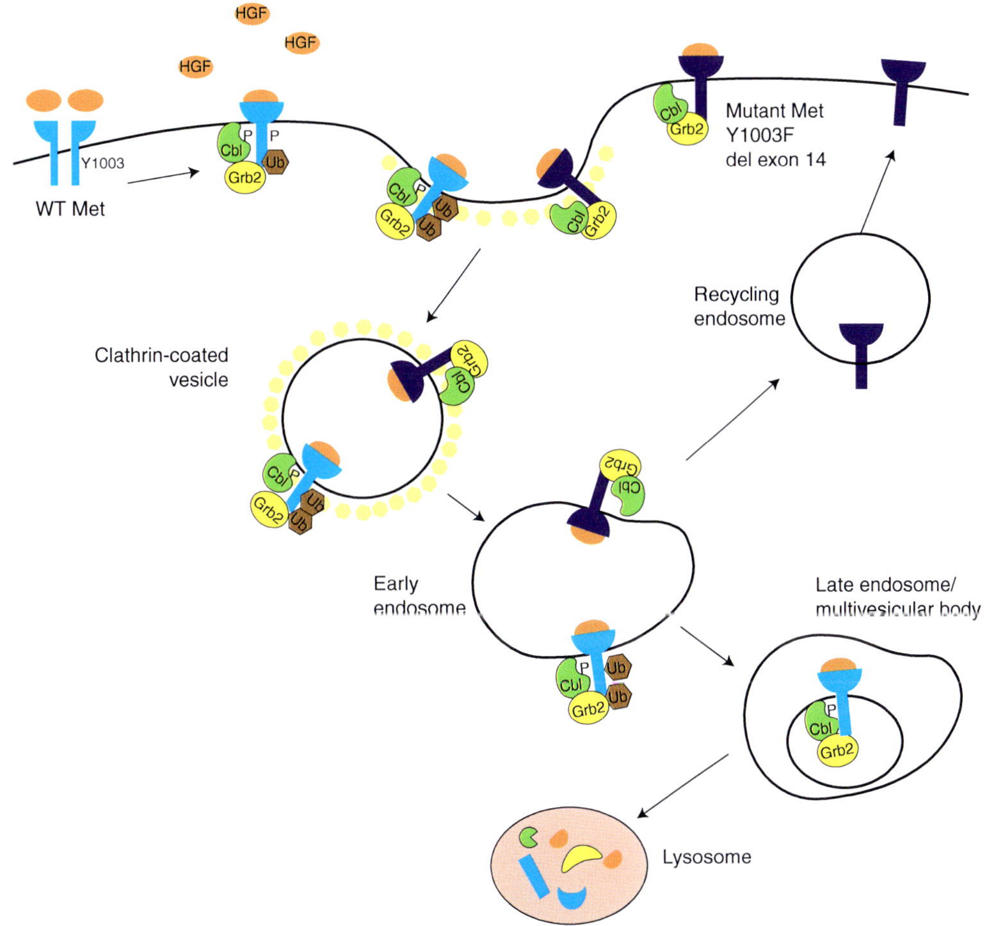

Figure 3. Down-regulation of the MET receptor. Following ligand binding and dimerization, MET Y1003 is phosphorylated and recruits the ligase Cbl. Cbl induces receptor ubiquitination and internalization. MET is processed through the endosomal pathway and is eventually degraded by lysosomal proteases. Receptors that are not ubiquitinated can be recycled back to the plasma membrane. Mutation or deletion of Y1003 allows MET to avoid lysosomal degradation and MET is recycled back to the plasma membrane.

scan of families with hereditary papillary renal carcinoma (HPRC) (Schmidt et al. 1997). This was the first genetic evidence demonstrating the oncogenicity of *MET* in humans. The missense mutations identified in HPRC patients flank the critical tyrosines Y1234 and Y1235 within the kinase domain (Fig. 4). Both germline and somatic mutations have been identified within the kinase domain, and several studies have shown that these mutations induce constitutive receptor activation (Jeffers et al. 1997, 1998; Schmidt et al. 1999). *MET* kinase domain mutations have also been observed in childhood hepatocellular carcinomas, metastatic head and neck cancers, gastric carcinomas, and squamous cell cancers (Park et al. 1999; Di Renzo et al. 2000; Lee et al. 2000; Aebersold et al. 2003). Numerous studies have delved into the mechanism by which mutational activation of *MET* is oncogenic, and it has been shown that mutationally activated MET can be ligand-dependent or -independent

(Jeffers et al. 1998; Michieli et al. 1999; Wang et al. 2001).

Screens for *MET* mutations in other solid cancers have shown that *MET* tyrosine kinase mutations are not a common event. However, several groups have looked outside the kinase domain and discovered mutations and deletions within the juxtamembrane and Sema domain of *MET* in gastric, small cell lung, and non-small cell lung cancers (Lee et al. 2000; Ma et al. 2003; Kong-Beltran et al. 2006). T1010I was identified in the juxtamembrane (JM) domain of a small cell lung cancer (SCLC) tumor sample (Fig. 4). Furthermore, one Sema domain missense mutation (E168D), two-base-pair insertional mutations (IVS13 [52−53]-insCT) within the pre-JM intron 13, and an alternative transcript involving exon 10 were identified in this study. The mutations within the juxtamembrane domain were found to alter cell adhesion and induce tumorigenicity by in vitro assays. The Sema

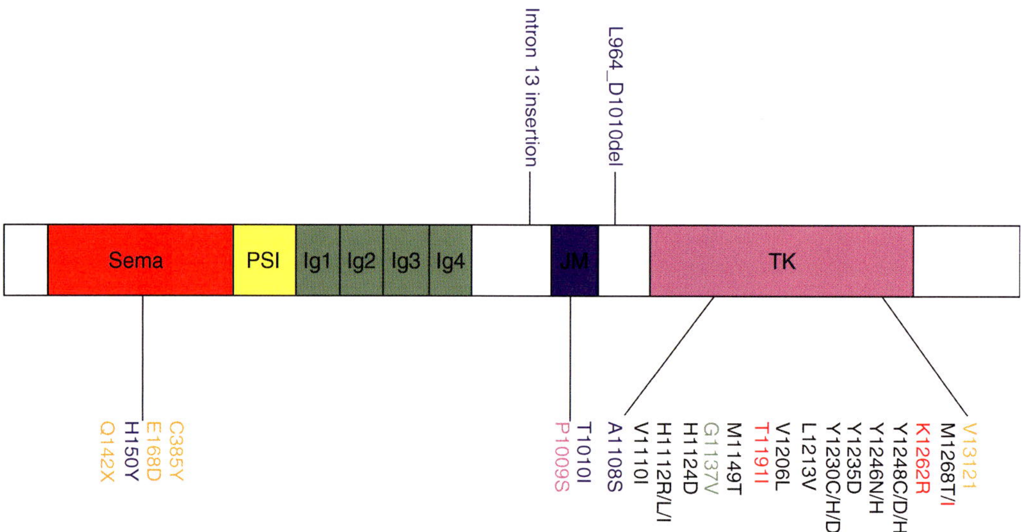

Figure 4. MET mutations identified in human cancers. Color coding of the mutation symbols above and below the MET protein diagram identifies the type of human tumor: black, papillary renal cell carcinoma; red, hepatocellular carcinoma; green, glioma; blue, lung carcinoma; purple, gastric carcinoma; orange, cancer unknown primary origin (CUP). Genetic alterations were found in the following tumors: L964_D1010del, lung adenocarcinoma; A1108S, squamous lung carcinoma; E168D, small cell lung carcinoma; intron 13 insertion 9IVS13-(52-52)insCT, small cell lung carcinoma; T1010I, small cell lung carcinoma. Note that T1010I was also found in a breast carcinoma and CUP; E168D was found in a CUP; Y1235D and Y1230C/D were identified in head and neck squamous cell carcinomas, respectively.

domain mutation has not been carefully evaluated but may affect the structure of the ligand-binding domain.

Another mutational analysis in lung and colon cancers identified several mutations that increased protein stability and MET receptor activation (Kong-Beltran et al. 2006). A mutation within a dinucleotide splice site resulted in deletion of exon 14, which decreased the binding of the Cbl E3-ligase and led to attenuated ubiquitination and degradation of the MET receptor. Treatment of cells containing the exon 14 deletion with an HGF-competitive MET antibody resulted in MET inhibition, suggesting that the exon 14 mutant is ligand independent and can be therapeutically targeted (Kong-Beltran et al. 2006). A recent study identified several novel mutations in cancers of unknown primary origin (CUP) including four somatic mutations clustered in the SEMA domain, one mutation in the juxtamembrane domain, and one near the active site of the tyrosine kinase domain (Stella et al. 2011). These mutations had a 30% incidence in the CUP cohort and were a negative prognostic marker.

Understanding how mutations activate the MET kinase is crucial to the development of effective cancer therapeutics. The efficacies of several promising kinase inhibitors, such as imatinib, have been overcome by tumors that acquired new point mutations. The efficacy of SU11274 (which targets the ATP-binding site of MET) was tested against the naturally occurring MET mutations H1112Y, L1213V, Y1248H, and M1268T in transformed NIH3T3 cells (Berthou et al. 2004). H1112Y and M1286T were both sensitive to SU11274 inhibition, but Y1248H and L1213V were resistant to treatment. Another study using the MET-dependent gastric carcinoma cell line SNU638 and two MET inhibitors (PHA-665752 and PF-2341066) observed resistance through acquisition of the Y1230H mutation in the activation loop (Qi et al. 2011). Structural analysis showed that Y1230H destabilizes the autoinhibitory conformation of MET and abrogates an important aromatic stacking interaction with the inhibitors. These results underscore the necessity for a clearer understanding of the three-dimensional structure of MET and how mutations alter the binding dynamics of therapeutic inhibitors.

MOUSE MODELS OF MET AND HGF/SF

Even though innumerable in vitro studies have shown the influence of MET activation on tumor development, mouse models have been invaluable for investigating the complexities of MET signaling in development and tumorigenesis. As shown in Table 1, knock-out, transgenic, inducible, and knock-in models have shown the diverse tissues in which MET signaling affects normal or tumor development.

Genetic studies in mice were the first to reveal the importance of MET and HGF/SF in embryonic development, during which MET and HGF/SF are expressed in close proximity to each other and receptor activation is regulated through paracrine signaling. When MET or HGF/SF is knocked out in mice, defective development of the liver and placenta results in embryonic death (Bladt et al. 1995; Schmidt et al. 1995; Uehara et al. 1995). The livers of MET$^{-/-}$ and HGF/SF$^{-/-}$ embryos are drastically smaller owing to decreased hepatocyte proliferation and increased apoptosis. Severely impaired placental development is a result of a reduction in labyrinthine trophoblasts. The identical phenotypes of the MET$^{-/-}$ and HGF/SF$^{-/-}$ embryos reaffirmed that MET and HGF/SF are an exclusive receptor/ligand pair. Furthermore, MET$^{-/-}$ and HGF/SF$^{-/-}$ embryos both lack skeletal muscles in the limb, tongue, and diaphragm, demonstrating the necessity of MET signaling for the migration of myogenic precursor cells. Interestingly, the mechanistic process by which myogenic precursors are released and migrate through the embryo is similar to the EMT observed during tumor invasion. In addition, placentation, liver development, and long-range migration of muscle progenitor cells are late evolutionary processes that confirm the emergence of MET signaling later in evolution (Birchmeier et al. 2003).

In 2004, three separate laboratories developed inducible mouse lines to further evaluate the regenerative abilities of MET and HGF/SF

Table 1. Mouse models of Met and HGF/SF

	Phenotype	References
Knock-out models		
HGF/SF	Embyronic lethality owing to placental and liver defects	Schmidt et al. 1995; Uehara et al. 1995
Met	Embryonic lethality owing to impaired migration of myogenic precursor cells	Bladt et al. 1995
Transgenic models		
MT-HGF	Diverse carcinomas, sarcomas, and melanomas	Takayama et al. 1997
	Melanomas	Otsuka et al. 1998
TRE-Met	Hepatocellular carcinomas	Wang et al. 2001
WAP-HGF	Mammary adenocarcinomas	Gallego et al. 2003
Tpr-Met	Mammary adenocarcinomas	Liang et al. 1996
HGF-scid	Enhanced growth of heterotopic xenografts	Zhang et al. 2005
GFP-Met	Adenomas, adenocarcinomas, and angiosarcomas in abdominal sebaceous glands	Moshitch-Moshkovitz et al. 2006
Met^{mt}	Mammary adenocarcinomas	Ponzo et al. 2009
Knock-in models		
Met^{mut}	Diverse carcinomas, sarcomas, and lymphomas	Graveel et al. 2004, 2009
Flox models		
Met	Mx-cre deletion of exon 15; impaired liver regeneration	Borowiak et al. 2004
HGF	Adeno-cre deletion of exon 5; impaired liver regeneration	Phaneuf et al. 2004
Met	Alb-cre deletion of exon 15; impaired liver regeneration	Huh et al. 2004

in the adult liver. Borowiak et al. developed MET^{flox} mice that were crossed to transgenic mice expressing *cre* under the control of the IFN-inducible *Mx* promoter (Borowiak et al. 2004). Liver regeneration after partial hepatectomy was greatly impaired in the MET^{flox} mice. Impaired regeneration was also observed in another MET^{flox} line in which Alb-cre mice were crossed to induce selective inactivation of MET within hepatocytes (Huh et al. 2004). MET^{flox} livers were found to have increased sensitivity to Fas-induced apoptosis and significantly decreased regeneration after treatment with the hepatocarcinogens phenobarbital and carbon tetrachloride. Another study created $HGF^{ex5-flox}$ mice that were administered a recombinant adenoviral vector coding for cre recombinase (Ad-Cre1) to selectively ablate HGF/SF (Phaneuf et al. 2004). $HGF^{ex5-flox}$ mice injected with carbon tetrachloride had significant reductions in regeneration compared with controls, reaffirming the observations made in the MET^{flox} models.

Several transgenic mouse models of MET and HGF have been created that illuminate the effect of MET activation on tumor development. Transgenic mice expressing HGF/SF under the metallothionein promoter develop a diverse array of tumors, many of which originate in tissues that exhibit abnormal development, including the mammary gland, skeletal muscle, and melanocytes (Takayama et al. 1997). Autocrine signaling of MET and HGF/SF had been implicated in several cell culture systems, but this was the first in vivo model to show the connection between autocrine MET-HGF/SF signaling and tumorigenesis. Further studies with MT-HGF/SF mice revealed that MET-HGF/SF autocrine signaling could induce metastatic melanomas (Otsuka et al. 1998). Other transgenic models have been created that target HGF/SF expression to the mammary epithelium under the control of the whey acidic protein (WAP) gene promoter (Gallego et al. 2003). WAP-HGF mice developed multiple metastatic adenocarcinomas within the mammary gland after pregnancy and lactation. These mammary tumors had high levels of MET activation and HGF expression, similar to what is observed in some human breast carcinomas. Earlier studies of TPR-MET transgenic mice (under the metallothionein promoter) also showed mammary adenocarcinoma development after pregnancy (Liang et al. 1996). These transgenic models supported the hypothesis that altered MET activation or signaling may play a role in a variety of tumor types.

To understand how ligand-independent activation of RTKs can affect tumorigenesis, Wang et al. created transgenic mice that express human *MET* in hepatocytes under the control of tetracycline (Wang et al. 2001). These mice developed hepatocellular carcinomas that regressed when the transgene was suppressed. Even though human MET cannot be activated by murine HGF/SF, MET activation was present in hepatocytes and was dependent on cell adherence. This was the first evidence that MET was tumorigenic in vivo through ligand-independent mechanisms. This study also shed light on how MET overexpression may be tumorigenic in tumors that do not express HGF/SF.

To further evaluate MET expression through imaging, Moshitch-Moshkovitz et al. (2006) developed a GFP-MET transgenic mouse that allowed for direct subcellular-resolution imaging of expression patterns during tumor progression. The GFP-MET mice developed sebaceous gland tumors that were imaged by confocal laser scanning microscopy (CLSM) and analyzed by Western blot analysis. A gradual increase in GFP-MET levels from normal skin, to adenoma and angiosarcoma, to adenocarcinoma, was observed. In addition, single cells expressing high levels of GFP-MET were observed spreading from the tumor, similar to micrometastases.

To investigate MET and HGF/SF in human cancers, human tumor xenografts are often grown in immunocompromised mice. Because human MET is not activated by murine HGF/SF, traditional xenograft models do not accurately represent paracrine signaling in human tumors. To solve this dilemma, Zhang et al. created a transgenic mouse expressing human HGF (designated hHGF-Tg) on a severe combined immunodeficiency (SCID) background (Zhang et al. 2005). The expression of hHGF significantly enhanced the growth of heterotopic sub-

cutaneous xenografts derived from human MET-expressing cancer cells. This model has been invaluable for the testing of therapeutic agents in human cancer cells in the context of MET signaling (Gao et al. 2006; Merchant et al. 2008; Zhang et al. 2010).

In 1997, activating mutations within the MET kinase domain were identified in families with HPRCs (Schmidt et al. 1997). The oncogenic potential of these mutations was confirmed through several in vitro, xenograft, and transgene experiments (Jeffers et al. 1997, 1998). To examine mutational activation of MET in vivo, knock-in mouse models were created with activating mutations (WT, D1226N, Y1228C, M1248T, and M1248T/L1193V) (Graveel et al. 2004). The different mutant MET lines developed unique tumor profiles including carcinomas, sarcomas, and lymphomas. It was also observed that the majority of tumors had nonrandom duplication of the mutant *MET* allele. This selective chromosomal amplification has been observed in patients with HPRC. Further studies have shown that when the M1248T/L1193V is congenically bred onto the FVB/N background, these mice develop a high incidence of aggressive mammary tumors (Graveel et al. 2009), which are histologically diverse and have several characteristics similar to those of human basal breast cancers. Similar mammary tumorigenesis also occurs in a transgenic model of the M1248T mutation (Ponzo et al. 2009), demonstrating that MET may induce progression of aggressive breast cancer subtypes. Overall, these germline and conditional mouse models have provided us with a greater understanding of the significant functions of MET signaling in development, tissue homeostasis, and tumorigenesis.

TARGETING MET-HGF/SF IN CANCER

Given that MET is involved in multiple stages of tumor progression in a variety of human cancers, it is a highly promising therapeutic target (Knudsen and Vande Woude 2008). In recent years, several approaches have been used to specifically target MET in neoplastic cells. Early studies of small molecule inhibitors showed that selective MET inhibition impeded tumor growth in mouse models (Christensen et al. 2003; Sattler et al. 2003; Wang et al. 2003; Knudsen and Vande Woude 2008). These compounds act by inhibiting ATP binding to the kinase domain. Based on the success of other receptor tyrosine kinase inhibitors such as gefitinib and imatinib, competitive inhibitors of MET are promising. Other approaches include small interfering RNA (siRNA) or ribozymes targeting MET and/or HGF/SF (Abounader et al. 1999; Shinomiya et al. 2004), neutralizing antibodies against MET and/or HGF/SF (Cao et al. 2001), and the HGF/SF antagonist NK4 (Date et al. 1997). Decoy receptors that inhibit MET activation by preventing both HGF binding and ligand-independent MET dimerization are able to inhibit MET signaling and MET-dependent tumor growth (Kong-Beltran et al. 2004; Michieli et al. 2004).

Recent work on drug-resistant lung cancers has shown that MET may be a critical player in developed resistance to targeted therapies. The epidermal growth factor receptor (EGFR) kinase inhibitors gefitinib and erlotinib are effective treatments for non-small cell lung cancers (NSCLC), but the majority of these tumors develop resistance with time. Approximately 50% of the tumors develop resistance owing to a secondary mutation in EGFR; however, focal amplification of MET was observed in 22% of the resistant tumors (Engelman et al. 2007). MET activation has also been associated with resistance to EGFR and ERBB2 inhibitors in colorectal cancer cells and breast cancer cells, respectively (Shattuck et al. 2008; Liska et al. 2011). Conversely, activation of ERBB family members mediates resistance to MET inhibition in gastric carcinoma cell lines (Corso et al. 2010). These studies and others indicate that signaling cross talk between RTKs may drive resistance to targeted therapies. Therefore, it is critical that the molecular profile of MET and other RTKs is understood to achieve clinical success with targeted inhibitors.

At this time, numerous MET-HGF/SF therapeutics are being evaluated in clinical trials. Several MET-specific inhibitors have shown promising results, including the monovalent antibody MetMAb (onartuzumab), which has shown activity in combination with erlotinib in NSCLC patients (Spigel et al. 2011). The HGF/SF monoclonal antibody AMG 102 (rilotumumab) improved overall survival of gastric adenocarcinoma (Oliner et al. 2012). In both of these studies, the best response was observed in patients with high MET expression levels. These clinical results underscore the necessity of patient stratification for targeted studies of MET and other molecular targets. Other therapeutic agents targeting MET include the noncompetitive inhibitor ARQ 197 (tivantinib), which has improved progression-free and overall survival in NSCLC and hepatocellular carcinoma (Sequist et al. 2011; Rimassa et al. 2012). Concurrent inhibition of angiogenesis and MET has been evaluated using the multi-target MET inhibitor XL184 (cabozantinib), which also targets VEGFR2. XL184 has been effective against several solid cancers including medullary thyroid cancer, breast, NSCLC, melanoma, and liver cancer (Gordon et al. 2012; Hellerstedt et al. 2012; Schoffski et al. 2012; Winer et al. 2012). The most significant results were observed in the reduction of bone metastatic lesions in castration-resistant prostate cancer (Smith et al. 2012). These results raise the question of using multi-targeting versus specific inhibitors in the clinic. The success of either approach requires balancing activity, toxicity, and resistance mechanisms in each cancer type.

CONCLUDING REMARKS

Since its discovery more than 25 years ago, numerous studies have established that MET is unique among RTKs and is critical for normal development and for the progression of a wide range of human cancers. MET activation has been shown to mediate numerous signaling pathways in both in vitro and in vivo models. Nevertheless, we are still unraveling how MET signaling can mediate such diverse cellular responses as motility, invasion, growth, and angiogenesis. A deeper understanding of how MET activation controls tumorigenesis will require further analysis using three-dimensional molecular structures, cell culture systems, human tumors, and animal models. Through these combined approaches, and with new drugs targeting MET, we will, in the near future, realize the influence MET has on tumorigenesis and how it may be controlled.

ACKNOWLEDGMENTS

The authors thank David Nadziejka for manuscript editing and Jack DeGroot and Yu-Wen Zhang for a critical reading of the manuscript. We apologize to authors whose studies we were unable to cite because of space restrictions.

REFERENCES

Abella JV, Peschard P, Naujokas MA, Lin T, Saucier C, Urbe S, Park M. 2005. Met/Hepatocyte growth factor receptor ubiquitination suppresses transformation and is required for Hrs phosphorylation. *Mol Cell Biol* **25:** 9632–9645.

Abounader R, Laterra J. 2005. Scatter factor/hepatocyte growth factor in brain tumor growth and angiogenesis. *Neuro Oncol* **7:** 436–451.

Abounader R, Ranganathan S, Lal B, Fielding K, Book A, Dietz H, Burger P, Laterra J. 1999. Reversion of human glioblastoma malignancy by U1 small nuclear RNA/ribozyme targeting of scatter factor/hepatocyte growth factor and c-met expression. *J Natl Cancer Inst* **91:** 1548–1556.

Aebersold DM, Landt O, Berthou S, Gruber G, Beer KT, Greiner RH, Zimmer Y. 2003. Prevalence and clinical impact of Met Y1253D-activating point mutation in radiotherapy-treated squamous cell cancer of the oropharynx. *Oncogene* **22:** 8519–8523.

Apte U, Zeng G, Muller P, Tan X, Micsenyi A, Cieply B, Dai C, Liu Y, Kaestner KH, Monga SP. 2006. Activation of Wnt/β-catenin pathway during hepatocyte growth factor-induced hepatomegaly in mice. *Hepatology* **44:** 992–1002.

Bardella C, Costa B, Maggiora P, Patane S, Olivero M, Ranzani GN, De Bortoli M, Comoglio PM, Di Renzo MF. 2004. Truncated RON tyrosine kinase drives tumor cell progression and abrogates cell–cell adhesion through E-cadherin transcriptional repression. *Cancer Res* **64:** 5154–5161.

Berthou S, Aebersold DM, Schmidt LS, Stroka D, Heigl C, Streit B, Stalder D, Gruber G, Liang C, Howlett AR, et al. 2004. The Met kinase inhibitor SU11274 exhibits a selective inhibition pattern toward different receptor mutated variants. *Oncogene* **23:** 5387–5393.

Bertotti A, Burbridge MF, Gastaldi S, Galimi F, Torti D, Medico E, Giordano S, Corso S, Rolland-Valognes G, Lockhart BP, et al. 2009. Only a subset of Met-activated pathways are required to sustain oncogene addiction. *Sci Signal* **2:** er11.

Birchmeier W, Brinkmann V, Niemann C, Meiners S, DiCesare S, Naundorf H, Sachs M. 1997. Role of HGF/SF and c-Met in morphogenesis and metastasis of epithelial cells. *Ciba Found Symp* **212:** 230–240.

Birchmeier C, Birchmeier W, Gherardi E, Vande Woude GF. 2003. Met, metastasis, motility and more. *Nat Rev Mol Cell Biol* **4:** 915–925.

Bladt F, Riethmacher D, Isenmann S, Aguzzi A, Birchmeier C. 1995. Essential role for the c-met receptor in the migration of myogenic precursor cells into the limb bud. *Nature* **376:** 768–771.

Blume-Jensen P, Hunter T. 2001. Oncogenic kinase signalling. *Nature* **411:** 355–365.

Boccaccio C, Ando M, Tamagnone L, Bardelli A, Michieli P, Battistini C, Comoglio PM. 1998. Induction of epithelial tubules by growth factor HGF depends on the STAT pathway. *Nature* **391:** 285–288.

Borowiak M, Garratt AN, Wustefeld T, Strehle M, Trautwein C, Birchmeier C. 2004. Met provides essential signals for liver regeneration. *Proc Natl Acad Sci* **101:** 10608–10613.

Bottaro DP, Rubin JS, Faletto DL, Chan AM, Kmiecik TE, Vande Woude GF, Aaronson SA. 1991. Identification of the hepatocyte growth factor receptor as the c-met proto-oncogene product. *Science* **251:** 802–804.

Bussolino F, Di Renzo MF, Ziche M, Bocchietto E, Olivero M, Naldini L, Gaudino G, Tamagnone L, Coffer A, Comoglio PM. 1992. Hepatocyte growth factor is a potent angiogenic factor which stimulates endothelial cell motility and growth. *J Cell Biol* **119:** 629–641.

Camp ER, Liu W, Fan F, Yang A, Somcio R, Ellis LM. 2005. RON, a tyrosine kinase receptor involved in tumor progression and metastasis. *Ann Surg Oncol* **12:** 273–281.

Cao B, Su Y, Oskarsson M, Zhao P, Kort EJ, Fisher RJ, Wang LM, Vande Woude GF. 2001. Neutralizing monoclonal antibodies to hepatocyte growth factor/scatter factor (HGF/SF) display antitumor activity in animal models. *Proc Natl Acad Sci* **98:** 7443–7448.

Carafoli F, Chirgadze DY, Blundell TL, Gherardi E. 2005. Crystal structure of the β-chain of human hepatocyte growth factor-like/macrophage stimulating protein. *FEBS J* **272:** 5799–5807.

Chirgadze DY, Hepple JP, Zhou H, Byrd RA, Blundell TL, Gherardi E. 1999. Crystal structure of the NK1 fragment of HGF/SF suggests a novel mode for growth factor dimerization and receptor binding. *Nat Struct Biol* **6:** 72–79.

Chmielowiec J, Borowiak M, Morkel M, Stradal T, Munz B, Werner S, Wehland J, Birchmeier C, Birchmeier W. 2007. c-Met is essential for wound healing in the skin. *J Cell Biol* **177:** 151–162.

Christensen JG, Schreck R, Burrows J, Kuruganti P, Chan E, Le P, Chen J, Wang X, Ruslim L, Blake R, et al. 2003. A selective small molecule inhibitor of c-Met kinase inhibits c-Met-dependent phenotypes in vitro and exhibits cytoreductive antitumor activity in vivo. *Cancer Res* **63:** 7345–7355.

Cooper CS, Park M, Blair DG, Tainsky MA, Huebner K, Croce CM, Vande Woude GF. 1984. Molecular cloning of a new transforming gene from a chemically transformed human cell line. *Nature* **311:** 29–33.

Corso S, Comoglio PM, Giordano S. 2005. Cancer therapy: Can the challenge be MET? *Trends Mol Med* **11:** 284–292.

Corso S, Ghiso E, Cepero V, Sierra JR, Migliore C, Bertotti A, Trusolino L, Comoglio PM, Giordano S. 2010. Activation of HER family members in gastric carcinoma cells mediates resistance to MET inhibition. *Mol Cancer* **9:** 121.

Date K, Matsumoto K, Shimura H, Tanaka M, Nakamura T. 1997. HGF/NK4 is a specific antagonist for pleiotrophic actions of hepatocyte growth factor. *FEBS Lett* **420:** 1–6.

Deakin JA, Lyon M. 1999a. Differential regulation of hepatocyte growth factor/scatter factor by cell surface proteoglycans and free glycosaminoglycan chains. *J Cell Sci* **112:** 1999–2009.

Deakin JA, Lyon M. 1999b. Differential regulation of hepatocyte growth factor/scatter factor by cell surface proteoglycans and free glycosaminoglycan chains. *J Cell Sci* **112:** 1999–2009.

Di Renzo MF, Olivero M, Martone T, Maffe A, Maggiora P, Stefani AD, Valente G, Giordano S, Cortesina G, Comoglio PM. 2000. Somatic mutations of the MET oncogene are selected during metastatic spread of human HNSC carcinomas. *Oncogene* **19:** 1547–1555.

Donate LE, Gherardi E, Srinivasan N, Sowdhamini R, Aparicio S, Blundell TL. 1994. Molecular evolution and domain structure of plasminogen-related growth factors (HGF/SF and HGF1/MSP). *Protein Sci* **3:** 2378–2394.

Engelman JA, Zejnullahu K, Mitsudomi T, Song Y, Hyland C, Park JO, Lindeman N, Gale CM, Zhao X, Christensen J, et al. 2007. MET amplification leads to gefitinib resistance in lung cancer by activating ERBB3 signaling. *Science* **316:** 1039–1043.

Fan S, Gao M, Meng Q, Laterra JJ, Symons MH, Coniglio S, Pestell RG, Goldberg ID, Rosen EM. 2005. Role of NF-κB signaling in hepatocyte growth factor/scatter factor-mediated cell protection. *Oncogene* **24:** 1749–1766.

Furge KA, Zhang YW, Vande Woude GF. 2000. Met receptor tyrosine kinase: Enhanced signaling through adapter proteins. *Oncogene* **19:** 5582–5589.

Gallego MI, Bierie B, Hennighausen L. 2003. Targeted expression of HGF/SF in mouse mammary epithelium leads to metastatic adenosquamous carcinomas through the activation of multiple signal transduction pathways. *Oncogene* **22:** 8498–8508.

Gao CF, Xie Q, Zhang YW, Su Y, Roys C, Zhao P, Cao B, Burgess T, Coxon A, Vande Woude G. 2006. Therapeutic potential of neutralizing antibodies to HGF/SF against human c-MET driven human tumors. In *American Association for Cancer Research Annual Meeting*, p. LB229. Washington, DC.

Gherardi E, Sandin S, Petoukhov MV, Finch J, Youles ME, Ofverstedt LG, Miguel RN, Blundell TL, Vande Woude GF, Skoglund U, et al. 2006. Structural basis of hepatocyte growth factor/scatter factor and MET signalling. *Proc Natl Acad Sci* **103:** 4046–4051.

Gherardi E, Birchmeier W, Birchmeier C, Vande Woude G. 2012. Targeting MET in cancer: Rationale and progress. *Nat Rev Cancer* **12:** 89–103.

Gonzatti-Haces M, Seth A, Park M, Copeland T, Oroszlan S, Vande Woude GF. 1988. Characterization of the TPR-MET oncogene p65 and the MET protooncogene p140 protein-tyrosine kinases. *Proc Natl Acad Sci* **85:** 21–25.

Gordon MS, Kluger HM, Shapiro G, Kurzrock R, Edelman G, Samuel TA, Moussa AH, Ramies DA, Laird AD, Schimmoller F, et al. 2012. Activity of cabozantinib (XL184) in metastatic melanoma: Results from a phase II randomized discontinuation trial (RDT). *J Clin Oncol Abstr* **30:** 8531.

Grant DS, Kleinman HK, Goldberg ID, Bhargava MM, Nickloff BJ, Kinsella JL, Polverini P, Rosen EM. 1993. Scatter factor induces blood vessel formation *in vivo*. *Proc Natl Acad Sci* **90:** 1937–1941.

Graveel C, Su Y, Koeman J, Wang LM, Tessarollo L, Fiscella M, Birchmeier C, Swiatek P, Bronson R, Vande Woude G. 2004. Activating Met mutations produce unique tumor profiles in mice with selective duplication of the mutant allele. *Proc Natl Acad Sci* **101:** 17198–17203.

Graveel CR, DeGroot JD, Su Y, Koeman J, Dykema K, Leung S, Snider J, Davies SR, Swiatek PJ, Cottingham S, et al. 2009. Met induces diverse mammary carcinomas in mice and is associated with human basal breast cancer. *Proc Natl Acad Sci* **106:** 12909–12914.

Gu H, Maeda H, Moon JJ, Lord JD, Yoakim M, Nelson BH, Neel BG. 2000. New role for Shc in activation of the phosphatidylinositol 3-kinase/Akt pathway. *Mol Cell Biol* **20:** 7109–7120.

Hellerstedt BA, Edelman G, Vogelzang NJ, Kluger HM, Yasenchak CA, Shen X, Ramies DA, Gordon MS. 2012. Activity of cabozantinib (XL184) in metastatic NSCLC: Results from a phase II randomized discontinuation trial (RDT). *J Clin Oncol Abstr* **30:** 7514.

Holmes O, Pillozzi S, Deakin JA, Carafoli F, Kemp L, Butler PJ, Lyon M, Gherardi E. 2007. Insights into the structure/function of hepatocyte growth factor/scatter factor from studies with individual domains. *J Mol Biol* **367:** 395–408.

Huh CG, Factor VM, Sanchez A, Uchida K, Conner EA, Thorgeirsson SS. 2004. Hepatocyte growth factor/c-met signaling pathway is required for efficient liver regeneration and repair. *Proc Natl Acad Sci* **101:** 4477–4482.

Itoh M, Yoshida Y, Nishida K, Narimatsu M, Hibi M, Hirano T. 2000. Role of Gab1 in heart, placenta, and skin development and growth factor- and cytokine-induced extracellular signal-regulated kinase mitogen-activated protein kinase activation. *Mol Cell Biol* **20:** 3695–3704.

Jeffers M, Rong S, Vande Woude GF. 1996. Enhanced tumorigenicity and invasion-metastasis by hepatocyte growth factor/scatter factor-Met signalling in human cells concomitant with induction of the urokinase proteolysis network. *Mol Cell Biol* **16:** 1115–1125.

Jeffers M, Schmidt L, Nakaigawa N, Webb CP, Weirich G, Kishida T, Zbar B, Vande Woude GF. 1997. Activating mutations for the Met tyrosine kinase receptor in human cancer. *Proc Natl Acad Sci* **94:** 11445–11450.

Jeffers M, Fiscella M, Webb CP, Anver M, Koochekpour S, Vande Woude GF. 1998. The mutationally activated Met receptor mediates motility and metastasis. *Proc Natl Acad Sci* **95:** 14417–14422.

Kirchhofer D, Yao X, Peek M, Eigenbrot C, Lipari MT, Billeci KL, Maun HR, Moran P, Santell L, Wiesmann C, et al. 2004. Structural and functional basis of the serine protease-like hepatocyte growth factor β-chain in Met binding and signaling. *J Biol Chem* **279:** 39915–39924.

Knudsen BS, Vande Woude G. 2008. Showering c-Met dependent cancers with drugs. *Curr Opin Genet Dev* **18:** 87–96.

     Cite this article as *Cold Spring Harb Perspect Biol* doi: 10.1101/cshperspect.a009209

Kong-Beltran M, Stamos J, Wickramasinghe D. 2004. The Sema domain of Met is necessary for receptor dimerization and activation. *Cancer Cell* **6:** 75–84.

Kong-Beltran M, Seshagiri S, Zha J, Zhu W, Bhawe K, Mendoza N, Holcomb T, Pujara K, Stinson J, Fu L, et al. 2006. Somatic mutations lead to an oncogenic deletion of Met in lung cancer. *Cancer Res* **66:** 283–289.

Lamorte L, Rodrigues S, Naujokas M, Park M. 2002a. Crk synergizes with epidermal growth factor for epithelial invasion and morphogenesis and is required for the Met morphogenic program. *J Biol Chem* **277:** 37904–37911.

Lamorte L, Royal I, Naujokas M, Park M. 2002b. Crk adapter proteins promote an epithelial-mesenchymal-like transition and are required for HGF-mediated cell spreading and breakdown of epithelial adherens junctions. *Mol Biol Cell* **13:** 1449–1461.

Lee JH, Han SU, Cho H, Jennings B, Gerrard B, Dean M, Schmidt L, Zbar B, Vande Woude GF. 2000. A novel germ line juxtamembrane Met mutation in human gastric cancer. *Oncogene* **19:** 4947–4953.

Liang TJ, Reid AE, Xavier R, Cardiff RD, Wang TC. 1996. Transgenic expression of tpr-met oncogene leads to development of mammary hyperplasia and tumors. *J Clin Invest* **97:** 2872–2877.

Lietha D, Chirgadze DY, Mulloy B, Blundell TL, Gherardi E. 2001. Crystal structures of NK1-heparin complexes reveal the basis for NK1 activity and enable engineering of potent agonists of the MET receptor. *EMBO J* **20:** 5543–5555.

Liska D, Chen CT, Bachleitner-Hofmann T, Christensen JG, Weiser MR. 2011. HGF rescues colorectal cancer cells from EGFR inhibition via MET activation. *Clin Cancer Res* **17:** 472–482.

Love CA, Harlos K, Mavaddat N, Davis SJ, Stuart DI, Jones EY, Esnouf RM. 2003. The ligand-binding face of the semaphorins revealed by the high-resolution crystal structure of SEMA4D. *Nat Struct Biol* **10:** 843–848.

Lowenstein EJ, Daly RJ, Batzer AG, Li W, Margolis B, Lammers R, Ullrich A, Skolnik EY, Bar-Sagi D, Schlessinger J. 1992. The SH2 and SH3 domain-containing protein GRB2 links receptor tyrosine kinases to ras signaling. *Cell* **70:** 431–442.

Lyon M, Deakin JA, Lietha D, Gherardi E, Gallagher JT. 2004. The interactions of hepatocyte growth factor/scatter factor and its NK1 and NK2 variants with glycosaminoglycans using a modified gel mobility shift assay. Elucidation of the minimal size of binding and activatory oligosaccharides. *J Biol Chem* **279:** 43560–43567.

Ma PC, Kijima T, Maulik G, Fox EA, Sattler M, Griffin JD, Johnson BE, Salgia R. 2003. c-MET mutational analysis in small cell lung cancer: Novel juxtamembrane domain mutations regulating cytoskeletal functions. *Cancer Res* **63:** 6272–6281.

Maroun CR, Holgado-Madruga M, Royal I, Naujokas MA, Fournier TM, Wong AJ, Park M. 1999. The Gab1 PH domain is required for localization of Gab1 at sites of cell-cell contact and epithelial morphogenesis downstream from the Met receptor tyrosine kinase. *Mol Cell Biol* **19:** 1784–1799.

Maroun CR, Naujokas MA, Holgado-Madruga M, Wong AJ, Park M. 2000. The tyrosine phosphatase SHP-2 is required for sustained activation of extracellular signal-regulated kinase and epithelial morphogenesis downstream from the met receptor tyrosine kinase. *Mol Cell Biol* **20:** 8513–8525.

Merchant M, Zhang Y, Su Y, Romero M, Louie S, Severin C, Mendoza N, Zheng Z, Dekoning T, DW K, et al. 2008. Combination efficacy with MetMAb and erlotinib in a NSCLC tumor model highlight therapeutic opportunities for c-Met inhibitors in combination with EGFR inhibitors. In *Proceedings of the 99th Annual Meeting of the American Association for Cancer Research*. San Diego, CA.

Michieli P, Basilico C, Pennacchietti S, Maffe A, Tamagnone L, Giordano S, Bardelli A, Comoglio PM. 1999. Mutant Met-mediated transformation is ligand-dependent and can be inhibited by HGF antagonists. *Oncogene* **18:** 5221–5231.

Michieli P, Mazzone M, Basilico C, Cavassa S, Sottile A, Naldini L, Comoglio PM. 2004. Targeting the tumor and its microenvironment by a dual-function decoy Met receptor. *Cancer Cell* **6:** 61–73.

Monga SP, Mars WM, Pediaditakis P, Bell A, Mule K, Bowen WC, Wang X, Zarnegar R, Michalopoulos GK. 2002. Hepatocyte growth factor induces Wnt-independent nuclear translocation of β-catenin after Met-β-catenin dissociation in hepatocytes. *Cancer Res* **62:** 2064–2071.

Montesano R, Soriano JV, Malinda KM, Ponce ML, Bafico A, Kleinman HK, Bottaro DP, Aaronson SA. 1998. Differential effects of hepatocyte growth factor isoforms on epithelial and endothelial tubulogenesis. *Cell Growth Differ* **9:** 355–365.

Mosesson Y, Mills GB, Yarden Y. 2008. Derailed endocytosis: An emerging feature of cancer. *Nat Rev Cancer* **8:** 835–850.

Moshitch-Moshkovitz S, Tsarfaty G, Kaufman DW, Stein GY, Shichrur K, Solomon E, Sigler RH, Resau JH, Vande Woude GY, Tsarfaty I. 2006. In vivo direct molecular imaging of early tumorigenesis and malignant progression induced by transgenic expression of GFP-Met. *Neoplasia* **8:** 353–363.

Muller M, Morotti A, Ponzetto C. 2002. Activation of NF-κB is essential for hepatocyte growth factor-mediated proliferation and tubulogenesis. *Mol Cell Biol* **22:** 1060–1072.

Nakamura T, Nishizawa T, Hagiya M, Seki T, Shimonishi M, Sugimura A, Tashiro K, Shimizu S. 1989. Molecular cloning and expression of human hepatocyte growth factor. *Nature* **342:** 440–443.

Oliner KS, Tang R, Anderson A, Lan Y, Iveson T, Donehower RC, Jiang Y, Dubey S, Loh E. 2012. Evaluation of MET pathway biomarkers in a phase II study of rilotumumab (R, AMG 102) or placebo (P) in combination with epirubicin, cisplatin, and capecitabine (ECX) in patients (pts) with locally advanced or metastatic gastric (G) or esophagogastric junction (EGJ) cancer. *J Clin Oncol* **30** (Abstr): 4005.

Ong SH, Dilworth S, Hauck-Schmalenberger I, Pawson T, Kiefer F. 2001. ShcA and Grb2 mediate polyoma middle T antigen-induced endothelial transformation and Gab1 tyrosine phosphorylation. *EMBO J* **20:** 6327–6336.

Orian-Rousseau V, Chen L, Sleeman JP, Herrlich P, Ponta H. 2002. CD44 is required for two consecutive steps in HGF/c-Met signaling. *Genes Dev* **16:** 3074–3086.

Otsuka T, Takayama H, Sharp R, Celli G, LaRochelle WJ, Bottaro DP, Ellmore N, Vieira W, Owens JW, Anver M, et al. 1998. c-Met autocrine activation induces development of malignant melanoma and acquisition of the metastatic phenotype. *Cancer Res* **58:** 5157–5167.

Park M, Dean M, Cooper CS, Schmidt M, O'Brien SJ, Blair DG, Vande Woude GF. 1986. Mechanism of met oncogene activation. *Cell* **45:** 895–904.

Park WS, Dong SM, Kim SY, Na EY, Shin MS, Pi JH, Kim BJ, Bae JH, Hong YK, Lee KS, et al. 1999. Somatic mutations in the kinase domain of the Met/hepatocyte growth factor receptor gene in childhood hepatocellular carcinomas. *Cancer Res* **59:** 307–310.

Pediaditakis P, Monga SP, Mars WM, Michalopoulos GK. 2002. Differential mitogenic effects of single chain hepatocyte growth factor (HGF)/scatter factor and HGF/NK1 following cleavage by factor Xa. *J Biol Chem* **277:** 14109–14115.

Peek M, Moran P, Mendoza N, Wickramasinghe D, Kirchhofer D. 2002. Unusual proteolytic activation of pro-hepatocyte growth factor by plasma kallikrein and coagulation factor XIa. *J Biol Chem* **277:** 47804–47809.

Pennacchietti S, Michieli P, Galluzzo M, Mazzone M, Giordano S, Comoglio PM. 2003. Hypoxia promotes invasive growth by transcriptional activation of the met protooncogene. *Cancer Cell* **3:** 347–361.

Peschard P, Park M. 2007. From Tpr-Met to Met, tumorigenesis and tubes. *Oncogene* **26:** 1276–1285.

Phaneuf D, Moscioni AD, LeClair C, Raper SE, Wilson JM. 2004. Generation of a mouse expressing a conditional knockout of the hepatocyte growth factor gene: Demonstration of impaired liver regeneration. *DNA Cell Biol* **23:** 592–603.

Ponzo MG, Lesurf R, Petkiewicz S, O'Malley FP, Pinnaduwage D, Andrulis IL, Bull SB, Chughtai N, Germain D, Omeroglu A, et al. 2009. Met induces mammary tumors with multiple pathologies and is associated with both poor outcome and basal-type breast cancers. *Proc Natl Acad Sci* **106:** 12903–12908.

Qi J, McTigue MA, Rogers A, Lifshits E, Christensen JG, Janne PA, Engelman JA. 2011. Multiple mutations and bypass mechanisms can contribute to development of acquired resistance to MET inhibitors. *Cancer Res* **71:** 1081–1091.

Reshetnikova G, Troyanovsky S, Rimm DL. 2007. Definition of a direct extracellular interaction between Met and E-cadherin. *Cell Biol Int* **31:** 366–373.

Rimassa L, Porta C, Borbath I, Daniele B, Salvagni S, Van Laethem JL, Van Vlieberghe H, Trojan J, Kolligs FT, Weiss A, et al. 2012. Tivantinib (ARQ 197) versus placebo in patients (Pts) with hepatocellular carcinoma (HCC) who failed one systemic therapy: Results of a randomized controlled phase II trial (RCT). *J Clin Oncol* **30** (Abstr): 4006.

Rodrigues SP, Fathers KE, Chan G, Zuo D, Halwani F, Meterissian S, Park M. 2005. CrkI and CrkII function as key signaling integrators for migration and invasion of cancer cells. *Mol Cancer Res* **3:** 183–194.

Ronsin C, Muscatelli F, Mattei MG, Breathnach R. 1993. A novel putative receptor protein tyrosine kinase of the met family. *Oncogene* **8:** 1195–1202.

Rosario M, Birchmeier W. 2003. How to make tubes: Signaling by the Met receptor tyrosine kinase. *Trends Cell Biol* **13:** 328–335.

Rozakis-Adcock M, McGlade J, Mbamalu G, Pelicci G, Daly R, Li W, Batzer A, Thomas S, Brugge J, Pelicci PG, et al. 1992. Association of the Shc and Grb2/Sem5 SH2-containing proteins is implicated in activation of the Ras pathway by tyrosine kinases. *Nature* **360:** 689–692.

Rubin JS, Day RM, Breckenridge D, Atabey N, Taylor WG, Stahl SJ, Wingfield PT, Kaufman JD, Schwall R, Bottaro DP. 2001. Dissociation of heparan sulfate and receptor binding domains of hepatocyte growth factor reveals that heparan sulfate-c-Met interaction facilitates signaling. *J Biol Chem* **276:** 32977–32983.

Sachs M, Brohmann H, Zechner D, Muller T, Hulsken J, Walther I, Schaeper U, Birchmeier C, Birchmeier W. 2000. Essential role of Gab1 for signaling by the c-Met receptor in vivo. *J Cell Biol* **150:** 1375–1384.

Sattler M, Pride YB, Ma P, Gramlich JL, Chu SC, Quinnan LA, Shirazian S, Liang C, Podar K, Christensen JG, et al. 2003. A novel small molecule Met inhibitor induces apoptosis in cells transformed by the oncogenic TPR-MET tyrosine kinase. *Cancer Res* **63:** 5462–5469.

Saucier C, Khoury H, Lai KM, Peschard P, Dankort D, Naujokas MA, Holash J, Yancopoulos GD, Muller WJ, Pawson T, et al. 2004. The Shc adaptor protein is critical for VEGF induction by Met/HGF and ErbB2 receptors and for early onset of tumor angiogenesis. *Proc Natl Acad Sci* **101:** 2345–2350.

Schaeper U, Gehring NH, Fuchs KP, Sachs M, Kempkes B, Birchmeier W. 2000. Coupling of Gab1 to c-Met, Grb2, and Shp2 mediates biological responses. *J Cell Biol* **149:** 1419–1432.

Schiering N, Knapp S, Marconi M, Flocco MM, Cui J, Perego R, Rusconi L, Cristiani C. 2003. Crystal structure of the tyrosine kinase domain of the hepatocyte growth factor receptor c-Met and its complex with the microbial alkaloid K-252a. *Proc Natl Acad Sci* **100:** 12654–12659.

Schmidt C, Bladt F, Goedecke S, Brinkmann V, Zschiesche W, Sharpe M, Gherardi E, Birchmeier C. 1995. Scatter factor/hepatocyte growth factor is essential for liver development. *Nature* **373:** 699–702.

Schmidt L, Duh F-M, Chen F, Kishida T, Glenn G, Choyke P, Scherer SW, Zhuang Z, Lubensky I, Dean M, et al. 1997. Germline and somatic mutations in the tyrosine kinase domain of the *Met* proto-oncogene in papillary renal carcinomas. *Nat Genet* **16:** 68–73.

Schmidt L, Junker K, Nakaigawa N, Kinjerski T, Weirich G, Miller M, Lubensky I, Neumann HP, Brauch H, Decker J, et al. 1999. Novel mutations of the MET proto-oncogene in papillary renal carcinomas. *Oncogene* **18:** 2343–2350.

Schoffski P, Elisei R, Müller S, Brose MS, Shah MH, Licitra LF, Jarzab B, Medvedev V, Kreissl M, Niederle B, et al. 2012. An international, double-blind, randomized, placebo-controlled phase III trial (EXAM) of cabozantinib (XL184) in medullary thyroid carcinoma (MTC) patients (pts) with documented RECIST progression at baseline. *J Clin Oncol Abstr* **30:** 5508.

Sequist LV, von Pawel J, Garmey EG, Akerley WL, Brugger W, Ferrari D, Chen Y, Costa DB, Gerber DE, Orlov S, et al. 2011. Randomized phase II study of erlotinib plus

     Cite this article as *Cold Spring Harb Perspect Biol* doi: 10.1101/cshperspect.a009209

tivantinib versus erlotinib plus placebo in previously treated non-small-cell lung cancer. *J Clin Oncol* **29:** 3307–3315.

Shattuck DL, Miller JK, Carraway KL III, Sweeney C. 2008. Met receptor contributes to trastuzumab resistance of Her2-overexpressing breast cancer cells. *Cancer Res* **68:** 1471–1477.

Shen Z, Yang ZF, Gao Y, Li JC, Chen HX, Liu CC, Poon RT, Fan ST, Luk JM, Sze KH, et al. 2008. The kringle 1 domain of hepatocyte growth factor has antiangiogenic and antitumor cell effects on hepatocellular carcinoma. *Cancer Res* **68:** 404–414.

Shimomura T, Kondo J, Ochiai M, Naka D, Miyazawa K, Morimoto Y, Kitamura N. 1993. Activation of the zymogen of hepatocyte growth factor activator by thrombin. *J Biol Chem* **268:** 22927–22932.

Shinomiya N, Gao CF, Xie Q, Gustafson M, Waters DJ, Zhang YW, Vande Woude GF. 2004. RNA interference reveals that ligand-independent Met activity is required for tumor cell signaling and survival. *Cancer Res* **64:** 7962–7970.

Smith MR, Sweeney C, Rathkopf DE, Scher HI, Logothetis C, George DJ, Higano CS, Yu EY, Harzstark AL, Small EJ, et al. 2012. Cabozantinib (XL184) in chemotherapy-pretreated metastatic castration resistant prostate cancer (mCRPC): Results from a phase II nonrandomized expansion cohort (NRE). *J Clin Oncol Abstr* **30:** 4513.

Spigel DR, Ervin TJ, Ramlau R, Daniel DB, Goldschmidt JH, Blumenschein GR, Krzakowski MJ, Robinet G, Clement-Duchene C, Barlesi F, et al. 2011. Final efficacy results from OAM4558 g, a randomized phase II study evaluating MetMAb or placebo in combination with erlotinib in advanced NSCLC. *J Clin Oncol Abstr* **29:** 7505.

Stahl SJ, Wingfield PT, Kaufman JD, Pannell LK, Cioce V, Sakata H, Taylor WG, Rubin JS, Bottaro DP. 1997. Functional and biophysical characterization of recombinant human hepatocyte growth factor isoforms produced in *Escherichia coli*. *Biochem J* **326:** 763–772.

Stamos J, Lazarus RA, Yao X, Kirchhofer D, Wiesmann C. 2004. Crystal structure of the HGF β-chain in complex with the Sema domain of the Met receptor. *EMBO J* **23:** 2325–2335.

Stella GM, Benvenuti S, Gramaglia D, Scarpa A, Tomezzoli A, Cassoni P, Senetta R, Venesio T, Pozzi E, Bardelli A, et al. 2011. MET mutations in cancers of unknown primary origin (CUPs). *Hum Mutat* **32:** 44–50.

Stoker M, Gherardi E, Perryman M, Gray J. 1987. Scatter factor is a fibroblast-derived modulator of epithelial cell mobility. *Nature* **327:** 239–242.

Takayama H, LaRochelle WJ, Sharp R, Otsuka T, Kriebel P, Anver M, Aaronson SA, Merlino G. 1997. Diverse tumorigenesis associated with aberrant development in mice overexpressing hepatocyte growth factor/scatter factor. *Proc Natl Acad Sci* **94:** 701–706.

Tolbert WD, Daugherty J, Gao C, Xie Q, Miranti C, Gherardi E, Woude GV, Xu HE. 2007. A mechanical basis for converting a receptor tyrosine kinase agonist to an antagonist. *Proc Natl Acad Sci* **104:** 14592–14597.

Uehara Y, Minowa O, Mori C, Shiota K, Kuno J, Noda T, Kitamura N. 1995. Placental defect and embryonic lethality in mice lacking hepatocyte growth factor/scatter factor. *Nature* **373:** 702–705.

Vermeulen L, De Sousa EMF, van der Heijden M, Cameron K, de Jong JH, Borovski T, Tuynman JB, Todaro M, Merz C, Rodermond H, et al. 2010. Wnt activity defines colon cancer stem cells and is regulated by the microenvironment. *Nat Cell Biol* **12:** 468–476.

Wang R, Ferrell LD, Faouzi S, Maher JJ, Bishop JM. 2001. Activation of the Met receptor by cell attachment induces and sustains hepatocellular carcinomas in transgenic mice. *J Cell Biol* **153:** 1023–1034.

Wang X, DeFrances MC, Dai Y, Pediaditakis P, Johnson C, Bell A, Michalopoulos GK, Zarnegar R. 2002. A mechanism of cell survival: Sequestration of Fas by the HGF receptor Met. *Mol Cell* **9:** 411–421.

Wang X, Le P, Liang C, Chan J, Kiewlich D, Miller T, Harris D, Sun L, Rice A, Vasile S, et al. 2003. Potent and selective inhibitors of the Met [hepatocyte growth factor/scatter factor (HGF/SF) receptor] tyrosine kinase block HGF/SF-induced tumor cell growth and invasion. *Mol Cancer Ther* **2:** 1085–1092.

Wang D, Li Z, Messing EM, Wu G. 2005. The SPRY domain-containing SOCS box protein 1 (SSB-1) interacts with MET and enhances the hepatocyte growth factor-induced Erk-Elk-1-serum response element pathway. *J Biol Chem* **280:** 16393–16401.

Weidner KM, Arakaki N, Hartmann G, Vandekerckhove J, Weingart S, Rieder H, Fonatsch C, Tsubouchi H, Hishida T, Daikuhara Y, et al. 1991. Evidence for the identity of human scatter factor and human hepatocyte growth factor. *Proc Natl Acad Sci* **88:** 7001–7005.

Winer EP, Tolaney S, Nechushtan H, Berger R, Kurzrock R, Ron I, Schoffski P, Awada A, Yasenchak CA, Burris HA, et al. 2012. Activity of cabozantinib (XL184) in metastatic breast cancer (MBC): Results from a phase II randomized discontinuation trial (RDT). *J Clin Oncol Abstr* **30:** 535.

Youles M, Holmes O, Petoukhov MV, Nessen MA, Stivala S, Svergun DI, Gherardi E. 2008. Engineering the NK1 fragment of hepatocyte growth factor/scatter factor as a MET receptor antagonist. *J Mol Biol* **377:** 616–622.

Zhang Y-W, Wang L-M, Jove R, Vande Woude GF. 2002. Requirement of Stat3 signaling for HGF/SF-Met mediated tumorigenesis. *Oncogene* **21:** 217–226.

Zhang Y-W, Su Y, Volpert OV, Vande Woude GF. 2003. Hepatocyte growth factor/scatter factor mediates angiogenesis through positive VEGF and negative thrombospondin-1 regulation. *Proc Natl Acad Sci* **100:** 12718–12723.

Zhang YW, Su Y, Lanning N, Gustafson M, Shinomiya N, Zhao P, Cao B, Tsarfaty G, Wang LM, Hay R, et al. 2005. Enhanced growth of human Met-expressing xenografts in a new strain of immunocompromised mice transgenic for human hepatocyte growth factor/scatter factor. *Oncogene* **24:** 101–106.

Zhang YW, Staal B, Essenburg C, Su Y, Kang L, West R, Kaufman D, Dekoning T, Eagleson B, Buchanan SG, et al. 2010. MET kinase inhibitor SGX523 synergizes with epidermal growth factor receptor inhibitor erlotinib in a hepatocyte growth factor-dependent fashion to suppress carcinoma growth. *Cancer Res* **70:** 6880–6890.

Zou C, Ma J, Wang X, Guo L, Zhu Z, Stoops J, Eaker AE, Johnson CJ, Strom S, Michalopoulos GK, et al. 2007. Lack of Fas antagonism by Met in human fatty liver disease. *Nat Med* **13:** 1078–1085.

Central Role of RET in Thyroid Cancer

Massimo Santoro and Francesca Carlomagno

Dipartimento di Medicina Molecolare e Biotecnologie Mediche, Universita' degli Studi di Napoli Federico II, 80131 Napoli, Italy

Correspondence: masantor@unina.it

RET (rearranged during transfection) is a receptor tyrosine kinase involved in the development of neural crest derived cell lineages, kidney, and male germ cells. Different human cancers, including papillary and medullary thyroid carcinomas, lung adenocarcinomas, and myeloproliferative disorders display gain-of-function mutations in *RET*. Accordingly, RET protein has become a promising molecular target for cancer treatment.

The human *RET* (rearranged during transfection) gene maps on 10q11.2 and is composed of 21 exons spanning a region of 55,000 bp. It encodes a single-pass *trans*-membrane protein, RET, that belongs to the receptor tyrosine kinase (RTK) family (Pasini et al. 1995). The RET extracellular segment contains four cadherin-like domains, followed by a domain containing cysteine residues involved in the formation of intramolecular disulfide bonds (Fig. 1A) (Anders et al. 2001; Airaksinen and Saarma 2002). RET protein is highly glycosylated and *N*-glycosylation is necessary for its transport to the cell surface. Only the fully mature glycosylated 170 kDa RET protein isoform is exposed to the extracellular compartment, whereas the mannose-rich 150 kDa isoform is confined to the Golgi (Takahashi et al. 1993; Carlomagno et al. 1996). The transmembrane segment is composed of 22 amino acids, among which S649 and S653 mediate self-association and dimerization of RET, possibly via formation of intermolecular hydrogen bonding (Kjaer et al. 2006). The intracellular portion of RET contains the tyrosine kinase domain split into two subdomains by the insertion of 27 amino acids. The RET COOH-terminal tail varies in length as a result of alternative splicing of the 3′ end (carboxy terminal with respect to glycine 1063), generating three different isoforms that contain 9 (RET9), 43 (RET43), or 51 (RET51) amino acids (Myers et al. 1995). RET9 and RET51 are the most abundant isoforms, and they activate similar signaling pathways through interaction with diverse protein complexes, and may exert a differential role in development (Fig. 1A) (de Graaff et al. 2001).

RET shows several autophosphorylation sites (Fig. 1A) (Liu et al. 1996; Kawamoto et al. 2004). RET tyrosine 1062 (Y1062) functions as a multidocking site for signaling molecules containing a phosphotyrosine-binding (PTB) domain (Asai et al. 1996). Phospho-Y1062 binding proteins include SHC, N-SHC (RAI), FRS2, IRS1/2, DOK1, and DOK4/5 that, in turn, contribute to the activation of RAS-MAPK (mitogen-activated protein kinases) and PI3K (phosphatidyl inositol 3 kinase)-AKT pathways.

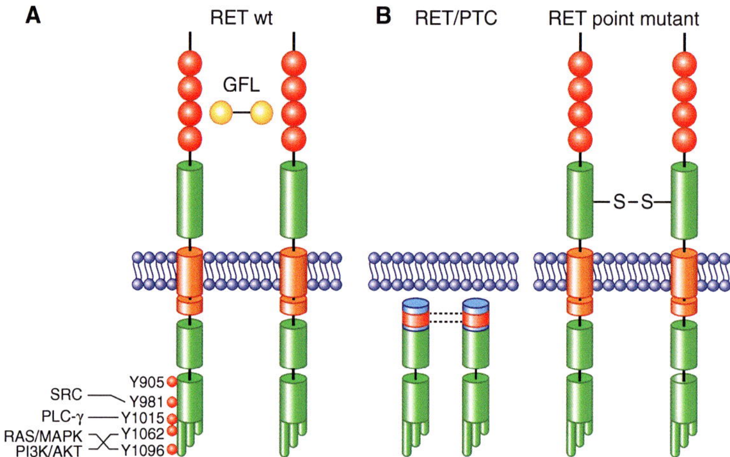

Figure 1. Illustration of the mechanisms of activation of wild-type (wt) RET and RET-derived oncoproteins. (*A*) Wild-type RET activation is mediated by ligand (GFL)-induced dimerization; ligand binding to RET is not direct and mediated by GFR-α coreceptors (not shown); major RET autophosphorylation sites and downstream signaling pathways are indicated. RET extracellular cadherin-like domains are represented in red. The split intracellular RET tyrosine kinase domain, as well as the three alternative carboxy-terminal RET tails, are also depicted. (*B*) RET/PTC activation is mediated by coiled-coil-induced dimerization (*left*); activation of RET cysteine mutants associated with MEN2A or FMTC is mediated by disulfide bonds-mediated dimerization (*right*).

Y1096, specific to the RET51 splicing variant, couples to the PI3K-AKT and RAS-MAPK pathways, as well. These signaling cascades mediate RET-dependent cell survival, proliferation, and motility (Alberti et al. 1998; Murakami et al. 1999; Segouffin-Cariou and Billaud 2000; Melillo et al. 2001a,b; Schuetz et al. 2004). Y905 is located in the activation loop of the RET kinase and its phosphorylation is associated with RET kinase activation (Knowles et al. 2006). Finally, Y981 and Y1015 have been shown to be coupled to important signaling molecules such as SRC and PLC-γ, respectively (Borrello et al. 1996; Encinas et al. 2004).

RET is the receptor for a group of neurotrophic growth factors that belong to the glial cell line-derived neurotrophic factor (GDNF) family (GFLs, GDNF family ligands), namely, GDNF, Neurturin (NRT), Artemin (ART), and Persephin (PSF) (Airaksinen and Saarma 2002). GFLs mediate RET protein dimerization and activation (Fig. 1A). GFLs are presented to RET by GPI (glycosylphosphatidylinositol)-anchored coreceptors, called GFR-α (GDNF family receptor α 1-4). Differential tissue expression dictates the specificity of action displayed by alternative GLF-GFR-α pairs during development and adult life (Baloh et al. 2000; Airaksinen and Saarma 2002).

Together with other membrane (DCC and p75NTR) or nuclear (androgen receptor, AR) receptors, RET belongs to the family of so-called "dependence" receptors (Mehlen and Bredesen 2011). In the absence of ligand, RET exerts a proapoptotic activity, that is blocked on ligand stimulation (Bordeaux et al. 2000). Such proapoptotic activity is RET kinase-independent and mediated by cleavage of RET cytosolic portion by caspase-3, which, in turn, releases a carboxy-terminal RET peptide that is able to induce cell death (Bordeaux et al. 2000). It is feasible that such activity is important for RET developmental function, because it may control migration of RET-expressing cells by limiting survival of cells that move beyond ligand availability (Bordeaux et al. 2000; Cañibano et al. 2007). Whether modulation of this function is also important for RET-associated diseases is still unknown. However, it is interesting to note that a cancer-associated RET mutant (RET-C634R, see below) does not exert cleavage-dependent proapoptotic effects, whereas RET mutants associated with defective development (Hirschsprung disease, see below) exert strong proapoptotic activity that is refractory to modulation by ligand (Bordeaux et al. 2000).

RET is expressed in enteric ganglia, adrenal medulla chromaffin cells, thyroid C cells, sensory and autonomic ganglia of the peripheral nervous system, a subset of central nervous system nuclei, developing kidney and testis germ cells (Manié et al. 2001; de Graaff et al. 2001). RET null mice display impaired development of superior cervical ganglia and enteric nervous system, kidney agenesia, reduction of thyroid C cells, and impaired spermatogenesis (Manié et al. 2001). Accordingly, individuals with germline loss-of-function mutations of *RET* are affected by intestinal aganglionosis causing congenital megacolon (Hirschsprung disease) (Brooks et al. 2005). *RET* loss-of-function mutations have also been identified in congenital anomalies of kidney and urinary tract (CAKUT), either isolated or in combination with Hirschsprung disease (Jain 2009).

Several genetic alterations convert *RET* into a dominantly transforming oncogene. This review will describe *RET*-derived oncogenes that are associated with different types of human neoplasia (Fig. 1B).

RET/PTC IN PAPILLARY THYROID CARCINOMA

RET/PTC Oncogenes

Papillary thyroid carcinoma (PTC) is the most frequent thyroid cancer and endocrine malignancy overall (Nikiforov and Nikiforova 2011). PTC originates from endodermal-derived thyroid follicular cells and is etiologically associated with exposure to ionizing radiation (Williams 2008). PTC features genetic lesions targeting the RTK-RAS-MAPK pathway. Roughly half of PTC cases display activation of *BRAF* oncogene (most commonly secondary to V600E mutation) and a small proportion of them, mainly belonging to the follicular variant-PTC, carry mutations of *RAS* genes (Xing 2005; Nikiforov and Nikiforova 2011). Although quite uncommon, rearrangements of

the *NTRK1* RTK are found in PTC as well (Greco et al. 2010).

In PTC cases that are negative for *BRAF, RAS,* or *NTRK1* mutations, chromosomal rearrangements targeting the long arm of chromosome 10 cause the disruption of *RET* gene and its fusion to various heterologous genes (Grieco et al. 1990). Such chromosomal aberrations give rise to chimeric oncogenes named *RET/PTC* (Fig. 1B). *RET/PTC* oncogenes are composed by the tyrosine kinase and COOH-tail encoding sequence of *RET* (from exon 12 to the 3′-end) and fused at the 5′ end to the promoter sequence and 5′-terminal exons of heterologous genes (Fig. 2) (Nikiforov and Nikiforova 2011). The fusion partner genes encode proteins that share the presence of protein–protein interaction domains, such as coiled-coil motifs able to mediate RET TK dimerization (Figs. 1B and 2). Most common *RET/PTC* rearrangements (90% of the cases) are *RET/PTC1* and *RET/PTC3*, the fusions between *RET* and *CCDC6* or *NCOA4* (*RFG, ARA70*) genes, respectively. *RET/PTC1* and *RET/PTC3* (and *RET/PTC4*, another *NCOA4-RET* fusion variant) are generated through a paracentric inversion of the long arm of chromosome 10, where *RET, CCDC6,* and *NCOA4* map (Grieco et al. 1990; Santoro et al. 1994). Instead, the other *RET/PTC* variants are generated by translocations between different chromosomes and are either rare (*RET/PTC2*) or identified only in single cases of radiation-induced PTC (Fig. 2).

RET/PTC Prevalence

The frequency of *RET/PTC* rearrangements (average 25% of the cases) varies considerably in different patient series (Nikiforov and Nikiforova 2011). This may depend on patients' exposure to different etiologic factors. As an example, in pediatric patients and in cases from areas contaminated by radioiodine isotopes, *RET/PTC* frequency can reach 50%–70% (Zhu et al. 2006). It is also possible that variable prevalence of *RET/PTC* may depend on the methodology used for the detection. Accordingly, the rearrangement can be present only in a subset of cancer or even benign cells (nonclonal *RET/*

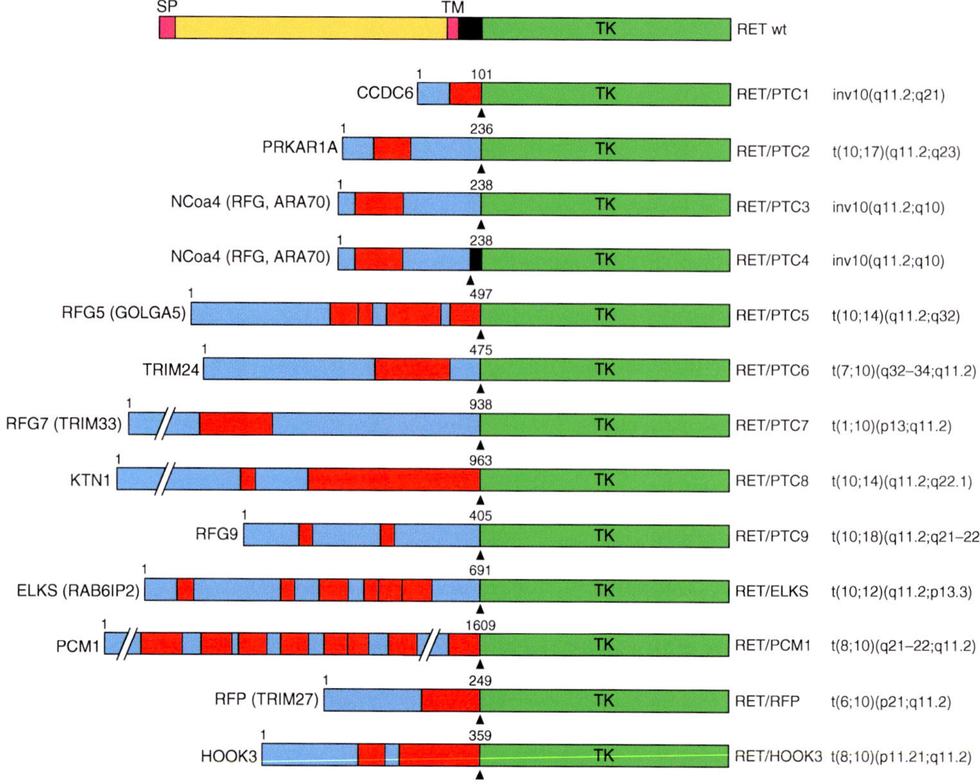

Figure 2. Schematic representation of RET/PTC oncoproteins. On the *top*, wild-type RET protein is illustrated. For each RET/PTC rearrangement, the name of the fusion partner is indicated on the *left* and the corresponding chromosomal alteration is indicated on the *right*. The fusion points are indicated by arrowheads. The length in amino acids of the partner protein portion is also indicated. Boxes in red indicate dimerization (coiled-coil) domains. SP, Signal peptide; TM, transmembrane domain; TK, tyrosine kinase domain; CCDC6, coiled-coil domain-containing protein 6; PRKAR1A, protein kinase, cAMP-dependent, regulatory, type I, alpha; NCOA4 (RFG, ARA70), nuclear coactivator 4 (RET-fused gene, androgen receptor-associated protein 70); RFG5 (GOLGA5), RET-fused gene 5 (Golgin A5); TRIM24, tripartite motif-containing 24; RFG7 (TRIM33), RET-fused gene 7 (tripartite motif-containing 33); KTN1, kinectin 1; RFG9, RET-fused gene 9; ELKS (RAB6IP2), glutamate, leucine, lysine, serine-rich sequence (RAB6-interacting protein 2); PCM1, pericentriolar material 1; RFP (TRIM27), RET finger protein (tripartite motif-containing 27); HOOK3, hook homolog 3.

PTC); and in these cases, it is detected only when highly sensitive techniques (such as nested reverse transcriptase-PCR or real-time reverse transcriptase PCR) or single-cell assays (such as FISH: fluorescent in situ hybridization) are used (Unger et al. 2004; Rhoden et al. 2006; Zhu et al. 2006).

RET/PTC Oncogenic Activity

Adoptive expression of *RET/PTC* oncogenes induces thyroid cell transformation in vitro (Santoro et al. 1993; Wang et al. 2003; Melillo et al. 2005). Moreover, targeted expression of *RET/PTC* in thyroid follicular cells induce thyroid hyperplasia or neoplasia in transgenic mice (Santoro et al. 1996; Powell et al. 1998). This evidence supports the causal contribution of *RET/PTC* formation to PTC development. Nevertheless, the low penetrance of the disease in transgenic animals, as well as the presence of *RET/PTC* rearrangements in papillary microcarcinoma, that may not progress to invasive cancer (Viglietto et al. 1995), suggests that additional oncogenic events should occur and cooperate with *RET/PTC* to generate an overt disease. Notably, the acute expression of *RET/PTC* in immortalized rat thyroid follicular cells in vitro activates a proapoptotic response because of unscheduled activation of the RAS-MAPK pathway (Castellone et al. 2003; Wang et al. 2003). Moreover, expression of *RET/PTC* in primary human thyrocytes induces oncogene-induced senescence (OIS) (Vizioli et al. 2011). It is not uncommon that normal cells oppose a barrier to neoplastic transformation by switching on suicidal or senescent programs in response to oncogene activation (Hanahan and Weinberg 2011). Thus, it is conceivable that *RET/PTC* rearrangements may be fairly frequent events in thyroid cells but insufficient alone to induce a full-blown cancer, unless further mutational events or epigenetic modifications occur to enable cells to escape cell death or growth arrest defenses.

Mechanisms of *RET/PTC* Formation

Breakage of *RET* and partner genes and their fusion are believed to result from unfaithful repair of DNA double-strand breaks (Ameziane-El-Hassani et al. 2010; Gandhi et al. 2010a). *RET* and its most common fusion partners (*CCDC6* and *NCOA4*) seem to be particularly susceptible to breakage because they map in DNA fragile sites (Gandhi et al. 2010b). Fragile sites are nonrandom DNA loci that are stable under normal conditions but become hot spots of chromosome breakage under exposure to different agents, such as ethanol, caffeine, and hypoxia (Durkin and Glover 2007).

In addition, *RET* gene disruption may be caused by genotoxic agents such as ionizing radiation and reactive oxygen species. *RET/PTCs* are enriched in patients with a known history of exposure to internal (because of thyrocyte ability to concentrate radioiodine) (Williams 2008) or external beam (Collins et al. 2002) radiation, as well as in atomic bomb survivors (Hamatani et al. 2008). Accordingly, *RET/PTC* formation can be experimentally induced by irradiation of cultured thyrocytes and thyroid tissue xenografts in SCID mice (Ito et al. 1993; Mizuno

et al. 1997, 2000; Caudill et al. 2005). However, it is important to note that most patients with *RET/PTC*-positive cancer do not have a documented exposure to radiation. In these cases, genotoxic agents other than radiation may cause DNA breakage (Gandhi et al. 2010b). Of note, H_2O_2, a potent DNA-damaging agent, is produced in large amounts by thyrocytes during the process of thyroid hormone biosynthesis and may represent one such agent (Ameziane-El-Hassani et al. 2010). Therefore, *RET/PTC*-specific occurrence in thyroid cancer might be explained by the fact that thyroid gland is commonly exposed to agents such as ionizing radiation or H_2O_2 that can disrupt *RET* and its fusion partners. In addition, in thyroid tissue, simultaneous breakage of *RET* and fusion partners, as well as their recombination, may be facilitated by thyroid cell-specific architecture of nuclear chromatin. It has been shown, indeed, that *CCDC6*, *NCOA4*, and *RET* loci display close proximity specifically in thyroid follicular cell chromatin (Nikiforova et al. 2000; Gandhi et al. 2006).

Mechanisms of *RET/PTC* Oncogenic Activation

Two major mechanisms underlie *RET* oncogenic conversion on *RET/PTC* formation. First, secondary to gene fusion, *RET* tyrosine kinase encoding domain is placed under the transcriptional control of the promoter and regulatory elements of *RET* fusion partners. Differently from *RET*, whose expression is restricted to neuroectoderm-derived cells, partner genes are ubiquitously expressed and able to drive *RET* expression in thyroid follicular cells. In addition, fusion to heterologous proteins containing protein homodimerization motifs results in constitutive RET kinase dimerization, ligand-independent activation, and autophosphorylation followed by continuous activation of downstream signaling pathways (Fig. 1B) (Bongarzone et al. 1993; Monaco et al. 2001).

The expression of a constitutively active RET kinase leads to chronic exposure of thyroid follicular cells to the activation of intracellular signaling pathways, such as RAS-MAPK, which is initiated at the level of RET tyrosine 1062 (Y1062) (Fig. 1A). Intriguingly, this pathway includes BRAF, the other oncogenic protein commonly activated in PTC (Fig. 3). Constitutive signaling is, in turn, responsible for the acquisition of several hallmarks of cancer cells including cell autonomy (independence from growth factors like TSH—thyroid-stimulating hormone), cell motility, and invasion (Melillo et al. 2005). RET/PTC signaling is also able to induce remodeling of the tumor stroma that may facilitate tumor growth. Accordingly, several reports have shown that RET/PTC via the RAS-MAPK cascade endorses an inflammatory-like response characterized by the production of several cytokines and chemokines that, in turn, recruit macrophages, lymphocytes, and mast cells within the tumor thereby promoting cell survival, invasion, and angiogenesis (Russell et al. 2003; Borrello et al. 2005; Melillo et al. 2005, 2010; Puxeddu et al. 2005).

RET Fusion Partners

It is possible that, besides causing activation of RET kinase, *RET/PTC* rearrangements affect also the function of *RET* fusion partners, this in turn contributing to thyroid tumorigenesis. According to this possibility, the rearrangement might cause a genetic double hit, inducing simultaneously the gain of *RET* oncogenic activity and the knockdown of the tumor suppressor function of *RET* partner gene. The *RET/PTC2* rearrangement nicely illustrates this possibility. In this case, the *RET* fusion partner is represented by the *PRKARIA* gene, which encodes the regulatory subunit RIα of protein kinase A. *PRKARIA* is a bona fide tumor suppressor gene that is targeted by germline inactivating mutations in patients affected by the Carney complex. This is a rare autosomal dominant cancer syndrome characterized by lentiginosis, atrial, and cutaneous myxoma,

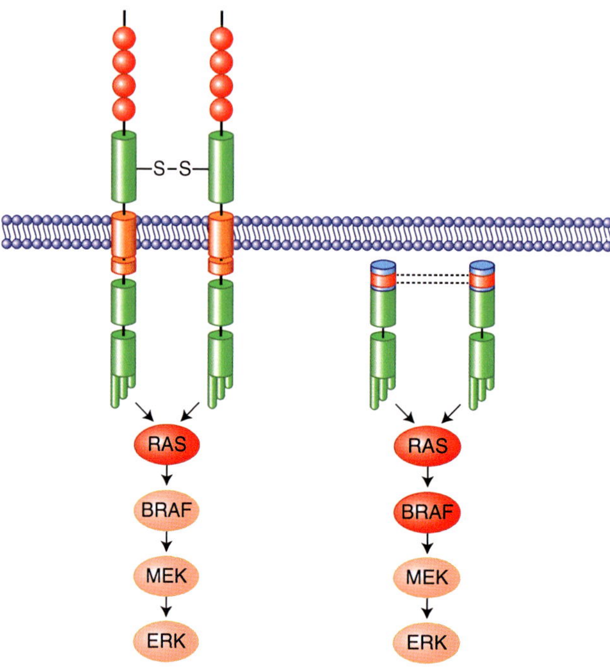

Figure 3. RAS-MAPK oncogenic signaling cascade activated by RET oncoproteins in MTC (*left*) and PTC (*right*). In red are indicated other mutational targets (RAS and BRAF) associated with MTC or PTC, respectively. In most of the cases, these mutations are alternative and not present simultaneously in a single tumor.

 Cite this article as *Cold Spring Harb Perspect Biol* doi: 10.1101/cshperspect.a009233

pituitary adenoma, testicular tumors, ovarian cysts, schwannoma, and thyroid neoplasia (Kirschner et al. 2000; Boikos and Stratakis 2006). Another example may be represented by *TRIM24* (also known as *HTIF1*), the *RET* fusion partner in *RET/PTC6* (Klugbauer and Rabes 1999). *TRIM24* null mice develop hepatic cell carcinoma, and *TRIM24* gene, on chromosome 7q32, shows frequent genetic aberrations in human hepatic cell carcinoma, strongly suggesting its role as a tumor suppressor gene (Wong et al. 1999; Khetchoumian et al. 2007). Finally, several studies also suggest that *CCDC6* and *NCOA4*, the genes involved in most common *RET/PTC* variants (*RET/PTC1* and *RET/PTC3*, respectively), might display tumor suppressor function. *CCDC6* gene product is a ubiquitously expressed 65 kDa protein that displays proapoptotic activity and is involved in ATM-mediated cellular response to DNA damage (Celetti et al. 2004; Merolla et al. 2007). A role of CCDC6 in the repression of CREB1, a transcriptional factor essential for thyroid cell growth and differentiation, has also been described (Leone et al. 2010). *NCOA4* gene encodes a 70 kDa protein that functions as a coactivator of PPARγ (peroxisome-proliferator-activated receptor γ) and AR (androgen receptor) (Yeh and Chang 1996; Heinlein et al. 1999). The ectopic overexpression of *NCOA4* in prostate cancer cells reduces cell proliferation and *NCOA4* expression is reduced in aggressive prostate and breast cancers, thus suggesting that this gene may function as a suppressor of tumorigenesis (Kollara et al. 2001; Li et al. 2002; Ligr et al. 2010).

RET MUTATIONS IN MEDULLARY THYROID CARCINOMA AND MEN2 SYNDROMES

Multiple Endocrine Neoplasia Type 2

Medullary thyroid carcinoma (MTC) arises from neural crest-derived calcitonin-producing thyroid parafollicular C cells and represents 5%–10% of all thyroid cancers. Although most MTCs are sporadic and affect adults, around 25% of cases are familial occurring in the frame of inherited cancer syndromes named multiple endocrine neoplasia type 2 (MEN2) syndromes

(online Mendelian inheritance in men, OMIM: #171400) (de Groot et al. 2006). MEN2 comprises three related disorders: MEN2A, MEN2B, and familial medullary thyroid carcinoma (FMTC). MEN2A, first described in 1961 by J.H. Sipple, is characterized by MTC associated with pheochromocytoma (a benign tumor of adrenal medulla) in 50% of cases and parathyroid hyperplasia or adenoma in 10%–30% of cases; more rarely, MEN2A patients show other disease features such as cutaneous lichen amyloidosis and congenital megacolon (see below). In MEN2B syndrome, MTC is associated with pheochromocytoma, ganglioneuromatosis of the intestine, thickening of corneal nerves, and marfanoid habitus (de Groot et al. 2006). Finally, MTC is the only disease phenotype of patients displaying FMTC. Recently, some investigators have suggested FMTC as a phenotypic variant of MEN2A with decreased expression.

RET and MEN2

Specific germline missense mutations of *RET* gene cause the MEN2 syndromes (Fig. 4). Most MEN2A and FMTC mutations affect cysteines in the extracellular cysteine-rich domain of RET. MEN2A is associated most frequently with mutations of cysteine 634 (85%), particularly C634R, whereas FMTC mutations are evenly distributed among the various cysteines (C609, C611, C618, C620, C630) (de Groot et al. 2006). Rare MEN2A or FMTC mutations in RET ectodomain do not target cysteine-rich domain (de Groot et al. 2006; Fazioli et al. 2008; Castellone et al. 2010). FMTC can also be associated with changes in the RET kinase domain (including E768D, L790F, V804L, and V804M).

Most MEN2B patients carry the M918T mutation in RET kinase domain, whereas only a small fraction of them harbor the A883F substitution (Fig. 4). Very rarely, the MEN2B phenotype is sustained by double mutations targeting either the same or two different *RET* alleles (de Groot et al. 2006).

Finally, several additional rare germline *RET* variants have been identified through the systematic screening of MTC patients. However, their pathological significance is not always ob-

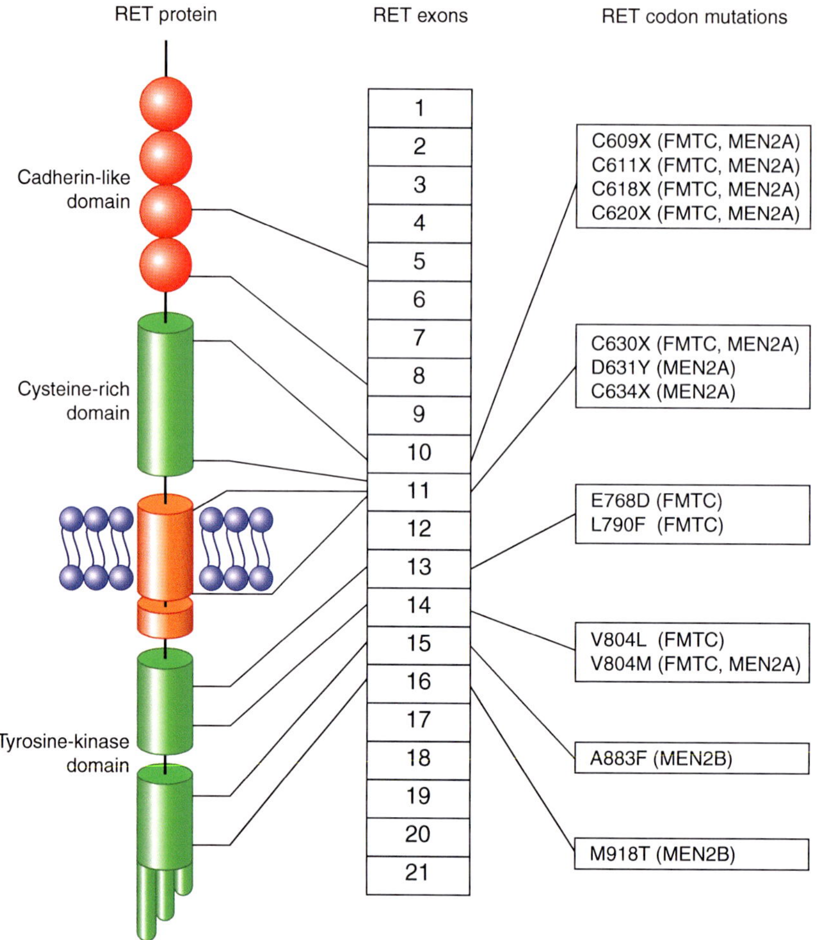

Figure 4. Representation of the most common germline missense mutations in *RET* gene found in MEN2 syndromes. The location of the mutation with respect to RET protein domains and RET exons is indicated. MEN2 phenotypes associated with the various mutations are indicated on the *right*.

vious, especially if data on cosegregation of the mutation with the disease and functional studies are not available. A comprehensive database annotating all *RET* variants and their pathogenetic relevance has been generated (Margraf et al. 2009).

Genetic testing in MTC patients is important not only to distinguish sporadic from familial cases (and thus to identify mutation carriers in the family at risk of developing the disease) but also to assess risk of developing aggressive MTC and MTC-associated neoplastic lesions, such as pheochromocytoma and parathyroid adenoma. Accordingly, most common

RET mutations have been classified to different disease risk levels, ranging from A (less severe) to D (most severe). Risk level classification guides decisions regarding timing of prophylactic thyroidectomy and intraoperative management of the parathyroid glands (Kloos et al. 2009).

RET and Sporadic MTC

MTC arises sporadically in about 75% of cases and *RET* somatic mutations, mainly M918T, occur in about 50% of sporadic MTC, but very rarely in sporadic pheochromocytoma (Beldjord et al. 1995; Lindor et al. 1995). The presence of somatic *RET* mutation correlates with an aggressive MTC disease phenotype (Romei et al. 1996; Schilling et al. 2001). Recently, sporadic MTC negative for *RET* mutations have been shown to frequently display mutation of *RAS* genes (Moura et al. 2011). Thus, similarly to follicular cell-derived thyroid tumors, thyroid C-cell-derived carcinomas commonly feature the activation of RET-RAS-MAPK pathway (Fig. 3).

MECHANISMS OF *RET* ONCOGENIC CONVERSION SECONDARY TO POINT MUTATIONS

The mechanisms leading to *RET* oncogenic conversion in MEN2 and MTC depend on the site of the amino-acid change. In the case of cysteine mutants, cysteine removal is believed to prevent the formation of intramolecular disulfide bonds, thus allowing free cysteine residues to form intermolecular bonds and mediate the formation of covalent RET dimers with constitutive kinase and signaling activity (Fig. 3) (Santoro et al. 1995). Moreover, differently from wild-type (wt) RET, C634 RET mutant is resistant to intracellular domain cleavage and does not show cytotoxic activity in the absence of ligand, thereby losing the "dependence" receptor feature. This implies that a single mutation (e.g., C634) may induce at the same time increased mitogenic signaling and reduced proapoptotic activity (Bordeaux et al. 2000). Mutations associated with FMTC, which generally target cysteines other than C634 (Fig. 4), are less potently transforming than MEN2A-associated C634 mutations because of their weaker ability to induce formation of RET dimers (Carlomagno et al. 1997; Ito et al. 1997; Chappuis-Flament et al. 1998).

A change in substrate specificity together with a ligand-independent activation of the enzymatic function has been implicated in the mechanism of activation induced by M918T mutation (Santoro et al. 1995; Songyang et al. 1995). In line with this model, M918T mutants differ from wild-type RET in the stoichiometry of phosphorylation of RET tyrosines and of various intracellular proteins (Santoro et al. 1995;

Salvatore et al. 2001). RET/M918T-expressing tumors have different gene expression profiles compared with RET/C634-expressing tumors (Jain et al. 2004). Moreover, X-ray crystallographic analysis of RET tyrosine kinase domain has shown that wild-type RET kinase adopts a head-to-tail autoinhibited dimeric state and that this inactive conformation is destabilized by M918T mutation (Knowles et al. 2006). The mechanism through which RET intracellular mutations (other than M918T) activate constitutively RET enzymatic function has not been clearly elucidated.

RET Gain- and Loss-of-Function in Disease

As mentioned above, germline mutations in *RET* have been implicated in both sporadic and familial cases of Hirschsprung disease (HSCR). HSCR is characterized by the congenital absence of enteric innervation, causing block of peristalsis and bowel obstruction (congenital megacolon) (Brooks et al. 2005). *RET* mutations found in HSCR patients are heterogeneous, ranging from gene deletions to nonsense and missense point mutations, and in most cases they cause a loss of RET signaling. Moreover, some HSCR mutations were found to cause a constitutive caspase-3-mediated RET cleavage and proapoptotic activity probably attributable to the "dependence receptor" features of RET (Bordeaux et al. 2000).

In a few cases, HSCR cosegregates with MEN2A/FMTC that instead, as described above, are associated with RET gain-of-function. These promiscuous HSCR-MEN2A/FMTC cases typically display mutations in RET cysteines other than C634. The most reasonable explanation of this paradox relies on the possibility that these particular mutations have a *Janus*-faced effect, as they feature both a decreased cell surface expression and a constitutive, although low-level, kinase activity. This may on the one hand be sufficient to cause thyroid C-cells transformation, but on the other hand not be sufficient to sustain correct enteric neurons development for the reduced ability of such mutants of interacting with the ligand on the cell surface (Carlomagno et al. 1997; Ito et al. 1997; Chappuis-Flament et al. 1998; Arighi et al. 2004).

RET IN MALIGNANCIES OTHER THAN THYROID CARCINOMA

RET in Lung Cancer

For many years, *RET* oncogenic conversion has been thought to be confined to thyroid cancer. More recently, structural *RET* alterations or changes of its expression have been described in neoplasms affecting organs other than thyroid. Systematic high-throughput sequencing screening identified the M918T *RET* mutation in one single case of non-small-cell lung cancer (NSCLC). In this case, *RET* mutation co-occurred with K-RAS mutation, whereas K-RAS is generally alternative to RTK mutation (Thomas et al. 2007). Very recently, *RET* has been shown to play an important role in a subset of NSCLC patients. In about 1% NSCLC, particularly in adenocarcinoma, chromosomal inversions cause the fusion of the *RET*-encoded TK domain (from exon 12 to the 3′-end of *RET*, as in RET/PTC rearrangements) to different 5′-terminal exons (15, 16, 22, 23, or 24 exons in different rearrangement variants) of the *KIF5B* (kinesin family member 5B) gene (Ju et al. 2012; Kohno et al. 2012; Li et al. 2012; Lipson et al. 2012; Takeuchi et al. 2012). Less commonly, the *RET*-encoded TK domain was found to be fused to *CCDC6, NCOA4,* or *TRIM33* genes (as in *RET/PTC1, RET/PTC3* and *RET/PTC7,* re-spectively) (Li et al. 2012; Wang et al. 2012; Drilon et al. 2013). Similarly to *RET/PTC* rearrangements, KIF5B-RET fusion proteins are likely to form homodimers through the coiled-coil domain present in the NH$_2$-terminal portion of KIF5B (Fig. 5). The coiled-coil domain is retained in all variants of KIF5B-RET rearranged proteins. Consistently, KIF5B-RET fusion proteins display ligand-independent activation of RET kinase and are able to transform fibroblasts in vitro (Kohno et al. 2012).

RET in Leukemia

RET gene was found up-regulated in a subtype of acute myeloid leukemia with myelomonocytic stage of differentiation (Camos et al. 2006). More recently, gene rearrangements causing the fusion of the *RET*-encoding TK domain (from exon 12 to the 3′-end of *RET,* as in *RET/PTC* and *KIF5B-RET*) in one case to the first 5′-terminal 4 exons of *BCR* (breakpoint cluster region) and in another case to the first 5′-terminal 12 exons of *FGFR1OP* (fibroblast growth factor receptor 1 oncogenic partner) genes have been cloned from two cases of chronic myelomonocytic leukemia (CMML) (Ballerini et al. 2012). BCR-RET and FGFR1OP-RET fusion proteins act as bona fide oncoproteins, display aberrant activation of RET kinase, and transform hematopoietic cells in vitro. Thus, although the prevalence of these rearrangements is still unknown, CMML appears to be another neoplasia that,

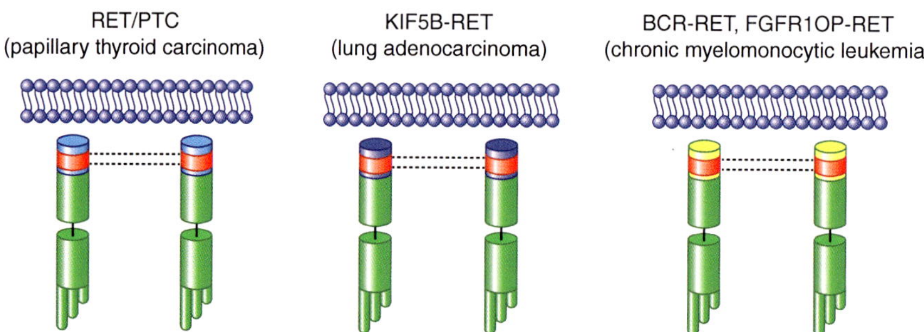

Figure 5. Schematic representation of *RET* gene rearrangements in human cancer. RET/PTC rearrangements in PTC are shown, together with KIF5B-RET (found in lung adenocarcinoma) and BCR-RET and FGFR1OP-RET (found in chronic myelomonocytic leukemia). All the *RET* fusion partners invariably code for protein–protein dimerization motifs (depicted as red boxes).

together with PTC and lung adenocarcinoma, is associated with *RET* activation through gene rearrangement (Fig. 5).

RET Role in Additional Cancer Types

Systematic DNA sequencing in a set of colon cancers has revealed the rare occurrence of somatic *RET* sequence variants (Wood et al. 2007). Similarly, sequencing of around 1500 cancer-related genes uncovered the presence of rare *RET* variants of unknown functional significance in hormone receptor positive breast cancer (Kan et al. 2010).

Apart from mutations, many reports have indicated a positive correlation between *RET* overexpression and ER (estrogen receptor)-positive breast carcinoma (Tozlu et al. 2006; Esseghir et al. 2007). *RET* transcriptional promoter was found to display three EREs (estrogen receptor elements) that may mediate *RET* up-regulation by estrogens (Boulay et al. 2008; Kang et al. 2010). A functional link between RET and ER is indicated by the observation that RET inhibition restored a hormone-sensitive phenotype of antiestrogen-resistant breast cancer cells (Plaza-Menacho et al. 2010).

Some frequent *RET* sequence variants (polymorphisms) have been correlated to specific tumor types. RET G691S polymorphism co-segregates with MTC, pancreatic cancer, and desmoplastic subtype of cutaneous malignant melanoma (Robledo et al. 2003; Cebrian et al. 2005; Sawai et al. 2005; Narita et al. 2009). G691S caused increased receptor-signaling responses to GDNF stimulation, with augmented cell proliferation, migration, and invasion (Narita et al. 2009).

RET protein was found overexpressed in pancreatic carcinoma and GDNF polarized migration of pancreatic cancer cells toward nerves mediating neural invasion (Veit et al. 2004; Gil et al. 2010). Finally, although the molecular basis was not elucidated, overactivation of RET protein has been observed in glioblastoma multiforme and involved in maintaining a robust downstream signaling able to limit the efficacy of therapies with kinase inhibitors (Stommel et al. 2007).

RET PROTEIN AS AN ANTINEOPLASTIC THERAPEUTIC TARGET

RET Kinase Inhibitors

Small molecule protein kinase inhibitors (PKI) are an important class of anticancer agents. In most of the cases, they compete with ATP, thereby obstructing autophosphorylation and signal transduction downstream from the targeted kinase (Zhang et al. 2009). Prominent examples of successful PKI are represented by imatinib (Gleevec) against BCR-ABL in chronic myeloid leukemia (CML), gefitinib (iressa), and erlotinib against EGFR in non-small-cell lung carcinoma (NSCLC), and vemurafenib (PLX4032) against BRAF in melanoma (Zhang et al. 2009).

Based on its involvement in cancer and particularly in MTC, a neoplasm that does not respond to conventional chemotherapy, RET has raised interest as a molecular target of kinase inhibitors (Schlumberger et al. 2008; Gild et al. 2011). Preclinical studies have shown that MTC cell lines are addicted to RET oncogenic signaling and that RET PKIs are able to block MTC cell proliferation (Schlumberger et al. 2008). Some compounds were isolated that exerted potent RET inhibition. They include vandetanib (ZD6474), sorafenib (BAY 43-9006), sunitinib (SU11248), cabozantinib (XL184), lenvatinib (E7080), and ponatinib (AP24534) (Carlomagno et al. 2002, 2006; Kim et al. 2006; Verbeek et al. 2011; De Falco et al. 2013). These compounds are multitargeted, being able to inhibit RET and also additional kinases.

Therapeutic Targeting of RET in Human Cancer

Most of RET PKIs have been or are being evaluated in clinical trials for MTC treatment (Schlumberger et al. 2008). In particular, vandetanib (ZD6474), an inhibitor of RET, EGFR (epidermal growth factor receptor) and VEGFR (vascular endothelial growth factor receptor), and cabozantinib, an inhibitor of RET, MET, and VEGFR, have been approved by the FDA for MTC treatment, based on significant progression-free survival prolongation in clinical trials (Kurzrock et al. 2011; Wells et al. 2012).

However, it should be noted that some specific RET mutations (codons 804 and 806) confer resistance to vandetanib and would require alternative inhibitors (Carlomagno et al. 2004, 2009).

Medullary thyroid cancer (MTC) may not remain the only cancer that benefits from RET kinase inhibition. Papillary thyroid carcinoma (PTC) in general responds well to adjuvant radioiodine treatment and therefore RET kinase inhibitors may not prove particularly useful for this tumor (Gild et al. 2011). Given the recent discovery of RET gain-of-function in lung adenocarcinoma and CMML, it is feasible that RET kinase inhibitors may also find applications in these cancers. Consistently, KIF5B-RET transformed fibroblasts growth was inhibited by vandetanib (ZD6474) (Kohno et al. 2012), treatment with cabozantinib induced objective response in some patients with *RET* mutant lung adenocarcinoma (Drilon et al. 2013), and treatment with sorafenib (BAY 43-9006), another RET kinase inhibitor, induced cytological, and clinical remission in a patient carrying the BCR-RET fusion (Ballerini et al. 2012).

CONCLUSIONS

After its initial isolation from patients affected by papillary thyroid carcinoma, RET receptor tyrosine kinase also proved to be a key player in the pathogenesis of medullary thyroid carcinoma (MTC). More recently, rare forms of leukemia and non-small-cell lung carcinoma have been associated with oncogenic conversion of RET, thus expanding the spectrum of cancer diseases associated with mutations of this gene. This knowledge will foster efforts to develop agents able to target RET oncoproteins for cancer treatment.

REFERENCES

Airaksinen MS, Saarma M. 2002. The GDNF family: Signalling, biological functions and therapeutic value. *Nat Rev Neurosci* 3: 383–394.

Alberti L, Borrello MG, Ghizzoni S, Torriti F, Rizzetti MG, Pierotti MA. 1998. Grb2 binding to the different isoforms of Ret tyrosine kinase. *Oncogene* 17: 1079–1087.

Ameziane-El-Hassani R, Boufraqech M, Lagente-Chevallier O, Weyemi U, Talbot M, Métivier D, Courtin F, Bidart JM, El Mzibri M, Schlumberger M, et al. 2010. Role of H2O2 in RET/PTC1 chromosomal rearrangement produced by ionizing radiation in human thyroid cells. *Cancer Res* 70: 4123–4132.

Anders J, Kjar S, Ibáñez CF. 2001. Molecular modeling of the extracellular domain of the RET receptor tyrosine kinase reveals multiple cadherin-like domains and a calcium-binding site. *J Biol Chem* 276: 35808–35817.

Arighi E, Popsueva A, Degl'Innocenti D, Borrello MG, Carniti C, Perälä NM, Pierotti MA, Sariola H. 2004. Biological effects of the dual phenotypic Janus mutation of ret cosegregating with both multiple endocrine neoplasia type 2 and Hirschsprung's disease. *Mol Endocrinol* 18: 1004–1017.

Asai N, Murakami H, Iwashita T, Takahashi M. 1996. A mutation at tyrosine 1062 in MEN2A-Ret and MEN2B-Ret impairs their transforming activity and association with shc adaptor proteins. *J Biol Chem* 271: 17644–17649.

Ballerini P, Struski S, Cresson C, Prade N, Toujani S, Deswarte C, Dobbelstein S, Petit A, Lapillonne H, Gautier EF, et al. 2012. RET fusion genes are associated with chronic myelomonocytic leukemia and enhance monocytic differentiation. *Leukemia* 26: 2384–2399.

Baloh RH, Enomoto H, Johnson EM Jr, Milbrandt J. 2000. The GDNF family ligands and receptors—Implications for neural development. *Curr Opin Neurobiol* 10: 103–110.

Beldjord C, Desclaux-Arramond F, Raffin-Sanson M, Corvol JC, De Keyzer Y, Luton JP, Plouin PF, Bertagna X. 1995. The RET protooncogene in sporadic pheochromocytomas: Frequent MEN 2-like mutations and new molecular defects. *J Clin Endocrinol Metab* 80: 2063–2068.

Boikos SA, Stratakis CA. 2006. Carney complex: Pathology and molecular genetics. *Neuroendocrinology* 83: 189–199.

Bongarzone I, Monzini N, Borrello MG, Carcano C, Ferraresi G, Arighi E, Mondellini P, Della Porta G, Pierotti MA. 1993. Molecular characterization of a thyroid tumor-specific transforming sequence formed by the fusion of ret tyrosine kinase and the regulatory subunit RI alpha of cyclic AMP-dependent protein kinase A. *Mol Cell Biol* 13: 358–366.

Bordeaux MC, Forcet C, Granger L, Corset V, Bidaud C, Billaud M, Bredesen DE, Edery P, Mehlen P. 2000. The RET proto-oncogene induces apoptosis: A novel mechanism for Hirschsprung disease. *EMBO J* 19: 4056–4063.

Borrello MG, Alberti L, Arighi E, Bongarzone I, Battistini C, Bardelli A, Pasini B, Piutti C, Rizzetti MG, Mondellini P, et al. 1996. The full oncogenic activity of Ret/ptc2 depends on tyrosine 539, a docking site for phospholipase Cγ. *Mol Cell Biol* 16: 2151–2163.

Borrello MG, Alberti L, Fischer A, Degl'innocenti D, Ferrario C, Gariboldi M, Marchesi F, Allavena P, Greco A, Collini P, et al. 2005. Induction of a proinflammatory program in normal human thyrocytes by the RET/PTC1 oncogene. *Proc Natl Acad Sci* 102: 14825–14830.

Boulay A, Breuleux M, Stephan C, Fux C, Brisken C, Fiche M, Wartmann M, Stumm M, Lane HA, Hynes NE. 2008. The Ret receptor tyrosine kinase pathway functionally

Cite this article as *Cold Spring Harb Perspect Biol* doi: 10.1101/cshperspect.a009233

interacts with the ERα pathway in breast cancer. *Cancer Res* **68**: 3743–3751.

Brooks AS, Oostra BA, Hofstra RM. 2005. Studying the genetics of Hirschsprung's disease: Unraveling an oligogenic disorder. *Clin Genet* **67**: 6–14.

Camós M, Esteve J, Jares P, Colomer D, Rozman M, Villamor N, Costa D, Carrió A, Nomdedéu J, Montserrat E, et al. 2006. Gene expression profiling of acute myeloid leukemia with translocation t(8;16)(p11;p13) and MYST3-CREBBP rearrangement reveals a distinctive signature with a specific pattern of HOX gene expression. *Cancer Res* **66**: 6947–6954.

Cañibano C, Rodriguez NL, Saez C, Tovar S, Garcia-Lavandeira M, Borrello MG, Vidal A, Costantini F, Japon M, Dieguez C, et al. 2007. The dependence receptor Ret induces apoptosis in somatotrophs through a Pit-1/p53 pathway, preventing tumor growth. *EMBO J* **26**: 2015–2028.

Carlomagno F, De Vita G, Berlingieri MT, de Franciscis V, Melillo RM, Colantuoni V, Kraus MH, Di Fiore PP, Fusco A, Santoro M. 1996. Molecular heterogeneity of RET loss of function in Hirschsprung's disease. *EMBO J* **15**: 2717–2725.

Carlomagno F, Salvatore G, Cirafici AM, De Vita G, Melillo RM, de Franciscis V, Billaud M, Fusco A, Santoro M. 1997. The different RET-activating capability of mutations of cysteine 620 or cysteine 634 correlates with the multiple endocrine neoplasia type 2 disease phenotype. *Cancer Res* **57**: 391–395.

Carlomagno F, Vitagliano D, Guida T, Ciardiello F, Tortora G, Vecchio G, Ryan AJ, Fontanini G, Fusco A, Santoro M. 2002. ZD6474, an orally available inhibitor of KDR tyrosine kinase activity, efficiently blocks oncogenic RET kinases. *Cancer Res* **62**: 7284–7290.

Carlomagno F, Guida T, Anaganti S, Vecchio G, Fusco A, Ryan AJ, Billaud M, Santoro M. 2004. Disease associated mutations at valine 804 in the RET receptor tyrosine kinase confer resistance to selective kinase inhibitors. *Oncogene* **23**: 6056–6063.

Carlomagno F, Anaganti S, Guida T, Salvatore G, Troncone G, Wilhelm SM, Santoro M. 2006. BAY 43–9006 inhibition of oncogenic RET mutants. *J Natl Cancer Inst* **98**: 326–334.

Carlomagno F, Guida T, Anaganti S, Provitera L, Kjaer S, McDonald NQ, Ryan AJ, Santoro M. 2009. Identification of tyrosine 806 as a molecular determinant of RET kinase sensitivity to ZD6474. *Endocr Relat Cancer* **16**: 233–241.

Castellone MD, Cirafici AM, De Vita G, De Falco V, Malorni L, Tallini G, Fagin JA, Fusco A, Melillo RM, Santoro M. 2003. Ras-mediated apoptosis of PC CL 3 rat thyroid cells induced by RET/PTC oncogenes. *Oncogene* **22**: 246–255.

Castellone MD, Verrienti A, Magendra Rao D, Sponziello M, Fabbro D, Muthu M, Durante C, Maranghi M, Damante G, Pizzolitto S, et al. 2010. A novel de novo germ-line V292M mutation in the extracellular region of RET in a patient with phaeochromocytoma and medullary thyroid carcinoma: functional characterization. *Clin Endocrinol (Oxf)* **73**: 529–534.

Caudill CM, Zhu Z, Ciampi R, Stringer JR, Nikiforov YE. 2005. Dose-dependent generation of RET/PTC in human thyroid cells after in vitro exposure to γ-radiation: A model of carcinogenic chromosomal rearrangement induced by ionizing radiation. *J Clin Endocrinol Metab* **90**: 2364–2369.

Cebrian A, Lesueur F, Martin S, Leyland J, Ahmed S, Luccarini C, Smith PL, Luben R, Whittaker J, Pharoah PD, et al. 2005. Polymorphisms in the initiators of RET (rearranged during transfection) signaling pathway and susceptibility to sporadic medullary thyroid carcinoma. *J Clin Endocrinol Metab* **90**: 6268–6274.

Celetti A, Cerrato A, Merolla F, Vitagliano D, Vecchio G, Grieco M. 2004. H4(D10S170), a gene frequently rearranged with RET in papillary thyroid carcinomas: Functional characterization. *Oncogene* **23**: 109–121.

Chappuis-Flament S, Pasini A, De Vita G, Ségouffin-Cariou C, Fusco A, Attié T, Lenoir GM, Santoro M, Billaud M. 1998. Dual effect on the RET receptor of MEN 2 mutations affecting specific extracytoplasmic cysteines. *Oncogene* **17**: 2851–2861.

Collins BJ, Chiappetta G, Schneider AB, Santoro M, Pentimalli F, Fogelfeld L, Gierlowski T, Shore-Freedman E, Jaffe G, Fusco A. 2002. RET expression in papillary thyroid cancer from patients irradiated in childhood for benign conditions. *J Clin Endocrinol Metab* **87**: 3941–3946.

De Falco V, Buonocore P, Muthu M, Torregrossa L, Basolo F, Billaud M, Gozgit JM, Carlomagno F, Santoro M. 2013. Ponatinib (AP24534) is a novel potent inhibitor of oncogenic RET mutants associated with thyroid cancer. *J Clin Endocrinol Metab* doi: 10.1210/jc.2012–2672.

de Graaff E, Srinivas S, Kilkenny C, D'Agati V, Mankoo BS, Costantini F, Pachnis V. 2001. Differential activities of the RET tyrosine kinase receptor isoforms during mammalian embryogenesis. *Genes Dev* **15**: 2433–2444.

de Groot JW, Links TP, Plukker JT, Lips CJ, Hofstra RM. 2006. RET as a diagnostic and therapeutic target in sporadic and hereditary endocrine tumors. *Endocr Rev* **27**: 535–560.

Drilon A, Wang L, Hasanovic A, Suehara Y, Lipson D, Stephens PJ, Ross J, Miller VA, Ginsberg MS, Zakowski MF, et al. 2013. Response to cabozantinib in patients with RET fusion-positive lung adenocarcinomas. *Cancer Discov* doi: 10.1158/2159–8290.CD-13–0035.

Durkin SG, Glover TW. 2007. Chromosome fragile sites. *Annu Rev Genet* **41**: 169–192.

Encinas M, Crowder RJ, Milbrandt J, Johnson EM Jr. 2004. Tyrosine 981, a novel ret autophosphorylation site, binds c-Src to mediate neuronal survival. *J Biol Chem* **279**: 18262–18269.

Esseghir S, Todd SK, Hunt T, Poulsom R, Plaza-Menacho I, Reis-Filho JS, Isacke CM. 2007. A role for glial cell derived neurotrophic factor induced expression by inflammatory cytokines and RET/GFR-α 1 receptor up-regulation in breast cancer. *Cancer Res* **67**: 11732–11741.

Fazioli F, Piccinini G, Appolloni G, Bacchiocchi R, Palmonella G, Recchioni R, Pierpaoli E, Silvetti F, Scarpelli M, Bruglia M, et al. 2008. A new germline point mutation in Ret exon 8 (cys515ser) in a family with medullary thyroid carcinoma. *Thyroid* **18**: 775–782.

Gandhi M, Medvedovic M, Stringer JR, Nikiforov YE. 2006. Interphase chromosome folding determines spatial proximity of genes participating in carcinogenic RET/PTC rearrangements. *Oncogene* **25**: 2360–2366.

Gandhi M, Evdokimova V, Nikiforov YE. 2010a. Mechanisms of chromosomal rearrangements in solid tumors: The model of papillary thyroid carcinoma. *Mol Cell Endocrinol* **321:** 36–43.

Gandhi M, Dillon LW, Pramanik S, Nikiforov YE, Wang YH. 2010b. DNA breaks at fragile sites generate oncogenic RET/PTC rearrangements in human thyroid cells. *Oncogene* **29:** 2272–2280.

Gil Z, Cavel O, Kelly K, Brader P, Rein A, Gao SP, Carlson DL, Shah JP, Fong Y, Wong RJ. 2010. Paracrine regulation of pancreatic cancer cell invasion by peripheral nerves. *J Natl Cancer Inst* **102:** 107–118.

Gild ML, Bullock M, Robinson BG, Clifton-Bligh R. 2011. Multikinase inhibitors: A new option for the treatment of thyroid cancer. *Nat Rev Endocrinol* **7:** 617–624.

Greco A, Miranda C, Pierotti MA. 2010. Rearrangements of NTRK1 gene in papillary thyroid carcinoma. *Mol Cell Endocrinol* **321:** 44–49.

Grieco M, Santoro M, Berlingieri MT, Melillo RM, Donghi R, Bongarzone I, Pierotti MA, Della Porta G, Fusco A, Vecchio G. 1990. PTC is a novel rearranged form of the ret proto-oncogene and is frequently detected in vivo in human thyroid papillary carcinomas. *Cell* **60:** 557–563.

Hamatani K, Eguchi H, Ito R, Mukai M, Takahashi K, Taga M, Imai K, Cologne J, Soda M, Arihiro K, et al. 2008. RET/PTC rearrangements preferentially occurred in papillary thyroid cancer among atomic bomb survivors exposed to high radiation dose. *Cancer Res* **68:** 7176–7182.

Hanahan D, Weinberg RA. 2011. Hallmarks of cancer: The next generation. *Cell* **144:** 646–674.

Heinlein CA, Ting HJ, Yeh S, Chang C. 1999. Identification of ARA70 as a ligand-enhanced coactivator for the peroxisome proliferator-activated receptor γ. *J Biol Chem* **274:** 16147–16152.

Ito T, Seyama T, Iwamoto KS, Hayashi T, Mizuno T, Tsuyama N, Dohi K, Nakamura N, Akiyama M. 1993. In vitro irradiation is able to cause RET oncogene rearrangement. *Cancer Res* **53:** 2940–2943.

Ito S, Iwashita T, Asai N, Murakami H, Iwata Y, Sobue G, Takahashi M. 1997. Biological properties of Ret with cysteine mutations correlate with multiple endocrine neoplasia type 2A, familial medullary thyroid carcinoma, and Hirschsprung's disease phenotype. *Cancer Res* **57:** 2870–2872.

Jain S. 2009. The many faces of RET dysfunction in kidney. *Organogenesis* **5:** 177–190.

Jain S, Watson MA, DeBenedetti MK, Hiraki Y, Moley JF, Milbrandt J. 2004. Expression profiles provide insights into early malignant potential and skeletal abnormalities in multiple endocrine neoplasia type 2B syndrome tumors. *Cancer Res* **64:** 3907–3913.

Ju YS, Lee WC, Shin JY, Lee S, Bleazard T, Won JK, Kim YT, Kim JI, Kang JH, Seo JS. 2012. A transforming KIF5B and RET gene fusion in lung adenocarcinoma revealed from whole-genome and transcriptome sequencing. *Genome Res* **22:** 436–445.

Kan Z, Jaiswal BS, Stinson J, Janakiraman V, Bhatt D, Stern HM, Yue P, Haverty PM, Bourgon R, Zheng J, et al. 2010. Diverse somatic mutation patterns and pathway alterations in human cancers. *Nature* **466:** 869–873.

Kang J, Qian PX, Pandey V, Perry JK, Miller LD, Liu ET, Zhu T, Liu DX, Lobie PE. 2010. Artemin is estrogen regulated and mediates antiestrogen resistance in mammary carcinoma. *Oncogene* **29:** 3228–3240.

Kawamoto Y, Takeda K, Okuno Y, Yamakawa Y, Ito Y, Taguchi R, Kato M, Suzuki H, Takahashi M, Nakashima I. 2004. Identification of RET autophosphorylation sites by mass spectrometry. *J Biol Chem* **279:** 14213–14224.

Khetchoumian K, Teletin M, Tisserand J, Mark M, Herquel B, Ignat M, Zucman-Rossi J, Cammas F, Lerouge T, Thibault C, et al. 2007. Loss of Trim24 (Tif1α) gene function confers oncogenic activity to retinoic acid receptor α. *Nat Genet* **39:** 1500–1506.

Kim DW, Jo YS, Jung HS, Chung HK, Song JH, Park KC, Park SH, Hwang JH, Rha SY, Kweon GR, et al. 2006. An orally administered multitarget tyrosine kinase inhibitor, SU11248, is a novel potent inhibitor of thyroid oncogenic RET/papillary thyroid cancer kinases. *J Clin Endocrinol Metab* **91:** 4070–4076.

Kirschner LS, Carney JA, Pack SD, Taymans SE, Giatzakis C, Cho YS, Cho-Chung YS, Stratakis CA. 2000. Mutations of the gene encoding the protein kinase A type I-α regulatory subunit in patients with the Carney complex. *Nat Genet* **26:** 89–92.

Kjaer S, Kurokawa K, Perrinjaquet M, Abrescia C, Ibáñez CF. 2006. Self-association of the transmembrane domain of RET underlies oncogenic activation by MEN2A mutations. *Oncogene* **25:** 7086–7095.

Kloos RT, Eng C, Evans DB, Francis GL, Gagel RF, Gharib H, Moley JF, Pacini F, Ringel MD, Schlumberger M, et al. 2009. Medullary thyroid cancer: Management guidelines of the American Thyroid Association. *Thyroid* **19:** 565–612.

Klugbauer S, Rabes HM. 1999. The transcription coactivator HTIF1 and a related protein are fused to the RET receptor tyrosine kinase in childhood papillary thyroid carcinomas. *Oncogene* **18:** 4388–4393.

Knowles PP, Murray-Rust J, Kjaer S, Scott RP, Hanrahan S, Santoro M, Ibáñez CF, McDonald NQ. 2006. Structure and chemical inhibition of the RET tyrosine kinase domain. *J Biol Chem* **281:** 33577–33587.

Kohno T, Ichikawa H, Totoki Y, Yasuda K, Hiramoto M, Nammo T, Sakamoto H, Tsuta K, Furuta K, Shimada Y, et al. 2012. KIF5B-RET fusions in lung adenocarcinoma. *Nat Med* **18:** 375–377.

Kollara A, Kahn HJ, Marks A, Brown TJ. 2001. Loss of androgen receptor associated protein 70 (ARA70) expression in a subset of HER2-positive breast cancers. *Breast Cancer Res Treat* **67:** 245–253.

Kurzrock R, Sherman SI, Ball DW, Forastiere AA, Cohen RB, Mehra R, Pfister DG, Cohen EE, Janisch L, Nauling F, et al. 2011. Activity of XL184 (Cabozantinib), an oral tyrosine kinase inhibitor, in patients with medullary thyroid cancer. *J Clin Oncol* **29:** 2660–2666.

Leone V, Mansueto G, Pierantoni GM, Tornincasa M, Merolla F, Cerrato A, Santoro M, Grieco M, Scaloni A, Celetti A, et al. 2010. CCDC6 represses CREB1 activity by recruiting histone deacetylase 1 and protein phosphatase 1. *Oncogene* **29:** 4341–4351.

Li P, Yu X, Ge K, Melamed J, Roeder RG, Wang Z. 2002. Heterogeneous expression and functions of androgen receptor co-factors in primary prostate cancer. *Am J Pathol* **161:** 1467–1474.

Li F, Feng Y, Fang R, Fang Z, Xia J, Han X, Liu XY, Chen H, Liu H, Ji H. 2012. Identification of RET gene fusion by exon array analyses in "pan-negative" lung cancer from never smokers. *Cell Res* **22:** 928–931.

Ligr M, Li Y, Zou X, Daniels G, Melamed J, Peng Y, Wang W, Wang J, Ostrer H, Pagano M, et al. 2010. Tumor suppressor function of androgen receptor coactivator ARA70α in prostate cancer. *Am J Pathol* **176:** 1891–1900.

Lindor NM, Honchel R, Khosla S, Thibodeau SN. 1995. Mutations in the RET protooncogene in sporadic pheochromocytomas. *J Clin Endocrinol Metab* **80:** 627–629.

Lipson D, Capelletti M, Yelensky R, Otto G, Parker A, Jarosz M, Curran JA, Balasubramanian S, Bloom T, Brennan KW, et al. 2012. Identification of new ALK and RET gene fusions from colorectal and lung cancer biopsies. *Nat Med* **18:** 382–384.

Liu X, Vega QC, Decker RA, Pandey A, Worby CA, Dixon JE. 1996. Oncogenic RET receptors display different autophosphorylation sites and substrate binding specificities. *J Biol Chem* **271:** 5309–5312.

Manié S, Santoro M, Fusco A, Billaud M. 2001. The RET receptor: Function in development and dysfunction in congenital malformation. *Trends Genet* **17:** 580–589.

Margraf RL, Crockett DK, Krautscheid PM, Seamons R, Calderon FR, Wittwer CT, Mao R. 2009. Multiple endocrine neoplasia type 2 RET protooncogene database: Repository of MEN2-associated RET sequence variation and reference for genotype/phenotype correlations. *Hum Mutat* **30:** 548–556.

Mehlen P, Bredesen DE. 2011. Dependence receptors: From basic research to drug development. *Sci Signal* **4:** mr2.

Melillo RM, Carlomagno F, De Vita G, Formisano P, Vecchio G, Fusco A, Billaud M, Santoro M. 2001a. The insulin receptor substrate (IRS)-1 recruits phosphatidylinositol 3-kinase to Ret: Evidence for a competition between Shc and IRS-1 for the binding to Ret. *Oncogene* **20:** 209–218.

Melillo RM, Santoro M, Ong SH, Billaud M, Fusco A, Hadari YR, Schlessinger J, Lax I. 2001b. Docking protein FRS2 links the protein tyrosine kinase RET and its oncogenic forms with the mitogen-activated protein kinase signaling cascade. *Mol Cell Biol* **21:** 4177–4187.

Melillo RM, Castellone MD, Guarino V, De Falco V, Cirafici AM, Salvatore G, Caiazzo F, Basolo F, Giannini R, Kruhoffer M, et al. 2005. The RET/PTC-RAS-BRAF linear signaling cascade mediates the motile and fitogeni phenotype of thyroid cancer cells. *J Clin Invest* **115:** 1068–1081.

Melillo RM, Guarino V, Avilla E, Galdiero MR, Liotti F, Prevete N, Rossi FW, Basolo F, Ugolini C, de Paulis A, et al. 2010. Mast cells have a protumorigenic role in human thyroid cancer. *Oncogene* **29:** 6203–6215.

Merolla F, Pentimalli F, Pacelli R, Vecchio G, Fusco A, Grieco M, Celetti A. 2007. Involvement of H4(D10S170) protein in ATM-dependent response to DNA damage. *Oncogene* **26:** 6167–6175.

Mizuno T, Kyoizumi S, Suzuki T, Iwamoto KS, Seyama T. 1997. Continued expression of a tissue specific activated oncogene in the early steps of radiation-induced human thyroid carcinogenesis. *Oncogene* **15:** 1455–1460.

Mizuno T, Iwamoto KS, Kyoizumi S, Nagamura H, Shinohara T, Koyama K, Seyama T, Hamatani K. 2000. Preferential induction of RET/PTC1 rearrangement by X-ray irradiation. *Oncogene* **19:** 438–443.

Monaco C, Visconti R, Barone MV, Pierantoni GM, Berlingieri MT, De Lorenzo C, Mineo A, Vecchio G, Fusco A, Santoro M. 2001. The RFG oligomerization domain mediates kinase activation and re-localization of the RET/PTC3 oncoprotein to the plasma membrane. *Oncogene* **20:** 599–608.

Moura MM, Cavaco BM, Pinto AE, Leite V. 2011. High prevalence of RAS mutations in RET-negative sporadic medullary thyroid carcinomas. *J Clin Endocrinol Metab* **96:** E863–E868.

Murakami H, Iwashita T, Asai N, Shimono Y, Iwata Y, Kawai K, Takahashi M. 1999. Enhanced phosphatidylinositol 3-kinase activity and high phosphorylation state of its downstream signalling molecules mediated by ret with the MEN 2B mutation. *Biochem Biophys Res Commun* **262:** 68–75.

Myers SM, Eng C, Ponder BA, Mulligan LM. 1995. Characterization of RET proto-oncogene 3′ splicing variants and polyadenylation sites: A novel C-terminus for RET. *Oncogene* **11:** 2039–2045.

Narita N, Tanemura A, Murali R, Scolyer RA, Huang S, Arigami T, Yanagita S, Chong KK, Thompson JF, Morton DL, et al. 2009. Functional RET G691S polymorphism in cutaneous malignant melanoma. *Oncogene* **28:** 3058–3068.

Nikiforov YE, Nikiforova MN. 2011. Molecular genetics and diagnosis of thyroid cancer. *Nat Rev Endocrinol* **7:** 569–580.

Nikiforova MN, Stringer JR, Blough R, Medvedovic M, Fagin JA, Nikiforov YE. 2000. Proximity of chromosomal loci that participate in radiation-induced rearrangements in human cells. *Science* **290:** 138–141.

Pasini B, Hofstra RM, Yin L, Bocciardi R, Santamaria G, Grootscholten PM, Ceccherini I, Patrone G, Priolo M, Buys CH, et al. 1995. The physical map of the human RET proto-oncogene. *Oncogene* **11:** 1737–1743.

Plaza-Menacho I, Morandi A, Robertson D, Pancholi S, Drury S, Dowsett M, Martin LA, Isacke CM. 2010. Targeting the receptor tyrosine kinase RET sensitizes breast cancer cells to tamoxifen treatment and reveals a role for RET in endocrine resistance. *Oncogene* **29:** 4648–4657.

Powell DJ Jr, Russell J, Nibu K, Li G, Rhee E, Liao M, Goldstein M, Keane WM, Santoro M, Fusco A, et al. 1998. The RET/PTC3 oncogene: Metastatic solid-type papillary carcinomas in murine thyroids. *Cancer Res* **58:** 5523–5528.

Puxeddu E, Knauf JA, Sartor MA, Mitsutake N, Smith EP, Medvedovic M, Tomlinson CR, Moretti S, Fagin JA. 2005. RET/PTC-induced gene expression in thyroid PCCL3 cells reveals early activation of genes involved in regulation of the immune response. *Endocr Relat Cancer* **12:** 319–334.

Rhoden KJ, Unger K, Salvatore G, Yilmaz Y, Vovk V, Chiappetta G, Qumsiyeh MB, Rothstein JL, Fusco A, Santoro M, et al. 2006. RET/papillary thyroid cancer rearrangement in nonneoplastic thyrocytes: Follicular cells of Hashimoto's thyroiditis share low-level recombination events with a subset of papillary carcinoma. *J Clin Endocrinol Metab* **91**: 2414–2423.

Robledo M, Gil L, Pollán M, Cebrián A, Ruíz S, Azañedo M, Benitez J, Menárguez J, Rojas JM. 2003. Polymorphisms G691S/S904S of RET as genetic modifiers of MEN 2A. *Cancer Res* **63**: 1814–1817.

Romei C, Elisei R, Pinchera A, Ceccherini I, Molinaro E, Mancusi F, Martino E, Romeo G, Pacini F. 1996. Somatic mutations of the ret protooncogene in sporadic medullary thyroid carcinoma are not restricted to exon 16 and are associated with tumor recurrence. *J Clin Endocrinol Metab* **81**: 1619–1622.

Russell JP, Shinohara S, Melillo RM, Castellone MD, Santoro M, Rothstein JL. 2003. Tyrosine kinase oncoprotein, RET/PTC3, induces the secretion of myeloid growth and chemotactic factors. *Oncogene* **22**: 4569–4577.

Salvatore D, Melillo RM, Monaco C, Visconti R, Fenzi G, Vecchio G, Fusco A, Santoro M. 2001. Increased in vivo phosphorylation of ret tyrosine 1062 is a potential pathogenetic mechanism of multiple endocrine neoplasia type 2B. *Cancer Res* **61**: 1426–1431.

Santoro M, Melillo RM, Grieco M, Berlingieri MT, Vecchio G, Fusco A. 1993. The TRK and RET tyrosine kinase oncogenes cooperate with ras in the neoplastic transformation of a rat thyroid epithelial cell line. *Cell Growth Differ* **4**: 77–84.

Santoro M, Dathan NA, Berlingieri MT, Bongarzone I, Paulin C, Grieco M, Pierotti MA, Vecchio G, Fusco A. 1994. Molecular characterization of RET/PTC3; a novel rearranged version of the RET proto-oncogene in a human thyroid papillary carcinoma. *Oncogene* **9**: 509–516.

Santoro M, Carlomagno F, Romano A, Bottaro DP, Dathan NA, Grieco M, Fusco A, Vecchio G, Matoskova B, Kraus MH, et al. 1995. Activation of RET as a dominant transforming gene by germline mutations of MEN2A and MEN2B. *Science* **267**: 381–383.

Santoro M, Chiappetta G, Cerrato A, Salvatore D, Zhang L, Manzo G, Picone A, Portella G, Santelli G, Vecchio G, et al. 1996. Development of thyroid papillary carcinomas secondary to tissue-specific expression of the RET/PTC1 oncogene in transgenic mice. *Oncogene* **12**: 1821–1826.

Sawai H, Okada Y, Kazanjian K, Kim J, Hasan S, Hines OJ, Reber HA, Hoon DS, Eibl G. 2005. The G691S RET polymorphism increases glial cell line-derived neurotrophic factor-induced pancreatic cancer cell invasion by amplifying mitogen-activated protein kinase signaling. *Cancer Res* **65**: 11536–11544.

Schilling T, Bürck J, Sinn HP, Clemens A, Otto HF, Höppner W, Herfarth C, Ziegler R, Schwab M, Raue F. 2001. Prognostic value of codon 918 (ATG→ACG) RET proto-oncogene mutations in sporadic medullary thyroid carcinoma. *Int J Cancer* **95**: 62–66.

Schlumberger M, Carlomagno F, Baudin E, Bidart JM, Santoro M. 2008. New therapeutic approaches to treat medullary thyroid carcinoma. *Nat Clin Pract Endocrinol Metab* **4**: 22–32.

Schuetz G, Rosário M, Grimm J, Boeckers TM, Gundelfinger ED, Birchmeier W. 2004. The neuronal scaffold protein Shank3 mediates signaling and biological function of the receptor tyrosine kinase Ret in epithelial cells. *J Cell Biol* **167**: 945–952.

Segouffin-Cariou C, Billaud M. 2000. Transforming ability of MEN2A-RET requires activation of the phosphatidylinositol 3-kinase/AKT signaling pathway. *J Biol Chem* **275**: 3568–3576.

Songyang Z, Carraway KL III, Eck MJ, Harrison SC, Feldman RA, Mohammadi M, Schlessinger J, Hubbard SR, Smith DP, Eng C, et al. 1995. Catalytic specificity of protein-tyrosine kinases is critical for selective signalling. *Nature* **373**: 536–539.

Stommel JM, Kimmelman AC, Ying H, Nabioullin R, Ponugoti AH, Wiedemeyer R, Stegh AH, Bradner JE, Ligon KL, Brennan C, et al. 2007. Coactivation of receptor tyrosine kinases affects the response of tumor cells to targeted therapies. *Science* **318**: 287–290.

Takahashi M, Asai N, Iwashita T, Isomura T, Miyazaki K, Matsuyama M. 1993. Characterization of the ret protooncogene products expressed in mouse L cells. *Oncogene* **8**: 2925–2929.

Takeuchi K, Soda M, Togashi Y, Suzuki R, Sakata S, Hatano S, Asaka R, Hamanaka W, Ninomiya H, Uehara H, et al. 2012. RET, ROS1 and ALK fusions in lung cancer. *Nat Med* **18**: 378–381.

Thomas RK, Baker AC, Debiasi RM, Winckler W, Laframboise T, Lin WM, Wang M, Feng W, Zander T, MacConaill L, et al. 2007. High-throughput oncogene mutation profiling in human cancer. *Nat Genet* **39**: 347–351.

Tozlu S, Girault I, Vacher S, Vendrell J, Andrieu C, Spyratos F, Cohen P, Lidereau R, Bieche I. 2006. Identification of novel genes that co-cluster with estrogen receptor α in breast tumor biopsy specimens, using a large-scale real-time reverse transcription-PCR approach. *Endocr Relat Cancer* **13**: 1109–1120.

Unger K, Zitzelsberger H, Salvatore G, Santoro M, Bogdanova T, Braselmann H, Kastner P, Zurnadzhy L, Tronko N, Hutzler P, et al. 2004. Heterogeneity in the distribution of RET/PTC rearrangements within individual post-Chernobyl papillary thyroid carcinomas. *J Clin Endocrinol Metab* **89**: 4272–4279.

Veit C, Genze F, Menke A, Hoeffert S, Gress TM, Gierschik P, Giehl K. 2004. Activation of phosphatidylinositol 3-kinase and extracellular signal-regulated kinase is required for glial cell line-derived neurotrophic factor-induced migration and invasion of pancreatic carcinoma cells. *Cancer Res* **64**: 5291–5300.

Verbeek HH, Alves MM, de Groot JW, Osinga J, Plukker JT, Links TP, Hofstra RM. 2011. The effects of four different tyrosine kinase inhibitors on medullary and papillary thyroid cancer cells. *J Clin Endocrinol Metab* **96**: E991–E995.

Viglietto G, Chiappetta G, Martinez-Tello FJ, Fukunaga FH, Tallini G, Rigopoulou D, Visconti R, Mastro A, Santoro M, Fusco A. 1995. RET/PTC oncogene activation is an early event in thyroid carcinogenesis. *Oncogene* **11**: 1207–1210.

Vizioli MG, Possik PA, Tarantino E, Meissl K, Borrello MG, Miranda C, Anania MC, Pagliardini S, Seregni E, Pierotti MA, et al. 2011. Evidence of oncogene-induced senescence in thyroid carcinogenesis. *Endocr Relat Cancer* **18**: 743–757.

Wang J, Knauf JA, Basu S, Puxeddu E, Kuroda H, Santoro M, Fusco A, Fagin JA. 2003. Conditional expression of RET/PTC induces a weak oncogenic drive in thyroid PCCL3 cells and inhibits thyrotropin action at multiple levels. *Mol Endocrinol* **17:** 1425–1436.

Wang R, Hu H, Pan Y, Li Y, Ye T, Li C, Luo X, Wang L, Li H, Zhang Y, et al. 2012. RET fusions define a unique molecular and clinicopathologic subtype of non-small-cell lung cancer. *J Clin Oncol* **30:** 4352–4359.

Wells SA Jr, Robinson BG, Gagel RF, Dralle H, Fagin JA, Santoro M, Baudin E, Elisei R, Jarzab B, Vasselli JR, et al. 2012. Vandetanib in patients with locally advanced or metastatic medullary thyroid cancer: A randomized, double-blind phase III trial. *J Clin Oncol* **30:** 134–141.

Williams D. 2008. Radiation carcinogenesis: Lessons from Chernobyl. *Oncogene* **27:** S9–S18.

Wong N, Lai P, Lee SW, Fan S, Pang E, Liew CT, Sheng Z, Lau JW, Johnson PJ. 1999. Assessment of genetic changes in hepatocellular carcinoma by comparative genomic hybridization analysis: Relationship to disease stage, tumor size, and cirrhosis. *Am J Pathol* **154:** 37–43.

Wood LD, Parsons DW, Jones S, Lin J, Sjöblom T, Leary RJ, Shen D, Boca SM, Barber T, Ptak J, et al. 2007. The genomic landscapes of human breast and colorectal cancers. *Science* **318:** 1108–1113.

Xing M. 2005. BRAF mutation in thyroid cancer. *Endocr Relat Cancer* **12:** 245–262.

Yeh S, Chang C. 1996. Cloning and characterization of a specific coactivator, ARA70, for the androgen receptor in human prostate cells. *Proc Natl Acad Sci* **93:** 5517–5521.

Zhang J, Yang PL, Gray NS. 2009. Targeting cancer with small molecule kinase inhibitors. *Nat Rev Cancer* **9:** 28–39.

Zhu Z, Ciampi R, Nikiforova MN, Gandhi M, Nikiforov YE. 2006. Prevalence of RET/PTC rearrangements in thyroid papillary carcinomas: Effects of the detection methods and genetic heterogeneity. *J Clin Endocrinol Metab* **91:** 3603–3610.

Index

Insulin receptor substrate 3 (IRS3),
tissue distribution, 411
Insulin receptor substrate 4 (IRS4),
tissue distribution, 411
Insulin resistance
gene defects
Akt, 420
insulin receptor, 418
insulin receptor substrate 1, 418
phosphatidylinositol 3-kinase, 418
PTEN, 419
hyperglycemia mechanisms, 421
lipotoxicity, 420
mitochondrial dysfunction, 421–422
obesity-associated inflammation, 421
prospects for study, 422
unfolded protein response, 422
Internalization. *See* Endocytosis, receptor
tyrosine kinases
Intracellular domain (ICD)
caspase cleavage from receptor tyrosine kinases, 55
γ-secretase cleavage from receptor tyrosine kinases
colony-stimulating factor-1 receptor, 53–54
Ephrin-B2 receptor, 53
ErbB-4, 50–53
fibroblast growth factor receptor 3, 55
insulin receptor, 54
insulin-like growth factor-1 receptor, 54
Met, 55
overview, 49–51
protein tyrosine kinase 7, 55
Ryk, 54–55
Tie1, 54
vascular endothelial growth factor receptor, 54
granzyme B cleavage from receptor
tyrosine kinases, 56
prospects for study, 59
rhomboid-dependent fragments, 56
iRhoms. *See* Rhomboids
IRS1. *See* Insulin receptor substrate 1
IRS2. *See* Insulin receptor substrate 2
IRS3. *See* Insulin receptor substrate 3
IRS4. *See* Insulin receptor substrate 4

J

JAK, TAM receptor signaling, 371

K

Kinome, human, 16
Kit
diseases
acute myeloid leukemia, 217
gastrointestinal stromal tumor, 217
lung cancer, 217–218

mastocytosis, 217
melanoma, 217
piebaldism, 218
down-regulation, 216
functions
fertility, 217
hematopoiesis, 216
neurons, 217
peristalsis, 217
pigmentation, 216–217
ligand dimers, 8
phosphorylation sites, 9
prospects for study, 218
signaling, 215–216
stem cell factor activation, 215
KLF2, 141–142
Klotho, fibroblast growth factor receptor signaling,
252–253

L

Leukemia, RET mutations, 460–461
Ligand shedding. *See also specific ligands and receptors*
ADAM proteases, 119–123
cancer dysregulation, 126–127
Drosophila studies, 116–118
ephrins, 126
epidermal growth factor receptor ligands, 118–119
rhomboids, 123–125
lin-18, Wnt signaling, 330–332
LRIG1, 140–141
LRP1, ephrin receptor association, 314–315
Lrp4
Agrin binding, 265, 269–270
myasthenia gravis autoantibodies, 271–272
neuromuscular synapse formation role, 265–268
structure and function, 269–270
LRP5, Wnt ligands, 325–326
LRP6
ligand concentration effects on entry routes and
intracellular fate, 98–99
Wnt ligands, 325–326
Lung cancer
epidermal growth factor receptor mutations and
therapeutic targeting, 166, 168
Kit role, 217–218
RET mutations, 460
Lymphangiogenesis. *See* Angiogenesis
Lymphedema, lymphangiogenesis defects, 375

M

MAPK. *See* Mitogen-activated protein kinase
Mass spectrometry (MS)
kinase assays, 23–24
phosphotyrosine proteomics, 22